UNITEXT - La Matematica per il 3+2

Volume 103

More information about this series at http://www.springer.com/series/5418

Carlo Mariconda · Alberto Tonolo

Discrete Calculus

Methods for Counting

 Springer

Carlo Mariconda
Dipartimento di Matematica
Università degli Studi di Padova
Padova
Italy

Alberto Tonolo
Dipartimento di Matematica
Università degli Studi di Padova
Padova
Italy

ISSN 2038-5722   ISSN 2038-5757   (electronic)
UNITEXT - La Matematica per il 3+2
ISBN 978-3-319-03037-1   ISBN 978-3-319-03038-8   (eBook)
DOI 10.1007/978-3-319-03038-8

Library of Congress Control Number: 2016946021

Cover illustration: A derangement of Venice (2016). © 2016, Graziella Giacobbe, Padova, Italy

Printed on acid-free paper

This Springer imprint is published by Springer Nature
The registered company is Springer International Publishing AG Switzerland

To Frank Sullivan

Preface

Several years ago, we were asked to give a second-year course for a Computer Science Engineering degree at the University of Padova, with some mathematical content about discrete mathematics: combinatorics, finite calculus, formal series, approximation of finite sums, etc. We were unable at that time to find a suitable book for our students: either the mathematical aspects were too profound and abstract, or else explanations were poor and each exercise was solved by means of some unexplainable tricks. That last aspect is particularly common in combinatorics, where, too often, the basic results on the subject are so distant from their applications that exercises often appear to have an "ad hoc" solution. Our first purpose was to develop a method for explaining combinatorics to our students, which would allow them to solve certain advanced problems with some facility. We achieved this goal after several attempts, ultimately meeting with the approval of our students, the involuntary guinea pigs in our experiment, to whom we express our thanks.

In the part of the book devoted to combinatorics, roughly a third of the volume, we begin by relating every application to a very few essential basic mathematical concepts. A key role in this is played by the sequences and collections, terms that we prefer to the more widely used, but more ambiguous terms of arrangements and combinations: indeed, thinking of real life (combination locks), why should a combination denote a non-ordered set of symbols? We spend some time on the Basic Principles of combinatorics, like the Multiplication and Division Principles: we strongly believe that passing over these subjects is the number one cause of errors in the applications. We focus on the occupancy problems, where one prescribes a fixed number of repetitions of the elements in a sequence or a collection, and we thoroughly discuss the Inclusion/Exclusion Principle and its consequences (derangements, partitions, etc.). In these first chapters, we encounter, of course, some famous stars such as the factorials, binomials, derangements, and the Bell, Catalan, Euler, and Stirling numbers: all of them are first defined via a combinatorial characteristic property and only afterward explicitly computed. This allows us to prove most of their properties—like the recurrence ones—with some simple combinatorial arguments instead of the more tedious inductive method.

Chapter 6 is devoted to the techniques for computing the sums of a finite number of consecutive terms of a sequence. They reproduce and actually can be an introduction to those of the differential calculus: derivatives, primitives, the fundamental theorem of calculus, and even the Taylor expansion find their discrete counterparts here. We also encounter the harmonic numbers.

A substantial part of the book is devoted to power series and generating formal series. In Chap. 7, we state the basic definitions and properties, though there are some delicate points like closed forms and compositions of formal power series that need more careful attention. The Basic Principle for occupancy problems explains how the tools introduced here are useful to solve combinatorial problems.

We made an effort to keep separate as much as possible the algebraic properties and the convergence of formal power series, which is introduced and used only in the subsequent Chap. 8. In this chapter, we compute the generating formal series of the sequences of famous numbers such as those mentioned above, or the Fibonacci and the Bernoulli numbers introduced here. As a by-product, we estimate easily the Bernoulli numbers via their relation with the famous Riemann zeta function.

Generating formal series return in the following chapters (though the reader can skip these without losing the main content), e.g., in Chap. 11 concerning symbolic calculus, where they play a prominent role in finding the number of sequences, of a prescribed length and alphabet, that do not contain a given pattern, or in computing the odds in favor of the appearing of a pattern before another one.

Chapters 9 and 10 are devoted to recurrence relations. In the first of these, we show how these relations arise. A section is devoted to discrete dynamical systems, i.e., recurrence relations of the form $x_{n+1} = f(x_n)$. We give a thorough analysis of the case where f is monotonic on an interval, and a simple proof of the famous Sarkovskii's theorem on the existence of orbits of any minimal period, once there is a point of minimal period 3: we did not find this material in other texts. In Chap. 10, we mainly deal with the classical theory of the linear recurrence relations. Here, the reader can find an alternative resolution method based on generating formal series, which turns out also to be useful in the proofs of the main results. We expect that the average reader will skip these, and therefore, they can be found in a separate section. The chapter ends with the divide and conquer relations and the estimates of the magnitude of their solutions, so useful in the analysis of algorithms.

Chapters 12 and 13 are devoted to the Euler–Maclaurin formula, which relates the sum of the values of a smooth function f on the integers of an interval with its integral on the same interval and its consequences like: the approximation of sums and sum of series, asymptotic estimates for the partial sums of a series, well-known and unusual integral criteria for the convergence of a series, and integral convergence, the trapezoidal methods, and the Hermite formula for the estimate of integrals. We believe that this material, so rich in applications and beauty, is not given its due place in courses in mathematics or computer science. In Chap. 12, we have made a strong effort to keep things as easy as possible: the chapter is devoted to the first- and second- order Euler–Maclaurin formulas, thus leaving matters such as the estimates of the Bernoulli polynomials, the Bernoulli numbers, and the proofs of every technical detail to the subsequent Chap. 13. It is rather surprising how even the first-order formula, whose proof is based on a simple integration by parts, leads to a primitive version of Stirling's formula for the approximation of the factorial.

Chapter 14 deals with the approximation of sums of binomials, like the Ramanujan Q-functions. The proofs here are based on some uniform estimates on families of sequences and may be skipped by the inexperienced reader.

Finally, a list of the main formulas and a detailed index can be found at the end of the book.

Just to give a taste of the book, here are some of the applications that we consider and discuss:

- The birthday problem (Example 2.30): what is the probability that two (or more) people randomly chosen from a group of 25 people have the same birthday?
- How to count card shuffles and to perform astonishing magic tricks by means of the Gilbreath Principle (Sect. 2.3).
- The hats problem (Example 4.25): each of the n diners entering a restaurant leaves his hat at the checkroom. In how many ways can the n hats be redistributed in such a way that no one receives his own hat? If the tipsy hatchecker randomly distributes the hats to the exiting diners, what is the probability that no one receives his own hat? Actually, we will be able to solve the more difficult variant of the problem which asks for the probability that no one receives a hat whose brand is the same as the original one.
- The Leibniz rule for the derivative of a product of functions (Theorem 3.26) and Faà di Bruno's formula for the derivatives of a composition of functions (Theorem 5.23) as an application of, respectively, the concept of sharing with a given occupancy and the concept of partition.
- The Smith College diploma problem (Example 5.68): at Smith College, diplomas are delivered as follows. The diplomas are randomly distributed to the graduating students. Those who do not receive their own diploma form a circle and pass the diploma received to their counterclockwise neighbor. Those then receiving their own diploma leave the circle, while the others form a smaller circle and repeat the procedure. Determine the probability that exactly $k \geq 0$ hand-offs are necessary before each graduate has his/her own diploma.
- The Latin teacher's random choice (Example 7.30): a teacher wants to select a student to test in the class. He randomly opens a book, sums up the digits of the number of the page, and chooses the corresponding student from the alphabetical list. What is the probability of being chosen for each student?
- The Titus Flavius Josephus problem (Example 9.14) is connected with an autobiographical episode recounted by the historian Titus Flavius Josephus, which we now translate into mathematical terms. The problem presents

n persons arranged in a circle. Having chosen an initial person and the direction of rotation, one moves multiple times along the circle, eliminating every second person one meets in the chosen direction until only a single person remains. Given $n \geq 1$, one seeks to determine the position of the remaining person if the circle is initially formed by n persons.

- Given any word in any language, how many sequences of letters of a prescribed length do not contain that word? We give a recursive answer in Corollary 11.39.

- The theorem of the monkey (Theorem 11.43): What is the average number of chance strokes on a keyboard with m keys necessary to make a given word appear?

- The Conway equation (Theorem 11.53): Two players select distinct sequences of equal length ℓ, with elements in a fixed set Γ. They toss a dice with as many faces as the cardinality of Γ, all labeled with the elements of Γ, until the last ℓ results match one player's pattern. What are the odds in favor of one or the other pattern?

- Evaluate the mysterious Euler constant $\gamma = \lim\limits_{n \to +\infty} H_n - \log n$ up to 12 decimal digits (Example 13.47) and Apéry's constant $\zeta(3) = \sum\limits_{k=1}^{\infty} 1/k^3$ up to 16 decimal digits (Example 13.51).

- Faulhaber's formula (Example 13.22) for the sum $\sum\limits_{k=1}^{n} k^m$ in terms of the first m Bernoulli numbers.

- We pick a card from a deck of 52 playing cards, and we reinsert it in the deck. After shuffling, we pick another card and we reinsert it again in the deck. How many times, on average, must we repeat this procedure in order for a repetition to appear?

The book includes an unusually large collection of examples and problems whose solutions, as well as corrections and updates, can be found at:

https://discretecalculus.wordpress.com

On many occasions, we have tried to give different proofs of the same result. This is the case, for instance, for the Inclusion/Exclusion Principle, Faulhaber's formula, many combinatorial results, and the estimates for the harmonic numbers and the binomials. Thus, several paths are possible, depending on the interests of the reader. Some independent blocks are as follows: the first five chapters on combinatorics, the sums and the finite calculus techniques (Chap. 6), the essentials of formal power series (Chap. 7) and their applications to the symbolic calculus (Chap. 11), the basic facts about recurrence relations with an insight into discrete dynamical systems including Sarkovskii's theorem (Chap. 9), and the classical theory on linear recurrences (Chap. 10): here, the alternative resolution method based on generating formal series needs the content of Chap. 8. Almost all the main tools on the approximation of sums appear in Chap. 12, whereas Chap. 13 is reserved for those who want to master Euler–Maclaurin formulas in the whole generality, or need more than a second-order expansion. Finally, for most readers, it will be sufficient in the final Chap. 14 to read the basic definitions and the claims of the theorems, without working through the proofs, which are the most difficult part of the book.

The many authors and books that inspired us are listed in the references. Among them, the books by Flajolet and Sedgewick [35] and by Graham et al. [20] were favorite sources of interest.

We are grateful to our readers for their comments that can be sent to:
discretecalculus@gmail.com

There are a couple of recurrent symbols in the book: ☞ denotes an important matter or comment, or even a warning; conversely, ♣ denotes a difficult proof, a boring computation, etc. These parts can be omitted, at least at first (and maybe second) reading.

With the acronym CAS, we indicate a generic computer algebra system, i.e., a software program that allows computation over mathematical expressions. The famous Maple, Mathematica, Maxima, and Pari GP software are among our preferred ones. We sometimes use them to make some hard computation and to evaluate the precision of our estimates.

A word about the authors: we are both Venetians, and this is probably the only point in common. Mathematically, this is perhaps justified by the fact that the field of Carlo Mariconda is analysis, whereas Alberto Tonolo's research topic is algebra; discrete mathematics is in any case just a hobby for both. The different points of view, even with respect to details such as fonts, symbols, and notation, explain why it took about 15 years to write this book. For the readers' sake, every single page initially proposed by one of us was then criticized and changed by the other (and then sometimes both agreed that the initial version was better...). We sincerely hope that this distillation process has been useful for the readers of our book.

We are indebted to our colleague Frank Sullivan, who gave us the taste for these kinds of argument, and who translated and revised with such great enthusiasm a large part of the book; sadly, he did not witness the final product.

Padova, Italy   Carlo Mariconda
April 2016   Alberto Tonolo

Acknowledgements

Many colleagues, friends, and students helped us during the long period of preparation of this book.

Frank Sullivan and then Thom Cuschieri worked on every line of the book, applying their skills in English, mathematics, and LaTeX.

Our former student Marco De Zotti carefully and critically read several parts of the text, suggesting some useful changes (e.g., regarding the use of Fourier analysis in the estimates on the Bernoulli polynomials).

Graziella Giacobbe drafted uncountable versions of the *derangement* of Venice in the cover.

To all of the above, we express our sincere gratitude.

Moreover, we warmly thank

- Anna Bohun, who reviewed the chapter devoted to linear recurrences.
- Francesca Bonadei and Francesca Ferrari of Springer for their help and their patience for our uncountable delays.
- Raphaël Cerf for his advice on combinatorics and the need of combinatorial formulas.
- Giuseppe De Marco for his comments and many answers to our questions.
- Alessandro Languasco for his insights into algebraic and analytic number theory.
- Luigi Provenzano for having carefully read the initial part of the book.
- Our students, who were the favorite test subjects for the book. In particular, we thank the students of the Engineering School, of the Galilean School of Higher Education, and of the degrees in Mathematics of the University of Padova. They stimulated the development of several sections of the book.
- Ashok Arumairaj and Sukanya Servai of Springer for their care on assisting us in proofreading the book.
- Francesca Zancan, who helped us with her precious legal advices.

Special thanks go to Francis Clarke, for his support during the preparation of the book and for his various suggestions: He was right when he bet that he would find something wrong, in the English language, in every part he read. Any mistake therefore belongs to the parts we have kept from him.

Finally, we are grateful to Francesca, Niccolò, Paola, and Tommaso for their constant encouragement.

Contents

Chapter 1
Let's Learn to Count

Abstract In this chapter, after a quick review of the basic concepts of set theory, we introduce the fundamental notions and principles of combinatorics. Even though its contents are elementary, we warmly suggest to take a look at the chapter. Our approach consists in trying to describe every combinatorial problem by means of sets of (ordered) sequences, or (unordered) collections, and their dual concepts of sharings and compositions. Computations are successively done via some basic fundamental tools like the Multiplication and the Division Principle. A rigorous and effective formulation of these principles, in particular of the Multiplication Principle, is of fundamental importance for their correct application. Indeed they constitute, at the same time, the royal way to solve combinatorial problems, and the main source of errors, when misused. We conclude the section with a brief discussion of uniform probability on finite sample spaces, which is here just a way to express combinatorial results in probabilistic terms.

1.1 Operations on Finite Sets

This section is dedicated to the basic operations and the most elementary techniques of counting. For most readers it will be a review of known concepts, though it will also serve to fix the notations which will be used throughout this book.

1.1.1 Review of Set Theory

We introduce the concept of set in a "naive" way: a rigorous axiomatic treatment of set theory is beyond the goals of the present text.

A **set** is defined when for any given *object* one can decide whether or not it belongs to the set. The objects belonging to a set are called **elements** of the set. They can be anything at all: numbers, people, other sets, etc..

The islands of the Mediterranean Sea constitute a set. Whatever object we might choose to consider, we know how to decide whether or not it is an element of that set: for example, the number 3, the United States, the Moon, and many other things do not belong to it, while the islands of Cyprus, Crete and Malta certainly do.

C. Mariconda and A. Tonolo, *Discrete Calculus*,
UNITEXT - La Matematica per il 3+2 103, DOI 10.1007/978-3-319-03038-8_1

✋   The *sets which do not belong to themselves* do not form a set: indeed, if they did form a set I, then given the element "the set I" one would not be able to decide whether or not it belonged to I. Indeed, if it belonged to I, then as an element of I it would not belong to itself, and so it would not belong to I, which gives a contradiction. If, on the contrary, it did not belong to I, then by the very definition of I it would have to belong to I, which again gives a contradiction.

Two sets are equal if and only if the elements lying in one also lie in the other, and vice versa; as a particular example, the sets $\{1, 1, 2\}$ and $\{1, 2\}$ are equal sets because both contain the numbers 1 and 2, and nothing else.

We write $a \in A$ to say that an element a belongs to the set A.

In giving a formal description of a set we either list its elements, for example $\{2, 4, 5, 7\}$, or, taking care to be precise, we describe the property that characterizes its elements, for example $\{n \in \mathbb{N} : 2 \text{ divides } n\}$.

In the remainder of this text we will use the symbols $\emptyset$, $\mathbb{N}$, $\mathbb{N}_{\geq 1}$, $\mathbb{N}_{\geq k}$, $\mathbb{Z}$, $\mathbb{Q}$, $\mathbb{R}$ and $\mathbb{C}$ to indicate the empty set, the sets of natural numbers, non-zero natural numbers, natural numbers greater or equal than k, integers, rational numbers, real numbers, and complex numbers respectively.

We now quickly review a few well-known notions of set theory.

Definition 1.1 Given two sets A and B, we say that A is a **subset** of B (and we will write $A \subseteq B$) if A is contained in B, that is, if

$$(a \in A) \Longrightarrow (a \in B).$$ □

Definition 1.2 Let X be a set. We use $\mathscr{P}(X)$ to denote the **set of parts of** X, that is, the set of subsets of X, including the empty subset and X itself.   □

Example 1.3 Let $X = \{1, 2\}$; then the set of parts of X is

$$\mathscr{P}(X) = \{\emptyset, \{1\}, \{2\}, X\}.$$

☞   Note that $\mathscr{P}(\emptyset) = \{\emptyset\} \neq \emptyset$: in fact, $\mathscr{P}(\emptyset)$ has one element (the empty set), while $\emptyset$ has no elements.   □

Definition 1.4 Let A, B be subsets of a set X; then the **union** $A \cup B$, the **intersection** $A \cap B$, the **difference** $A \backslash B$, the **symmetric difference** $A \triangle B$ and the **complement** A^c are defined as follows:

- $A \cup B = \{x \in X : x \in A \text{ or } x \in B\}$;
- $A \cap B = \{x \in X : x \in A \text{ and } x \in B\}$;
- $A \backslash B = \{x \in X : x \in A \text{ and } x \notin B\}$;
- $A \triangle B = \{x \in X : x \in A \cup B \text{ and } x \notin A \cap B\}$;
- $A^c = \{x \in X : x \notin A\}$.

If $A \cap B = \emptyset$, we say that A and B are **disjoint** subsets of X.   □

A useful tool to verify the equality between different descriptions of the same subset is the notion of characteristic function.

Definition 1.5 Let X be a set and A a subset of X. The **characteristic function** of A is defined setting

$$\chi_A : X \to \{0, 1\}, \quad \chi_A(x) = \begin{cases} 1 & \text{if } x \in A, \\ 0 & \text{otherwise.} \end{cases}$$ □

Clearly, if A, B are two subsets of a set X, the characteristic functions χ_A of A and χ_B of B coincide if and only if $A = B$. Moreover one easily gets $\chi_X(x) = 1$ and $\chi_\emptyset(x) = 0$ for each $x \in X$.

Lemma 1.6 *Let A, B be two subsets of X and $x \in X$. Then:*

1. $\chi_{A \cap B}(x) = \chi_A(x)\chi_B(x)$;
2. $\chi_{A \cup B}(x) = \chi_A(x) + \chi_B(x) - \chi_A(x)\chi_B(x)$;
3. *If $A \cap B = \emptyset$, then $\chi_{A \cup B}(x) = \chi_A(x) + \chi_B(x)$;*
4. $\chi_{A \setminus B}(x) = \chi_A(x) - \chi_A(x)\chi_B(x)$;
5. $\chi_{A \triangle B}(x) = \chi_A(x) + \chi_B(x) - 2\chi_A(x)\chi_B(x)$;
6. $\chi_{A^c}(x) = 1 - \chi_A(x)$.

Proof. 1. The element x belongs to $A \cap B$ if and only if $x \in A$ and $x \in B$; therefore $\chi_{A \cap B}(x) = 1$ if and only if $\chi_A(x) = 1 = \chi_B(x)$, or equivalently $\chi_A(x)\chi_B(x) = 1$.
2. The element x belongs to $A \cup B$ if and only if either $x \in A$ and $x \notin B$, $x \notin A$ and $x \notin B$, or $x \in A$ and $x \in B$. We conclude since

$$\chi_A(x) + \chi_B(x) - \chi_A(x)\chi_B(x) = \begin{cases} 1 + 0 - 0 = 1 & \text{if } x \in A \text{ and } x \notin B, \\ 0 + 1 - 0 = 1 & \text{if } x \notin A \text{ and } x \in B, \\ 1 + 1 - 1 = 1 & \text{if } x \in A \text{ and } x \in B, \\ 0 & \text{otherwise.} \end{cases}$$

3. It follows by Point 2: indeed, if A and B are disjoint, then $\chi_A(x)\chi_B(x) = 0$.
4. We have $A \cup B = (A \setminus B) \cup B$; since $A \setminus B$ and B are disjoint, by Points 2 and 3 we have

$$\chi_A(x) + \chi_B(x) - \chi_A(x)\chi_B(x) = \chi_{A \cup B}(x) = \chi_{(A \setminus B) \cup B}(x) = \chi_{A \setminus B}(x) + \chi_B(x).$$

Therefore $\chi_{A \setminus B}(x) = \chi_A(x) - \chi_A(x)\chi_B(x)$.
5. We have $A \triangle B = (A \cup B) \setminus (A \cap B)$; by Points 4, 1 and 2 we have

$$\chi_{A \triangle B}(x) = \chi_{A \cup B}(x) - \chi_{A \cup B}(x)\chi_{A \cap B}(x)$$
$$= \chi_A(x) + \chi_B(x) - \chi_A(x)\chi_B(x) - \big(\chi_A(x) + \chi_B(x) - \chi_A(x)\chi_B(x)\big)\big(\chi_A(x)\chi_B(x)\big)$$
$$= \chi_A(x) + \chi_B(x) - \chi_A(x)\chi_B(x) - \chi_A(x)\chi_B(x) = \chi_A(x) + \chi_B(x) - 2\chi_A(x)\chi_B(x).$$

6. We have $X = A \cup A^c$; since A and A^c are disjoint, by Point 3 we have

$$1 = \chi_X(x) = \chi_A(x) + \chi_{A^c}(x),$$

and hence $\chi_{A^c}(x) = 1 - \chi_A(x)$. $\square$

Using the characteristic functions and Lemma 1.6 it is easy to verify the following simple properties.

Proposition 1.7 *Let X be a set and A, B and C three of its subsets. One then has*

1. $A \cup (B \cap C) = (A \cup B) \cap (A \cup C)$;
2. $A \cap (B \cup C) = (A \cap B) \cup (A \cap C)$;
3. $A \backslash (B \cap C) = (A \backslash B) \cup (A \backslash C)$;
4. $A \backslash (B \cup C) = (A \backslash B) \cap (A \backslash C)$;
5. $A \cap (B \bigtriangleup C) = (A \cap B) \bigtriangleup (A \cap C)$;
6. $(A \cap B)^c = A^c \cup B^c$;
7. $(A \cup B)^c = A^c \cap B^c$;
8. $A \bigtriangleup B = A^c \bigtriangleup B^c$;
9. $(A \bigtriangleup B) \bigtriangleup B = A$.

Proof. Let us prove the identities 2, 6, 8 and 9 proving that the characteristic functions of the involved sets coincide; the other identities can be obtained in a similar way. By Lemma 1.6, for any $x \in X$ we get:

2. $\chi_{A \cap (B \cup C)}(x) = \chi_A(x) \chi_{B \cup C}(x) = \chi_A(x) \left(\chi_B(x) + \chi_C(x) - \chi_B(x) \chi_C(x) \right)$ and

$$\chi_{(A \cap B) \cup (A \cap C)}(x) = \chi_{A \cap B}(x) + \chi_{A \cap C}(x) - \chi_{A \cap B \cap C}(x)$$
$$= \chi_A(x) \chi_B(x) + \chi_A(x) \chi_C(x) - \chi_A(x) \chi_B(x) \chi_C(x).$$

6. $\chi_{(A \cap B)^c}(x) = 1 - \chi_{A \cap B}(x) = 1 - \chi_A(x) \chi_B(x)$ and

$$\chi_{A^c \cup B^c}(x) = \chi_{A^c}(x) + \chi_{B^c}(x) - \chi_{A^c}(x) \chi_{B^c}(x)$$
$$= 1 - \chi_A(x) + 1 - \chi_B(x) - (1 - \chi_A(x))(1 - \chi_B(x))$$
$$= 1 - \chi_A(x) \chi_B(x).$$

8. $\chi_{A^c \bigtriangleup B^c}(x) = \chi_{A^c}(x) + \chi_{B^c}(x) - 2\chi_{A^c}(x) \chi_{B^c}(x)$

$$= 1 - \chi_A(x) + 1 - \chi_B(x) - 2(1 - \chi_A(x))(1 - \chi_B(x))$$
$$= \chi_A(x) + \chi_B(x) - 2\chi_A(x) \chi_B(x)$$
$$= \chi_{A \bigtriangleup B}(x).$$

9. $\chi_{(A \bigtriangleup B) \bigtriangleup B}(x) = \chi_{A \bigtriangleup B}(x) + \chi_B(x) - 2\chi_{A \bigtriangleup B}(x) \chi_B(x) = \chi_A(x) + \chi_B(x)+$

$-2\chi_A(x) \chi_B(x) + \chi_B(x) - 2(\chi_A(x) + \chi_B(x) - 2\chi_A(x)\chi_B(x))\chi_B(x) = \chi_A(x).$
                                                                                            $\square$

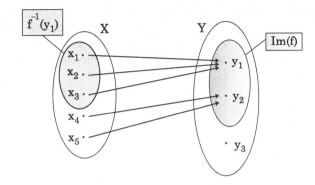

Fig. 1.1 An example of a function

The following notion allows one to "connect" two sets.

Definition 1.8 A **function**

$$f : X \to Y \quad \text{consists of:}$$

1. A set X called the **domain** of f;
2. A set Y called the **codomain** of f;
3. A law associating each element x in X to one and only one corresponding element y in Y. The element of Y associated to $x \in X$ is called the **image** of x under f and is denoted by $f(x)$.

Given $y \in Y$, the set $\{x \in X : f(x) = y\}$ is called the **counterimage** or **inverse image** of y under f and is denoted by $f^{-1}(y)$ (see, e.g., Fig. 1.1). The subset $\{f(x) : x \in X\}$ of Y is called the **image of** f and is denoted by $\mathrm{Im}(f)$. $\qquad\square$

We also recall the following classification of functions.

Definition 1.9 A function $f : X \to Y$ is said to be:

- **Injective** if $(f(x) = f(y)) \implies (x = y)$, that is, when distinct elements of the domain X have distinct images in the codomain Y;
- **Surjective** if $\mathrm{Im}(f) = Y$, that is, if each element of the codomain is the image under f of at least one element of the domain;
- **Bijective** if it is both injective and surjective; in this case we say that X and Y are in **one to one correspondence**. $\qquad\square$

Given two sets A and B we can consider the set of all ordered pairs realized by taking (in order) an element of A and one from B.

Definition 1.10 Given n sets $A_1, \ldots, A_n$, their **cartesian product** is the set $A_1 \times \cdots \times A_n$ of n-tuples $(a_1, \ldots, a_n)$ with $a_i \in A_i$, for $i = 1, \ldots, n$. The cartesian product $\underbrace{A \times \cdots \times A}_{n \text{ times}}$ is denoted by A^n. $\qquad\square$

The cartesian product is a formal construction frequently used in mathematics to construct complex sets starting from simple ones. One notes that, given two distinct sets A and B, the sets $A \times B$ and $B \times A$ are different. It is, however, possible to construct a one to one correspondence (bijection) between them. Indeed,

$$f : A \times B \to B \times A, \quad f(a, b) = (b, a)$$

is obviously a bijective function.

1.1.2 Cardinality of a Finite Set

A fundamental characteristic of a finite set is its *size*.

Definition 1.11 The **cardinality** of a finite set X is the number of its distinct elements. The symbol $|X|$ is used to denote the cardinality of X.   □

The cardinality of a subset of a finite set can be recovered from its characteristic function.

Proposition 1.12 *Let A be a subset of a finite set X. Then*

$$|A| = \sum_{x \in X} \chi_A(x).$$

Proof. The characteristic function $\chi_A(x)$ equals 1 in every point of A and 0 elsewhere.   □

Remark 1.13 The above easy result is the discrete analogous of the fact that the integral of the characteristic function of an interval is equal to its length.

To count the number of elements of a set, it is often convenient to interpret the set as either a union, an intersection or a cartesian product (or some other operation) obtained from simpler sets. Hence it is important to know how to calculate the cardinality of a set obtained through such operations.

Proposition 1.14 *Let A, B be two finite subsets of a set X. Then:*

1. *If A and B are disjoint, namely, if $A \cap B = \emptyset$, then $|A \cup B| = |A| + |B|$;*
2. *$|A \cup B| = |A| + |B| - |A \cap B|$;*
3. *$|A \times B| = |A| \times |B|$;*
4. *If X is finite then $|A^c| = |X| - |A|$.*

Proof. Let $A = \{a_1, \ldots, a_m\}$ and $B = \{b_1, \ldots, b_n\}$ be two sets of cardinality $|A| = m$ and $|B| = n$ respectively.
1. If A and B are disjoint then $A \cup B = \{a_1, \ldots, a_m, b_1, \ldots, b_n\}$ has $m + n$ elements.
2. $A \cup B$ is the union of the pairwise disjoint subsets $A \setminus B$ and B; since A is the disjoint union of $A \setminus B$ and $A \cap B$, one has $|A \setminus B| = |A| - |A \cap B|$. Then

$$|A \cup B| = |A| - |A \cap B| + |B|.$$

Alternatively, using Proposition 1.12 together with Lemma 1.6, we get

$$|A \cup B| = \sum_{x \in X} \chi_{A \cup B}(x) = \sum_{x \in X} \left(\chi_A(x) + \chi_B(x) - \chi_A(x)\chi_B(x) \right)$$

$$= \sum_{x \in X} \left(\chi_A(x) + \chi_B(x) - \chi_{A \cap B}(x) \right) = |A| + |B| - |A \cap B|.$$

3. Clearly

$$A \times B = (\{a_1\} \times B) \cup (\{a_2\} \times B) \cup \cdots \cup (\{a_m\} \times B),$$

with the $\{a_i\} \times B$ pairwise disjoint. The set $\{a_1\} \times B$ has n elements:

$$(a_1, b_1), (a_1, b_2), \ldots, (a_1, b_n).$$

Similarly, $|\{a_i\} \times B| = n$ for each element a_i of A. In view of Point 1 one has

$$|A \times B| = |\{a_1\} \times B| + \cdots + |\{a_m\} \times B| = \underbrace{n + n + \cdots + n}_{m \text{ times}} = m \times n.$$

4. Since $X = A \cup A^c$ and $A \cap A^c = \emptyset$, it follows that $|X| = |A| + |A^c|$, and hence $|A^c| = |X| - |A|$. $\qquad\square$

Proposition 1.15 *Two finite sets X and Y have the same cardinality if and only if there is a one to one correspondence between X and Y.*

Proof. If $|X| = |Y| = n$, we can denote all the elements of X by $x_1, \ldots, x_n$ and all the elements of Y by $y_1, \ldots, y_n$. Then the function that sends x_i to y_i, $i = 1, \ldots, n$, is a bijection. Conversely, if $f : X \to Y$ is a bijection, and $|X| = n$, let $x_1, \ldots, x_n$ be the list of all the elements of X. Then $f(x_1), \ldots, f(x_n)$ are distinct (since f is injective) and they constitute the entire set Y (since f is surjective). Therefore $Y = \{f(x_1), \ldots, f(x_n)\}$ has cardinality n. $\qquad\square$

Example 1.16 Count the number of possible committees composed of a man and a woman selected from a set M of 8 men and a set W of 10 women respectively.
Solution. Let Y be the set of such committees. Every element of Y is uniquely represented by an ordered pair (m, w) where m is one of the men and w one of the women. Since Y is in one to one correspondence with $M \times W$, we have

$$|Y| = 8 \times 10 = 80.$$
$\qquad\square$

1.2 Sequences, Collections, Sharings, Compositions and Partitions

We now introduce the terminology which will allow us to give precise descriptions of the various situations that we will encounter.

In what follows, we will deal with finite sets; for simplicity, we associate a positive integer to each element of such a finite set X, that is, we enumerate the elements of X so that if $|X| = n$ we can in this way identify X with *the set I_n* of numbers from 1 to n:

$$\forall n \in \mathbb{N}_{\geq 1} \quad I_n = \{1, \ldots, n\}, \quad I_0 = \emptyset.$$

We will call such an identification a **labeling** of X. Once labeled, the set X, inter alia, acquires an ordering: we will call the element of X with label 1 the *first element* of X, that with label 2 the *second element* of X, and so on. By convention, writing $I_n = \{*, !, ?, \%, \ldots, \#\}$ we mean that we label 1 the element $*$, we label 2 the element $!, \ldots$, we label n the element $\#$.

We now define the key concepts for the entire book. The reason for the exchange of roles between k and n in the first four points of the following definition will be clarified by Theorem 1.28.

Definition 1.17 Let $n, k \in \mathbb{N}$. We distinguish two classes of objects:

1. Ordered:

   a. A k-**sequence** of I_n is an ordered k-*tuple* $(a_1, \ldots, a_k)$ of not necessarily distinct elements of I_n, that is, an element of the cartesian product I_n^k.
   b. An n-**sharing** of I_k is an ordered n-*tuple* $(C_1, \ldots, C_n)$ of pairwise disjoint (and possibly *empty*) subsets of I_k, whose union $C_1 \cup \cdots \cup C_n$ is I_k.
   c. An n-**composition** of k is an ordered n-*tuple* $(k_1, \ldots, k_n)$ of natural numbers (possibly equal to 0) such that $k_1 + \cdots + k_n = k$; one also says that $(k_1, \ldots, k_n)$ is a **natural solution** or **natural number solution** of $x_1 + \cdots + x_n = k$.

2. Non ordered:

   a. A k-**collection** of I_n is a *non ordered family* of k (not necessarily distinct) elements of I_n. A k-collection containing k_1 copies of 1, k_2 copies of 2, $\ldots, k_n$ copies of n will be denoted by inserting those elements, in any order, each with its multiplicity, between square brackets:

   $$[\underbrace{1, \ldots, 1}_{k_1}, \ldots, \underbrace{n, \ldots, n}_{k_n}].$$

☞   Note that a k-collection of non repeated elements from I_n is nothing but a subset of I_n of cardinality k.

    b. An n-**partition** of I_k is a *non ordered family* of n *non empty* disjoint subsets of I_k whose union is I_k.

    c. An n-**partition** of k is a *non ordered family* of $[k_1, \ldots, k_n]$ of natural numbers in $\mathbb{N}_{\geq 1}$ such that $k_1 + \cdots + k_n = k$.

We use the terms **sequence, sharing, composition, collection, partition**, respectively to describe any k-sequence, n-sharing, n-composition, k-collection, or n-partition.[1] We define a **binary sequence** to be a sequence of $\{0, 1\}$.   □

Remark 1.18 An n-sharing of I_k is a particular n-sequence of $\mathscr{P}(I_k)$, while an n-partition of I_k is a particular n-collection of $\mathscr{P}(I_k) \setminus \{\emptyset\}$. An n-composition of k is a particular n-sequence of the set $\mathbb{N}$ of natural numbers, while an n-partition of k is a particular n-collection of the set $\mathbb{N}_{\geq 1}$ of positive natural numbers.

Remark 1.19 (*Extreme cases*) With regard to the notions introduced in the Definition 1.17, we now clarify the following cases, which are, however, of little significance when it comes to applications:

1a. For every $n \in \mathbb{N}$ there exists *one and only one* 0-sequence of I_n: the *empty sequence*. For each $k \geq 1$, *there are no k-sequences of I_0*.

1b. For every $n \in \mathbb{N}$ there is *one and only one* n-sharing of I_0: $(\underbrace{\emptyset, \ldots, \emptyset}_{n\text{-times}})$. For every $k \geq 1$, *there are no 0-sharings of I_k*.

1c. For every $n \in \mathbb{N}$ there is *one and only one* n-composition of 0: the n-sequence $(\underbrace{0, \ldots, 0}_{n\text{-times}})$. For every $k \geq 1$, *there are no 0-compositions of k*.

2a. For every $n \in \mathbb{N}$ there is *one and only one* 0-collection of I_n: the *empty collection*. For each $k \geq 1$, *there are no k-collections of I_0*.

2b. There is one and only one 0-partition of I_0: the *empty partition*. For each $n \in \mathbb{N}_{\geq 1}$ there exist neither 0-partitions of I_n, nor n-partitions of I_0.

2c. There is one and only one 0-partition of 0: the *empty partition*. For each $n \in \mathbb{N}_{\geq 1}$ there exist neither 0-partitions of n, nor n-partitions of 0.

☞   One should bear in mind the difference between sharings and compositions on one side, and partitions of sets and numbers on the other one:

- The sets forming a **sharing** can be *empty*, and are listed *in order*;
- The numbers forming a **composition** can be 0, and are listed *in order*;
- The sets forming a **partition** of a set must be *non empty*, and their *order does not matter*;
- The numbers forming a **partition** of a number must be *positive*, and their *order does not matter*.

---

[1]In many textbooks, the k-sequences of I_n are called k-fold *dispositions* of n elements, the k-collections of I_n are referred to as *combinations* of n elements taken k at a time, while generally sharings into non empty sets are called *ordered partitions*.

We could have called *composition of a set* the sharing to have a better correspondence with the notions of partition of a set and partition of a number. We have preferred the present notation to avoid the risk of confusion. Indeed, the sharing of a set will play an important role in the sequel, while we will say only few words about partitions of numbers in Sect. 7.10.

Definition 1.20 Given a k-sequence $(a_1, \ldots, a_k)$ of I_n, we say that $(b_1, \ldots, b_k)$ is a **permutation** of $(a_1, \ldots, a_k)$ if $[a_1, \ldots, a_k] = [b_1, \ldots, b_k]$, namely when it is obtained by re-ordering the elements $a_1, \ldots, a_n$ in any manner whatsoever. □

We have borrowed the terms sequence, collection, sharing, composition partition, and permutation from ordinary language, and we have transformed them into technical terms by assigning them a precise meaning. The meanings taken from a dictionary, cited here below, seem to us to allow one to recall the technical definitions which will be used from here on in this text:

Sequence: an ordered series of things or successive facts. A sequence of accidents. A series of successive frames.

Sharing: the act of distributing or spreading or apportioning. A distribution in shares. Sharing an inheritance.

Composition: the way in which something is made up of different parts, things, or members.

Collection: systematic collection of objects of the same species that have historical, artistic or scientific interest. Collection of coins, stamps, trading cards.

Partition: to divide into parts.

Permutation: to interchange one thing with another.

The notions just introduced turn out to be particularly useful for giving precise descriptions of various situations which arise frequently in combinatorial analysis. For example, sequences and collections allow for good descriptions of various types of lottery draws; the notions of sharing, composition and partition of numbers, not widely used in the combinatorial literature, turn out to be useful, for example, in the description of the distribution of objects (e.g., jelly beans to children, books to students, ...); partitions of sets allow one to discuss the division into subsets of sets of objects (e.g., formation of squads, distribution of a group of people to different tables at a restaurant, ...); finally, permutations arise in dealing with problems of shuffling or rearranging (e.g., anagrams or codes).

Example 1.21 A few examples of the objects defined in Definition 1.17.

- $(1, 1, 3, 2, 2)$ and $(1, 3, 1, 2, 2)$ are two distinct 5-*sequences* of I_4.
- The sequence $(\{1, 4\}, \emptyset, \{3\}, \{2\}, \emptyset)$ is a 5-*sharing* of I_4.
- The sequence $(1, 0, 1, 3)$ is a 4-*composition* of 5.
- $[1, 1, 3, 2, 2]$ and also $[1, 3, 2, 1, 2]$ represent the 5-*collection* of I_4 containing two 1, two 2, one 3 and zero 4.
- The collection $[\{2\}, \{1, 3, 4\}]$ is a 2-*partition* of I_4.

- The collection $[1, 1, 1, 2]$ is the only 4-*partition* of 5.
- The sequences $(2, 2, 3, 5)$, $(2, 2, 5, 3)$, $(2, 3, 2, 5)$, $(2, 5, 2, 3)$, $(2, 3, 5, 2)$, $(2, 5, 3, 2)$, $(3, 2, 2, 5)$, $(5, 2, 2, 3)$, $(3, 2, 5, 2)$, $(5, 2, 3, 2)$, $(3, 5, 2, 2)$, $(5, 3, 2, 2)$ are precisely the *permutations* of the 4-sequence $(2, 3, 2, 5)$ of I_5. $\qquad\square$

Example 1.22 (Extractions of numbers) Let $\ell, m \in \mathbb{N}_{\geq 1}$. The extractions of ℓ numbers of I_m can be either *ordered* or *non-ordered*: in the first case, to every extraction there corresponds an ℓ-*sequence* of I_m, while in the second case each extraction corresponds to an ℓ-*collection* of I_m. Depending on whether the extractions take place with or without replacement of the objects extracted, the resulting sequences and collections are with or without possible repetitions.

For example, we draw 3 numbers with replacement from I_4. The 3-sequence $(2, 4, 2)$ represents an ordered extraction of the numbers 2, 4, 2. The 3-collection $[2, 2, 4]$ represents an extraction in which, without keeping track of the order, 2 was extracted twice, and 4 once. $\qquad\square$

Example 1.23 (Distribution of objects) Let $\ell, m \in \mathbb{N}_{\geq 1}$. The distribution of ℓ jelly beans to m children, labeled respectively by I_ℓ and I_m, may be distinguished according to whether:

- *The jelly beans are all different (distinguishable) from one another:* in this case every distribution uniquely determines (and is determined by) the *m-sharing* $(C_1, \ldots, C_m)$ of the set I_ℓ of the jelly beans, where C_i is the subset of jelly beans given to date child i.
  For example, we distribute 3 jelly beans to two children. If the jelly beans are all distinguishable (different) then the distribution of jelly bean 1 to child 2 and the remaining jelly beans 2 and 3 to child 1 is described by the 2-sharing $(\{2, 3\}, \{1\})$ of I_3. Similarly, the 2-sharing $(\{1\}, \{2, 3\})$ describes the distribution of jelly bean 1 to child 1 and the remaining two jelly beans 2 and 3 to child 2.
- *The jelly beans are all alike (indistinguishable):* in this case every distribution uniquely determines (and is uniquely determined by) the *m-composition* $(n_1, \ldots, n_m)$ of ℓ, where n_i is the number of jelly beans given to child i.
  For example, if the jelly beans are indistinguishable and one gives two of them to child 1 and one of them to child 2, then the corresponding 2-composition of 3 is $(2, 1)$. Similarly, the 2-composition $(1, 2)$ of 3 describes the case in which child 1 receives a single jelly bean, while child 2 gets two of them. $\qquad\square$

Example 1.24 (Functions between finite sets) Let $\ell, m \in \mathbb{N}_{\geq 1}$. Any function $f : I_\ell \to I_m$ uniquely determines and is determined by each of the following objects:

1. The ℓ-*sequence* $(f(1), \ldots, f(\ell))$ of I_m;
2. The *m-sharing* $(f^{-1}(1), \ldots, f^{-1}(m))$ of I_ℓ.

The function f is:

- *Injective* if and only if the sequence $(f(1), \ldots, f(\ell))$ has no repetitions or equivalently, each component of the sharing $(f^{-1}(1), \ldots, f^{-1}(m))$ has cardinality at most 1; in this case one necessarily has $\ell \leq m$.

- *Surjective* if and only if in the sequence $(f(1), \ldots, f(\ell))$ every element of I_m appears at least once, or, equivalently, if and only if the empty set does never appear in the sharing $(f^{-1}(1), \ldots, f^{-1}(m))$ of the inverse images. In this case one necessarily has $m \leq \ell$.

The *permutations* of $(1, \ldots, n)$ can be identified with the one to one correspondences $I_n \to I_n$: each permutation $\mathfrak{a} = (a_1, \ldots, a_n)$ uniquely determines the one to one correspondences $f_{\mathfrak{a}} : I_n \to I_n$ defined by setting $f_{\mathfrak{a}}(i) = a_i$ for $i = 1, \ldots, n$. □

Remark 1.25 Every *permutation* of the sequence $(f(1), \ldots, f(\ell))$ determined by a function $f : I_\ell \to I_m$ again determines a function $I_\ell \to I_m$. The new function so obtained is injective (resp. surjective) if and only if the original function f was so.

Example 1.26 Let $f : I_5 \to I_3$ be the function defined by setting

$$f(1) = f(3) = 1, \quad f(2) = f(4) = f(5) = 2.$$

The sequence determined by the function f is the 5-sequence $(1, 2, 1, 2, 2)$ of I_3 whose components are the images of the elements of I_5 endowed with its natural ordering. The sharing determined by f is the 3-sharing $(\{1, 3\}, \{2, 4, 5\}, \emptyset)$ of the inverse images of the elements of I_3 endowed with its natural ordering (see Fig. 1.2). The function f is not injective: in the sequence $(1, 2, 1, 2, 2)$ there are repetitions, just as, analogously, in the sharing $(\{1, 3\}, \{2, 4, 5\}, \emptyset)$ there are components with cardinality greater than 1. The function f is not surjective: 3 does not appear in the 5-sequence $(1, 2, 1, 2, 2)$ of I_3, just as, equivalently, the empty set appears in the 3-sharing $(\{1, 3\}, \{2, 4, 5\}, \emptyset)$ of the inverse images of I_5. Every permutation of the sequence $(1, 2, 1, 2, 2)$ determines a function $I_5 \to I_3$ which is certainly neither surjective nor injective. □

Example 1.27 There is a sort of duality between the notions of sequence and sharing, and between the notions of collection and composition.

1. Consider the 8-sequence $(5, 5, 1, 1, 4, 2, 4, 1)$ of I_5; let $C_1 = \{3, 4, 8\}$ denote the set of positions where 1 appears, $C_2 = \{6\}$ the set of positions where 2 appears, $C_3 = \emptyset$ the set of positions where 3 appears, $C_4 = \{5, 7\}$ the set of positions where 4 appears, and finally let $C_5 = \{1, 2\}$ be the set of positions where 5 appears. The sequence $(C_1, \ldots, C_5)$ is a 5-sharing of I_8. Conversely, the 5-sharing $(\{3, 4, 8\}, \{6\}, \emptyset, \{5, 7\}, \{1, 2\})$ of I_8 determines the original 8-sequence of I_5 in

Fig. 1.2 Function $f : I_5 \to I_3$ and diagram of the corresponding 3-sharing associated to f

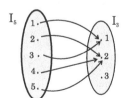

which 1 appears in positions three, four, and eight; 2 appears in position six, 3 does not appear, 4 appears in positions five and seven, and finally 5 appears in positions one and two.

2. Consider the 8-collection [5, 5, 1, 1, 4, 2, 4, 1] of I_5; let $k_1 = 3$ denote the number of repetitions of 1, $k_2 = 1$ the number of repetitions of 2, $k_3 = 0$ the number of repetitions of 3, $k_4 = 2$ the number of repetitions of 4, and $k_5 = 2$ the number of repetitions of 5. The sequence $(k_1, \ldots, k_5)$ is a 5-composition of 8. Conversely, the 5-composition $(3, 1, 0, 2, 2)$ of 8 determines the original 8-collection of I_5 in which 1 appears three times, 2 appears once, 3 does not appear, 4 appears twice and 5 appears twice. □

In general, one has the following elementary but illuminating result, which establishes a correspondence between the k-sequences of I_n and the n-sharings of I_k, as well as between k-collections of I_n and n-compositions of k. Figure 1.3 shows a simple example of these correspondences.

Theorem 1.28 (Sequences-sharings and collections-compositions) *Let $n, k \in \mathbb{N}$.*

1. *By associating to each n-sharing $(C_1, \ldots, C_n)$ of I_k the k-sequence $(a_1, \ldots, a_k)$ of I_n, defined by setting a_i equal to the index of the subset that contains i (equivalently $a_i = j$ if i lies in C_j) one obtains a one to one correspondence between the n-sharings of I_k and the k-sequences of I_n.*

2. *By associating to each n-composition $(k_1, \ldots, k_n)$ of k the k-collection of I_n*

$$[\underbrace{1, \ldots, 1}_{k_1}, \ldots, \underbrace{n, \ldots, n}_{k_n}]$$

formed by setting k_1 terms equal to $1, \ldots, k_n$ terms equal to n, one obtains a one to one correspondence between the n-compositions of k and the k-collections of I_n.

Proof. 1. Any sequence $(a_1, \ldots, a_k)$ of I_n arises uniquely from the sequence of sets $(C_1, \ldots, C_n)$ where $C_i = \{j \in I_k : a_j = i\}$ for $i = 1, \ldots, n$. The sets C_i's are clearly disjoint and their union is I_k: thus $(C_1, \ldots, C_n)$ is a n-sharing of I_k.

2. Any k-collection of I_n with k_1 terms equal to $1, \ldots, k_n$ terms equal to n arises uniquely from the n-composition $(k_1, \ldots, k_n)$ of k. □

Fig. 1.3 *Left*: diagram of a 3-sharing of I_5 and the corresponding 5-sequence of I_3. *Right*: a 3-composition of 8 and the diagram of the corresponding 8-collection of I_3

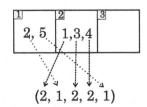

3 +	1		1	1
1 +		2		
4	3	3	3	3

☞ *Remark 1.29* In the first correspondence established by Theorem 1.28 we have that:

- The k-sequences of I_n in which there appear elements all of which are distinct correspond to the n-sharings of I_k formed by sets of cardinality ≤ 1;
- The k-sequences of I_n for which *all* the elements of I_n appear at least once correspond to the n-sharings of I_k formed by non-empty subsets.

Similarly, under the second correspondence of Theorem 1.28, we have that:

- The k-collections of I_n in which the elements appearing are all distinct correspond to the n-compositions of k by natural numbers ≤ 1;
- The k-collections of I_n in which *all* the elements of I_n appear at least once correspond to the n-compositions of k by natural numbers ≥ 1.

1.3 Fundamental Principles

In this section we will discuss two principles which will be of constant use in problems involving counting.

1.3.1 Multiplication Principle

It often happens that one must find the cardinality of certain sets whose structure generalizes that of the cartesian product.

Example 1.30 We extract two numbers, one after the other, from an urn containing 10 balls numbered from 1 to 10. We consider the set X of all the ordered pairs that can be obtained in this way. The set X consists of the pairs (a, b) with $a \in X_1 = \{1, \ldots, 10\}$ and $b \in X_2(a) = \{1, \ldots, 10\}\setminus\{a\}$. Bear in mind that the set of numbers from which one selects the second component depends on the selection made for the first component, but its cardinality is independent of that choice and always equal to 9. □

Definition 1.31 Let $k \in \mathbb{N}_{\geq 1}$. Given natural numbers $m_1, \ldots, m_k$, we define **conditional product of multiplicities** $(m_1, \ldots, m_k)$ to be a set X of k-sequences $(x_1, \ldots, x_k)$ where:

- x_1 belongs to a set X_1 having exactly m_1 elements;
- For every choice of x_1, the component x_2 lies in a set $X_2(x_1)$, which may depend on x_1, and in any case has exactly m_2 elements;
- ...
- For every choice of $x_1, \ldots, x_{k-1}$, the component x_k lies in a set $X_k(x_1, \ldots, x_{k-1})$, which may depend on $x_1, \ldots, x_{k-1}$, and in any case has exactly m_k elements. □

Example 1.32   1. The set X of Example 1.30 is a conditional product of multiplicities $(10, 9)$.

2. The cartesian product $X_1 \times \cdots \times X_k$ of sets of respective cardinalities $m_1, \ldots,$
m_k is a particular case of a conditional product with multiplicities $(m_1, \ldots, m_k)$;
in this case the set X_i from which the i-th component is chosen does not depend
on the choices of the previous $i - 1$ components.

3. The set X of the order pairs of integers between 1 and 20 with neither both
even nor both odd is a conditional product of multiplicities $(20, 10)$: indeed, X
consists of the pairs (x_1, x_2) with $x_1 \in X_1 = I_{20}$ and $x_2 \in X_2(x_1)$ where, if x_1
is even, then $X_2(x_1) = \{1, 3, 5, 7, 9, 11, 13, 15, 17, 19\}$, while if x_1 is odd, then
$X_2(x_1) = \{2, 4, 6, 8, 10, 12, 14, 16, 18, 20\}$. The set X_1 has cardinality 20, while
$X_2(x_1)$ has cardinality 10 for every choice of $x_1 \in X_1$.   □

Remark 1.33 If X is a conditional product with multiplicities $(m_1, m_2, \ldots, m_k)$ as
in Definition 1.31, then for every $\mathbf{x_1} \in X_1$ the set

$$\{(x_2, \ldots, x_k) : (\mathbf{x_1}, x_2, \ldots, x_k) \in X\}$$

is a conditional product with multiplicities sequence $(m_2, \ldots, m_k)$.

The number of elements of a conditional product depends only on its multiplicities.

Theorem 1.34 (Multiplication Principle) *A conditional product of multiplicities*
$(m_1, \ldots, m_k)$ *has* $m_1 \times \cdots \times m_k$ *elements.*

Proof. We give the proof via induction on k. The case $k = 1$ is clear, since a conditional product of multiplicity m_1 is just a set with m_1 elements. Now let $k > 1$ and
suppose that the statement hold for $k - 1$. Let X be a conditional product of multiplicities $(m_1, \ldots, m_k)$. We use $\{x_1, \ldots, x_{m_1}\}$ to denote the set from which we choose
the first component of the sequences constituting the conditional product X. For each
$i = 1, \ldots, m_1$, the sequences of X that start with x_i are in number equal to the cardinality of a conditional product of multiplicities $(m_2, \ldots, m_k)$ (see Remark 1.33). By
the inductive hypothesis, for each $i = 1, \ldots, m_1$ there are therefore $m_2 \times \cdots \times m_k$
sequences in X that start with x_i. The set X is the union of the m_1 disjoint subsets
formed by the sequences that start with x_i, $i = 1, \ldots, m_1$. By Proposition 1.14 one
concludes that $|X| = m_1 \times m_2 \times \cdots \times m_k$.   □

Example 1.35 As a byproduct of the Multiplication Principle 1.34 it follows immediately that in particular the cardinality of a cartesian product $X_1 \times \cdots \times X_k$ of finite
sets is equal to $|X_1| \times \cdots \times |X_k|$.   □

Remark 1.36 A set A is in one to one correspondence with a conditional product
of multiplicities $(m_1, \ldots, m_k)$, and so has $m_1 \times \cdots \times m_k$ elements, if its elements
are the results of a procedure which is decomposable into k successive phases, for
which:

1. The first phase has m_1 different possible outcomes, and, no matter what the outcomes of the first i phases are, $1 \leq i \leq k - 1$, there are m_{i+1} different possible outcomes for the $(i + 1)$-st.
2. Distinct partial outcomes of the k phases lead to distinct elements of A (or, equivalently, every element of A uniquely determines the partial outcomes of the k phases that determine that element).

Indeed, Point 1 guarantees the existence of a surjective map between a conditional product with multiplicities $(m_1, \ldots, m_k)$ and the set A, while Point 2 ensures the injectivity of that map.

Example 1.37 Count the number of ways in which it is possible to choose a president and a vice president from an assembly of 15 people.
Solution. Every possible choice may be obtained in *two phases*: in the first phase one chooses the president, and in the second phase, the vice president. The first phase has 15 possible outcomes, while the second has 14. Given an arbitrary pair of people, chosen respectively as president and vice president, it is clear who was chosen in the first phase (the president) and who was chosen in the second (the vice president). Therefore, by Remark 1.36 the set of possible choices is in one to one correspondence with a conditional product of multiplicities $(15, 14)$ and so by the Multiplication Principle 1.34 that set has $15 \times 14 = 210$ elements.   □

Example 1.38 Three balls are successively extracted (without replacement) from an urn containing 10 balls numbered from 1 to 10. The extraction may be naturally divided into three phases:

(1st extraction, 2nd extraction, 3rd extraction).

The two conditions of Remark 1.36 hold; therefore the set of possible extractions is in one to one correspondence with a conditional product with multiplicities $(10, 9, 8)$. Indeed, for each of the 10 possible outcomes of the first extraction, there remain 9 possible outcomes for the second and 8 for the third: the possible outcomes for the extraction taking place depends on the outcomes of the previous extractions, but the number of its possible outcomes does not. By the Multiplication Principle 1.34 the set of triples obtained has $10 \times 9 \times 8$ elements.   □

Proposition 1.39 *If X is a finite set, then the set $\mathscr{P}(X)$ of parts of X has cardinality*

$$|\mathscr{P}(X)| = 2^{|X|}.$$

Proof. Set $|X| = n$, and label the elements of the set X with I_n. We construct a subset of I_n by deciding whether or not to include each element i in the subset. This is a process made up of n phases (one phase for each element of I_n), and at each phase we have two possible outcomes (either to include the element or to exclude it).

Obviously changing the outcome at any phase leads to the construction of a different subset of I_n. Hence by Remark 1.36 the set $\mathscr{P}(I_n)$ of subsets of I_n is in one to one correspondence with a conditional product of multiplicities $(\underbrace{2, 2, \ldots, 2}_{n})$ and thus by the Multiplication Principle 1.34 has $2^n = 2^{|X|}$ elements.   □

☞ A first source of errors in combinatorics is due to the misuse of the Multiplication Principle 1.34 by overlooking the importance of checking both conditions stated in Remark 1.36: the following two examples show how such an error can occur.

Example 1.40 A coin is tossed; if it comes up Heads a die having 6 faces numbered from 1 to 6 is cast; if the coin comes up Tails, then the coin is tossed again. Let us consider the set of 2-sequences that describe the possible outcomes. Each element of this set is obtained in two phases: first the tossing of the coin, and then the tossing of, according to the pertinent case, the die or the coin. The number of outcomes of the second toss varies according to the outcome of the first, and so this subdivision into phases does not respect the first condition of Remark 1.36. Therefore, the subdivision in phases under consideration does not determine a one to one correspondence with a conditional product.

In this case we can divide the set of 2-sequences that describe the possible outcomes into two disjoint subsets: the 2-sequences whose first element is Heads (and thus whose second element is an integer between 1 and 6) consisting of the pairs in $\{H\} \times \{1, 2, 3, 4, 5, 6\}$, and the 2-sequences whose first element is Tails (and whose second element is "Heads" or "Tails") consisting of the pairs in $\{T\} \times \{H, T\}$. The first set has 6 elements, the second has 2, and so there are in all $6 + 2 = 8$ possible overall outcomes.   □

Example 1.41 We wish to form a committee of two people, containing at least one woman, with choices to be made among 5 women and 7 men. One might use the following procedure to create a committee: first, choose a woman, and then choose a person among the remaining people. This procedure might seem to suggest that the committees sought form a conditional product of multiplicities $(5, 11)$. But be careful: this is NOT the case!

Indeed, from a committee consisting of Ellen and Paula it is not possible to recover the choices made in the two phases into which we have divided the procedure. We could, in fact, have first selected Paula from among the 5 women, and then have selected Ellen from the remaining 11 people, or, vice versa, we could have first chosen Ellen from among the 5 women and then Paula from among the remaining 11 people. In other words, the procedure followed does not respect the second condition of Remark 1.36.

We could however proceed by distinguishing two cases: the committees containing exactly one woman, and those consisting of two women. These are two disjoint subsets whose union gives the set of all possible committees. The committees containing exactly one woman can be formed by first choosing a woman and then a man: this corresponds to a conditional product of multiplicities $(5, 7)$ and so gives 35 such committees. The committees with two women are 10 in number: indeed, the ordered pairs of women form a conditional product of multiplicities $(5, 4)$ and so by the Multiplication Principle 1.34 there are 20 such ordered pairs. However, every committee with two women determines two different ordered pairs, and so the number of committees with only women is $20/2 = 10$ (see Sect. 1.3.2 below). In all there are therefore 45 possible committees with at least one woman.   □

We now consider some more involved examples of a correct application of the Multiplication Principle 1.34.

Example 1.42 Count the number of automobiles that can be registered using the license plates used in Italy, which consist of two letters followed by three numbers followed by another pair of letters. How many of these license plates are there in which there are neither repeated letters or repeated numbers?
Solution. Let L denote the set of the 26 letters of the English alphabet, and let D denote the set of digits $\{0, 1, 2, \ldots, 9\}$. We must calculate the cardinality of the product $L \times L \times D \times D \times D \times L \times L$. This is a conditional product of multiplicities $(26, 26, 10, 10, 10, 26, 26)$ and so by the Multiplication Principle 1.34 we have

$$|L \times L \times D \times D \times D \times L \times L| = 26^4 \times 10^3 = 456\,976\,000.$$

The license plates without repetitions form a conditional product of multiplicities $(26, 25, 10, 9, 8, 24, 23)$, and thus there are

$$26 \times 25 \times 10 \times 9 \times 8 \times 24 \times 23 = 258\,336\,000$$

of them.   □

Example 1.43 In how many ways can one form a three letter word using the alphabet $\{a, b, c, d, e, f\}$:

1. With possible repetitions of the letters?
2. Without repetitions of the letters?
3. Without repetitions and containing the letter e?
4. With possible repetitions and containing the letter e?

Solution. 1. Let $X = \{a, b, c, d, e, f\}$. The possible words are in one to one correspondence with the cartesian product $X \times X \times X$, which is a conditional product with multiplicities $(6, 6, 6)$; thus, by the Multiplication Principle 1.34 we have in all $6 \times 6 \times 6 = 216$ different words of three letters.
2. The possible words without repetitions are in one to one correspondence with a conditional product of multiplicities $(6, 5, 4)$: indeed, there are 6 choices for the first letter, 5 for the second, and 4 for the third. Thus, by the Multiplication Principle 1.34, there are in all $6 \times 5 \times 4 = 120$ three letter words without repetitions.
3. We can use the following procedure: in a first phase decide the position of the letter e (there are 3 possibilities); in the second phase choose a letter different from e to be placed in the first free position from the left (there are 5 possibilities); finally choose a letter different from the two used in the preceding steps (there are 4 possibilities) and

put it in the remaining position. The set of the words specified by this third condition is thereby placed into one to one correspondence with a conditional product whose multiplicities are $(3, 4, 5)$. The Multiplication Principle 1.34 then tells us that there are $3 \times 5 \times 4 = 60$ words of three distinct letters and containing the letter e.

4. Let Ω be the set of all possible three letter words that can be formed using the letters a, b, c, d, e, f and A be the subset of those words containing the letter e. By Point 1 we know that $|\Omega| = 6^3$ and $|A^c| = 5^3$; in fact, A^c is the set of all possible three letter words that can be formed using the letters a, b, c, d, f. Hence, $|A| = 6^3 - 5^3 = 91$. $\square$

Example 1.44 On a bookshelf there are five different books by Spanish authors, six different books by French authors, and eight different books by English authors. In how many ways is it possible to choose two books by authors of different nationalities?

Solution. The set of possible choices of a book by a Spaniard and one by a French author, is a conditional product of multiplicities $(5, 6)$ and so by the Multiplication Principle 1.34 has $5 \times 6 = 30$ elements. Similarly, a book by a Spanish author and one by an English author can be chosen in $5 \times 8 = 40$ ways, while a book by a French author and one by an English author can be chosen in $6 \times 8 = 48$ ways. The sets formed by these three types of choices are disjoint, and so there are $30 + 40 + 48 = 118$ ways in all to choose two books by authors of different nationalities. $\square$

1.3.2 Partition and Division Principles

We illustrate here some other fundamental tools for computing the cardinality of a set.

Theorem 1.45 (Partition Principle) *Let $f : X \to Y$ be a surjective function between two finite sets X and Y. Then*

$$|X| = \sum_{y \in Y} |f^{-1}(y)|. \tag{1.45.a}$$

Proof. The family $\{f^{-1}(y) : y \in Y\}$ is a partition of X. Indeed the sets of the family are pairwise disjoint and, of course, $X = \bigcup_{y \in Y} f^{-1}(y)$. The conclusion is straightforward. $\square$

Grouping together the counter images with the same cardinality one gets:

Corollary 1.46 *Let X and Y be two finite sets and $[Y_1, \ldots, Y_k]$ a k-partition of Y. Suppose that, for $i = 1, \ldots, k$, each element y_i of Y_i corresponds to m_i elements of X via a function $f : X \to Y$. Then $|X| = m_1|Y_1| + \cdots + m_k|Y_k|$.*

Proof. Grouping together the elements of the sets $Y_1, \ldots, Y_k$ in (1.45.a) we get

$$|X| = \sum_{y \in Y} |f^{-1}(y)| = \sum_{i=1}^{k} \left(\sum_{y \in Y_i} |f^{-1}(y)| \right)$$

$$= \sum_{i=1}^{k} m_i |Y_i|.$$

$\square$

The celebrated Division Principle is an immediate consequence of the Partition Principle 1.45. It simply states that, if you need 5 g of ground coffee to prepare a traditional 3-cup Moka pot, then with a packet of 250 g of ground coffee you are allowed to prepare fifty 3-cup Moka pots.

Theorem 1.47 (Division Principle) *Let X and Y be two finite sets. Suppose that each element y of Y corresponds to m elements of X via a function $f : X \to Y$. Then $|Y| = |X|/m$.*

Proof. We directly apply the Partition Principle 1.45: from (1.45.a) we get

$$|X| = \sum_{y \in Y} |f^{-1}(y)| = m|Y|.$$

$\square$

Figure 1.4 exemplifies the Division Principle 1.47. Let us examine some other instances of its possible applications.

Example 1.48 We wish to count the number of committees of three people chosen from a group of 27. Let X be the set of 3-sequences of I_{27} without repeated elements and Y the set of committees of three people. Each committee corresponds to 6 different 3-sequences without repetitions of I_{27}. By the Multiplication Principle 1.34 the 3-sequences without repetitions of I_{27} are $27 \times 26 \times 25$. Then, by the Division Principle 1.47 there are $(27 \times 26 \times 25)/6$ different committees. $\square$

Example 1.49 How many hands of two cards can be formed from a deck of 52 cards? *Solution.* Each hand $[a, b]$ corresponds to the two 2-sequences (a, b), and (b, a). By the Multiplication Principle 1.34 the number of 2-sequences of distinct cards is

Fig. 1.4 An application of the Division Principle

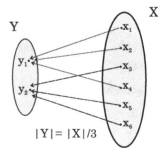

Fig. 1.5 An arrangement and the corresponding 8-sequences

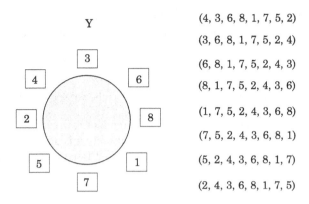

(4, 3, 6, 8, 1, 7, 5, 2)

(3, 6, 8, 1, 7, 5, 2, 4)

(6, 8, 1, 7, 5, 2, 4, 3)

(8, 1, 7, 5, 2, 4, 3, 6)

(1, 7, 5, 2, 4, 3, 6, 8)

(7, 5, 2, 4, 3, 6, 8, 1)

(5, 2, 4, 3, 6, 8, 1, 7)

(2, 4, 3, 6, 8, 1, 7, 5)

52×51, and therefore by the Division Principle 1.47 the number of two card hands is $26 \times 51 = 326$. □

Example 1.50 Determine the number of anagrams of the word *ALA*.
Solution. Here it is a question of calculating the permutations of the sequence (A, L, A). If we artificially distinguish the two *A*'s by labeling them A_1 and A_2, then to each anagram of the word *ALA* there are associated two anagrams of the word $A_1 L A_2$. The anagrams of $A_1 L A_2$ form a conditional product with multiplicities $(3, 2, 1)$, and so in view of the Multiplication Principle 1.34 there are a total of 6 such anagrams. Then, by the Division Principle 1.47 one concludes that *ALA* has $6/2 = 3$ anagrams. □

Example 1.51 In how many ways can one arrange 8 people around a round table?
Solution. Let *Y* be the set of arrangements of the 8 people around the table. To each arrangement there are associated the eight 8-sequences obtained by proceeding clockwise starting from any one of the people seated at the table (see, e.g., Fig. 1.5). By the Multiplication Principle 1.34 the set of 8-sequences of the eight people has $8 \times 7 \times \cdots \times 2 \times 1$ elements. Therefore, by the Division Principle 1.47 one has

$$|Y| = \frac{8 \times 7 \times \cdots \times 2 \times 1}{8} = 7 \times \cdots \times 2 \times 1.$$ □

1.4 Sample Spaces and Uniform Probability

Counting problems are closely connected to the frequency of certain related phenomena.

Let Ω be a set. A **random experiment with sample space** Ω is a procedure which randomly determines the choice of an element of Ω. The elements of Ω are called **elementary events** or **outcomes** of the experiment; more generally, any subset of Ω defines (and is defined to be) an **event** of the experiment.

Example 1.52 Tossing a die is a procedure for choosing an integer between 1 and 6 by chance. It is therefore a random experiment with sample space

$$\Omega = \{1, 2, 3, 4, 5, 6\}.$$

The event "the die comes up even" is the subset $A = \{2, 4, 6\}$.   □

☞   Often the sample space of a random experiment is implicitly understood. For example, when one tosses a coin, usually it is understood that one is dealing with a procedure for choosing either Heads or Tails, in other words, that the sample space is $\Omega = \{H, T\}$.

Throughout the text we will, for simplicity of notation, identify an elementary event $\omega \in \Omega$ with the event $\{\omega\} \subseteq \Omega$ consisting of that single element.

Definition 1.53 Let Ω be a finite set. The **uniform probability** on Ω is the function

$$P : \mathscr{P}(\Omega) \rightarrow [0, 1]$$

which associates to each subset A of Ω the number

$$P(A) = \frac{|A|}{|\Omega|}.$$ □

Thus the uniform probability of an event is the quotient of the number of *favorable cases* (the number of elements lying in the event) by the number of *possible cases* (the number of elements in the sample space). Two events A and B (elementary or not) are said to be equi-probable if $P(A) = P(B)$.

Theorem 1.54 *Let A, B be two events of a finite sample space Ω which is endowed with the uniform probability P. Then:*

1. For each elementary event $\omega \in \Omega$ one has

$$P(\omega) = \frac{1}{|\Omega|};$$

2. $P(\emptyset) = 0$ and $P(\Omega) = 1$;
3. $P(A \cup B) = P(A) + P(B) - P(A \cap B)$;
4. If A and B are disjoint, then $P(A \cup B) = P(A) + P(B)$;
5. $P(A^c) = 1 - P(A)$.

Proof. Points 1 and 2 are immediate consequences of the definition of the uniform probability.
3. By definition one has

$$P(A \cup B) = \frac{|A \cup B|}{|\Omega|} = \frac{|A| + |B| - |A \cap B|}{|\Omega|} = P(A) + P(B) - P(A \cap B).$$

4. This follows immediately from Points 2 and 3.
5. Since A^c and A are disjoint, by Point 4 one has

$$1 = P(\Omega) = P(A^c \cup A) = P(A^c) + P(A).$$ □

Remark 1.55 The uniform probability furnishes a correct indication of the frequency with which a certain event takes place only when the sample space consists of equi-probable elements. For example, the outcomes of tossing a fair die can be considered equi-probable. On the other hand, an automobile race or a football game generally do not have equi-probable outcomes. When one encounters an exercise in probability to be resolved using the uniform probability, the first thing to be done is to formalize the problem as an experiment with a sample space made up of equi-probable elements.

Example 1.56 One extracts two integers between 1 and 50 from an urn. Calculate the probability that the larger number is equal to twice the smaller.
Solution. This may be viewed as an experiment with sample space Ω_1 consisting of the 2-sequences of I_{50} without repetitions, or as one having sample space Ω_2 consisting of the 2-collections of I_{50} without repetitions. In the first instance we bear in mind the order in which the two choices are made, while in the second approach the order is irrelevant. If the choices are made randomly, the elementary events of both the spaces should be considered equi-probable. In the first case the event in which we are interested is

$$A_1 = \{(x, 2x) : 1 \le x \le 25\} \cup \{(2x, x) : 1 \le x \le 25\}.$$

Clearly A_1 has $25 + 25 = 50$ elements, while by the Multiplication Principle 1.34 the space Ω_1 has 50×49 elements; therefore, the probability sought is

$$P(A_1) = \frac{|A_1|}{|\Omega_1|} = \frac{50}{50 \times 49} = \frac{1}{49}.$$

In the second instance, the event in which we are interested is

$$A_2 = \{[x, 2x] : 1 \le x \le 25\}.$$

Clearly A_2 has 25 elements, while Ω_2 has $(50 \times 49)/2$ elements (see Example 1.49); therefore the probability sought is

$$P(A_2) = \frac{|A_2|}{|\Omega_2|} = \frac{25}{(50 \times 49)/2} = \frac{1}{49}.$$

Obviously the probability that in this extraction the larger number is twice the smaller number does not depend on the sample space chosen.   □

Example 1.57 One extracts (with replacement) two integers between 1 and 50 from an urn. Calculate the probability that the two numbers extracted are equal.

Solution. This can be viewed as a random experiment with sample space the set Ω_1 of 2-sequences of I_{50}, or as one with sample space Ω_2 consisting of the 2-collections of I_{50}. While in the former case the elements of Ω_1 should be considered equi-probable, in the second case the elements of Ω_2 are not equi-probable: indeed, the 2-collection [1, 1] is realized only if the 1 is chosen in both the first and second extractions, whereas the 2-collection [1, 2] may be realized both when the first extraction gives 1 and the second yields 2, or also vice versa. Since we wish to use the uniform probability, we choose the sample space Ω_1. The requested probability is then

$$\frac{|\{(x, x); 1 \leq x \leq 50\}|}{|\Omega_1|} = \frac{50}{50^2} = \frac{1}{50}. \qquad \square$$

Example 1.58 What is the probability that a four digit number contains one or more repeated digits?

Solution. The four digit numbers are the integers from 1000 to 9999, and thus there are 9000 such numbers in all. The numbers without repeated digits form a conditional product of multiplicities $(9, 9, 8, 7)$. Indeed, one notes that the first digit (the thousands digit) must be between 1 and 9, while the others are allowed to be between 0 and 9, but different from those already chosen. Hence, the set of such numbers that contain one or more repeated digits has cardinality

$$9000 - 9 \times 9 \times 8 \times 7 = 9000 - 4536 = 4464.$$

The probability requested is therefore equal to $\dfrac{4464}{9000} = 0.496.$ $\qquad \square$

1.5 Problems

Problem 1.1 Prove the identities of Proposition 1.7 without using the characteristic functions.

Problem 1.2 A store carries 8 different brands of pants. For every brand there are 10 sizes, 6 lengths, and 4 colors. How many different types of pants are there in the store?

Problem 1.3 How many four letter words can be formed with an alphabet of 26 letters? How many of these are without a repeated letter?

Problem 1.4 Give 8 different books in English, 7 different books in French, and 5 different books in German: in how many ways can one choose three books, one for each language?

Problem 1.5 In how many ways can one pick two cards from a deck of 52 playing cards so that:

1. The first card is an ace and the second is not a queen?
2. One is an ace and the other is not a queen?
3. The first card is a spade and the second is not a queen?
4. One is a spade and the other is not a queen?

Problem 1.6 In how many ways can one toss two dice, one red and one green, so as to obtain a sum divisible by 3?

Problem 1.7 Consider the set X of 5 digit numbers, in other words, the numbers between 10000 and 99999.

1. Determine the cardinality of X.
2. How many even numbers are there in X?
3. In how many numbers of X does the digit 3 appear exactly once?
4. How many 5 digit palindromic numbers are there (in other words, how many 5 digit numbers are there which remain unchanged if one inverts the order of its digits, e.g., 15251)?

Problem 1.8 What is the probability that the two top cards in a deck of 52 cards do not form a pair, that is, are not two cards with the same value (from different suits)?

Problem 1.9 A message is spread in a group of 10 people in the following way: the first person telephones a second who in turn telephones a third, and so on in a random way. A person of the group can pass the message to any other member of the group, except the person whose call has just been received.

1. In how many different ways can the message be spread via three phone calls? And via n calls?
2. What is the probability that A receives the third call, if it is known that A made the initial call?
3. What is the probability that A receives the third call, if it is known that A did not make the initial call?

Problem 1.10 How many three letter words without repetition of letters can be made by using the letters a, b, c, d, e, f in such a way that either the letter e or the letter f or both appear?

Problem 1.11 What is the probability that a natural number between 1 and 10000 contains both the digits 8 and 9 exactly once?

Problem 1.12 An assembly of 20 people must vote by raising hands to choose a president from among 7 candidates A, B, C, D, E, F, G.

1. In how many different ways can the votes of the assembly be cast?
2. How many outcomes of the voting are there in which A and D receive exactly one vote?

Problem 1.13 How many 4 digit numbers divisible by 4 may be formed using the digits 1, 2, 3, 4, 5 (with possible repetitions)?

Problem 1.14 In how many ways can one place two identical rooks in the same row or column of an 8×8 chessboard? What is the result in the case of a chessboard with n rows and m columns?

Problem 1.15 In how many ways can one place two identical queens on an 8×8 chessboard in such a way that the two queens do not lie in the same row, column, or diagonal?

Problem 1.16 In how many ways can one invite friends (at least one!) chosen from 10 people?

Problem 1.17 Following the rules of 'Checkers', in how many ways can one put a white pawn and a black pawn in two black squares of a checkerboard in such a way that the white pawn can jump the black one? Recall that a pawn jumps diagonally, and jumps over the pawn to be taken, and also that pawns can not move backwards.

Chapter 2
Counting Sequences and Collections

Abstract In this chapter we count sequences and sharings, collections and compositions, furnishing many applications and examples. Factorials and binomial coefficients are, on the one hand, indispensable tools for such counting problems, and, on the other hand, their combinatorial interpretation gives a valuable contribution in suggesting and proving many useful identities both concerning sums or alternating sums of binomials and their products.

2.1 Sequences and Collections Having Distinct Elements

In this section we shall learn how to count sequences and collections that do not contain repeated elements.

2.1.1 Sequences Without Repetitions

The notion of the *factorial* of a natural number arises very frequently in counting problems.

Definition 2.1 Let $n \geq 0$ be a natural number. The **factorial of** n is

$$n! = \begin{cases} n \times (n-1) \times \cdots \times 2 \times 1 & \text{for } n \geq 1, \\ 1 & \text{for } n = 0. \end{cases} \qquad \square$$

Remark 2.2 (*Stirling's formula*) Calculating $n!$ for large values of n is rather tedious. However one can prove (see Corollary 12.73) that $n!$ is asymptotic to $\sqrt{2\pi n}(n/e)^n$ for $n \to \infty$, that is,

© Springer International Publishing Switzerland 2016

C. Mariconda and A. Tonolo, *Discrete Calculus*,

UNITEXT - La Matematica per il 3+2 103, DOI 10.1007/978-3-319-03038-8_2

☞

$$\lim_{n \to \infty} \frac{n!}{\sqrt{2\pi n}\,(n/e)^n} = 1.$$

The following table shows the accuracy of this approximation, called **Stirling**[1] **approximation**, already for rather small values of n.

n	$n!$	$\approx \sqrt{2\pi n}(n/e)^n$	$\frac{n!-\sqrt{2\pi n}(n/e)^n}{n!}$ in percent
1	1	0.92	7.79 %
2	2	1.92	4.05 %
3	6	5.84	2.73 %
4	24	23.51	2.06 %
5	120	118.02	1.65 %
6	720	710.08	1.38 %
7	5 040	4980.4	1.18 %
8	40 320	39902.4	1.04 %
9	362 880	359536.87	0.92 %
10	3.6288×10^6	3.5987×10^6	0.83 %
11	3.99168×10^7	3.9616×10^7	0.75 %
12	$\approx 4.79002 \times 10^8$	4.7569×10^8	0.69 %
13	$\approx 6.22702 \times 10^9$	6.1872×10^9	0.64 %
14	$\approx 8.7178 \times 10^{10}$	8.6661×10^{10}	0.59 %
15	$\approx 1.3077 \times 10^{12}$	1.3004×10^{12}	0.55 %
16	$\approx 2.0923 \times 10^{13}$	2.0814×10^{13}	0.52 %
17	$\approx 3.5569 \times 10^{14}$	3.5395×10^{14}	0.49 %
18	$\approx 6.4024 \times 10^{15}$	6.3728×10^{15}	0.46 %
19	$\approx 1.2165 \times 10^{17}$	1.2111×10^{17}	0.44 %
20	$\approx 2.4329 \times 10^{18}$	2.4228×10^{18}	0.42 %

Example 2.3 Using Stirling's formula it is possible to obtain an estimate of the number of decimal digits in 100!. The number of decimal digits of a natural number k is equal to $[\log_{10} k]+1$ where $[x]$ denotes the *integer part* or *floor* of x. By Stirling's formula[2] one has

---

[1] James Stirling (1692–1770). We Venetians can hardly neglect to mention that he was also known as "*the Venetian*" after his sojourn in Venice from 1715 to 1724. Stirling had to flee from Venice with his life at risk after being accused of having stolen the secret method used for the industrial production of Murano glass.

[2] Here one uses the fact that if a_n and b_n are two sequences diverging to $+\infty$ with $a_n \sim b_n$ ($\sim$ stands for "asymptotic to") for $n \to +\infty$, then $\log_{10} a_n \sim \log_{10} b_n$. One should bear in mind that in general for a function f if $a_n \sim b_n$ it does not necessarily follow that $f(a_n) \sim f(b_n)$ for $n \to +\infty$.

$$\log_{10} 100! \approx \log_{10}(\sqrt{2\pi \times 100}(100/e)^{100}) =$$

$$= \frac{1}{2}(\log_{10}(2\pi) + 2) + 100 \times \log_{10}(100/e)$$

$$= \frac{1}{2}\log_{10}(2\pi) + 1 + 200 - 100\log_{10} e = 157.97\ldots.$$

Thus the number of decimal digits in 100! is approximately equal to 158. One can in fact verify that 100! has exactly 158 decimal digits.   □

How many sequences of prescribed length with terms from a given finite set, and with no repetitions, are there?

Definition 2.4 Let $n, k \in \mathbb{N}$. We use $S(n, k)$ to denote the **number of k-sequences of I_n without repetitions**, that is, the number of k-sequences of distinct elements of I_n.   □

Remark 2.5 In Theorem 1.28 we have proved the existence of a one to one correspondence between the k-sequences of I_n and the n-sharings of I_k, associating to each k-sequence of I_n the n-sharing of I_k whose elements are the subsets of the positions of $1, 2, \ldots, n$ in the k-sequence. Therefore $S(n, k)$ is also the **number of n-sharings** $(C_1, \ldots, C_n)$ of I_k **with at most one element** in each $C_i, i = 1, \ldots, n$.

The value of $S(n, k)$ is easily found using the notion of conditional product and the Multiplication Principle 1.34 introduced in the preceding chapter.

Theorem 2.6 *Let $n, k \in \mathbb{N}$. Then*

$$S(n, k) = \begin{cases} \dfrac{n!}{(n-k)!} & \text{if } k \leq n, \\ 0 & \text{otherwise.} \end{cases}$$

In particular $S(n, 0) = 1$ and $S(n, n) = n!$ for each $n \in \mathbb{N}$.

Proof. If $k > n$, there are no k-sequences of I_n without repetitions, and so $S(n, k) = 0$. Now let $k \leq n$. If $k = 0$ one has $S(n, 0) = 1$: indeed the empty sequence is the unique 0-sequence of I_n. Suppose that $k \geq 1$. The set of such k-sequences of I_n is a conditional product with multiplicities $(n, n-1, \ldots, n-(k-1))$: indeed, we can choose any one of the n elements of I_n as the first component of the sequence; for the second component we have $n-1$ choices, since we can not repeat the choice made for the first component; we have $n-2$ choices for the third component, and so on. Thus by the Multiplication Principle 1.34

$$S(n, k) = n \times (n-1) \times \cdots \times (n-(k-1)) = \frac{n!}{(n-k)!}.$$ □

Corollary 2.7 (Number of injective functions) *Let $k, n \in \mathbb{N}_{\geq 1}$. The number of injective functions $I_k \to I_n$ equals $S(n, k)$.*

Proof. In Example 1.24 we have seen how every function $f : I_k \to I_n$ is determined by a k-sequence of I_n; the function f is injective if and only if the associated sequence has no repetitions. The result follows from Theorem 2.6. □

Example 2.8 In how many ways is it possible to make a list of n people? What is the probability that Mr. Caruso is in the second position on the list?

Solution. Such a list is just an n-sequence without repetitions of the set of n people: hence there are $S(n, n) = n!$ possible lists. Therefore

$$P(\text{Caruso second}) = \frac{\text{number of lists with Caruso second}}{\text{total number of lists}}.$$

Every list with Caruso second is determined by the sequence of the other $n - 1$ people making up the list; thus there are $(n - 1)!$ such sublists, and so the desired probability is $(n - 1)!/n! = 1/n$. □

We end this section giving a combinatorial proof of a nice identity:

Proposition 2.9 *For each natural number $n \geq 2$ we have*

$$n! = 1 + \sum_{k=1}^{n-1} k\, k!.$$

Proof. The number $n! = S(n, n)$ counts the n-sequences without repetitions of I_n, i.e., the permutations of $(1, \ldots, n)$. Let us count in a different manner the cardinality of the same set of n-sequences. Let $A_0 = \{(1, \ldots, n)\}$ and A_k, $1 \leq k \leq n - 1$, be the set of permutations $(a_1, \ldots, a_n)$ of $(1, \ldots, n)$ such that $a_i = i$ for $i > k + 1$ and $a_{k+1} \neq k + 1$. Clearly $|A_0| = 1$. Let us compute $|A_k|$ for $1 \leq k \leq n - 1$. If $(a_1, \ldots, a_n)$ belongs to A_k, then $(a_{k+2}, \ldots, a_n) = (k+2, \ldots, n)$, $a_{k+1} \in \{1, \ldots, k\}$, and $(a_1, \ldots, a_k)$ is a permutation of the elements of $\{1, \ldots, k+1\}\setminus\{a_{k+1}\}$. There are k possibilities for the choice of a_{k+1} and $k!$ possibilities for the choice of $(a_1, \ldots, a_k)$; by the Multiplication Principle 1.34, one gets $|A_k| = kk!$ for each $1 \leq k \leq n - 1$. Since the sets A_k, $k = 0, \ldots, n - 1$, form a partition of the set of the n-sequences without repetitions of I_n, we get $n! = |A_0| + \sum_{k=1}^{n-1} |A_k| = 1 + \sum_{k=1}^{n-1} k\, k!$. □

2.1.2 Collections Without Repetitions, or Subsets

In addition to the notion of factorials, the concept of the *binomial coefficient* is also very important.

Definition 2.10 Let $k, n \in \mathbb{N}$. The **binomial coefficient** "n over k" is the number

$$\binom{n}{k} = \begin{cases} \dfrac{n!}{k!(n-k)!} & \text{if } 0 \le k \le n, \\ 0 & \text{otherwise.} \end{cases}$$

Remark 2.11 Note that if $n \in \mathbb{N}$ and $k \ge 1$ one has

$$\binom{n}{k} = \frac{n(n-1)\cdots(n-k+1)}{k!}; \tag{2.11.a}$$

in the numerator k consecutive factors from n to $n - k + 1$ appear. Bear in mind that for each $n \in \mathbb{N}$

$$\binom{n}{0} = 1, \quad \binom{n}{1} = n, \quad \binom{n}{n} = 1, \quad \binom{n}{n-1} = n.$$

☞ We shall soon see that $\binom{n}{k}$ is an integer for each pair $k, n \in \mathbb{N}$.

The following symmetry formula is straightforward, but very important.

Proposition 2.12 (Symmetry of binomial coefficients) *Let $n \in \mathbb{N}$. Then*

$$\binom{n}{k} = \binom{n}{n-k} \qquad \forall k \in \{0, \ldots, n\}.$$

Proof. One has

$$\binom{n}{k} = \frac{n!}{k!(n-k)!} = \frac{n!}{(n-k)!k!} = \frac{n!}{(n-k)!(n-(n-k))!} = \binom{n}{n-k}. \qquad \square$$

It is useful to have an idea of the behavior of the function $k \mapsto \binom{n}{k}$ for a given $n \in \mathbb{N}$.

Proposition 2.13 *For fixed n, the binomial coefficient $\binom{n}{k}$ grows as k varies between 0 and the integer part $[n/2]$ of $n/2$, and decreases as k varies from $[n/2] + 1$ to n.*

Proof. Indeed, for each $1 \leq k \leq n$ one has $\binom{n}{k-1} \leq \binom{n}{k}$ if and only if

$$\frac{\binom{n}{k-1}}{\binom{n}{k}} = \frac{k}{n-k+1} \leq 1.$$

i.e., if and only if $k \leq \dfrac{n+1}{2}$. □

Remark 2.14 For n fixed, the binomial coefficient $\binom{n}{k}$ assumes its maximum value at $k = [n/2]$ (see Fig. 2.1 for $n = 10$ and $n = 100$). An application of Stirling's formula (Remark 2.2) shows that this maximum value is asymptotic to

$$\frac{2^{n+\frac{1}{2}}}{\sqrt{\pi n}} \quad \text{for } n \to +\infty.$$

Definition 2.15 Let $n, k \in \mathbb{N}$. We use $C(n, k)$ to denote the **number of k-collection without repetition** of I_n, or equivalently the **number of subsets of I_n of cardinality** k. □

Fig. 2.1 Graph of the values assumed by the binomial coefficients $\binom{10}{k}$ and $\binom{100}{k}$ as k varies

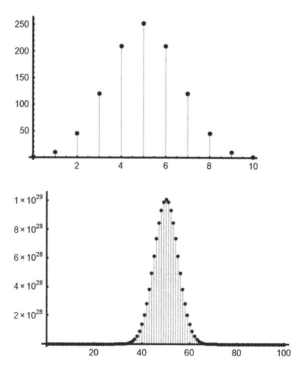

Remark 2.16 In Theorem 1.28 we have proved the existence of a one to one correspondence between the k-collections of I_n and the n-compositions of k, associating to each k-collection of I_n the n-composition of k whose terms are the number of 1, 2, ..., n in the collection. Therefore $C(n, k)$ is also the **number of n-compositions** $(k_1, \ldots, k_n)$ of k **with** $k_i \leq 1$ for each $i = 1, \ldots, n$.

The binomial coefficients have an important combinatorial interpretation.

Theorem 2.17 *Let $n, k \in \mathbb{N}$. Then*

$$C(n, k) = \frac{S(n, k)}{k!} = \binom{n}{k}.$$

In particular $\binom{n}{k}$ *is a natural number.*

Proof. If $k > n$, there are no k-collections without repetitions in I_n, and so $C(n, k) = 0 = \binom{n}{k}$. Now let $k \leq n$. If $k = 0$, one has $C(n, 0) = 1 = \binom{n}{0}$: indeed, the empty collection is the unique 0-collection of I_n. Suppose that $k \geq 1$. Consider the function that associates to each k-sequence of I_n without repetitions the k-collection obtained by *forgetting* the order of the elements (see, e.g., Fig. 2.2).

By way of this function each k-collection without repetitions is the image of exactly $k!$ sequences without repetitions, to wit, the k-sequences that one obtains by ordering its elements in all possible ways. By the Division Principle 1.47 it follows that the number of k-collections of I_n without repetitions is $S(n, k)/k!$.   □

Example 2.18 Let $0 \leq k \leq n$ be integers. Using the definition of the binomial coefficient we have seen that

Fig. 2.2 The function that associates to each 2-sequence of I_3 without repetitions the 2-collection obtained by *forgetting* the order of the elements

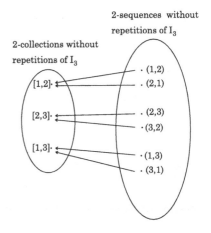

$$\binom{n}{k} = \binom{n}{n-k}.$$

This formula can also be found, in the light of Theorem 2.17, by noting that **to choose** k objects from a set of n is equivalent **to excluding** $n-k$ objects from the same set of n. Thus the number $\binom{n}{k} = C(n,k)$ of the possible choices of k objects is equal to the number $\binom{n}{n-k} = C(n, n-k)$ of the possible ways of excluding $n-k$ objects. $\qquad\square$

The following clarifies the reason for the terminology "binomial coefficient".

Proposition 2.19 (Binomial formula) *Let* $0 \le k \le n$. *The coefficient of* $x^k y^{n-k}$ *in the expansion of* $(x+y)^n$ *is* $\binom{n}{k}$. *More, precisely, the expansion of* $(x+y)^n$ *is given by*

$$\binom{n}{0}x^0 y^n + \binom{n}{1}x y^{n-1} + \binom{n}{2}x^2 y^{n-2} + \cdots + \binom{n}{n-1}x^{n-1} y + \binom{n}{n}x^n y^0.$$

Proof. The formula is trivially true for $n=0$. Now let $n \ge 1$. Multiplying $x+y$ by itself n times one obtains a sum of monomials of the type $x^k y^{n-k}$; the term $x^k y^{n-k}$ appears every time one chooses the x term k times and the y term $n-k$ times in the n-fold product. The k factors in which one chooses x constitute a k-collection of the set of n factors, and so they can be chosen in $C(n,k) = \binom{n}{k}$ ways. Hence the summand $x^k y^{n-k}$ is repeated $C(n,k) = \binom{n}{k}$ times, and so this integer is the coefficient of the monomial $x^k y^{n-k}$ in the expansion of the binomial. $\qquad\square$

Corollary 2.20 *Let* n, k *and* $n_1, \ldots, n_k$ *be natural numbers. Then we have:*

1. $\displaystyle\sum_{i=0}^{n}\binom{n}{i} = 2^n$;

2. $\displaystyle\sum_{\substack{0\le \ell_1\le n_1 \\ 0\le \ell_k\le n_k}}\binom{n_1}{\ell_1}\cdots\binom{n_k}{\ell_k} = 2^{n_1+\cdots+n_k}$.

Proof. 1. By Proposition 2.19 one has

$$2^n = (1+1)^n = \sum_{i=0}^{n} \binom{n}{i} 1^{n-i} 1^i = \sum_{i=0}^{n} \binom{n}{i}.$$

2. In view of Point 1 one gets

$$\sum_{\substack{0 \leq \ell_1 \leq n_1 \\ \overset{\cdots}{0 \leq \ell_k \leq n_k}}} \binom{n_1}{\ell_1} \cdots \binom{n_k}{\ell_k} = \prod_{i=1}^{k} \sum_{0 \leq \ell_i \leq n_i} \binom{n_i}{\ell_i} = \prod_{i=1}^{k} 2^{n_i} = 2^{n_1 + \cdots + n_k}. \qquad \square$$

The following identity is another easy consequence of the binomial formula.

Corollary 2.21 *Let* $n \in \mathbb{N}_{\geq 1}$. *Then*

$$\sum_{k=0}^{n} (-1)^k \binom{n}{k} = 0.$$

Proof. Indeed, by Proposition 2.19 one has

$$0 = (1-1)^n = \sum_{k=0}^{n} \binom{n}{k} 1^{n-k} (-1)^k = \sum_{k=0}^{n} (-1)^k \binom{n}{k}. \qquad \square$$

The following recursive formula, known also as **Stifel**[3] **formula,** is frequently used.

Proposition 2.22 (Stifel Recursive Formula) *Let* $k, n \in \mathbb{N}_{\geq 1}$. *Then*

$$\binom{n-1}{k-1} + \binom{n-1}{k} = \binom{n}{k}.$$

Proof. Note that $\binom{n}{k} = C(n, k)$ is the number of k-element subsets of I_n. Such subsets may be divided into two disjoint classes: those containing 1 and those that do not. The first class has just as many elements as there are $(k-1)$-element subsets of $\{2, \ldots, n\}$: it suffices to add the number 1 to each of the latter $(k-1)$-element subsets to obtain a set of the first class. The second class consists of the subsets of k

---

[3] Michael Stifel (1487–1567).

elements chosen from $\{2, \ldots, n\}$. Thus the first class has $C(n-1, k-1) = \binom{n-1}{k-1}$ elements, while the second has $C(n-1, k) = \binom{n-1}{k}$ elements.   □

Example 2.23 (*The triangle of Tartaglia–Pascal*) The preceding proposition shows that to calculate the binomial coefficients with first argument ("numerator") equal to n, it is sufficient to know those with first argument equal to $n-1$. This observation underlies the construction of the Tartaglia[4]–Pascal[5] triangle[6]

$$\binom{0}{0}$$
$$\binom{1}{0} \quad \binom{1}{1}$$
$$\binom{2}{0} \quad \binom{2}{1} \quad \binom{2}{2}$$
$$\binom{3}{0} \quad \binom{3}{1} \quad \binom{3}{2} \quad \binom{3}{3}$$
$$\binom{4}{0} \quad \binom{4}{1} \quad \binom{4}{2} \quad \binom{4}{3} \quad \binom{4}{4}$$
$$\ldots \quad \ldots \quad \ldots \quad \ldots \quad \ldots \quad \ldots \quad \ldots \quad \ldots \quad \ldots \quad \ldots \quad \ldots$$

Since $\binom{n}{k} = \binom{n}{n-k}$, the Tartaglia–Pascal triangle is symmetric with respect to the vertical line passing through its vertex. The values $\binom{n}{0}$ and $\binom{n}{n}$ on its sides are all equal to one. Using the Stifel recursive formula (see Proposition 2.22)

$$\binom{n}{k} = \binom{n-1}{k-1} + \binom{n-1}{k}, \quad 1 \le k \le n-1,$$

one can easily **find the internal values of any row when one knows the values of the preceding row**; they are, in fact, equal to the sum of the two adjacent values of the preceding row:

$$
\begin{array}{ccc}
 & & k \\
n-1 \cdots\cdots a & & b \\
n \cdots\cdots\cdots a+b & &
\end{array}
$$

---

[4]Niccolò Tartaglia (1499–1557).

[5]Blaise Pascal (1623–1662).

[6]Actually, this triangle was known around the year 1000 by Indian, Persian and Chinese mathematicians.

Fig. 2.3 "Enological application" of the Tartaglia–Pascal triangle. Reproduced with the permission of Perrone's family

Using the recursion we can construct the initial rows of the Tartaglia–Pascal triangle:

$$
\begin{array}{ccccccccc}
 & & & & 1 & & & & \\
 & & & 1 & & 1 & & & \\
 & & 1 & & 2 & & 1 & & \\
 & 1 & & 3 & & 3 & & 1 & \\
1 & & 4 & & 6 & & 4 & & 1 \\
\end{array}
$$
$$\dots \dots \dots \dots \dots \dots \dots \dots \dots \dots \dots$$

$\square$

Figure 2.3 shows the popularity of this triangle among people with taste.

Example 2.24 (G. Pólya[7]'s walk) Imagine that the following diagram describes the network of streets of a city. Each vertex is determined by an ordered pair (n, k) where $n \geq 0$ is the number of the row and $k \geq 0$ is the number of the column. We wish to determine how many different "paths" lead from the vertex $(0, 0)$ to the generic vertex (n, k).

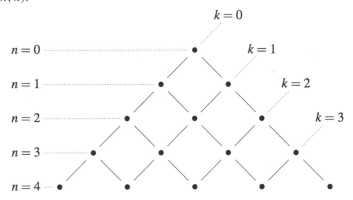

[7]György Pólya (1887–1985).

We may describe any path joining $(0, 0)$ to (n, k) by way of a sequence of Lefts and Rights: for example a path connecting $(0, 0)$ to $(3, 1)$ is given by (L, L, R); another path joining the same vertices is (L, R, L) or also (R, L, L). In general, in order to arrive at the vertex (n, k) one must turn right exactly k times, and consequently also make exactly $n - k$ left turns. Every path is uniquely determined by an n-sequence in $\{L, R\}$ in which the symbol "R" appears exactly k times. Such a sequence is determined by the k positions in which R appears: therefore, there are $\binom{n}{k}$ possible paths leading from $(0, 0)$ to (n, k).   □

Example 2.25   1. How many 5 card hands can be formed from a standard 52 card deck?
2. If one randomly chooses a 5 card hand, what is the probability that all the cards belong to the same suit?
3. If one randomly chooses a 5 card hand, what is the probability of holding exactly 3 Aces?

Solution. 1. A 5 card hand is a 5-collection without repetitions of I_{52}, and so there are $C(52, 5) = 52!/(47!5!) = 2\,598\,960$ hands consisting of 5 cards.

2. We must first find the number of 5-collections without repetitions made up of cards of the same suit. The set of such collections constitutes a conditional product. Indeed, such a collection can be obtained by the following two steps procedure: first choose a suit, and then pick 5 cards of that suit. The first step has 4 possible outcomes, clubs, diamonds, hearts, or spades. The second step consists in choosing a 5-collection without repetitions from the set of the 13 cards of the suit chosen in phase 1: this may be done in $C(13, 5) = 13!/(5!8!) = 1\,287$ ways. Thus, by the Multiplication Principle 1.34 there are $4 \times 1\,287 = 5\,148$ possible hands composed of cards all of the same suit. Hence one has

$$P(5 \text{ cards of the same suit}) = \frac{4 \times C(13, 5)}{C(52, 5)} = 0.00198(\approx 0.2\,\%).$$

3. The set of hands with three Aces is a conditional product. Indeed, a hand with three Aces can be obtained in two phases using the following procedure: first we choose three of the four Aces, and this may be done in $C(4, 3) = 4$ ways. Then we must complete the hand by using two cards chosen from the remaining 48 cards (non-Aces), and this may be done in $C(48, 2) = 1\,128$ ways. By the Multiplication Principle 1.34 there are in all $4 \times 1\,128 = 4\,512$ hands with exactly three Aces, and thus the desired probability is

$$\frac{C(4, 3) \times C(48, 2)}{C(52, 5)} = 0.00174(\approx 0.17\,\%).$$ □

Example 2.26 We must choose a committee of k people from a set formed by 7 women and 4 men. In how many ways can the choice be made if:

1. The committee consists of 5 people, 3 women and 2 men?

2. The committee can have any (strictly positive) number of members but must have an equal number of men and women?
3. The committee is composed of 4 people, one of whom is Mr. Jones?
4. The committee has 4 people, of which at least two are women?
5. The committee has 4 people, two men and two women, and Mr. and Mrs. Jones can not both be members?

Solution. 1. The set of committees under consideration is a conditional product. In fact, such a committee is obtained by first choosing the 3 women, and then the 2 men. Thus it will suffice to multiply the number of 3-collections without repetitions of women with that of the 2-collections without repetitions of men. Thus in all one has $C(7, 3) \times C(4, 2) = 35 \times 6 = 210$ ways of choosing the committee.

2. To count the number of committees that have the same number of men and women we divide the problem into 4 distinct sub-cases: committees composed of 1 woman and 1 man, those having 2 women and 2 men, those with 3 women and 3 men, and finally those with 4 women and 4 men (since there are only 4 men in all). The total number of committees is then the sum of the numbers of committees obtained in these 4 subcases: $[C(7, 1) \times C(4, 1)] + [C(7, 2) \times C(4, 2)] + [C(7, 3) \times C(4, 3)] + [C(7, 4) \times C(4, 4)] = 7 \times 4 + 21 \times 6 + 35 \times 4 + 35 \times 1 = 329$.

3. If Mr. Jones must be a member of the committee, the problem reduces to choosing the 3 other members from among the remaining 10 people (7 women and 3 men). We may choose these three people in any way we desire, both for the women and the men. Thus the number sought is $C(10, 3) = 120$.

4. We can split the problem into subcases by considering the exact number of men and women composing the committee: 2 women and 2 men, 3 women and 1 man, or 4 women. Thus the number of committees satisfying this condition is $[C(7, 2) \times C(4, 2)] + [C(7, 3) \times C(4, 1)] + C(7, 4) = 21 \times 6 + 35 \times 4 + 35 = 301$.

5. It is useful to split the condition "Mr. and Mrs. Jones can not both be members of the committee" into a series of sub-cases in which we specify the membership of Mr. Jones or Mrs. Jones. Note that the committees from which both are absent are allowed under the stated condition. Thus there are 3 sub-cases to consider: the first case is that in which Mrs. Jones is on the committee but Mr. Jones is not. We can then choose a second woman among the remaining 6, and two men from the remaining 3 (after excluding Mr. Jones). This can be done in $C(6, 1) \times C(3, 2) = 6 \times 3 = 18$ ways. The other two cases are the one in which Mr. Jones is a member but not Mrs. Jones, and which similar reasoning shows to give rise to $C(6, 2) \times C(3, 1) = 15 \times 3 = 45$ different committees, and that in which neither Mr. Jones nor Mrs. Jones are present, which produces $C(6, 2) \times C(3, 2) = 15 \times 3 = 45$ different committees. Therefore, the total number of committees of the desired type is $18 + 45 + 45 = 108$.

☞ A simpler solution of this last point can be obtained by subtracting the number of committees in which both spouses are present from the total number of committees. The committees formed by 2 women and 2 men are in number $C(7, 2) \times C(4, 2)$. The committees in which both spouses are present are formed by choosing a second woman and a second man to complete the committee, and this may be done in $C(6, 1) \times C(3, 1)$ ways. Therefore there are

$$C(7, 2) \times C(4, 2) - C(6, 1) \times C(3, 1) = 21 \times 6 - 6 \times 3 = 108$$

committees in which Mr. and Mrs. Jones are not simultaneously members.   □

2.2 Arbitrary Sequences and Collections

We now consider the sequences and collections of I_n without any restriction regarding possible repetitions.

2.2.1 Sequences and Sharings

In this section we will learn how to count the k-sequences of I_n, without any restriction on their particular nature.

Definition 2.27 Let $n, k \in \mathbb{N}$. The symbol $S((n, k))$ will be used to denote the **number of k-sequences of I_n**, or equivalently (see Theorem 1.28) **the number of n-sharings of I_k**.   □

☞   One should beware of the fact that, unlike $S(n, k)$, the symbol $S((n, k))$ is not zero when $k > n$: indeed there are sequences of I_n of arbitrary length.

Theorem 2.28 *Let $n, k \in \mathbb{N}$. Then*

$$S((n, k)) = n^k.$$

Proof. If $k = 0$, the unique 0-sequence of I_n is the empty sequence: therefore, $S((n, 0)) = 1 = n^0$ for every $n \in \mathbb{N}$. Now let $k \geq 1$. The k-sequences in I_n are exactly the elements of the cartesian product I_n^k, and so in all they number n^k.   □

Corollary 2.29 (Number of functions) *Let $k, n \in \mathbb{N}_{\geq 1}$. The number of functions $I_k \rightarrow I_n$ equals $S((n, k)) = n^k$.*

Proof. In Example 1.24 we have seen how each function $f : I_k \rightarrow I_n$ is determined by a k-sequence of I_n. The conclusion follows from Theorem 2.28.   □

Example 2.30 (Birthday's problem) What is the probability that two (or more) people randomly chosen from a group of 25 people have the same birthday?

Solution. We label the 25 people with I_{25}. Their birthdays constitute a 25-sequence of I_{365}. The probability that the 25 people were all born on different days is

$$Q(365, 25) := \frac{S(365, 25)}{S((365, 25))} = \frac{365 \times 364 \times \cdots \times 341}{365^{25}} \approx 0{,}431.$$

Therefore the desired probability is $1 - Q(365, 25) \approx 0{,}569$. The numbers

$$Q(n, k) := \frac{S(n, k)}{S((n, k))} = \frac{n!}{(n - k)!n^k}$$

are called Ramanujan numbers; they will be studied in Sect. 14.6, where we will give
an estimate for them for large values of k and n.   □

Example 2.31 Consider an urn containing 10 red balls and 4 black ones. Determine
the probability that the third ball extracted is red, and then the probability that the
first two extracted are both black, for extractions *with* replacement as well as those
without replacement.

Solution. Let A and B be the events: "$A =$ the third ball extracted is red", "$B =$ the
first two balls extracted are black". We number the red and black balls and consider
the sets $R = \{R_1, \ldots, R_{10}\}$ of red balls, and the set $N = \{N_1, \ldots, N_4\}$ of black
balls.
(1) The case of extractions with replacement: we choose the sample space Ω consisting of the 3-sequences in $R \cup N$; indeed possible further extractions after the
third are of no interest for us. Clearly the elements in Ω are equi-probable and
$|\Omega| = S((14, 3)) = 14^3$. The event A coincides with the cartesian product

$$(R \cup N) \times (R \cup N) \times R = \{(x, y, z) : x, y \in R \cup N, z \in R\}.$$

Hence, $|A| = 14^2 \times 10$ elements and so

$$P(A) = \frac{|A|}{|\Omega|} = 14^2 \times 10/14^3 = 10/14 = 5/7 \approx 71.43\,\%.$$

Similarly, the event B coincides with the cartesian product

$$N \times N \times (R \cup N) = \{(x, y, z) : x, y \in N, z \in R \cup N\}.$$

Therefore, $|B| = 4^2 \times 14$, and so

$$P(B) = \frac{|B|}{|\Omega|} = 4^2 \times 14/14^3 = 4^2/14^2 = 2^2/7^2 \approx 8.16\,\%.$$

(2) The case of extractions without replacement: we choose the sample space Ω'
consisting of 3-sequences without repetition in $R \cup N$. The elements of Ω' are equi-
probable and by the Multiplication Principle 1.34 we have $|\Omega'| = 14 \times 13 \times 12$.
We calculate the cardinality of the event $A = \{(x, y, z) : x, y \in R \cup N, z \in$
R with x, y, z distinct$\}$: there are 10 choices for z, 13 for x and 12 for y; hence

A is a conditional product with multiplicities $(10, 13, 12)$ and consequently $|A| = 10 \times 13 \times 12$. The probability sought is

$$P(A) = \frac{|A|}{|\Omega'|} = (10 \times 13 \times 12)/(14 \times 13 \times 12) = 10/14 = 5/7 \approx 71.43\,\%.$$

Here one should note that by dividing the construction of the triples of A starting from the first component x it would not have been possible to apply the Multiplication Principle 1.34: for example if x and y are both red there are 8 choices for z, whereas if they are both black the number of choices available for z becomes 10.
In a similar fashion,

$$B = \{(x, y, z) : x, y \in N, \ z \in R \cup N \text{ with } x, y, z \text{ distinct}\}$$

has $4 \times 3 \times 12$ elements, from which one has

$$P(B) = \frac{|B|}{|\Omega'|} = 4 \times 3 \times 12/(14 \times 13 \times 12) = 4 \times 3/14 \times 13$$

$$= 6/(7 \times 13) \approx 6.59\,\%. \qquad \Box$$

Example 2.32 In Proposition 1.39 we have proved that the set of parts $\mathscr{P}(I_k)$ has 2^k elements. We obtain the same result observing that, mapping each subset A of I_k into the 2-sharing $(A, I_k \setminus A)$ of I_k, we get a bijective correspondence between $\mathscr{P}(I_k)$ and the 2-sharings of I_k. Therefore

$$|\mathscr{P}(I_k)| = S((2, k)) = 2^k. \qquad \Box$$

2.2.2 Collections and Compositions

Here we will consider the collections of finite sets in which repetitions are possible, or equivalently, the compositions of a natural number.

Definition 2.33 Let $n, k \in \mathbb{N}$. We use $C((n, k))$ to denote the **number of k-collections** of I_n, or equivalently (see Theorem 1.28) the **number of n-compositions** of k. $\qquad \Box$

☞   Bear in mind that unlike the situation for $C(n, k)$, the symbol $C((n, k))$ is not zero for $k > n$: indeed there are collections of I_n of arbitrary length.

Example 2.34 (*Roman Numerals*) Every 4-composition (k_1, k_2, k_3, k_4) of 7 may be represented by a $(7 + 4 - 1)$-sequence of $\{I, +\}$ formed in order by k_1 "I", a "$+$", k_2 "I", a "$+$", k_3 "I", a "$+$", k_4 "I": for example, the 4-composition $(2, 1, 0, 4)$ is represented by the sequence $(I, I, +, I, +, +, I, I, I, I)$, while the 4-composition

$(1, 1, 3, 2)$ is represented by the sequence $(I, +, I, +, I, I, I, +, I, I)$. Essentially it is a question of writing the sums

$$2 + 1 + 0 + 4 \quad \text{and} \quad 1 + 1 + 3 + 2$$

in the corresponding "roman numeral notation"

$$II + I + +IIII \quad \text{and} \quad I + I + III + II. \qquad \square$$

Using the "roman numeral representation" of the n-compositions of k, one easily obtains the value of $C((n, k))$.

Theorem 2.35 *Let* $n, k \in \mathbb{N}$*. Then one has*

$$C((n, k)) = \begin{cases} C(n - 1 + k, k) & \text{if } n + k \geq 1, \\ 1 & \text{if } n = 0 = k. \end{cases}$$

Proof. If $k = 0$, the only 0-collection of I_n is the empty one: hence, $C((n, 0)) = 1$ which is exactly the value indicated in the statement of the theorem both for $n = 0$ and for $n \geq 1$. Now let $k \geq 1$. If $n = 0$ there are no k-collections of $I_0 = \emptyset$: therefore $C((0, k)) = 0 = C(k - 1, k)$. If $n \geq 1$, we count the number of n-compositions of k. Every such composition $(k_1, \ldots, k_n)$ is uniquely determined by the corresponding representation as "roman numerals", that is, by the $(n + k - 1)$-sequence of $\{I, +\}$ formed in order by k_1 "I", a "$+$", k_2 "I", a "$+$", $\ldots$, a "$+$", k_n "I". Therefore, the number of n-compositions of k coincides with the number of $(n - 1 + k)$-sequences of $\{I, +\}$ with k appearances of "I" and $n - 1$ of "$+$". Note that in order to determine such a sequence it is sufficient to indicate the k positions where "I" appears, and so we have $C(n - 1 + k, k)$ possible n-compositions of k. $\qquad \square$

Example 2.36 In how many ways is it possible to distribute 20 identical bars of white chocolate and 15 identical bars of dark chocolate among 5 children?

Solution. The number of possible distributions of 20 white chocolate bars among five children is equal to the number of 5-compositions of 20, namely $C((5, 20)) = C(20 + 5 - 1, 20) = 10\,626$: indeed, denoted by k_i, $i = 1, \ldots, 5$, the unknown number of white chocolate bars we give to the i-th child, our problem reduces to finding the number of natural numbers solutions of the equation $k_1 + \cdots + k_5 = 20$, i.e., the number of 5-compositions of 20. Similarly, the 15 identical bars of dark chocolate can be distributed in $C((5, 15)) = C(15 + 5 - 1, 15) = 3\,876$ different ways. By the Multiplication Principle 1.34, we therefore have $10\,626 \times 3\,876 = 41\,186\,376$ ways to carry out the overall distribution. $\qquad \square$

Example 2.37 Grandma Emily goes to the drugstore with 10 euros to buy decorations for a cake. The drugstore sells packages of sugar rosettes, of glazed coffee beans, and of birthday cake candles for 50 cents per package, and jelly beans for 2 euros per bag. In how many ways can she spend all of the money she brings to the store?

Solution. The first step consists of taking 50 cents as the unit of measure. Thus, Grandma Emily must spend a total of twenty of these monetary units. Bearing in mind that a bag of jelly beans costs 4 units, our problem reduces to counting the number of natural number solutions of the following equation:

$$x_1 + x_2 + x_3 + 4x_4 = 20.$$

Perhaps the simplest way to solve the problem consists in specifying right from the start how many bags of jelly beans to buy. If we buy one bag, then the equation to resolve becomes $x_1 + x_2 + x_3 = 16$, which is an equation with

$$C((3, 16)) = C(16 + 2, 2) = 153$$

solutions in natural numbers. In general, if $x_4 = i$ we obtain the equation $x_1 + x_2 + x_3 = 20 - 4i$, which has $C((3, 20 - 4i)) = C(22 - 4i, 2)$ solutions, for $i = 0, 1, 2, 3, 4, 5$. Summing these, we obtain in all

$$\binom{22}{2} + \binom{18}{2} + \binom{14}{2} + \binom{10}{2} + \binom{6}{2} + \binom{2}{2} = 536$$

possible different purchases.   □

Example 2.38 In how many different ways can one choose six bottles of wine of three different types?

Solution. Suppose that the three types of wine are Barbera, Merlot and Sauvignon. Each possible choice corresponds to a solution in natural numbers of the equation

$$x_B + x_M + x_S = 6,$$

where x_B, x_M, x_S represent respectively the number of bottles of Barbera, Merlot and Sauvignon. In all there are $C((3, 6)) = C(8, 6) = \binom{8}{6} = \dfrac{8!}{6!2!}$ possible choices for the bottles of wine.   □

Here are a few properties of the function $C((n, k))$.

Proposition 2.39 *Let $n \in \mathbb{N}_{\geq 1}$ and $k \in \mathbb{N}$. Then:*

1. $C((n, k)) = C((k + 1, n - 1))$;
2. *If $k \geq 1$, then $C((n, k - 1)) + C((n - 1, k)) = C((n, k))$;*
3. $C((n - 1, 0)) + \cdots + C((n - 1, k)) = C((n, k))$.

Proof. 1. By Theorem 2.35 we have (see Problem 2.52 for a combinatorial proof of this identity)

$$C((n, k)) = C(n + k - 1, k) = C(n + k - 1, n - 1) = C((k + 1, n - 1)).$$

2. Among the k-collections of I_n, there are those that contain at least a 1 and those that do not contain any 1. The first type are obtained by adding a 1 to an arbitrary $(k - 1)$-collection of I_n, and so there are $C((n, k - 1))$ of them. The second type are the k-collections in $\{2, \ldots, n\}$ and so are $C((n - 1, k))$ in number.

3. Among the k-collections of I_n, there are those that contain 1 repeated exactly k times, those containing 1 repeated exactly $k - 1$ times, $\ldots$, those containing 1 repeated exactly once, and those that do not contain 1. In general, the k-collections of I_n with 1 repeated j times number $C((n - 1, k - j))$, since it is a question of counting the number of ways in which the elements different from 1 may be chosen. Therefore, summing for $j = k, k - 1, \ldots, 1, 0$ one obtains the desired equality. $\square$

2.2.3 Collections and Compositions with Constraints

We now wish to count collections that contain at least a certain number of elements of each type, or equivalently the compositions with some constraints on its elements. In the following result, the used strategy is more important to fix than the result itself.

Corollary 2.40 *Let $n \in \mathbb{N}_{\geq 1}$ and $\ell_1, \ldots, \ell_n \in \mathbb{N}$. If $k \geq \ell_1 + \cdots + \ell_n$, then the number of solutions in natural numbers to*

$$x_1 + \cdots + x_n = k, \quad \text{with } x_i \geq \ell_i, \ i = 1, \ldots, n,$$

is equal to the number $C((n, k - (\ell_1 + \cdots + \ell_n)))$ of n-compositions of $k - (\ell_1 + \cdots + \ell_n)$.

Proof. Set $x_i = y_i + \ell_i$. The equation $x_1 + \cdots + x_n = k$ then becomes

$$(y_1 + \ell_1) + \cdots + (y_n + \ell_n) = k,$$

that is, $y_1 + \cdots + y_n = k - (\ell_1 + \cdots + \ell_n)$. By associating the n-tuple $(b_1 - \ell_1, \ldots, b_n - \ell_n)$ to the n-tuple $(b_1, \ldots, b_n)$ we obtain a one to one correspondence between the natural number solutions of the equation $x_1 + \cdots + x_n = k$ satisfying $x_i \geq \ell_i$ with the natural number solutions of

$$y_1 + \cdots + y_n = k - (\ell_1 + \cdots + \ell_n),$$

that is, with the n-compositions of $k - (\ell_1 + \cdots + \ell_n)$. $\square$

Example 2.41 In how many ways is it possible to fill a box with 12 chocolates of 5 different types under the restriction that there must be at least one chocolate of each type?

Solution. By Corollary 2.40, this reduces to the problem of counting the natural number solutions of the equation

$$y_1 + \cdots + y_5 = 12 - (1 + 1 + 1 + 1 + 1) = 7.$$

By Theorem 2.35 that number is $C((5, 7)) = C(7 + 5 - 1, 7) = 330.$ □

Example 2.42 How many possible choices of 10 balls from a basket containing red, blue, and green balls (at least 10 of each color) are there if the choice must include at least 5 red balls? What if there must be at most 5?

Solution. The first question is similar to that posed in Example 2.41: by Corollary 2.40, it reduces to counting the natural number solutions to the equation

$$y_1 + y_2 + y_3 = 10 - 5 = 5.$$

By Theorem 2.35, the result is $C((3, 5)) = C(5 + 3 - 1, 5) = 21.$

To handle the case in which there are at most 5 red balls, we count the number of elements in the complementary subset; thus, we count the possible collections in which there are at least 6 red balls. Among all the $C((3, 10)) = C(10 + 3 - 1, 10) = 66$ possible ways of choosing 10 balls from among those of three colors without any restriction, there are $C((3, 4)) = C(4 + 3 - 1, 4) = 15$ ways to choose a collection with at least 6 red balls. Thus there are $66 - 15 = 51$ possible choices of 10 balls with at most 5 red balls in the selection.   □

Example 2.43 Consider the equation $x_1 + x_2 + x_3 + x_4 = 12.$

1. How many solutions does it have in natural numbers?
2. How many solutions does it have with natural numbers satisfying $x_i \geq 1$, for $i = 1, \ldots, 4$?
3. How many natural number solutions does it have which satisfy $x_1 \geq 2, x_2 \geq 2, x_3 \geq 4, x_4 \geq 0$?

Solution. 1. The number of solutions is $C((4, 12)) = C(12 + 4 - 1, 12) = 455$, by Theorem 2.35.
2. By Corollary 2.40, the question reduces to counting how many natural number solutions of the equation

$$y_1 + y_2 + y_3 + y_4 = 12 - 4 = 8$$

there are. By Theorem 2.35 that number is $C((4, 8)) = C(8 + 4 - 1, 8) = 165.$
3. By Corollary 2.40, one need only count the natural number solutions of the equation

$$y_1 + y_2 + y_3 + y_4 = 12 - (2 + 2 + 4) = 4.$$

By Theorem 2.35 the answer is $C((4, 4)) = C(7, 4) = 35.$ □

Example 2.44 How many sequences formed with the letters $a, e, i, o, u, x, x, x, x,$ $x, x, x, x,$ (8 x's) are there if one requires that no two vowels appear consecutively?

Solution. One can construct the words requested in two phases: in the first phase we choose a 5-sequence of vowels, and in the second we insert the x's respecting the restriction imposed. For the first phase we have 5! possible outcomes. We now pass to the calculation of the outcomes of the second phase. For an arbitrary given sequence of vowels, we write y_1 for the number of x's to the left of the first vowel, then y_2 for the number of x's between the first and second vowels, ..., y_5 for the number of x's between the fourth and fifth vowel, and finally y_6 for the number of x's appearing after the last vowel. Our restriction amounts to solving the equation

$$y_1 + \cdots + y_6 = 8, \qquad y_1, y_6 \geq 0, \qquad y_2, y_3, y_4, y_5 \geq 1.$$

By Corollary 2.40 such an equation has $C((6, 8 - 4)) = C(6 - 1 + 4, 4) = 126$ solutions. By the Multiplication Principle 1.34, there are a total of $5! \times 126 = 15\,120$ possible words of the type required. □

Many processes involving automatic calculations are controlled by input consisting of a sequence of zeroes and ones. Certain types of sequences can be read very quickly. Hence, it is useful to know how often these special types of sequences appear. The sequences that appear in the next example are among those that admit fast reading.

Example 2.45 How many binary sequences of length 10 are there if the sequence is required to consist of a sequence of 1's followed by a sequence of 0's followed by a sequence of 1's followed by a sequence of 0's (for example, 1110111000)?

Solution. Let x_1 denote the number of 1's in the first block, x_2 the number of 0's in the second block, x_3 the number of 1's in the third block and x_4 the number of 0's in the fourth block. This problem then corresponds to determining how many solutions there are to the equation $x_1 + \cdots + x_4 = 10$ with each $x_i \geq 1$. By Corollary 2.40 the number is $C((4, 6)) = C(4 + 6 - 1, 6) = 84$. □

We now count the natural number solutions of an inequality in several variables with unitary coefficients.

Proposition 2.46 *Let $n \in \mathbb{N}_{\geq 1}$ and $k \in \mathbb{N}$. The natural number solutions of the inequality*

$$x_1 + \cdots + x_n \leq k$$

is equal to the number $C((n + 1, k))$ of $(n + 1)$-compositions of k.

Proof. By associating to every natural number solution $(k_1, \ldots, k_n)$ of the inequality under consideration the $(n + 1)$-tuple $(k_1, \ldots, k_n, k - (k_1 + \cdots + k_n))$, one obtains a one to one correspondence with the set of natural number solutions of the equation $x_1 + \cdots + x_n + x_{n+1} = k$. □

Example 2.47 How many words can be formed using three A's and no more than seven B's?

Solution. We use x_1 to denote the number of B's which precede the first A, x_2 for the number of B's between the first and second A, x_3 for the number of B's between the second and third A and x_4 for the number of B's after the third A. Clearly we have

$$x_1 + x_2 + x_3 + x_4 \leq 7,$$

and so there are $C(7 + 4, 4)$ ways to create such words. □

2.3 The Gilbreath Principle ☕

The following trick is a very simplified version of the beautiful *Schizoid Rosary* performed by the famous magician Phil Goldstein (also called Max Maven): we refer to [19] for a suitable presentation of the effect.

The effect. Ask two students A and B to participate to the experiment: you first give to B a closed envelope, containing a prediction. Then give a deck of 20 cards to A; ask him to think of a number between 1 and 19, and to deal, from the deck face-down in one hand, one by one the thought number of cards, face down on the table. You then ask him to riffle shuffle[8] the pile on the table with the pile in his hands. Now the first 10 cards of the deck are dealt off on the table, the other pile of 10 cards remains on A's hands. At this stage spectator B is asked to make the *most important* choice: which pile he has to take, and which will be kept aside. Spectator B is asked to look at the cards of his pile: it turns out that there are exactly two spades, and that this corresponds to what is predicted in the envelope, once opened.

Preparation. It is enough to give to A a prearranged deck, in a repeated sequence of 1 spade and 4 non-spades, from top to bottom (face down) as in the following:

$$(\spadesuit, \diamondsuit, \diamondsuit, \clubsuit, \heartsuit, \spadesuit, \heartsuit, \clubsuit, \heartsuit, \diamondsuit, \spadesuit, \clubsuit, \diamondsuit, \diamondsuit, \clubsuit, \spadesuit, \heartsuit, \clubsuit, \heartsuit, \diamondsuit).$$

Explanation. Everything relies on the Gilbreath Principle, from the magician and mathematician Norman L. Gilbreath, that appeared in its first version in the magic magazine [18] in 1958. We will give all the details in Example 2.54.

In the next definition, an *interval of* $\mathbb{N}$ is a set of consecutive natural numbers.

Definition 2.48 (*Gilbreath permutation*) A permutation $(a_1, \ldots, a_n)$ of $(1, \ldots, n)$ is said to be a **Gilbreath permutation** if $\{a_1, \ldots, a_\ell\}$ is an interval of $\mathbb{N}$ for each $\ell \in I_n$. □

---

[8] A riffle shuffle is obtained by holding one deck in each hand with the thumbs inward, then releasing the cards by the thumbs so that they fall to the table interleaved; alternatively the two decks can be put on the table and gently pushed one into the other.

Example 2.49 The sequence $(3, 4, 2, 1)$ is a Gilbreath permutation of $(1, 2, 3, 4)$. Indeed the sets

$$\{3\}, \quad \{3, 4\}, \quad \{3, 4, 2\}, \quad \{3, 4, 2, 1\}$$

are all intervals of $\mathbb{N}$. Instead, $(3, 4, 1, 2)$ is not a Gilbreath permutation of $(1, 2, 3, 4)$ due to the fact that $\{3, 4, 1\}$ is not an interval of $\mathbb{N}$. Observe that the only Gilbreath permutation of $(1, 2, 3, 4)$ starting with 1 is necessarily $(1, 2, 3, 4)$ itself.   □

Let us recall that $(a_{i_1}, \ldots, a_{i_k})$ is said to be a subsequence of $(a_1, \ldots, a_n)$ if we have $1 \leq i_1 < \cdots < i_k \leq n$. For instance $(2, 4, 1)$ is a subsequence of $(5, 2, 4, 3, 1)$, whereas $(4, 5, 1)$ is not a subsequence of $(5, 2, 4, 3, 1)$.

Remark 2.50 Consider the Gilbreath permutation $(4, 3, 2, 5, 6, 1)$ of $(1, 2, 3, 4, 5, 6)$. Every term of $(4, 3, 2, 5, 6, 1)$ belongs either to an increasing subsequence of consecutive numbers from 4 up to 6, or to a decreasing subsequence of consecutive numbers from 4 down to 1. We will show that this is a general fact.

The next result sounds as a new characterisation of the Gilbreath permutations: some others may be found in [13, Main Theorem, Chap. 5].

Proposition 2.51 (Characterisation of the Gilbreath permutations) *A permutation $(a_1, \ldots, a_n)$ of $(1, \ldots, n)$ is a Gilbreath permutation if and only if every a_i, $i = 2, \ldots, n$, belongs either to an increasing subsequence of consecutive numbers from a_1 up to n, or to a decreasing subsequence of consecutive numbers from a_1 down to 1.*

Proof. We begin remarking the obvious fact that whenever I is an interval of $\mathbb{N}$ and $x \in \mathbb{N}\backslash I$, then $I \cup \{x\}$ is an interval of $\mathbb{N}$ if and only if either $x = \max I + 1$ or $x = \min I - 1$. Let us assume that $n \geq 2$, the result being trivial if $n = 1$.

($\Leftarrow$). Clearly the set $\{a_1\}$ is an interval of $\mathbb{N}$. If a_2 belongs to an increasing subsequence of consecutive numbers from a_1 up to n, then necessarily $a_2 = a_1 + 1$; if it belongs to a decreasing subsequence of consecutive numbers from a_1 down to 1, then necessarily $a_2 = a_1 - 1$. In both the cases $\{a_1, a_2\}$ is an interval of $\mathbb{N}$. Given $2 \leq i < n$, let us assume that $I := \{a_1, \ldots, a_i\}$ is an interval of $\mathbb{N}$. Then if a_{i+1} belongs to an increasing subsequence of consecutive numbers from a_1 up to n, then necessarily $a_{i+1} = \max I + 1$; if it belongs to a decreasing subsequence of consecutive numbers from a_1 down to 1, then necessarily $a_{i+1} = \min I - 1$. In both the cases $\{a_1, a_2, \ldots, a_{i+1}\}$ is an interval of $\mathbb{N}$.

($\Rightarrow$). We build a pair of subsequences of $(a_1, \ldots, a_n)$ starting from a_1 as follows: for each $i \in \{1, \ldots, n-1\}$ consider the set $I := \{a_1, \ldots, a_i\}$; since $\{a_1, \ldots, a_i, a_{i+1}\}$ is an interval of $\mathbb{N}$, then either $a_{i+1} = \max I + 1$ or $a_{i+1} = \min I - 1$. In the first case we keep a_{i+1} for the first subsequence, otherwise we keep it for the second subsequence. This leads to an increasing and to a decreasing subsequences of consecutive terms starting from a_1. Necessarily these subsequences end, respectively, with n and 1.   □

50

Remark 2.52 By Proposition 2.51, a Gilbreath permutation of $(1, \ldots, n)$ is uniquely determined either by the choice in $\{2, \ldots, n\}$ of the indexes $i_2 < \cdots < i_k$ such that $(a_1, a_{i_2}, \ldots, a_{i_k})$ is the subsequence of increasing consecutive terms from a_1 to n or the indexes $j_2 < \cdots < j_h$ such that $(a_1, a_{j_2}, \ldots, a_{j_h})$ is the subsequence of decreasing consecutive terms from a_1 to 1. For example, let $(a_1, \ldots, a_8)$ be a Gilbreath permutation of $(1, \ldots, 8)$. If the increasing subsequence of consecutive terms of $(a_1, \ldots, a_8)$ up to 8 is (a_1, a_4, a_7, a_8) then $(a_1, a_2, a_3, a_5, a_6)$ is the decreasing subsequence of consecutive terms down to 1. Hence necessarily $a_8 = 8, a_7 = 7, a_4 = 6$, $a_1 = 5$ and then $a_2 = 4, a_3 = 3, a_5 = 2, a_6 = 1$:

$$(a_1, a_2, a_3, a_4, a_5, a_6, a_7, a_8) = (5, 4, 3, 6, 2, 1, 7, 8).$$

Example 2.53 (*The Gilbreath Principle*) Proposition 2.51 shows how to obtain a Gilbreath permutation of a deck of cards, numerated from 1 to n (1 on top, n on bottom). A *Gilbreath shuffle* is performed as follows: fix $j \in \{0, 1, \ldots, n - 1, n\}$, and deal the top j cards face-down on a pile, thus reversing their initial order forming the sequence $(j, j - 1, \ldots, 1)$; these j cards are then riffle shuffled with the other pile of $n - j$ cards which forms the sequence $(j + 1, \ldots, n)$. The resulting sequence, from top to bottom, forms by Proposition 2.51 a Gilbreath permutation. □

Figure 2.4 illustrates a Gilbreath shuffle.

We are now able to explain the trick described at the beginning of this section.[9]

Example 2.54 (*Trick's explanation*) Image to enumerate from 1 to 20 the cards of the sequence

$$\begin{matrix} 1 & 2 & 3 & 4 & 5 & 6 & 7 & 8 & 9 & 10 & 11 & 12 & 13 & 14 & 15 & 16 & 17 & 18 & 19 & 20 \\ (\spadesuit, & \diamondsuit, & \diamondsuit, & \clubsuit, & \heartsuit, & \spadesuit, & \heartsuit, & \clubsuit, & \heartsuit, & \diamondsuit, & \spadesuit, & \clubsuit, & \diamondsuit, & \diamondsuit, & \clubsuit, & \spadesuit, & \heartsuit, & \clubsuit, & \heartsuit, & \diamondsuit) \end{matrix}$$

given to the student A; first he performs a Gilbreath shuffle of the deck. Hence he dealt off on the table the first 10 cards of the deck, holding the other pile of 10 cards. The 10 cards of the pile dealt off on the table forms an interval of $\mathbb{N}$, i.e., are the cards $i, i + 1, \ldots, i + 9$ for a suitable $1 \le i \le 11$. It is easy to check that any subset of 10 consecutive elements of the original sequence of cards and its complementary subset both contain exactly two spades. Whatever pile B chooses, the prediction is satisfied.

Notice that, knowing that the initial deck consists of four spades, six diamonds, five clubs and five hearts, the probability that B has a hand with exactly two spades is

$$\frac{\binom{4}{2}\binom{18}{8}}{\binom{20}{10}} \approx 0.42.$$

□

---

[9]See https://discretecalculus.wordpress.com for more details on the nice periodicity properties of a Gilbreath permutation, and the explanation of more complicated tricks.

Fig. 2.4 A Gilbreath shuffle (cards face up for the convenience of the reader): the cards from 1 to 6 are dealt off on the *right-side* thus reversing their order; the other cards from 6 to 13 remain in their initial order on the *left-side* deck. The two decks are riffle shuffled. The result yields a Gilbreath permutation of $(1, \ldots, 13)$

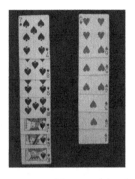

Remark 2.55 (Counting the Gilbreath shuffles) How frequent are the Gilbreath shuffles among shuffles? A deck of n cards can be obviously shuffled in $n!$ ways, the number of permutations of $(1, \ldots, n)$. Among these shuffles there are however only 2^{n-1} Gilbreath shuffles, or equivalently Gilbreath permutations of $(1, \ldots, n)$. In order to prove this claim, let us remember that, as observed in Remark 2.52, a Gilbreath shuffle is uniquely determined by the choice in $\{2, \ldots, n\}$ of the indexes of the increasing (or decreasing) subsequence of consecutive terms. Since these indexes could form any subset of $\{2, \ldots, n\}$, the Gilbreath shuffles are 2^{n-1} (see Proposition 1.39). For instance, with four cards, the Gilbreath permutations of $(1, 2, 3, 4)$ are

$$(1, 2, 3, 4), \quad (2, 1, 3, 4), \quad (2, 3, 1, 4), \quad (2, 3, 4, 1),$$

$$(3, 2, 1, 4), \quad (3, 2, 4, 1), \quad (3, 4, 2, 1), \quad (4, 3, 2, 1).$$

It follows that the probability that a shuffle of n cards is a Gilbreath one is $2^{n-1}/n!$: for instance, with $n = 20$, this probability is approximately equal to 2.16×10^{-13}.

Gilbreath was proud to have created more than 150 tricks based on his Principle, which was also applied in scientific fields: for instance Knuth used it in [24, 5.4.9] for an "improved superblock striping" technique that allows two or more files, distributed on discs, to be merged without possible conflicts.

2.4 Binomial Identities

In this section we study some formulas regarding binomial coefficients. It is convenient to consider separately the formulas that involve alternating signs from those that contains only positive summands.

2.4.1 Elementary Binomial Identities

In Corollary 2.20 we have proved that summing on the lower index:

$$\binom{n}{0} + \cdots + \binom{n}{n} = 2^n.$$

The next proposition provides some other interesting combinatorial identities regarding sums of binomial coefficients or squares of binomial coefficients.

Proposition 2.56 *Let n, k be two natural numbers. The following identities hold:*

1. *Sum of squares:*

$$\binom{n}{0}^2 + \binom{n}{1}^2 + \binom{n}{2}^2 + \cdots + \binom{n}{n-1}^2 + \binom{n}{n}^2 = \binom{2n}{n};$$

2. *Summing on two indices:*

$$\binom{n}{0} + \binom{n+1}{1} + \binom{n+2}{2} + \cdots + \binom{n+k}{k} = \binom{n+k+1}{k};$$

3. *Sum on the upper index:*

$$\binom{0}{k} + \binom{1}{k} + \cdots + \binom{n}{k} = \binom{n+1}{k+1}.$$

Proof. The three identities hold trivially for $n = 0$. Now let $n \geq 1$.

1. The number $C(2n, n) = \binom{2n}{n}$ represents the number of subsets of cardinality n of I_{2n}. Any subset of this type is formed by choosing k elements from the set $\{1, \ldots, n\}$ and $n - k$ elements from the set $\{n + 1, \ldots, 2n\}$, for the appropriate value of $k \in \{0, 1, \ldots, n\}$. The first choice can be made in $C(n, k) = \binom{n}{k}$ ways, while the second choice can be made in $C(n, n - k) = \binom{n}{n - k}$ ways. Thus one obtains that the total number of ways to construct an n element subset of I_{2n} exactly k of whose elements lie in I_n is

$$\binom{n}{k} \binom{n}{n - k} = \binom{n}{k}^2.$$

The total number of n element subsets of I_{2n} is therefore given by the sum of these terms as k varies from 0 to n.

2. The binomial coefficient $\binom{n + k + 1}{k}$ is the number of binary $(n + k + 1)$-sequences with exactly k copies of "0" and $n + 1$ copies of "1": indeed, such a sequence is determined when one knows the positions, for example, of the k zeroes. Fix such a sequence. Then the last 1 can be in position $n + 1$ (if the all 1's appear at the beginning of the sequence), $n + 2, \ldots, n + k + 1$ (if the sequence ends with a 1). In general, the last 1 is necessarily in position $n + 1 + j$ for a suitable j satisfying $0 \leq j \leq k$; it is followed by a sequence of zeroes, and is preceded by a binary $(n + j)$-sequence that contains "1" exactly n times. The sequences that have their last 1 in position $n + 1 + j$ are equal in number to the binary $(n + j)$-sequences in which "1" appears n times. Since the latter are $\binom{n + j}{n} = \binom{n + j}{j}$, there are therefore

$$\binom{n + k + 1}{k} = \sum_{j=0}^{k} \binom{n + j}{j}$$

binary $(n + k + 1)$-sequences in which "1" appears $n + 1$ times.

3. Suppose that we wish to construct a $(k + 1)$-collection of I_{n+1}: we can do this in $\binom{n + 1}{k + 1}$ ways. If $n \geq i \geq k$ there are $\binom{i}{k}$ different $(k + 1)$-collection of I_{n+1} having $i + 1$ as the maximum number. Thus one has the relation

$$\binom{n + 1}{k + 1} = \sum_{i=k}^{n} \binom{i}{k} = \sum_{i=0}^{n} \binom{i}{k}. \qquad \square$$

The following formula, attributed to Vandermonde,[10] was well known to Chinese mathematicians of the fourteenth century.

Proposition 2.57 (Vandermonde Convolution) *Let $k, m, n \in \mathbb{N}$. Then*

$$\sum_{j=0}^{k} \binom{m}{j}\binom{n}{k-j} = \binom{m+n}{k}. \qquad (2.57.a)$$

Proof. The binomial $\binom{m+n}{k}$ corresponds to the number of subsets of I_{m+n} of cardinality k. One can get such a subset by first choosing $0 \le j \le k$ elements from $\{1, \ldots, m\}$, and then taking the remaining $k-j$ elements from $\{m+1, \ldots, m+n\}$. For the first phase we have $\binom{m}{j}$ possible outcomes, while for the second the possible outcomes are $\binom{n}{k-j}$. By the Multiplication Principle 1.34, summing on j one yields easily the conclusion.   □

2.4.2 Alternating Sign Binomial Identities

We consider here identities of the form

$$\sum_{i\in I}(-1)^i a_i = b, \qquad (2.57.b)$$

where I is a finite subset of $\mathbb{N}$, $b \in \mathbb{R}$ and, for each $i \in I$, $a_i \in \mathbb{R}$.

Following [5], the main argument in the combinatorial proofs of the identities of this section is the obvious *very important* fact that (2.57.b) is equivalent to the following equality:

$$\sum_{\substack{i\in I \\ i \text{ even}}} a_i = \sum_{\substack{i\in I \\ i \text{ odd}}} a_i + b.$$

Example 2.58 In Corollary 2.21 we have proved that if $n \in \mathbb{N}_{\ge 1}$, then

$$\sum_{k=0}^{n}(-1)^k \binom{n}{k} = 0. \qquad (2.58.a)$$

---

[10] Alexandre-Théophile Vandermonde (1735–1796).

The latter identity is equivalent to

$$\sum_{\substack{k \in \mathbb{N} \\ k \text{ even}}} \binom{n}{k} = \sum_{\substack{k \in \mathbb{N} \\ k \text{ odd}}} \binom{n}{k}.$$

This means that, among the subsets of I_n, the number of subsets of even cardinality equals the number of subsets of odd cardinality. This can be seen by considering the map Φ defined on the subsets of even cardinality of I_n by

$$\Phi(A) = A \triangle \{1\} \qquad \forall A \subseteq I_n.$$

Then Φ changes the parity of the cardinality of the subsets of I_n: indeed

$$\Phi(A) = \begin{cases} A \setminus \{1\} & \text{if } 1 \in A, \\ A \cup \{1\} & \text{if } 1 \notin A. \end{cases}$$

It is easy to see that Φ is a bijection: indeed by Point 9 of Proposition 1.7, $\Phi(A) = B$ if and only if $A = B \triangle \{1\}$. The subsets of I_n of even cardinality thus correspond one to one, via Φ, to the subsets of I_n of odd cardinality. It follows that the two families of subsets have the same cardinality. $\qquad\qquad\qquad\qquad\qquad\qquad\qquad\qquad\qquad\qquad$ $\square$

The next result generalizes (2.58.a).

Proposition 2.59 *Let $0 \le i \le n$ be two integers. One has*

$$\sum_{k=i}^{n} (-1)^k \binom{n}{k} \binom{k}{i} = (-1)^n \delta_{i,n}, \quad \text{where} \quad \delta_{i,n} := \begin{cases} 1 & \text{if } i = n, \\ 0 & \text{otherwise.} \end{cases} \qquad (2.59.a)$$

Proof. Notice that, for $i \le k \le n$, $\binom{n}{k}\binom{k}{i}$ counts the number of 2-sequences of sets (S, T) with $T \subseteq S \subseteq I_n$, $|T| = i$, $|S| = k$. For $0 \le i \le n$ we thus define the families

$$\mathscr{E} = \{(S, T) : T \subseteq S \subseteq I_n, \ |T| = i, \ |S| \text{ even}\},$$

$$\mathscr{O} = \{(S, T) : T \subseteq S \subseteq I_n, \ |T| = i, \ |S| \text{ odd}\}.$$

The claim is equivalent to

$$|\mathscr{E}| = |\mathscr{O}| + (-1)^n \delta_{i,n}. \qquad (2.59.b)$$

We consider the following cases:

- If $i = n$ is even then $\mathscr{E} = \{(I_n, I_n)\}$ and $\mathscr{O} = \emptyset$ so that $|\mathscr{E}| = 1 = |\mathscr{O}| + 1$;

- If $i = n$ is odd then $\mathscr{E} = \emptyset$ and $\mathscr{O} = \{(I_n, I_n)\}$ so that $|\mathscr{E}| = 0 = |\mathscr{O}| - 1$;
- Assume that $i < n$. If $(S, T) \in \mathscr{E}$ then $T \neq I_n$: let $\xi(T) = \max I_n \setminus T$. We set

$$\Phi(S, T) = (S \triangle \{\xi(T)\}, T) \qquad (S, T) \in \mathscr{E}.$$

Just to clarify, assume for instance that $n = 5, i = 2, T = \{1, 3\}$ and $S = \{1, 3, 5\}$. Then $\xi(T) = \max I_5 \setminus \{1, 3\} = \max\{2, 4, 5\} = 5$ and $\Phi(S, T) = (\{1, 3\}, \{1, 3\})$. Notice that $|S \triangle \{\xi(T)\}| = |S| \pm 1$ depending whether $\xi(T) \in S$ or $\xi(T) \notin S$; thus $\Phi(S, T) \subseteq \mathscr{O}$ for all $(S, T) \in \mathscr{E}$. Then $\Phi : \mathscr{E} \to \mathscr{O}$ is a bijective map, proving that $|\mathscr{E}| = |\mathscr{O}|$: indeed, by Point 9 of Proposition 1.7, one has $\Phi(S_0, T) = (S_1, T)$ if and only if $S_0 = S_1 \triangle \{\xi(T)\}$.   □

Remark 2.60 Notice that, by choosing $i = 0 < n$ in (2.59.a), we get (2.58.a).

Example 2.61 Niccolò and Tommaso have money to buy a tub with b scoops from an ice-cream shop that sells a number of different flavours of ice cream, and b number of different flavours of sorbets. They agree that Niccolò will choose different scoops of ice-cream and sorbets, and Tommaso will choose only ice-cream scoops. How many different tubs can they buy? Are there more tubs in which Tommaso chooses an even number of scoops, or an odd number of scoops?

Solution. If Tommaso takes c scoops, k_1 of the first flavour, k_2 of the second one, $\dots$, k_a of the a-th, then Niccolò can choose $b - c$ different scoops of ice-cream and sorbets among the $a + b$ possible ones. For each $0 \leq c \leq b$, Tommaso and Niccolò have $C((a, c))$ and $\binom{a + b}{b - c}$ different possibilities, respectively. Therefore, by the Multiplication Principle 1.34 and Theorem 2.35 the possible tubs are

$$\sum_{c=0}^{b} \binom{a + b}{b - c} C((a, c)).$$

The next proposition will permit us to prove that the tubs in which Tommaso chooses an even number of scoops are one more than the tubs in which he chooses an odd number of scoops.   □

Proposition 2.62 *Let $a, b \in \mathbb{N}$. Then*

$$\sum_{c=0}^{b} (-1)^c \binom{a + b}{b - c} C((a, c)) = 1. \qquad (2.62.\text{a})$$

Proof. Consider the sets

$$\mathscr{E} := \{(S, (k_1, \dots, k_a)) : S \subseteq I_{a+b}, k_1, \dots, k_a \in \mathbb{N}, k_1 + \dots + k_a = b - |S| \text{ is even}\},$$

$$\mathscr{O} := \{(S, (k_1, \ldots, k_a)) : S \subseteq I_{a+b}, k_1, \ldots, k_a \in \mathbb{N}, k_1 + \cdots + k_a = b - |S| \text{ is odd}\}.$$

For instance, if $a = 4$ and $b = 5$, then $(\{2, 5, 8\}, (0, 0, 0, 2))$ belongs to $\mathscr{E}$, while $(\{2, 8\}, (0, 1, 0, 2))$ belongs to $\mathscr{O}$. Any element $(S, (k_1, \ldots, k_a))$ of $\mathscr{E} \cup \mathscr{O}$ with $k_1 + \cdots + k_a = c$ can be obtained in two phases: first one chooses a subset S of I_{a+b} of cardinality $b - c$, and then one consider an a-composition of c. Since there are $C(a + b, b - c)$ ways to choose S, and $C((a, c))$ ways to choose the a-composition of c, by the Multiplication Principle 1.34 one gets easily that the set $\mathscr{E}$ has cardinality

$$|\mathscr{E}| = \sum_{\substack{0 \le c \le b \\ c \text{ even}}} C(a + b, b - c) C((a, c)),$$

and the set $\mathscr{O}$ has cardinality

$$|\mathscr{O}| = \sum_{\substack{0 \le c \le b \\ c \text{ odd}}} C(a + b, b - c) C((a, c)).$$

We denote by $\mathscr{E}^*$ the set $\mathscr{E} \setminus \{(\{a + 1, \ldots, a + b\}, \underbrace{(0, \ldots, 0)}_{a})\}$; for instance, again with $a = 4$ and $b = 5$, $\mathscr{E}^* = \mathscr{E} \setminus \{(\{5\}, (0, 0, 0, 0))\}$. Given $\mathfrak{s} := (S, (k_1, \ldots, k_a)) \in \mathscr{O} \cup \mathscr{E}^*$, the set $(S \cap I_a) \cup \{1 \le i \le a : k_i \ne 0\}$ is not empty; let

$$i(\mathfrak{s}) = \max \left((S \cap I_a) \cup \{1 \le i \le a : k_i \ne 0\}\right).$$

Then we define

$$\Psi(\mathfrak{s}) = \begin{cases} (S \setminus \{i(\mathfrak{s})\}, (k_1, \ldots, k_{i(\mathfrak{s})-1}, k_{i(\mathfrak{s})} + 1, \ldots, k_a)) & \text{if } i(\mathfrak{s}) \in S, \\ (S \cup \{i(\mathfrak{s})\}, (k_1, \ldots, k_{i(\mathfrak{s})-1}, k_{i(\mathfrak{s})} - 1, \ldots, k_a)) & \text{if } i(\mathfrak{s}) \notin S. \end{cases}$$

Observe that if $i(\mathfrak{s}) \notin S$, then $k_{i(\mathfrak{s})} \ge 1$ and hence $(k_1, \ldots, k_{i(\mathfrak{s})-1}, k_{i(\mathfrak{s})} - 1, \ldots, k_a)$ is an a-composition of $b - |S|$. If $\mathfrak{s} := (S, (k_1, \ldots, k_a))$ and $\Psi(\mathfrak{s}) = \mathfrak{t} := (T, (h_1, \ldots, h_a))$, then it is easy to check that the following conditions are satisfied:

1. $i(\mathfrak{s}) = i(\mathfrak{t})$;
2. $T = S \triangle \{i(\mathfrak{s})\}$;
3. $(k_1, \ldots, k_{i(\mathfrak{s})-1}, k_{i(\mathfrak{s})+1}, \ldots, k_a) = (h_1, \ldots, h_{i(\mathfrak{s})-1}, h_{i(\mathfrak{s})+1}, \ldots, h_a)$ and

$$k_{i(\mathfrak{s})} = \begin{cases} h_{i(\mathfrak{s})} - 1 & \text{if } i(\mathfrak{s}) \in S \text{ (or equiv. } i(\mathfrak{s}) \notin T), \\ h_{i(\mathfrak{s})} + 1 & \text{if } i(\mathfrak{s}) \notin S \text{ (or equiv. } i(\mathfrak{s}) \in T). \end{cases}$$

Therefore $\Psi(\Psi(\mathfrak{s})) = ((S \triangle \{i(\mathfrak{s})\}) \triangle \{i(\mathfrak{t})\}, (k_1, \ldots, k_{i(\mathfrak{s})-1}, k_{i(\mathfrak{s})} + 1 - 1, \ldots, k_a)) = \mathfrak{s}$. For instance, always with $a = 4$ and $b = 5$, we have

$$\Psi(\Psi(\{2, 8\}, (0, 1, 0, 2))) = \Psi(\{2, 4, 8\}, (0, 1, 0, 1)) = (\{2, 8\}, (0, 1, 0, 2)).$$

Hence Ψ coincides with its inverse map and then it is a bijective map of $(\mathscr{O} \cup \mathscr{E}^*)$ in itself, sending elements of $\mathscr{O}$ to elements of $\mathscr{E}^*$, and vice versa. Hence

$$\sum_{c=0}^{b}(-1)^c C(a+b, b-c)C((a, c)) = |\mathscr{E}| - |\mathscr{O}| = |\mathscr{E}^*| + 1 - |\mathscr{O}| = 1. \qquad \square$$

Surprisingly, the following application of Proposition 2.62 will be used to prove the formula for the n-th derivative of functions of the form $1/g(x)$ in Sect. 5.1.4.

Corollary 2.63 *Let $0 \le j \le n$ be natural numbers. Then*

$$\sum_{k=j}^{n}(-1)^k \binom{n+1}{k+1}\binom{k}{j} = (-1)^j.$$

Proof. Indeed, applying Proposition 2.62 with $a = j + 1$, $b = n - j$ we get

$$1 = \sum_{c=0}^{n-j}(-1)^c \binom{n+1}{n-j-c} C((j+1, c)),$$

so that, by Theorem 2.35, $1 = \displaystyle\sum_{c=0}^{n-j}(-1)^c \binom{n+1}{n-j-c}\binom{j+c}{c}$. By setting $k = j+c$ we obtain

$$1 = \sum_{k=j}^{n}(-1)^{k-j} \binom{n+1}{n-k}\binom{k}{k-j} = \sum_{k=j}^{n}(-1)^{k-j} \binom{n+1}{k+1}\binom{k}{j}.$$

We conclude, multiplying on both sides by $(-1)^j$. $\square$

2.5 Problems

Problem 2.1 Prove by induction Stifel recursive formula

$$\binom{n}{k} = \binom{n-1}{k-1} + \binom{n-1}{k} \quad \text{for } k, n \in \mathbb{N}_{\geq 1}.$$

Problem 2.2 How many ways are there for ordering the 52 cards of a deck?

Problem 2.3 How many ways are there to distribute 9 different books among 15 students, if no student receives more than one book?

Problem 2.4 How many possible anagrams are there for the word INFINITY?

Problem 2.5 If a coin is tossed 10 times, what is the probability that Heads comes up at least 8 times?

Problem 2.6 In a weekly lottery, five balls are selected (without replacement) from an urn containing balls numbered from 1 to 90. Calculate the probability that in a given week:

1. The first number chosen is 37;
2. The second number chosen is 37;
3. The first and second numbers chosen are respectively 37 and 51.

Problem 2.7 How many sequences of 4 numbers are there with only one 8, and without any digit repeated exactly twice? (Sequences starting with 0 are allowed.)

Problem 2.8 If one writes all the numbers from 1 to 10^5, how many times does one write the digit 5?

Problem 2.9 If one throws three distinct dice what is the probability that the highest number is twice the lowest?

Problem 2.10 How many n-sequences of I_3 have exactly 9 digits equal to 1?

Problem 2.11 How many possible committees can be formed from a set of 4 men and 6 women if:

1. There are at least two men and twice as many women as men?
2. There are 4 members in all, at least two of whom are women, and Mr. and Mrs. Jones can not simultaneously be members?

Problem 2.12 There are 6 different books in English, 8 in Russian and 5 in Spanish. In how many ways can one arrange the books in a row on a shelf with all books in the same language grouped together?

Problem 2.13 How many words of 10 different letters can be formed using the 5 vowels and 5 consonants chosen from among the 21 possible consonants of the English alphabet? What is the probability that one of these words does not contain two consecutive consonants?

Problem 2.14 In how many ways can one distribute 40 identical jelly beans to 4 children in the following cases:

1. Without any restrictions?
2. If each child receives 10 jelly beans?
3. If each child receives at least one jelly bean?

Problem 2.15 What is the probability that in a sequence without repetitions of $\{a, b, c, d, e, f\}$ one has:

1. a, b in consecutive positions?
2. a appearing before b?

Problem 2.16 A man has n friends and each evening for a year (365 evenings) he invites a different group of 4 of them to his home. How large must n be in order for this to be possible?

Problem 2.17 In the first round of a tournament involving $n = 2^m$ players, the n players are divided into $n/2$ pairs each of which then plays a game. The losers are eliminated and the winners participate in the second round, and so on, until there remains a single player, the winner of the tournament.

1. How many outcomes are possible for the first round?
2. How many possible outcomes can the tournament have, if by "outcome of the tournament" we mean complete information on all the rounds?

Problem 2.18 Suppose that a subset of 60 different days of the year is chosen by extraction. What is the probability that there are 5 days for each month in the subset? (For simplicity, assume that there are 12 months of 30 days each.)

Problem 2.19 In how many bridge hands do the players North and South have all the spades?

Problem 2.20 What is the probability of choosing at random a 10-sequence of I_{10} without repetitions such that:

1. In the first position there is an odd digit and one of 1, 2, 3, 4, 5 occupies the final position?
2. 5 is not in the first position and 9 is not in the last?

Problem 2.21 What is the probability that in a hand of 5 cards taken from a deck of 52 there is:

1. At least one of each of the following cards: Ace, King, Queen, Jack?
2. At least one of the following cards: Ace, King, Queen, Jack?
3. The same number of hearts and spades?

Problem 2.22 In how many ways can one form a group (not ordered) of four couples chosen from a set of 30 people?

Problem 2.23 Let k be a prescribed natural number satisfying $1 \le k \le 17$; then fix 4 numbers chosen between 1 and 20.

1. What is the probability that k appears among the four numbers chosen, and is the smallest of the four?
2. What is the probability that k appears among the four numbers chosen and is the second largest of them?

Problem 2.24 What is the probability that in five tosses of a die only two different numbers come up?

Problem 2.25 From a set of $2n$ objects, of which n are identical and the other n all different from each other, how many possible selections of n objects are there?

Problem 2.26 In a lake 10 fish are tagged from a population of k. Twenty fish are caught. What is the probability that two of them are tagged?

Problem 2.27 We wish to organize three dinners on three consecutive evenings to each of which will be invited three friends chosen from among the n schoolmates with whom we are still in contact. In how many ways can the guests for the three evenings be chosen?

Problem 2.28 We have organized ten dinners on 10 consecutive evenings. To these dinners we wish to invite, among others, the 8 schoolmates with whom we are still in contact, but we are uncertain whether to invite them all for the first evening, or to not invite more than one friend each evening, or to make invitations in some other way (but with no friend invited more than once). How many possible choices are there?

Problem 2.29 What is the probability that in a hand of 5 cards from a deck of 52:

1. There is exactly one pair (not two pairs or three of a kind)?
2. There are at least two cards with the same value?
3. There is at least one spade, one heart, no club or diamond, and the face values of the spade cards are all strictly higher than the face values of the heart cards?

Problem 2.30 How many subsets consisting of three distinct natural numbers between 1 and 90 (extremes included) are there if the sum of the three is:

1. An even number?
2. Divisible by 3?
3. Divisible by 4?

Problem 2.31 In how many ways can one choose 10 coins from a pile of euro-coins consisting of 1 cent, 2 cent, 5 cent, and 10 cent coins?

Problem 2.32 We must establish how many places on a commission of 15 congressmen will be awarded to Democrats, Republicans, and Independents. How many possibilities are there if each party must have at least two members of the commission? What if, moreover, no party should by itself comprise the majority of the commission?

Problem 2.33 In how many ways can one distribute 18 chocolate donuts, 12 cinnamon donuts and 14 honey-dip donuts to 4 pupils if each of these requires at least two donuts of each type?

Problem 2.34 How many integer solutions of $x_1 + x_2 + x_3 = 0$ are there with each $x_i \geq -5$?

Problem 2.35 How many electoral results are possible (number of votes for each candidate) if there are 3 candidates and 30 voters? What if, moreover, some candidate obtains an absolute majority?

Problem 2.36 How many numbers between 0 and 10 000 are such that the sum of their digits is:

1. Equal to 7?
2. Less than or equal to 7?

Problem 2.37 How many natural number solutions are there for the equation

$$2x_1 + 2x_2 + x_3 + x_4 = 12 \quad ?$$

Problem 2.38 How many natural number solutions are there to the system of inequalities

$$\begin{cases} x_1 + \cdots + x_6 \le 20 \\ x_1 + x_2 + x_3 \le 7 \end{cases} ?$$

Problem 2.39 How many binary sequences are there in which 0 appears n times and 1 appears m times, and having k groups of consecutive 0's?

Problem 2.40 How many binary sequences of n terms contain the *pattern* 01 exactly m times?

Problem 2.41 Let $m \le n$ and $s \le r$ be natural numbers. How many ways are there to distribute r identical balls in n distinct boxes in such a way that the first m boxes contain a total of at least s balls?

Problem 2.42   1. In how many ways can one seat 8 people in a row of 15 seats of a cinema?
2. In how many of the preceding seating arrangements, do 3 given friends receive adjacent seats?

Problem 2.43 If a coin is tossed n times, what is the probability that:

1. The first "Heads" appears after exactly m "Tails";
2. The i-th "Heads" appears after "Tails" has come up m times?

Problem 2.44 In how many ways can one distribute 3 different teddy bears and 9 identical lollipops to four children:

1. Without restrictions?
2. Without having any child receive two or more teddy bears?
3. With each child receiving 3 "gifts"?

Problem 2.45 Find the number of binary 20-sequences with exactly 15 terms equal to 0 and 5 terms equal to 1. How many sequences with 15 terms of one type and 5 of the other?

Problem 2.46 If n different objects are distributed randomly into n different boxes what is the probability that:

1. No box is empty?
2. Exactly one box is empty?
3. Exactly two boxes are empty?

Problem 2.47 In how many ways can one distribute 4 red balls, 5 blue ones, and 7 black balls into:

1. Two boxes?
2. Two boxes neither of which is empty?
3. In how many ways can one place 4 red balls, 6 blue, and 8 black ones into two boxes?

Discuss the case in which the boxes are distinct separately from that in which they are indistinguishable.

Problem 2.48 In a 4 story house (in addition to the ground floor) an elevator leaves the ground floor with 5 people aboard. No one else gets on, and every person gets off randomly at one of the four (upper) stories. Calculate the probability that the elevator:

1. Arrives empty at the fourth (top) floor;
2. Arrives empty at the third floor;
3. Becomes empty at exactly the third floor;
4. Arrives at the fourth floor carrying 2 people.

Problem 2.49 We wish to open a locked door. We have a ring of 100 keys, of which exactly 2 open the door in question. The keys are tried successively one by one

1. What is the probability that the 56th key opens the door?
2. What is the probability that the 56th key is the second key that opens the door?

Problem 2.50 Five marbles are extracted *simultaneously* from an urn containing 10 red marbles and 20 blue ones. Find the probability that only one blue marble is extracted.

Problem 2.51 How many committees of 5 people with at least two women and at least one man can be formed by choosing from a group of 6 women and 8 men?

Problem 2.52 Give a combinatorial proof of the equality (see Point 1 of Proposition 2.39)

$$C((n, k)) = C((k + 1, n - 1)) \quad \forall k \in \mathbb{N}, \ n \in \mathbb{N}_{\geq 1}.$$

Problem 2.53 By using the Binomial Theorem, one sees that $\binom{2n}{n}$ is the coefficient of $x^n y^n$ in the sum giving $(x + y)^{2n}$. Write $(x + y)^{2n}$ in the form $(x + y)^n (x + y)^n$, expand both factors $(x + y)^n$ using the Binomial Theorem, and look for the coefficient of $x^n y^n$ that arises from expanding this product. Show that this procedure leads to an alternative proof for Identity 2 of Proposition 2.56.

Problem 2.54 (*Magnetic colors*) Explain the following magic trick, published by N. Gilbreath in [18]:

Take a complete deck of cards out of the case. Ask a spectator to give a few straight cuts, deal off any number of cards into a pile on the table, and then riffle shuffle the pile on the table with the pile still in his hand. You say something like "as you know, red and black cards are magnetically attracted one to each other". Then, ask the spectator to pick the deck up into dealing position, and deal off the top two cards. They will definitely be one red and one black. Deal off the next two cards. Again, one red/one black. Keep going. You'll find that each consecutive pair alternates in color!

Problem 2.55 Let k, m, n be natural numbers. Prove the following identity:

$$\binom{n}{0}\binom{m}{k} + \binom{n}{1}\binom{m}{k-1} + \binom{n}{2}\binom{m}{k-2} + \cdots$$

$$\cdots + \binom{n}{k-1}\binom{m}{1} + \binom{n}{k}\binom{m}{0} = \binom{n+m}{k}.$$

[Hint: one can proceed in a manner similar to the proof of Identity 2 of Proposition 2.56, or use the Binomial Theorem as in Problem 2.53.]

Problem 2.56 Prove by induction on k the Point 3 of Proposition 2.56.

Problem 2.57 Prove Proposition 2.59 by means of Corollary 2.21.

$\left[\text{Hint: for } i < n \text{ multiply } \binom{n}{k}\binom{k}{i} \text{ by } \dfrac{(n-i)!}{(n-i)!}.\right]$

Problem 2.58 Let $k, n \in \mathbb{N}_{\geq 1}$.

1. Given an n-sharing $(C_1, \ldots, C_n)$ of I_k prove that there is a one to one correspondence between $\mathcal{P}(I_k)$ and the n-sharings $(B_1, \ldots, B_n)$ of I_k with

$$B_1 \subseteq C_1, \ldots, B_n \subseteq C_n.$$

2. Let $(k_1, \ldots, k_n)$ be a given n-composition of k. Deduce from Point 1 the value of the sum

$$\sum_{\substack{(j_1,\ldots,j_n)\in\mathbb{N}^n \\ j_1 \leq k_1,\ldots,j_n \leq k_n}} \binom{k_1}{j_1} \cdots \binom{k_n}{j_n}.$$

3. Prove that

$$\sum_{\substack{(k_1,\ldots,k_n)\in\mathbb{N}^n \\ k_1+\cdots+k_n=k}} \sum_{\substack{(j_1,\ldots,j_n)\in\mathbb{N}^n \\ j_1 \leq k_1,\ldots,j_n \leq k_n}} \binom{k_1}{j_1} \cdots \binom{k_n}{j_n} = 2^k \binom{k+n-1}{k}. \tag{2.58a}$$

Chapter 3
Occupancy Constraints

Abstract This chapter introduces a finer level of analysis for counting sequences or collections that are subject to some occupancy constraint, namely a constraint on the number of repetitions of its elements. Several problems are considered. As more unusual application in this framework, we prove the Leibniz rule for the derivatives of a product of functions, and count, in terms of the Catalan numbers, the Dyck sequences, i.e., the binary sequences of even length with equal number of 0's and 1's where, at each position, the number of 1's does not exceed the number of 0's.

3.1 Sequences and Sharings with Occupancy Constraints

In this section we discuss sequences and sharings subjected to specific constraints regarding repetitions of their elements.

3.1.1 Sequences and Sharings with Occupancy Sequences: Permutations and Anagrams

We wish to count the k-sequences of I_n with a specified number of copies of the numbers $1, \ldots, n$, or equivalently the n-sharings of I_k into subsets of prescribed cardinalities.

Definition 3.1 Let $n \in \mathbb{N}_{\geq 1}$ and $k_1, \ldots, k_n, k \in \mathbb{N}$ be such that $k_1 + \cdots + k_n = k$. We call:

1. **k-sequence of I_n with occupancy sequence** $(k_1, \ldots, k_n)$ any k-sequence of I_n with k_1 repetitions of $1, \ldots, k_n$ repetitions of n;
2. **n-sharing of I_k with occupancy sequence** $(k_1, \ldots, k_n)$ any n-sharing $(C_1, \ldots, C_n)$ of I_k such that $|C_1| = k_1, \ldots, |C_n| = k_n$.

We will denote **the number of k-sequences of I_n, or equivalently of n-sharings of I_k, with occupancy sequence** $(k_1, \ldots, k_n)$ by $S(n, k; (k_1, \ldots, k_n))$. □

© Springer International Publishing Switzerland 2016   65
C. Mariconda and A. Tonolo, *Discrete Calculus*,
UNITEXT - La Matematica per il 3+2 103, DOI 10.1007/978-3-319-03038-8_3

Remark 3.2 In view of Theorem 1.28 and Remark 1.29 the number of k-sequences of I_n with occupancy $(k_1, \ldots, k_n)$ is the same as that of the number of n-sharings of I_k with occupancy $(k_1, \ldots, k_n)$. Note that the first n in the symbol $S(n, k; (k_1, \ldots, k_n))$ is redundant since the value of n may be inferred from the length of the occupancy sequence.

Example 3.3 The 9-sequence $(3, 5, 5, 7, 7, 2, 2, 2, 2)$ of I_7 is an example of a *sequence with occupancy* $(0, 4, 1, 0, 2, 0, 2)$. Such a sequence corresponds to the 7-sharing $(C_1, C_2, C_3, C_4, C_5, C_6, C_7)$ of I_9 defined by $C_1 = \emptyset, C_2 = \{6, 7, 8, 9\}$, $C_3 = \{1\}, C_4 = \emptyset, C_5 = \{2, 3\}, C_6 = \emptyset, C_7 = \{4, 5\}$. Since $|C_1| = |C_4| = |C_6| = 0$, $|C_2| = 4, |C_3| = 1, |C_5| = |C_7| = 2$ this is a *sharing with occupancy* $(0, 4, 1, 0, 2, 0, 2)$. □

Example 3.4 The possible ways of distributing 10 different prizes (labeled by the set I_{10}) among four people in such a way that the first person receives 3 prizes, the second 2, the third 1 and the fourth 4 can be represented via the 4-sharings of I_{10} with occupancy $(3, 2, 1, 4)$. □

Remark 3.5 If $(a_1, \ldots, a_k)$ is a k-sequence of I_n with occupancy $(k_1, \ldots, k_n)$, the set of all k-sequences of I_n with that same occupancy is given by the set of permutations (see Definition 1.20) of the sequence $(a_1, \ldots, a_k)$.

Definition 3.6 Let $(a_1, \ldots, a_k)$ be a k-sequence of I_n. We use $P(a_1, \ldots, a_k)$ to denote the **number of permutations of** $(a_1, \ldots, a_k)$. □

Theorem 3.7 *Let* $(a_1, \ldots, a_k)$ *be a k-sequence of I_n with occupancy* $(k_1, \ldots, k_n)$. *Then*

$$S(n, k; (k_1, \ldots, k_n)) = P(a_1, \ldots, a_k) = \frac{k!}{k_1! k_2! \cdots k_n!}.$$

Proof. The permutations of the sequence $(a_1, \ldots, a_k)$ consist of precisely the k-sequences of I_n with occupancy $(k_1, \ldots, k_n)$. Thus one certainly has $P(a_1, \ldots, a_k)$ equals $S(n, k; (k_1, \ldots, k_n))$. The construction of a k-sequence of I_n with occupancy $(k_1, \ldots, k_n)$ can then be carried out in n distinct steps (see Fig. 3.1): in the first step we choose the k_1 positions where 1 appears, in the second step, the k_2 positions for $2, \ldots,$ and in the last step we choose the k_n positions for n. In the first step we have $C(k, k_1)$ choices for the position of 1, in the second step we have $C(k - k_1, k_2)$ choices for placement of $2, \ldots,$ in the last step we have $C(k - (k_1 + \cdots + k_{n-1}), k_n)$ choices for the positions of n. Different choices at any step lead necessarily to distinct final results. By the Multiplication Principle 1.34 the total number of possible choices is

$$\binom{k}{k_1}\binom{k-k_1}{k_2}\binom{k-(k_1+k_2)}{k_3}\cdots\binom{k-(k_1+k_2+\cdots+k_{n-1})}{k_n}=$$

$$=\frac{k!}{k_1!(k-k_1)!}\frac{(k-k_1)!}{k_2!(k-(k_1+k_2))!}\cdots\frac{(k-(k_1+\cdots+k_{n-1}))!}{k_n!0!}=\frac{k!}{k_1!k_2!\cdots k_n!}.$$

□

Remark 3.8 One sees immediately from Theorem 3.7 that $S(n, k; (k_1, \ldots, k_n))$ is invariant under permutations of $(k_1, \ldots, k_n)$.

One should be careful not to confuse $S(n, k; (k_1, \ldots, k_n))$ with $P(k_1, \ldots, k_n)$. For example,

$$S(4, 8; (2, 2, 3, 1)) = \frac{8!}{2!2!3!1!}, \quad P(2, 2, 3, 1) = S(3, 4; (1, 2, 1)) = \frac{4!}{2!1!1!}.$$

One gets easily the following identity:

Corollary 3.9 *Let* $n, k \in \mathbb{N}$. *Then*

$$n^k = \sum_{k_1+\cdots+k_n=k} \frac{k!}{k_1!\cdots k_n!}.$$

Proof. Any k-sequence of I_n has a certain number k_1 of 1, k_2 of 2, $\ldots, k_n$ of n with $k_1 + \cdots + k_n = k$. Therefore the number $S((n, k)) = n^k$ of all k-sequences is equal to the sum of the numbers of k-sequences of I_n with all possible occupancies $(k_1, \ldots, k_n)$. □

Example 3.10 Calculate the number of anagrams of BANANA.
Solution. Label the letters B, A, and N respectively with 1, 2, and 3. Then the number of anagrams sought coincides with the number of 6-sequences of I_3 with occupancy $(3, 1, 2)$. By Theorem 3.7 that number is

$$S(3, 6; (3, 1, 2)) = 6!/(3! \times 2!) = 60.$$

□

Fig. 3.1 Construction of a sequence with 1 repeated k_1 times, 2 repeated k_2 times,...

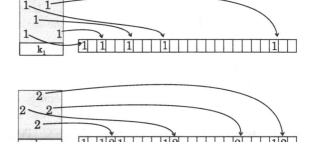

Example 3.11 How many different eight-digit binary sequences are there with six digits equal to 1 and two equal to 0?

Solution. One need only count the number of permutations of $(0, 0, 1, 1, 1, 1, 1, 1)$. The number of such permutations is $S(2, 8; (2, 6)) = 28$. One notes that an alternative approach would be to count the number $C(8, 6) = 28$ of possible placements for the digit 1.   □

Example 3.12 A factory produces kitchen stoves. At the last stop on the assembly line, a quality controller, after checking the appliance, marks each stove either with the number 1 (acceptable) or with the number 0 (unacceptable). Every day the quality controller examines 15 stoves. Supposing that the controller does not really verify the presence of defects, but rather chooses his rating by chance, calculate the probability that the 12-th stove is the third one to be marked unacceptable.

Solution. At the end of the day, the quality controller produces a binary 15-sequence. The number of such sequences is $S((2, 15)) = 2^{15}$. If the 12-th stove is the third one to be marked unacceptable, then the sequence produced by the controller must have exactly two 0's (and nine 1's) in the first 11 positions, and the number of such binary 11-sequences is $S(2, 11; (9, 2)) = C(11, 2) = 55$; in the 12-th position there must be a 0 (1 possibility), and the remaining three positions can be either 0 or 1 randomly ($2^3 = 8$ possibilities). By the Multiplication Principle 1.34 there are $55 \times 1 \times 8 = 440$ 15-sequences with the third 0 occurring at the 12-th position, and so the probability sought is $440/2^{15} \approx 0.013 = 1.3\,\%$.   □

Example 3.13 Nine students, three from class A, three from class B and three from class C must seat themselves in a row with 9 seats. If they take their seats randomly, what is the probability that the three students of class A, the three from class B, and the three from class C all find themselves in three consecutive seats?

Solution. From the terms of the problem, we need only to take into account the class to which each student belongs. Therefore we can consider the sample space of the 3-sequences of $I_3 = \{A, B, C\}$ with occupancy $(3, 3, 3)$. We seek the number of favorable outcomes, and will divide that number by the total number of possible outcomes to obtain the desired probability. The first question is: what is the number of all possible outcomes here? There are $S(3, 9; (3, 3, 3)) = 9!/3!3!3! = 1\,680$ ways of arranging the three students of each of the three classes A, B, C in a row of nine places. If the three students of each class must be together, we can imagine that we have to place the three blocks AAA, BBB, CCC. Thus instead of dealing with 9 letters we really have to deal with three blocks of letters. There are $3! = 6$ ways to place these blocks. Hence the probability that the students of each class sit together is $6/1\,680 (\approx 0.32\,\%)$.   □

Example 3.14 How many ways are there to make 10 letter words using the alphabet $\{a, b, c, d\}$ if each letter must appear at least twice but not more than four times?

Solution. We divide the problem into sub-problems by fixing exactly how many a's, how many b's, how many c's and how many d's there are in the word. There are two possibilities for having a total of 10 letters with each appearing at least twice. The first is that some letter appears four times, and the other 3 all appear twice; the second is that two letters appear 3 times, and the other two appear twice. In the first situation, there are 4 choices for the letter appearing 4 times, and $S(4, 10; (4, 2, 2, 2)) = 18\,900$ ways to place one letter 4 times and each of the other three twice. In the second instance, there are $C(4, 2) = 6$ choices for the two letters that appear 3 times and $S(4, 10; (3, 3, 2, 2)) = 25\,200$ ways of placing the two letters 3 times and two other letters twice. Hence the solution to the problem is that there are $4 \times 18\,900 + 6 \times 25\,200 = 226\,800$ possible words of the specified type.   □

Example 3.15 In how many ways can one assign 100 different diplomats to 5 embassies? In how many ways can one carry out these assignments if each embassy must have 20 diplomats assigned to it?
Solution. The first part of the problem is equivalent to calculating the number of 5-sharings of I_{100}. By Proposition 2.28 there are $S((100, 5)) = 5^{100}$ possible assignments. The second part of the problem amounts to finding the number of 5-sharings of I_{100} with occupancy $(20, 20, 20, 20, 20)$, and here the number is

$$S(5, 100; (20, 20, 20, 20, 20)) = 100!/(20!)^5. \qquad\qquad □$$

Example 3.16 In a bridge hand, a deck of 52 cards is randomly distributed to four players, North, South, East and West with each player receiving 13 cards. What is the probability that West has all 13 spades? What is the probability that each player holds an Ace?
Solution. There are $S(4, 52; (13, 13, 13, 13))$ ways of dealing the 52 cards to the four players N, S, E, and O. The cardinality of the event "West has all the spades" can be calculated by multiplying the number of ways in which West can hold all the spades (1 way) by the number of ways of distributing all the remaining 39 non-spade cards in the deck to the other three players (which may be done in $S(3, 39; (13, 13, 13))$ ways). Hence the probability that West has all the spades is

$$\frac{S(3, 39; (13, 13, 13))}{S(4, 52; (13, 13, 13, 13))} = \frac{39!/(13!)^3}{52!/(13!)^4}$$

$$= \frac{13!39!}{52!} = \binom{52}{13}^{-1} \approx 0.000000000002.$$

One could also give a more direct solution to this problem by considering only the cards given to West, without regard to the cards dealt to the other players. Indeed, if West is dealt 13 cards at random, there are $C(52, 13)$ possible different hands, and so the unique hand with 13 spades has probability $1/C(52, 13)$.

    In order to analyze the deals in which every player receives an Ace, we can adopt a two step procedure. The first step deals with the Aces: in all we have 4! ways to distribute the four Aces among the 4 players. In the second step we distribute the remaining 48 non-Aces to the 4 players: there are in all $S(4, 48; (12, 12, 12, 12))$ ways to do so. Hence the probability that every player holds an Ace is

$$\frac{4! \times S(4, 48; (12, 12, 12, 12))}{S(4, 52; (13, 13, 13, 13))} = \frac{4!48!/(12!)^4}{52!/(13!)^4}$$

$$= (13)^4 \times \binom{52}{4}^{-1} \approx 0.105. \qquad \square$$

Example 3.17 A boat has 3 cabins, each of which can contain at most 4 people. The various cabins are very different: cabin A has portholes and a bath, cabin B has no porthole but does have a bath, cabin C is the worst: it has neither portholes nor a bath, and moreover it smells of kerosene. Eight people take a trip on this boat. How many ways can cabin assignments be made if the captain wants only one person in cabin A, 3 people in cabin B and 4 in cabin C? What if he wants 3 people in A, 3 in B and 2 in C?

Solution. We must count the number of 3-sharings of I_8, in the first case with occupancy $(1, 3, 4)$, and in the second with occupancy $(3, 3, 2)$. Therefore there are $S(3, 8; (1, 3, 4)) = \dfrac{8!}{3!4!}$ cabin assignments in the first instance and $S(3, 8; (3, 3, 2)) = \dfrac{8!}{3!3!2!}$ in the second. $\qquad \square$

Example 3.18 In how many ways can one distribute 4 identical oranges and 6 distinguishable apples (for example, of different type) into 5 different containers? In how many ways is it possible to carry out the same distribution if it is required that every container should have exactly two pieces of fruit?

Solution. The possible distributions of the oranges correspond to the 5-compositions of 4, and the possible distributions of the apples correspond to the 5-sharings of I_6. Then, there are $C(4 + 5 - 1, 4) = 70$ ways to distribute 4 identical oranges in 5 distinct jars, and $5^6 = 15\,625$ ways for placing the 6 apples. By the Multiplication Principle 1.34 there are in all $70 \times 15\,625 = 1\,093\,750$ ways of distributing the 4 oranges and 6 apples. The requirement that there should be exactly two pieces of fruit in each jar complicates the matter. We begin by dealing with the oranges. There are three possible ways to distribute the 4 oranges in the 5 jars without having more than 2 oranges for each jar:

1. Two oranges in two jars, and no oranges in the remaining three jars. The two jars getting the two pairs of oranges can be chosen in $C(5, 2) = 10$ ways. The six apples can then be placed in the three jars left, two per jar, in $S(3, 6; (2, 2, 2)) = 90$ ways. Thus the first case consists in all of $10 \times 90 = 900$ possible distributions.
2. Two oranges in one jar, and the other two oranges in different jars. The jar with two oranges can be chosen in $C(5, 1) = 5$ ways, while the two containers with a single orange can be chosen in $C(4, 2) = 6$ ways (one could also consider, combining these two steps, of arranging the numbers 2,1,1,0,0 in the 5 jars in $S(3, 5; (1, 2, 2)) = 30$ ways). Now the two empty jars must each get two apples, while the two jars that already have an orange must each be given an apple. Hence the 6 apples can be distributed in $S(4, 6; (2, 2, 1, 1)) = 180$ ways, giving a total of $180 \times 5 \times 6 = 5400$ distributions in this second case.

3. Four oranges distributed in four different jars. This may be done in $C(5, 4) = 5$ ways, and the apples can be distributed in $S(5, 6; (2, 1, 1, 1, 1)) = 360$ ways, giving a total of $5 \times 360 = 1\,800$ distributions in this case.

Summing the various cases we have in all $900 + 5\,400 + 1\,800 = 8\,100$ distributions with 2 pieces of fruit in each jar.   □

3.1.2 Sequences and Sharings with Occupancy Collection

We now discuss a different type of constraint on the repetitions of the elements of a sequence and the cardinalities of the sets that comprise a sharing.

Definition 3.19 Let $n \in \mathbb{N}_{\geq 1}$ and $k, k_1, \ldots, k_n \in \mathbb{N}$ satisfy $k = k_1 + \cdots + k_n$. We define:

1. **k-sequence of I_n with occupancy collection** $[k_1, \ldots, k_n]$ a k-sequence of I_n whose occupancy sequence is any permutation of $(k_1, \ldots, k_n)$;
2. **n-sharing of I_k with occupancy collection** $[k_1, \ldots, k_n]$ an n-sharing of I_k whose occupancy sequence is any permutation of $(k_1, \ldots, k_n)$.

We use $S(n, k; [k_1, \ldots, k_n])$ to denote **the number of k-sequences of I_n**, or equivalently **of n-sharings of I_k, with occupancy** $[k_1, \ldots, k_n]$.   □

Example 3.20 The 9-sequences $(3, 5, 5, 7, 7, 2, 2, 2, 2)$ and $(4, 4, 4, 4, 1, 3, 1, 6, 3)$ of I_7 have occupancy $[1, 2, 2, 4, 0, 0, 0]$; indeed, in both one has one number repeated once, two numbers repeated twice, one number repeated 4 times and three numbers which do not appear.   □

Example 3.21 The words of 10 letters constructed using the alphabet $\{A, B, C\}$ and in which one letter is repeated twice, another letter repeated three times, and still another letter with five repetitions are precisely the set of 10-sequences of $\{A, B, C\}$ with occupancy $[2, 3, 5]$.   □

Example 3.22 The possible distributions of 10 different pieces of candy to three children such that one receives 2, another receives 3, and still another receives 5 can be represented via the 3-sharings (C_1, C_2, C_3) with occupancy $[2, 3, 5]$ of the set of pieces of candy.   □

Theorem 3.23 *Let $n \in \mathbb{N}_{\geq 1}$ and $k, k_1, \ldots, k_n \in \mathbb{N}$ satisfy $k = k_1 + \cdots + k_n$. Then*

$$S(n, k; [k_1, \ldots, k_n]) = S(n, k; (k_1, \ldots, k_n)) \times P(k_1, \ldots, k_n).$$

Proof. We must count the k-sequences of I_n whose occupancy sequence is a permutation of $(k_1, \ldots, k_n)$. We proceed in two steps: first we fix a permutation of $(k_1, \ldots, k_n)$, and then we choose a k-sequence of I_n whose occupancy sequence is given by that permutation. By the Multiplication Principle 1.34, we have in all $S(n, k; (k_1, \ldots, k_n)) \times P(k_1, \ldots, k_n)$ sequences. $\square$

Example 3.24 Let us reconsider Example 3.17 of the preceding section. A boat has three different cabins, each of which can accommodate 4 passengers. How many possible cabin assignments are there if the captain requires that one cabin has a single occupant, another cabin 3 occupants, and the remaining cabin 4 occupants? What if the captain requires that two cabins have 3 occupants each, and the third cabin has the remaining two passengers?

Solution. Use I_8 to label the set of people to be assigned cabins. The question then reduces to calculating the number of 3-sharings of I_8 with occupancy $[1, 3, 4]$ in the first instance, and $[3, 3, 2]$ in the second. By Theorem 3.23 we find that the number of possible assignments in the two cases are respectively

$$S(3, 8; (1, 3, 4)) \times P(1, 3, 4) = S(3, 8; (1, 3, 4)) \times S(3, 3; (1, 1, 1)) = \frac{8!}{3!4!} \times 3!,$$

$$S(3, 8; (3, 3, 2)) \times P(3, 3, 2) = S(3, 8; (3, 3, 2)) \times S(2, 3; (1, 2)) = \frac{8!}{3!3!2!} \times 3.$$
$\square$

Example 3.25 A doctor sees 5 patients in a given week.

1. In how many ways can the doctor see 2 patients on Monday, 2 patients on Tuesday, and 1 patient on Thursday?
2. In how many ways can the doctor see patients on Monday, Tuesday, and Thursday so as to see 2 on one day, 2 on another day and 1 on a third day?
3. In how many ways can he see the patients in a seven-day week in such a way as to see 2 one day, 2 another day, and 1 on a third day?

Solution. 1. The question reduces to counting the 3-sharings of I_5 with occupancy $(2, 2, 1)$, and these number $S(3, 5; (2, 2, 1)) = 30$.
2. The question reduces to counting the 3-sharings of I_5 with occupancy $[2, 2, 1]$, and in view of Theorem 3.23 in all these number

$$S(3, 5; [2, 2, 1]) = S(3, 5; (2, 2, 1)) \times P(2, 2, 1) =$$

$$= S(3, 5; (2, 2, 1)) \times S(2, 3; (1, 2)) = 90.$$

3. Here we must count the 7-sharings of I_5 with occupancy $[2, 2, 1, 0, 0, 0, 0]$. In all they are

$$S(7, 5; [2, 2, 1, 0, 0, 0, 0]) = S(7, 5; (2, 2, 1, 0, 0, 0, 0)) \times P(2, 2, 1, 0, 0, 0, 0) =$$

$$= S(7, 5; (2, 2, 1, 0, 0, 0, 0)) \times S(3, 7; (4, 1, 2)) = 3\,150.$$
$\square$

3.1.3 The Leibniz Rule for the Derivative of a Product ☕

A remarkable application of the results on the sharings with occupancy is the Leibniz formula for the n-th derivative of a product of functions, which generalizes the most known rule for the derivative of the product of two functions.

For each $n \in \mathbb{N}$, let $f^{(n)}$ denote the n-th derivative of a function f, with $f^{(0)} = f$.

Theorem 3.26 (Leibniz rule) *Let $k, n \geq 1$ and $f_1, \ldots, f_k$ be functions which are n-times differentiable on an open interval J of $\mathbb{R}$. The n-th derivative of the product $f_1 \cdots f_k$ of the functions $f_1, \ldots, f_k$ is*

$$(f_1 \cdots f_k)^{(n)} = \sum_{\substack{(C_1, \ldots, C_k) \\ k\text{-sharing of } I_n}} f_1^{(|C_1|)} \cdots f_k^{(|C_k|)}. \qquad (3.26.a)$$

Therefore one has

$$(f_1 \cdots f_k)^{(n)} = \sum_{\substack{(n_1, \ldots, n_k) \in \mathbb{N}^k \\ n_1 + \cdots + n_k = n}} \frac{n!}{n_1! \cdots n_k!} f_1^{(n_1)} \cdots f_k^{(n_k)}. \qquad (3.26.b)$$

Proof. Denote by $\mathscr{S}_k(I_n)$ the set of the k-sharings of I_n. Let us prove (3.26.a) by induction on n. If $n = 1$, one has

$$(f_1 \cdots f_k)' = f_1' f_2 \cdots f_k + \cdots + f_1 \cdots f_{k-1} f_k'.$$

Now, the k-sharings of I_1 are $(\{1\}, \underbrace{\emptyset, \ldots, \emptyset}_{k-1}), \ldots, (\underbrace{\emptyset, \ldots, \emptyset}_{k-1}, \{1\})$, and hence

$$\sum_{(C_1, \ldots, C_k) \in \mathscr{S}_k(I_1)} f_1^{(|C_1|)} \cdots f_k^{(|C_k|)} = f_1' f_2 \cdots f_k + \cdots + f_1 \cdots f_{k-1} f_k'.$$

The statement is then true for $n = 1$. Suppose (3.26.a) is true for a given $n \geq 1$ and let us prove it for $n + 1$.

First observe that each k-sharing $(B_1, \ldots, B_k)$ of I_n can be obtained from the k different k-sharings of I_{n+1}

$$(B_1 \cup \{n + 1\}, B_2, \ldots, B_k), \ldots, (B_1, \ldots, B_{k-1}, B_k \cup \{n + 1\})$$

eliminating the element $n+1$. Differentiating further the n-th derivative of the product $f_1 \cdots f_k$, by the inductive hypothesis we get

$$
\begin{aligned}
(f_1 \cdots f_k)^{(n+1)} &= \left((f_1 \cdots f_k)^{(n)}\right)' \\
&= \sum_{(B_1,\ldots,B_k)\in\mathscr{S}_k(I_n)} (f_1^{(|B_1|)} \cdots f_k^{(|B_k|)})'.
\end{aligned}
$$

Now, the derivative $(f_1^{(|B_1|)} \cdots f_k^{(|B_k|)})'$ consists of

$$
\sum_{j=1}^{k} f_1^{(|B_1|)} \cdots f_j^{(|B_j|+1)} \cdots f_k^{(|B_k|)} = \sum_{j=1}^{k} f_1^{(|B_1|)} \cdots f_j^{(|B_j\cup\{n+1\}|)} \cdots f_k^{(|B_k|)}.
$$

Therefore

$$
\sum_{(B_1,\ldots,B_k)\in\mathscr{S}_k(I_n)} (f_1^{(|B_1|)} \cdots f_k^{(|B_k|)})' = \sum_{(C_1,\ldots,C_k)\in\mathscr{S}_k(I_{n+1})} f_1^{(|C_1|)} \cdots f_k^{(|C_k|)}.
$$

To prove (3.26.b), it is sufficient to group the sharings in the sum in (3.26.a), according to their occupancy:

$$
\begin{aligned}
(f_1 \cdots f_k)^{(n)} &= \sum_{(C_1,\ldots,C_k)\in\mathscr{S}_k(I_n)} f_1^{(|C_1|)} \cdots f_k^{(|C_k|)} \\
&= \sum_{\substack{(n_1,\ldots,n_k)\,\in\,\mathbb{N}^k \\ n_1+\cdots+n_k=n}} \ \sum_{\substack{(C_1,\ldots,C_k)\,\in\,\mathscr{S}_k(I_n) \\ |C_1|=n_1,\ldots,|C_k|=n_k}} f_1^{(|C_1|)} \cdots f_k^{(|C_k|)}. \qquad (3.26.c)
\end{aligned}
$$

Now, if $(C_1,\ldots,C_k)$ is a k-sharing of I_n with occupancy $(n_1,\ldots,n_k)$, one has

$$
f_1^{(|C_1|)} \cdots f_k^{(|C_k|)} = f_1^{(n_1)} \cdots f_k^{(n_k)}.
$$

Moreover, by Theorem 3.7, there are exactly $S(k,n;(n_1,\ldots,n_k)) = \dfrac{n!}{n_1!\cdots n_k!}$ such sharings: therefore

$$
\sum_{\substack{(C_1,\ldots,C_k)\,\in\,\mathscr{S}_k(I_n) \\ |C_1|=n_1,\ldots,|C_k|=n_k}} f_1^{(|C_1|)} \cdots f_k^{(|C_k|)} = \frac{n!}{n_1!\cdots n_k!} f_1^{(n_1)} \cdots f_k^{(n_k)}.
$$

The formula (3.26.b) follows now easily from (3.26.c).   □

3.2 Collections and Compositions with Occupancy Constraints

In this section we discuss collections and compositions subject to certain constraints.

3.2.1 Collections and Compositions with Occupancy Sequence

The situation that we consider in this section will turn out to be rather trivial.

Definition 3.27 Let $k, k_1, \ldots, k_n \in \mathbb{N}$ with $k = k_1 + \cdots + k_n$. We call:

1. **k-collection of I_n with occupancy sequence** $(k_1, \ldots, k_n)$ any k-collection of I_n consisting of k_1 repetitions of 1, k_2 repetitions of 2, $\ldots$, k_n repetitions of n;
2. **n-composition of I_k with occupancy sequence** $(k_1, \ldots, k_n)$ the n-composition $k_1 + \cdots + k_n = k$. □

It follows immediately from the definitions that there is a unique k-collection of I_n or n-composition of k with a prescribed occupancy sequence.

3.2.2 Collections and Compositions with Occupancy Collection

We now deal with collections or compositions in which there appears a prescribed number of repetitions.

Definition 3.28 Let $k, k_1, \ldots, k_n \in \mathbb{N}$ with $k = k_1 + \cdots + k_n$. We define:

1. **k-collection of I_n with occupancy collection** $[k_1, \ldots, k_n]$ any k-collection of I_n with occupancy sequence equal to a permutation of $(k_1, \ldots, k_n)$;
2. **n-composition of k with occupancy collection** $[k_1, \ldots, k_n]$ any n-composition $(m_1, \ldots, m_n)$ of k with $(m_1, \ldots, m_n)$ a permutation of $(k_1, \ldots, k_n)$.

We will use $C(n, k; [k_1, \ldots, k_n])$ for **the number of k-collections of I_n**, or equivalently, **the number of n-compositions of k, with occupancy** $[k_1, \ldots, k_n]$. □

Remark 3.29 Note that the first n in the symbol $C(n, k; [k_1, \ldots, k_n])$ is redundant since it may be deduced from the length of the occupancy collection.

Example 3.30 Paula has asked Albert to buy packets of butter, bottles of milk, and bars of cooking chocolate so she can do some baking. Albert, however, does not recall the quantities of each of these to be purchased, but only that he should get 3 things of one type, 2 of another, and 5 of yet another of the types. The possible shopping lists from which he must choose one can be represented as the 10-collections of I_3 with occupancy $[2, 3, 5]$. Alternatively, the possible outcomes of his shopping trip can be represented by the 3-compositions (m_B, m_L, m_C) of 10 formed respectively by the quantities of Butter, Milk, and Chocolate, such that $[m_B, m_L, m_C] = [3, 2, 5]$.   □

Theorem 3.31 *Let* $n \in \mathbb{N}_{\geq 1}$ *and* $k, k_1, \ldots, k_n \in \mathbb{N}$ *with* $k = k_1 + \cdots + k_n$. *Then*

$$C(n, k; [k_1, \ldots, k_n]) = P(k_1, \ldots, k_n),$$

that is, the number of permutations of $(k_1, \ldots, k_n)$.

Proof. Since $C(n, k; (h_1, \ldots, h_n)) = 1$ for every permutation $(h_1, \ldots, h_n)$ of $(k_1, \ldots, k_n)$, the k-collections of I_n with occupancy $[k_1, \ldots, k_n]$ are by definition equal in number to the permutations of $(k_1, \ldots, k_n)$.   □

☞   Be careful not to confuse $C(n, k; [k_1, \ldots, k_n])$ with $S(n, k; (k_1, \ldots, k_n))$: the first is the number $P(k_1, \ldots, k_n)$ of permutations of $(k_1, \ldots, k_n)$ while the second equals the number $P(a_1, \ldots, a_k)$ of permutations of any k-sequence $(a_1, \ldots, a_k)$ with occupancy $(k_1, \ldots, k_n)$.

Example 3.32 With reference to Example 3.30, the number of possible shopping lists among which Albert must choose is

$$C(3, 10; [2, 3, 5]) = P(2, 3, 5) = S(3, 3; (1, 1, 1)) = 3!.$$

Therefore, there is a probability of $1 - 1/3! \approx 0.83 = 83\%$ that Paula will get angry.   □

The following example summarizes what we have seen regarding sequences and collections with occupancy constraints.

Example 3.33 We distribute 5 red balls amongst 8 different boxes. Determine:

1. The total number of possible distributions;
2. The number of distributions such that no box contains more than one ball;
3. The probability that no box contains more than one ball;
4. The probability that no box contains more than 2 balls.

Solution. We label the set of boxes with I_8.
1. Every distribution of the balls is specified by an 8-composition $(k_1, \ldots, k_8)$ of 5, where k_i is the number of balls in box i: in all there are

$$C((8, 5)) = C(7 + 5, 7) = C(12, 7) = 792$$

such distributions.

2. The distributions that place at most one red ball into each box correspond to the 8-compositions of 5 with occupancy $[1, 1, 1, 1, 1, 0, 0, 0]$. There are

$$C(8, 5; [1, 1, 1, 1, 1, 0, 0, 0]) = P(1, 1, 1, 1, 1, 0, 0, 0) = S(2, 8; (5, 3)) = 56$$

such 8-compositions. Alternatively, each such distribution corresponds to a choice of the boxes into which a red ball is to be placed, and so such distributions can be made in $C(8, 5) = \binom{8}{5} = 56$ ways.

3. The 8-compositions of 5 are not equiprobable: for instance, to realize the 8-composition $(5, 0, 0, \ldots, 0)$ we are forced to choose box 1 for five times among the eight available boxes; on the other hand, if one has to realize the 8-composition $(1, 1, 1, 1, 1, 0, 0, 0)$, he can place the first ball in one of the first five boxes (5 choices) then the second ball in one of the four remaining boxes (4 choices), and so on!

In order to have equi-probable distributions it is necessary to distinguish between the 5 balls. Let us label the balls with I_5. Then each distribution becomes equivalent to an 8-sharing of I_5, and these are equal in number to the 5-sequences of I_8, namely 8^5. The distributions with at most one ball per box correspond to the 8-sharings of I_5 with occupancy $[1, 1, 1, 1, 1, 0, 0, 0]$, and in all they number

$$S(8, 5; [1, 1, 1, 1, 1, 0, 0, 0]) = S(8, 5; (1, 1, 1, 1, 1, 0, 0, 0)) \times P(1, 1, 1, 1, 1, 0, 0, 0) =$$

$$= S(8, 5; (1, 1, 1, 1, 1, 0, 0, 0)) \times S(2, 8; (3, 5)) = 5! \times \frac{8!}{5!3!} = 8!/3!.$$

Therefore, the probability sought is $\dfrac{8!}{3!8^5} \approx 0.205 = 20.5\,\%$.

4. The admissible distributions in this case are the 8-sharings of I_5 with occupancy $[1, 1, 1, 1, 1, 0, 0, 0], [2, 1, 1, 1, 0, 0, 0, 0]$ or $[2, 2, 1, 0, 0, 0, 0, 0]$. The total number of such 8-sharings is

$$S(8, 5; (1, 1, 1, 1, 1, 0, 0, 0)) \times P(1, 1, 1, 1, 1, 0, 0, 0)+$$

$$+S(8, 5; (2, 1, 1, 1, 0, 0, 0, 0)) \times P(2, 1, 1, 1, 0, 0, 0, 0)+$$

$$+S(8, 5; (2, 2, 1, 0, 0, 0, 0, 0)) \times P(2, 2, 1, 0, 0, 0, 0, 0) =$$

$$= 5! \frac{8!}{5!3!} + \frac{5!}{2!} \frac{8!}{3!4!} + \frac{5!}{2!2!} \frac{8!}{2!5!},$$

and so the desired probability is therefore

$$\frac{5!8!}{8^5} \left(\frac{1}{5!3!} + \frac{1}{2!3!4!} + \frac{1}{(2!)^3 5!} \right) \approx 0.872 = 87.2\,\%. \qquad \square$$

3.3 Catalan Numbers and Dyck Sequences ☕

In this section we discuss the sequence of *Catalan*[1] *numbers*. There are many counting
problems in combinatorics in which the solution is given by the Catalan numbers: in
[36] there are descriptions of 66 different interpretations for them.

Definition 3.34 The **Catalan numbers** are defined by

$$\mathrm{Cat}_n = \frac{1}{n+1}\binom{2n}{n}, \ n \in \mathbb{N}.$$

$\Box$

The first 10 Catalan numbers from Cat_0 to Cat_9 are

$$1, \ 1, \ 2, \ 5, \ 14, \ 42, \ 132, \ 429, \ 1\,430, \ 4\,862.$$

The 25-th Catalan number is $\mathrm{Cat}_{24} = 1\,289\,904\,147\,324$. Using Stirling's formula
(see Remark 2.2) one finds that

$$\mathrm{Cat}_n \sim \frac{2^{2n}}{n\sqrt{n\pi}} \ \text{for } n \to \infty.$$

Here we relate the Catalan numbers to *Dyck sequences*; we shall see other their
interpretations in Chap. 11.

Definition 3.35 We say that a binary $2n$-sequence with occupancy (n, n) is a **Dyck**[2]
$2n$-**sequence** if, for each $i, \ 1 \le i \le 2n$, within the first i components of the $2n$-
sequence, the number of 1's is less or equal than the number of 0's. $\Box$

Example 3.36 The following are Dyck 8-sequences:

$$(0, 1, 0, 1, 0, 1, 0, 1), \quad (0, 1, 0, 0, 0, 1, 1, 1), \quad (0, 0, 1, 0, 1, 1, 0, 1).$$

The sequences $(0, 0, 1, 1)$, $(0, 1, 0, 1)$ are both Dyck 4-sequences, and there are no
others. $\Box$

☕ **Proposition 3.37** *The number of Dyck $2n$-sequences coincides with the n-th Catalan number* Cat_n.

---

[1] Eugène Charles Catalan (1814–1884)

[2] Walther Franz Anton von Dyck (1856–1934).

Proof. Given any binary $2n$-sequence $\mathfrak{a} := (a_1, \ldots, a_{2n})$, we define $i(\mathfrak{a})$ to be either the minimal index j for which $(a_1, \ldots, a_j)$ contains more 1's than 0's, if such index exists, or $2n$ if $\mathfrak{a}$ is a Dyck sequence. We use $(a_1, \ldots, a_{2n})^*$ to denote the $2n$-sequence $(a_1, \ldots, a_{i(\mathfrak{a})}, 1 - a_{i(\mathfrak{a})+1}, \ldots, 1 - a_{2n})$. Clearly one has

$$(a_1, \ldots, a_{2n})^{**} = (a_1, \ldots, a_{2n});$$

and therefore * is a one to one correspondence of the set of binary $2n$-sequences with itself. In this bijection the binary $2n$-sequences with occupancy (n, n) which are not Dyck correspond to the binary $2n$-sequences with occupancy $(n - 1, n + 1)$. Indeed, if $(a_1, \ldots, a_{2n})$ is a non Dyck $2n$-sequence of $\{0, 1\}$ with occupancy (n, n), then $(a_1, \ldots, a_{i(\mathfrak{a})})$ contains more 1's than 0's, and, more precisely, exactly one more. The sequence $(a_{i(\mathfrak{a})+1}, \ldots, a_{2n})$ will then contain more 0's than 1's (indeed precisely one more); therefore, $(a_1, \ldots, a_{2n})^*$ is a binary $2n$-sequence with occupancy $(n - 1, n + 1)$. If instead $\mathfrak{b} := (b_1, \ldots, b_{2n})$ is a $2n$-sequence of $\{0, 1\}$ with occupancy $(n - 1, n + 1)$, then certainly $i(\mathfrak{b})$ is strictly less than $2n$. Since there are two more 1's in $(b_1, \ldots, b_{2n})$, the sequence $(b_{i(\mathfrak{b})+1}, \ldots, b_{2n})$ will again contain another 1, and so $(b_1, \ldots, b_{2n})^*$ is a binary $2n$-sequence with occupancy (n, n) and is not a Dyck sequence since $(b_1, \ldots, b_{i(\mathfrak{b})})$ contains more 1's than 0's. Therefore, the number of Dyck $2n$-sequences is obtained by subtracting the number of binary $2n$-sequences with occupancy $(n - 1, n + 1)$ from the total number of binary $2n$-sequences with occupancy (n, n):

$$\binom{2n}{n} - \binom{2n}{n-1} = \frac{2n!}{n!n!} - \frac{2n!}{(n+1)!(n-1)!} = \frac{2n!}{(n-1)!n!}\left(\frac{1}{n} - \frac{1}{n+1}\right)$$

$$= \frac{2n!}{(n-1)!n!}\frac{n+1-n}{n(n+1)} = \frac{1}{n+1}\binom{2n}{n}. \qquad \square$$

Example 3.38 A DJ has to play 20 songs chosen randomly without repetitions from a set of 10 House and 10 Electronic. What is the probability that, at every stage in the evening, the number of soundtracks of House music does not outnumber the Electronic ones?

Solution. The possible sequences of House and Electronic songs are $\binom{20}{10}$. Among these, the wished ones are exactly the number of Dyck 20-sequences. By Proposition 3.37 the wanted probability is

$$\frac{\text{Cat}_{10}}{\binom{20}{10}} = \frac{1}{11} \sim 0.09 = 9\%. \qquad \square$$

Example 3.39 The operation of subtraction in the set of integers $\mathbb{Z}$ does not enjoy the associative property. Indeed, for $c \neq 0$

$$(a - b) - c \neq a - (b - c).$$

The order in which the two subtractions are performed changes, depending on the manner in which the difference is written. To distinguish them, one could keep in mind when the two subtractions *begin* and when they are *completed*. For example, if we enclose the two expressions between parentheses

$$((a - b) - c) \qquad (a - (b - c)),$$

we can distinguish them by bearing in mind the sequence of "−'s" and ")'s": the minus sign represents the beginning of a subtraction while the right parenthesis represents the completion of a subtraction. For example, to "$((a - b) - c)$" we can associate the sequence −)−), while "$(a - (b - c))$" is associated with the sequence −−)). Representing the "−" with 0's and the ")" with 1's, we obtain the Dyck 4-sequences $(0, 1, 0, 1)$ and $(0, 0, 1, 1)$. Proceeding in a similar fashion, in the general situation it is possible to set up a one to one correspondence between the possible orders for carrying out the subtractions in

$$a_1 - a_2 - \cdots - a_{n+1}$$

and the Dyck $2n$-sequences. For example, the Dyck 6-sequence $(0, 0, 1, 0, 1, 1)$ corresponds to the sequence −−)−)) and thus to the subtraction $(a_1 - ((a_2 - a_3) - a_4))$, while the subtraction $((a_1 - (a_2 - a_3)) - a_4)$ corresponds to the Dyck 6-sequence $(0, 0, 1, 1, 0, 1)$. $\square$

3.4 Problems

Problem 3.1 A die is tossed six times with the outcome of each toss being recorded. Determine the probability that the six outcomes consist of one 1, three 5's and two 6's.

Problem 3.2 1. How many 6 digit numbers can be formed with the numbers 3, 5, and 7?
2. How many of the numbers considered above contain two 3's, two 5's, and two 7's?

Problem 3.3 Assuming that passwords are generated randomly, what is the probability that a password of 8 digits from 0 to 9 contains two 5's and two 8's, three 2's and a 4?

Problem 3.4 What is the probability that in randomly distributing 6 indistinguishable objects amongst 9 distinct boxes one obtains a collection with occupancy $[2, 2, 1, 1, 0, 0, 0, 0, 0]$?

Problem 3.5 A waiter has taken orders for four types of drink: Martinis, Manhattans, White Wines, and Ginger Ales. The waiter remembers only that he must serve 3 drinks for two of these types, and 2 drinks for the other two types. What is the probability the he correctly guesses the drinks to be served?

Problem 3.6 In how many ways can one distribute 20 distinct objects amongst 3 distinct boxes with 6 objects in one box and 7 in the two other boxes? What if the objects are indistinguishable?

Problem 3.7 A program downloads and randomly distributes 15 different videos in mp4 format in the folders of 5 different users. What is the probability that one folder remains empty, 3 folders have 4 videos, and one folder contains 3?

Problem 3.8 The components of a group of 30 people travel in a railway car in such a way as to have three of them in compartment 1, six travellers in compartments 2, 4, and 6, five travellers in compartment 3, and four in compartment 5. Represent the outcome of such an arrangement in terms of sequences or collections, and then determine the number of such possible seating arrangements.

Problem 3.9 How many permutations of the digits in $1\,224\,666$ produce numbers less than $3\,000\,000$?

Problem 3.10 How many 8 digit numbers are there in which all six of the digits 1, 2, 3, 4, 5, 6 and no others appear? How many 8 digit numbers are there with six different (unspecified) digits?

Problem 3.11 Show that
$$\sum_{\substack{k_1 + k_2 + k_3 = 10 \\ k_i \in \mathbb{N}}} S(3, 10; (k_1, k_2, k_3)) = 3^{10}.$$

Problem 3.12 How many words can be formed using seven A's, eight B's, three C's, and six D's if the pairs CC and CA do not appear in immediate succession.

Problem 3.13 In a bridge hand (a card game with 4 players each receiving 13 cards) what is the probability that:

1. Player West holds 4 spades, 3 hearts, 3 diamonds and 3 clubs?
2. Players North and South each hold 5 spades, West has 2 spades, and East holds 1 spade?
3. A player holds all the Aces?
4. All the players hold 4 cards of one suit and 3 of all the other suits?

Problem 3.14 If one tosses a coin 20 times, getting 14 Heads and 6 Tails, what is the probability that two consecutive Tails do not come up?

Problem 3.15 How many anagrams of MATHEMATICIAN are there in which two A's do not appear consecutively?

Problem 3.16 How many anagrams are there of ANCESTORS in which each S is followed by a vowel?

Problem 3.17 How many anagrams of MISSISSIPPI have no two consecutive S's?

Problem 3.18 Among all the anagrams of UNABRIDGED how many have:

1. Four consecutive vowels?
2. At least three consecutive vowels?
3. No two consecutive vowels?

Problem 3.19 A bartender must serve drinks to 14 people seated at his bar.

1. The bartender does not remember who ordered a given drink: in how many ways could he serve the drinks if 3 people ordered a Martini, 2 a Manhattan, 2 a beer, 3 ordered a glass of Chablis, and 4 did not order anything?
2. The bartender is even more forgetful: he remembers only that Martinis, Manhattans, beers and Chablis have been ordered, and that someone made no order, and also that he must serve 3 drinks of one of these types, 2 of another type, 2 of yet another type, 3 of another type, and 4 of still another type (and here the "types of drink ordered" includes the order for no drink at all). In how many ways can these orders be served?

Problem 3.20 Nine people arrive at a restaurant with three empty dining rooms, and each person randomly chooses a dining room. Find the probability that:

1. There are exactly three people in the first dining room;
2. There are three people in each dining room;
3. There is a dining room with 2 people, one with 3 people, and 4 people in the remaining room.

Problem 3.21 Determine the probability that in a binary 10-sequence with four 1's and six 0's there are 2 adjacent 1's and the other 1's are not adjacent (for example, $(1, 1, 0, 1, 0, 1, 0, 0, 0, 0)$ fits our requirement).

Problem 3.22 A teacher has decided to quiz 8 of his 27 students in English, Latin, and History. In how many ways can he do so if he wishes to quiz 3 students in one subject, 3 in a second subject, and 2 in yet another subject?

Problem 3.23 A class of 18 students goes on a trip accompanied by two teachers. The evening accommodation consists of 5 rooms with 4 beds each. The two teachers do not want to sleep in the same room. In how many ways can room assignments satisfying this condition be made?

Chapter 4
Inclusion/Exclusion

Abstract The Inclusion/Exclusion Principle is a formula that allows us to compute the cardinality of a finite union, or intersection, of finite sets. We present the most popular applications of the Principle, like finding the number of surjective applications between two finite sets, or the number of derangements, i.e., point free permutations, of $(1, \ldots, n)$: this leads us to show that, collecting at random a hat at the wardrobe, the probability that nobody recovers their own hat tends to $1/e$ as the number of people grows. The curious reader will find some more special results, like the computation of the number of derangements of a sequence with repetitions.

4.1 Inclusion/Exclusion Principle

The material discussed in this section is indispensable in understanding how to count the number of elements in a union of sets without missing any such element or counting it several times.

4.1.1 Cardinality of a Union of Sets

We have seen in Proposition 1.14 that in order to calculate the cardinality of the union of two finite subsets A, B of a subset X, one must subtract the cardinality of $A \cap B$ from $|A| + |B|$, since otherwise the elements of the intersection would be counted twice. Thus one has the formula

$$|A \cup B| = |A| + |B| - |A \cap B|,$$

which is the Inclusion/Exclusion Principle for two sets.

C. Mariconda and A. Tonolo, *Discrete Calculus*,
UNITEXT - La Matematica per il 3+2 103, DOI 10.1007/978-3-319-03038-8_4

Example 4.1 (*Inclusion/Exclusion Principle for three sets*) Let A, B, C be finite subsets of a set X. Let $D = B \cup C$, so that one has

$$|A \cup B \cup C| = |A \cup D| = |A| + |D| - |A \cap D|,$$

$$|D| = |B \cup C| = |B| + |C| - |B \cap C| \quad \text{and}$$

$$\begin{aligned}
|A \cap D| = |A \cap (B \cup C)| &= |(A \cap B) \cup (A \cap C)| \\
&= |A \cap B| + |A \cap C| - |(A \cap B) \cap (A \cap C)| \\
&= |A \cap B| + |A \cap C| - |A \cap B \cap C|.
\end{aligned}$$

Thus,

$$|A \cup B \cup C| = |A| + |B| + |C| - (|B \cap C| + |A \cap B| + |A \cap C|) + |A \cap B \cap C|. \qquad \square$$

Let us now see how to extend the result of Example 4.1 to the general case of n sets. The main tool is to deal with linear combinations of characteristic functions of sets.

Lemma 4.2 *Let $B_1, \dots, B_n$ be subsets of a finite set X and $a_1, \dots, a_n$ real numbers. Then*

$$\sum_{x \in X} \left(\sum_{i=1}^{n} a_i \chi_{B_i}(x) \right) = \sum_{i=1}^{n} a_i |B_i|.$$

Proof. By changing the order of summation, one has

$$\sum_{x \in X} \left(\sum_{i=1}^{n} a_i \chi_{B_i}(x) \right) = \sum_{i=1}^{n} a_i \left(\sum_{x \in X} \chi_{B_i}(x) \right) = \sum_{i=1}^{n} a_i |B_i|. \qquad \square$$

If $A_1, A_2, \dots, A_n$ are subsets of a set X, we write $\mathfrak{S}_k(A_1, \dots, A_n)$ to denote the **sum of the cardinalities of all the possible intersections of k of the subsets** $A_1, A_2, \dots, A_n$:

$$\begin{aligned}
\mathfrak{S}_1(A_1, \dots, A_n) &= |A_1| + |A_2| + \cdots + |A_n|, \\
\mathfrak{S}_2(A_1, \dots, A_n) &= |A_1 \cap A_2| + |A_1 \cap A_3| + \cdots + |A_{n-1} \cap A_n|,
\end{aligned}$$

$$\dots \dots$$

$$\mathfrak{S}_k(A_1, \ldots, A_n) = \underbrace{\sum_{1 \le i_1 < \cdots < i_k \le n} |A_{i_1} \cap \cdots \cap A_{i_k}|,}_{C(n,k) \text{ summands}}$$

$$\ldots \ldots$$

$$\mathfrak{S}_n(A_1, \ldots, A_n) = |A_1 \cap \cdots \cap A_n|.$$

☞   In the sum defining $\mathfrak{S}_k(A_1, \ldots, A_n)$ there are $C(n, k) = \binom{n}{k}$ **summands:** one for each k-collection without repetitions of the set $\{A_1, \ldots, A_n\}$. When it is clear from the context that we are dealing with a prescribed family of sets $\{A_1, A_2, \ldots, A_n\}$, we will write $\mathfrak{S}_k$ rather than $\mathfrak{S}_k(A_1, \ldots, A_n)$.

Theorem 4.3 (Inclusion/Exclusion Principle for a union of sets) *Let $A_1, \ldots, A_n$ be subsets of X. Then*

$$|A_1 \cup \cdots \cup A_n| = \mathfrak{S}_1 - \mathfrak{S}_2 + \cdots + (-1)^{n-1} \mathfrak{S}_n.$$

Proof. For every $k = 1, \ldots, n$ and $x \in X$ let

$$\chi_k(x) := \sum_{1 \le i_1 < \cdots < i_k \le n} \chi_{A_{i_1} \cap \cdots \cap A_{i_k}}(x).$$

Notice that by Lemma 4.2 we have

$$\mathfrak{S}_k = \sum_{x \in X} \chi_k(x). \qquad (4.3.a)$$

Let us prove that $\chi_{A_1 \cup \cdots \cup A_n}(x) = \chi_1(x) - \chi_2(x) + \cdots + (-1)^{n-1} \chi_n(x)$ for any $x \in X$.

• If $x \notin A_1 \cup \cdots \cup A_n$, obviously $\chi_k(x) = 0$ for all k, so that

$$\chi_1(x) - \chi_2(x) + \cdots + (-1)^{n-1} \chi_n(x) = 0 = \chi_{A_1 \cup \cdots \cup A_n}(x).$$

• If $x \in A_1 \cup \cdots \cup A_n$, one may (after relabelling the A_i's) suppose without loss of generality that $x \in A_1 \cap \cdots \cap A_m$ for some $m \le n$ and that $x \notin A_i$ for $i \ge m+1$. The intersection $A_{i_1} \cap \cdots \cap A_{i_k}$ contains x if and only if $i_1, \ldots, i_k \le m$. Thus $\chi_k(x)$ equals the number $\binom{m}{k}$ of k-collections without repetitions of $\{A_1, \ldots, A_m\}$: it follows by Corollary 2.21 that

$$\chi_1(x) - \chi_2(x) + \cdots + (-1)^{n-1} \chi_n(x) = \binom{m}{1} - \binom{m}{2} + \cdots + (-1)^{m-1} \binom{m}{m}$$

$$= 1 - (1-1)^m = 1 = \chi_{A_1 \cup \cdots \cup A_n}(x).$$

Then Lemma 4.2 and (4.3.a) yield

$$|A_1 \cup \cdots \cup A_n| = \sum_{x \in X} \chi_{A_1 \cup \cdots \cup A_n}(x) = \sum_{x \in X} \left(\chi_1(x) - \chi_2(x) + \cdots + (-1)^{n-1} \chi_n(x) \right)$$

$$= \mathfrak{S}_1 - \mathfrak{S}_2 + \cdots + (-1)^{n-1} \mathfrak{S}_n . \qquad \square$$

Problems asking one to count in how many ways at least one of the events $A_1, \ldots, A_n$ may occur, can be solved by using the Inclusion/Exclusion Principle 4.3 to count the number of elements in $A_1 \cup \cdots \cup A_n$.

Example 4.4 Each student in a freshman dormitory must take at least one of the introductory courses in Biology (B), English (E), History (H), and Mathematics (M). There are 6 students who take all 4 courses. Moreover each course has 25 students, and for each pair of courses 15 students take that pair, and for each triple of courses 10 students take that triple. How many students are there in the dormitory?
Solution. Let $\mathfrak{S}_i := \mathfrak{S}_i(B, I, S, M)$, $i = 1, 2, 3, 4$. The given data state that

$$\mathfrak{S}_1 = 25 \times 4 = 100, \quad \mathfrak{S}_2 = 15 \times C(4, 2) = 90, \quad \mathfrak{S}_3 = 10 \times C(4, 3) \text{ and}$$

$$\mathfrak{S}_4 = 6 \times C(4, 4) = 6.$$

Since each student takes at least one course, the number of students in the dormitory is $|B \cup I \cup S \cup M|$. By the Inclusion/Exclusion Principle 4.3, we find

$$|B \cup I \cup S \cup M| = \mathfrak{S}_1 - \mathfrak{S}_2 + \mathfrak{S}_3 - \mathfrak{S}_4 = 100 - 90 + 40 - 6 = 44. \qquad \square$$

Example 4.5 In how many ways can one form a hand of 10 cards from a deck of 52 so that the hand contains all four suits of at least one card (value)?
Solution. Consider the set Ω consisting of the possible hands, that is, of 10 card subsets of the deck. Let

$$A_i = \{10 \text{ card hands with 4 cards of face value } i\};$$

the problem amounts to counting the cardinality of $A_1 \cup \cdots \cup A_{13}$ via the Inclusion/Exclusion Principle 4.3. If $1 \le i, j, k \le 13$ are distinct numbers, then

$$|A_i| = C(48, 6), \ |A_i \cap A_j| = C(44, 2) \text{ and } |A_i \cap A_j \cap A_k| = 0.$$

Indeed, to realise an element in A_i one has to add 6 cards to the four i; analogously an element in $A_i \cap A_j$ is obtained adding two cards to the four i and the four j. Finally, there are no 10 card subsets with four i, four j and four k. Hence, $\mathfrak{S}_1(A_1, \ldots, A_{13}) = 13 \times C(48, 6)$, $\mathfrak{S}_2(A_1, \ldots, A_{13}) = C(13, 2) \times C(44, 2)$ and $\mathfrak{S}_3(A_1, \ldots, A_{13}) = 0$. Therefore one can form a hand of the type described in

$$13 \times C(48, 6) - C(13, 2) \times C(44, 2) = 159\,455\,868 \text{ ways.} \qquad \square$$

Using the same method involved in the proof of the Inclusion/Exclusion Principle 4.3, one gets the following generalisation.

Proposition 4.6 (Inclusion/Exclusion Principle for the number of elements in at least m among n sets) *Let $A_1, \ldots, A_n$ be subsets of X. For $1 \le m \le n$, let $\mathfrak{U}_m(A_1, \ldots, A_n)$ be the subset of X of the elements belonging to at least m among the sets $A_1, \ldots, A_n$. Then*

$$|\mathfrak{U}_m(A_1, \ldots, A_n)| = \sum_{k=m}^{n} (-1)^{k-m} \binom{k-1}{m-1} \mathfrak{S}_k.$$

Remark 4.7 For $m = 1$, Proposition 4.6 gives Theorem 4.3: indeed

$$\mathfrak{U}_1(A_1, \ldots, A_n) = A_1 \cup \cdots \cup A_n.$$

Proof (of Proposition 4.6). For every $k = 1, \ldots, n$ and $x \in X$ let, as in the proof of Theorem 4.3,

$$\chi_k(x) := \sum_{1 \le i_1 < \cdots < i_k \le n} \chi_{A_{i_1} \cap \cdots \cap A_{i_k}}(x).$$

Let us prove that $\chi_{\mathfrak{U}_m(A_1, \ldots, A_n)}(x) = \sum_{k=m}^{n} (-1)^{k-m} \binom{k-1}{m-1} \chi_k(x)$ for any $x \in X$.

- If $x \notin \mathfrak{U}_m(A_1, \ldots, A_n)$, then x does not belong to any intersection of $k \ge m$ sets among $A_1, \ldots, A_n$; therefore $\chi_k(x) = 0$ for all $k \ge m$, and hence

$$\sum_{k=m}^{n} (-1)^{k-m} \binom{k-1}{m-1} \chi_k(x) = 0 = \chi_{\mathfrak{U}_m(A_1, \ldots, A_n)}(x).$$

- If, instead, $x \in \mathfrak{U}_m(A_1, \ldots, A_n)$, one may (after relabelling the A_i's) suppose without loss of generality that $x \in A_1 \cap \cdots \cap A_s$ for some $m \le s \le n$ and, if $s < n$, that $x \notin A_i$ for $i \ge s + 1$. The intersection $A_{i_1} \cap \cdots \cap A_{i_k}$ contains x if and only if each of $A_{i_1}, \ldots, A_{i_k}$ contains x, and this is the case if and only if $i_1, \ldots, i_k \le s$. Thus, for $k \ge m$, $\chi_k(x)$ equals the number $\binom{s}{k}$ of k-collections without repetitions of $\{A_1, \ldots, A_s\}$: it follows that

$$\sum_{k=m}^{n}(-1)^{k-m}\binom{k-1}{m-1}\chi_k(x)=\sum_{k=m}^{s}(-1)^{k-m}\binom{k-1}{m-1}\binom{s}{k}.$$

Now, if we set

$$a:=m,\quad b:=s-m,\quad c=k-m,$$

then

$$\binom{s}{k}=\binom{s}{s-k}=\binom{a+b}{b-c},\quad \binom{k-1}{m-1}=\binom{k-1}{k-m}=\binom{a+c-1}{c},$$

and hence by (2.62.a)

$$\sum_{k=m}^{n}(-1)^{k-m}\binom{k-1}{m-1}\chi_k(x)=\sum_{c=0}^{b}(-1)^c\binom{a+b}{b-c}\binom{a+c-1}{c}$$

$$=1=\chi_{\mathfrak{U}_m(A_1,\ldots,A_n)}(x).$$

Then by Lemma 4.2 and (4.3.a) we get

$$|\mathfrak{U}_m(A_1,\ldots,A_n)|=\sum_{x\in X}\chi_{\mathfrak{U}_m(A_1,\ldots,A_n)}(x)=\sum_{x\in X}\left(\sum_{k=m}^{n}(-1)^{k-m}\binom{k-1}{m-1}\chi_k(x)\right)$$

$$=\sum_{k=m}^{n}\left((-1)^{k-m}\binom{k-1}{m-1}\sum_{x\in X}\chi_k(x)\right)=\sum_{k=m}^{n}(-1)^{k-m}\binom{k-1}{m-1}\mathfrak{S}_k.\ \square$$

4.1.2 Cardinality of Intersections

Suppose now that we wish to count in how many ways several events $B_1,\ldots,B_n$ can take place simultaneously, that is, the number of elements in $B_1\cap\cdots\cap B_n$. The general method to use here is illustrated by the following example, and then formalized in the subsequent theorem.

Example 4.8 There are 15 students taking Mathematical Analysis, 12 taking Discrete Mathematics, and 9 taking both courses. If there are a total of 30 students, how many of them are not taking either of these courses? The number in question is $|A^c\cap D^c|$ where A is the set of students taking Analysis and D the set of those taking Discrete Mathematics. Now, if X is the set of the 30 students, one has

$$A^c\cap D^c=(A\cup D)^c=X\setminus(A\cup D)$$

and so

$$|A^c \cap D^c| = |X| - |A \cup D| = |X| - (|A| + |D| - |A \cap D|)$$
$$= 30 - (15 + 12 - 9) = 12. \qquad \square$$

In the general case one has the following result:

Corollary 4.9 (Inclusion/Exclusion Principle for an intersection of sets) *Let* $A_1, \ldots, A_n$ *be subsets of a finite set* X. *Then*

$$|A_1^c \cap A_2^c \cap \cdots \cap A_n^c| = |X| - \mathfrak{S}_1 + \mathfrak{S}_2 + \cdots + (-1)^n \, \mathfrak{S}_n.$$

Proof. Since $A_1^c \cap A_2^c \cap \cdots \cap A_n^c = (A_1 \cup A_2 \cup \cdots \cup A_n)^c$, one has

$$|A_1^c \cap \cdots \cap A_n^c| = |X| - |A_1 \cup \cdots \cup A_n|.$$

The result now follows by the Inclusion/Exclusion Principle 4.3. $\qquad \square$

As a first application of the Inclusion/Exclusion Principle 4.9 for an intersection of sets we are now going to calculate the number of sequences in which certain elements appear at least once.

Proposition 4.10 (Number of k-sequences of I_q containing $1, \ldots, n$ $(q \geq n)$) *Let* $k, n, q \in \mathbb{N}$ *with* $q \geq n \geq 1$. *The number of* k-*sequences of* I_q *in which* all *the elements* $1, \ldots, n$ *appear at least once, or, equivalently, the number of* q-*sharings* $(C_1, \ldots, C_q)$ *of* I_k *with the first* n *subsets* $C_1, \ldots, C_n$ *non-empty, is given by*

$$\sum_{i=0}^{n} (-1)^i \binom{n}{i} (q - i)^k.$$

In particular, for $q \geq n \geq 1$, *and* $0 \leq k < n$ *one gets*

$$\sum_{i=0}^{n} (-1)^i \binom{n}{i} (q - i)^k = 0. \tag{4.10.a}$$

Proof. If $k = 0$ there are no such 0-sequences, and in fact one has

$$\sum_{i=0}^{n} (-1)^i \binom{n}{i} (q - i)^0 = \sum_{i=0}^{n} (-1)^i \binom{n}{i} = (1 - 1)^n = 0.$$

Suppose that $k \geq 1$. The correspondence between sharings and sequences described in Theorem 1.28 induces a bijection between the q-sharings of I_k whose first n subsets are non-empty and the k-sequences of I_q in which $1, \ldots, n$ appear at least once. We calculate the number of the latter. Let A_ℓ, $\ell = 1, \ldots, n$ be the set of k-sequences of I_q which do not contain ℓ. We wish to determine the cardinality of $A_1^c \cap A_2^c \cap \cdots \cap A_n^c$. By Corollary 4.9 one has

$$|A_1^c \cap A_2^c \cap \cdots \cap A_n^c| = S((q,k)) - \mathfrak{S}_1(A_1, \ldots, A_n) + \cdots + (-1)^n \, \mathfrak{S}_n(A_1, \ldots, A_n)$$

$$= q^k - \sum_{i=1}^n (-1)^{i-1} C(n,i)(q-i)^k = \sum_{i=0}^n (-1)^i \binom{n}{i} (q-i)^k.$$

Indeed, for each i with $1 \leq i \leq n$, and for any choice of $1 \leq j_1 < \cdots < j_i \leq n$, the cardinality of $A_{j_1} \cap \cdots \cap A_{j_i}$ is equal to $(q-i)^k$.

If $k < n$ it is evident that there are no k-sequences of I_q in which $1, \ldots, n$ appear at least once, from which (4.10.a) follows.   □

There are several consequences of Proposition 4.10.

Corollary 4.11 *Let $m \geq 1$ and $Q(X)$ be a polynomial of degree $\deg Q(X) < m$. Then*

$$\sum_{i=0}^m (-1)^i \binom{m}{i} Q(i) = 0. \qquad (4.11.a)$$

Proof. Since $Q(X)$ is a sum of monomials of degree less than $\deg Q(X)$, it is enough to prove the claim for $Q(X) = X^k$, with $k < m$. Now

$$\sum_{i=0}^m (-1)^i \binom{m}{i} i^k = \sum_{i=0}^m (-1)^i \binom{m}{m-i} i^k$$

$$= \sum_{j=0}^m (-1)^{m-j} \binom{m}{j} (m-j)^k = \sum_{j=0}^m (-1)^j \binom{m}{j} (m-j)^k.$$

The conclusion follows directly from (4.10.a), since $k < m$.   □

Corollary 4.12 (Number of surjective functions) *Let $k, n \in \mathbb{N}_{\geq 1}$. The number of surjective functions $I_k \to I_n$ equals*

$$\sum_{i=0}^{n}(-1)^i\binom{n}{i}(n-i)^k = \sum_{i=0}^{n}(-1)^{n-i}\binom{n}{i}i^k.$$

Proof. As we noticed in Example 1.24, the surjective functions $I_k \to I_n$ are in one to one correspondence with the k-sequences of I_n for which each element of I_n appears at least once. The conclusion follows from Proposition 4.10 with $q = n$. □

We now give some other examples showing how to apply the Inclusion/Exclusion Principles 4.3 and 4.9.

Example 4.13 Each of the 32 offices in the Mathematics Department is equipped with a personal computer. Of these, 15 are laptops, 10 have a scanner, and 8 have a wireless mouse. Moreover, 2 offices have a laptop computer with printer and wireless mouse. Are there any offices in which the computer does not have any of these properties? If so, give an estimate for the number of such offices.
Solution. Let us denote the set of offices having laptops, scanner and wireless mouse by L, S and M. If r is the number of offices in which the computer does not have any of the three properties, then one has

$$r = |L^c \cap S^c \cap M^c| =$$

$$= 32 - (|L| + |S| + |M| - (|L \cap S| + |M \cap S| + |L \cap M|) + |L \cap S \cap M|).$$

Since $|M \cap S|, |L \cap S|, |P \cap M| \geq 2$, one has

$$r = 32 - [15 + 10 + 8] + (|L \cap S| + |M \cap S| + |L \cap M|) - 2 \geq 3. \quad □$$

Example 4.14 Find the number of natural number solutions to

$$a+b+c+d = 19 \quad 2 \leq a \leq 4,\ 3 \leq b \leq 6,\ 4 \leq c \leq 6,\ d \geq 2.$$

Solution. Put $x = a - 2$, $y = b - 3$, $z = c - 4$, and $w = d - 2$. Then the desired number of solutions coincides with the number of natural number solutions of $x + y + z + w = 8$, where x can be at most 2, y at most 3 and z at most 2. Let:

$$N := \{(x, y, z, w) \in \mathbb{N}^4 : x+y+z+w = 8\},$$

$$X := \{(x, y, z, w) \in N : x \geq 3\}, \quad Y := \{(x, y, z, w) \in N : y \geq 4\}, \quad \text{and}$$

$$Z := \{(x, y, z, w) \in N : z \geq 3\}.$$

We must calculate the cardinality of $X^c \cap Y^c \cap Z^c$. We have

$$|N| = C(11,3), \; |X| = C(8,3), \; |Y| = C(7,3), \; |Z| = C(8,3),$$

$$|X \cap Y| = C(4,3), \; |X \cap Z| = C(5,3), \; |Y \cap Z| = C(4,3) \text{ and } |X \cap Y \cap Z| = 0.$$

By the Inclusion/Exclusion Principle 4.9 the number of solutions to our problem is:

$$C(11,3) - (C(8,3) + C(7,3) + C(8,3)) + (C(4,3) + C(5,3) + C(4,3)) - 0 =$$

$$= 165 - (56 + 35 + 56) + (4 + 10 + 4) - 0 = 36. \qquad \square$$

Example 4.15 In how many ways can a president[1] choose 12 jelly beans from 4 different types if he does not take exactly 2 jelly beans of the same type?
Solution. If one uses I_4 to label the different types of jelly beans, the problem amounts to determining the cardinality of $B_1 \cap B_2 \cap B_3 \cap B_4$ where:

$$B_i = \{12\text{-collections of } I_4 \text{ not containing exactly 2 copies of } i\}.$$

Let A_i be the complement of B_i; by the Inclusion/Exclusion Principle 4.9, one has

$$|B_1 \cap B_2 \cap B_3 \cap B_4| = |A_1^c \cap A_2^c \cap A_3^c \cap A_4^c| =$$

$$= |X| - \mathfrak{S}_1(A_1, \ldots, A_4) + \cdots + \mathfrak{S}_4(A_1, \ldots, A_4).$$

Now $|X|$ is equal to the number of natural number solutions to the equation $x_1 + x_2 + x_3 + x_4 = 12$, and by Proposition 2.35 this number is $C(15,3)$. Clearly $|A_1|$ is equal to the number of natural number solutions of $2 + x_2 + x_3 + x_4 = 12$, that is, of $x_2 + x_3 + x_4 = 10$; therefore, one has $|A_1| = C(12,2)$. Analogously one proves $|A_2| = |A_3| = |A_4| = C(12,2)$ and hence $\mathfrak{S}_1 = 4 \times C(12,2)$. Similarly one finds that $\mathfrak{S}_2 = C(4,2) \times C(9,1)$, $\mathfrak{S}_3 = C(4,3)$ and $\mathfrak{S}_4 = 0$. Hence $|B_1 \cap B_2 \cap B_3 \cap B_4|$ is equal to

$$C(15,3) - 4 \times C(12,2) + C(4,2) \times C(9,1) - C(4,3) + 0 = 241. \qquad \square$$

Example 4.16 Count the number of anagrams of TAMTAM in which no two consecutive letters are equal.
Solution. Let Ω be the set of anagrams of TAMTAM. Put

$$A_M = \{\text{Anagrams of TAMTAM with two consecutive M's}\},$$

---

[1] Jelly Belly beans were a favorite of U.S. President Ronald Reagan, who kept a jar of them on his desk in the White House, at Blair House and on Air Force One. Reagan also made them the first jelly beans in outer space, sending them on the Shuttle Orbiter Challenger during the STS-7 mission in 1983, as a surprise for the astronauts.

and define A_T and A_A analogously. The question amounts to calculating $|A_M^c \cap A_T^c \cap A_A^c|$. For example, to calculate $|A_M|$ one must count the anagrams of $(MM)TATA$ considering (MM) as a unique letter: thus $|A_M| = 5!/(2!2!))$. To calculate $|A_M \cap A_T|$ one must count the anagrams of $(MM)(TT)AA$: in all there are $4!/2!$. Finally $|A_M \cap A_T \cap A_A| = 3!$, and so by the Inclusion/Exclusion Principle 4.9 the specified type of anagrams of TAMTAM number

$$|\Omega| - \mathfrak{S}_1 + \mathfrak{S}_2 - \mathfrak{S}_3 = S(3, 6; 2, 2, 2) - 3 \times \frac{5!}{2!2!} + \frac{C(3, 2)4!}{2!} - 3! = 30. \quad \square$$

Example 4.17 Find the number of integers x, $1 \le x \le 600$, which are not divisible either by 3, by 5, or by 7.

Solution. If $1 \le x \le n$, the number of multiples of x that do not exceed n is the integer part $[n/x]$ of n/x. If we let A, B, C respectively be the sets of integers x with $1 \le x \le 600$ which are divisible by 3, 5, and 7, we must then calculate $|A^c \cap B^c \cap C^c|$. The cardinality of A coincides with $[600/3] = 200$, that of B with $[600/5] = 120$ and that of C with $[600/7] = 85$. The set $A \cap B$ then contains the integers $1 \le x \le 600$ which are divisible by 15, and so $|A \cap B| = [600/15] = 40$. Similarly one calculates that $|A \cap C| = [600/21] = 28$ and $|B \cap C| = [600/35] = 17$. Finally, $|A \cap B \cap C| = [600/105] = 5$. By the Inclusion/Exclusion Principle 4.9 we find that

$$|A^c \cap B^c \cap C^c| = 600 - \mathfrak{S}_1 + \mathfrak{S}_2 - \mathfrak{S}_3 = 600 - 405 + 85 - 5 = 275.$$

There are 275 numbers between 1 and 600 which are not divisible by 3, 5 or 7. $\quad \square$

Example 4.18 Two natural numbers m and n are said to be **relatively prime (or prime to one another)** if the only positive divisor which they have in common is 1. The cardinality of the set formed by the strictly positive integers less than n and relatively prime to n is the **Euler function** evaluated at n, denoted by $\phi(n)$. For example, $\phi(8)$ is the cardinality of the set $\{1, 3, 5, 7\}$, and so $\phi(8) = 4$. Use the Inclusion/Exclusion Principle 4.9 to calculate $\phi(60)$.

Solution. The distinct primes dividing 60 are 2, 3, and 5. Let A, B, and C be respectively the sets of integers between 1 and 60 that are divisible by 2, 3 and 5. Then, $|A| = 60/2, |B| = 60/3, |C| = 60/5$. Moreover, $|A \cap B| = 60/(2 \times 3), |A \cap C| = 69/(2 \times 5), |B \cap C| = 60/(3 \times 5)$ and finally $|A \cap B \cap C| = 60/(2 \times 3 \times 5)$. By the Inclusion/Exclusion Principle 4.9 one has

$$\phi(60) = |A^c \cap B^c \cap C^c| =$$

$$= 60 - \left(\frac{60}{2} + \frac{60}{3} + \frac{60}{5}\right) + \left(\frac{60}{2 \times 3} + \frac{60}{2 \times 5} + \frac{60}{3 \times 5}\right) - \frac{60}{2 \times 3 \times 5} =$$

$$= 60 \left(1 - \frac{1}{2}\right) \left(1 - \frac{1}{3}\right) \left(1 - \frac{1}{5}\right) = 16. \quad \square$$

Remark 4.19 The preceding example may be generalized to an arbitrary positive integer. Let $n \geq 1$ be a positive integer, and let p_i for $i = 1, 2, \ldots, k$ be the distinct prime factors of n. Then

$$\phi(n) = n \left(1 - \frac{1}{p_1}\right) \left(1 - \frac{1}{p_2}\right) \cdots \left(1 - \frac{1}{p_k}\right).$$

4.2 Derangements (Fixed Point Free Permutations)

The permutations of a sequence which "move" all of its terms are of considerable interest.

Definition 4.20 Let $n, k \in \mathbb{N}_{\geq 1}$. A **derangement** of an n-sequence $(b_1, \ldots, b_n)$ of I_k is a permutation $(a_1, \ldots, a_n)$ of $(b_1, \ldots, b_n)$ such that $a_i \neq b_i$ for all $i = 1, \ldots, n$. If $n_1, \ldots, n_k$ are non-zero natural numbers such that $n_1 + \cdots + n_k = n$, we use $D_n(n_1, \ldots, n_k)$ to denote the number of derangements of an n-sequence of I_k with occupancy $(n_1, \ldots, n_k)$. If $n = k$ and $n_1 = \cdots = n_n = 1$ then for simplicity we write D_n to denote the number $D_n(\underbrace{1, \ldots, 1}_{n})$. $\qquad\square$

Example 4.21 • The sequence $(1, 2, 4, 3)$ is not a derangement of $(1, 2, 3, 4)$, since 1 and 2 are in their original positions. In contrast, the sequence $(2, 1, 4, 3)$ is a derangement of $(1, 2, 3, 4)$.
• The sequences $(2, 3, 1, 1)$ and $(3, 2, 1, 1)$ are the only derangements of $(1, 1, 2, 3)$. $\qquad\square$

Remark 4.22 It is easy to verify that two n-sequences of I_k both with occupancy $(n_1, \ldots, n_k)$ have the same number of derangements, and so $D_n(n_1, \ldots, n_k)$ is well defined. In particular, D_n counts the number of derangements of the sequence $(1, 2, \ldots, n)$. Note further that the number $D_n(n_1, \ldots, n_k)$ is invariant under permutations of the occupancy $(n_1, \ldots, n_k)$.

Example 4.23 A new deck has 52 cards of which the first 13 are hearts, the next 13 are diamonds, the next 13 clubs, and, finally, the last 13 are spades. To deal the cards in such a way that person 1 receives no hearts, person 2 receives no diamonds, person 3 receives no clubs, and person 4 receives no spades corresponds to carrying out a derangement of the 52-sequence of $I_4 = \{H, D, C, S\}$ composed of 13 hearts, followed successively by 13 diamonds, 13 clubs, and 13 spades. $\qquad\square$

4.2.1 Calculation of D_n

Calculating the number D_n of derangements of the n-sequence $(1, 2, \ldots, n)$ is an easy application of the Inclusion/Exclusion Principle 4.9.

Theorem 4.24 *Let $n \geq 1$. Then*

$$D_n = n! \left(1 - \frac{1}{1!} + \frac{1}{2!} - \frac{1}{3!} + \cdots + \frac{(-1)^n}{n!} \right).$$

Proof. Let A_i, for $i = 1, \ldots, n$, be the set of permutations of $(1, 2, \ldots, n)$ under which the number i remains in the i-th position. The derangements of $(1, 2, \ldots, n)$ are the elements of $A_1^c \cap \cdots \cap A_n^c$; the total number of permutations of $(1, 2, \ldots, n)$ is $n!$. By the Inclusion/Exclusion Principle 4.9 one therefore has

$$D_n = n! - \mathfrak{S}_1(A_1, \ldots, A_n) + \mathfrak{S}_2(A_1, \ldots, A_n) - \cdots + (-1)^n \, \mathfrak{S}_n(A_1, \ldots, A_n).$$

For any $1 \leq i_1 < \cdots < i_j \leq n$ and $1 \leq j \leq n$, the intersection $A_{i_1} \cap \cdots \cap A_{i_j}$ contains the permutations of $(1, 2, \ldots, n)$ under which the numbers $i_1, \ldots, i_j$ remain in their positions; therefore $|A_{i_1} \cap \cdots \cap A_{i_j}| = (n - j)!$. Then one has

$$\mathfrak{S}_1(A_1, \ldots, A_n) = C(n, 1)(n - 1)! = n!,$$

$$\mathfrak{S}_2(A_1, \ldots, A_n) = C(n, 2)(n - 2)! = \frac{n!}{2!},$$

$$\cdots \cdots \cdots$$

$$\mathfrak{S}_k(A_1, \ldots, A_n) = C(n, k)(n - k)! = \frac{n!}{k!},$$

$$\cdots \cdots \cdots$$

$$\mathfrak{S}_n(A_1, \ldots, A_n) = C(n, n)(n - n)! = \frac{n!}{n!} = 1.$$

Summing the various terms one obtains the desired result. □

Example 4.25 (*The hats problem*) Each of the n diners entering a restaurant leaves his hat at the check-room. In how many ways can the n hats be redistributed in such a way that no one receives his own hat? If the tipsy hat-checker randomly distributes the hats to the exiting diners, what is the probability that no one receives his own hat?

Solution. We label the clients and hats with I_n, assuming that hat i belongs to person i. We denote the outcome of a distribution of hats by an n-sequence of I_n, where the term in the i-th place is the hat received by the i-th person. The event that no one receives his own hat consists of the set of all derangements of $(1, \ldots, n)$, and there are D_n such derangements. If the distribution of hats takes place in random fashion, then every n-sequence of I_n is equally probable: thus the probability of our event is

$$\frac{D_n}{n!} = 1 - \frac{1}{1!} + \frac{1}{2!} - \frac{1}{3!} + \cdots + \frac{(-1)^n}{n!}.$$

One notes that since

$$e^{-1} = \lim_{n \to \infty} \sum_{k=0}^{n} (-1)^k \frac{1}{k!},$$

for large values of n the probability of this event approaches $1/e$.

The problem can become more complex if one introduces the brands of the different hats. If we assume that, among the n hats, there are n_1 hats of brand $1, \ldots, n_k$ hats of brand k, it turns out easily that the probability that no one receives a hat whose brand is the same as the original one is

$$\frac{D_n(n_1, \ldots, n_k)}{S(k, n; (n_1, \ldots, n_k))} = \frac{n!}{n_1! \cdots n_k!} D_n(n_1, \ldots, n_k).$$

Next section is devoted to the computation of $D_n(n_1, \ldots, n_k)$.   □

4.2.2 Calculation of $D_n(n_1, \ldots, n_k)$ ☕

The calculation of the number $D_n(n_1, \ldots, n_k)$ of derangements of an n-sequence of I_k with occupancy $(n_1, \ldots, n_k)$ is a bit more complicated. The general case is obtained in an analogous fashion to that used above for D_n via an application of the Inclusion/Exclusion Principle 4.9. Problem 4.26 suggests an *ad hoc* method for treating the case $k = 3$.

Theorem 4.26 *Let* $k, n_1, \ldots, n_k \in \mathbb{N}_{\geq 1}$, $n = n_1 + \cdots + n_k$. *Then*

$$D_n(n_1, \ldots, n_k) = \sum_{m=0}^{n} (-1)^m \sum_{\substack{0 \leq j_1 \leq n_1 \\ 0 \leq j_k \leq n_k \\ j_1 + \cdots + j_k = m}} \binom{n_1}{j_1} \cdots \binom{n_k}{j_k} \frac{(n-m)!}{(n_1 - j_1)! \cdots (n_k - j_k)!}.$$

$$(4.26.a)$$

Proof. We count the derangements of the following n-sequence of I_k:

$$\mathfrak{a} := (\underbrace{1, \ldots, 1}_{n_1}, \underbrace{2, \ldots, 2}_{n_2}, \ldots, \underbrace{k, \ldots, k}_{n_k}).$$

Denote by Ω the set of all permutations of $\mathfrak{a}$ and by A_j^i the subset of Ω consisting of the permutations of $\mathfrak{a}$ in which the j-th element equal to i in the sequence $\mathfrak{a}$ remains in the same position:

$$A_j^i = \{(b_1^1, \ldots, b_{n_1}^1, b_1^2, \ldots, b_{n_2}^2, \ldots, b_1^k, \ldots, b_{n_k}^k) \in \Omega : b_j^i = i\}.$$

The set of derangements of $\mathfrak{a}$ is the intersection of the complements in Ω of the n sets A_j^i, $1 \leq i \leq k$, $1 \leq j \leq n_i$. Thus, by the Inclusion/Exclusion Principle 4.9 one has

$$D_n(n_1, \ldots, n_k) = |\Omega| - \mathfrak{S}_1\left(\{A_j^i : i \in I_k, 1 \leq j \leq n_i\}\right) +$$
$$+ \mathfrak{S}_2\left(\{A_j^i : i \in I_k, 1 \leq j \leq n_i\}\right) - \cdots +$$
$$+ (-1)^n \mathfrak{S}_n\left(\{A_j^i : i \in I_k, 1 \leq j \leq n_i\}\right).$$

Of course $|\Omega| = \dfrac{n!}{n_1! \cdots n_k!}$. For a given m with $1 \leq m \leq n$, we have that the term $\mathfrak{S}_m\left(\{A_j^i : i \in I_k, 1 \leq j \leq n_i\}\right)$ is equal to the sum of the cardinalities of the sets

$$A_{(\Sigma_1, \ldots, \Sigma_k)} := \left(\bigcap_{j \in \Sigma_1} A_j^1\right) \cap \cdots \cap \left(\bigcap_{j \in \Sigma_k} A_j^k\right)$$

as $\Sigma_1, \ldots, \Sigma_k$ vary respectively over the subsets of $\{1, \ldots, n_1\}, \ldots, \{1, \ldots, n_k\}$ with $|\Sigma_1| + \cdots + |\Sigma_k| = m$. Now, if

$$|\Sigma_1| = j_1, \ldots, |\Sigma_k| = j_k, \quad j_1 + \cdots + j_k = m,$$

one has

$$|A_{(\Sigma_1, \ldots, \Sigma_k)}| = \frac{(n-m)!}{(n_1 - j_1)! \cdots (n_k - j_k)!}.$$

Indeed, one must count the permutations $(b_1^1, \ldots, b_{n_1}^1, b_1^2, \ldots, b_{n_2}^2, \ldots, b_1^k, \ldots, b_{n_k}^k)$ of $\mathfrak{a}$ with $b_\ell^1 = 1$ for each $\ell \in \Sigma_1, \ldots, b_\ell^k = k$ for each $\ell \in \Sigma_k$, or equivalently, considering only the terms that one may permute, the permutations of the $(n-m)$-sequence

$$(\underbrace{1, \ldots, 1}_{n_1 - j_1}, \underbrace{2, \ldots, 2}_{n_2 - j_2}, \ldots, \underbrace{k, \ldots, k}_{n_k - j_k}).$$

For given $j_1, \ldots, j_k$ there are $\binom{n_1}{j_1}$ ways to choose $\Sigma_1 \subseteq \{1, \ldots, n_1\}$ with $|\Sigma_1| = j_1, \ldots, \binom{n_k}{j_k}$ ways to choose $\Sigma_k \subseteq \{1, \ldots, n_k\}$ with $|\Sigma_k| = j_k$. Therefore one has

$$\mathfrak{S}_m\left(\{A_j^i : i \in I_k, 1 \leq j \leq n_i\}\right) = \sum_{\substack{0 \leq j_1 \leq n_1 \\ 0 \leq j_k \leq n_k \\ j_1 + \cdots + j_k = m}} \binom{n_1}{j_1} \cdots \binom{n_k}{j_k} \frac{(n-m)!}{(n_1 - j_1)! \cdots (n_k - j_k)!}.$$

The conclusion now immediately follows. $\qquad\square$

Example 4.27 If $k = n$ and $n_1 = \cdots = n_n = 1$ one has $D_n = D_n(1, \ldots, 1)$; indeed, Eq. (4.26.a) becomes

$$D_n(1, \ldots, 1) = \sum_{m=0}^{n}(-1)^m \sum_{\substack{0 \le j_1 \le 1 \\ 0 \le j_n \le 1 \\ j_1 + \cdots + j_n = m}} \binom{1}{j_1} \cdots \binom{1}{j_n} \frac{(n-m)!}{(1-j_1)! \cdots (1-j_n)!}$$

$$= \sum_{m=0}^{n}(-1)^m \sum_{\substack{0 \le j_1 \le 1 \\ 0 \le j_n \le 1 \\ j_1 + \cdots + j_n = m}} (n-m)! \,.$$

Since there are $\binom{n}{m}$ solutions of $j_1 + \cdots + j_n = m$ with $0 \le j_1 \le 1, \ldots, 0 \le j_n \le 1$ it follows that

$$D_n(1, \ldots, 1) = \sum_{m=0}^{n}(-1)^m (n-m)! \binom{n}{m} = \sum_{m=0}^{n}(-1)^m \frac{n!}{m!},$$

which corresponds to the result of Theorem 4.24.   □

Example 4.28 Let us calculate the probability that in dealing 52 cards to 4 people after having shuffled the deck, each of the four hands dealt lacks a different suit (i.e., one hand has no Hearts, a second has no Diamonds, a third has no Clubs, and a fourth no Spades). Each deal of the cards produces a 52-sequence of $I_4 = \{H, D, C, S\}$ with occupancy $(13, 13, 13, 13)$: these are $52!/(13!)^4$. The sequences that correspond to deals of the required type are precisely the derangements of the 4! different 52-sequences of $I_4 = \{H, D, C, S\}$ consisting only of blocks with 13 elements equal to a H, 13 elements equal to a D, 13 elements equal to a C, and 13 elements equal to a S. The desired probability is thus equal to

$$\frac{4! D_{52}(13, 13, 13, 13)}{52!/(13!)^4}.$$

By Theorem 4.26 one has

$$D_{52}(13, 13, 13, 13) = \sum_{m=0}^{52}(-1)^m \sum_{\substack{0 \le j_1 \le 13 \\ 0 \le j_4 \le 13 \\ j_1 + \cdots + j_4 = m}} \binom{13}{j_1} \cdots \binom{13}{j_4} \frac{(52-m)!}{(13-j_1)! \cdots (13-j_4)!};$$

using a CAS one finds

$$D_{52}(13, 13, 13, 13) = 20\,342\,533\,966\,643\,026\,042\,641,$$

while

$$\frac{52!}{(13!)^4} = 53\,644\,737\,765\,488\,792\,839\,237\,440\,000.$$

The desired probability is approximately equal to 0.0009 %. □

Example 4.29 A deck of 52 cards is shuffled, the cards are then dealt face up one-by-one. One wins if when counting the 52 cards as they are dealt by calling aloud the consecutive numbers from 1 to 13 for four times, one never encounters a card whose face value (Jack = 11, Queen = 12, King = 13) coincides with the number being called out. We can describe the possible outcomes of the experiment via 52-sequences of I_{13} with occupancy $(\underbrace{4, 4, \ldots, 4}_{13})$: in all there are $\dfrac{52!}{(4!)^{13}}$ such sequences.

One wins the game if after the shuffling of the deck one obtains a derangement of the 52-sequence consisting of four repeated blocks of the form 1, 2, 3, ..., 13: in all these are $D_{52}(\underbrace{4, 4, \ldots, 4}_{13})$. Hence the probability of winning this game is

$$\frac{D_{52}(4, 4, \ldots, 4)}{52!/(4!)^{13}}.$$

As usual, with the assistance of a CAS we find

$$D_{52}(4, 4, \ldots, 4) = 1\,493\,804\,444\,499\,093\,354\,916\,284\,290\,188\,948\,031\,229\,880\,469\,556$$

while

$$\frac{52!}{(4!)^{13}} = 92\,024\,242\,230\,271\,040\,357\,108\,320\,801\,872\,044\,844\,750\,000\,000\,000.$$

Thus the desired probability is approximately 1.62 %. □

In Example 4.29 the CAS we used did not succeed, on the authors computer, in calculating the sum present in the formula (4.26.a) within a reasonable period of time. The explicit calculation of $D_{52}(\underbrace{4, 4, \ldots, 4}_{13})$ was actually done by using an alternative formula involving the so-called Laguerre[2] polynomials.

Definition 4.30 Let $n \in \mathbb{N}_{\geq 1}$. The *n*-th **Laguerre polynomial** is

---

[2]Edmond Nicolas Laguerre (1834–1886).

$$L_n(X) := \sum_{\ell=0}^{n} (-1)^{\ell} \binom{n}{\ell} \frac{X^{\ell}}{\ell!}.$$ □

Corollary 4.31 [14] *Let* $k, n_1, \ldots, n_k \in \mathbb{N}_{\geq 1}$, $n = n_1 + \cdots + n_k$. *Then*

$$D_n(n_1, \ldots, n_k) = (-1)^n \int_0^{+\infty} e^{-x} \prod_{i=1}^{k} L_{n_i}(x)\, dx. \tag{4.31.a}$$

Proof. We calculate the product of the polynomial functions $L_{n_1}(x), \ldots, L_{n_k}(x)$. Using the well-known formulas for the product of polynomials, one has

$$\prod_{i=1}^{k} L_{n_i}(x) = \prod_{i=1}^{k} \left(\sum_{\ell_i=0}^{n_i} (-1)^{\ell_i} \binom{n}{\ell_i} \frac{x^{\ell_i}}{\ell_i!} \right)$$

$$= \sum_{p=0}^{n} \sum_{\substack{0 \leq \ell_1 \leq n_1 \\ \cdots \\ 0 \leq \ell_k \leq n_k \\ \ell_1 + \cdots + \ell_k = p}} (-1)^p \binom{n_1}{\ell_1} \cdots \binom{n_k}{\ell_k} \frac{x^p}{\ell_1! \cdots \ell_k!},$$

from which, bearing in mind[3] that $\int_0^{+\infty} x^p e^{-x}\, dx = p!$ for all $p \in \mathbb{N}$, one has

$$\int_0^{+\infty} e^{-x} \prod_{i=1}^{k} L_{n_i}(x)\, dx = \sum_{p=0}^{n} \sum_{\substack{0 \leq \ell_1 \leq n_1 \\ \cdots \\ 0 \leq \ell_k \leq n_k \\ \ell_1 + \cdots + \ell_k = p}} (-1)^p \binom{n_1}{\ell_1} \cdots \binom{n_k}{\ell_k} \frac{p!}{\ell_1! \cdots \ell_k!}. $$

$$\tag{4.31.b}$$

On replacing each index ℓ_i in the sum with $n_i - j_i$ one obtains

$$(-1)^n \int_0^{+\infty} e^{-x} \prod_{i=1}^{k} L_{n_i}(x)\, dx =$$

$$= (-1)^n \sum_{p=0}^{n} \sum_{\substack{0 \leq j_1 \leq n_1 \\ \cdots \\ 0 \leq j_k \leq n_k \\ j_1 + \cdots + j_k = n-p}} (-1)^p \binom{n_1}{j_1} \cdots \binom{n_k}{j_k} \frac{p!}{(n_1 - j_1)! \cdots (n_k - j_k)!}.$$

---

[3] This may be seen immediately with a repeated integrating by parts or by using the basic properties of Euler Γ function [33,Theorem 8.18].

The change of variable $m = n - p$ yields

$$(-1)^n \int_0^{+\infty} e^{-x} \prod_{i=1}^{k} L_{n_i}(x)\, dx =$$

$$= (-1)^n \sum_{m=0}^{n} \sum_{\substack{0 \le j_1 \le n_1 \\ 0 \le j_k \le n_k \\ j_1 + \cdots + j_k = m}} (-1)^{n-m} \binom{n_1}{j_1} \cdots \binom{n_k}{j_k} \frac{(n-m)!}{(n_1 - j_1)! \cdots (n_k - j_k)!}.$$

Since $(-1)^n (-1)^{n-m} = (-1)^{-m} = (-1)^m$, by Theorem 4.26 the latter coincides with $D_n(n_1, \ldots, n_k)$. $\quad\square$

Remark 4.32 Formula (4.31.a) applied to the calculation of the number $D_{52}(\underbrace{4, 4, \ldots, 4}_{13})$ in Example 4.29 yields

$$D_{52}(\underbrace{4, 4, \ldots, 4}_{13}) = \int_0^{+\infty} e^{-x} \left(1 - 4x + 3x^2 - \frac{2x^3}{3} + \frac{x^4}{24} \right)^{13} dx.$$

Using a CAS one obtains the result in the blink of an eye.

4.3 Problems

Problem 4.1 A pastry chef prepares baskets with 6 chocolate eggs; the eggs can be wrapped in tin-foil of any one of 5 different colors: Blue, Green, Red, White, and Yellow. The order in which the eggs are placed in the basket does not matter.

1. What is the maximum number of distinct baskets that can be prepared under these assumptions?
2. The chef has prepared one basket for each of the possible types. A client buys all the baskets in which there is at least one Blue egg or exactly 2 Yellow eggs. How many baskets does the client buy?

Problem 4.2 In how many ways is it possible to give a child 16 jelly beans when choosing from a large bin containing (at least 13 of each type) lemon, mint, and raspberry jelly beans if the child is to receive exactly 3 of at least one of the flavors?

Problem 4.3 Determine the number of 13 card hands that one can get from a deck of 52 cards, if the hand has 4 Kings, or 4 Aces, or exactly four spades.

Problem 4.4 How many 5 letter words can be formed using an alphabet of 26 letters (with repetitions allowed) if every word must begin or end with a vowel?

Problem 4.5 In how many ways can one form a sequence of length 5 using an alphabet of 3 letters if the sequence is to have at least two consecutive equal letters?

Problem 4.6 Suppose that in a bookstore there are 200 books, 70 in French and 100 dealing with a mathematical topic. How many books are there not written in French and not dealing with mathematics if there are 30 books in French dealing with mathematics?

Problem 4.7 A group of 200 students are eligible to take three Mathematics courses: Discrete Mathematics, Analysis, and Geometry. Each course has 80 students. Each pair of courses has 30 students in common, and 15 students are taking all three courses.

1. How many students are not taking any of the three mathematics courses?
2. How many students are taking only Discrete Mathematics?

Problem 4.8 How many numbers between 1 and 30 are relatively prime to 30?

Problem 4.9 How many 10-sequences of I_9 are there in which the digits 1, 2, and 3 all appear?

Problem 4.10 In how many ways can 20 different people be assigned to 3 rooms if each room must receive at least one person?

Problem 4.11 How many anagrams of SINGER are there in which at least one of the following three conditions holds: (i) S precedes I, (ii) I precedes N, (iii) N precedes G? Here "precedes" means "occurs earlier than", but not necessarily "immediately before".

Problem 4.12 The Bakers, the Vinsons and the Caseys each have 5 children. If the 15 youngsters camp out in 5 different tents, with three in each tent, and are assigned randomly to the 5 tents, what is the probability that each family has at least two of its children in the same tent?

Problem 4.13 What is the probability that a hand of 13 cards taken from a deck of 52 has:

1. At least one missing suit?
2. At least one card of each suit?
3. At least one of each type of face card and at least one Ace (that is, at least one Ace, at least one Jack, at least one Queen, and at least one King)?

Problem 4.14 How many 9-sequences of I_3 are there in which there appear three 1's, three 2's, and three 3's, but without three consecutive equal numbers?

Problem 4.15 How many permutations of the 26 letters of the English alphabet are there which do not contain any of the words SOAP, FLY, LENS, GIN?

Problem 4.16 In how many ways can one distribute 25 identical balls in 6 distinct containers so as to have a maximum of 6 balls in any one of the first 3 containers?

Problem 4.17 A witch doctor has 5 friends. During a long convention on black magic he went to lunch with each friend 10 times, with every pair of friends 5 times, with every triple of friends 3 times, with every quadruple of friends twice, and only once with all 5 friends. If, moreover, the witch doctor lunched alone 6 times, how many days did the convention last?

Problem 4.18 Suppose that in a mathematics department there are 10 courses to be assigned to 5 different professors. In how many ways can one assign the 5 professors two courses per year in two successive academic years in such a way that no professor teaches the same 2 courses both years?

Problem 4.19 How many permutations of $(1, 2, \ldots, n)$ are there in which 1 is not immediately followed by 2, 2 is not immediately followed by 3, $\ldots$, n is not immediately followed by 1?

Problem 4.20 In how many ways can one distribute 10 books to 10 boys (one book to each boy), and then collect the books and redistribute them in such a way that every boy gets a new book?

Problem 4.21 In a city 3 newspapers (A, B, and C) are sold. A survey reveals that 47 % of the inhabitants read newspaper A, 34 % read newspaper B, 12 % read newspaper C. Moreover, 8 % read both A and B, 5 % both A and C, and 4 % both B and C. Finally, 4 % read all three papers. If one picks an inhabitant of the city at random, find the probability that:

1. She/he does not read any paper;
2. She/he reads only one paper.

Problem 4.22 We must insert 9 distinct numbers between 1 and 90 (including the extreme values) into a table of 4 rows and 6 columns.

1. How many different tables can be created?
2. How many tables have an empty row or 90 in the first row?

Problem 4.23 Consider an alphabet composed of 13 symbols.

1. How many 8 letter words containing at least one symbol repeated three times is it possible to write?
2. How many 8 letter words containing at least two distinct symbols repeated exactly 3 times is it possible to write?

Problem 4.24 Consider the red cards (13 hearts and 13 diamonds) from a poker deck of 52 cards. The 13 heart cards are distributed to 13 people, and then the diamonds, with one card of each type to each person.

1. How many possible outcomes are there for such a distribution?
2. What is the probability that at least one person receives a pair (two cards with the same face value)?

Problem 4.25 Prove that the factorial $n!$ and the number of derangements D_n satisfy for $n \geq 3$ the following recurrence formulas:

1. $n! = (n-1)\big((n-1)! + (n-2)!\big)$;
2. $D_n = (n-1)\big(D_{n-1} + D_{n-2}\big)$.

[Hint for Part 2: Carry out every permutation of $(1, \ldots, n)$ in two steps. The first step consists of choosing the position i to which 1 is moved; the second step is to carry out the repositioning of the other digits, distinguishing between the case where i is moved to position 1, and when it is not …]

Problem 4.26 Let $n_1, n_2, n_3 \in \mathbb{N}_{\geq 1}$ and let $n = n_1 + n_2 + n_3$. Show, without using Theorem 4.26, that

$$D_n(n_1, n_2, n_3) = \sum_{k=1}^{n_1} \binom{n_2}{k} \binom{n_3}{n_1 - k} \binom{n_1}{n_3 - n_1 + k}.$$

[Hint: perform a derangement of the n-sequence

$$(\underbrace{1, \ldots, 1}_{n_1}, \underbrace{2, \ldots, 2}_{n_2}, \underbrace{3, \ldots, 3}_{n_3}).$$

To obtain a derangement, k elements equal to 1 are placed where the 2's were, the other $n_1 - k$ elements equal to 1 are placed where the 3's were. Finally, the remaining free positions where there were 2's are filled by $n_3 - (n_1 - k)$ elements equal to 3.]

Chapter 5
Stirling Numbers and Eulerian Numbers

Abstract This chapter is dedicated to counting partitions of sets and partitions of sets into cycles, and also introduces Stirling numbers and Bell numbers. As an application of the concepts discussed here we state Faà di Bruno chain rule for the n-th derivative of a composite of n-times differentiable functions on $\mathbb{R}$. In the last section we discuss Eulerian numbers and as an application we solve the famous problem of the Smith College diplomas, and we establish some notable identities like Worpitzky's formula.

5.1 Partitions of Sets

In this section we deal with partitions of sets, that is, with dividing a set into non-empty disjoint subsets whose union is the whole set.

5.1.1 Stirling Numbers of the Second Kind

We now introduce Stirling numbers of the second kind; just as in Stirling's original work [37] we find it convenient to present the Stirling numbers of the second kind before those of the first kind, strangely enough.

Definition 5.1 (*Stirling Numbers of the second kind*) Let n, $k \in \mathbb{N}$. The **Stirling number of the second kind n over k**, denoted by $\left\{ {n \atop k} \right\}$, is the **number of k-partitions of I_n**. □

Remark 5.2 There is still no standard notation for the Stirling numbers. Here we follow the most frequent one in use at the present time,[1] and which is due to Karamata.[2]

---

[1] "par cette notation les formules deviennent plus symétriques"': with this notation the formulas become more symmetric [23].

[2] Jovan Karamata (1902–1967).

© Springer International Publishing Switzerland 2016
C. Mariconda and A. Tonolo, *Discrete Calculus*,
UNITEXT - La Matematica per il 3+2 103, DOI 10.1007/978-3-319-03038-8_5

Example 5.3 Bearing in mind that the elements of a partition of a set must be non-empty disjoint subsets with union equal to the whole set, one has:

- $\left\{ {n \atop n} \right\} = 1$ for all $n \geq 0$: if $n = 0$ the empty collection is the unique 0-partition of I_0, while if $n > 0$ the unique n-partition of I_n is $[\{1\}, \ldots, \{i\}, \ldots, \{n\}]$;

- $\left\{ {n \atop 0} \right\} = 0$ for all $n > 0$: there is no 0-partition of I_n;

- $\left\{ {0 \atop n} \right\} = 0$ for all $n > 0$: there is no n-partition of $I_0 = \emptyset$;

- $\left\{ {n \atop k} \right\} = 0$ for all $n < k$: there is no k-partition of I_n;

- $\left\{ {n \atop 1} \right\} = 1$ for all $n > 0$: the unique 1-partition of I_n is $[I_n]$.   □

Example 5.4 Let $n \geq 2$. Then

$$\left\{ {n \atop n-1} \right\} = \binom{n}{2}.$$

Indeed, an $(n-1)$-partition of I_n is necessarily made up of $n-2$ subsets with 1 element and a single subset with 2 elements; it is, therefore, determined by the choice of the two elements for the subset of cardinality 2: this choice can be made in $\binom{n}{2}$ ways.   □

Proposition 5.5 *Let $k, n \in \mathbb{N}$. The number of k-sharings of I_n into non-empty subsets equals $k! \left\{ {n \atop k} \right\}$. In particular*

$$k! \left\{ {n \atop k} \right\} = \sum_{\substack{(n_1,\ldots,n_k)\in(\mathbb{N}_{\geq 1})^k \\ n_1+\cdots+n_k=n}} \frac{n!}{n_1! \cdots n_k!}. \tag{5.5.a}$$

Proof. Every k-sharing of I_n into non-empty subsets is obtained by ordering in the $k!$ possible ways the elements of a k-partition of I_n. By the Division Principle 1.47, the number of k-sharings of I_n into non-empty subsets is $k!$ times the number of k-partitions of I_n.

Now, each k-sharing of I_n into non-empty subsets has an occupancy sequence of the form $(n_1, \ldots, n_k)$ for some $n_1, \ldots, n_k \in \mathbb{N}_{\geq 1}$ and $n_1 + \cdots + n_k = n$. Thus

$$k! \left\{ {n \atop k} \right\} = \sum_{\substack{(n_1,\ldots,n_k)\in(\mathbb{N}_{\geq 1})^k \\ n_1+\cdots+n_k=n}} S(k, n; (n_1, \ldots, n_k)).$$

The Identity (5.5.a) follows then directly from Theorem 3.7.   □

We now provide an explicit general formula for $\left\{{n \atop k}\right\}$; an alternative version is given in Theorem 5.12.

Theorem 5.6 (Formula for the Stirling numbers of the second kind) *Let $k, n \in \mathbb{N}$. Then*

$$\left\{{n \atop k}\right\} = \begin{cases} 1 & \text{if } k = 0 = n, \\ 0 & \text{if } k > n, \\ \dfrac{1}{k!} \displaystyle\sum_{i=0}^{k} (-1)^i \binom{k}{i} (k-i)^n & \text{otherwise.} \end{cases}$$

Proof. By Proposition 5.5 the number $\left\{{n \atop k}\right\}$ counts the k-sharings of I_n into non-empty subsets divided by $k!$. The result follows from Proposition 4.10 with $q = n$.   □

Example 5.7 In how many ways can 14 students be divided into at most 3 groups?
Solution. The students can all be inserted into a single group, or divided into two groups, or divided into three groups, and so there are

$$N = \left\{{14 \atop 1}\right\} + \left\{{14 \atop 2}\right\} + \left\{{14 \atop 3}\right\} = 1 + \left\{{14 \atop 2}\right\} + \left\{{14 \atop 3}\right\}$$

ways to divide them into groups. Moreover, one has

$$\left\{{14 \atop 2}\right\} = \frac{1}{2!}\left(\binom{2}{0} 2^{14} - \binom{2}{1} 1^{14} + \binom{2}{2} 0^{14}\right) = 2^{13} - 1$$

$$\left\{{14 \atop 3}\right\} = \frac{1}{3!}\left(\binom{3}{0} 3^{14} - \binom{3}{1} 2^{14} + \binom{3}{2} 1^{14} - \binom{3}{3} 0^{14}\right) = \frac{1}{2}(3^{13} - 2^{14} + 1)$$

and so $N = (3^{13} + 1)/2$.   □

As a consequence of Corollary 4.12 we get the number of surjective functions between two finite sets.

Corollary 5.8 (Number of surjective functions) *Let $m \leq \ell \in \mathbb{N}_{\geq 1}$. The number of ℓ-sequences of I_m in which each element of I_m appears at least once, or equivalently, the number of surjective functions $I_\ell \to I_m$, equals*

$$m! \left\{ {\ell \atop m} \right\} = \sum_{i=0}^{m} (-1)^i \binom{m}{i} (m-i)^\ell. \tag{5.8.a}$$

More generally, one has the following result:

Proposition 5.9 (Number of sharings with a prescribed number of non empty sets)
Given $\ell, m \in \mathbb{N}_{\geq 1}$ and an integer $1 \leq i \leq m$, the

- *m-sharings of I_ℓ with exactly i non empty sets,*
- *ℓ-sequences of I_m in which there appear exactly i elements of I_m,*
- *functions $f : I_\ell \to I_m$ whose image has cardinality i,*

have all the same cardinality given by

$$\left\{ {\ell \atop i} \right\} \frac{m!}{(m-i)!}.$$

Proof. We can construct an m-sharing $(C_1, \ldots, C_m)$ of I_ℓ with exactly i non empty sets using the following two step procedure:

(1) We choose the i positions $j_1 < \cdots < j_i$ in I_m corresponding to a non empty set of the m-sharing: this can be done in $\binom{m}{i}$ ways;
(2) We choose the i-sharing $(C_{j_1}, \ldots, C_{j_i})$ of I_ℓ into non empty sets: by Proposition 5.5, this can be done in $i! \left\{ {\ell \atop i} \right\}$ ways.

By the Multiplication Principle 1.34 there are $\binom{m}{i} i! \left\{ {\ell \atop i} \right\} = \frac{m!}{(m-i)!} \left\{ {\ell \atop i} \right\}$ such m-sharings. $\qquad \square$

The Stirling numbers of the second kind may be calculated via a recursive procedure.

Proposition 5.10 (Recursive formula for the Stirling numbers of the second kind)
Let $k, n \in \mathbb{N}_{\geq 1}$. Then

$$\left\{ {n-1 \atop k-1} \right\} + k \left\{ {n-1 \atop k} \right\} = \left\{ {n \atop k} \right\}.$$

Proof. We treat the cases $n - 1 < k - 1$, $n - 1 = k - 1$ and $n - 1 > k - 1$ in different ways. If $n - 1 < k - 1$ or $n - 1 = k - 1$, by what we have seen in Example 5.3 the equalities to be proved become respectively the identities $0 + k \times 0 = 0$ and $1 + k \times 0 = 1$. Suppose now that $n - 1 > k - 1$, that is, $n > k$. Among the k-partitions of I_n there are those that contain the singleton subset $\{n\}$ consisting of n alone, and those in which n belongs to a subset with at least two elements. The partitions of the first class are equal in number to the $(k - 1)$-partitions of I_{n-1}, that is, to $\begin{Bmatrix} n-1 \\ k-1 \end{Bmatrix}$: indeed, it suffices to add the subset $\{n\}$ to an arbitrary $(k - 1)$-partition of I_{n-1} to obtain such a partition of I_n. The partitions of the second class may be obtained by first considering a k-partition of I_{n-1}, and then adding the element n to one of the k subsets of that partition. By the Multiplication Principle 1.34 there are $k \begin{Bmatrix} n-1 \\ k \end{Bmatrix}$ such partitions. $\qquad\square$

Example 5.11 (The triangle of the Stirling numbers of the second kind) The preceding proposition shows that in order to calculate the Stirling numbers of the second kind with first argument equal to n, it suffices to know those with first argument equal to $n - 1$. This fact constitutes the basis for the construction of the triangle of the Stirling numbers of the second kind, an analogue of the Tartaglia–Pascal triangle.

In view of Example 5.3 the values $\begin{Bmatrix} n \\ 1 \end{Bmatrix}$ and $\begin{Bmatrix} n \\ n \end{Bmatrix}$ along the edges are all equal to 1. Using the formula

$$\begin{Bmatrix} n \\ k \end{Bmatrix} = \begin{Bmatrix} n-1 \\ k-1 \end{Bmatrix} + k \begin{Bmatrix} n-1 \\ k \end{Bmatrix}$$

proved in Proposition 5.10, we can easily **obtain the internal values along a row if we know the values of the preceding row**:

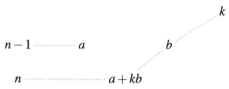

Thus, one obtains the triangle

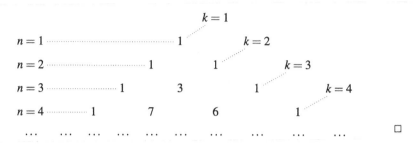

□

As an application of the recursive formula found in Proposition 5.10, we propose an explicit formula for the Stirling numbers of the second kind less known than the one shown in Proposition 5.9 but useful in applications to the generating formal series.

Theorem 5.12 (Formula for the Stirling numbers of the second kind) *Let k, n in $\mathbb{N}$. Then*

$$\begin{Bmatrix} n \\ k \end{Bmatrix} = \sum_{\substack{(n_1,\dots,n_k)\in\mathbb{N}^k \\ n_1+\cdots+n_k=n-k}} 1^{n_1} 2^{n_2} \cdots k^{n_k}. \tag{5.12.a}$$

Proof. The formula is valid for $k > n$ or $k = 0$: in such cases both the members of the equality vanish. Next, the formula is true for

- $k = n$: the only n-composition of $n - k = 0$ is the n-tuple $(0, 0, \dots, 0)$ and it is $1^0 \cdots n^0 = 1$;
- $k = 1 \le n$: indeed, $\begin{Bmatrix} n \\ 1 \end{Bmatrix} = 1$, the only 1-composition of $n - 1$ is the 1-tuple $(n - 1)$ and it is $1^{n-1} = 1$.

Let us prove now (5.12.a) by induction on $n \geq 1$. The case $n = 1$ has already been considered. Suppose $n \geq 2$ and that the assertion holds for $n - 1$. We show that it then holds for n as well. The cases $k = 1$ and $k = n$ have been proved; assume $2 \leq k \leq n - 1$. By Proposition 5.10 one has

$$\begin{Bmatrix} n \\ k \end{Bmatrix} = \begin{Bmatrix} n - 1 \\ k - 1 \end{Bmatrix} + k \begin{Bmatrix} n - 1 \\ k \end{Bmatrix}. \tag{5.12.b}$$

The inductive hypothesis gives

$$\begin{Bmatrix} n - 1 \\ k - 1 \end{Bmatrix} = \sum_{\substack{(n_1,\dots,n_{k-1}) \in \mathbb{N}^{k-1} \\ n_1 + \cdots + n_{k-1} = (n-1)-(k-1)}} 1^{n_1} 2^{n_2} \cdots (k-1)^{n_{k-1}}$$

$$= \sum_{\substack{(n_1,\dots,n_k) \in \mathbb{N}^k \\ n_1 + \cdots + n_k = n-k \\ n_k = 0}} 1^{n_1} 2^{n_2} \cdots (k-1)^{n_{k-1}} k^{n_k},$$

and

$$k \begin{Bmatrix} n - 1 \\ k \end{Bmatrix} = \sum_{\substack{(m_1,\dots,m_k) \in \mathbb{N}^k \\ m_1 + \cdots + m_k = (n-1)-k}} 1^{m_1} 2^{m_2} \cdots k^{m_k+1}$$

$$= \sum_{\substack{(n_1,\dots,n_k) \in \mathbb{N}^k \\ n_1 + \cdots + n_k = n-k \\ n_k \geq 1}} 1^{n_1} 2^{n_2} \cdots k^{n_k}, \quad (n_k := m_k + 1).$$

We conclude by (5.12.b). $\qquad\qquad\qquad\qquad\qquad\qquad\qquad\qquad\qquad\qquad\qquad\qquad\qquad\square$

Remark 5.13 The formula shown in Theorem 5.6 is often more convenient of the formula (5.12.a) for the explicit computation of the Stirling numbers of the second kind, since the first generally involves a significantly lower number of summands.

Example 5.14 Let us compute again the numbers $\begin{Bmatrix} 14 \\ 2 \end{Bmatrix}$ and $\begin{Bmatrix} 14 \\ 3 \end{Bmatrix}$ considered in the previous Example 5.7 by means of Theorem 5.12. We use the well-known fact that $\sum_{k=0}^{m} a^k = \dfrac{a^{m+1} - 1}{a - 1}$ for any $a \neq 1$.

$$\begin{Bmatrix} 14 \\ 2 \end{Bmatrix} = \sum_{\substack{(n_1,n_2) \in \mathbb{N}^2 \\ n_1 + n_2 = 12}} 1^{n_1} 2^{n_2} = \sum_{m=0}^{12} 2^m = 2^{13} - 1;$$

$$
\left\{ {14 \atop 3} \right\} = \sum_{\substack{(n_1,n_2,n_3)\in\mathbb{N}^3 \\ n_1+n_2+n_3=11}} 1^{n_1} 2^{n_2} 3^{n_3} = \sum_{\substack{(m_1,m_2)\in\mathbb{N}^2 \\ m_1+m_2\leq 11}} 2^{m_1} 3^{m_2}
$$

$$
= \sum_{m=0}^{11} \sum_{\substack{(m_1,m_2)\in\mathbb{N}^2 \\ m_1+m_2=m}} 2^{m_1} 3^{m_2} = \sum_{m=0}^{11} \sum_{k=0}^{m} 2^k 3^{m-k} = \sum_{m=0}^{11} 3^m \sum_{k=0}^{m} \left(\frac{2}{3}\right)^k
$$

$$
= \sum_{m=0}^{11} 3^m \frac{1 - (2/3)^{m+1}}{1 - 2/3} = \sum_{m=0}^{11} 3^{m+1} \left(1 - (2/3)^{m+1}\right)
$$

$$
= \sum_{m=0}^{11} \left(3^{m+1} - 2^{m+1}\right) = 3\frac{3^{12} - 1}{3 - 1} - 2\frac{2^{12} - 1}{2 - 1}
$$

$$
= \frac{3^{13}}{2} - \frac{3}{2} - 2^{13} + 2 = \frac{1}{2}\left(3^{13} - 2^{14} + 1\right). \qquad \square
$$

5.1.2   Bell Numbers

In this section we count the total number of partitions of a set with n elements. In Fig. 5.1 the partitions of a set with 5 elements are described.

Definition 5.15 The number of all possible partitions of I_n, $n \in \mathbb{N}$, is said to be the n-**th Bell**[3] **number**, and is denoted by the symbol $\mathfrak{B}_n$. $\square$

The first eleven Bell numbers are

$$
\mathfrak{B}_0 = 1, \ \mathfrak{B}_1 = 1, \ \mathfrak{B}_2 = 2, \ \mathfrak{B}_3 = 5, \ \mathfrak{B}_4 = 15, \ \mathfrak{B}_5 = 52, \ \mathfrak{B}_6 = 203,
$$

$$
\mathfrak{B}_7 = 877, \ \mathfrak{B}_8 = 4\,140, \ \mathfrak{B}_9 = 21\,147, \ \mathfrak{B}_{10} = 115\,975.
$$

Clearly the n-th Bell number may be expressed as a function of the Stirling numbers of the second kind:

$$
\mathfrak{B}_n = \sum_{k=0}^{n} \left\{ {n \atop k} \right\}.
$$

Proposition 5.16 (Recursive formula for the Bell numbers) *Let* $n \in \mathbb{N}$. *Then*

---

[3]Eric Temple Bell (1883–1960).

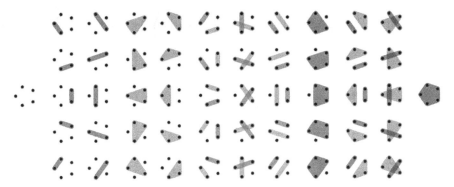

Fig. 5.1 The 52 partitions of a set with 5 elements. (*Source* commons wikimedia, https://commons.wikimedia.org/wiki/File:Set_partitions_5;_circles.svg. Author Watchduck (a.k.a. Tilman Piesk) (Own work), Creative Commons Attribution CC BY 3.0 Unported license)

$$\mathfrak{B}_{n+1} = \sum_{i=0}^{n} \binom{n}{i} \mathfrak{B}_i.$$

Proof. Let j be a natural number with $1 \leq j \leq n+1$. A partition of I_{n+1} in which $n+1$ is contained in a subset of cardinality j may be constructed in two steps: first one chooses $j-1$ elements of I_n to which $n+1$ is then adjoined; in the second step one chooses a partition of the remaining $n+1-j$ elements. There are, therefore, in all a total of $\binom{n}{j-1} \times \mathfrak{B}_{n+1-j}$ such partitions. Then, by summing on the index j one has

$$\mathfrak{B}_{n+1} = \sum_{j=1}^{n+1} \binom{n}{j-1} \times \mathfrak{B}_{n+1-j} = \sum_{j=1}^{n+1} \binom{n}{n-(j-1)} \times \mathfrak{B}_{n+1-j} = \sum_{i=0}^{n} \binom{n}{i} \mathfrak{B}_i. \quad \square$$

It is possible to explicitly calculate the Bell numbers $\mathfrak{B}_n$. We warn the reader that the result is expressed in terms of the sum of a series.

Theorem 5.17 *Let $n \in \mathbb{N}$. The n-th Bell number is given by the sum of the series*

$$\mathfrak{B}_n = \frac{1}{e} \sum_{j=0}^{\infty} \frac{j^n}{j!}.$$

Proof. By Theorem 5.6 one has

$$\mathcal{B}_n = \sum_{k=0}^{n} \begin{Bmatrix} n \\ k \end{Bmatrix} = \sum_{k=0}^{n} \frac{1}{k!} \sum_{i=0}^{k} (-1)^i \binom{k}{i} (k-i)^n.$$

Now, by (4.10.a), for each $m < k$ one has

$$\sum_{i=0}^{k} (-1)^i \binom{k}{i} (k-i)^m = 0,$$

from which it follows that

$$\mathcal{B}_n = \sum_{k=0}^{n} \begin{Bmatrix} n \\ k \end{Bmatrix} = \sum_{k=0}^{n} \frac{1}{k!} \sum_{i=0}^{k} (-1)^i \binom{k}{i} (k-i)^n = \sum_{k=0}^{\infty} \frac{1}{k!} \sum_{i=0}^{k} (-1)^i \binom{k}{i} (k-i)^n$$

$$= \sum_{k=0}^{\infty} \sum_{i=0}^{k} (-1)^i \frac{1}{i!} \frac{(k-i)^n}{(k-i)!}.$$

Let $a_{ki} := (-1)^i \frac{1}{i!} \frac{(k-i)^n}{(k-i)!}$. Then one has

$$\mathcal{B}_n = \sum_{k=0}^{\infty} \sum_{i=0}^{k} a_{ki} = (a_{00}) + (a_{10} + a_{11}) + (a_{20} + a_{21} + a_{22}) + \dots$$

$$= (a_{00} + a_{10} + a_{20} + \dots) + (a_{11} + a_{21} + a_{31} + \dots) + \dots$$

$$= \sum_{i=0}^{\infty} \sum_{k=i}^{\infty} a_{ki} = \sum_{i=0}^{\infty} \sum_{k=i}^{\infty} (-1)^i \frac{1}{i!} \frac{(k-i)^n}{(k-i)!}$$

$$= \sum_{i=0}^{\infty} (-1)^i \frac{1}{i!} \sum_{k=i}^{\infty} \frac{(k-i)^n}{(k-i)!}.$$

The interchange of the order of summation is justified since the series

$$\sum_{i=0}^{\infty} \frac{1}{i!} \sum_{k=i}^{\infty} \frac{(k-i)^n}{(k-i)!}$$

converges.[4] Indeed, by [33, Theorem 3.55], on rearranging the order in which one sums the terms of an absolutely convergent series, one again obtains an absolutely convergent series which converges to the same value as the original series. On putting $j = k - i$ we obtain

---

[4]For all i the series $\displaystyle\sum_{k=i}^{\infty} \frac{(k-i)^n}{(k-i)!} = \sum_{r=0}^{\infty} \frac{r^n}{r!}$ converges. Let τ be its sum. One then has $\displaystyle\sum_{i=0}^{\infty} \frac{\tau}{i!} = \tau e$.

$$\mathfrak{B}_n = \sum_{i=0}^{\infty} \frac{(-1)^i}{i!} \sum_{k=i}^{\infty} \frac{(k-i)^n}{(k-i)!} = \sum_{i=0}^{\infty} \frac{(-1)^i}{i!} \sum_{j=0}^{\infty} \frac{j^n}{j!} = \frac{1}{e} \sum_{j=0}^{\infty} \frac{j^n}{j!},$$

since $\sum_{i=0}^{\infty} \frac{(-1)^i}{i!} = e^{-1} = \frac{1}{e}$. $\qquad\qquad\square$

5.1.3 Partitions of Sets with Occupancy Constraints

We now consider partitions of sets formed by subsets of assigned cardinality, which obviously must be strictly positive.

Definition 5.18 Let $n, k, n_1, \ldots, n_k \in \mathbb{N}_{\geq 1}$ satisfy $n_1 + \cdots + n_k = n$. A k-**partition** of I_n **with occupancy collection** $[n_1, \ldots, n_k]$ is a k-partition of I_n consisting of subsets, one of which has cardinality n_1, a second of cardinality n_2, $\ldots$, and one of cardinality n_k. We use the symbol $\left\{ \begin{matrix} n \\ k \end{matrix}; [n_1, \ldots, n_k] \right\}$ to denote **the number of** k-**partitions of** I_n **with occupancy** $[n_1, \ldots, n_k]$. $\qquad\square$

Example 5.19 The 2-partitions of I_3 with occupancy collection $[1, 2]$ are

$$[\{1\}, \{2, 3\}], \quad [\{2\}, \{1, 3\}], \quad [\{3\}, \{1, 2\}]. \qquad\square$$

Example 5.20 The division of a group of 20 people into 3 groups consisting of 7, 5, and 8 people may be represented by a 3-partition of the set of people with occupancy collection $[5, 7, 8]$. $\qquad\square$

Theorem 5.21 *Let* $n, k, n_1, \ldots, n_k \in \mathbb{N}_{\geq 1}$ *satisfy* $n_1 + \cdots + n_k = n$. *Then*

$$\left\{ \begin{matrix} n \\ k \end{matrix}; [n_1, \ldots, n_k] \right\} = \frac{1}{k!} S(k, n; [n_1, \ldots, n_k])$$

$$= \frac{1}{k!} \times \frac{n!}{n_1! n_2! \cdots n_k!} \times P(n_1, \ldots, n_k).$$

Proof. To each k-sharing $(C_1, \ldots, C_k)$ of I_n with occupancy collection $[n_1, \ldots, n_k]$ we associate the k-partition $[C_1, \ldots, C_k]$. Obviously $[C_1, \ldots, C_k]$ also has occupancy $[n_1, \ldots, n_k]$. Starting from the $k!$-sharings that one obtains by permuting the sequence $(C_1, \ldots, C_k)$ all the associated partitions are the same. By the Division Principle 1.47, therefore the number of partitions with occupancy $[n_1, \ldots, n_k]$ is $\frac{1}{k!}$ times the number of sharings with the same occupancy. In view of Theorem 3.23 this concludes the proof. $\qquad\square$

Example 5.22 In how many ways is it possible to divide 5 youngsters into three groups, one group consisting of only one person, and the two other groups each having two people?

Solution. We need to count the number of 3-partitions of the set I_5 with occupancy collection $[2, 2, 1]$:

$$\left\{ \begin{matrix} 5 \\ 3 \end{matrix}; [2, 2, 1] \right\} = \frac{1}{3!} S(3, 5; [2, 2, 1]) = \frac{1}{3!} S(3, 5; (2, 2, 1)) \times P(2, 2, 1) = 15. \quad \square$$

5.1.4 Faà di Bruno Formula ☕

We conclude this section with an application of the concepts concerning partitions to the chain-rule formula for the n-th derivative of the composition of two functions, known also as *Faà di Bruno*[5] formula.

Theorem 5.23 (Faà di Bruno chain rule) *Let* $n \in \mathbb{N}_{\geq 1}$, *I and J be open intervals in* $\mathbb{R}$, $f : J \rightarrow \mathbb{R}$ *and* $g : I \rightarrow J$ *be n-times differentiable functions. For each* $x \in I$ *one has*

$$(f \circ g)^{(n)}(x) = \sum_{\substack{\mathscr{P} \text{ partition} \\ \text{of } I_n}} f^{(|\mathscr{P}|)}(g(x)) \prod_{B \in \mathscr{P}} g^{(|B|)}(x). \tag{5.23.a}$$

Example 5.24 In the 4-partition of I_{11}

$$\mathscr{P}_0 = [\{1\}, \{2, 3, 4\}, \{5, 6, 7\}, \{8, 9, 10, 11\}]$$

there are one set of cardinality 1, two sets of cardinality 3, and one set of cardinality 4: thus, one has

$$f^{(|\mathscr{P}_0|)}(g(x)) \prod_{B \in \mathscr{P}_0} g^{(|B|)}(x) = f^{(4)}(g(x))(g'(x))(g^{(3)}(x))^2(g^{(4)}(x)). \quad \square$$

Proof (of Theorem 5.23). For each partition $\mathscr{P}$ of I_n, $n \in \mathbb{N}_{\geq 1}$, we set

$$\varphi_{\mathscr{P}}(x) := f^{(|\mathscr{P}|)}(g(x)) \prod_{B \in \mathscr{P}} g^{(|B|)}(x).$$

We proceed by induction on n. For $n = 1$ one has

$$(g \circ f)'(x) = g'(f(x)) f'(x).$$

---

[5]Francesco Faà di Bruno (1825–1888).

The unique partition of I_1 consists of the 1-partition $\mathscr{P}_1 = [\{1\}]$ and

$$\varphi_{\mathscr{P}_1} = f'(g(x))g'(x).$$

Thus in the case $n = 1$ the formula holds. Suppose now that the assertion holds for a given natural number $n \geq 1$. We show that it then holds for $n + 1$ as well.

Note first that from every partition of I_{n+1} one obtains a partition of I_n by eliminating the element $n + 1$ (if the set containing $n + 1$ is the singleton $\{n + 1\}$, we eliminate the whole set), and conversely, every partition $\mathscr{R}$ of $\{1, \ldots, n + 1\}$ arises from a k-partition $\mathscr{P} = [B_1, \ldots, B_k]$ of $\{1, \ldots, n\}$ by adding $n + 1$ according to one of the following methods (we will suppose that in $\mathscr{P}$ there are b_i sets with i elements, for $i = 1, \ldots, n$):

- Method 1: One adds the singleton set $\{n + 1\}$ to the sets of the k-partition $\mathscr{P}$ obtaining the $(k + 1)$-partition $\mathscr{R} = [B_1, \ldots, B_k, \{n + 1\}]$: in this case

$$\varphi_{\mathscr{R}}(x) = f^{(k+1)}(g(x))(g'(x))^{b_1+1} \cdots (g^{(n)}(x))^{b_n};$$

  this may be done in only one way.
- Method 2: One adds the element $n + 1$ to one of the sets of the partition $\mathscr{P}$ consisting of i elements, $1 \leq i \leq n - 1$: in this case $\mathscr{R}$ is a k-partition and, with respect to $\mathscr{P}$, in $\mathscr{R}$ there is one less set with i elements and one more set with $i + 1$ elements. Therefore,

$$\varphi_{\mathscr{R}}(x) = f^{(k)}(g(x))(g'(x))^{b_1} \cdots (g^{(i)}(x))^{b_i-1}(g^{(i+1)}(x))^{b_{i+1}+1} \cdots (g^{(n)}(x))^{b_n};$$

  this may be done in b_i ways (as many ways as there are subsets with i elements).
- Method 3: In the case $\mathscr{P} = [I_n]$, one adds the element $n + 1$ to the unique subset of the 1-partition $\mathscr{P}$ thus obtaining the 1-partition $\mathscr{R} = [I_{n+1}]$: in this case

$$\varphi_{\mathscr{R}}(x) = f'(g(x))g^{(n+1)}(x);$$

  and this may be done in only one way.

Now we take the derivative of $(f \circ g)^{(n)}$, and invoke the induction hypothesis:

$$(f \circ g)^{(n+1)}(x) = \left((f \circ g)^{(n)}\right)'(x) = \sum_{\substack{\mathscr{P} \text{ partition} \\ \text{of } I_n}} \varphi'_{\mathscr{P}}(x).$$

Fix a k-partition $\mathscr{P}$ of I_n, and suppose, as usual, that in it there are b_i subsets consisting of i elements, for $i = 1, \ldots, n$, so that

$$\varphi_{\mathscr{P}}(x) = f^{(k)}(g(x))(g'(x))^{b_1} \cdots (g^{(n)}(x))^{b_n}.$$

The derivative of $\varphi_{\mathscr{P}}$ consists of the following summands:

1. $f^{(k+1)}(g(x))(g'(x))^{b_1+1} \cdots (g^{(n)}(x))^{b_n}$: this is just the derivative of $f^{(k)}(g(x))$ multiplied by $(g'(x))^{b_1} \cdots (g^{(n)}(x))^{b_n}$. We recognize the function $\varphi_{\mathscr{R}}$ with $\mathscr{R}$ obtained from $\mathscr{P}$ according to Method 1.

2. $b_i f^{(k)}(g(x))(g'(x))^{b_1} \cdots (g^{(i)}(x))^{b_i-1}(g^{(i+1)}(x))^{b_{i+1}+1} \cdots (g^{(n)}(x))^{b_n}, i = 1, \ldots, n-1$: this is just the derivative of $(g^{(i)}(x))^{b_i}$ multiplied by

$$f^{(k)}(g(x))(g'(x))^{b_1} \cdots (g^{(i-1)}(x))^{b_{i-1}}(g^{(i+1)}(x))^{b_{i+1}} \cdots (g^{(n)}(x))^{b_n}.$$

Here we recognize the function $b_i\varphi_{\mathscr{R}}$ with $\mathscr{R}$ obtained from $\mathscr{P}$ via Method 2; there were b_i possibilities for obtaining such a $\mathscr{R}$ using Method 2.

3. $f'(g(x))g^{(n+1)}(x)$: this is the derivative of $g^{(n)}(x)$ multiplied by $f'(g(x))$. Here we recognize the function $\varphi_{\mathscr{R}}$ with $\mathscr{R}$ obtained from $\mathscr{P}$ via Method 3.

Therefore,

$$(f \circ g)^{(n+1)}(x) = \sum_{\substack{\mathscr{P} \text{ partition} \\ \text{of } I_n}} \varphi'_{\mathscr{P}}(x) = \sum_{\substack{\mathscr{R} \text{ partition} \\ \text{of } I_{n+1}}} \varphi_{\mathscr{R}}(x). \qquad \square$$

In practical applications it can be more convenient to use the following reformulations of Faà di Bruno Theorem 5.23.

Corollary 5.25 (Faà di Bruno chain rule, explicit version) *Let $n \in \mathbb{N}_{\geq 1}$, I and J be open intervals in $\mathbb{R}$, $f : J \to \mathbb{R}$ and $g : I \to J$ be n-times differentiable functions. For each $x \in I$ one has*

$$(f \circ g)^{(n)}(x) = \sum_{\substack{(b_1, \ldots, b_n) \in \mathbb{N}^n \\ b_1 + 2b_2 + \cdots + nb_n = n}} \frac{n!}{b_1! \cdots b_n!} f^{(b_1+\cdots+b_n)}(g(x)) \left(\frac{g'(x)}{1!}\right)^{b_1} \cdots \left(\frac{g^{(n)}(x)}{n!}\right)^{b_n}$$

$$(5.25.\text{a})$$

$$= n! \sum_{k=1}^{n} \frac{f^{(k)}(g(x))}{k!} \sum_{\substack{(n_1, \ldots, n_k) \in (\mathbb{N}_{\geq 1})^k \\ n_1 + \cdots + n_k = n}} \frac{g^{(n_1)}(x)}{n_1!} \cdots \frac{g^{(n_k)}(x)}{n_k!}. \quad (5.25.\text{b})$$

☞ *Remark 5.26* The reader should bear in mind that while the indices $b_1, \ldots, b_n$ appearing in (5.25.a) vary over sets of natural numbers, the indices n_i appearing in (5.25.b) are required to satisfy $n_i \geq 1$ for $i = 1, \ldots, k$.

Proof (of Corollary 5.25). Fix an occupancy $[\underbrace{1, \ldots, 1}_{b_1}, \ldots, \underbrace{n, \ldots, n}_{b_n}]$ with a certain

number $b_1 \geq 0$ of 1's, ..., a certain number $b_n \geq 0$ of n's. A partition with occupancy
$[\underbrace{1, \ldots, 1}_{b_1}, \ldots,$

$\underbrace{n, \ldots, n}_{b_n}]$ is a $(b_1 + \cdots + b_n)$-partition of I_n containing b_1 subsets of cardinality

$1, \ldots, b_n$ subsets of cardinality n. Since the union of the sets making up a partition is equal to I_n, one must have $b_1 + 2b_2 + \cdots + nb_n = n$. By Theorem 5.21 the
$(b_1 + \cdots + b_n)$-partitions of I_n with occupancy $[\underbrace{1, \ldots, 1}_{b_1}, \ldots, \underbrace{n, \ldots, n}_{b_n}]$ are

$$\frac{1}{(b_1 + \cdots + b_n)!\,(1!)^{b_1} \cdots (n!)^{b_n}} \frac{n!}{} \frac{(b_1 + \cdots + b_n)!}{b_1! \cdots b_n!} = \frac{n!}{b_1! \cdots b_n!} \frac{1}{(1!)^{b_1} \cdots (n!)^{b_n}}.$$

Thus, by grouping the partitions of I_n according to the number of times $1, \ldots, n$ are
present in their occupancy, by formula (5.23.a) one finds that $(f \circ g)^{(n)}(x)$ coincides
with

$$\sum_{\substack{(b_1, \ldots, b_n) \in \mathbb{N}^n \\ b_1 + 2b_2 + \cdots + nb_n = n}} \sum_{\substack{\text{partitions of } I_n \text{ with} \\ \text{occupancy } [\underbrace{1, \ldots, 1}_{b_1}, \ldots, \underbrace{n, \ldots, n}_{b_n}]}} f^{(b_1 + \cdots + b_n)}(g(x))(g'(x))^{b_1} \cdots (g^{(n)}(x))^{b_n} =$$

$$= \sum_{\substack{(b_1, \ldots, b_n) \in \mathbb{N}^n \\ b_1 + 2b_2 + \cdots + nb_n = n}} \frac{n!}{b_1! \cdots b_n!} \frac{1}{(1!)^{b_1} \cdots (n!)^{b_n}} f^{(b_1 + \cdots + b_n)}(g(x))(g'(x))^{b_1} \cdots (g^{(n)}(x))^{b_n}.$$

In order to prove (5.25.b), we have $(f \circ g)^{(n)}(x) =$

$$= \sum_{\substack{(b_1, \ldots, b_n) \in \mathbb{N}^n \\ b_1 + 2b_2 + \cdots + nb_n = n}} \frac{n!}{b_1! \cdots b_n!} \frac{1}{(1!)^{b_1} \cdots (n!)^{b_n}} f^{(b_1 + \cdots + b_n)}(g(x))(g'(x))^{b_1} \cdots (g^{(n)}(x))^{b_n}$$

$$= n! \sum_{k=1}^{n} \frac{f^{(k)}(g(x))}{k!} \sum_{\substack{(b_1, \ldots, b_n) \in \mathbb{N}^n \\ b_1 + \cdots + b_n = k \\ b_1 + 2b_2 + \cdots + nb_n = n}} \frac{k!}{b_1! \cdots b_n!} \left(\frac{g'(x)}{1!}\right)^{b_1} \cdots \left(\frac{g^{(n)}(x)}{n!}\right)^{b_n}.$$

Let $(b_1, \ldots, b_n) \in \mathbb{N}^n$ be such that $b_1 + \cdots + b_n = k$ and $b_1 + 2b_2 + \cdots + nb_n = n$; taking b_1 copies of 1's, ..., b_n copies of n's we can form $\dfrac{k!}{b_1! \cdots b_n!}$ different k-sequences $(n_1, \ldots, n_k)$ of $\mathbb{N}_{\geq 1}$ with $n_1 + \cdots + n_k = b_1 + 2b_2 + \cdots + nb_n = n$. Therefore

$$\sum_{\substack{(b_1,\ldots,b_n)\in\mathbb{N}^n \\ b_1+\cdots+b_n=k \\ b_1+2b_2+\cdots+nb_n=n}} \frac{k!}{b_1!\cdots b_n!}\left(\frac{g'(x)}{1!}\right)^{b_1}\cdots\left(\frac{g^{(n)}(x)}{n!}\right)^{b_n}=$$

$$=\sum_{\substack{(n_1,\ldots,n_k)\in\mathbb{N}^k_{\geq 1} \\ n_1+\cdots+n_k=n}}\left(\frac{g^{(n_1)}(x)}{n_1!}\right)\cdots\left(\frac{g^{(n_k)}(x)}{n_k!}\right),$$

and hence

$$(f\circ g)^{(n)}(x)=n!\sum_{k=1}^n\frac{f^{(k)}(g(x))}{k!}\sum_{\substack{(n_1,\ldots,n_k)\in\mathbb{N}^k_{\geq 1} \\ n_1+\cdots+n_k=n}}\left(\frac{g^{(n_1)}(x)}{n_1!}\right)\cdots\left(\frac{g^{(n_k)}(x)}{n_k!}\right). \quad\square$$

We now give an application of Corollary 5.25.

Example 5.27 For $n=3$ the formula (5.25.a) yields

$$(f\circ g)^{(3)}(x)=\sum_{\substack{(b_1,b_2,b_3) \\ b_1+2b_2+3b_3=3}}\frac{3!}{b_1!b_2!b_3!}f^{(b_1+b_2+b_3)}(g(x))\left(\frac{g'(x)}{1!}\right)^{b_1}\left(\frac{g''(x)}{2!}\right)^{b_2}\left(\frac{g^{(3)}(x)}{3!}\right)^{b_3}.$$

Now, the triples (b_1,b_2,b_3) of natural numbers such that $b_1+2b_2+3b_3=3$ are precisely

$$(0,0,1),\quad(1,1,0),\quad(3,0,0);$$

to these triples there correspond the summands

$$(0,0,1):\quad\frac{3!}{1!}f'(g(x))\frac{g^{(3)}(x)}{3!}=f'(g(x))g^{(3)}(x),$$

$$(1,1,0):\quad\frac{3!}{1!}f''(g(x))\frac{g'(x)}{1!}\frac{g''(x)}{2!}=3f''(x)g'(x)g''(x),$$

$$(3,0,0):\quad\frac{3!}{3!}f^{(3)}(g(x))\left(\frac{g'(x)}{1!}\right)^3=f^{(3)}(g(x))\left(g'(x)\right)^3.$$

Thus one has

$$(f\circ g)^{(3)}(x)=f'(g(x))g^{(3)}(x)+3f''(g(x))g'(x)g''(x)+f^{(3)}(g(x))\left(g'(x)\right)^3.$$

Applying the formula (5.25.b) one gets

$$(f \circ g)^{(3)}(x) = 3! \sum_{k=1}^{3} \frac{f^{(k)}(g(x))}{k!} \sum_{\substack{(n_1, \ldots, n_k) \in \mathbb{N}_{\geq 1}^{k} \\ n_1 + \cdots + n_k = 3}} \left(\frac{g^{(n_1)}(x)}{n_1!} \right) \cdots \left(\frac{g^{(n_k)}(x)}{n_k!} \right).$$

For $k = 1$ one has

$$\sum_{\substack{(n_1) \in \mathbb{N}_{\geq 1} \\ n_1 = 3}} \left(\frac{g^{(n_1)}(x)}{n_1!} \right) = \frac{g^{(3)}(x)}{3!};$$

for $k = 2$ one has

$$\sum_{\substack{(n_1, n_2) \in \mathbb{N}_{\geq 1} \\ n_1 + n_2 = 3}} \left(\frac{g^{(n_1)}(x)}{n_1!} \right) \left(\frac{g^{(n_2)}(x)}{n_2!} \right) = \frac{g'(x)}{1!} \frac{g''(x)}{2!} + \frac{g''(x)}{2!} \frac{g'(x)}{1!};$$

for $k = 3$ one has

$$\sum_{\substack{(n_1, n_2, n_3) \in \mathbb{N}_{\geq 1} \\ n_1 + n_2 + n_3 = 3}} \left(\frac{g^{(n_1)}(x)}{n_1!} \right) \left(\frac{g^{(n_2)}(x)}{n_2!} \right) \left(\frac{g^{(n_3)}(x)}{n_3!} \right) = \frac{g'(x)}{1!} \frac{g'(x)}{1!} \frac{g'(x)}{1!} = (g'(x))^3.$$

Therefore

$$(f \circ g)^{(3)}(x) = 3! \left(\frac{f'(g(x))}{1!} \frac{g^{(3)}(x)}{3!} + \frac{f''(g(x))}{2!} g'(x) g''(x) + \frac{f^{(3)}(g(x))}{3!} (g'(x))^3 \right)$$

$$= f'(g(x)) g^{(3)}(x) + 3 f''(g(x)) g'(x) g''(x) + f^{(3)}(g(x)) (g'(x))^3. \qquad \square$$

Let us use now (5.25.b) to obtain the formula for the n-th derivative of $1/g$.

Proposition 5.28 (Derivatives of $1/g$) *Let $n \in \mathbb{N}_{\geq 1}$, J be an open interval in $\mathbb{R}$, $g : J \to \mathbb{R}$ be an n-times differentiable functions such that $g(x) \neq 0$ for each $x \in J$. Then for all $x \in J$ one has*

$$\left(\frac{1}{g} \right)^{(n)} (x) = \sum_{k=1}^{n} (-1)^k \binom{n+1}{k+1} g^{-k-1}(x) \left(g^k \right)^{(n)} (x). \tag{5.28.a}$$

Proof. Since g is continuous and $0 \notin g(J)$, we have that either $g(J)$ is contained in $\mathbb{R}_{>0}$, or in $\mathbb{R}_{<0}$; let us assume that $g(J) \subseteq \mathbb{R}_{>0}$. The function $\frac{1}{g} : J \to \mathbb{R}$ is obtained by composing the function $f : \mathbb{R}_{>0} \to \mathbb{R}$, $f(x) = \frac{1}{x}$, with the function $g : J \to \mathbb{R}_{>0}$. It is easy to see that the successive derivatives of $f(x)$ are given by

$$f'(x) = -\frac{1}{x^2}, \ f''(x) = \frac{2}{x^3}, \ f^{(3)}(x) = -\frac{3!}{x^4}, \ \ldots, f^{(k)}(x) = (-1)^k \frac{k!}{x^{k+1}}, \ldots.$$

Then (5.25.b) yields

$$\left(\frac{1}{g}\right)^{(n)}(x) = n! \sum_{j=1}^{n} (-1)^j g^{-j-1}(x) \sum_{\substack{(n_1, \ldots, n_j) \in (\mathbb{N}_{\geq 1})^j \\ n_1 + \cdots + n_j = n}} \frac{1}{n_1! \cdots n_j!} g^{(n_1)}(x) \cdots g^{(n_j)}(x).$$

$$(5.28.b)$$

We wish to prove that the right hand members of (5.28.a) and (5.28.b) coincide. By Theorem 3.26 for $x \in J$ one has

$$\left(g^k\right)^{(n)}(x) = \sum_{\substack{(m_1, \ldots, m_k) \in \mathbb{N}^k \\ m_1 + \cdots + m_k = n}} \frac{n!}{m_1! \cdots m_k!} g^{(m_1)}(x) \cdots g^{(m_k)}(x). \qquad (5.28.c)$$

☞   The reader should beware the fact that in the last formula the indices m_j can assume the value 0, whereas the indices n_j in (5.28.b) vary in $\mathbb{N}_{\geq 1}$.

Now, after removing possible zero terms from a k-sequence $(m_1, \ldots, m_k) \in \mathbb{N}^k$ with $m_1 + \cdots + m_k = n$ one obtains, for some $1 \leq j \leq k$ a j-sequence $(n_1, \ldots, n_j)$ in $\mathbb{N}_{\geq 1}$ with

$$[n_1, \ldots, n_j, \underbrace{0, \ldots, 0}_{k-j}] = [m_1, \ldots, m_k], \quad n_1 + \cdots + n_j = n \text{ and} \qquad (5.28.d)$$

$$\forall 1 \leq i \leq j-1 \ \exists 1 \leq p < q \leq k: n_i = m_p, n_{i+1} = m_q, m_\ell = 0 \text{ if } p < \ell < q. \qquad (5.28.e)$$

Note, moreover, that conversely a j-sequence $(n_1, \ldots, n_j)$ in $\mathbb{N}_{\geq 1}$ with $j \leq k$, arises from exactly $\binom{k}{j}$ different k-sequences $(m_1, \ldots, m_k)$ that satisfy (5.28.d) and (5.28.e): there are in fact precisely this number of ways to choose the $k-j$ terms to be set equal to zero out of the values $m_1, \ldots, m_k$ of the k-sequence. Since in (5.28.c), in correspondence with each such index $m_i = 0$, one has $g^{(m_i)} = g$ and $m_i! = 0! = 1$, one finds that for each $k = 1, \ldots, n$

$$\left(g^k\right)^{(n)}(x) = \sum_{j=1}^{k} \binom{k}{j} \sum_{\substack{(n_1, \ldots, n_j) \in (\mathbb{N}_{\geq 1})^k \\ n_1 + \cdots + n_j = n}} \frac{n!}{n_1! \cdots n_j!} g^{k-j}(x) g^{(n_1)}(x) \cdots g^{(n_j)}(x).$$

Therefore, changing the order of summation one gets

$$\sum_{k=1}^{n} (-1)^k \binom{n+1}{k+1} g^{-k-1}(x) \left(g^k\right)^{(n)}(x) =$$

$$= \sum_{k=1}^{n} (-1)^k \binom{n+1}{k+1} \sum_{j=1}^{k} \binom{k}{j} g^{-j-1}(x) \sum_{\substack{(n_1,\ldots,n_j) \in (\mathbb{N}_{\geq 1})^k \\ n_1 + \cdots + n_j = n}} \frac{n!}{n_1! \cdots n_j!} g^{(n_1)}(x) \cdots g^{(n_j)}(x)$$

$$= \sum_{k=1}^{n} (-1)^k \binom{n+1}{k+1} \sum_{j=1}^{n} \binom{k}{j} g^{-j-1}(x) \sum_{\substack{(n_1,\ldots,n_j) \in (\mathbb{N}_{\geq 1})^k \\ n_1 + \cdots + n_j = n}} \frac{n!}{n_1! \cdots n_j!} g^{(n_1)}(x) \cdots g^{(n_j)}(x)$$

$$= \sum_{j=1}^{n} \left(\sum_{k=1}^{n} (-1)^k \binom{n+1}{k+1} \binom{k}{j} \right) g^{-j-1}(x) \sum_{\substack{(n_1,\ldots,n_j) \in (\mathbb{N}_{\geq 1})^k \\ n_1 + \cdots + n_j = n}} \frac{n!}{n_1! \cdots n_j!} g^{(n_1)}(x) \cdots g^{(n_j)}(x)$$

$$= \sum_{j=1}^{n} \left(\sum_{k=j}^{n} (-1)^k \binom{n+1}{k+1} \binom{k}{j} \right) g^{-j-1}(x) \sum_{\substack{(n_1,\ldots,n_j) \in (\mathbb{N}_{\geq 1})^k \\ n_1 + \cdots + n_j = n}} \frac{n!}{n_1! \cdots n_j!} g^{(n_1)}(x) \cdots g^{(n_j)}(x).$$

Now, (2.63) yields

$$\sum_{k=j}^{n} (-1)^k \binom{n+1}{k+1} \binom{k}{j} = (-1)^j.$$

Thus one has

$$\sum_{k=1}^{n} (-1)^k \binom{n+1}{k+1} g^{-k-1}(x) \left(g^k \right)^{(n)} (x) =$$

$$= \sum_{j=1}^{n} (-1)^j g^{-j-1}(x) \sum_{\substack{(n_1,\ldots,n_j) \in (\mathbb{N}_{\geq 1})^k \\ n_1 + \cdots + n_j = n}} \frac{n!}{n_1! \cdots n_j!} g^{(n_1)}(x) \cdots g^{(n_j)}(x).$$

The conclusion then follows from (5.28.b). □

5.2 Cycles and Partitions into Cycles

Rather than partitioning a set into subsets, one is sometimes interested in dividing it into *cyclical sequences*.

5.2.1 Cycles

How can one formalize the placement of a certain number of people around a circular table, or the bus stops along a circular route (see Fig. 5.2)?

Definition 5.29 Given two natural numbers $k \leq n \in \mathbb{N}_{\geq 1}$, and k distinct elements $a_1, \ldots, a_k$ of I_n, we use $\langle a_1, a_2, \ldots, a_k \rangle$ to denote the bijective function

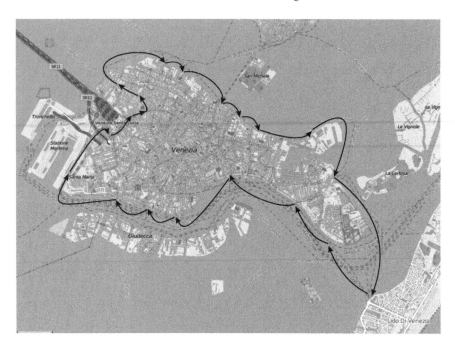

Fig. 5.2 A cycle of water-bus stops: line 5.2 makes a clockwise cyclical run starting from the Lido, and with stops at S. Elena, Giardini, San Zaccaria, Spirito Santo, Zattere, San Basilio, Santa Marta, Piazzale Roma, Ferrovia, Riva de Biasio, and then (passing along the Cannaregio canal) with further stops at Guglie, Tre Archi, Sant'Alvise, Madonna dell'Orto, Fondamente Nuove, Ospedale, Celestia, Bacini and San Pietro di Castello, and finally returning to the Lido. (©OpenStreetMap contributors, the data is available under the Open Database License)

$$\langle a_1, a_2, \ldots, a_k \rangle : I_n \to I_n$$

defined by setting $a_1 \mapsto a_2$, $a_2 \mapsto a_3$, ..., $a_{k-1} \mapsto a_k$, $a_k \mapsto a_1$ and

$$i \mapsto i \quad \text{for all } i \in I_n \setminus \{a_1, a_2, \ldots, a_k\}.$$

The bijective function $\langle a_1, a_2, \ldots, a_k \rangle$ is called a k-**cycle** of I_n and the set $\{a_1, \ldots, a_k\}$ is called the **support** of the cycle.   □

Remark 5.30 For each $j \in I_n$ the 1-cycle $\langle j \rangle$ is the identity function $I_n \to I_n$. The 3-cycles $\langle 1, 3, 2 \rangle$, $\langle 3, 2, 1 \rangle$ and $\langle 2, 1, 3 \rangle$ of I_4 coincide: indeed, in each of the three cases $1 \mapsto 3, 2 \mapsto 1, 3 \mapsto 2$ and $4 \mapsto 4$. By contrast, $\langle 1, 3, 2 \rangle \neq \langle 1, 2, 3 \rangle$: in fact, in the first case $1 \mapsto 3$, whilst in the second $1 \mapsto 2$. In general, given k distinct elements $a_1, \ldots, a_k$ of I_n one has

$$\langle a_1, a_2, \ldots, a_k \rangle = \langle a_2, a_3, \ldots, a_k, a_1 \rangle = \cdots = \langle a_k, a_1, a_2, \ldots, a_{k-1} \rangle.$$

Every k-cycle may thus be written in exactly k different ways, as many as the **cyclic permutations**

$$(a_1, a_2, \ldots, a_k), (a_2, a_3, \ldots, a_k, a_1), \ldots, (a_k, a_1, a_2, \ldots, a_{k-1})$$

of $(a_1, \ldots, a_k)$.

Among the k possible representations of a k-cycle, we call *canonical* the one beginning with the minimal element of its support.

Definition 5.31 Let σ be a cycle. We say that $\langle a_1, a_2, \ldots, a_k \rangle$ is the **canonical representation** of σ if $\sigma = \langle a_1, a_2, \ldots, a_k \rangle$ and $a_1 = \min\{a_1, \ldots, a_k\}$.   □

If we fix an element of the support, for example a_k, then every k-cycle admits a unique representation starting from a_k; in particular, every cycle has a unique canonical representation.

Example 5.32 The number of ways in which k people out of n may be placed around a circular table with k places corresponds to the number of k-cycles of I_n. Indeed, if we label the n people with I_n, we can associate to each placement around the table the function f that associates person i to the person on his right if i is among those seated, and otherwise associates i to himself:

$$f(i) = \begin{cases} j & \text{if } j \text{ is seated next to } i \text{ on the right,} \\ i & \text{if } i \text{ is not among those seated at the table.} \end{cases}$$

This describes a one-to-one correspondence between the set of possible placements of k people (out of n) around the table and the set of k-cycles of I_n. The cycle $\langle 3, 5, 2 \rangle$ of I_5 describes the situation in which 1 and 4 remain standing, while 2, 3 and 5 are seated at the table with 5 placed to the right of 3, 2 to the right of 5 and 3 to the right of 2.   □

We now give an explicit calculation for the number of k-cycles of I_n.

Theorem 5.33 *Let $k \leq n \in \mathbb{N}_{\geq 1}$. The number of k-cycles of I_n equals*

$$\frac{1}{k} S(n, k) = \frac{n!}{k \times (n - k)!}.$$

In particular, the number of n-cycles of I_n is $\dfrac{n!}{n} = (n - 1)!$.

Proof. In order to determine a k-cycle of I_n we first fix its support $\{a_1, a_2, \ldots, a_k\}$; this may be done in $C(n, k)$ ways. Without loss of generality, we may suppose that a_1 is the minimal element of $\{a_1, a_2, \ldots, a_k\}$. The canonical representations of the

k-cycles with support $\{a_1, a_2, \ldots, a_k\}$ are in one-to-one correspondence with the k-sequences of $\{a_1, a_2, \ldots, a_k\}$ that begin with a_1. The latter are obviously equal in number to the permutations of $(a_2, \ldots, a_k)$, namely, $(k-1)!$. Therefore, there are

$$\binom{n}{k} \times (k-1)! = \frac{n!}{k \times (n-k)!}$$

k-cycles of I_n.   □

5.2.2 Stirling Numbers of the First Kind ☙

In how many ways is it possible to place people at round tables? How many circular rings of children can be realized in a kindergarten?

Definition 5.34 Let k, n in $\mathbb{N}$. A k-**partition into cycles** of I_n is *a collection of cycles whose supports constitute a k-partition of I_n.* A **partition into cycles** is an arbitrary k-partition into cycles as k varies in $\mathbb{N}$.   □

Example 5.35 Tommy, Francie, Jack, Emma and Katy want to play ring around the rosy. If the boys and girls decide to split up into separate groups, we necessarily obtain one of the following 2-partitions into cycles of $\{T, F, J, E, K\}$:

$$[\langle J, T\rangle, \langle F, E, K\rangle], \quad [\langle J, T\rangle, \langle F, K, E\rangle].$$ □

Definition 5.36 We say that the k-sequence $(\langle a_1^1, \ldots, a_{n_1}^1\rangle, \ldots, \langle a_1^k, \ldots, a_{n_k}^k\rangle)$ is the **canonical representation** of the k-partition into cycles of I_n

$$[\langle a_1^1, \ldots, a_{n_1}^1\rangle, \ldots, \langle a_1^k, \ldots, a_{n_k}^k\rangle],$$

if the cycles $\langle a_1^i, \ldots, a_{n_i}^i\rangle$, $i = 1, \ldots, k$, are in canonical form (that is, if a_1^i is equal to $\min\{a_1^i, \ldots, a_{n_i}^i\}$) and $a_1^1 > a_1^2 > \cdots > a_1^k$.   □

Remark 5.37 Obviously every k-partition into cycles of I_n has a unique canonical representation. Moreover, the last cycle in a canonical representation of a partition into cycles always begins with 1.

Remark 5.38 If $\sigma_1, \ldots, \sigma_k$ are cycles of I_n with disjoint support, then $\sigma_1 \circ \cdots \circ \sigma_k = \sigma_{i_1} \circ \cdots \circ \sigma_{i_k}$ for every permutation $(i_1, \ldots, i_k)$ of $(1, \ldots, k)$. Indeed for each $i \in I_n$ one has

$$\sigma_{i_1} \circ \cdots \circ \sigma_{i_k}(i) = \begin{cases} \sigma_j(i) & \text{if } i \text{ belongs to the support of} \sigma_j, \\ i & \text{if } i \text{ does not belong to the support of any} \sigma_j. \end{cases}$$

Definition 5.39 The **composition of a k-partition into cycles** $[\sigma_1, \ldots, \sigma_k]$ of I_n is the function $f = \sigma_1 \circ \cdots \circ \sigma_k$.   □

Example 5.40 Consider the following 3-partition into cycles of I_7:

$$[\langle 6, 3, 5\rangle, \langle 1, 2, 7\rangle, \langle 4\rangle].$$

Its composition is the bijective function $f : I_7 \to I_7$ defined by setting

$$f(1) = 2, \ f(2) = 7, \ f(3) = 5, \ f(4) = 4, \ f(5) = 6, \ f(6) = 3, \ f(7) = 1.$$

Its canonical representation is the 3-sequence

$$(\langle 4\rangle, \langle 3, 5, 6\rangle, \langle 1, 2, 7\rangle). \qquad \qquad \square$$

Obviously, since cycles are bijective functions, also the composition of a partition into cycles is bijective. Let us prove that, conversely, any bijective function of a finite set onto itself is a composition of a partition into cycles.

Theorem 5.41 *For $n \in \mathbb{N}_{\geq 1}$, every bijective function $f : I_n \to I_n$ uniquely determines a partition into cycles $[\sigma_1, \ldots, \sigma_k]$ of I_n of which it is the composition.*

Proof. Put $f^i = \underbrace{f \circ f \circ \cdots \circ f}_{i}$, and consider the $(n + 1)$-sequence of I_n

$$(1, f(1), f^2(1), \ldots, f^n(1)).$$

In this sequence, consisting of $n + 1$ elements, there is certainly at least one repeated element of I_n. If $f^\ell(1)$, $\ell \geq 1$, is the first repetition, then $f^\ell(1) = 1$; indeed, if one had $f^\ell(1) = f^j(1)$ with $1 \leq j < \ell$, then one would also have

$$f(f^{\ell-1}(1)) = f^\ell(1) = f^j(1) = f(f^{j-1}(1))$$

and so, by the injectivity of f, it would follow that $f^{\ell-1}(1) = f^{j-1}(1)$, contradicting the choice of $f^\ell(1)$. Consider then the ℓ-cycle $\sigma_1 = \langle 1, f(1), \ldots, f^{\ell-1}(1)\rangle$. If the support of σ_1 does not coincide with I_n, then we take an element x of $I_n \setminus \{1, f(1), \ldots, f^{\ell-1}(1)\}$, that is from I_n without the support of σ_1, and we repeat the procedure just carried out: the $(n + 1)$-sequence $(x, f(x), \ldots, f^n(x))$ consists of $n + 1$ elements of I_n and hence there is certainly at least one repeated element of I_n. If $f^m(x)$, $m \geq 1$, is the first repetition, then as above one has $f^m(x) = x$. In this way we get a cycle $\sigma_2 = \langle x, \ldots, f^{m-1}(x)\rangle$, whose support is disjoint from the support of σ_1; indeed, arguing by contradiction, let $r \geq 0$ be the smallest natural number such that $f^r(x) = f^s(1)$ for a suitable $0 \leq s \leq \ell - 1$. By the way in which x was chosen, one necessarily has $r \geq 1$. If $s \geq 1$, then one would have $f(f^{r-1}(x)) = f(f^{s-1}(1))$ and so, by the injectivity of f, one would have $f^{r-1}(x) = f^{s-1}(1)$ which contradicts the choice of r. Therefore, one must have $f^r(x) = 1 = f(f^{\ell-1}(1))$ whence $f^{r-1}(x) = f^{\ell-1}(1)$, again contradicting our choice of r. If the union of the supports of σ_1 and σ_2 were not all of I_n, we could continue with the same procedure and construct cycles σ_j, $j = 3, \ldots, k$, until arriving at cycles $\sigma_1, \ldots, \sigma_k$ the union of whose supports coincides with I_n. The collection $[\sigma_1, \ldots, \sigma_k]$ is a partition of I_n into cycles whose composition coincides with f.   □

Example 5.42 Consider the function $f : I_7 \to I_7$ defined by setting:

$$f(1) = 7,\ f(2) = 4,\ f(3) = 5,\ f(4) = 2,\ f(5) = 3,\ f(6) = 1,\ f(7) = 6.$$

To the function f there is associated the partition into cycles $[\langle 1, 7, 6 \rangle, \langle 2, 4 \rangle, \langle 3, 5 \rangle]$. Indeed, considering $1, f(1), f(f(1)) = f^2(1), f^3(1), \ldots, f^8(1)$ the first repetition is $1 = f^3(1)$: thus one has the cycle $\langle 1, f(1), f^2(1) \rangle = \langle 1, 7, 6 \rangle$. The first element of I_7 not belonging to the support of the cycle $\langle 1, 7, 6 \rangle$ is 2; considering $2, f(2), \ldots$ the first repetition is $2 = f^2(2)$: thus one has the cycle $\langle 2, f(2) \rangle = \langle 2, 4 \rangle$. The first element of I_7 not belonging to the supports of $\langle 1, 7, 6 \rangle$ and $\langle 2, 4 \rangle$ is 3; considering $3, f(3), \ldots$ the first repetition is $3 = f^2(3)$: thus one has the cycle $\langle 3, f(3) \rangle = \langle 3, 5 \rangle$. The canonical representation of the partition into cycles associated to f is the sequence

$$(\langle 3, 5 \rangle, \langle 2, 4 \rangle, \langle 1, 7, 6 \rangle).$$ □

The time has now come to introduce the Stirling numbers of the first kind.

Definition 5.43 (*Stirling numbers of the first kind*) Let $k, n \in \mathbb{N}$. The **Stirling number of the first kind n over k**, denoted by $\begin{bmatrix} n \\ k \end{bmatrix}$, is the **number of k-partitions into cycles of I_n**.   □

Example 5.44 Bearing in mind the very definition of partitions into cycles one sees easily that:

- $\begin{bmatrix} n \\ n \end{bmatrix} = 1$ for all $n \geq 0$: if $n = 0$ the *empty collection* is the unique 0-partition into cycles of I_0, while if $n > 0$ the unique n-partition into cycles of I_n is $[\langle 1 \rangle, \ldots, \langle i \rangle, \ldots, \langle n \rangle]$;

- $\begin{bmatrix} n \\ 0 \end{bmatrix} = 0$ for all $n > 0$: there is no 0-partition into cycles of I_n;

- $\begin{bmatrix} 0 \\ k \end{bmatrix} = 0$ for all $k > 0$: there is no k-partition into cycles of I_0;

- $\begin{bmatrix} n \\ k \end{bmatrix} = 0$ for all $k > n$: there is no k-partition into cycles of I_n;

- $\begin{bmatrix} n \\ 1 \end{bmatrix} = (n - 1)!$ for all $n > 0$: indeed, this symbol denotes the number of n-cycles of I_n (see Theorem 5.33).   □

Remark 5.45 The map that associates to each k-partition into cycles

$$[\langle a_1^1, \ldots, a_{n_1}^1 \rangle, \ldots \langle a_1^k, \ldots, a_{n_k}^k \rangle]$$

the k-partition

$$[\{a_1^1, \ldots, a_{n_1}^1\}, \ldots \{a_1^k, \ldots, a_{n_k}^k\}]$$

is a surjective function from the set of k-partitions into cycles of I_n to the set of k-partitions of I_n. Therefore, the number of k-partitions into cycles of I_n is always greater than or equal to the number of k-partitions of I_n:

$$\begin{bmatrix} n \\ k \end{bmatrix} \geq \begin{Bmatrix} n \\ k \end{Bmatrix}.$$

A non-trivial case in which the Stirling numbers of the first and second kinds coincide is illustrated by the following example.

Example 5.46 Let $n \geq 2$. Then

$$\begin{bmatrix} n \\ n-1 \end{bmatrix} = \binom{n}{2} = \begin{Bmatrix} n \\ n-1 \end{Bmatrix}.$$

Indeed, an $(n-1)$-partition into cycles of I_n necessarily consists of $n-2$ cycles with 1 element and a single cycle with 2 elements; it is therefore determined by the choice of the two elements in the cycle with 2 elements, and that choice may be made in $\binom{n}{2}$ ways. The second equality was verified in Example 5.4.   □

The following identities have a nice combinatorial description.

Corollary 5.47 *For every $n \in \mathbb{N}$ and $m \in \mathbb{N}_{\geq 2}$ one has:*

1. $\displaystyle\sum_{0 \leq k \leq n} \begin{bmatrix} n \\ k \end{bmatrix} = n!;$

2. $\displaystyle\sum_{1 \leq k \leq m} (-1)^k \begin{bmatrix} m \\ k \end{bmatrix} = 0.$

Proof. 1. If $n = 0$ the statement becomes the identity $1 = \begin{bmatrix} 0 \\ 0 \end{bmatrix} = 0! = 1$. Now let $n \geq 1$; by Theorem 5.41 there is a one-to-one correspondence between the $n!$ bijective functions $I_n \to I_n$ and the partitions into cycles of I_n. Therefore $n!$ coincides with the total number of k-partitions into cycles of I_n as k varies from 0 to n.
2. Let us prove that there exists a bijective correspondence between the even partitions into cycles and the odd partitions into cycles of I_m; it will follow

$$\sum_{\substack{1 \leq k \leq m \\ k \text{ even}}} \begin{bmatrix} m \\ k \end{bmatrix} = \sum_{\substack{1 \leq k \leq m \\ k \text{ odd}}} \begin{bmatrix} m \\ k \end{bmatrix},$$

and hence the thesis. In the sequel we write the partitions into cycles in their canonical representation. Let

$$(\sigma_1, \ldots, \sigma_k) = (\langle a_1^1, \ldots, a_{m_1}^1 \rangle, \ldots, \langle a_1^k, \ldots, a_{m_k}^k \rangle)$$

be the canonical representation of a k-partition into cycles. By Remark 5.37 it is $a_1^k = 1$. The element 2 either belongs to the support $\{a_1^k = 1, \ldots, a_{m_k}^k\}$ of the last cycle, or is the first element of the penultimate cycle $\langle a_1^{k-1}, \ldots, a_{m_{k-1}}^{k-1} \rangle$. In the first case, if

$$\sigma_k = \langle a_1^k = 1, \ldots, a_j^k, a_{j+1}^k = 2, a_{j+2}^k, \ldots, a_{m_k}^k \rangle,$$

we split σ_k and set

$$\Psi(\sigma_1, \ldots, \sigma_k) = (\langle a_1^1, \ldots, a_{m_1}^1 \rangle, \ldots, \langle a_{j+1}^k = 2, a_{j+2}^k, \ldots, a_{m_k}^k \rangle, \langle a_1^k = 1, \ldots, a_j^k \rangle).$$

In the second case it is $\sigma_{k-1} = \langle a_1^{k-1} = 2, \ldots, a_{m_{k-1}}^{k-1} \rangle$ and $\sigma_k = \langle a_1^k = 1, \ldots, a_{m_k}^k \rangle$; then we join σ_k with σ_{k-1} and set

$$\Psi(\sigma_1, \ldots, \sigma_k) = (\langle a_1^1, \ldots, a_{m_1}^1 \rangle, \ldots, \langle a_1^{k-2}, \ldots, a_{m_{k-2}}^{k-2} \rangle, \langle a_1^k = 1, \ldots, a_{m_k}^k, a_1^{k-1} = 2, \ldots, a_{m_{k-1}}^{k-1} \rangle).$$

It is easy to see that Ψ coincides with its inverse map and then it is a bijective map of the set of the k-partition into cycles of I_m in itself. Since Ψ changes the parity of the number of cycles in the partitions, it induces a bijection between the even partitions into cycles and the odd partitions into cycles of I_m.   □

It is not difficult to give an explicit formula for the Stirling numbers of the first kind.

Theorem 5.48 (Formula for the Stirling numbers of the first kind) *If $k \leq n \in \mathbb{N}_{\geq 1}$, one has*

$$\begin{bmatrix} n \\ k \end{bmatrix} = \frac{n!}{k!} \sum_{\substack{n_1, \ldots, n_k \in \mathbb{N}_{\geq 1} \\ n_1 + \cdots + n_k = n}} \frac{1}{n_1 \cdots n_k}.$$

Proof. We carry out a combinatorial proof. We have to show that

$$k! \begin{bmatrix} n \\ k \end{bmatrix} = \sum_{\substack{n_1, \ldots, n_k \in \mathbb{N}_{\geq 1} \\ n_1 + \cdots + n_k = n}} \frac{n!}{n_1 \cdots n_k}.$$

By the Multiplication Principle 1.34, the number $k! \begin{bmatrix} n \\ k \end{bmatrix}$ is equal to the number of k-sequences $(\sigma_1, \ldots, \sigma_k)$ of cycles whose supports constitute a k-partition of I_n. Let

$n_1, \ldots, n_k \in \mathbb{N}_{\geq 1}$ satisfy the equality $n_1 + \cdots + n_k = n$. We can associate with any permutation $(b_1, \ldots, b_n)$ of $(1, \ldots, n)$ the k-sequence of cycles

$$(\langle b_1, \ldots, b_{n_1} \rangle, \langle b_{n_1+1}, \ldots, b_{n_1+n_2} \rangle, \ldots, \langle b_{n_1+\cdots+n_{k-1}+1}, \ldots, b_n \rangle).$$

By Remark 5.30, there are $n_1 \cdots n_k$ different permutations of $(1, \ldots, n)$ with associated the same k-sequence of cycles. By the Division Principle 1.47, the sequences of cycles of length $n_1, \ldots, n_k$ are $\dfrac{n!}{n_1 \cdots n_k}$. Therefore the total number $k! \begin{bmatrix} n \\ k \end{bmatrix}$ of k-sequences of cycles whose supports constitute a partition of I_n is equal to

$$\sum_{\substack{n_1, \ldots, n_k \in \mathbb{N}_{\geq 1} \\ n_1 + \cdots + n_k = n}} \frac{n!}{n_1 \cdots n_k}. \qquad \square$$

Example 5.49 For $n = 3$ and $k = 1, 2$ the formula found in Theorem 5.48 gives

$$\begin{bmatrix} 3 \\ 1 \end{bmatrix} = \frac{3!}{1!} \sum_{\substack{n_1 \in \mathbb{N}_{\geq 1} \\ n_1 = 3}} \frac{1}{n_1} = 2,$$

$$\begin{bmatrix} 3 \\ 2 \end{bmatrix} = \frac{3!}{2!} \sum_{\substack{n_1, n_2 \in \mathbb{N}_{\geq 1} \\ n_1 + n_2 = 3}} \frac{1}{n_1 n_2} = 3 \left(\frac{1}{1 \times 2} + \frac{1}{2 \times 1} \right) = 3. \qquad \square$$

Just as for Stirling numbers of the second kind, there is a recursive formula for the Stirling numbers of the first kind.

Proposition 5.50 (Recursive formula for the Stirling numbers of the first kind) *Let $k, n \in \mathbb{N}_{\geq 1}$. Then*

$$\begin{bmatrix} n-1 \\ k-1 \end{bmatrix} + (n-1) \begin{bmatrix} n-1 \\ k \end{bmatrix} = \begin{bmatrix} n \\ k \end{bmatrix}.$$

Proof. We treat the cases $n < k$, $n = k$ and $n > k$ separately. If $n < k$ or $n = k$, the equalities to be proved become respectively the identities $0 + (n-1) \times 0 = 0$ and $1 + (n-1) \times 0 = 1$.

Suppose now that $n > k$. We divide the k-partitions into cycles of I_n into two classes: those that contain the cycle $\langle n \rangle$ and those that do not. The k-partitions into cycles of the first type are formed by $\langle n \rangle$ and by a $(k-1)$-partition into cycles of I_{n-1}, and so they number $\begin{bmatrix} n-1 \\ k-1 \end{bmatrix}$. Those of the second type are the k-partitions into cycles, whose cycle containing n has support of cardinality greater than or equal to 2. Such k-partitions into cycles may be obtained from a k-partition into cycles of I_{n-1} by inserting n into one of the cycles. The number n may be inserted into an m-cycle $\langle a_1, \ldots, a_m \rangle$ of I_{n-1} so as to generate distinct $(m+1)$-cycles in m ways:

$$\langle n, a_1, \ldots, a_m \rangle, \ \langle a_1, n, \ldots, a_m \rangle, \ \ldots, \ \langle a_1, \ldots, n, a_m \rangle.$$

Thus, starting from a k-partition into cycles of I_{n-1} formed by cycles of lengths $m_1, \ldots, m_k$, on inserting the element n one obtains $m_1 + \cdots + m_k = n - 1$ different k-partitions into cycles of I_n. Thus there are $(n-1) \times \begin{bmatrix} n-1 \\ k \end{bmatrix}$ different k-partitions of the second type. Summing the number of elements of the two classes one obtains the desired result.   □

Example 5.51 (The triangle of the Stirling numbers of the first kind) The proposition proved above shows that in order to calculate the Stirling numbers of the first kind whose first argument is equal to n, it suffices to know the Stirling numbers of the first kind with first argument equal to $n - 1$. This leads to the construction of the triangle of the Stirling numbers of the first kind:

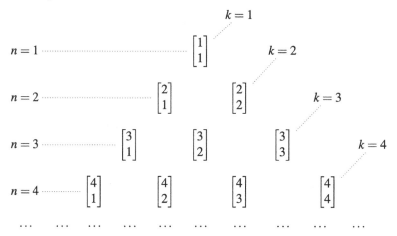

In view of Example 5.44 the values $\begin{bmatrix} n \\ 1 \end{bmatrix}$ and $\begin{bmatrix} n \\ n \end{bmatrix}$ along the edges are respectively equal to $(n-1)!$ and to 1; using the formula

$$\begin{bmatrix} n \\ k \end{bmatrix} = \begin{bmatrix} n-1 \\ k-1 \end{bmatrix} + (n-1) \times \begin{bmatrix} n-1 \\ k \end{bmatrix}$$

established in Proposition 5.50, we can easily find the internal values of a row when the values of the preceding row are known:

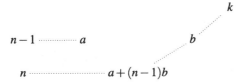

Thus one obtains the triangle

$$
\begin{array}{cccccccc}
 & & & & k=1 & & & \\
n=1 & \cdots\cdots\cdots & 1 & & & k=2 & & \\
n=2 & \cdots\cdots\cdots & 1 & & 1 & & k=3 & \\
n=3 & \cdots\cdots\cdots & 2 & & 3 & & 1 & k=4 \\
n=4 & \cdots\cdots & 6 & 11 & & 6 & & 1 \\
\cdots & \cdots & \cdots & \cdots & \cdots & \cdots & \cdots & \cdots
\end{array}
$$ □

By Proposition 5.50 we can also deduce an explicit alternative formula for $\begin{bmatrix} n \\ k \end{bmatrix}$.
If $I \neq \emptyset$, we are used to denote the sum and the product of the elements of I by $\sum_{i \in I} i$ and $\prod_{i \in I} i$. Furthermore, let us recall that always

$$
\sum_{i \in \emptyset} a_i = 0 \quad \text{and} \quad \prod_{i \in \emptyset} a_i = 1.
$$

Theorem 5.52 (Formula for the Stirling numbers of the first kind) *If $n \geq k \geq 1$, one has*

$$
\begin{bmatrix} n \\ k \end{bmatrix} = \sum_{\substack{J \subseteq I_{n-1} \\ |J| = n-k}} \prod_{i \in J} i.
$$

Proof. First note that for each $n \geq 1$ one has

$$\begin{bmatrix} n \\ n \end{bmatrix} = 1 = \prod_{i \in \emptyset} i = \sum_{\substack{J \subseteq I_{n-1} \\ |J| = 0}} \prod_{i \in J} i.$$

We now proceed by induction on n. For $n = 1$ one has only the case $\begin{bmatrix} 1 \\ 1 \end{bmatrix}$ which has already been discussed. Now let $n > 1$ and $1 \leq k \leq n$. Again, the case $n = k$ is already in hand; suppose, then, that $1 \leq k < n$. By Proposition 5.50 one has

$$\begin{bmatrix} n \\ k \end{bmatrix} = (n - 1) \begin{bmatrix} n - 1 \\ k \end{bmatrix} + \begin{bmatrix} n - 1 \\ k - 1 \end{bmatrix}.$$

The inductive hypothesis gives

$$\begin{bmatrix} n - 1 \\ k \end{bmatrix} = \sum_{\substack{J \subseteq I_{n-2} \\ |J| = n - 1 - k}} \prod_{i \in J} i$$

and therefore

$$(n - 1) \times \begin{bmatrix} n - 1 \\ k \end{bmatrix} = \sum_{\substack{J \subseteq I_{n-2} \\ |J| = n - 1 - k}} (n - 1) \times \prod_{i \in J} i = \sum_{\substack{J \subseteq I_{n-1},\, n - 1 \in J \\ |J| = n - k}} \prod_{i \in J} i.$$

We now deal with the summand $\begin{bmatrix} n - 1 \\ k - 1 \end{bmatrix}$. If $k = 1$, since there are no subsets of I_{n-2} having $n - 1$ elements, it follows that

$$\begin{bmatrix} n - 1 \\ k - 1 \end{bmatrix} = \begin{bmatrix} n - 1 \\ 0 \end{bmatrix} = 0 = \sum_{\substack{J \subseteq I_{n-2} \\ |J| = n - 1}} \prod_{i \in J} i.$$

If $k > 1$ the inductive hypothesis now yields

$$\begin{bmatrix} n - 1 \\ k - 1 \end{bmatrix} = \sum_{\substack{J \subseteq I_{n-2} \\ |J| = n - k}} \prod_{i \in J} i = \sum_{\substack{J \subseteq I_{n-1},\, n - 1 \notin J \\ |J| = n - k}} \prod_{i \in J} i.$$

Hence it follows that

$$\begin{bmatrix} n \\ k \end{bmatrix} = (n-1) \times \begin{bmatrix} n-1 \\ k \end{bmatrix} + \begin{bmatrix} n-1 \\ k-1 \end{bmatrix}$$

$$= \sum_{\substack{J \subseteq I_{n-1},\, n-1 \in J \\ |J| = n-k}} \prod_{i \in J} i + \sum_{\substack{J \subseteq I_{n-1},\, n-1 \notin J \\ |J| = n-k}} \prod_{i \in J} i$$

$$= \sum_{\substack{J \subseteq I_{n-1} \\ |J| = n-k}} \prod_{i \in J} i. \qquad \square$$

Example 5.53 We compute again, as in Example 5.49, the numbers $\begin{bmatrix} 3 \\ 2 \end{bmatrix}$ and $\begin{bmatrix} 3 \\ 1 \end{bmatrix}$, following now the formula found in Theorem 5.52. We have

$$\begin{bmatrix} 3 \\ 2 \end{bmatrix} = \sum_{\substack{J \subseteq I_2 \\ |J| = 1}} \prod_{i \in J} i = 1 + 2 = 3,$$

while for $k = 1$ we obtain

$$\begin{bmatrix} 3 \\ 1 \end{bmatrix} = \sum_{\substack{J \subseteq I_2 \\ |J| = 2}} \prod_{i \in J} i = 1 \times 2 = 2. \qquad \square$$

The formula stated in Theorem 5.52 has the disadvantage of being rather tedious for calculation; it has a more convenient reformulation which allows one to find the Stirling numbers of the first kind as the coefficients of an easily remembered polynomial.

Corollary 5.54 *Let $n \in \mathbb{N}_{\geq 1}$. The Stirling number of the first kind $\begin{bmatrix} n \\ k \end{bmatrix}$ is the coefficient of the term of degree k in the polynomial $X(X+1)(X+2)\cdots(X+(n-1))$:*

$$\sum_{k=1}^{n} \begin{bmatrix} n \\ k \end{bmatrix} X^k = X(X+1)(X+2)\cdots(X+(n-1)).$$

Proof. Multiplying $X, X+1, \ldots, X+(n-1)$ one gets:

$$X(X+1)\cdots(X+(n-1)) = X(a_1 + a_2 X + \cdots + a_n X^{n-1})$$

where each a_k, $k = 1, \ldots, n$, is the sum of the products of all possible choices of $(n-1) - (k-1) = n - k$ terms of I_{n-1} (with the convention that, as above, an empty product equals 1). By Theorem 5.52, one has

$$a_k = \sum_{\substack{J \subseteq I_{n-1} \\ |J| = n - k}} \prod_{i \in J} i = \begin{bmatrix} n \\ k \end{bmatrix}. \qquad \square$$

Corollary 5.54 permits one to calculate the Stirling numbers of the first kind by carrying out a simple product of polynomials.

Example 5.55 In Example 5.53 we used the formula from Theorem 5.52 to calculate the Stirling numbers $\begin{bmatrix} 3 \\ k \end{bmatrix}$, $k = 1, 2$. Alternatively, one can obtain the same result more easily in the light of Corollary 5.54, since these numbers are the coefficients of the terms of degree $k = 1, 2$ in the polynomial

$$X(X + 1)(X + 2) = X^3 + 3X^2 + 2X.$$

We can appreciate better the new formulation determining the Stirling numbers $\begin{bmatrix} 5 \\ k \end{bmatrix}$, for $k = 0, \ldots, 5$. By Corollary 5.54, the Stirling number $\begin{bmatrix} 5 \\ k \end{bmatrix}$ is the coefficient of the term of degree k in

$$X(X + 1)(X + 2)(X + 3)(X + 4) = 24X + 50X^2 + 35X^3 + 10X^4 + X^5.$$

Hence,

$$\begin{bmatrix} 5 \\ 0 \end{bmatrix} = 0, \quad \begin{bmatrix} 5 \\ 1 \end{bmatrix} = 24, \quad \begin{bmatrix} 5 \\ 2 \end{bmatrix} = 50, \quad \begin{bmatrix} 5 \\ 3 \end{bmatrix} = 35, \quad \begin{bmatrix} 5 \\ 4 \end{bmatrix} = 10, \quad \begin{bmatrix} 5 \\ 5 \end{bmatrix} = 1. \qquad \square$$

5.2.3   Partitions into Cycles with Occupancy Constraints ☕

We now consider partitions into cycles with a prescribed occupancy collection.

Definition 5.56 Let $n, k, n_1, \ldots, n_k \in \mathbb{N}_{\geq 1}$ with $n_1 + \cdots + n_k = n$. A k-**partition into cycles** of I_n **with occupancy collection** $[n_1, \ldots, n_k]$ is a k-partition into cycles of I_n consisting of cycles whose supports constitute a k-partition of I_n with occupancy collection $[n_1, \ldots, n_k]$. We use the symbol $\begin{bmatrix} n \\ k \end{bmatrix}; [n_1, \ldots, n_k]$ to denote **the number of k-partitions into cycles of I_n with occupancy** $[n_1, \ldots, n_k]$. $\qquad \square$

It is rather easy to calculate the number $\begin{bmatrix} n \\ k \end{bmatrix}; [n_1, \ldots, n_k]$.

Theorem 5.57 *Let $k, n, n_1, \ldots, n_k \in \mathbb{N}_{\geq 1}$ with $n_1 + \cdots + n_k = n$. Then*

$$\left[\begin{matrix} n \\ k \end{matrix}; [n_1, \ldots, n_k]\right] = \frac{1}{k!} \times \frac{n!}{n_1 \cdots n_k} \times P(n_1, \ldots, n_k).$$

Proof. By Theorem 5.21, there are

$$\left\{\begin{matrix} n \\ k \end{matrix}; [n_1, \ldots, n_k]\right\} = \frac{1}{k!} \times \frac{n!}{n_1! \cdots n_k!} \times P(n_1, \ldots, n_k)$$

k-partitions of I_n with occupancy collection $[n_1, \ldots, n_k]$. By Theorem 5.33, every subset of I_n having cardinality n_i is the support of $(n_i - 1)!$ different n_i-cycles. Thus each k-partition of I_n with occupancy collection $[n_1, \ldots, n_k]$ produces exactly $(n_1 - 1)! \cdots (n_k - 1)!$ different k-partitions into cycles of I_n with occupancy collection $[n_1, \ldots, n_k]$. Hence, the total number of k-partitions into cycles of I_n with occupancy collection $[n_1, \ldots, n_k]$ is

$$\left[\begin{matrix} n \\ k \end{matrix}; [n_1, \ldots, n_k]\right] = \left\{\begin{matrix} n \\ k \end{matrix}; [n_1, \ldots, n_k]\right\} \times ((n_1 - 1)! \times \cdots \times (n_k - 1)!)$$

$$= \frac{1}{k!} \times \frac{n!}{n_1 \cdots n_k} \times P(n_1, \ldots, n_k). \qquad \square$$

Example 5.58 In how many ways can one seat 6 people at 3 equally-sized round tables with at least one person per table? What if one would like that there be 1 table with two people, 1 table with one and 1 table with three?

Solution. Two placements of the 6 people are different if they place different groups of people at some table, or if at some table they assign some person a different neighbor (for example, proceeding clockwise). Label the 6 people with I_6. In the first case we must count the 3-partitions into cycles of I_6, that is, calculate $\left[\begin{matrix} 6 \\ 3 \end{matrix}\right]$.

Applying Theorem 5.52 one obtains

$$\left[\begin{matrix} 6 \\ 3 \end{matrix}\right] = \sum_{\substack{I \subseteq I_5 \\ |I| = 3}} \prod_{i \in I} i =$$

$$= 1 \times 2 \times 3 + 1 \times 2 \times 4 + 1 \times 2 \times 5 + 1 \times 3 \times 4 + 1 \times 3 \times 5 + 1 \times 4 \times 5 +$$

$$+ 2 \times 3 \times 4 + 2 \times 3 \times 5 + 2 \times 4 \times 5 + 3 \times 4 \times 5 = 225.$$

In the second case we must count the 3-partitions into cycles of I_6 with occupancy collection $[1, 2, 3]$. By Theorem 5.57 the number of such 3-partitions is

$$\left[\begin{matrix} 6 \\ 3 \end{matrix}; [1, 2, 3]\right] = \frac{1}{3!} \times \frac{6!}{1 \times 2 \times 3} \times P(1, 2, 3) = 120. \qquad \square$$

5.3 Eulerian Numbers: Excesses and Ascending Couples in a Permutation

Eulerian numbers count both the number of permutations of $(1, \ldots, n)$ with a given number of elements larger than the position they occupy (*excesses*) and the number of permutations of $(1, \ldots, n)$ in which a given number of elements are larger than their predecessor (*ascending couples*, or *ascending pairs*). They will both be used in solving some classical problems such as the Smith College diploma problem.

5.3.1 Eulerian Numbers and Excesses: The Smith College Diplomas

We begin by formalizing what it means to say that an *element is larger than the position it occupies* in a permutation:

Definition 5.59 Let $a = (a_1, \ldots, a_n)$ be a permutation of $(1, \ldots, n)$. An **excess** is an element i of I_n such that $a_i \geq i$. A **strict excess** is an element i of I_n such that $a_i > i$. □

Example 5.60 The permutation $a = (4, 2, 3, 1)$ has three excesses and only one strict excess: in fact $a_1 = 4 > 1$, $a_2 = 2$, $a_3 = 3$ and $a_4 = 1 < 4$. □

Remark 5.61 A permutation a of $(1, \ldots, n)$ has at least one excess: necessarily $a_1 \geq 1$. The permutation a has no strict excesses if and only if it has n excesses: in both cases one must have $a_i = i$ for all $i \in I_n$ and so $a = (1, \ldots, n)$.

Definition 5.62 (*Eulerian numbers*) Let $k, n \in \mathbb{N}$ with $n \geq 1$. The **Eulerian number** n **over** k, denoted by $\left\langle {n \atop k} \right\rangle$, is the **number of permutations of** $(1, \ldots, n)$ **with exactly** k **strict excesses**. □

Example 5.63 Bearing in mind that n will never be a strict excess in a permutation of $(1, \ldots, n)$, for each $n \geq 1$ one has:

- $\left\langle {n \atop 0} \right\rangle = 1$: $(1, \ldots, n)$ is the unique permutation of $(1, \ldots, n)$ without strict excesses;

- $\left\langle {n \atop k} \right\rangle = 0$ for each $k \geq n$: there is no permutation of $(1, \ldots, n)$ with $k \geq n$ strict excesses;

- $\left\langle {n \atop n-1} \right\rangle = 1$: $(2, 3, \ldots, n, 1)$ is the unique permutation of $(1, \ldots, n)$ with $n - 1$ strict excesses. □

Next we introduce the *shift* of a permutation of $(1, \ldots, n)$.

Definition 5.64 The **shift** $\Theta(a_1, \ldots, a_n)$ of a permutation $(a_1, \ldots, a_n)$ of $(1, \ldots, n)$ is defined by

$$\Theta(a_1, \ldots, a_n) := (a_2, \ldots, a_n, a_1). \qquad \Box$$

Example 5.65 The shift of $(3, 5, 2, 4, 1)$ is $\Theta(3, 5, 2, 4, 1) = (5, 2, 4, 1, 3)$. $\qquad \Box$

There is a useful and interesting relation between the number of excesses of a permutation and the number of strict excesses of its shift.

Lemma 5.66 *The permutation* $\mathfrak{a} = (a_1, \ldots, a_n)$ *of* $(1, \ldots, n)$ *has* $k \geq 1$ *excesses if and only if the shift* $\Theta(\mathfrak{a})$ *of* $\mathfrak{a}$ *has* $k - 1$ *strict excesses.*

Proof. Let $i \in I_n$; there are two possible cases:

$i = 1$: clearly $1 \leq a_1 \leq n$; therefore 1 is an excess for $\mathfrak{a}$ and n is not a strict excess for $\Theta(\mathfrak{a})$.

$i > 1$: in this case $i \leq a_i$ if and only if $i - 1 < a_i$; hence i is an excess for $\mathfrak{a}$ if and only if $i - 1$ is a strict excess for $\Theta(\mathfrak{a})$. $\qquad \Box$

Example 5.67 The emperor distributes twelve bags containing respectively $1, 2, \ldots,$ 12 gold coins to 12 of his knights. Having numbered the 12 places at a round table in the clockwise sense, he then seats the knights at the table. We use a_i to denote the number of gold coins in the bag received by the knight seated at position i. Suppose that only 5 knights have a number of coins greater than or equal to the number of the place they occupy. If each knight passes his bag of coins to his counterclockwise neighbor, there will be exactly 4 knights whose money bag strictly exceeds his place number. Indeed, if the initial distribution of money bags corresponds to the sequence $(a_1, \ldots, a_{12})$, then the new distribution of bags corresponds to the sequence $(a_2, a_3, \ldots, a_{12}, a_1) = \Theta(a_1, \ldots, a_{12})$; thus, by Lemma 5.66 one obtains the desired conclusion. $\qquad \Box$

Example 5.68 (*The Smith College diploma problem* [25]) At Smith College diplomas are conferred as follows. The diplomas are randomly distributed to the graduating students. Those who do not receive their own diploma form a circle and pass the diploma received to their counterclockwise neighbor: those then receiving their own diploma leave the circle, while the others form a smaller circle and repeat the procedure. Determine the probability that exactly $k \geq 0$ hand-offs are necessary before each graduate has her/his own diploma.
Solution. Label the graduating students with I_n. There are $n!$ possible distributions of the diplomas. We now count the initial distributions for which after exactly $k \geq 0$ hand-offs the procedure terminates. Suppose that graduate i has been given the diploma of graduate a_i. We eliminate from $(a_1, \ldots, a_n)$ the ℓ graduates who have received their own diplomas; let $i_1 < \cdots < i_{n-\ell}$ be the remaining graduates. Relabeling i_1 with $1, i_2$ with $2, \ldots,$ and $i_{n-\ell}$ with $n - \ell$, the distribution of the diplomas is now represented by a permutation $(b_1, \ldots, b_{n-\ell})$ of $(1, \ldots, n - \ell)$. By construction $j \neq b_j$ for $j = 1, \ldots, n - \ell$ and so the excesses of the permutation $(b_1, \ldots, b_{n-\ell})$

are all strict; moreover $(b_1, \ldots, b_{n-\ell})$ has the same number of strict excesses as $(a_1, \ldots, a_n)$: indeed, $i_h < a_{i_h}$ if and only if $h < b_h$. At the first hand-off the distribution of the diplomas is represented by the permutation $\Theta(b_1, \ldots, b_{n-\ell})$: in view of Lemma 5.66 the number of strict excesses decreases by one. We repeat the same procedure eliminating the graduates who have received their own diplomas and relabelling the remaining ones. Proceeding in this way, in order that k hand-offs should be required before every graduate has her own diploma, it is necessary and sufficient that the permutation $(a_1, \ldots, a_n)$ of $(1, \ldots, n)$ which represented the initial distribution of the diplomas should have exactly k strict excesses. The number of favorable cases is therefore $\left\langle {n \atop k} \right\rangle$, and so the desired probability is $\dfrac{1}{n!} \left\langle {n \atop k} \right\rangle$.   □

5.3.2 Eulerian Numbers and Ascending Couples: Recursive and Explicit Formulas

We introduce the notion of an *ascending couple* (sometimes also called an *ascending pair*) of a permutation.

Definition 5.69 Let $n \in \mathbb{N}_{\geq 1}$. If $(a_1, \ldots, a_n)$ is a permutation of $(1, \ldots, n)$, we say that (a_i, a_{i+1}), $1 \leq i < n$, is an **ascending couple** if $a_i < a_{i+1}$.   □

Example 5.70 The permutation $(4, 6, 1, 3, 5, 2)$ of $(1, 2, 3, 4, 5, 6)$ has three ascending couples: $(4, 6)$, $(1, 3)$ and $(3, 5)$. The permutations of $(1, 2, 3, 4)$ that have 2 ascending couples are

$$(1, 2, 4, 3), \ (1, 3, 4, 2), \ (1, 3, 2, 4), \ (1, 4, 2, 3), \ (2, 1, 3, 4), \ (2, 3, 4, 1),$$

$$(2, 3, 1, 4), \ (2, 4, 1, 3), \ (3, 4, 1, 2), \ (3, 1, 2, 4), \ (4, 1, 2, 3).$$ □

The Eulerian number $\left\langle {n \atop k} \right\rangle$ also counts the permutations of $(1, \ldots, n)$ with k ascending couples.

Theorem 5.71 *Let $k, n \in \mathbb{N}$ with $n \geq 1$. Then the number of permutations of $(1, \ldots, n)$ with k ascending couples coincides with the number $\left\langle {n \atop k} \right\rangle$ of permutations of $(1, \ldots, n)$ with k strict excesses.*

Proof. In every permutation of $(1, \ldots, n)$ both the number of strict excesses and the number of ascending couples is greater than or equal to 0 and less than or equal to $n - 1$. Now let $0 \leq k \leq n - 1$. We construct a bijection T from the set of permutations of $(1, \ldots, n)$ to itself in such a way as to induce a one-to-one correspondence between the permutations with k strict excesses and the permutations with k ascending couples. As we have seen in Example 1.24, to a permutation $\mathfrak{a} = (a_1, \ldots, a_n)$ of $(1, \ldots, n)$ we can associate the bijective function $f_{\mathfrak{a}} : I_n \to I_n$

defined by setting $f_\mathfrak{a}(i) = a_i$ for $i = 1, \ldots, n$. By Theorem 5.41, the bijective function $f_\mathfrak{a}$ determines a partition into cycles of I_n of which it is the composition: let $\left(\langle a_1^1, \ldots, a_{k_1}^1 \rangle, \ldots \langle a_1^m, \ldots, a_{k_m}^m \rangle \right)$ be its canonical representation (see Definition 5.36). We then set

$$T(\mathfrak{a}) := (a_1^1, \ldots, a_{k_1}^1, a_1^2, \ldots, a_{k_2}^2, \ldots, a_1^m, \ldots, a_{k_m}^m).$$

We first verify that T is a surjective function: let $\mathfrak{b} = (b_1, \ldots, b_n)$ be a permutation of $(1, \ldots, n)$; there are $m \geq 0$ and integers $k_0, k_1, \ldots, k_m$ such that

$$k_0 = 1, \; k_1 = \min\{i > k_0 : b_i < b_{k_0}\}, \; k_2 = \min\{i > k_1 : b_i < b_{k_1}\}, \ldots,$$

$$k_m = \min\{i > k_{m-1} : b_i < b_{k_{m-1}}\} \quad \text{and} \quad b_{k_m} < b_i \; \forall i : k_m < i \leq n.$$

Consider the cycles $\langle b_1 = b_{k_0}, \ldots, b_{k_1-1} \rangle, \ldots, \langle b_{k_m}, \ldots, b_n \rangle$; by construction

$$(\langle b_1 = b_{k_0}, \ldots, b_{k_1-1} \rangle, \ldots, \langle b_{k_m}, \ldots, b_n \rangle)$$

is the canonical representation of a partition into cycles of I_n. Let $f : I_n \to I_n$ be the composition of that partition into cycles; clearly one has $T(f(1), \ldots, f(n)) = \mathfrak{b}$. Since the domain and codomain of T have the same cardinality, T is therefore a bijective function. We now verify that a permutation $\mathfrak{a}$ of $(1, \ldots, n)$ has the same number of strict excesses as there are ascending couples in the permutation $T(\mathfrak{a})$. Let $\mathfrak{a} = (a_1, \ldots, a_n)$ e $T(\mathfrak{a}) = (x_1, \ldots, x_n)$. By construction one has

$$(x_1, \ldots, x_n) = (a_1^1, \ldots, a_{k_1}^1, a_1^2, \ldots, a_{k_2}^2, \ldots, a_1^m, \ldots, a_{k_m}^m)$$

with

- $a_j = f_\mathfrak{a}(j) = \langle a_1^1, \ldots, a_{k_1}^1 \rangle \circ \cdots \circ \langle a_1^m, \ldots, a_{k_m}^m \rangle(j)$ for each $j = 1, \ldots, n$;
- $a_1^s = \min\{a_j^s : j = 1, \ldots, k_s\}$ for each $s = 1, \ldots, m$;
- $a_1^1 > a_1^2 > \cdots > a_1^m = 1$.

Given $i \in I_n$, let $\langle a_1^\ell, \ldots, a_{k_\ell}^\ell \rangle$ be that cycle of the partition into cycles determined by the function $f_\mathfrak{a}$ which contains x_i. There are two possible cases:

(a) If $x_i = a_h^\ell$ with $1 \leq h < k_\ell$, then $i < n$ and $x_{i+1} = a_{h+1}^\ell = f_\mathfrak{a}(a_h^\ell) = f_\mathfrak{a}(x_i) = a_{x_i}$.

(b) If $x_i = a_{k_\ell}^\ell$ then

$$a_{x_i} = f_\mathfrak{a}(x_i) = f_\mathfrak{a}(a_{k_\ell}^\ell) = a_1^\ell < a_{k_\ell}^\ell = x_i$$

and either $i = n$ or

$$x_{i+1} = a_1^{\ell+1} < a_1^\ell < a_{k_\ell}^\ell = x_i.$$

If x_i is a strict excess for $\mathfrak{a}$, then $a_{x_i} > x_i$ and necessarily we are in case (a) thus, one has $x_i < a_{x_i} = x_{i+1}$ and so (x_i, x_{i+1}) is an ascending couple in $T(\mathfrak{a})$. Conversely, if (x_i, x_{i+1}) is an ascending couple in $T(\mathfrak{a})$, then since $i < n$ and $x_i < x_{i+1}$, we again find that we are necessarily in case (a) hence one has $x_i < x_{i+1} = a_{x_i}$ and so x_i is a strict excess for $\mathfrak{a}$. It follows that the number of permutations of $(1, \ldots, n)$ with k ascending couples coincides with that of permutations with k strict excesses.   □

☞ *Example 5.72* One should beware that the preceding result does *not* state that a permutation having k ascending couples has k strict excesses: for example, the permutation $(2, 3, 4, 1)$ has 2 ascending couples but has 3 strict excesses! The permutation $(2, 3, 4, 1)$ determines the partition into cycles $[\langle 1, 2, 3, 4\rangle]$ consisting of a single cycle; the image of $(2, 3, 4, 1)$ under the map T constructed in the preceding proof is therefore the permutation $(1, 2, 3, 4)$ which does indeed have 3 ascending couples.

□

Theorem 5.71 allows us to obtain the following useful relations rather easily.

Proposition 5.73 *Let $n \in \mathbb{N}_{\geq 1}$; then one has:*

1. $\left\langle\!\!{n \atop k}\!\!\right\rangle = \left\langle\!\!{n \atop n-1-k}\!\!\right\rangle$ *for* $0 \leq k < n$;

2. $\left\langle\!\!{n \atop k}\!\!\right\rangle = (n-k)\left\langle\!\!{n-1 \atop k-1}\!\!\right\rangle + (k+1)\left\langle\!\!{n-1 \atop k}\!\!\right\rangle$ *for* $1 \leq k < n$.

Proof. 1. To each permutation $\mathfrak{a} = (a_1, \ldots, a_{n-1}, a_n)$ of $(1, \ldots, n)$ we can associate the permutation $\mathfrak{b}$ obtained by reading the components of the permutation $\mathfrak{a}$ from right to left:

$$\mathfrak{b} = (a_n, a_{n-1}, \ldots, a_1).$$

Obviously the ascending couples in $\mathfrak{a}$ become descending in $\mathfrak{b}$ and vice versa; therefore, if there are k ascending couples in $\mathfrak{a}$, then in $\mathfrak{b}$ there are $(n-1) - k$. Thus we have obtained a one-to-one correspondence between the permutations with k ascending couples and those with $(n-1) - k$ ascending couples.

2. To each permutation $\mathfrak{a}$ of $(1, \ldots, n)$ we associate the permutation $\pi_n(\mathfrak{a})$ of the sequence $(1, \ldots, n-1)$ obtained from $\mathfrak{a}$ by making the element n "disappear": thus, for example $\pi_3(2, 3, 1) = (2, 1)$. To a permutation $\mathfrak{a}$ of $(1, \ldots, n)$ with k ascending couples the map π_n associates the permutation $\pi_n(\mathfrak{a}) = (b_1, \ldots, b_{n-1})$ of $(1, \ldots, n-1)$ with a number of ascending couples that varies according to the position of n in the permutation $\mathfrak{a}$. There are four possible cases to be considered:

(a) n occupies the first position of $\mathfrak{a}$, that is $\mathfrak{a} = (n, b_1, \ldots, b_{n-1})$; since surely $n > b_1$, the permutation $\pi_n(\mathfrak{a})$ has k ascending couples.

(b) n occupies the last position of $\mathfrak{a}$, that is $\mathfrak{a} = (b_1, \ldots, b_{n-1}, n)$; certainly $b_{n-1} < n$, and so in $\pi_n(\mathfrak{a})$ there are $k - 1$ ascending couples, one less than in $\mathfrak{a}$.

(c) n is located "inside" $\mathfrak{a}$, that is $\mathfrak{a} = (b_1, \ldots, b_i, n, b_{i+1}, \ldots, b_n)$ for some $1 \leq i < n - 1$.

(c1) If $b_i < b_{i+1}$, then since the couple (b_i, n) is ascending, while (n, b_{i+1}) is not, it follows that in $\pi_n(\mathfrak{a})$ there are k ascending couples, just as many as there are in $\mathfrak{a}$.

(c2) If $b_i > b_{i+1}$, then since the couple (b_i, n) is ascending while (n, b_{i+1}) is not, in $\pi_n(\mathfrak{a})$ there are $k - 1$ ascending couples, one less than in $\mathfrak{a}$.

Every permutation $(b_1, \ldots, b_{n-1})$ of $(1, \ldots, n-1)$ with k ascending couples is the image under π_n of $1 + k$ permutations of $(1, \ldots, n)$ with k ascending couples: one of these is the permutation $(n, b_1, \ldots, b_{n-1})$ (case (a)) and there are k permutations of $(1, \ldots, n)$ obtained by inserting n between two elements b_i and b_{i+1} of an ascending couple of $(b_1, \ldots, b_{n-1})$ (case (c1)). Every permutation $(b_1, \ldots, b_{n-1})$ of $(1, \ldots, n-1)$ with $k - 1$ ascending couples is the image under π_n of $1 + n - 1 - k = n - k$ permutations of $(1, \ldots, n)$ with k ascending couples: one is the permutation $(b_1, \ldots, b_{n-1}, n)$ (case (b)) and the $n - 1 - k$ others are the permutations of $(1, \ldots, n)$ obtained by inserting n between two elements b_i and b_{i+1} of a non-ascending couple of $(b_1, \ldots, b_{n-1})$ (case (c2)). Therefore, by Corollary 1.46 one has

$$\left\langle \begin{matrix} n \\ k \end{matrix} \right\rangle = (k+1) \left\langle \begin{matrix} n-1 \\ k \end{matrix} \right\rangle + (n-k) \left\langle \begin{matrix} n-1 \\ k-1 \end{matrix} \right\rangle. \qquad \square$$

Example 5.74 (The triangle of the Eulerian numbers) Proposition 5.73, 2, shows that in order to calculate the Eulerian numbers with first argument equal to n it suffices to know those with first argument equal to $n - 1$. This observation permits one to construct the triangle of the Eulerian numbers.

For all n one has $\left\langle \begin{matrix} n \\ 0 \end{matrix} \right\rangle = \left\langle \begin{matrix} n \\ n-1 \end{matrix} \right\rangle = 1$: thus, along non-horizontal sides of the Eulerian triangle all elements are equal to one. Using the formula

$$\left\langle \begin{matrix} n \\ k \end{matrix} \right\rangle = (n-k) \left\langle \begin{matrix} n-1 \\ k-1 \end{matrix} \right\rangle + (k+1) \left\langle \begin{matrix} n-1 \\ k \end{matrix} \right\rangle$$

proved in Point 2 of Proposition 5.73, we can easily obtain the internal values of a row once the values of the preceding row are known:

$$n-1 \quad\text{......}\quad a \qquad\qquad\qquad\qquad\qquad b \qquad\qquad\qquad k$$

$$n \quad\text{..............}\quad (n-k)a+(k+1)b$$

Thus one obtains the triangle

$$k=0$$

$$n=1 \quad\text{.......................................}\quad 1 \qquad\qquad k=1$$

$$n=2 \quad\text{..................................}\quad 1 \qquad\qquad 1 \qquad\qquad k=2$$

$$n=3 \quad\text{.............................}\quad 1 \qquad\qquad 4 \qquad\qquad 1 \qquad\qquad k=3$$

$$n=4 \quad\text{...................}\quad 1 \qquad\qquad 11 \qquad\qquad 11 \qquad\qquad 1$$

...   .

Note that in view of the property

$$\left\langle {n \atop k} \right\rangle = \left\langle {n \atop n-1-k} \right\rangle$$

it is sufficient to complete the first half of each row.   □

We conclude this section by giving an explicit calculation of the Eulerian number $\left\langle {n \atop k} \right\rangle$. Here we give an elementary but rather long and tedious proof; a quick proof, but making use of more elaborate techniques like generating formal series and Worpitzky formula (Proposition 5.84), is given in Corollary 8.45.

Proposition 5.75 *Let* $k, n \in \mathbb{N}$ *with* $n \geq 1$. *Then*

$$\left\langle {n \atop k} \right\rangle = \begin{cases} \sum_{i=0}^{k}(-1)^i \binom{n+1}{i}(k+1-i)^n & \text{if } 0 \leq k < n, \\ 0 & \text{otherwise.} \end{cases}$$

Proof. The equality is obvious if $k \geq n$; for $k = 0$ one has

$$\left\langle {n \atop 0} \right\rangle = 1 = \binom{n+1}{0}.$$

The case $n = 1$ is part of that with $k = 0$. We now prove the statement for $1 \leq k \leq n - 1$ using induction on n. Suppose that the equality holds for a given $n \geq 1$. We prove that it then also holds for $n + 1$. By Point 2 of Proposition 5.73 one has

$$\left\langle {n+1 \atop k} \right\rangle = (k+1)\left\langle {n \atop k} \right\rangle + (n+1-k)\left\langle {n \atop k-1} \right\rangle.$$

The induction hypothesis gives

$$\left\langle {n+1 \atop k} \right\rangle = \underbrace{(k+1)\sum_{i=0}^{k}(-1)^i\binom{n+1}{i}(k+1-i)^n}_{A} +$$

$$+ \underbrace{(n+1-k)\sum_{i=0}^{k-1}(-1)^i\binom{n+1}{i}(k-i)^n}_{B}.$$

Now one has

$$B = \sum_{i=0}^{k-1}(-1)^i(n+1-i-k+i)\binom{n+1}{i}(k-i)^n$$

$$= \sum_{i=0}^{k}(-1)^i(n+1-i-k+i)\binom{n+1}{i}(k-i)^n$$

$$= \sum_{i=0}^{k}(-1)^i(n+1-i)\binom{n+1}{i}(k-i)^n - \sum_{i=0}^{k}(-1)^i(k-i)\binom{n+1}{i}(k-i)^n.$$

Since

$$(n+1-i)\binom{n+1}{i} = (n+1-i)\frac{(n+1)!}{i!(n+1-i)!} = \frac{(n+1)!}{i!(n-i)!}$$

$$= (i+1)\frac{(n+1)!}{(i+1)!(n-i)!} = (i+1)\binom{n+1}{i+1},$$

it follows that

$$B = \sum_{i=0}^{k}(-1)^i(i+1)\binom{n+1}{i+1}(k-i)^n - \sum_{i=0}^{k}(-1)^i\binom{n+1}{i}(k-i)^{n+1}$$

$$= -\sum_{i=1}^{k+1}(-1)^i i\binom{n+1}{i}(k-(i-1))^n - \sum_{i=0}^{k}(-1)^i\binom{n+1}{i}(k-i)^{n+1}$$

$$= -\sum_{i=0}^{k}(-1)^i i\binom{n+1}{i}(k+1-i)^n - \sum_{i=0}^{k}(-1)^i\binom{n+1}{i}(k-i)^{n+1}.$$

Therefore we have

$$\left\langle{n+1\atop k}\right\rangle = A + B = \sum_{i=0}^{k}(-1)^i(k+1)\binom{n+1}{i}(k+1-i)^n +$$

$$-\sum_{i=0}^{k}(-1)^i i\binom{n+1}{i}(k+1-i)^n - \sum_{i=0}^{k}(-1)^i\binom{n+1}{i}(k-i)^{n+1} =$$

$$= \sum_{i=0}^{k}(-1)^i(k+1-i)\binom{n+1}{i}(k+1-i)^n - \sum_{i=0}^{k}(-1)^i\binom{n+1}{i}(k-i)^{n+1}$$

$$= \sum_{i=0}^{k}(-1)^i\binom{n+1}{i}(k+1-i)^{n+1} + \sum_{i=1}^{k}(-1)^i\binom{n+1}{i-1}(k-(i-1))^{n+1}$$

$$= (k+1)^{n+1} + \sum_{i=1}^{k}(-1)^i\left[\binom{n+1}{i}+\binom{n+1}{i-1}\right](k+1-i)^{n+1}$$

$$= (k+1)^{n+1} + \sum_{i=1}^{k}(-1)^i\binom{n+2}{i}(k+1-i)^{n+1}$$

$$= \sum_{i=0}^{k}(-1)^i\binom{n+2}{i}(k+1-i)^{n+1}. \qquad \square$$

5.3.3 The Key of a Sequence and Some Notable Identities ☕

The Eulerian numbers appear in a variety of notable identities. A *key role* in the verification of such identities is played by the concept of *key* of a sequence: it is simply the sequence of the positions of the elements taken in order of increasing, respecting the initial order in case of repetitions.

Definition 5.76 Let $n \in \mathbb{N}_{\geq 1}$ and $(x_1, \ldots, x_n)$ be an n-sequence of natural numbers. There exists a unique permutation $\sigma = (\sigma_1, \ldots, \sigma_n)$ of $(1, \ldots, n)$, called the **key** of the sequence $(x_1, \ldots, x_n)$, such that

$$x_{\sigma_1} \leq \cdots \leq x_{\sigma_n} \quad \text{and} \quad x_{\sigma_i} = x_{\sigma_{i+1}} \implies \sigma_i < \sigma_{i+1}.$$

In this case we write

$$\sigma = \text{key}(x_1, \ldots, x_n). \qquad \qquad \square$$

More formally, the key σ of a sequence $(x_1, \ldots, x_n)$ is defined recursively by setting

$$\sigma_1 = \min\{j \in I_n : x_j = \min\{x_i : i \in I_n\}\}$$

and, for $k \geq 2$,

$$\sigma_k = \min\{j \in I_n \setminus \{\sigma_1, \ldots, \sigma_{k-1}\} : x_j = \min\{x_i : i \in I_n \setminus \{\sigma_1, \ldots, \sigma_{k-1}\}\}.$$

☞ *Remark 5.77* The key σ of a sequence $(x_1, \ldots, x_n)$ is therefore the permutation whose values are given by the list of the positions of the elements of the sequence in increasing order; in the case of equal values in different positions by convention the key lists the lowest position first. Thus one has $\sigma_i = j$ if, under the foregoing convention, x_j is the i-th element in increasing order of the sequence. Note that while an n-sequence can very well have various repeated elements, its key is an n-sequence without repetitions of I_n, that is to say, a permutation of $(1, \ldots, n)$.

Example 5.78 The key of the 6-sequence $(4, 2, 5, 3, 1, 3)$ of I_5 is the permutation $(5, 2, 4, 6, 1, 3)$ of $(1, \ldots, 6)$. In fact, in the given sequence:

- 1 is in position 5;
- 2 is in position 2;
- 3 is in positions 4 and 6;
- 4 is in position 1;
- 5 is in position 3.

Obviously various sequences can have the same key: for example $(5, 2, 4, 6, 1, 3)$ is also the key of the 6-sequence $(4, 2, 4, 2, 1, 2)$ of I_5. $\qquad \square$

Remark 5.79 The permutation $\sigma = (\sigma_1, \ldots, \sigma_n)$ is the key of the n-sequence of I_n that has 1 in position σ_1, 2 in position σ_2, $\ldots$, n in position σ_n. For example the permutation $\sigma = (4, 2, 3, 5, 1)$ is the key of $(5, 2, 3, 1, 4)$. Thus, every permutation of $(1, \ldots, n)$ is the key of at least one n-sequence of I_n.

Example 5.80 We seek a 5-sequence of I_5 whose key is $\sigma = (3, 2, 5, 4, 1)$. Proceeding as indicated in Remark 5.79, it is easy to obtain the 5-sequence $(5, 2, 1, 4, 3)$. Given that in σ we have the ascending couple $(2, 5)$, repeating in position 5 the value located in position 2 we obtain another 5-sequence $(5, 2, 1, 4, 2)$ of I_5 with key σ. $\qquad \square$

We now count the number of sequences that have a given key and that are composed of a given number of distinct elements: they are obtained once one knows the number of ascending couples of the key.

Proposition 5.81 *Let $n, m \in \mathbb{N}_{\geq 1}$, and σ be a permutation of $(1, \ldots, n)$ with $k \leq n - 1$ ascending couples.*

1. *An n-sequence $(x_1, \ldots, x_n)$ of I_m such that $\sigma = \mathrm{key}(x_1, \ldots, x_n)$ has at least $n - k$ distinct elements; in particular $k \geq n - m$.*
2. *For each $\ell \leq n$, there are $\binom{m}{\ell} \times \binom{k}{n-\ell}$ different n-sequences of I_m composed of exactly ℓ distinct values with key σ.*

Proof. 1. Let $\sigma = (\sigma_1, \ldots, \sigma_n)$; by hypothesis k out of the $n - 1$ couples $(\sigma_1, \sigma_2), \ldots, (\sigma_{n-1}, \sigma_n)$ are ascending. In correspondence to the $n - 1 - k$ values $1 \leq j_1 < \cdots < j_{n-k-1} \leq n - 1$ for which $\sigma_{j_i} > \sigma_{j_i+1}$ one must have $n - k$ distinct values

$$x_{\sigma_{j_1}} < x_{\sigma_{j_2}} < \cdots < x_{\sigma_{j_{n-k-1}}} < x_{\sigma_{j_{n-k-1}+1}}.$$

Indeed, for each i one necessarily has $x_{\sigma(j_i)} \leq x_{\sigma(j_i+1)}$; if then the equality were to hold, one would have $x_{\sigma(j_i)} = x_{\sigma(j_i+1)} = \cdots = x_{\sigma(j_i+1)}$ and so $\sigma_{j_i} < \sigma_{j_i+1}$. Therefore $(x_1, \ldots, x_n)$ has at least $n - k$ distinct elements, and thus, since $n - k \leq m$, one has $k \geq n - m$.

2. By Point 1, if $\ell < n - k$ or $\ell > m$ the claim is true: indeed $\binom{m}{\ell} \times \binom{k}{n-\ell} = 0$ and there are no n-sequences of I_m composed of ℓ distinct elements and having key σ. Now let $n - k \leq \ell \leq m$; we wish to construct an n-sequence $(x_1, \ldots, x_n)$ of I_m with ℓ distinct elements and key σ. We do so in three steps. In the first step we choose ℓ distinct elements of I_m: clearly, there are $\binom{m}{\ell}$ ways to carry out this step. In the second step we decide for which of the k ascending couples $\sigma_i < \sigma_{i+1}$ we will set $x_{\sigma_i} < x_{\sigma_{i+1}}$ and for which we will set $x_{\sigma_i} = x_{\sigma_{i+1}}$. As noted in the proof of part (1), corresponding to the $n - k$ values $j_1 < \cdots < j_{n-k-1}$ for which $\sigma_{j_i} > \sigma_{j_i+1}$, we have $n - k$ distinct elements

$$x_{\sigma_{j_1}} < x_{\sigma_{j_2}} < \cdots < x_{\sigma_{j_{n-k}}}.$$

So we must choose $\ell - (n - k)$ ascending couples $\sigma_i < \sigma_{i+1}$ for which we set $x_{\sigma_i} < x_{\sigma_{i+1}}$: this may be done in $\binom{k}{\ell-(n-k)} = \binom{k}{k-(n-\ell)}$ ways. In the third step we construct the sequence $(x_1, \ldots, x_n)$ using the ℓ elements chosen from I_m and the instructions furnished by the key σ, with the specifications established in the second step. This can be done in only one way. By the Multiplication Principle 1.34, the number of n-sequences of I_m consisting of ℓ distinct elements and having key σ is $\binom{m}{\ell} \times \binom{k}{n-\ell}$. $\qquad \square$

Example 5.82 The permutation $\sigma = (2, 5, 3, 1, 4)$ of $(1, 2, 3, 4, 5)$ has 2 ascending couples. Each 5-sequence of I_m, $m \geq 1$, having σ as its key has at least $5 - 2 = 3$ distinct elements. Indeed, if $(x_1, x_2, x_3, x_4, x_5)$ is a sequence with key σ one must have

$$x_2 \leq x_5 < x_3 < x_1 \leq x_4.$$

For each $\ell \leq 5$ with $3 \leq \ell \leq m$ there are $\binom{m}{\ell} \times \binom{2}{5 - \ell}$ different 5-sequences of I_m with ℓ distinct elements and with key σ. For example, there are $\binom{4}{3} \times \binom{2}{5 - 3} =$ 4 different 5-sequences of I_4 with 3 distinct elements and key σ:

$$(3, 1, 2, 3, 1), (4, 1, 2, 4, 1), (4, 1, 3, 4, 1), (4, 2, 3, 4, 2). \qquad \square$$

One has the following relation between the Stirling numbers of the second kind and the Eulerian numbers:

Proposition 5.83 *Let $n \geq m \geq 1$. Then*

$$m! \left\{ {n \atop m} \right\} = \sum_{k \in \mathbb{N}} \left\langle {n \atop k} \right\rangle \binom{k}{n - m} = \sum_{k = n - m}^{n-1} \left\langle {n \atop k} \right\rangle \binom{k}{n - m}. \qquad (5.83.a)$$

Proof. Since $\left\langle {n \atop k} \right\rangle = 0$ for $k \geq n$ and $\binom{k}{n - m} = 0$ for $k < n - m$, one certainly has

$$\sum_{k \in \mathbb{N}} \left\langle {n \atop k} \right\rangle \binom{k}{n - m} = \sum_{k = n - m}^{n-1} \left\langle {n \atop k} \right\rangle \binom{k}{n - m}.$$

We now give a combinatorial demonstration of the first equality. An alternative one is suggested in Problem 6.10. In Corollary 5.8 we have verified that there are $m! \left\{ {n \atop m} \right\}$ different n-sequences of I_m composed of m distinct elements. By Point 1 of Proposition 5.81, each such sequence has a key whose number of ascending couples lies between $n - m$ and $n - 1$. Since there are $\left\langle {n \atop k} \right\rangle$ keys with k ascending couples, and for each of these there are $\binom{m}{m} \binom{k}{n - m} = \binom{k}{n - m}$ different n-sequences of I_m with m distinct values (see Point 2 of Proposition 5.81), there are

$$\sum_{k = n - m}^{n-1} \left\langle {n \atop k} \right\rangle \binom{k}{n - m}$$

different n-sequences of I_m composed of m distinct elements.   □

The same combinatorial technique used to prove Proposition 5.83 permits us to obtain an easy proof of the so-called **Worpitzky**[6] **Identity**.

Proposition 5.84 (Worpitzky Identity) *Let $m, n \in \mathbb{N}$ and $n \geq 1$; then*

$$m^n = \sum_{k=0}^{n-1} \left\langle {n \atop k} \right\rangle \binom{m+k}{n}. \tag{5.84.a}$$

Proof. The equality holds easily for $m = 0$. In case $m \geq 1$, we carry out a combinatorial proof of the identity, leaving the proof by induction as an exercise (Problem 5.12). The number m^n coincides with the number of n-sequences of I_m. By Proposition 5.81 there are

$$\sum_{\ell=0}^{n} \binom{m}{\ell} \binom{k}{n-\ell}$$

different n-sequences of I_m with a given key, having $k \leq n-1$ ascending couples; in view of the Vandermonde convolution (Proposition 2.57) that number is equal to $\binom{m+k}{n}$. Since there are $\left\langle {n \atop k} \right\rangle$ permutations of $(1, \ldots, n)$ with $k \leq n-1$ ascending couples, one immediately obtains the identity.   □

Remark 5.85 Observe that (5.83.a) counts the number of the surjective functions $I_n \to I_m$, whereas (5.84.a) counts the number of all functions $I_n \to I_m$.

5.4 Problems

Problem 5.1 In how many ways can one distribute 23 different objects into 8 identical boxes, in such a way as to have no box left empty?

Problem 5.2 The members of a group of 30 tourists decide to go for jeep rides; in how many ways can they split up into 6 groups in such a way as to have 3 groups consisting of 6 people, one group of 3 people, one of 4, and one of 5 people?

---

[6]Julius Worpitzky (1835–1895).

Problem 5.3 A rescue team of 14 people is searching for a missing person. They decide to split into 5 squads: 3 squads with 2 people and 2 squads with four people. In how many ways is it possible to perform this division into squads?

Problem 5.4 In how many ways can a deck of 52 cards be divided into:

(*a*) 4 decks of 13 cards?
(*b*) 3 decks of 8 cards and 4 decks of 7 cards?

Problem 5.5 A class of 18 students goes on a trip accompanied by two teachers. They will spend the night in 5 quadruple rooms; the two teachers do not wish to sleep in the same room. Under this condition, in how many ways can the entire group be split up into 5 groups of exactly 4 people?

Problem 5.6 In how many ways is it possible to split up 20 people into 5 squads, of which two consist of 5 people, two of three people, and one of 4 people?

Problem 5.7 In how many ways is it possible to fill a Ferris wheel with 30 one-place seats from a school group of 100 people?

Problem 5.8 Twenty guests are to be seated around a large circular table. Find the probability that Carlo, Alberto, Elena and Paola are neighbors.

Problem 5.9 In how many ways can one seat 40 people in a banquet hall containing 8 equally sized round tables if no table is to be left empty?

Problem 5.10 We wish to place 100 students on 4 Ferris wheels with one-place seats, two of the wheels having 25 seats, one with 20 seats and one with 30 seats. Calculate the probability that Nicky, Tommy, and Francies sit in three consecutive places on one of the Ferris wheels.

Problem 5.11 Let $k \in \mathbb{N}_{\geq 1}$ and $b_1, \ldots, b_k$ be natural numbers such that

$$b_1 + 2b_2 + \cdots + kb_k = k.$$

Prove that the number of partitions into cycles of I_k with occupancy collection $[\underbrace{1, \ldots, 1}_{b_1}, \ldots, \underbrace{k, \ldots, k}_{b_k}]$ is equal to

$$\frac{k!}{(1!)^{b_1} \cdots (k!)^{b_k}} \frac{1}{b_1! \cdots b_k!}.$$

Problem 5.12 Prove the Worpitzky Identity (see Proposition 5.84) by induction on n.

Chapter 6
Manipulation of Sums

Abstract This chapter deals with the calculus of finite sums: after examining some special techniques, we develop the general theory of finite calculus, the discrete analogue of differential calculus. The discrete primitives are the tool that enable to compute finite sums. We examine in detail the case of the sums of powers of consecutive natural numbers: quite surprisingly this leads to the Stirling numbers of second kind. A section is devoted to the inversion formula, a powerful tool in many mathematical fields: we use it here to obtain the discrete analogue of the Taylor expansion, an alternative short proof of both the number of derangements of a sequence and of surjective functions between two finite sets, and, finally, a more general version of the inclusion/exclusion principle.

6.1 Some Techniques

Explicitly determining the sum of a given number of summands is not always a simple matter; sometimes, however, some simple but astute adjustments allow one to solve the problem rapidly.

6.1.1 Gauss Method

The story goes that when Gauss's[1] elementary school teacher punished the class of the 9-year old Gauss by requiring the pupils to find the sum of the first 100 natural numbers, the future great mathematician easily, rapidly (and disdainfully) solved the problem. Rather than summing $1 + 2$ and then adding 3 to the result, and so on, Gauss observed that if S denoted the desired sum, then S could be expressed in two ways:

$$S = 100 \ + 99 \ + 98 \ + \ \cdots \ + \ 3 + \ 2 + \ 1$$
$$S = 1 \ \ + 2 \ \ + 3 \ \ + \ \cdots \ + \ 98 + \ 99 + \ 100.$$

---

[1]Carl Friedrich Gauss (1777–1855).

© Springer International Publishing Switzerland 2016

C. Mariconda and A. Tonolo, *Discrete Calculus*,

UNITEXT - La Matematica per il 3+2 103, DOI 10.1007/978-3-319-03038-8_6

Adding the terms column by column one obtains

$$2S = (100 + 1) + (99 + 2) + \cdots + (2 + 99) + (1 + 100) = 101 \times (100) = 10\,100,$$

whence $S = 5\,050$. With the same method one sees that more generally one has

$$1 + 2 + \cdots + n = \frac{n(n+1)}{2}.$$

6.1.2 Perturbation Technique

Given a sequence $(a_k)_{k \in \mathbb{N}}$ of real numbers, suppose that we wish to calculate the n-th term of the sequence of **partial sums**

$$s_n = \sum_{k=0}^{n} a_k.$$

Adding a_{n+1} to both the terms of the equality we get

$$s_n + a_{n+1} = a_0 + \sum_{k=0}^{n} a_{k+1};$$

now one attempts to express the right hand side of the equation as a function of s_n, so as to obtain an equation for s_n.

Example 6.1 Let $r \in \mathbb{R}$, and suppose that we wish to calculate

$$s_n = \sum_{k=0}^{n} r^k.$$

Directly applying the method introduced above one has

$$s_n + r^{n+1} = 1 + \sum_{k=0}^{n} r^{k+1} = 1 + rs_n.$$

Therefore, $(1 - r)s_n = 1 - r^{n+1}$ and so, for $r \neq 1$, one has $s_n = (1 - r^{n+1})/(1 - r)$. Note that when $r = 1$ one sees immediately that $s_n = n$.   □

Example 6.2 We now use the method of perturbation to calculate the value of

$$s_n = \sum_{k=0}^{n} k2^k.$$

Again, applying the above reasoning, one has

$$s_n + (n+1)2^{n+1} = 0 + \sum_{k=0}^{n}(k+1)2^{k+1} = 2\sum_{k=0}^{n}(k2^k + 2^k) = 2s_n + 2\sum_{k=0}^{n}2^k.$$

Since $\sum_{k=0}^{n} 2^k = 2^{n+1} - 1$ by Example 6.1, we get

$$s_n + (n+1)2^{n+1} = 2s_n + 2(2^{n+1} - 1),$$

and hence $s_n = 2^{n+1}(n-1) + 2.$ □

Example 6.3 We calculate the sum

$$s_n = \sum_{k=1}^{n} kr^k$$

for any given real number r. If $r = 1$ one has $s_n = \sum_{k=1}^{n} k = \dfrac{n(n+1)}{2}$ as seen in Sect. 6.1.1. If $r \neq 1$, then applying the method of perturbation we find that

$$s_n + (n+1)r^{n+1} = r + \sum_{k=1}^{n}(k+1)r^{k+1} = r + rs_n + r\sum_{k=1}^{n}r^k$$

$$= r + rs_n + r^2\sum_{k=0}^{n-1}r^k = r + rs_n + r^2\frac{1 - r^n}{1 - r}$$

$$= rs_n + r\frac{(1 - r) + r - r^{n+1}}{1 - r} = rs_n + r\frac{1 - r^{n+1}}{1 - r}.$$

Therefore one has

$$s_n = r\frac{1 - r^{n+1}}{(1 - r)^2} - (n+1)\frac{r^{n+1}}{1 - r}.$$ □

6.1.3 Derivative Method

Let $r \in \mathbb{R}$ and $(a_k)_k$ be a sequence of real numbers. We observe that if $p(r)$ is the polynomial function $p(r) = a_0 + a_1 r + \cdots + a_n r^n$, then its derivative is

$$p'(r) = \left(\sum_{k=0}^{n} a_k r^k\right)' = a_1 + 2a_2 r + \cdots + na_n r^{n-1} = \sum_{k=1}^{n} ka_k r^{k-1}.$$

We now apply this elementary remark to the calculation of the sum $\sum_{k=1}^{n} k r^k$, which has already been discussed in Example 6.3 where the perturbation technique was employed. Factoring r from the sum gives

$$\sum_{k=1}^{n} k r^k = r \sum_{k=1}^{n} k r^{k-1} = r \left(\sum_{k=1}^{n} r^k \right)'.$$

For $r \neq 1$ we have

$$\sum_{k=1}^{n} r^k = r \sum_{k=0}^{n-1} r^k = r \frac{1 - r^n}{1 - r} = \frac{r - r^{n+1}}{1 - r};$$

and so

$$\sum_{k=1}^{n} k r^k = r \left(\sum_{k=1}^{n} r^k \right)' = r \left(\frac{r - r^{n+1}}{1 - r} \right)'$$

$$= r \frac{(1 - (n+1)r^n)(1 - r) - (-1)(r - r^{n+1})}{(1 - r)^2}$$

$$= \frac{r + n r^{n+2} - (n+1) r^{n+1}}{(1 - r)^2}.$$

6.1.4 Changing the Order of Summation

In discussing finite sums it is often useful to be able to interchange the order of the summation; this is a simple consequence of the commutative and associative laws for addition. Nevertheless, it is worthwhile to dedicate a few lines to a definitive confirmation of this fact.

Proposition 6.4 (Interchanging the order of summation) *Let m, n be two natural numbers, and as i varies in $\{0, \ldots, m\}$ and k varies in $\{0, \ldots, n\}$, let $a_{i,k}$ be real numbers. Then*

$$\sum_{i=0}^{m} \left(\sum_{k=0}^{n} a_{i,k} \right) = \sum_{k=0}^{n} \left(\sum_{i=0}^{m} a_{i,k} \right). \qquad (6.4.a)$$

In particular

$$\sum_{k=0}^{n}\left(\sum_{i=0}^{k}a_{i,k}\right)=\sum_{i=0}^{n}\left(\sum_{k=i}^{n}a_{i,k}\right). \tag{6.4.b}$$

Proof. Grouping together the terms with second index equal to 0, then those with second index equal to 1,..., then those with second index equal to n one has:

$$\sum_{i=0}^{m}\left(\sum_{k=0}^{n}a_{i,k}\right)=(a_{0,0}+\cdots+a_{0,n})+\cdots+(a_{m,0}+\cdots+a_{m,n})$$

$$=(a_{0,0}+\cdots+a_{m,0})+\cdots+(a_{0,n}+\cdots+a_{m,n})=\sum_{k=0}^{n}\left(\sum_{i=0}^{m}a_{i,k}\right).$$

To establish the second identity we put $a_{i,k}=0$ if $i>k$; applying (6.4.a) one obtains

$$\sum_{k=0}^{n}\left(\sum_{i=0}^{k}a_{i,k}\right)=\sum_{k=0}^{n}\left(\sum_{i=0}^{n}a_{i,k}\right)$$

$$=\sum_{i=0}^{n}\left(\sum_{k=0}^{n}a_{i,k}\right)=\sum_{i=0}^{n}\left(\sum_{k=i}^{n}a_{i,k}\right). \qquad \square$$

6.2   Finite Calculus

In the preceding section we have seen some techniques that yielded explicit calculations for certain sums. However, these were particular techniques that can be fruitfully applied only in special situations. We now attempt a more systematic approach which introduces the discrete analogue of differential calculus.

6.2.1   Shift and Difference Operators

It is convenient here to represent any sequence of real numbers $(a_k)_{k\in\mathbb{N}}$ as the function $f:\mathbb{N}\to\mathbb{R}$ defined by

$$f(k)=a_k \quad \forall k\in\mathbb{N}.$$

Given two functions $f, g : \mathbb{N} \to \mathbb{R}$ and $r \in \mathbb{R}$ we consider the functions

$$(f + g)(k) = f(k) + g(k), \quad \text{and} \quad (rf)(k) = rf(k) \quad \forall k \in \mathbb{N}.$$

Endowed with these operations, the set of functions from $\mathbb{N}$ to $\mathbb{R}$ is an $\mathbb{R}$-vector space. We will also consider the function

$$(fg)(k) = f(k)g(k) \quad \forall k \in \mathbb{N}.$$

A linear map from the space of functions from $\mathbb{N}$ to $\mathbb{R}$ in itself is an *operator*.

Definition 6.5 Consider the space of the functions from $\mathbb{N}$ to $\mathbb{R}$. For any function $f : \mathbb{N} \to \mathbb{R}$ the **identity operator** $\mathbb{1}$ and the **shift operator** θ are defined by

$$\mathbb{1}(f) = f \quad \text{and} \quad \theta(f)(k) = f(k + 1) \quad \forall k \in \mathbb{N}. \qquad \square$$

One verifies immediately that the identity and the shift operator are indeed linear.

Proposition 6.6 (Linearity of the identity and shift operators) *Let $f, g : \mathbb{N} \to \mathbb{R}$ and $c \in \mathbb{R}$. Then one has:*

1. $\mathbb{1}(f + g)(k) = \mathbb{1}(f)(k) + \mathbb{1}(g)(k)$ *and* $\theta(f + g)(k) = \theta(f)(k) + \theta(g)(k)$;
2. $\mathbb{1}(cf)(k) = c\mathbb{1}(f)(k)$ *and* $\theta(cf)(k) = c\theta(f)(k)$.

Proof. Let $k \in \mathbb{N}$; then

$$\theta(f + g)(k) = (f + g)(k + 1) = f(k + 1) + g(k + 1) = \theta(f)(k) + \theta(g)(k);$$

$$\theta(cf)(k) = (cf)(k + 1) = cf(k + 1) = c\theta(f)(k).$$

Checking the linearity of $\mathbb{1}$ is immediate. $\qquad \square$

☞   For any operator T, it will be convenient, at the price of a slight abuse of notation, to write $Tf(k)$ instead of $T(f)(k)$. Moreover in some cases, for example when f depends on other parameters, one writes $T_k f(k)$ rather than $Tf(k)$ to avoid ambiguity. Thus, for instance, denoted by $\mathbb{1}_{\mathbb{N}} : \mathbb{N} \to \mathbb{N}$ the function defined by $\mathbb{1}_{\mathbb{N}}(k) = k$ for each $k \in \mathbb{N}$, we will write $\theta k = k + 1$ rather than $\theta(\mathbb{1}_{\mathbb{N}})(k) = k + 1$. Analogously $\theta k^2 = (k + 1)^2$, $\theta_k k^a = (k + 1)^a$ and $\theta_k a^k = a^{k+1}$ for each $a \in \mathbb{R}$.

For real valued functions of a *natural number variable* we now introduce the analogue of the usual derivative for real valued functions of a real variable:

Definition 6.7 The **difference operator** is the operator Δ which to each function $f : \mathbb{N} \to \mathbb{R}$ assigns the function $\Delta f : \mathbb{N} \to \mathbb{R}$ defined by setting

$$\Delta f(k) = f(k+1) - f(k) \quad \forall k \in \mathbb{N}.$$

Remark 6.8 Using the shift operator, one has $\Delta = \theta - \mathbb{1}$, i.e.

$$\Delta f = \theta f - f, \quad \forall f : \mathbb{N} \to \mathbb{R}.$$

Remark 6.9 Clearly, for a function $f : \mathbb{N} \to \mathbb{R}$, one has

$$\Delta f(k) = \frac{f(k+1) - f(k)}{1}$$

and so $\Delta f : \mathbb{N} \to \mathbb{R}$ is a function that measures the difference quotient of f over the smallest possible interval of natural numbers, namely, an interval of length one. In this sense, the difference operator constitutes the discrete analogue of the notion of derivative for functions of a real variable. In what follows, the reader will have occasion to note analogies and contrasts between these two notions.

Just like the derivative, the difference operator is linear: indeed, it is a difference of two linear operators.

Proposition 6.10 (Linearity of the difference) *Let $f, g : \mathbb{N} \to \mathbb{R}$ and $c \in \mathbb{R}$. Then one has:*

1. $\Delta(f + g) = \Delta f + \Delta g;$
2. $\Delta(cf) = c \, \Delta f.$

Proof. Since $\Delta = \theta - \mathbb{1}$ one gets:

1. $\Delta(f + g) = (\theta - \mathbb{1})(f + g) = \theta(f + g) - \mathbb{1}(f + g) = \theta(f) - f + \theta(g) - g = \Delta(f) + \Delta(g);$
2. $\Delta(cf) = (\theta - \mathbb{1})(cf) = \theta(cf) - \mathbb{1}(cf) = c\,\theta(f) - cf = c(\theta - \mathbb{1})(f) = c\,\Delta(f).$ $\quad\square$

We now see how the difference operator acts on some simple functions with domain $\mathbb{N}$.

Example 6.11 1. *Constant Functions*: Just as in the case of the derivative of constant functions with domains in $\mathbb{R}$, here too we have that the difference of a constant function (with domain $\mathbb{N}$) is equal to the zero function: indeed, if $f(k) = c \in \mathbb{R}$ for every k in $\mathbb{N}$, then

$$\Delta f(k) = f(k+1) - f(k) = c - c = 0.$$

2. *Identity Function on the natural numbers*: Just as in the continuous case, the difference of the identity function $\mathbb{1}_{\mathbb{N}} : \mathbb{N} \to \mathbb{N}$ is the constant function $k \mapsto 1$ for all k in $\mathbb{N}$: indeed

$$\Delta \mathbb{1}_{\mathbb{N}}(k) = \mathbb{1}_{\mathbb{N}}(k+1) - \mathbb{1}_{\mathbb{N}}(k) = k + 1 - k = 1.$$

3. *Powers*: (a) Given $m \in \mathbb{N}_{\geq 1}$, consider the function

$$\mathbb{N} \to \mathbb{R}, \quad k \mapsto k^m.$$

In view of the binomial formula (2.19), one has that for each $k \in \mathbb{N}$

$$\Delta_k k^m = (k+1)^m - k^m = \sum_{i=0}^{m-1} \binom{m}{i} k^i \quad \forall k \in \mathbb{N}.$$

(b) Given $k \in \mathbb{N}$, consider the function

$$\mathbb{N} \to \mathbb{R}, \quad m \mapsto k^m.$$

One has

$$\Delta_m k^m = k^{m+1} - k^m \quad \forall m \in \mathbb{N}.$$

The differences of $k \mapsto k^m$ and $m \mapsto k^m$ do not at all resemble the formulas for the derivatives of a power and of an exponential in the continuous case. □

In Sect. 6.5 we will use the following simple commutation property between the shift and difference operators to derive the discrete analogue of Taylor formula.

Example 6.12 The shift and difference operators commute. More explicitly, one has

$$\Delta \circ \theta = \theta \circ \Delta.$$

Indeed for each $k \in \mathbb{N}$ and each function $f : \mathbb{N} \to \mathbb{R}$ one has

$$\Delta(\theta f)(k) = \theta f(k+1) - \theta f(k) = f(k+2) - f(k+1),$$

while

$$\theta(\Delta f)(k) = \Delta f(k+1) = f(k+2) - f(k+1).$$

Thus one has $\Delta(\theta f)(k) = \theta(\Delta f)(k)$. □

The formula for the difference of a product resembles that for the derivative of a product, except for the introduction of the shift operator:

Proposition 6.13 (Difference of a product) *If $f, g : \mathbb{N} \to \mathbb{R}$ then*

$$\Delta(fg) = \Delta f \, \theta \, g + f \, \Delta g.$$

Remark 6.14 One should note the obvious fact that despite the apparent lack of symmetry, one has $\Delta(fg) = \Delta(gf)$.

Proof. Given $k \in \mathbb{N}$, one obtains

$$
\begin{aligned}
\Delta(f(k)g(k)) &= f(k+1)g(k+1) - f(k)g(k) \\
&= f(k+1)g(k+1) - f(k)g(k+1) + f(k)g(k+1) - f(k)g(k) \\
&= (f(k+1) - f(k))g(k+1) + f(k)(g(k+1) - g(k)) \\
&= \Delta f(k) \, \theta \, g(k) + f(k) \, \Delta g(k).
\end{aligned}
$$
$\square$

Example 6.15 Let us calculate $\Delta \, 3^k$ and then, using the product formula, $\Delta(k3^k)$. One easily calculates that

$$
\Delta \, 3^k = 3^{k+1} - 3^k = 3^k(3-1) = 2 \times 3^k.
$$

By Proposition 6.13 one has

$$
\Delta(k3^k) = \Delta \, k \, \theta \, 3^k + k \, \Delta \, 3^k = 3^{k+1} + k(3^{k+1} - 3^k) = 3^k(3+2k). \qquad \square
$$

As we have seen also in Point 3 of Example 6.11, the presence of the shift operator in the formula for the difference of a product makes the calculation of the difference of a power considerably less straightforward than in differential calculus.

6.2.2 Descending Factorial Powers

In the finite calculus the "good" analogue for the ordinary powers $x \mapsto x^n$ of differential calculus is given by the notion of a descending factorial power.

Definition 6.16 (*Descending factorial power*) For each $k \in \mathbb{N}$ and $m \in \mathbb{Z}$, we define the **descending factorial power** $k^{\underline{m}}$ by setting:

$$
k^{\underline{m}} = \begin{cases} k(k-1)\cdots(k-m+1) & \text{if } m > 0, \\ 1 & \text{if } m = 0, \\ \dfrac{1}{(k+1)(k+2)\cdots(k+|m|)} & \text{if } m < 0. \end{cases}
$$
$\square$

The term *descending powers* derives from the expression of $k^{\underline{m}}$ when $m > 0$; note that in that case $k^{\underline{m}}$ coincides with the number $S(k, m)$ of m-sequences of I_k without repetition (see Theorem 2.6).

Remark 6.17 It is easy to verify that

$$k^{\underline{m}} = \begin{cases} \dfrac{k!}{(k-m)!} & \text{if } k \geq m, \\ 0 & \text{otherwise.} \end{cases}$$

In the case $m > 0$ the equality is obvious if $k \geq m$, while if $k < m$, then

$$k(k-1)\cdots(k-m+1) = 0$$

since one of the terms of the product under consideration is zero. If, on the other hand, $m < 0$ one has

$$\frac{k!}{(k-m)!} = \frac{k!}{(k+|m|)!} = \frac{1}{(k+1)(k+2)\cdots(k+|m|)} \quad \forall k \in \mathbb{N}.$$

Finally, the case $m = 0$ is obvious.

Example 6.18 A factorial is a particular case of a descending power. Indeed, if $k \in \mathbb{N}$ one has

$$k^{\underline{k}} = \frac{k!}{0!} = k! \,.$$

Notice also that

$$0^{\underline{n}} = \begin{cases} 0 & \text{if } n > 0, \\ 1 & \text{if } n = 0, \\ \frac{1}{|n|!} & \text{if } n < 0. \end{cases}$$ □

☞ *Remark 6.19* Bear in mind the fact that in general $k^{\underline{m+n}}$ does not coincide with $k^{\underline{m}} k^{\underline{n}}$; actually, for $m, n \in \mathbb{Z}$ and $k \geq m$, one has (see Problem 6.3)

$$k^{\underline{m+n}} = k^{\underline{m}}(k-m)^{\underline{n}}.$$

The difference operator acts on descending powers of natural numbers in a manner analogous to the derivative on ordinary powers.

Proposition 6.20 *For any given $m \in \mathbb{Z}$ one has*

$$\Delta_k \, k^{\underline{m}} = m k^{\underline{m-1}} \quad \forall k \in \mathbb{N}.$$

Proof. One has $\Delta_k k^{\underline{m}} = (k+1)^{\underline{m}} - k^{\underline{m}}$. We consider the following three cases:

$$m > 0: (k+1)^{\underline{m}} - k^{\underline{m}} = (k+1)\cdots(k+1-m+1) - k\cdots(k-m+1)$$
$$= k\cdots(k-m+2)\big((k+1)-(k-m+1)\big)$$
$$= mk^{\underline{m-1}}.$$
$$m = 0: (k+1)^{\underline{0}} - k^{\underline{0}} = 1 - 1 = 0 = 0k^{\underline{-1}}.$$
$$m < 0: (k+1)^{\underline{m}} - k^{\underline{m}} = \frac{1}{(k+2)\cdots(k+1+|m|)} - \frac{1}{(k+1)\cdots(k+|m|)}$$
$$= \frac{1}{(k+2)\cdots(k+|m|)}\left(\frac{1}{k+1+|m|} - \frac{1}{k+1}\right)$$
$$= \frac{1}{(k+2)\cdots(k+|m|)}\frac{-|m|}{(k+1)(k+|m|+1)}$$
$$= mk^{\underline{m-1}}. \qquad \square$$

6.2.3 Discrete Primitives and Fundamental Theorem of Finite Calculus

We now investigate the discrete analogue of the concept of primitive (or antiderivative) for a function.

Definition 6.21 If $f : \mathbb{N} \to \mathbb{R}$ is a function defined on the natural numbers, we say that $F : \mathbb{N} \to \mathbb{R}$ is a **discrete primitive** of f if $\Delta F = f$. $\qquad \square$

Example 6.22 (Discrete primitive of 0) The discrete primitive of 0 are the constant functions. Indeed, clearly the difference of a constant is zero. Conversely, let $F : \mathbb{N} \to \mathbb{R}$ be such that $\Delta F = 0$. Then, one has

$$0 = F(1) - F(0) = \cdots = F(n+1) - F(n) \quad \forall n \in \mathbb{N};$$

so that $F(0) = F(1) = \cdots = F(n)$ for all $n \in \mathbb{N}$. $\qquad \square$

Remark 6.23 The linearity of the operator Δ, proved in Proposition 6.10, yields the linearity of the primitives: let $f, g, F, G : \mathbb{N} \to \mathbb{R}$ with F, G discrete primitives of f, g, respectively, and $c \in \mathbb{R}$. Then $F + G$ is a discrete primitive of $f + g$ and cF is a discrete primitive of cf.

Every function defined on the natural numbers admits discrete primitives.

Proposition 6.24 (Existence of discrete primitives) *Given any* $f : \mathbb{N} \to \mathbb{R}$*, the function* $F : \mathbb{N} \to \mathbb{R}$ *defined by*

$$F(k) = \begin{cases} 0 & \text{if } k = 0, \\ f(0) + \cdots + f(k-1) & \text{if } k \geq 1 \end{cases}$$

is a discrete primitive for f.

Proof. Indeed, $\Delta F(0) = F(1) - F(0) = f(0)$ and, for $k \geq 1$,

$$\Delta F(k) = F(k+1) - F(k) = (f(0) + \cdots + f(k)) - (f(0) + \cdots + f(k-1)) = f(k). \quad \square$$

In complete analogy with the known result from differential calculus, two functions are discrete primitives of the same function if and only if they differ by a constant.

Proposition 6.25 *Let $f, F : \mathbb{N} \to \mathbb{R}$ be functions such that F is a discrete primitive for f. Then the function $G : \mathbb{N} \to \mathbb{R}$ is a discrete primitive for f if and only if $G = F + c$ for some constant $c \in \mathbb{R}$.*

Proof. The function G is a discrete primitive of f if and only if $\Delta G = \Delta F$; by the linearity of the operator Δ, this occurs precisely when $\Delta(G - F) = 0$. We conclude by Example 6.22. $\quad \square$

Example 6.26 Let us calculate the difference of $f(k) = k^2$ and the primitives of $g(k) = 2k$. Clearly,

$$\Delta f(k) = \Delta k^2 = (k+1)^2 - k^2 = 2k + 1.$$

Therefore, $f(k) = k^2$ is a primitive of $2k + 1$; since $1 = \Delta k$, we have

$$2k = \Delta k^2 - \Delta k = \Delta(k^2 - k).$$

Hence the discrete primitives of $g(k) = 2k$ are the functions $G_c(k) = k^2 - k + c$, with c constant. $\quad \square$

The following result is the discrete analogue of the Fundamental Theorem of Differential Calculus: discrete primitives allow to compute sums.

Theorem 6.27 (Fundamental Theorem of Finite Calculus) *Let $F, f : \mathbb{N} \to \mathbb{R}$. The function F is a discrete primitive of f if and only if*

$$\sum_{m \leq k < n} f(k) = F(n) - F(m) =: [F(k)]_{k=m}^n \qquad \forall m < n \in \mathbb{N}. \qquad (6.27.a)$$

Proof. Let F be a discrete primitive for f. If $m < n \in \mathbb{N}$ one has

$$[F(k)]_{k=m}^n = F(n) - F(m) =$$

$$= (F(n) - F(n-1)) + (F(n-1) - F(n-2)) + \cdots + (F(m+1) - F(m))$$

$$= \Delta F(n-1) + \cdots + \Delta F(m) = f(n-1) + f(n-2) + \cdots + f(m) = \sum_{m \leq k < n} f(k).$$

Conversely, if (6.27.a) holds, then for all natural numbers n we get

$$f(n) = \sum_{n \le k < n+1} f(k) = F(n+1) - F(n) = \Delta F(n),$$

proving that F is a discrete primitive of f. □

Definition 6.28 *(The symbol $\sum f$)* Let $f : \mathbb{N} \to \mathbb{R}$. We use the symbol $\sum f$, or $\sum_k f(k)$ if we wish to emphasize the variable k, to denote the set of the discrete primitives of f:

$$\sum f := \{F : \mathbb{N} \to \mathbb{R} : \Delta F = f\}.$$

□

Clearly this notation is the discrete analogue of the usual symbol $\int f(x)\,dx$ for the indefinite integral of classical analysis.

Remark 6.29 In light of the notation just introduced, Proposition 6.25 can be stated in the following concise form:

$$\Delta F = f \iff \sum f = F + \mathbb{R}.$$

6.3 Discrete Primitives of Some Important Functions

In order to make the Fundamental Theorem of Finite Calculus 6.27 an effective tool, it is now useful to understand how one finds the discrete primitives of a function whose domain is the set of natural numbers.

6.3.1 Discrete Primitive of the Descending Factorial Powers with Exponent $\ne -1$

The case of the exponent equal to -1 will be discussed in Sect. 6.3.4. In Proposition 6.20 we have seen that for each $m \in \mathbb{Z}$ one has

$$\Delta_k k^{\underline{m}} = m k^{\underline{m-1}}.$$

As a consequence, one has

$$\forall m \neq -1 \qquad \sum_k k^{\underline{m}} = \frac{1}{m+1} k^{\underline{m+1}} + \mathbb{R}. \qquad (6.29.a)$$

Example 6.30 Let us calculate $\displaystyle\sum_{1 \le k < n} k^{\underline{4}}$. In view of (6.29.a) one gets

$$\sum k^{\underline{4}} = \frac{1}{5} k^{\underline{5}} + \mathbb{R}.$$

By the Fundamental Theorem of Finite Calculus 6.27 we have

$$\sum_{1 \le k < n} k^{\underline{4}} = \left[\frac{1}{5} k^{\underline{5}}\right]_{k=1}^n = \frac{1}{5} n^{\underline{5}} - \frac{1}{5} 1^{\underline{5}} = \frac{1}{5} n^{\underline{5}} = \frac{1}{5} n(n-1)\cdots(n-4). \qquad \square$$

Example 6.31 (Telescopic sums) Consider the sum $\displaystyle\sum_{0 \le k < n} \frac{1}{(k+1)(k+2)}$. The usual trick to compute it is to use the fact that

$$\frac{1}{(k+1)(k+2)} = \frac{1}{k+1} - \frac{1}{k+2},$$

and hence

$$\sum_{0 \le k < n} \frac{1}{(k+1)(k+2)} = \left(1 - \frac{1}{2}\right) + \left(\frac{1}{2} - \frac{1}{3}\right) + \cdots + \left(\frac{1}{n} - \frac{1}{n+1}\right) = 1 - \frac{1}{n+1}.$$

We recognize that $\dfrac{1}{(k+1)(k+2)} = k^{\underline{-2}}$. Using the Fundamental Theorem of Finite Calculus 6.27 one can more generally compute, for any $m \ge 2$,

$$\sum_{0 \le k < n} \frac{1}{(k+1)\cdots(k+m)} = \sum_{0 \le k < n} k^{\underline{-m}} = \left[\frac{1}{-m+1} k^{\underline{-m+1}}\right]_{k=0}^n$$

$$= \frac{1}{-m+1}\left(n^{\underline{-m+1}} - 0^{\underline{-m+1}}\right)$$

$$= \frac{1}{-m+1}\left(\frac{1}{(n+1)\cdots(n+m-1)} - \frac{1}{(m-1)!}\right).$$

Notice in particular that we immediately get the sum of the series

$$\sum_{k=0}^{\infty} \frac{1}{(k+1)\cdots(k+m)} = \frac{1}{(m-1)(m-1)!}.$$ □

6.3.2 Discrete Primitive of Positive Integer Powers

The first powers k^0 and k^1 coincide, respectively, with the descending powers $k^{\underline{0}}$ and $k^{\underline{1}}$.

Example 6.32 Calculate the sum

$$\sum_{0\le k<n} k.$$

Solution. Since $k = k^{\underline{1}}$, the function $\frac{1}{2}k^{\underline{2}}$ is a discrete primitive for k. By the Fundamental Theorem of Finite Calculus 6.27 one has

$$\sum_{0\le k<n} k = \left[\frac{k^{\underline{2}}}{2}\right]_{k=0}^{n} = \frac{n^{\underline{2}}}{2} = \frac{n(n-1)}{2}.$$ □

Fix now $\ell \in \mathbb{N}_{\ge 1}$. There are a couple of methods to find the discrete primitive of $k \mapsto k^\ell$, $k \in \mathbb{N}$. The first one is recursive and is more suitable for lower values of the exponent ℓ: it allows to find a discrete primitive of $k \mapsto k^\ell$ once one has the primitives of the lower powers. Indeed, since

$$\Delta k^{\underline{\ell+1}} = (\ell+1)k^{\underline{\ell}} = (\ell+1)k(k-1)\cdots(k-\ell+1) = (\ell+1)k^\ell + p(k),$$

where $p(k)$ is a polynomial function of degree $\ell - 1$, the knowledge of the discrete primitives of $k \mapsto k^i$, $i = 0, \ldots, \ell - 1$, provides a discrete primitive $P(k)$ of $p(k)$, so that $\frac{1}{\ell+1}\left(k^{\underline{\ell+1}} - P(k)\right)$ turns out to be a discrete primitive of $k \mapsto k^\ell$. We illustrate this method with the computation of the discrete primitives of k^2 and of k^3.

Example 6.33 (The recursive method) We know that k is a discrete primitive of 1. We look for a discrete primitive of k^2. Since

$$\Delta k^{\underline{3}} = 3k^{\underline{2}} = 3k(k-1) = 3k^2 - 3k = 3k^2 - 3\Delta\left(\frac{k^{\underline{2}}}{2}\right),$$

we deduce that $\Delta\left(k^{\underline{3}} + 3\frac{k^{\underline{2}}}{2}\right) = 3k^2$, so that $\frac{1}{3}k^{\underline{3}} + \frac{1}{2}k^{\underline{2}}$ is a primitive of k^2. We now look for a discrete primitive of k^3. Since

$$\Delta\, k^{\underline{4}} = 4k^{\underline{3}} = 4k(k-1)(k-2) = 4k^3 - 12k^2 + 8k$$

$$= 4k^3 - 12\,\Delta\left(\frac{1}{3}k^{\underline{3}} + \frac{1}{2}k^{\underline{2}}\right) + 8\,\Delta\left(\frac{k^{\underline{2}}}{2}\right) \quad,$$

$$= 4k^3 - \Delta\left(4k^{\underline{3}} + 2k^{\underline{2}}\right)$$

we deduce that $\Delta\left(k^{\underline{4}} + 4k^{\underline{3}} + 2k^{\underline{2}}\right) = 4k^3$, thus $\frac{1}{4}k^{\underline{4}} + k^{\underline{3}} + \frac{1}{2}k^{\underline{2}}$ is a primitive of k^3. $\qquad\square$

The other method is based on the fact that the usual powers k^{ℓ} can be written as a linear combination of descending factorial powers. To that end, the Stirling numbers of the second kind (introduced in Sect. 5.1) prove to be very useful. Recall that given $\ell, i \in \mathbb{N}$, $\left\{{\ell \atop i}\right\}$ denotes the number of i-partitions of I_{ℓ}.

Proposition 6.34 (Positive powers in term of descending powers) *Fix $\ell \in \mathbb{N}_{\geq 1}$. For all $k \in \mathbb{N}$ one has*

$$k^{\ell} = \sum_{i=1}^{\ell} \left\{{\ell \atop i}\right\} k^{\underline{i}} = \sum_{i=0}^{\ell} \left\{{\ell \atop i}\right\} k^{\underline{i}}. \tag{6.34.a}$$

Hence the discrete primitives of $k \mapsto k^{\ell}$ are

$$\sum_{k} k^{\ell} = \sum_{i=1}^{\ell} \left\{{\ell \atop i}\right\} \frac{1}{i+1} k^{\underline{i+1}} + \mathbb{R} = \sum_{i=0}^{\ell} \left\{{\ell \atop i}\right\} \frac{1}{i+1} k^{\underline{i+1}} + \mathbb{R}.$$

Proof. If $k = 0$ the Identity (6.34.a) is trivial, so we suppose that $k \geq 1$. The integer $k^{\ell} = S((k, \ell))$ represents the number of ℓ-sequences of I_k. Now in every such sequence, for some i between 1 and k there appear exactly i elements of I_k. In Proposition 5.9 we verified that if $1 \leq i \leq k$, there are

$$\left\{{\ell \atop i}\right\} \frac{k!}{(k-i)!} = \left\{{\ell \atop i}\right\} k^{\underline{i}}$$

ℓ-sequences of I_k in which there appear exactly i elements of I_k. Thus

$$S((k, \ell)) = k^{\ell} = \sum_{i=1}^{k} \left\{{\ell \atop i}\right\} k^{\underline{i}}. \tag{6.34.b}$$

Since $k^{\underline{i}} = 0$ for $i > k$ and $\begin{Bmatrix} \ell \\ i \end{Bmatrix} = 0$ for $i > \ell$ then $\begin{Bmatrix} \ell \\ i \end{Bmatrix} k^{\underline{i}} = 0$ if $i > k$ or $i > \ell$, so that (6.34.b) may be rewritten as

$$S((k, \ell)) = k^\ell = \sum_{i=1}^{\ell} \begin{Bmatrix} \ell \\ i \end{Bmatrix} k^{\underline{i}}.$$

By linearity and (6.29.a) we get

$$\sum_k k^\ell = \sum_{i=1}^{\ell} \begin{Bmatrix} \ell \\ i \end{Bmatrix} \sum_k k^{\underline{i}} = \sum_{i=1}^{\ell} \begin{Bmatrix} \ell \\ i \end{Bmatrix} \left(\frac{1}{i+1} k^{\underline{i+1}} + \mathbb{R} \right) = \sum_{i=1}^{\ell} \begin{Bmatrix} \ell \\ i \end{Bmatrix} \frac{1}{i+1} k^{\underline{i+1}} + \mathbb{R}. \quad \Box$$

Example 6.35 In Example 6.33 we found the primitives of $k \mapsto k^2$ and $k \mapsto k^3$ via the recursive method. We use here instead Proposition 6.34. We have

$$k^2 = \sum_{i=1}^{2} \begin{Bmatrix} 2 \\ i \end{Bmatrix} k^{\underline{i}} = k^{\underline{2}} + k^{\underline{1}} \quad \text{and} \quad k^3 = \sum_{i=1}^{3} \begin{Bmatrix} 3 \\ i \end{Bmatrix} k^{\underline{i}} = k^{\underline{3}} + 3k^{\underline{2}} + k^{\underline{1}}.$$

In particular we get the following celebrated identities:

$$\sum_{k=0}^{n} k^2 = \sum_{0 \le k < n+1} k^2 = \sum_{0 \le k < n+1} (k^{\underline{2}} + k^{\underline{1}})$$

$$= \left[\frac{1}{3} k^{\underline{3}} + \frac{1}{2} k^{\underline{2}} \right]_{k=0}^{n+1} = \frac{(n+1)^{\underline{3}}}{3} + \frac{(n+1)^{\underline{2}}}{2}$$

$$= \frac{(n+1)n(2(n-1)+3)}{6} = \frac{n(n+1)(2n+1)}{6}$$

and

$$\sum_{k=0}^{n} k^3 = \sum_{0 \le k < n+1} k^3 = \sum_{0 \le k < n+1} (k^{\underline{3}} + 3k^{\underline{2}} + k^{\underline{1}})$$

$$= \left[\frac{1}{4} k^{\underline{4}} + k^{\underline{3}} + \frac{1}{2} k^{\underline{2}} \right]_{k=0}^{n+1} = \frac{(n+1)^{\underline{4}}}{4} + (n+1)^{\underline{3}} + \frac{(n+1)^{\underline{2}}}{2}$$

$$= \frac{(n+1)n((n-1)(n-2)+4(n-1)+2)}{4} = \frac{n^2(n+1)^2}{4}. \quad \Box$$

Example 6.36 It is easy to find a discrete primitive of k^4. Indeed, by Proposition 6.34 we have

$$k^4 = \sum_{i=1}^{4} \begin{Bmatrix} 4 \\ i \end{Bmatrix} k^{\underline{i}}$$

$$= k^{\underline{1}} + 7k^{\underline{2}} + 6k^{\underline{3}} + k^{\underline{4}}.$$

It follows from (6.29.a) that

$$\sum_k k^4 = \frac{1}{2}k^{\underline{2}} + \frac{7}{3}k^{\underline{3}} + \frac{3}{2}k^{\underline{4}} + \frac{1}{5}k^{\underline{5}} + \mathbb{R}.$$

In particular we get the sum of the first n fourth powers

$$\sum_{k=0}^{n} k^4 = \sum_{0 \le k < n+1} k^4$$

$$= \left[\frac{1}{2}k^{\underline{2}} + \frac{7}{3}k^{\underline{3}} + \frac{3}{2}k^{\underline{4}} + \frac{1}{5}k^{\underline{5}} \right]_{k=0}^{n+1}$$

$$= \frac{1}{30}(n+1)n\,[15 + 70(n-1) + 45(n-1)(n-2) +$$

$$+ 6(n-1)(n-2)(n-3)]$$

$$= \frac{n^2 + n}{30}\left[-1 + n + 9n^2 + 6n^3\right]$$

$$= \frac{6n^5 + 15n^4 + 10n^3 - n}{30}. \qquad\qquad \square$$

We have seen in Proposition 6.34 that every (ordinary) power is a linear combination of descending factorial powers. It is remarkable that, conversely, the descending factorial powers may also be expressed as linear combinations of ordinary powers, and that the coefficients appearing in those combinations are, up to their sign, the Stirling numbers of the first kind (see Sect. 5.2.2).

Proposition 6.37 (Descending powers in terms of natural powers) *Fix $\ell \in \mathbb{N}_{\ge 1}$. Then for each $k \in \mathbb{N}$ one has*

$$k^{\underline{\ell}} = \sum_{i=1}^{\ell} \begin{bmatrix} \ell \\ i \end{bmatrix} (-1)^{\ell+i} k^i.$$

Proof. In view of Corollary 5.54 we know that

$$X(X+1)\cdots(X+\ell-1) = \sum_{i=1}^{\ell} \begin{bmatrix} \ell \\ i \end{bmatrix} X^i.$$

Let $k \in \mathbb{N}$; evaluating both sides of this identity at $-k$ yields

$$(-1)^{\ell}k(k-1)\cdots(k-\ell+1) = (-k)(-k+1)\cdots(-k+\ell-1) = \sum_{i=1}^{\ell}(-1)^i \begin{bmatrix} \ell \\ i \end{bmatrix} k^i.$$

The desired conclusion is now immediate on multiplying both sides of the preceding equality by $(-1)^\ell$.   □

6.3.3   Discrete Primitive of the Discrete Exponential Function

Given a real number $a > 0$, the function $\mathbb{N} \to \mathbb{R}$, $k \mapsto a^k$, is called **discrete exponential**. One has

$$\Delta_k a^k = a^{k+1} - a^k = (a - 1)a^k,$$

and so

$$\forall a \neq 1 \quad \sum_k a^k = \frac{1}{a - 1} a^k + \mathbb{R}.$$

It is well known that the differentiable functions f of a real variable whose derivative $\frac{d}{dx} f$ coincides with f are given by constant multiples of the exponential function $x \mapsto e^x$. In the discrete case we have:

Example 6.38 ($\Delta f = f$) We seek the functions $f : \mathbb{N} \to \mathbb{R}$ that satisfy $\Delta f = f$, i.e.,

$$\Delta f(k) = f(k + 1) - f(k) = f(k).$$

One easily deduces that necessarily $f(k + 1) = 2f(k)$, and hence, setting $c := f(0)$, one has $f(k) = 2^k c$, for all $k \in \mathbb{N}$. Such functions are solutions of $\Delta f = f$:

$$\Delta 2^k c = 2^{k+1} c - 2^k c = 2^k c.$$

Thus, the functions $\mathbb{N} \to \mathbb{R}$ that coincide with their difference are given by constant multiples of the base 2 exponential function.   □

Example 6.39 For given natural numbers $m < n$ and $a \in \mathbb{R}$, calculate the value of the sum $\sum_{m \leq k < n} a^k$.

Solution. If $a = 1$ one has $\sum_{m \leq k < n} a^k = n - m$. If $a \neq 1$, the function $k \mapsto \frac{a^k}{a - 1}$ is a discrete primitive of $k \mapsto a^k$; hence

$$\sum_{m \leq k < n} a^k = \left[\frac{a^k}{a - 1} \right]_{k=m}^{n} = \frac{a^n - a^m}{a - 1}.$$ □

6.3.4 Harmonic Numbers and Discrete Primitives
of the Descending Factorial Power with Exponent = −1

The function that associates each natural number $k \in \mathbb{N}$ with the harmonic number H_k will turn out to be the discrete analogue of the usual logarithm function.

Definition 6.40 The **harmonic numbers** H_k, as k varies in $\mathbb{N}$, are defined by

$$H_0 = 0, \quad H_k = \frac{1}{1} + \frac{1}{2} + \frac{1}{3} + \cdots + \frac{1}{k}, \quad k \in \mathbb{N}_{\geq 1}. \qquad \Box$$

Example 6.41 The first eleven harmonic numbers (see also Fig. 6.1) are:

$$H_0 = 0, \ H_1 = 1, \ H_2 = \frac{3}{2}, \ H_3 = \frac{11}{6}, \ H_4 = \frac{25}{12}, \ H_5 = \frac{137}{60},$$

$$H_6 = \frac{49}{20}, \ H_7 = \frac{363}{140}, \ H_8 = \frac{761}{280}, \ H_9 = \frac{7129}{2520}, \ H_{10} = \frac{7381}{2520}. \qquad \Box$$

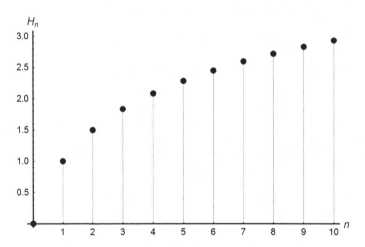

Fig. 6.1 Plotting the first eleven harmonic numbers

Proposition 6.42 *A discrete primitive of the function* $k \mapsto k^{\underline{-1}} = \dfrac{1}{k+1}$ *is given by the function* F *defined by* $F(k) := H_k$ *for all* $k \in \mathbb{N}$, *that is,*

$$\sum k^{\underline{-1}} = H_k + \mathbb{R}.$$

Proof. For each $k \in \mathbb{N}$ one has

$$\Delta F(k) = H_{k+1} - H_k = \frac{1}{k+1} = k^{\underline{-1}};$$

Proposition 6.25 then provides the desired conclusion.   □

Example 6.43 Let us calculate

$$\sum_{1 \le k < n} \frac{1}{k(k+5)}.$$

In view of the decomposition of rational functions into simple fractions (see, for example, Theorem 7.110) we know that there exist real numbers a, b such that

$$\frac{1}{k(k+5)} = \frac{a}{k} + \frac{b}{k+5}.$$

From this equation it is easy to calculate that $a = 1/5$ and $b = -1/5$. Hence one has

$$\begin{aligned}
\sum_{1 \le k < n} \frac{1}{k(k+5)} &= \frac{1}{5}\left(\sum_{1 \le k < n} \frac{1}{k} - \sum_{1 \le k < n} \frac{1}{k+5} \right) \\
&= \frac{1}{5}\left(\sum_{1 \le k < n} (k-1)^{\underline{-1}} - \sum_{1 \le k < n} (k+4)^{\underline{-1}} \right) \\
&= \frac{1}{5}\left(\sum_{0 \le k < n-1} k^{\underline{-1}} - \sum_{5 \le k < n+4} k^{\underline{-1}} \right) \\
&= \frac{1}{5}\left([H_k]_{k=0}^{n-1} - [H_k]_{k=5}^{n+4} \right) = \frac{1}{5}(H_{n-1} - H_{n+4} + H_5). \quad □
\end{aligned}$$

The function $k \mapsto H_k$ and the function $x \mapsto \log x$, respectively the discrete primitive of the function $k \mapsto k^{\underline{-1}}$ (defined on the natural numbers) and the classical primitive of the function $x \mapsto x^{-1}$ (defined on reals $x > 0$), have the same asymptotic behavior (see also Fig. 6.2).

Fig. 6.2 Comparing H_n and $\log n$

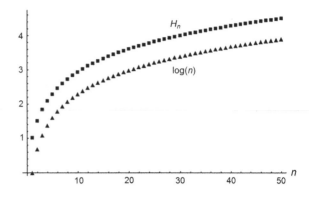

Proposition 6.44 (Harmonic numbers and logarithms are asymptotic) *One has*

$$\log n \le H_n \le \log n + 1 \qquad \forall n \in \mathbb{N}_{\ge 1} \tag{6.44.a}$$

whence

$$H_n \sim \log n \ \ for \ n \to \infty: \ \ \lim_{n \to +\infty} \frac{H_n}{\log n} = 1.$$

Proof. The harmonic number H_{n-1} is the sum of the values assumed by the function $1/x$ on the natural numbers $1, 2, \ldots, n-1$ or, equivalently, the sum of the areas of the rectangles with base $[k, k+1]$ and height $1/k$, for $k = 1, \ldots, n-1$. Since for $x \in [k, k+1]$ one has $1/x \le 1/k$ it follows that H_{n-1} is greater than or equal to the area of the trapezoid on the function $1/x$ between 1 and n, and thus one has

$$H_n \ge H_n - \frac{1}{n} = H_{n-1} = 1 + \frac{1}{2} + \cdots + \frac{1}{n-1} \ge \int_1^n \frac{1}{x}\,dx = \log n.$$

In particular, it follows that $\lim_{n \to \infty} H_n = \infty$. Similarly, for each $x \in [k, k+1]$ one has $1/x \ge 1/(k+1)$ and so

$$H_n - 1 = \frac{1}{2} + \cdots + \frac{1}{n}$$

is less than or equal to the trapezoid on $1/x$ between 1 and n, so that (see Fig. 6.3)

$$H_n - 1 = \frac{1}{2} + \cdots + \frac{1}{n} \le \int_1^n \frac{1}{x}\,dx = \log n.$$

Hence we obtain

$$H_n - 1 \le \log n \le H_n,$$

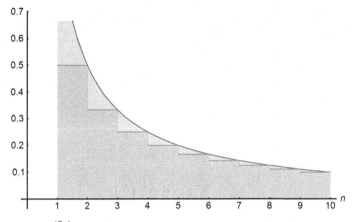

Fig. 6.3 $H_n - 1 \leq \int_1^n \frac{1}{x}\, dx$

implying (6.44.a). Dividing the three terms of the inequality by $\log n$ one yields

$$1 \leq \frac{H_n}{\log n} \leq 1 + \frac{1}{\log n},$$

and hence $H_n \sim \log n$. □

Remark 6.45 One of the fundamental inequalities on logarithm is

$$\frac{x}{1+x} \leq \log(1+x) \leq x \quad \forall x > -1.$$

Setting $x = 1/n$, $n \in \mathbb{N}_{\geq 1}$, one gets

$$\frac{1}{n+1} \leq \log\left(1 + \frac{1}{n}\right) \leq \frac{1}{n}$$

and hence

$$H_{n+1} - H_n \leq \log(n+1) - \log n \leq H_n - H_{n-1}.$$

We end this section with a couple of representation formulas of the harmonic numbers. The first one, an integral formula due to Euler, leads to express H_n in terms of a sum of binomials. The second one links the harmonic numbers and the Stirling numbers of the first kind.

Proposition 6.46 (Euler formula for harmonic numbers) *For all $n \in \mathbb{N}_{\geq 1}$ one has the following integral formula for H_n:*

$$H_n = \int_0^1 \frac{1 - x^n}{1 - x} \, dx. \tag{6.46.a}$$

As a consequence the following formula holds:

$$H_n = \sum_{k=1}^n (-1)^{k-1} \frac{1}{k} \binom{n}{k}. \tag{6.46.b}$$

Proof. For each $x \in \mathbb{R}$ one has

$$(1 + x + \cdots + x^{n-1})(1 - x) = 1 - x^n$$

and so, for $x \neq 1$,

$$1 + x + \cdots + x^{n-1} = \frac{1 - x^n}{1 - x}.$$

Note that in (6.46.a) the integrand may be extended continuously to 1 by setting it equal to $\lim_{x \to 1} \dfrac{1 - x^n}{1 - x} = \lim_{x \to 1} (1 + x + \cdots + x^{n-1}) = n$. One sees immediately that:

$$\int_0^1 \frac{1 - x^n}{1 - x} \, dx = \int_0^1 (1 + x + \cdots + x^{n-1}) \, dx = 1 + \frac{1}{2} + \cdots + \frac{1}{n} = H_n.$$

On setting $x = 1 - u$ in the integral formula just obtained one finds that

$$H_n = \int_0^1 \frac{1 - x^n}{1 - x} \, dx = \int_1^0 \frac{1 - (1 - u)^n}{u} (-du) = \int_0^1 \frac{1 - (1 - u)^n}{u} \, du.$$

Now

$$1 - (1 - u)^n = 1 - \sum_{k=0}^n \binom{n}{k} (-u)^k = -\sum_{k=1}^n \binom{n}{k} (-1)^k u^k$$

and therefore

$$\mathrm{H}_n = \int_0^1 \sum_{k=1}^n (-1)^{k-1} \binom{n}{k} u^{k-1}\, du$$

$$= \sum_{k=1}^n (-1)^{k-1} \binom{n}{k} \int_0^1 u^{k-1}\, du = \sum_{k=1}^n (-1)^{k-1} \frac{1}{k} \binom{n}{k}. \qquad \square$$

Proposition 6.47 *For each $n \in \mathbb{N}_{\geq 1}$ one has*

$$\mathrm{H}_n = \frac{1}{n!} \begin{bmatrix} n+1 \\ 2 \end{bmatrix}.$$

Proof. By Theorem 5.52 one has

$$\mathrm{H}_n = 1 + \frac{1}{2} + \cdots + \frac{1}{n} = \frac{\frac{n!}{1} + \frac{n!}{2} + \cdots + \frac{n!}{n}}{n!}$$

$$= \frac{1}{n!} \sum_{\substack{J \subseteq I_n \\ |J| = n-1}} \prod_{i \in J} i = \frac{1}{n!} \begin{bmatrix} n+1 \\ 2 \end{bmatrix}. \qquad \square$$

6.4 Formula for Summation by Parts

In this section we introduce the technique of summation by parts which is a direct consequence of the formula for the difference of a product proved in Proposition 6.13.

Theorem 6.48 (Summation by parts formula) *Let F, G be discrete primitives of $f, g : \mathbb{N} \to \mathbb{R}$, respectively. Then*

$$\sum Fg = FG - \sum f\,\theta\,G, \quad \text{that is for } m < n \text{ in } \mathbb{N}:$$

$$\sum_{m \leq k < n} F(k)g(k) = [F(k)G(k)]_{k=m}^n - \sum_{m \leq k < n} f(k)G(k+1).$$

Proof. By Proposition 6.13 one has $\Delta(FG) = Fg + f\,\theta\,G$; therefore FG is a discrete primitive of $Fg + f\,\theta\,G$. In particular by the Fundamental Theorem of Finite Calculus 6.27 one gets

$$\sum_{m \leq k < n} \bigl(F(k)g(k) + f(k)(\theta\,G)(k) \bigr) = [F(k)G(k)]_{k=m}^n. \qquad \square$$

☞   The formula just proved resembles the well know integration by parts formula. One should, however, beware the presence of the *shift* operator in the sum on the right side of the equation evaluated on the primitive G.

Example 6.49 Calculate $\displaystyle\sum_{0 \le k < n} k 2^k$.

Solution. Since $\Delta 2^k = 2^k$ and $\Delta k = 1$, summation by parts with $F(k) = k$ and $g(k) = 2^k$ yields

$$\sum_{0 \le k < n} k 2^k = \sum_{0 \le k < n} k \, \Delta \, 2^k = \left[k 2^k \right]_{k=0}^{n} - \sum_{0 \le k < n} 2^{k+1}$$

$$= n 2^n - \left[2^{k+1} \right]_{k=0}^{n} = n 2^n - (2^{n+1} - 2) = 2^n (n-2) + 2. \qquad \square$$

Example 6.50 Calculate $\displaystyle\sum_{0 \le k < n} \mathrm{H}_k$ and $\displaystyle\sum_{0 \le k < n} k \, \mathrm{H}_k$.

Solution. Since $\Delta \mathrm{H}_k = k^{\underline{-1}}$ and $\Delta k = 1$, summation by parts with $F(k) = \mathrm{H}_k$ and $g(k) = 1$ gives

$$\sum_{0 \le k < n} \mathrm{H}_k = \sum_{0 \le k < n} \mathrm{H}_k \, \Delta \, k = [k \, \mathrm{H}_k]_{k=0}^{n} - \sum_{0 \le k < n} \Delta \, \mathrm{H}_k (k+1)$$

$$= [k \, \mathrm{H}_k]_{k=0}^{n} - \sum_{0 \le k < n} k^{\underline{-1}}(k+1) = n \, \mathrm{H}_n - \sum_{0 \le k < n} 1 = n(\mathrm{H}_n - 1).$$

Since $n(\mathrm{H}_n - 1) = [h \, \mathrm{H}_k - k]_{k=0}^{n}$, by the Fundamental Theorem of Finite Calculus 6.27, from the previous equality we have that $k \, \mathrm{H}_k - k$ is a discrete primitive of H_k. Since $\Delta k^{\underline{2}} = 2k^{\underline{1}} = 2k$, summation by parts then yields

$$\sum_{0 \le k < n} k \, \mathrm{H}_k = \sum_{0 \le k < n} \mathrm{H}_k \, \Delta \left(\frac{1}{2} k^{\underline{2}} \right)$$

$$= \left[\mathrm{H}_k \frac{1}{2} k^{\underline{2}} \right]_{k=0}^{n} - \sum_{0 \le k < n} k^{\underline{-1}} \frac{1}{2} (k+1)^{\underline{2}}$$

$$= \frac{1}{2} n^{\underline{2}} \mathrm{H}_n - \frac{1}{2} \sum_{0 \le k < n} k = \frac{1}{2} \left(n^{\underline{2}} \mathrm{H}_n - \left[\frac{1}{2} k^{\underline{2}} \right]_{k=0}^{n} \right)$$

$$= \frac{n^{\underline{2}}}{2} \left(\mathrm{H}_n - \frac{1}{2} \right). \qquad \square$$

6.5 Discrete Taylor Series ☕

Many readers will certainly be aware of representations of functions of a real variable using Taylor or Maclaurin series. In this section we shall see that there are analogous representations for functions whose domain is $\mathbb{N}$.

If α is an operator, then for all $n \in \mathbb{N}$ we set

$$\alpha^n = \begin{cases} \mathbb{1} & \text{if } n = 0, \\ \underbrace{\alpha \circ \alpha \circ \cdots \circ \alpha}_{n} & \text{otherwise,} \end{cases}$$

where as usual $\mathbb{1}$ is the identity operator.

Example 6.51 Let $f : \mathbb{N} \to \mathbb{R}$. For all $a \in \mathbb{N}$ one has

$$\theta^2 f(a) = (\theta \circ \theta)f(a) = \theta f(a+1) = f(a+2).$$

More generally, for all $n \in \mathbb{N}$ one has

$$\theta^n f(a) = f(a+n). \qquad \square$$

In order to prove the discrete version of Taylor formula we need a couple of technical results.

Lemma 6.52 *Let α, β be two operators on functions with domain $\mathbb{N}$. If α and β commute, i.e., $\alpha \circ \beta = \beta \circ \alpha$, then:*

1. *If $m, n \in \mathbb{N}$ one has $\alpha^m \circ \beta^n = \beta^n \circ \alpha^m$;*
2. $(\alpha + \beta)^n = \sum_{k=0}^{n} \binom{n}{k} \alpha^k \circ \beta^{n-k}.$

Proof. 1. For $n = 0$ the identity is equivalent to $\alpha^m = \alpha^m$ and similarly for $m = 0$ it amounts to $\beta^n = \beta^n$. Fix $m \geq 1$. We prove the statement by induction on n. If $n = 1$, using several times that $\alpha \circ \beta = \beta \circ \alpha$, one gets

$$\alpha^m \circ \beta = \underbrace{\alpha \circ \cdots \circ \alpha}_{m} \circ \beta = \underbrace{\alpha \circ \cdots \circ \alpha}_{m-1} \circ \beta \circ \alpha = \cdots = \alpha \circ \beta \circ \underbrace{\alpha \circ \cdots \circ \alpha}_{m-1} = \beta \circ \alpha^m.$$

Let $n \geq 1$; by the inductive hypothesis one has

$$\begin{aligned} \alpha^m \circ \beta^{n+1} &= (\alpha^m \circ \beta^n) \circ \beta \\ &= (\beta^n \circ \alpha^m) \circ \beta = \beta^n \circ (\alpha^m \circ \beta) \\ &= \beta^n \circ (\beta \circ \alpha^m) = (\beta^n \circ \beta) \circ \alpha^m = \beta^{n+1} \circ \alpha^m, \end{aligned}$$

and this proves the desired identity.

2. The proof of this point parallels that of the proof of the binomial formula in Proposition 2.19. Composing $\alpha + \beta$ with itself n times we obtain a sum of 2^n operators of the type

$$\gamma_1 \circ \gamma_2 \circ \cdots \circ \gamma_n, \quad \gamma_i \in \{\alpha, \beta\}.$$

In view of the fact that the operators α and β commute, if in the composition considered above the operator α appears k times then one has

$$\gamma_1 \circ \gamma_2 \circ \cdots \circ \gamma_n = \alpha^k \circ \beta^{n-k}.$$

Such a term appears whenever the operator α is chosen k times and the operator β is chosen $n - k$ times from the n factors $(\alpha + \beta)$. The k terms $(\alpha + \beta)$ in which α is chosen form a k-collection of the set of n factors, and so they may be selected in $C(n, k) = \binom{n}{k}$ ways. The term $\alpha^k \circ \beta^{n-k}$ is therefore repeated $C(n, k) = \binom{n}{k}$ times and so this is the coefficient of the operator $\alpha^k \circ \beta^{n-k}$ in the expansion of $(\alpha + \beta)^n$.   □

Theorem 6.53 (Discrete Taylor formula) *Let $f : \mathbb{N} \to \mathbb{R}$ be a function, and $a \in \mathbb{N}$. Then*

$$f(a + n) = \sum_{k=0}^{n} \frac{\Delta^k f(a)}{k!} n^{\underline{k}} \qquad \forall n \in \mathbb{N}.$$

Proof. Since the operators Δ and $\mathbb{1}$ commute, by Lemma 6.52 one has

$$\theta^n = (\Delta + \mathbb{1})^n = \sum_{k=0}^{n} \binom{n}{k} \Delta^k \circ \mathbb{1}^{n-k} = \sum_{k=0}^{n} \binom{n}{k} \Delta^k$$

from which it follows that

$$f(a + n) = \theta^n f(a) = \sum_{k=0}^{n} \binom{n}{k} \Delta^k f(a) = \sum_{k=0}^{n} \frac{\Delta^k f(a)}{k!} n^{\underline{k}}. \qquad □$$

To make full use of the discrete Taylor formula one need to understand how the operator Δ^k acts.

Example 6.54 Let us calculate Δ^2. If $f : \mathbb{N} \to \mathbb{R}$, then for each $k \in \mathbb{N}$ one has

$$\begin{aligned}
\Delta^2 f(k) &= \Delta(\Delta f(k)) = \Delta f(k + 1) - \Delta f(k) \\
&= (f(k + 2) - f(k + 1)) - (f(k + 1) - f(k)) \\
&= f(k + 2) - 2f(k + 1) + f(k).
\end{aligned}$$
□

In general, the iterations of the operator Δ act as follows.

Proposition 6.55 (The operator Δ^n) *For every function f with domain $\mathbb{N}$ one has*

$$\Delta^n f(a) = \sum_{k=0}^n (-1)^{n-k} \binom{n}{k} f(a+k) \qquad \forall a, n \in \mathbb{N}.$$

Proof. Since $\Delta = \theta - \mathbb{1}$ and the two operators θ and $\mathbb{1}$ commute, by Lemma 6.52 one has

$$\Delta^n = (\theta - \mathbb{1})^n = \sum_{k=0}^n (-1)^{n-k} \binom{n}{k} \theta^k \circ \mathbb{1}^{n-k} = \sum_{k=0}^n (-1)^{n-k} \binom{n}{k} \theta^k.$$

Therefore,

$$\Delta^n f(a) = \sum_{k=0}^n (-1)^{n-k} \binom{n}{k} \theta^k f(a) = \sum_{k=0}^n (-1)^{n-k} \binom{n}{k} f(a+k). \qquad \square$$

Remark 6.56 Note that, on one hand to compute $f(a+n)$ in the Taylor formula (Theorem 6.53) it is necessary to know $\Delta^k f(a)$ for each $k = 1, \ldots, n$; on the other hand in Proposition 6.55 to compute $\Delta^n f$ one has to know $f(a+n)$, going round in circles. The interest of the discrete Taylor formula resides mainly in its formal analogy with the similar formula of differential calculus.

6.6 An Inversion Formula

The inversion formulas are very useful in many different fields. In this section we present one version for the partially ordered set of the finite subsets of a set, ordered by inclusion and some interesting consequences.

Proposition 6.57 (Inversion formula for set functions) *Let ϕ, ψ be two functions defined on the subsets of I_n, for some $n \in \mathbb{N}_{\geq 1}$. Assume that*

$$\phi(T) = \sum_{S \subseteq T} \psi(S) \qquad \forall T \subseteq I_n.$$

Then

$$\psi(T) = \sum_{S \subseteq T} (-1)^{|T|-|S|} \phi(S) \qquad \forall T \subseteq I_n. \tag{6.57.a}$$

Proof. Let $T \subseteq I_n$. By changing the summation order, we get

$$\sum_{S \subseteq T} (-1)^{|T|-|S|} \phi(S) = \sum_{S \subseteq T} (-1)^{|T \setminus S|} \sum_{R \subseteq S} \psi(R)$$

$$= \sum_{R \subseteq T} \psi(R) \sum_{R \subseteq S \subseteq T} (-1)^{|T \setminus S|}. \tag{6.57.b}$$

Now, as S varies among the subsets of T that contain R, $T \setminus S$ varies among the subsets of $T \setminus R$. It follows that

$$\sum_{R \subseteq S \subseteq T} (-1)^{|T \setminus S|} = \sum_{S' \subseteq T \setminus R} (-1)^{|S'|}.$$

For each $0 \le k \le m := |T \setminus R|$ there are $\binom{m}{k}$ subsets of cardinality k of $T \setminus R$, and therefore

$$\sum_{S' \subseteq T \setminus R} (-1)^{|S'|} = \sum_{k=0}^{m} \left(\sum_{\{S' \subseteq T \setminus R, \, |S'|=k\}} (-1)^k \right)$$

$$= \sum_{k=0}^{m} \binom{m}{k} (-1)^k = \begin{cases} (1-1)^m = 0 & \text{if } m \ge 1 \ (R \ne T), \\ 1 & \text{if } m = 0 \ (R = T). \end{cases}$$

By (6.57.b) one has

$$\sum_{S \subseteq T} (-1)^{|T|-|S|} \phi(S) = \sum_{R \subseteq T} \psi(R) \sum_{R \subseteq S \subseteq T} (-1)^{|T \setminus S|}$$

$$= \sum_{R \subseteq T} \psi(R) \sum_{S' \subseteq T \setminus R} (-1)^{|S'|} = \psi(T). \qquad \square$$

Corollary 6.58 (Inversion formula for natural functions) *Let f and g be two functions defined on $\mathbb{N}$. Assume that*

$$f(n) = \sum_{k=0}^{n} \binom{n}{k} g(k) \qquad \forall n \in \mathbb{N}.$$

Then

$$g(n) = \sum_{k=0}^{n} (-1)^{n-k} \binom{n}{k} f(k) \qquad \forall n \in \mathbb{N}. \tag{6.58.a}$$

Proof. Let $n \in \mathbb{N}$; for any subset S of I_n we define $\psi(S) = g(|S|)$. We set $\phi(T) = \sum_{S \subseteq T} \psi(S)$ for every subset T of I_n. Grouping together the sets of same cardinality, we get

$$\phi(T) = \sum_{S \subseteq T} \psi(S) = \sum_{S \subseteq T} g(|S|) = \sum_{k=0}^{|T|} \sum_{\{S \subseteq T, \, |S|=k\}} g(k) = \sum_{k=0}^{|T|} \binom{|T|}{k} g(k) = f(|T|).$$

Then by Proposition 6.57 one has

$$g(n) = \psi(I_n) = \sum_{S \subseteq I_n} (-1)^{n-|S|} \phi(S) = \sum_{S \subseteq I_n} (-1)^{n-|S|} f(|S|)$$

$$= \sum_{k=0}^{n} \sum_{\{S \subseteq I_n, \, |S|=k\}} (-1)^{n-k} f(k) = \sum_{k=0}^{n} (-1)^{n-k} \binom{n}{k} f(k). \qquad \square$$

We now give some applications of the inversion formula. Let us start proving that Theorem 6.53 and Proposition 6.55 are one consequence of the other through the *inversion formula*.

Example 6.59 Let $f : \mathbb{N} \to \mathbb{R}$ and $a \in \mathbb{N}$. Set $\widetilde{f}(k) = f(a+k)$ and $g(k) = \Delta^k f(a)$. Then Taylor formula from Theorem 6.53 yields

$$\widetilde{f}(n) = \sum_{k=0}^{n} \binom{n}{k} g(k) \qquad \forall n \in \mathbb{N}.$$

Applying the inversion formula (6.58.a) one obtains

$$\Delta^n f(a) = g(n) = \sum_{k=0}^{n} (-1)^{n-k} \binom{n}{k} \widetilde{f}(k) = \sum_{k=0}^{n} (-1)^{n-k} \binom{n}{k} f(a+k) \qquad \forall n \in \mathbb{N},$$

thus re-establishing the formula of Proposition 6.55. $\qquad \square$

We now apply the inversion formula to calculate the number of derangements (fixed point free permutations) on n elements. In Theorem 4.24 we have already calculated this number using the Inclusion/Exclusion Principle 4.9.

Example 6.60 (*Number of derangements*) Let $n \in \mathbb{N}$ and consider the $n!$ permutations of $(1, \ldots, n)$. In each permutation there are elements that are moved and others that remain fixed. Fix an integer k between 0 and n. Then there are $\binom{n}{k}$ ways of choosing $n - k$ elements which are not moved, and D_k ways of moving all of the remaining k elements ($D_0 := 1$). Thus one has

$$n! = \sum_{k=0}^{n} \binom{n}{k} D_k.$$

Set $f(n) = n!$ and $g(n) = D_n$ for each $n \in \mathbb{N}$. Then from the inversion formula (6.58.a) one obtains

$$D_n = \sum_{k=0}^{n} \binom{n}{k} (-1)^{n-k} k! = n! \left(\sum_{k=0}^{n} (-1)^{n-k} \frac{1}{(n-k)!} \right)$$
$$= n! \left(\sum_{k=0}^{n} (-1)^k \frac{1}{k!} \right),$$

just as we found in Theorem 4.24.   □

Example 6.61 (*Number of surjective functions*) Let $n \in \mathbb{N}$ and $\ell \geq 1$. In Corollary 5.8 we found the number $n! \begin{Bmatrix} \ell \\ n \end{Bmatrix}$ of surjective functions from I_ℓ onto I_n as a consequence of the inclusion-exclusion principle: we get it here by means of the inversion formula. Remind that, from (6.34.a), we have

$$n^\ell = \sum_{k=0}^{n} \begin{Bmatrix} \ell \\ k \end{Bmatrix} n^{\underline{k}}. \tag{6.61.a}$$

Since

$$\begin{Bmatrix} \ell \\ k \end{Bmatrix} n^{\underline{k}} = k! \begin{Bmatrix} \ell \\ k \end{Bmatrix} \binom{n}{k}$$

then (6.61.a) may be rewritten as

$$n^\ell = \sum_{k=0}^{n} \left(k! \begin{Bmatrix} \ell \\ k \end{Bmatrix} \right) \binom{n}{k}.$$

Set $f(n) = n^\ell$ and $g(n) = n! \begin{Bmatrix} \ell \\ n \end{Bmatrix}$. Then from the inversion formula (6.58.a) one obtains

$$n! \begin{Bmatrix} \ell \\ n \end{Bmatrix} = \sum_{k=0}^{n} \binom{n}{k} (-1)^{n-k} k^\ell = \sum_{i=0}^{n} \binom{n}{i} (-1)^i (n-i)^\ell,$$

which is exactly (5.8.a).   □

The Inclusion/Exclusion Principle 4.3 can be read as a formula for the *uniform probability* $P(A) := \dfrac{|A|}{|\Omega|}$ on the subsets A of a finite set Ω; namely if $A_1, \ldots, A_n$ are subsets of a finite set Ω then

$$P(A_1 \cup \cdots \cup A_n) = \frac{|A_1 \cup \cdots \cup A_n|}{|\Omega|} = \frac{\mathfrak{S}_1 - \mathfrak{S}_2 + \cdots + (-1)^{n-1} \mathfrak{S}_n}{|\Omega|}$$

$$= \mathfrak{S}_1' - \mathfrak{S}_2' + \cdots + (-1)^{n-1} \mathfrak{S}_n'$$

where

$$\mathfrak{S}_1' = P(A_1) + \cdots + P(A_n),$$
$$\mathfrak{S}_2' = P(A_1 \cap A_2) + P(A_1 \cap A_3) + \cdots + P(A_{n-1} \cap A_n),$$

......

$$\mathfrak{S}_k' = \underbrace{\sum_{1 \le i_1 < \cdots < i_k \le n} P(A_{i_1} \cap \cdots \cap A_{i_k})}_{C(n,k) \text{ summands}},$$

......

$$\mathfrak{S}_n' = P(A_1 \cap \cdots \cap A_n).$$

Given an arbitrary set Ω, an *algebra* $\mathscr{F}$ of subsets of Ω is a subset $\mathscr{F} \subset \mathscr{P}(\Omega)$ containing the empty set $\emptyset$, and closed under the set-theoretic operations of union and intersection. A real function μ defined on the algebra $\mathscr{F}$ is *additive* if $\mu(\emptyset) = 0$, and for any finite number of disjoint sets $A_1, \ldots, A_n$

$$\mu\left(\bigcup_{i=1}^{n} A_i\right) = \sum_{i=1}^{n} \mu(A_i).$$

All *probability functions* are possible examples of additive real functions on an algebra of subsets.

Let us show as the inversion formula yields a proof of the Inclusion-Exclusion formula for any additive function.

Proposition 6.62 (Inclusion/Exclusion Principle for finitely subadditive functions) *Let μ be an additive function defined on an algebra $\mathscr{F}$ of subsets of a space Ω. Let $A_1, \ldots, A_n$ be in $\mathscr{F}$. Then*

$$\mu(A_1 \cup \cdots \cup A_n) = \sum_{\emptyset \neq J \subseteq I_n} (-1)^{|J|-1} \mu \left(\bigcap_{i \in J} A_i \right)$$

$$= \sum_{k=1}^{n} (-1)^{k-1} \sum_{1 \leq i_1 < \cdots < i_k \leq n} \mu(A_{i_1} \cap \cdots \cap A_{i_n}).$$

(6.62.b)

Proof. For every subset S of I_n let

$$A_S := \begin{cases} \displaystyle\bigcap_{i \in I_n \setminus S} A_i \setminus \bigcup_{i \in S} A_i & \text{if } S \neq I_n, \\ \emptyset & \text{if } S = I_n. \end{cases}$$

Notice that we have:

- $A_\emptyset = A_1 \cap \cdots \cap A_n$;
- If $S_1 \neq S_2$ are two different subsets of I_n then $A_{S_1} \cap A_{S_2} = \emptyset$. Indeed, assume $S_1 \not\subseteq S_2$; if $i \in S_1 \setminus S_2$, then $A_{S_1} \cap A_i = \emptyset$ whereas $A_{S_2} \subseteq A_i$.

For any $T \subseteq I_n$ let

$$\psi(T) := \mu(A_T), \qquad \phi(T) := \sum_{S \subseteq T} \psi(S).$$

The sets $(A_S)_{S \subseteq I_n}$ being disjoint, we have

$$\phi(T) = \sum_{S \subseteq T} \mu(A_S) = \mu \left(\bigcup_{S \subseteq T} A_S \right).$$

Now

$$\bigcup_{S \subseteq T} A_S = \begin{cases} \displaystyle\bigcap_{i \in I_n \setminus T} A_i & \text{if } T \neq I_n, \\ A_1 \cup \cdots \cup A_n & \text{if } T = I_n. \end{cases}$$

Indeed if $S \subseteq T \subsetneq I_n$ then

$$A_S \subseteq \bigcap_{i \in I_n \setminus S} A_i \subseteq \bigcap_{i \in I_n \setminus T} A_i;$$

conversely, if $x \in \bigcap_{i \in I_n \setminus T} A_i$ and $S = \{i \in I_n : x \notin A_i\}$, then $S \subseteq T$ and $x \in A_S$. On the other side, if $S \subseteq T = I_n$ then $A_S \subseteq A_1 \cup \cdots \cup A_n$; conversely, if $x \in A_1 \cup \cdots \cup A_n$ and $S = \{i \in I_n : x \notin A_i\}$, then x belongs to A_S. It follows that, for any subset T of I_n,

$$\phi(T) = \begin{cases} \mu\left(\bigcap_{i \in I_n \setminus T} A_i\right) & \text{if } T \neq I_n, \\ \mu(A_1 \cup \cdots \cup A_n) & \text{if } T = I_n. \end{cases}$$

The inversion formula (6.57.a) now yields

$$0 = \mu(\emptyset) = \psi(I_n) = \sum_{S \subseteq I_n} (-1)^{n-|S|} \phi(S)$$

$$= \sum_{S \subsetneq I_n} (-1)^{n-|S|} \mu\left(\bigcap_{i \in I_n \setminus S} A_i\right) + \mu(A_1 \cup \cdots \cup A_n)$$

$$= \sum_{\emptyset \neq J \subseteq I_n} (-1)^{|J|} \mu\left(\bigcap_{i \in J} A_i\right) + \mu(A_1 \cup \cdots \cup A_n),$$

thus proving the validity of (6.62.b). $\square$

6.7 Abel–Dirichlet Convergence Test ☕

In this last section we abandon finite sums, and consider an application of the techniques of finite calculus to the study of the convergence of a class of numerical series. It is not surprising that the techniques employed to calculate the partial sums

$$s_n = a_0 + \cdots + a_n$$

of a given sequence $(a_k)_k$ can also provide convergence criteria for series: indeed, by definition the series $\sum_{k=0}^{\infty} a_k$ converges if the limit $\lim_{n \to \infty} s_n$ exists and is finite.

Most of the best known convergence criteria for series concern series with positive terms; thus in general they furnish criteria for absolute convergence. A notable exception is provided by the Leibniz[2] test for convergence of an alternating series, which is, indeed, often the only criterion given for non-absolute convergence. The summation by parts formula given in Theorem 6.48 allows us to obtain a more general test for non-absolute convergence, from which the Leibniz one may be deduced as a special case (see Corollary 6.67).

Definition 6.63 We say that a sequence $(b_k)_k$ has **bounded partial sums** if there exists an $M > 0$ such that

$$|b_0 + \cdots + b_k| \leq M \quad \forall k \in \mathbb{N}.$$ $\square$

---

[2]Gottfried Wilhelm von Leibniz (1646–1716).

Remark 6.64 If the series $\sum_{k=0}^{\infty} b_k$ converges, then the sequence $(b_k)_k$ has bounded partial sums (in fact, the sequence of partial sums $(b_0 + \cdots + b_k)_k$ converges, and is therefore bounded). However the sequence $(b_k)_k$ can very well have bounded partial sums without being convergent, as is illustrated by the sequence $(b_k)_k = \left((-1)^k\right)_k$.

Proposition 6.65 (Convergence test of Abel[3]–Dirichlet[4]) *Let $(a_k)_k$ be a decreasing sequence in $\mathbb{R}_{\geq 0}$ with $\lim_{k\to\infty} a_k = 0$, and let $(b_k)_k$ be a sequence whose partial sums $s_k = b_0 + \cdots + b_k$ are bounded. Then the series*

$$\sum_{k=0}^{\infty} a_k b_k$$

converges.

Proof. Let $a(k) = a_k$ and $B(k) = b_0 + \cdots + b_{k-1}$ (with $B(0) = 0$). Then one has $\Delta B(k) = b_k$. The summation by parts formula (Theorem 6.48) yields

$$\sum_{0\leq k<n} a_k b_k = \sum_{0\leq k<n} a(k)\,\Delta B(k) = [a(k)B(k)]_{k=0}^{n} - \sum_{0\leq k<n} \Delta a(k)\theta\, B(k)$$

$$= a_n B(n) - \sum_{0\leq k<n} (a_{k+1} - a_k)B(k+1). \tag{6.65.a}$$

Let $M > 0$ be such that $|B(k)| \leq M$ for all $k \in \mathbb{N}$. Since $|a_n B(n)| \leq M a_n$ and $(a_k)_k$ has limit zero, one has

$$\lim_{n\to\infty} a_n B(n) = 0.$$

Moreover, since $(a_k)_k$ is decreasing, one obtains

$$\sum_{0\leq k<n} |(a_{k+1} - a_k)B(k+1)| \leq M \sum_{0\leq k<n} |(a_{k+1} - a_k)|$$

$$= M \sum_{0\leq k<n} (a_k - a_{k+1}) = M(a_0 - a_n) \leq M a_0.$$

If follows that the series $\sum_{k=0}^{\infty} |(a_{k+1} - a_k)B(k+1)|$ converges; indeed, we recall that a series with positive terms either converges or diverges to $+\infty$. But then the series

$$\sum_{k=0}^{\infty} (a_{k+1} - a_k)B(k+1)$$

---

[3] Niels Henrik Abel (1802–1829).
[4] Johann Peter Gustav Lejeune Dirichlet (1805–1859).

converges (absolute convergence implies convergence). The conclusion now follows immediately from (6.65.a). □

Remark 6.66 Rereading the proof, one notes that Proposition 6.65 continues to hold if, instead of supposing that $(a_k)_k$ is decreasing with limit zero, one supposes only that $(a_k)_k$ has limit zero, and that it is of *bounded variation*, that is,

$$\sum_{k=0}^{\infty} |a_{k+1} - a_k|$$

converges.

The well-known Leibniz test for alternating series may be deduced as a special case of Proposition 6.65.

Corollary 6.67 (Leibniz test) *Let $(a_k)_k$ be a decreasing sequence of positive terms with $\lim_{k \to +\infty} a_k = 0$. Then the series*

$$\sum_{k=0}^{\infty} (-1)^k a_k$$

converges.

Proof. It suffices to set $b_k = (-1)^k$ and apply the Abel–Dirichlet test. □

The summation by parts formula, and most of the material expounded in this chapter, continues to hold in the case of functions with domain $\mathbb{N}$ and codomain $\mathbb{C}$. In particular, the Abel–Dirichlet test also holds for sequences $(b_k)_k$ of complex values.

Example 6.68 Let $t \in \mathbb{R}$ with $t \notin 2\pi \mathbb{Z}$. Verify that the series

$$\sum_{k=1}^{\infty} \frac{\cos(kt)}{k} \quad \text{and} \quad \sum_{k=1}^{\infty} \frac{\sin(kt)}{k}$$

are convergent.

Solution. Using the complex exponential function one has

$$\sum_{k=1}^{\infty} \frac{e^{ikt}}{k} = \sum_{k=1}^{\infty} \frac{\cos(kt) + i\sin(kt)}{k} = \sum_{k=1}^{\infty} \frac{\cos(kt)}{k} + i\sum_{k=1}^{\infty} \frac{\sin(kt)}{k};$$

hence it is equivalent to verify the convergence in $\mathbb{C}$ of the series

$$\sum_{k=1}^{\infty} \frac{e^{ikt}}{k}.$$

The sequence $1/k$ is decreasing and has limit 0. For a given $t \notin 2\pi \mathbb{Z}$, the sequence of partial sums of $(e^{ikt})_k$ is bounded: indeed for each $m \in \mathbb{N}$ one has

$$s_m := \sum_{k=1}^{m} e^{ikt} = -1 + \sum_{k=0}^{m} (e^{it})^k = \frac{1 - e^{i(m+1)t}}{1 - e^{it}} - 1$$

and thus

$$|s_m| \le \left| \frac{1 - e^{i(m+1)t}}{1 - e^{it}} \right| + 1 \le \frac{2}{|1 - e^{it}|} + 1.$$

An application of Proposition 6.65 then gives the desired conclusion.   □

6.8 Problems

Problem 6.1 Let $a, b \in \mathbb{R}$. Use Gauss method to calculate

$$S = \sum_{k=0}^{n} (a + bk).$$

Problem 6.2 Let x be a given real number. Calculate, using the perturbation method and the result of Example 6.3, the sum

$$T_n = \sum_{k=0}^{n} k^2 x^k.$$

From this deduce, for $|x| < 1$, the sum of the series

$$\sum_{k=0}^{\infty} k^2 x^k.$$

Problem 6.3 Let $k \in \mathbb{N}$ and $m, n \in \mathbb{Z}$ with $k \ge m$. Prove that

$$k^{\underline{m+n}} = k^{\underline{m}} (k - m)^{\underline{n}}.$$

Problem 6.4 Find a discrete primitive of $k \longmapsto \dfrac{k}{(k+1)(k+2)(k+3)}$.

Problem 6.5 Calculate

$$\sum_{0 \le k < n} k^2 2^k.$$

Problem 6.6 Calculate

$$\sum_{0 \le k < n} k^2 5^k.$$

Problem 6.7 Verify that $\Delta \log k! = \log(k + 1)$, and then calculate

$$\sum_{0 \le k < n} k \log(k + 1).$$

Problem 6.8 Prove the following equality for all $n \ge 1$:

$$\sum_{k=0}^{n} (-1)^k \frac{1}{(k + 1)^2} \binom{n}{k} = \frac{1}{n + 1} H_{n+1}.$$

Problem 6.9 We propose an alternative proof for Proposition 5.83. Let $n \ge 1$ and $k \in \mathbb{N}$. Consider the functions

$$m \mapsto m^n, \quad m \mapsto \binom{m + k}{n}.$$

1. For each $i \in \mathbb{N}$ calculate $\Delta^i m^n := (\underbrace{\Delta \circ \cdots \circ \Delta}_{i \text{ times}}) m^n$. [Hint: use the representation formula for ordinary powers in terms of descending factorial powers stated in Proposition 6.34.]
2. Calculate $\Delta^i \binom{m + k}{n}$.
3. Use the Worpistzky Identity (Proposition 5.84) to conclude.

Problem 6.10 For every $n \in \mathbb{N}$ and $s \in \mathbb{R}$, $s > 1$ let

$$H_n^{(s)} = 1 + \frac{1}{2^s} + \cdots + \frac{1}{n^s}.$$

Prove that

$$H_n^{(s)} - 1 \le \int_1^n \frac{1}{x^s} \, dx \le H_n^{(s)} - \frac{1}{n^s};$$

deduce that

$$1 + \frac{1}{2^s} + \cdots + \frac{1}{n^s} \le \frac{s}{s - 1} \qquad \forall s > 1. \tag{6.10.a}$$

Chapter 7
Formal Power Series

Abstract We begin here the subject of formal power series, objects of the form $\sum_{n=0}^{\infty} a_n X^n$ ($a_n \in \mathbb{R}$ or $\mathbb{C}$) which can be thought as a generalization of polynomials. We focus here on their algebraic properties and basic applications to combinatorics. The reader must not be confused by the many technical, though simple, details that are needed in a book to justify rigorously every step. In Chap. 3, we have learned how to count sequences and collections with occupancy constraints: the number of possible codes of 10 digits that use only four 1's, five 2's and one 3 is easily obtained: $\frac{10!}{4!5!}$. What about counting the possible codes of 10 digits that use an even number of 1's, an odd number of 2's and any number of 3's? What about the validity of the "Latin teacher's random choice" which selects the students to test by opening randomly a book and summing up the digits of the page? Many counting problems can be solved using the formal power series! These are extremely useful in studying recurrences, notable sequences, probability and many other arguments. Moreover, they constitute a rich and interesting environment in their own right. The chapter ends with a combinatorial proof of the celebrated Euler pentagonal theorem.

7.1 Formal Power Series: Basic Definitions

Formal power series generalize the notion of polynomial in one indeterminate.

Definition 7.1 A **formal power series** in the indeterminate X with real coefficients is an expression of the form

$$A(X) = a_0 + a_1 X + a_2 X^2 + \cdots + a_n X^n + \cdots = \sum_{n=0}^{\infty} a_n X^n$$

© Springer International Publishing Switzerland 2016
C. Mariconda and A. Tonolo, *Discrete Calculus*,
UNITEXT - La Matematica per il 3+2 103, DOI 10.1007/978-3-319-03038-8_7

where the **coefficients** a_n, $n \in \mathbb{N}$, **of the series** belong to $\mathbb{R}$. We use:

- $[X^n]\, A(X)$ to denote the **coefficient** a_n of X^n in $A(X)$;
- $\left[X^{\leq n}\right] A(X)$ to denote the polynomial $a_0 + a_1 X + a_2 X^2 + \cdots + a_n X^n$;
- $\left[X^{>n}\right] A(X)$ to denote $A(X) - \left[X^{\leq n}\right] A(X)$.

The element $a_0 = \left[X^0\right] A(X)$ is called the **constant term** of $A(X)$. Two formal power series $A(X)$ and $B(X)$ are **equal** if and only if $[X^n]\, A(X) = [X^n]\, B(X)$ for all $n \in \mathbb{N}$. □

Remark 7.2 Extending the notation introduced in Definition 7.1, for each $0 \neq a, b \in \mathbb{R}$, it will be useful to denote by $\left[\dfrac{aX^n}{b}\right] A(X)$ the coefficient of $\dfrac{aX^n}{b}$ in the formal series $A(X)$. Observe that

$$\left[\frac{aX^n}{b}\right] A(X) = [X^n]\,\frac{bA(X)}{a}.$$

The set of all the formal power series with real coefficients is denoted by the symbol $\mathbb{R}[[X]]$. Similarly, the symbol $\mathbb{C}[[X]]$ denotes the set of formal power series with complex coefficients. Most of the results do not depend on choosing $\mathbb{R}$ or $\mathbb{C}$ as set of coefficients. Nevertheless we will concentrate on formal power series with real coefficients.

Example 7.3 The polynomials in the indeterminate X are particular examples of formal power series: indeed, they are precisely the formal power series having only a finite number of non-zero coefficients. In particular, the real numbers themselves are polynomials, and thus *constant* formal power series in $\mathbb{R}[[X]]$. □

Remark 7.4 It will occasionally be useful to note the obvious fact that if $A(X)$ is a formal power series, then for each $0 \leq i \leq n$ one has

$$\left[X^i\right] A(X) = \left[X^i\right] \left(\left[X^{\leq n}\right] A(X)\right).$$

This reduces the calculation of the coefficient of X^i in formal power series to that of the degree i term in the polynomial obtained by truncating the formal power series at a degree $n \geq i$.

7.1.1 Sums, Products, and Derivatives

It is possible to define operations of *sum* and *product* on $\mathbb{R}[[X]]$ which are natural extensions of the corresponding operations for polynomials.

Definition 7.5 (*Sum and product of two formal power series*) Given $A(X) = \sum_{n=0}^{\infty} a_n X^n$ and $B(X) = \sum_{n=0}^{\infty} b_n X^n$ in $\mathbb{R}[[X]]$, the **sum** and the **product** of $A(X)$ and $B(X)$ are defined respectively by

$$A(X) + B(X) := \sum_{n=0}^{\infty} (a_n + b_n) X^n$$

$$A(X)B(X) := a_0 b_0 + (a_0 b_1 + a_1 b_0) X + (a_0 b_2 + a_1 b_1 + a_2 b_0) X^2 + \cdots$$

$$= \sum_{n=0}^{\infty} \left(\sum_{i=0}^{n} a_i b_{n-i} \right) X^n.$$

The **additive inverse** of $A(X) = \sum_{n=0}^{\infty} a_n X^n$ is the formal power series

$$-A(X) := \sum_{n=0}^{\infty} (-a_n) X^n$$

which satisfies $A(X) + \left(- A(X) \right) = 0$.   □

It is evident that the sum and product of formal power series with real coefficients are *commutative operations*, in other words, that for all $A(X), B(X) \in \mathbb{R}[[X]]$ one has

$$A(X) + B(X) = B(X) + A(X) \quad \text{and} \quad A(X)B(X) = B(X)A(X).$$

☞ *Remark 7.6* If $A(X)$ and $B(X)$ are two formal power series, when there is no risk of confusion, we will write $[X^n]A(X)B(X)$, rather than the less ambiguous $[X^n]\left(A(X)B(X)\right)$, to denote the coefficient of X^n in the product $A(X)B(X)$.

The sum and product may be extended inductively in the obvious way to sums and products of a finite number of formal power series.

Definition 7.7 (*Sum and product of several formal power series*) Let $m \in \mathbb{N}$ and let

$$\sum_{n=0}^{\infty} a_n^{(1)} X^n, \ldots, \sum_{n=0}^{\infty} a_n^{(m)} X^n$$

be m formal power series. One sets

$$\sum_{n=0}^{\infty} a_n^{(1)} X^n + \cdots + \sum_{n=0}^{\infty} a_n^{(m)} X^n := \sum_{n=0}^{\infty} \left(a_n^{(1)} + \cdots + a_n^{(m)} \right) X^n \qquad (7.7.a)$$

$$\left(\sum_{n=0}^{\infty} a_n^{(1)} X^n \right) \cdots \left(\sum_{n=0}^{\infty} a_n^{(m)} X^n \right) := \sum_{n=0}^{\infty} c_n X^n \qquad (7.7.b)$$

$$\text{with } c_n = \sum_{\substack{k_1, \ldots, k_m \in \mathbb{N} \\ k_1 + \cdots + k_m = n}} a_{k_1}^{(1)} \cdots a_{k_m}^{(m)}, \quad n \in \mathbb{N}.$$

If $m = 1, 2, 3, \ldots$ we will denote the m **times iterated product** of $A(X)$ with itself by $A^m(X)$. We say that $A(X)$ is an m-th **root** of $B(X) \in \mathbb{R}[[X]]$ if $A^m(X) = B(X)$.
□

Remark 7.8 (**Changing the order of summation**) In a more concise fashion, one can express (7.7.a) as

$$\sum_{k=1}^{m} \left(\sum_{n=0}^{\infty} a_n^{(k)} X^n \right) = \sum_{n=0}^{\infty} \left(\sum_{k=1}^{m} a_n^{(k)} \right) X^n$$

and one speaks of a *change in the order of summation*, by analogy with the language used for finite sums. One should note that here we are dealing with a definition, whereas Proposition 6.4 regarding finite sums was a consequence of the properties of sums of real numbers.

The notion of the derivative of a formal power series will play an important role in what follows, and it is inspired by the derivative of polynomials too.

Definition 7.9 (*Derivative of a formal power series*) Given $A(X) = \sum_{n=0}^{\infty} a_n X^n$, we use $A'(X)$ (or $A(X)'$) to denote the series

$$A'(X) = \sum_{n=1}^{\infty} (n a_n) X^{n-1}$$

called the **derivative formal power series** of $A(X)$. We then use $A^{(2)}(X)$ (or $A''(X)$), $A^{(3)}(X), \ldots$ to denote the derivatives of the formal power series $A'(X), A''(X), \ldots$.
□

Example 7.10 One sees easily that for all $\ell, m \in \mathbb{N}$,

$$[X^m]A^{(\ell)}(X) = (m + \ell) \cdots (m + 1)a_{\ell+m} = \frac{(m + \ell)!}{m!}[X^{m+\ell}]A(X). \qquad (7.10.\text{a})$$

$\square$

The linearity of the derivative for formal power series is a straightforward verification.

Proposition 7.11 (Linearity of the derivative for formal power series) *Let $A(X)$, $B(X)$ be two formal power series. Then:*

1. $(A(X) + B(X))' = A'(X) + B'(X)$;
2. *For every $c \in \mathbb{R}$ one has* $(cA(X))' = cA'(X)$.

Proof. Set

$$A(X) = \sum_{n=0}^{\infty} a_n X^n, \quad B(X) = \sum_{n=0}^{\infty} b_n X^n.$$

Then one has

$$(A(X) + B(X))' = \sum_{n=1}^{\infty} n(a_n + b_n)X^{n-1}$$

$$= \sum_{n=1}^{\infty} na_n X^{n-1} + \sum_{n=1}^{\infty} nb_n X^{n-1}$$

$$= A'(X) + B'(X).$$

Moreover,

$$(cA(X))' = \sum_{n=1}^{\infty} n(ca_n)X^{n-1} = c\sum_{n=1}^{\infty} na_n X^{n-1} = cA'(X). \qquad \square$$

7.1.2 The Codegree of a Formal Power Series

In the setting of formal power series, the following notion, dual to that of the degree of a polynomial, is of particular interest. It indicates the first term of a formal power series which is *effectively present*.

Definition 7.12 The **codegree** of a non-zero power series $A(X)$ is the degree of the first non-zero term:

$$\mathrm{codeg}\,A(X) = \min\{m : [X^m]\,A(X) \neq 0\}.$$

We set, for convenience, $\mathrm{codeg}\,0 = +\infty$.   □

Example 7.13 The codegree of $\displaystyle\sum_{n=1}^{\infty} X^{3n}$ is 3.   □

Analogously to the degree for polynomials, the codegree of a product of formal power series is the sum of their codegrees; here, and in what follows, we set

$$a + \infty = +\infty \text{ for every } a \in \mathbb{N}.$$

Proposition 7.14 *Let $A_1(X)$ and $A_2(X)$ be two formal power series. Then one has*

$$\mathrm{codeg}\,\big(A_1(X)A_2(X)\big) = \mathrm{codeg}\,A_1(X) + \mathrm{codeg}\,A_2(X).$$

More generally, if $A_1(X), \ldots, A_n(X)$ are formal power series, then

$$\mathrm{codeg}\,\big(A_1(X)\cdots A_n(X)\big) = \mathrm{codeg}\,A_1(X) + \cdots + \mathrm{codeg}\,A_n(X).$$

In particular, for all $n \in \mathbb{N}$ one has $\mathrm{codeg}\,A^n(X) = n\,\mathrm{codeg}\,A(X)$.

Proof. The equalities are trivial if one of the power series is 0; we thus assume $A_i(X) \neq 0$ for all i. If $\mathrm{codeg}\,A_1(X) = \ell$ and $\mathrm{codeg}\,A_2(X) = m$ then

$$A_1(X) = \sum_{k=\ell}^{\infty} a_k^{(1)} X^k \quad \text{and} \quad A_2(X) = \sum_{k=m}^{\infty} a_k^{(2)} X^k$$

with $a_\ell^{(1)} \neq 0$ and $a_m^{(2)} \neq 0$. The first non-zero term in the product $A_1(X)A_2(X)$ is $a_\ell^{(1)} a_m^{(2)} X^{\ell+m}$. The remaining assertions are obtained immediately by induction on n.   □

The following properties are an easy consequence of Proposition 7.14.

Corollary 7.15 *If $A(X)$ and $B(X)$ are two formal power series, then:*

1. $\mathrm{codeg}\,A(X)B(X) = 0$ *if and only if* $\mathrm{codeg}\,A(X) = 0 = \mathrm{codeg}\,B(X)$;
2. $\mathrm{codeg}\,B^{(m)}(X) = \mathrm{codeg}\,B(X) - m$ *for all* $m \leq \mathrm{codeg}\,B(X)$;
3. $\mathrm{codeg}\,B(X) > i$ *if and only if* $[X^{\leq i}]\,B(X) = 0$.

7.2 Generating Series of a Sequence

In the sequel a sequence $(a_n)_{n \in \mathbb{N}}$ of real numbers will be denoted simply by $(a_n)_n$. Therefore $(a_n)_n$ with a double n will denote a sequence, whereas a_n with a single n will denote its n-th term.

Every formal power series with real coefficients $A(X) = \sum_{n=0}^{\infty} a_n X^n$ determines the sequence $(a_n)_n$ of its coefficients. To each sequence $(a_n)_n$ we can, however, associate several formal series.

7.2.1 OGF and EGF of a Sequence

There are various types of formal series associated to a sequence. In this section we study those which appear most frequently in applications.

Definition 7.16 Let $(a_n)_n$ be a sequence of real numbers.

1. The **ordinary generating formal series** or **OGF of the sequence** $(a_n)_n$ is the formal power series

$$\mathrm{OGF}(a_n)_n = \sum_{n=0}^{\infty} a_n X^n.$$

2. The **exponential generating formal series** or **EGF of the sequence** $(a_n)_n$ is the formal power series

$$\mathrm{EGF}(a_n)_n = \sum_{n=0}^{\infty} a_n \frac{X^n}{n!}.$$

Of course $a_n = [X^n] \, \mathrm{OGF}(a_n)_n$ and $\left[\dfrac{X^n}{n!} \right] \mathrm{EGF}(a_n)_n = a_n$. □

Remark 7.17 Note that EGF's are particular types of OGF's and vice versa:

$$\mathrm{EGF}(a_n)_n = \mathrm{OGF} \left(\frac{a_n}{n!} \right)_n, \quad \mathrm{OGF}(a_n)_n = \mathrm{EGF}(n! a_n)_n.$$

Indeed

$$\mathrm{OGF}\left(\frac{a_n}{n!}\right)_n = \sum_{n=0}^{\infty} \frac{a_n}{n!} X^n = \sum_{n=0}^{\infty} a_n \frac{X^n}{n!} = \mathrm{EGF}(a_n)_n \; ;$$

$$\mathrm{EGF}(n!a_n)_n = \sum_{n=0}^{\infty} n! a_n \frac{X^n}{n!} = \sum_{n=0}^{\infty} a_n X^n = \mathrm{OGF}(a_n)_n \; .$$

Example 7.18 Consider the sequence of factorials $(n!)_n$. Its ordinary generating formal series is

$$\mathrm{OGF}(n!)_n = \sum_{n=0}^{\infty} n! X^n,$$

while its exponential generating formal series is

$$\mathrm{EGF}(n!)_n = \sum_{n=0}^{\infty} n! \frac{X^n}{n!} = \sum_{n=0}^{\infty} X^n. \qquad \square$$

When applied to generating formal series, the operations of sum and product produce generating formal series associated to new sequences of real numbers.

Definition 7.19 Let $a^{(1)} = (a_n^{(1)})_n, \ldots, a^{(m)} = (a_n^{(m)})_n$ be sequences of real numbers.

1. The **convolution product** of $a^{(1)}, \ldots, a^{(m)}$ is the sequence $a^{(1)} \star \cdots \star a^{(m)}$ whose n-th term is defined by

$$\left(a^{(1)} \star \cdots \star a^{(m)}\right)_n = \sum_{\substack{k_1, \ldots, k_m \in \mathbb{N} \\ k_1 + \cdots + k_m = n}} a_{k_1}^{(1)} \cdots a_{k_m}^{(m)}, \quad n \in \mathbb{N}.$$

2. The **binomial convolution product** of $a^{(1)}, \ldots, a^{(m)}$ is the sequence $a^{(1)} \diamond \cdots \diamond a^{(m)}$ whose n-th term is defined by

$$\left(a^{(1)} \diamond \cdots \diamond a^{(m)}\right)_n = \sum_{\substack{k_1, \ldots, k_m \in \mathbb{N} \\ k_1 + \cdots + k_m = n}} \frac{n!}{k_1! \cdots k_m!} a_{k_1}^{(1)} \cdots a_{k_m}^{(m)}, \quad n \in \mathbb{N}. \qquad \square$$

Example 7.20 For $m = 2$, Definition 7.19 yields

$$\left(a^{(1)} \star a^{(2)}\right)_n = a_0^{(1)} a_n^{(2)} + a_1^{(1)} a_{n-1}^{(2)} + \cdots + a_n^{(1)} a_0^{(2)} = \sum_{i=0}^{n} a_i^{(1)} a_{n-i}^{(2)}$$

$$\left(a^{(1)} \diamond a^{(2)}\right)_n = \binom{n}{0} a_0^{(1)} a_n^{(2)} + \binom{n}{1} a_1^{(1)} a_{n-1}^{(2)} + \cdots + \binom{n}{n} a_n^{(1)} a_0^{(2)}$$

$$= \sum_{i=0}^{n} \binom{n}{i} a_i^{(1)} a_{n-i}^{(2)}.$$

For example, if $a^{(1)}$ and $a^{(2)}$ are both the constant sequence equal to 1, the n-th terms of $a^{(1)} \star a^{(2)}$ and $a^{(1)} \diamond a^{(2)}$ are

$$\left(a^{(1)} \star a^{(2)}\right)_n = n + 1 \quad \text{and} \quad (a^{(1)} \diamond a^{(2)})_n = \sum_{i=0}^{n} \binom{n}{i} = 2^n (\text{see Proposition 2.56}).$$

$\square$

The operations on sequences described above have their counterparts on the corresponding ordinary and exponential generating formal series.

Proposition 7.21 (Operations on OGF's and EGF's) *Let* $(a_n)_n$ *and* $a^{(1)} := \left(a_n^{(1)}\right)_n$, $\ldots, a^{(m)} := \left(a_n^{(m)}\right)_n$ *be sequences. Then the following properties hold:*

Sums:

1. $\mathrm{OGF}(a^{(1)}) + \cdots + \mathrm{OGF}(a^{(m)}) = \mathrm{OGF}\left(a^{(1)} + \cdots + a^{(m)}\right);$
2. $\mathrm{EGF}(a^{(1)}) + \cdots + \mathrm{EGF}(a^{(m)}) = \mathrm{EGF}\left(a^{(1)} + \cdots + a^{(m)}\right).$

Products:

3. $\mathrm{OGF}(a^{(1)}) \cdots \mathrm{OGF}(a^{(m)}) = \mathrm{OGF}\left(a^{(1)} \star \cdots \star a^{(m)}\right);$
4. $\mathrm{EGF}(a^{(1)}) \cdots \mathrm{EGF}(a^{(m)}) = \mathrm{EGF}\left(a^{(1)} \diamond \cdots \diamond a^{(m)}\right).$

Derivatives:

5. $(\mathrm{OGF}(a_n)_n)' = \mathrm{OGF}\left((n+1)a_{n+1}\right)_n;$
6. $(\mathrm{EGF}(a_n)_n)' = \mathrm{EGF}\left(a_{n+1}\right)_n.$

Proof. Points 1 and 2 are immediate consequences of the definition of sum for formal power series.

3. One has

$$\left(\sum_{n=0}^{\infty} a_n^{(1)} X^n\right) \cdots \left(\sum_{n=0}^{\infty} a_n^{(m)} X^n\right) = \sum_{n=0}^{\infty} c_n X^n = \mathrm{OGF}(c_n)_n$$

where $c_n = \displaystyle\sum_{\substack{k_1, \ldots, k_m \in \mathbb{N} \\ k_1 + \cdots + k_m = n}} a_{k_1}^{(1)} \cdots a_{k_m}^{(m)} = \left(a^{(1)} \star \cdots \star a^{(m)}\right)_n.$

4. By Point 3 and Remark 7.17 one has

$$\left(\sum_{n=0}^{\infty} a_n^{(1)} \frac{X^n}{n!}\right) \cdots \left(\sum_{n=0}^{\infty} a_n^{(m)} \frac{X^n}{n!}\right) = \sum_{n=0}^{\infty} c_n X^n = \mathrm{OGF}(c_n)_n = \mathrm{EGF}(n!c_n)_n$$

where $(c_n)_n = \left(\dfrac{a_n^{(1)}}{n!}\right)_n \star \cdots \star \left(\dfrac{a_n^{(m)}}{n!}\right)_n$. Therefore, for each $n \in \mathbb{N}$, one has

$$n!c_n = n! \sum_{\substack{k_1, \ldots, k_m \in \mathbb{N} \\ k_1 + \cdots + k_m = n}} \frac{a_{k_1}^{(1)}}{k_1!} \cdots \frac{a_{k_m}^{(m)}}{k_m!} = \sum_{\substack{k_1, \ldots, k_m \in \mathbb{N} \\ k_1 + \cdots + k_m = n}} \frac{n!}{k_1! \cdots k_m!} a_{k_1}^{(1)} \cdots a_{k_m}^{(m)}$$

$$= \left(a^{(1)} \diamond \cdots \diamond a^{(m)}\right)_n .$$

5. The derivative of $\mathrm{OGF}(a_n)_n$ is the formal power series

$$\left(\mathrm{OGF}(a_n)_n\right)' = \left(\sum_{n=0}^{\infty} a_n X^n\right)' = \sum_{n=1}^{\infty} n a_n X^{n-1} = \sum_{n=0}^{\infty} (n+1) a_{n+1} X^n.$$

6. The derivative of $\mathrm{EGF}(a_n)_n$ is the formal power series

$$\left(\mathrm{EGF}(a_n)_n\right)' = \left(\sum_{n=0}^{\infty} a_n \frac{X^n}{n!}\right)' = \sum_{n=1}^{\infty} n a_n \frac{X^{n-1}}{n!} = \sum_{n=0}^{\infty} a_{n+1} \frac{X^n}{n!}. \qquad \square$$

We now see a first application of formal power series to combinatorics.

Example 7.22   1. Consider the formal power series

$$A_1(X) = X^2 + X^3 + X^4, \quad A_2(X) = X^3 + X^4 + X^5 + X^6,$$

$$A_3(X) = X^4 + X^5 + X^6, \quad A_4(X) = X^2 + X^3 + X^4 + \cdots .$$

These are the OGF's of the sequences

$$a^{(1)} = (0, 0, 1, 1, 1, 0, 0, \ldots), \quad a^{(2)} = (0, 0, 0, 1, 1, 1, 1, 0, 0, \ldots),$$

$$a^{(3)} = (0, 0, 0, 0, 1, 1, 1, 0, 0, \ldots), \quad a^{(4)} = (0, 0, 1, 1, 1, 1, 1, \ldots).$$

By Point 3 of Proposition 7.21, the product $A_1(X) A_2(X) A_3(X) A_4(X)$ is the OGF of the convolution product $a^{(1)} \star a^{(2)} \star a^{(3)} \star a^{(4)}$ where

$$\left(a^{(1)} \star a^{(2)} \star a^{(3)} \star a^{(4)}\right)_n = \sum_{\substack{k_1, \ldots, k_4 \in \mathbb{N} \\ k_1 + \cdots + k_4 = n}} a_{k_1}^{(1)} a_{k_2}^{(2)} a_{k_3}^{(3)} a_{k_4}^{(4)}.$$

Now, since

$$a_{k_1}^{(1)} a_{k_2}^{(2)} a_{k_3}^{(3)} a_{k_4}^{(4)} = \begin{cases} 1 & \text{if } k_1 \in \{2, 3, 4\}, \ k_2 \in \{3, 4, 5, 6\}, \\ & \quad k_3 \in \{4, 5, 6\}, \ k_4 \geq 2 \\ 0 & \text{otherwise,} \end{cases}$$

the n-th term $(a^{(1)} \star a^{(2)} \star a^{(3)} \star a^{(4)})_n$ coincides with the sum of as many 1's as there are natural number solutions to the equation

$$k_1 + k_2 + k_3 + k_4 = n \text{ with } 2 \leq k_1 \leq 4, \ 3 \leq k_2 \leq 6, \ 4 \leq k_3 \leq 6, \ k_4 \geq 2.$$

In Example 4.14 the reader was asked to find the number of natural solutions of

$$a + b + c + d = 19 \text{ with } 2 \leq a \leq 4, \ 3 \leq b \leq 6, \ 4 \leq c \leq 6, \ d \geq 2.$$

By what we have noted above, that number is precisely the coefficient of X^{19} in the product $A_1(X)A_2(X)A_3(X)A_4(X)$:

$$[X^{19}](X^2 + X^3 + X^4)(X^3 + X^4 + X^5 + X^6)(X^4 + X^5 + X^6)(X^2 + X^3 + \cdots) =$$
$$= [X^{19}]X^{11}(1 + X + X^2)(1 + X + X^2 + X^3)(1 + X + X^2)(1 + X + X^2 + \cdots)$$
$$= [X^8](1 + X + X^2)(1 + X + X^2 + X^3)(1 + X + X^2)(1 + X + X^2 + \cdots)$$
$$= [X^8](1 + 4X + 10X^2 + 18X^3 + 26X^4 + 32X^5 + 35X^6 + 36X^7 + 36X^8 + \cdots).$$

Thus the desired number of solutions is 36.

2. Consider the formal power series

$$B(X) = \frac{X^2}{2!} + \frac{X^3}{3!} + \frac{X^4}{4!}.$$

Here one is dealing with the exponential generating formal series of the sequence $b = (0, 0, 1, 1, 1, 0, 0, \ldots)$. By Point 4 of Proposition 7.21, $B^4(X)$ is the EGF of the binomial convolution product $b \diamond b \diamond b \diamond b$ where

$$(b \diamond b \diamond b \diamond b)_n = \sum_{\substack{k_1, \ldots, k_4 \in \mathbb{N} \\ k_1 + \cdots + k_4 = n}} \frac{n!}{k_1! k_2! k_3! k_4!} b_{k_1} b_{k_2} b_{k_3} b_{k_4}.$$

Now, since

$$b_{k_1} b_{k_2} b_{k_3} b_{k_4} = \begin{cases} 1 & \text{if } k_1, \ k_2, \ k_3, \ k_4 \in \{2, 3, 4\} \\ 0 & \text{otherwise,} \end{cases}$$

by Theorem 3.7 the n-th term $(b \diamond b \diamond b \diamond b)_n$ coincides with the number of n-sequences in I_4 with occupancy sequence (k_1, k_2, k_3, k_4), where k_1, k_2, k_3, k_4 vary in $\{2, 3, 4\}$. In Example 3.14 the reader was asked to calculate in how many ways one could form a word of 10 letters using an alphabet $\{a, b, c, d\}$ if each letter is required to appear at least twice and at most 4 times. Here the answer to the question amounts precisely to the coefficient of $\dfrac{X^{10}}{10!}$ in $B^4(X)$:

$$
\left[\frac{X^{10}}{10!}\right]\left(\frac{X^2}{2!} + \frac{X^3}{3!} + \frac{X^4}{4!}\right)^4 = \left[\frac{X^{10}}{10!}\right] X^8 \left(\frac{1}{2!} + \frac{X}{3!} + \frac{X^2}{4!}\right)^4
$$

$$
= \left[\frac{X^2}{10!}\right]\left(\frac{1}{2!} + \frac{X}{3!} + \frac{X^2}{4!}\right)^4
$$

$$
= \left[\frac{X^2}{10!}\right]\left(4\left(\frac{1}{2!}\right)^3 \frac{X^2}{4!} + \binom{4}{2}\left(\frac{1}{2!}\right)^2 \left(\frac{X}{3!}\right)^2\right)
$$

$$
= [X^2]\, 10!\left(4\left(\frac{1}{2!}\right)^3 \frac{X^2}{4!} + \binom{4}{2}\left(\frac{1}{2!}\right)^2 \frac{X^2}{(3!)^2}\right)
$$

$$
= 4\frac{10!}{2!2!2!4!} + \binom{4}{2}\frac{10!}{2!2!3!3!} = 226\,800.
$$

3. We apply generating formal series to rediscover a formula proved in Proposition 2.56. Given $n \in \mathbb{N}$, we calculate the sum $\sum_{k=0}^{n}\binom{n}{k}^2$. Put $a_k = \binom{n}{k}$. Since $a_k = a_{n-k}$, one has

$$
\sum_{k=0}^{n}\binom{n}{k}^2 = \sum_{k=0}^{n} a_k a_{n-k};
$$

hence, setting $a := (a_n)_n$, one has $\sum_{k=0}^{n}\binom{n}{k}^2 = (a \star a)_n$. By Proposition 7.21, if $A(X)$ denotes the OGF of the sequence $a = (a_n)_n$, one has $(a \star a)_n = [X^n]A^2(X)$. Since

$$
A(X) = 1 + \binom{n}{1} X + \cdots + \binom{n}{n} X^n = (1 + X)^n,
$$

one finds that $A^2(X) = (1 + X)^{2n}$, and therefore $(a \star a)_n = [X^n]A^2(X) = \binom{2n}{n}$.

□

7.2.2 Basic Principle for Occupancy Problems

We seek to formalize what is suggested by Example 7.22 regarding the possible use of formal power series in combinatorics.

Definition 7.23 (*Characteristic OGF and EGF of a set*) Let E be a subset of $\mathbb{N}$.

1. The **characteristic OGF** of E is the formal power series $I_E^{\text{OGF}}(X)$ defined by

$$I_E^{\text{OGF}}(X) = \sum_{n \in E} X^n.$$

2. The **characteristic EGF** of E is the formal power series $I_E^{\text{EGF}}(X)$ defined by

$$I_E^{\text{EGF}}(X) = \sum_{n \in E} \frac{X^n}{n!}. \qquad \square$$

In Point 1 of Example 7.22 we made use of the characteristic OGF of the sets $\{2, 3, 4\}$, $\{3, 4, 5, 6\}$, $\{4, 5, 6\}$ and $\{2, 3, 4, 5, \ldots\}$, while in Point 2 we used the characteristic EGF of the set $\{2, 3, 4\}$.

Remark 7.24 For each subset $E \subseteq \mathbb{N}$, denoted by χ_E the characteristic function of E (see Definition 1.5), one has

$$I_E^{\text{OGF}}(X) = \sum_{n \in \mathbb{N}} \chi_E(n) X^n \quad \text{and} \quad I_E^{\text{EGF}}(X) = \sum_{n \in \mathbb{N}} \chi_E(n) \frac{X^n}{n!}.$$

We formulate now one of the fundamental links between generating formal series and combinatorics: occupancy problems with multiplicities in prescribed sets.

Example 7.25 Let $E_1 = \{0, 2, 4, \ldots\}$, $E_2 = \{3, 6, 9, \ldots\}$, $E_3 = \{0, 1, 10\}$. Then

$$(2, 2, 1, 2, 3, 1) \quad \text{and} \quad [1, 1, 2, 2, 2, 3]$$

are, respectively, a 6-sequence and a 6-collection of I_3 with $2 \in E_1$ repetitions of 1, $3 \in E_2$ repetitions of 2, $1 \in E_3$ repetition of 3. $\qquad \square$

Definition 7.26 Given n subsets $E_1, \ldots, E_n \subseteq \mathbb{N}$ and $k \geq 1$ we denote by

- $C(n, k; (E_1, \ldots, E_n))$: the number of k-collections of I_n with $k_1 \in E_1$ repetitions of $1, \ldots, k_n \in E_n$ repetitions of n;
- $S(n, k; (E_1, \ldots, E_n))$: the number of k-sequences of I_n with $k_1 \in E_1$ repetitions of $1, \ldots, k_n \in E_n$ repetitions of n.

We will often use the fact that, in view of Theorem 1.28, $C(n, k; (E_1, \ldots, E_n))$ coincides also with the n-compositions $(k_1, \ldots, k_n)$ of k with $k_1 \in E_1, \ldots, k_n \in E_n$. $\qquad \square$

Theorem 7.27 (Basic Principle for occupancy problems) *Let $k, n \in \mathbb{N}$ and $E_1, \ldots, E_n$ be subsets of $\mathbb{N}$.*

1. $\mathrm{OGF}\left(C(n, k; (E_1, \ldots, E_n))_k\right) = I_{E_1}^{\mathrm{OGF}}(X) \cdots I_{E_n}^{\mathrm{OGF}}(X)$; *more precisely one has*

$$C(n, k; (E_1, \ldots, E_n)) = \left[X^k\right] I_{E_1}^{\mathrm{OGF}}(X) \cdots I_{E_n}^{\mathrm{OGF}}(X) \quad \forall k \geq 0.$$

2. $\mathrm{EGF}\left(S(n, k; (E_1, \ldots, E_n))_k\right) = I_{E_1}^{\mathrm{EGF}}(X) \cdots I_{E_n}^{\mathrm{EGF}}(X)$; *more precisely one has*

$$S(n, k; (E_1, \ldots, E_n)) = \left[\frac{X^k}{k!}\right] I_{E_1}^{\mathrm{EGF}}(X) \cdots I_{E_n}^{\mathrm{EGF}}(X) \quad \forall k \geq 0.$$

Proof. 1. By Sect. 3.2.1 one has

$$C(n, k; (E_1, \ldots, E_n)) = \sum_{\substack{k_1 \in E_1, \ldots, k_n \in E_n \\ k_1 + \cdots + k_n = k}} C(n, k; (k_1, \ldots, k_n))$$

$$= \sum_{\substack{k_1 \in E_1, \ldots, k_n \in E_n \\ k_1 + \cdots + k_n = k}} 1.$$

In view of Definition 7.19 and Proposition 7.21 this number is the coefficient of X^k in the product $I_{E_1}^{\mathrm{OGF}}(X) \cdots I_{E_n}^{\mathrm{OGF}}(X)$ of the characteristic OGF's of the sets $E_1, \ldots, E_n$.
2. By Theorem 3.7 one has

$$S(n, k; (E_1, \ldots, E_n)) = \sum_{\substack{k_1 \in E_1, \ldots, k_n \in E_n \\ k_1 + \cdots + k_n = k}} S(n, k; (k_1, \ldots, k_n))$$

$$= \sum_{\substack{k_1 \in E_1, \ldots, k_n \in E_n \\ k_1 + \cdots + k_n = k}} \frac{k!}{k_1! \cdots k_n!}.$$

In view of Definition 7.19 and Proposition 7.21 this number is the coefficient of $\dfrac{X^k}{k!}$ in the product $I_{E_1}^{\mathrm{EGF}}(X) \cdots I_{E_n}^{\mathrm{EGF}}(X)$ of the characteristic EGF's of $E_1, \ldots, E_n$. $\qquad\square$

Example 7.28 In how many ways is it possible to distribute 18 mint gum drops to 6 youngsters, if Charlie must receive no more than 5 (≤ 5) and Al must receive at least 4 (≥ 4)?

Solution. Consider the sets $E_C = \{0, 1, \ldots, 5\}$, $E_A = \{4, 5, \ldots\}$, $\mathbb{N}$. The number of gum drops that can be given to Charlie belongs to E_C; the number that can be given to Al belongs to E_A; the other 4 youngsters can receive any number of gum drops. Each distribution of the gum drops thus corresponds to a 6-composition $(x_C, x_A, x_3, x_4, x_5, x_6)$ of 28 with $x_C \in E_C$, $x_A \in E_A$ and $x_i \in \mathbb{N}$, $i = 3, \ldots, 6$. The product

$$I_{E_C}^{\text{OGF}}(X) I_{E_A}^{\text{OGF}}(X) I_{\mathbb{N}}^{\text{OGF}}(X) I_{\mathbb{N}}^{\text{OGF}}(X) I_{\mathbb{N}}^{\text{OGF}}(X) I_{\mathbb{N}}^{\text{OGF}}(X) =$$

$$= I_{E_C}^{\text{OGF}}(X) I_{E_A}^{\text{OGF}}(X) I_{\mathbb{N}}^{\text{OGF}}(X)^4$$

is, by the Basic Principle 7.27, the OGF of the sequence $(C(6, k; (E_C, E_A, \mathbb{N}, \ldots, \mathbb{N},)))_k$ of 6-compositions of k with the restrictions that have been imposed. We must therefore determine

$$[X^{28}] I_{E_C}^{\text{OGF}}(X) I_{E_A}^{\text{OGF}}(X) I_{\mathbb{N}}^{\text{OGF}}(X)^4 =$$

$$= [X^{28}](1 + X + X^2 + \cdots + X^5)(X^4 + X^5 + X^6 + \cdots)(1 + X + X^2 + X^3 + \cdots)^4$$

$$= [X^{28}](1 + X + X^2 + \cdots + X^5)X^4(1 + X + X^2 + X^3 + \cdots)(1 + X + X^2 + X^3 + \cdots)^4$$

$$= [X^{24}](1 + X + X^2 + \cdots + X^5)(1 + X + X^2 + X^3 + \cdots)^5.$$

In Example 7.91 we will see how to complete the calculation by means of closed forms. For now, using a CAS for calculating the product of polynomials, one finds

$$[X^{24}](1 + X + X^2 + \cdots + X^5)(1 + X + X^2 + X^3 + \cdots)^5 =$$

$$= [X^{24}](1 + X + X^2 + \cdots + X^5)(1 + X + X^2 + X^3 + \cdots + X^{24})^5$$

$$= [X^{24}](1 + 6X + 21X^2 + \cdots + 85\,106X^{24} + \cdots + X^{125}).$$

There are, therefore, 85 106 ways to distribute the gum drops according to the given instructions.   □

Example 7.29 Calculate the number of possible codes of 10 digits that use only the digits 1, 2, and 3, and in which 1 appears at least 4 times but not more than 7 times, while 3 appears an even number of times.

Solution. Consider the sets $E_1 = \{4, 5, 6, 7\}$, $E_2 = \mathbb{N}$, $E_3 = \{0, 2, 4, 6, \ldots\}$. The number of 1's present in the code belongs to E_1; the number of 3's present in the code belongs to E_3; there can be an arbitrary number of 2's. Thus each such code corresponds to a 10-sequence of I_3 with occupancy (k_1, k_2, k_3) with $k_i \in E_i$, $i = 1, 2, 3$. By the Basic Principle 7.27 the product

$$I_{E_1}^{\text{EGF}}(X) I_{\mathbb{N}}^{\text{EGF}}(X) I_{E_3}^{\text{EGF}}(X)$$

is the EGF of the sequence $S(3, k; E_1, E_2, E_3)_k$. We must therefore determine

$$\left[\frac{X^{10}}{10!}\right] I_{E_1}^{\text{EGF}}(X) I_{\mathbb{N}}^{\text{EGF}}(X) I_{E_3}^{\text{EGF}}(X) =$$

$$= \left[\frac{X^{10}}{10!}\right]\left(\frac{X^4}{4!} + \cdots + \frac{X^7}{7!}\right)\left(1 + X + \frac{X^2}{2!} + \cdots\right)\left(1 + \frac{X^2}{2!} + \frac{X^4}{4!} + \cdots\right).$$

In Example 7.83, we will see how to complete the calculation. For the moment, using again a CAS for calculating the product of polynomials, one finds that

$$\left[\frac{X^{10}}{10!}\right]\left(\frac{X^4}{4!} + \cdots + \frac{X^7}{7!}\right)\left(1 + X + \frac{X^2}{2!} + \cdots\right)\left(1 + \frac{X^2}{2!} + \frac{X^4}{4!} + \cdots\right) =$$

$$= [X^{10}]10!\left(\frac{X^4}{4!} + \cdots + \frac{X^7}{7!}\right)\left(1 + X + \frac{X^2}{2!} + \cdots + \frac{X^{10}}{10!}\right)\left(1 + \frac{X^2}{2!} + \frac{X^4}{4!} + \cdots + \frac{X^{10}}{10!}\right)$$

$$= [X^{10}]10!\left(\frac{X^4}{4!} + 6\frac{X^5}{5!} + 37\frac{X^6}{6!} + \cdots + 12\,912\frac{X^{10}}{10!} + \cdots\right) = 12\,912.$$

There are, therefore, 12 912 codes of the desired type.   □

Example 7.30 (*The Latin teacher's random choice*) The Latin teacher wants to select a student to test in the class. The teacher randomly opens the book dedicated to the philosophical writings of Cicero,[1] sums up the digits of the number of the page and chooses the corresponding student from the alphabetical list. If the book has 999 pages, and the class has 27 students, determine for each student the probability of being chosen.

Solution. The pages of the book correspond to the 3-sequences of $I_{10} = \{0, 1, 2, \ldots,$ $9\}$, except the sequence $(0, 0, 0)$. The number of pages corresponding to the student k, $1 \le k \le 27$, coincides with the number $C(3, k; I_{10}, I_{10}, I_{10})$ of 3-compositions (k_1, k_2, k_3) of k with $k_1, k_2, k_3 \in I_{10} = \{0, 1, 2, \ldots, 9\}$. Consider the sample space $\Omega = I_{10}^3 \setminus \{(0, 0, 0)\}$. For each $k \in \{1, \ldots, 27\}$, the probability that student k will be the *unlucky* one equals

$$\frac{C(3, k; (I_{10}, I_{10}, I_{10}))}{|\Omega|} = \frac{C(3, k; (I_{10}, I_{10}, I_{10}))}{999}.$$

Thanks to the Basic Principle 7.27, we have

$$C(3, k; (I_{10}, I_{10}, I_{10})) = [X^k]\left(I_{I_{10}}^{\text{OGF}}(X)\right)^3.$$

Now

---

[1] Marcus Tullius Cicero (106 BC–43 BC).

$$(1 + X + X^2 + X^3 + X^4 + X^5 + X^6 + X^7 + X^8 + X^9)^3 =$$
$$= X^{27} + 3X^{26} + 6X^{25} + 10X^{24} + 15X^{23} + 21X^{22} + 28X^{21} + 36X^{20} + 45X^{19} +$$
$$+ 55X^{18} + 63X^{17} + 69X^{16} + 73X^{15} + 75X^{14} + 75X^{13} + 73X^{12} + 69X^{11} + 63X^{10} +$$
$$+ 55X^9 + 45X^8 + 36X^7 + 28X^6 + 21X^5 + 15X^4 + 10X^3 + 6X^2 + 3X + 1.$$

We see how unjust this method is: for instance student 11 is expected to be chosen 23 more times than student 1!   □

7.3 Infinite Sums of Formal Power Series

As surprising as it may sound, we will sometimes need to consider infinite sums of formal series.

7.3.1 Locally Finite Families of Formal Power Series

Let us recall how we sum formal power series. If $A_0(X), \ldots, A_N(X) \in \mathbb{R}[[X]]$, then the formal power series $A_0(X) + \cdots + A_N(X)$ is defined by setting

$$[X^n](A_0(X) + \cdots + A_N(X)) = [X^n]A_0(X) + \cdots + [X^n]A_N(X)$$

for all $n \in \mathbb{N}$. The coefficient of X^n in a *finite* sum of formal power series is thus the sum of their coefficients of X^n. It is precisely the *finiteness* of the sum of the coefficients of X^n (rather than the finiteness of the number of formal power series) that makes it possible to define $A_0(X) + \cdots + A_N(X)$.

Definition 7.31 We say that a countable family $\{A_i(X) : i \in \mathbb{N}\}$ of formal power series is **locally finite** if for each $n \in \mathbb{N}$ one has codeg $A_i(X) > n$ for all but a finite number of i, or equivalently,

$$\{i \in \mathbb{N} : [X^{\leq n}]A_i(X) \neq 0\} \text{ is finite } \forall n \in \mathbb{N}.$$ □

☞ *Remark 7.32* It is worth noticing that a countable family of formal power series is locally finite if and only if for each $n \in \mathbb{N}$ all but a finite number of power series have the term of degree n equal to zero. This is also equivalent to

$$\lim_{i \to +\infty} \text{codeg } A_i(X) = +\infty. \tag{7.32.a}$$

Indeed, (7.32.a) holds if and only if for each $n \in \mathbb{N}$ there exists $\mu_n \in \mathbb{N}$ such that codeg $A_i(X) > n$ for each $i \geq \mu_n$.

A particularly interesting case of a locally finite but infinite family of formal power series is illustrated in the following result.

Proposition 7.33 *Let $B(X)$ be a non-zero formal power series in $\mathbb{R}[[X]]$. The family $\{B^i(X) : i \in \mathbb{N}\}$ of the powers of the formal power series $B(X)$ is locally finite if and only if $B(X)$ has constant term equal to zero, that is, $[X^0] B(X) = 0$.*

Proof. Suppose that $[X^0] B(X) = 0$, i.e., codeg $B(X) > 0$. For each $i \in \mathbb{N}$, by Proposition 7.14, one has

$$\text{codeg } B^i(X) = i \text{ codeg } B(X) \geq i$$

for each $i \in \mathbb{N}$, and hence

$$\lim_{i \to +\infty} \text{codeg } B^i(X) = +\infty.$$

Then, (7.32.a) shows that the family $\{B^i(X) : i \in \mathbb{N}\}$ is locally finite.
Suppose now that $[X^0] B(X) \neq 0$. Then

$$\text{codeg } B^i(X) = i \text{ codeg } B(X) = 0 \quad \forall i,$$

therefore the family $\{B^i(X) : i \in \mathbb{N}\}$ is not locally finite.   □

Corollary 7.34 *Let $(a_i)_i$ be a sequence of real numbers and $B(X) \in \mathbb{R}[[X]]$. The family $\{a_i B^i(X) : i \in \mathbb{N}\}$ of formal power series is locally finite if and only if one of the following conditions holds:*

1. *All but a finite number of the a_i's are equal zero;*
2. *$[X^0]B(X) = 0$.*

Proof. If $a_i \neq 0$ for a finite number of $i \in \mathbb{N}$, then $\{a_i B^i(X) : i \in \mathbb{N}\}$ is a finite family of formal power series and hence clearly locally finite. Otherwise, $\{a_i B^i(X) : i \in \mathbb{N}\}$ is locally finite if and only if $\{B^i(X) : i \in \mathbb{N}\}$ is locally finite: we conclude by Proposition 7.33.   □

7.3.2 Infinite Sums of Formal Power Series

Using the notion of a locally finite family of formal power series, we can define the concept of an infinite sum of formal power series.

Definition 7.35 Let $\{A_i(X) : i \in \mathbb{N}\}$ be a locally finite family of formal power series. The **sum of the infinitely many formal power series** $A_i(X), i \in \mathbb{N}$, is defined to be the formal power series

$$\sum_{i=0}^{\infty} A_i(X) := \sum_{n=0}^{\infty}\left(\sum_{i=0}^{\infty}[X^n]A_i(X)\right)X^n.$$

In other words, the formal power series $\sum\limits_{i=0}^{\infty} A_i(X)$ is the formal power series whose coefficient of X^n is equal to $\sum\limits_{i=0}^{\infty}[X^n]A_i(X)$ (which is a finite sum by Definition 7.31). □

Remark 7.36 It is a simple exercise to check that the sum of a finite number $A_0(X), \ldots, A_N(X)$ of power series coincides with the infinite sum of the locally finite family of power series $\{A_0(X), \ldots, A_N(X), 0, 0, 0, \ldots\}$.

Remark 7.37 Let $0 \neq B(X) \in \mathbb{R}[[X]]$. If $[X^0]B(X) = 0$, and $(a_i)_i$ is a sequence of real numbers, then for each $i \in \mathbb{N}$, one has $\operatorname{codeg} a_i B^i(X) \geq i$ and so the coefficient of X^n of the formal power series $\sum\limits_{i=0}^{\infty} a_i B^i(X)$ is given by

$$[X^n]\left(\sum_{i=0}^{\infty} a_i B^i(X)\right) = \sum_{i=0}^{n} a_i\,[X^n]\,B^i(X).$$

Remark 7.38 The following *interchange of order of summations* for a locally finite family $\{A_i(X) : i \in \mathbb{N}\}$ is a consequence of the given definition:

$$\sum_{i=0}^{\infty}\sum_{n=0}^{\infty}\left([X^n]A_i(X)\right)X^n = \sum_{i=0}^{\infty} A_i(X) = \sum_{n=0}^{\infty}\sum_{i=0}^{\infty}\left([X^n]A_i(X)\right)X^n.$$

☞ *Remark 7.39* If the family $\{A_i(X) : i \in \mathbb{N}\}$ is locally finite, for each $n \in \mathbb{N}$ the sum of real numbers $\sum\limits_{i=0}^{\infty}[X^n]\,A_i(X)$ is indeed well defined, because it has a finite number of summands. One could, in fact, extend the definition of an infinite sum of formal power series from the case of a locally finite family to that of a family $\{A_i(X) : i \in \mathbb{N}\}$ of formal power series in which for each $n \in \mathbb{N}$ the numerical series $\sum\limits_{i=0}^{\infty}[X^n]\,A_i(X)$ converges (absolutely); here we choose to forego a deeper discussion of this approach since in the present text we shall have no need for such generality.

Example 7.40 Every formal power series $A(X) = a_0 + a_1 X + \cdots + a_n X^n + \cdots$ is the infinite sum of the monomials

$$A_0(X) = a_0, \ A_1(X) = a_1 X, \ A_2(X) = a_2 X^2, \ldots, \ A_n(X) = a_n X^n, \ldots.$$

The family $\{A_i(X) : i \in \mathbb{N}\}$ is obviously locally finite since $\operatorname{codeg} A_i(X) \geq i$ for all i and hence $\lim\limits_{i \to +\infty} \operatorname{codeg} A_i(X) = +\infty$.   □

7.4 Composition of Formal Power Series

The result obtained in Proposition 7.33 allows us to define the *composite of two formal power series*:

Definition 7.41 Let $A(X) = \sum\limits_{n=0}^{\infty} a_n X^n$ and $B(X)$ be two formal power series. If $A(X)$ is a polynomial or if the constant term $\left[X^0\right] B(X)$ of $B(X)$ is equal to zero, we define the **composite formal power series** of $A(X)$ with $B(X)$ to be the formal power series obtained by *formally* replacing each appearance of X in $A(X)$ with the formal power series $B(X)$:

$$A(B(X)) := \sum_{i=0}^{\infty} a_i B^i(X). \qquad\qquad □$$

Example 7.42 Let

$$A(X) = 1 + X + X^2 + \cdots + X^n + \cdots, \quad B(X) = X + X^3, \quad C(X) = 1 + X.$$

Since $B(X)$ and $C(X)$ are polynomials and $[X^0]B(X) = 0$, we can certainly define $B(A(X))$, $B(C(X))$, $C(A(X))$, $C(B(X))$ and $A(B(X))$. By contrast, it makes no sense to consider $A(C(X))$. Indeed, formally replacing each appearance of X in $A(X)$ with $C(X)$, one would obtain

$$1 + (1 + X) + (1 + X)^2 + \cdots$$

in which, for example, the "constant term" would involve summing an infinite number of 1's. Let us rather explicitly calculate

$$A(B(X)) = 1 + (X + X^3) + (X + X^3)^2 + \cdots + (X + X^3)^n + \cdots.$$

By definition this amounts to the formal power series $D(X) := d_0 + d_1 X + \cdots + d_n X^n + \cdots$ where for each n one has

$$d_n = [X^n] \left(1 + (X + X^3) + \cdots + (X + X^3)^n + \cdots \right)$$
$$= [X^n] \left(1 + (X + X^3) + \cdots + (X + X^3)^n \right).$$

☕ Fix n. For each $k \in \mathbb{N}$ one has

$$(X + X^3)^k = \sum_{i=0}^{k} \binom{k}{i} X^i X^{3(k-i)} = \sum_{i=0}^{k} \binom{k}{i} X^{3k-2i};$$

therefore $[X^n](X + X^3)^k$ is different from zero if and only if $3k - 2i = n$ for a suitable integer $i = 0, \ldots, k$. This means that $3k - n$ is even, and $0 \leq 3k - n \leq 2k$, so that $n/3 \leq k \leq n$. Specifically, one has

$$[X^n](X + X^3)^k = \begin{cases} \binom{k}{\dfrac{3k-n}{2}} & \text{if } 3k - n \text{ is even and } n/3 \leq k \leq n, \\ 0 & \text{otherwise.} \end{cases}$$

Therefore

$$d_n = \sum_{k=0}^{n} [X^n](X + X^3)^k = \sum_{\substack{n/3 \leq k \leq n \\ 3k - n \text{ even}}} [X^n](X + X^3)^k = \sum_{\substack{n/3 \leq k \leq n \\ 3k - n \text{ even}}} \binom{k}{\dfrac{3k-n}{2}}.$$

Thus, for example, one has

$$d_0 = \binom{0}{0} = 1, \ d_1 = \binom{1}{1} = 1, \ d_2 = \binom{2}{2} = 1, \ d_3 = \binom{1}{0} + \binom{3}{3} = 2, \ \ldots. \ \square$$

Example 7.43 (*The formal power series $A(cX)$*) Let $A(X) = \displaystyle\sum_{n=0}^{\infty} a_n X^n \in \mathbb{R}[[X]]$ and $c \in \mathbb{R}$. Since $[X^0](cX) = 0$ we may compose $A(X)$ with cX and thus obtain the formal power series

$$A(cX) = \sum_{n=0}^{\infty} a_n (cX)^n = \sum_{n=0}^{\infty} a_n c^n X^n.$$

In particular, for $c = -1$, one has

$$A(-X) := \sum_{n=0}^{\infty} (-1)^n a_n X^n. \qquad \square$$

We can now give the definition of odd and even formal power series.

Definition 7.44 (*Odd and even formal power series*) A formal power series $A(X)$ is said to be **even** if $A(-X) = A(X)$ and **odd** if $A(-X) = -A(X)$.   □

The even (resp. odd) formal power series are those that have non-zero coefficients at most for the terms of even (resp. odd) degree.

Proposition 7.45 (Coefficients of even/odd formal power series) *Let $A(X)$ be a formal power series.*

1. *$A(X)$ is even if and only if $[X^n]\,A(X) = 0$ for all odd n.*
2. *$A(X)$ is odd if and only if $[X^n]\,A(X) = 0$ for all even n.*

Proof. Put $A(X) = \displaystyle\sum_{n=0}^{\infty} a_n X^n$. One then has

$$A(-X) = \sum_{n=0}^{\infty}(-1)^n a_n X^n.$$

Consequently $A(X)$ is even if and only if $(-1)^n a_n = a_n$ for each n, or equivalently if and only if $a_n = 0$ for all odd n; $A(X)$ is odd if and only if $(-1)^n a_n = -a_n$ for all n, or equivalently if and only if $a_n = 0$ for all even n.   □

The following rule for "substitution of the X" in a product of formal power series holds:

Proposition 7.46 (Change of variable) *Let $A(X)$, $B(X)$, $C_1(X)$, $C_2(X)$ and $D(X)$ in $\mathbb{R}[[X]]$ with*

$$A(X) + B(X) = C_1(X) \quad and \quad A(X)B(X) = C_2(X).$$

If $A(X)$, $B(X)$ are polynomials or $[X^0]\,D(X) = 0$, then it is possible to replace X with $D(X)$ in the preceding equations, thereby obtaining

$$A(D(X)) + B(D(X)) = C_1(D(X)) \quad and \quad A(D(X))B(D(X)) = C_2(D(X)).$$

Proof. If $A(X)$, $B(X)$ are polynomials, then $A(D(X)) + B(D(X)) = C_1(D(X))$ and $A(D(X))B(D(X)) = C_2(D(X))$ are consequences of the properties of sums and products in $\mathbb{R}[[X]]$.

Suppose now that $[X^0]\,D(X) = 0$: we set

$$A(X) = \sum_{i=0}^{\infty} a_i X^i, \quad B(X) = \sum_{j=0}^{\infty} b_j X^j, \quad C_1(X) = \sum_{m=0}^{\infty} c_m^{(1)} X^m, \quad C_2(X) = \sum_{m=0}^{\infty} c_m^{(2)} X^m.$$

By hypothesis one has $c_m^{(1)} = a_m + b_m$ and $c_m^{(2)} = \displaystyle\sum_{i=0}^{m} a_i b_{m-i}$. It is easy to check the first equality:

$$C_1(D(X)) = \sum_{m=0}^{\infty} c_m^{(1)} D^m(X) = \sum_{m=0}^{\infty} (a_m + b_m) D^m(X) = A(D(X)) + B(D(X)).$$

For the second, let us fix n in $\mathbb{N}$; we wish to show that

$$[X^n] (A(D(X))B(D(X))) = [X^n] (C_2(D(X))).$$

By Remark 7.37 one has

$$[X^n] C_2(D(X)) = \sum_{m=0}^{n} c_m^{(2)} [X^n] D^m(X) = \sum_{m=0}^{n} \sum_{i=0}^{m} a_i b_{m-i} [X^n] D^m(X). \quad (7.46.a)$$

We now examine $[X^n] (A(D(X))B(D(X)))$:

$$[X^n] (A(D(X))B(D(X))) = \sum_{k=0}^{n} [X^k] A(D(X)) [X^{n-k}] B(D(X)).$$

Since $k, n - k \leq n$ one has

$$[X^k] A(D(X)) = \sum_{i=0}^{n} a_i [X^k] D^i(X) \quad \text{and}$$

$$[X^{n-k}] B(D(X)) = \sum_{j=0}^{n} b_j [X^{n-k}] D^j(X).$$

Hence

$$[X^n] A(D(X))B(D(X)) = \sum_{k=0}^{n} \left(\sum_{i=0}^{n} a_i [X^k] D^i(X) \right) \left(\sum_{j=0}^{n} b_j [X^{n-k}] D^j(X) \right)$$

$$= \sum_{i=0}^{n} \sum_{j=0}^{n} a_i b_j \sum_{k=0}^{n} [X^k] D^i(X) [X^{n-k}] D^j(X)$$

$$= \sum_{i=0}^{n} \sum_{j=0}^{n} a_i b_j [X^n] D^{i+j}(X).$$

Bearing in mind that whenever $i + j$ is greater than n one has $[X^n] D^{i+j}(X) = 0$ and setting $m = i + j$, the latter equals

$$\sum_{i+j=0}^{n} a_i b_j [X^n] D^{i+j}(X) = \sum_{m=0}^{n} \sum_{i=0}^{m} a_i b_{m-i} [X^n] D^m(X).$$

We have thus proved that $A(D(X))B(D(X)) = C_2(D(X))$.   □

Locally finite families of formal power series behave well with respect to sums and products.

Proposition 7.47 *Let $C(X)$ be a formal power series, and $\{A_i(X) : i \in \mathbb{N}\}$, $\{B_i(X) : i \in \mathbb{N}\}$ two locally finite families of formal power series. Then $\{A_i(X) + B_i(X) : i \in \mathbb{N}\}$ and $\{C(X)A_i(X) : i \in \mathbb{N}\}$ are two locally finite families of power series and one has*

$$\sum_{i=0}^{\infty} A_i(X) + \sum_{i=0}^{\infty} B_i(X) = \sum_{i=0}^{\infty}(A_i(X) + B_i(X)) \qquad (7.47.a)$$

$$C(X)\left(\sum_{i=0}^{\infty} A_i(X)\right) = \sum_{i=0}^{\infty} C(X)A_i(X). \qquad (7.47.b)$$

Proof. Since for each $n \in \mathbb{N}$

$$[X^n](A_i(X) + B_i(X)) = [X^n]A_i(X) + [X^n]B_i(X)$$

one sees easily from the definition that $\{A_i(X) + B_i(X) : i \in \mathbb{N}\}$ is a locally finite family of formal power series. Next one has

$$\lim_{j \to +\infty} \mathrm{codeg}(C(X)A_j(X)) = \mathrm{codeg}\, C(X) + \lim_{j \to +\infty} \mathrm{codeg}\, A_j(X) = +\infty.$$

Therefore, by (7.32.a) $\{C(X)A_i(X) : i \in \mathbb{N}\}$ is also a locally finite family of formal power series. Finally, since for each $n \in \mathbb{N}$

$$[X^n]\left(\sum_{i=0}^{\infty} A_i(X)\right) + [X^n]\left(\sum_{i=0}^{\infty} B_i(X)\right) = \sum_{i=0}^{\infty}[X^n]A_i(X) + \sum_{i=0}^{\infty}[X^n]B_i(X)$$

$$= \sum_{i=0}^{\infty}[X^n](A_i(X) + B_i(X))$$

$$= [X^n]\sum_{i=0}^{\infty}(A_i(X) + B_i(X)) \quad \text{and}$$

$$[X^n]C(X)\left(\sum_{i=0}^{\infty}A_i(X)\right) = \sum_{k=0}^{n}[X^k]C(X)[X^{n-k}]\left(\sum_{i=0}^{\infty}A_i(X)\right)$$

$$= \sum_{k=0}^{n}[X^k]C(X)\sum_{i=0}^{\infty}[X^{n-k}]A_i(X)$$

$$= \sum_{i=0}^{\infty}\sum_{k=0}^{n}[X^k]C(X)[X^{n-k}]A_i(X)$$

$$= \sum_{i=0}^{\infty}[X^n]C(X)A_i(X) = [X^n]\sum_{i=0}^{\infty}C(X)A_i(X),$$

one obtains (7.47.a) and (7.47.b). □

7.5 Invertible Formal Power Series

Consider the formal power series $A(X) = 1 - X$ and $B(X) = \sum_{n=0}^{\infty}X^n$. One checks immediately that

$$A(X)B(X) = (1 - X)(1 + X + X^2 + \cdots)$$
$$= 1 + (1 - 1)X + (1 - 1)X^2 + \cdots = 1.$$

We will say that $A(X)$ is *invertible*, and that $B(X)$ is its *inverse*.

Definition 7.48 A formal power series $A(X)$ is said to be **invertible** if there exists $B(X)$ in $\mathbb{R}[[X]]$ such that $A(X)B(X) = 1$; if such a formal power series $B(X)$ exists, it is unique and is called **the inverse** of $A(X)$; we will denote it by $A^{-1}(X)$. □

Example 7.49 Let $A(X)$ be the formal power series

$$1 - X + X^2 - \cdots + (-1)^n X^n + \cdots.$$

Prove that $A(X) = 1 - XA(X)$; deduce that $A(X)$ is the inverse of $1 + X$. Use this to calculate the inverse of $1 - 3X$ as well.

Solution. One has

$$A(X) = 1 - X(1 - X + \cdots + (-1)^n X^n + \cdots) = 1 - XA(X).$$

Therefore $A(X)(1 + X) = 1$ from which it follows that $A(X) = (1 + X)^{-1}$. Since $[X^0](-3X) = 0$, on substituting $-3X$ for X in $A(X)(1 + X) = 1$ by Proposition 7.46 one obtains

$$1 = A(-3X)(1 + (-3X)) = A(-3X)(1 - 3X)$$

and thus $(1 - 3X)^{-1} = A(-3X) = 1 + 3X + 9X^2 + \cdots + 3^n X^n + \cdots$.   □

It is very easy to recognize invertible formal power series.

Proposition 7.50 *A formal power series in $\mathbb{R}[[X]]$ is invertible if and only if it has non-zero constant term, that is, if and only if its codegree is zero.*

Proof. The constant term of a product of two formal power series is the product of the constant terms of the factors; therefore, if a formal power series is invertible, then necessarily its constant term multiplied by the constant term of its inverse must give 1, and so certainly must be different from 0.

Conversely, let $A(X) = \sum_{n=0}^{\infty} a_n X^n \in \mathbb{R}[[X]]$ with $a_0 \neq 0$. We must establish the existence of another formal power series $B(X)$ which gives 1 when multiplied by $A(X)$. Put

$$B(X) = \sum_{n=0}^{\infty} b_n X^n.$$

The equation $A(X)B(X) = 1$ is then equivalent to the infinite set of equations

$$\begin{cases} a_0 b_0 = 1 \\ a_0 b_1 + a_1 b_0 = 0 \\ \cdots\cdots\cdots \\ a_0 b_n + a_1 b_{n-1} + \cdots + a_n b_0 = 0 \\ \cdots\cdots\cdots . \end{cases}$$

Now the first of these yields $b_0 = a_0^{-1}$; the second then yields

$$b_1 = -a_0^{-1}(a_1 b_0) = -a_0^{-1}(a_1 a_0^{-1}),$$

and so on: in general, once one has found $b_0, b_1, \ldots, b_{n-1}$ using the first n equations, from the $(n+1)$-th equation $a_0 b_n + a_1 b_{n-1} + \cdots + a_n b_0 = 0$ one immediately calculates the value of b_n:

$$b_n = -a_0^{-1}(a_1 b_{n-1} + \cdots + a_n b_0).$$

Therefore the equation $A(X)B(X) = 1$ has a unique solution $B(X)$, which is precisely the inverse of $A(X)$.   □

One obtains the following interesting result as an immediate consequence:

Corollary 7.51 *Every non-zero formal power series $A(X)$ is the product of a power of X times an invertible formal power series.*

Proof. Let $m = \operatorname{codeg} A(X)$. Then $A(X) = X^m C(X)$ where $C(X)$ is a formal power series with $\operatorname{codeg} C(X) = 0$. Proposition 7.50 then gives the desired conclusion.   □

Example 7.52 The formal power series $A(X) = 3 - X - X^2$ is clearly invertible in view of $[X^0] A(X) = 3 \neq 0$. Let us calculate its inverse $A^{-1}(X) = \sum_{n=0}^{\infty} b_n X^n$ using the method developed in the proof of Proposition 7.50. We must solve the following system of equations

$$\begin{cases} 3b_0 = 1 \\ 3b_1 - b_0 = 0 \\ 3b_2 - b_1 - b_0 = 0 \\ \quad \cdots\cdots \\ 3b_n - b_{n-1} - b_{n-2} = 0 \\ \quad \cdots\cdots \end{cases}$$

or equivalently

$$\begin{cases} b_0 = 1/3 \\ b_1 = b_0/3 \\ b_2 = (b_1 + b_0)/3 \\ \quad \cdots\cdots \\ b_n = (b_{n-1} + b_{n-2})/3 \\ \quad \cdots\cdots \end{cases}$$

It is easy to get $b_0 = 1/3$, $b_1 = 1/3^2$, $b_2 = (1/3^2 + 1/3)/3 = 4/3^3$, $b_3 = (3 + 4)/3^4 = 7/3^4$, $b_4 = (3 \times 4 + 7)/3^5 = 19/3^5$, $b_5 = (3 \times 7 + 19)/3^6 = 40/3^6, \ldots$ We will get a complete description of the coefficients of $A^{-1}(X)$ further ahead in Example 7.55. $\qquad\qquad\qquad\qquad\qquad\qquad\qquad\qquad\qquad\qquad\qquad\qquad\quad\square$

The existence of the inverse allows one to obtain the following important property with ease:

Proposition 7.53 (Vanishing of a product) *Let $A(X)$, $B(X)$ be two formal power series. Then $A(X)B(X) = 0$ if and only if $A(X) = 0$ or $B(X) = 0$.*

Proof. Suppose that $A(X) \neq 0$; by Corollary 7.51 we may write $A(X) = X^m C(X)$ for some $m \in \mathbb{N}$ and invertible $C(X)$. Therefore one has

$$0 = A(X)B(X) = X^m B(X)C(X);$$

multiplying both terms by the inverse of $C(X)$ one obtains $0 = X^m B(X)$, whence $[X^n] B(X) = [X^{n+m}] X^m B(X) = 0$ for all $n \in \mathbb{N}$ and hence $B(X) = 0$. $\qquad\square$

The following result furnishes an explicit formula for the inverse of a formal power series with non-zero constant term: observe that any invertible formal power series $A(X) = a_0 + a_1 X + \cdots$ can be written in the form $A(X) = a_0(1 - B(X))$ where $B(X) = 1 - a_0^{-1} A(X)$.

Proposition 7.54 (Inverse of a formal power series) *Let $B(X)$ be a formal power series with $[X^0]B(X) = 0$. Then*

$$(1 - B(X))^{-1} = \sum_{n=0}^{\infty} B^n(X).$$

Proof. The formal power series $B(X)$ has constant term $[X^0] B(X) = 0$; thus by Proposition 7.33 the family $\{B^i(X) : i \geq 0\}$ is locally finite. In Example 7.49 we obtained

$$(1 - X)^{-1} = 1 + X + X^2 + \cdots + X^n + \cdots;$$

replacing X with $B(X)$, in view of Proposition 7.46, we obtain:

$$(1 - B(X))^{-1} = 1 + B(X) + B^2(X) + \cdots + B^n(X) + \cdots. \qquad \square$$

Example 7.55 We again consider the formal power series $A(X) = 3 - X - X^2$ examined in Example 7.52. We now calculate its inverse using the method described in Proposition 7.54. Isolating the constant term 3 one has

$$A(X) = 3\left(1 - \frac{1}{3}(X + X^2)\right) = 3(1 - B(X))$$

with $B(X) = \dfrac{1}{3}(X + X^2)$. Hence one obtains

$$A^{-1}(X) = \frac{1}{3}(1 - B(X))^{-1} = \frac{1}{3} \sum_{n=0}^{\infty} B^n(X).$$

We now explicitly determine the coefficients of $A^{-1}(X)$. For each $n \in \mathbb{N}$ we have

$$B^n(X) = \frac{1}{3^n}(X + X^2)^n = \frac{1}{3^n} \sum_{k=0}^{n} \binom{n}{k} X^k (X^2)^{n-k}$$

$$= \frac{1}{3^n} \sum_{k=0}^{n} \binom{n}{k} X^{2n-k}.$$

Fix $m \in \mathbb{N}$. Then $[X^m] B^n(X) = 0$ for each $m < n$, and, if $m \geq n$,

$$[X^m] B^n(X) = [X^m] \frac{1}{3^n} \sum_{k=0}^{n} \binom{n}{k} X^{2n-k} = \frac{1}{3^n} \sum_{k=0}^{n} [X^m] \binom{n}{k} X^{2n-k}$$

$$= \frac{1}{3^n} \binom{n}{2n - m} = \frac{1}{3^n} \binom{n}{m - n}.$$

Therefore, we have

$$[X^m]A^{-1}(X) = [X^m]\frac{1}{3}\sum_{n=0}^{\infty}B^n(X) = \frac{1}{3}\sum_{n=0}^{m}[X^m]B^n(X)$$

$$= \frac{1}{3}\sum_{n=0}^{m}\frac{1}{3^n}\binom{n}{m-n} = \frac{1}{3^{m+1}}\sum_{n=0}^{m}3^{m-n}\binom{n}{m-n}.$$

Thus, for example,

$$[X^0]A^{-1}(X) = \frac{1}{3}\left(3^0\binom{0}{0}\right) = \frac{1}{3}$$

$$[X^1]A^{-1}(X) = \frac{1}{3^2}\left(3^1\binom{0}{1} + 3^0\binom{1}{0}\right) = \frac{1}{3^2}$$

$$[X^2]A^{-1}(X) = \frac{1}{3^3}\left(3^2\binom{0}{2} + 3^1\binom{1}{1} + 3^0\binom{2}{0}\right) = \frac{4}{3^3}. \qquad \square$$

It is natural to introduce the following notion in $\mathbb{R}[[X]]$.

Definition 7.56 (*Divisibility between formal power series*) A formal power series $B(X)$ **divides** a formal power series $A(X)$ in $\mathbb{R}[[X]]$ if and only if there exists a formal power series $C(X)$ such that $A(X) = B(X)C(X)$. $\qquad \square$

Remark 7.57 An invertible formal power series $B(X)$ divides any other formal power series. Indeed, given $A(X) \in \mathbb{R}[[X]]$ one has $A(X) = B(X)(B^{-1}(X)A(X))$.

Proposition 7.58 *Let $A(X)$ and $B(X)$ be two non-zero formal series. Then $B(X)$ divides $A(X)$ if and only if*

$$\text{codeg } B(X) \leq \text{codeg } A(X).$$

Proof. If $A(X) = B(X)C(X)$ for some formal power series $C(X)$, then

$$\text{codeg } B(X) \leq \text{codeg } B(X) + \text{codeg } C(X) = \text{codeg } A(X).$$

Conversely, assume $t := \text{codeg } B(X) \leq \text{codeg } A(X) := s$; by Corollary 7.51 one has $A(X) = X^s D(X)$ and $B(X) = X^t F(X)$ with $D(X)$ and $F(X)$ invertible formal power series. Since $s \geq t$, one has

$$A(X) = X^s D(X) = X^t X^{s-t} D(X) F(X) F^{-1}(X) = B(X)(X^{s-t}D(X)F^{-1}(X)). \square$$

7.6 Fractions of Formal Power Series and Applications

A rational fraction is given by the quotient of two rational integers, a numerator $a \in \mathbb{Z}$ and a denominator $b \in \mathbb{Z}\backslash\{0\}$. Two such fractions, for example $1/2$ and $2/4$, can represent the same rational number even though they have different numerators and denominators.

One easily convinces himself that two fractions a/b and c/d represent the same rational number if and only if $ad = bc$.

In a similar way it is possible to define *fractions* of formal power series.

7.6.1 Quotients of Formal Power Series

Let us extend the set $\mathbb{R}[[X]]$ of formal power series to the *larger* set of quotients of formal power series.

Definition 7.59 Given two formal power series $A(X)$ and $B(X) \neq 0$, we consider the symbol $\dfrac{A(X)}{B(X)}$. If $A_1(X), A_2(X), 0 \neq B_1(X), 0 \neq B_2(X)$ belong to $\mathbb{R}[[X]]$, we **identify** the symbols $\dfrac{A_1(X)}{B_1(X)}$ and $\dfrac{A_2(X)}{B_2(X)}$ whenever $A_1(X)B_2(X) = A_2(X)B_1(X)$ in $\mathbb{R}[[X]]$: we will say that they represent the same **fraction of formal power series**. The set of all fractions of formal power series is denoted by $\mathbb{R}((X))$.   □

Identifying each formal power series $A(X)$ with the fraction $\dfrac{A(X)}{1}$, the set $\mathbb{R}[[X]]$ of formal power series becomes a subset of $\mathbb{R}((X))$. We call **proper fractions of formal power series** the elements of $\mathbb{R}((X))$ which do not belong to $\mathbb{R}[[X]]$.

We can extend the operations of sum and product defined on $\mathbb{R}[[X]]$ to all of $\mathbb{R}((X))$ as follows:

Definition 7.60 (*Sum and product in* $\mathbb{R}((X))$) The **sum** and the **product** of two fractions of formal power series $\dfrac{A(X)}{B(X)}$ and $\dfrac{C(X)}{D(X)}$ are defined respectively by

$$\frac{A(X)}{B(X)} + \frac{C(X)}{D(X)} := \frac{A(X)D(X) + C(X)B(X)}{B(X)D(X)}$$
$$\frac{A(X)}{B(X)} \cdot \frac{C(X)}{D(X)} := \frac{A(X)C(X)}{B(X)D(X)}.$$

□

Note that the product of fractions of formal power series is well defined in view of Proposition 7.53: indeed, it guarantees that the product of the two denominators is never zero.

Remark 7.61 These definitions of the sum and product of fractions of formal power series effectively extend the corresponding definitions for the sum and product of power series: indeed,

$$A(X) + C(X) = \frac{A(X)}{1} + \frac{C(X)}{1} = \frac{A(X) \cdot 1 + C(X) \cdot 1}{1 \cdot 1} \quad \text{and}$$

$$A(X)C(X) = \frac{A(X)}{1} \cdot \frac{C(X)}{1} = \frac{A(X)C(X)}{1 \cdot 1}.$$

Remark 7.62 The motivation behind the construction of the fractions of formal power series is that in this larger setting every non-zero element is invertible: if $\dfrac{A(X)}{B(X)} \neq 0$, then $A(X) \neq 0$ and

$$\frac{A(X)}{B(X)} \frac{B(X)}{A(X)} = \frac{1}{1} = 1.$$

Thus X is not an invertible formal power series, but $X = \dfrac{X}{1}$ becomes invertible in $\mathbb{R}((X))$:

$$\frac{X}{1} \frac{1}{X} = \frac{1}{1} = 1.$$

On the other hand, if $A(X)$ is an invertible formal power series, then it is easy to verify that $\dfrac{A^{-1}(X)}{1}$ and $\dfrac{1}{A(X)}$ are the same fraction, and hence

$$A^{-1}(X) = \frac{1}{A(X)}.$$

The following result allows us to determine and give a clear description of the boundary between the set of formal power series and that of the proper fractions of power series.

Proposition 7.63 *A fraction* $\dfrac{A(X)}{B(X)}$ *of formal power series belongs to* $\mathbb{R}[[X]]$ *if and only if* $B(X)$ *divides* $A(X)$ *in* $\mathbb{R}[[X]]$, *i.e., if and only if*

$$\mathrm{codeg}\, B(X) \leq \mathrm{codeg}\, A(X).$$

If instead $\mathrm{codeg}\, B(X) > \mathrm{codeg}\, A(X)$, *then* $\dfrac{A(X)}{B(X)}$ *is a proper fraction of formal power series and one has*

$$\frac{A(X)}{B(X)} = \frac{C(X)}{X^{\mathrm{codeg}\, B(X) - \mathrm{codeg}\, A(X)}}$$

for a suitable invertible formal power series $C(X)$.

Proof. The fraction $\dfrac{A(X)}{B(X)}$ equals $C(X) \in \mathbb{R}[[X]]$ if and only if $A(X) = B(X)C(X)$, i.e., if and only if $B(X)$ divides $A(X)$; by Proposition 7.58 this is equivalent to the condition on the codegrees. On the other hand, if $t := \mathrm{codeg}\, B(X) > \mathrm{codeg}\, A(X) := s$, then for $r = t - s$ one has, for suitable invertible series D(X) and F(X),

$$\frac{A(X)}{B(X)} = \frac{X^s D(X)}{X^t F(X)} = \frac{C(X)}{X^r},$$

where $C(X) := D(X)F^{-1}(X)$ is invertible in as much as it is a product of two invertible formal power series. □

Example 7.64 By Preposition 7.63, for each $n \in \mathbb{N}$ the fraction $\dfrac{1 - X^{n+1}}{1 - X}$ belongs to $\mathbb{R}[[X]]$. It is a polynomial. Indeed, since $(1 + X + \cdots + X^n)(1 - X) = 1 - X^{n+1}$ one has □

$$\frac{1 - X^{n+1}}{1 - X} = \frac{1 + X + \cdots + X^n}{1} = 1 + X + \cdots + X^n.$$

Example 7.65 By Preposition 7.63, the fraction $\dfrac{1 - 2X - 3X^2}{1 + X}$ is a formal power series; let us find it. We have

$$
\begin{aligned}
\frac{1 - 2X - 3X^2}{1 + X} &= (1 - 2X - 3X^2)\frac{1}{1 + X} \\
&= (1 - 2X - 3X^2)\left(1 - X + X^2 - X^3 + \cdots\right) \\
&= \left(1 + (-2 - 1)X + (-3 + 2 + 1)X^2 + (3 - 2 - 1)X^3 + \cdots\right) \\
&= 1 - 3X.
\end{aligned}
$$
□

7.7 The Closed Forms of a Formal Power Series

In this section we introduce and develop the concept of a closed form for (or of) a formal power series. This notion will allow us to make significant simplifications in our calculations with series.

7.7.1 Maclaurin Formal Power Series and Closed Forms for Formal Power Series

We recall that a real-valued function f is said to be of class $\mathscr{C}^\infty$ around zero (in brief $f \in \mathscr{C}^\infty(0)$), if there exists a neighborhood of 0 in $\mathbb{R}$ on which the function is defined and admits derivatives of all orders.

Definition 7.66 Let f be a function in $\mathscr{C}^\infty(0)$. We use the symbol $f(X)$ to denote the **Maclaurin**[2] **series** of the function f, more specifically, $f(X)$ is the formal power series

---

[2]Colin Maclaurin (1698–1746).

$$f(X) := f(0) + \frac{f'(0)}{1!}X + \frac{f''(0)}{2!}X^2 + \cdots + \frac{f^{(n)}(0)}{n!}X^n + \cdots. \qquad \square$$

Remark 7.67 The Maclaurin formal power series $f(X)$ of a function $f \in \mathscr{C}^\infty(0)$ is the EGF of the sequence of the derivatives of f calculated at 0.

☞   One should avoid confusion of the symbol $f(X)$ which denotes the Maclaurin series of the function f with the symbol $f(x)$ which, as usual, denotes the value of the function f calculated at the real (or complex) number x.

Example 7.68   1. The Maclaurin formal power series of the polynomial function $f(x) = a_0 + a_1 x + \cdots + a_k x^k$ is the polynomial

$$f(X) = a_0 + a_1 X + \cdots + a_k X^k.$$

Indeed, one easily checks that

$$f(0) = a_0, \quad f'(0) = a_1, \quad \ldots, f^{(k)}(0) = k! a_k \text{ and } f^{(n)}(0) = 0 \; \forall n > k.$$

2. The Maclaurin formal power series of $f(x) = e^x$ is

$$e^X := 1 + X + \frac{X^2}{2!} + \cdots + \frac{X^n}{n!} + \cdots.$$

In fact, for each $n \in \mathbb{N}$ one has $f^{(n)}(x) = e^x$, and consequently $f^{(n)}(0) = 1$.

3. The Maclaurin formal power series of $f(x) = \dfrac{1}{1-x}$ is

$$\frac{1}{1-X} := 1 + X + X^2 + \cdots + X^n + \cdots.$$

Here, for each $n \in \mathbb{N}$ one has $f^{(n)}(x) = \dfrac{n!}{(1-x)^{n+1}}$ from which it follows that $f^{(n)}(0) = n!$. $\square$

Remark 7.69 The equality $\dfrac{1}{1-X} = 1 + X + X^2 + \cdots + X^n + \cdots$ can be read in several ways:

- $\displaystyle\sum_{i=0}^{\infty} X^i$ is the inverse of the formal power series $1 - X$;

- $\displaystyle\sum_{i=0}^{\infty} X^i$ is the Maclaurin formal power series of the polynomial function

$$x \mapsto \frac{1}{1-x};$$

- The fraction of formal power series $\dfrac{1}{1-X}$ coincides with $\displaystyle\sum_{i=0}^{\infty} X^i$.

Definition 7.70 We say that f in $\mathscr{C}^{\infty}(0)$ is a **closed form** for (or of) the power series $A(X)$ if $f(X) = A(X)$, that is, if $A(X)$ coincides with the Maclaurin formal power series of f.   □

☞ *Remark 7.71* A closed form of a formal power series is a function, NOT a formal power series! To say that a function f in $\mathscr{C}^{\infty}(0)$ is a closed form of $a_0 + a_1 X + \cdots + a_n X^n + \cdots$, that is, to write

$$f(X) = a_0 + a_1 X + \cdots + a_n X^n + \cdots$$

means precisely that

$$\forall n \in \mathbb{N} \quad a_n = [X^n](a_0 + a_1 X + \cdots + a_n X^n + \cdots) = \frac{f^{(n)}(0)}{n!}.$$

Example 7.72 (*Notable closed forms*) Here is a list of some important closed forms; they arise from the well-known Maclaurin series expansions of some elementary functions.

$$\frac{1}{1-X} = 1 + X + X^2 + \cdots + X^n + \cdots$$

$$e^X = 1 + X + \frac{X^2}{2!} + \cdots + \frac{X^n}{n!} + \cdots$$

$$\cos X = 1 - \frac{X^2}{2!} + \cdots + (-1)^n \frac{X^{2n}}{(2n)!} + \cdots$$

$$\sin X = X - \frac{X^3}{3!} + \cdots + (-1)^n \frac{X^{2n+1}}{(2n+1)!} + \cdots$$

$$\cosh X = \frac{e^X + e^{-X}}{2} = 1 + \frac{X^2}{2!} + \cdots + \frac{X^{2n}}{(2n)!} + \cdots$$

$$\sinh X = \frac{e^X - e^{-X}}{2} = X + \frac{X^3}{3!} + \cdots + \frac{X^{2n+1}}{(2n+1)!} + \cdots$$ □

Example 7.73 Consider the function $\log(1 + x)$, $x > -1$; for each $n \in \mathbb{N}_{\geq 1}$ its n-th derivative is

$$(-1)^{n+1} \frac{(n-1)!}{(1+x)^n}.$$

Thus one has

$$\log(1 + X) = \sum_{n=1}^{\infty} (-1)^{n+1} \frac{X^n}{n}. \qquad \square$$

Example 7.74 Determine a closed form for

$$X + \frac{X^2}{2} + \cdots + \frac{X^n}{n} + \cdots .$$

Solution. For $|x| < 1$ the function $\log(1 + x)$ is of class $\mathscr{C}^\infty$; one has

$$\log(1 + X) = X - \frac{X^2}{2} + \cdots + (-1)^{n+1} \frac{X^n}{n} + \cdots$$

from which it follows that

$$- \log(1 - X) = X + \frac{X^2}{2} + \cdots + \frac{X^n}{n} + \cdots$$

and so $x \mapsto -\log(1 - x)$ is a closed form for the formal power series under consideration. $\qquad \square$

7.7.2 Properties of Closed Forms

Knowing a closed form for a formal power series is important in applications of the theory since the derivatives of the closed form furnish its coefficients. It is clear that with a CAS it is possible to find, if not a general formula, at least any desired number of coefficients of the series.

It follows from the definition that, given a function $f \in \mathscr{C}^\infty(0)$, there is exactly one formal power series of which f is a closed form.

☞ *Example 7.75* One should beware, however, the fact that, conversely, *a formal power series does not uniquely determine a closed form.* Two functions in $\mathscr{C}^\infty(0)$ all of whose corresponding derivatives have the same value at 0 yield the same Maclaurin formal power series, and so are closed forms of the same formal power series. For example, the function $f : \mathbb{R} \to \mathbb{R}$ defined by setting

$$f(x) = \begin{cases} e^{-1/x^2} & \text{if } x \neq 0, \\ 0 & \text{otherwise} \end{cases}$$

is of class $\mathscr{C}^\infty$ and all its derivatives at 0 are equal to 0. Therefore, not only the identically zero function, but the function f also are closed forms for the zero power series. $\qquad \square$

In the foregoing example we have seen that the zero power series admits a non-zero closed form; the presence of such "spurious" closed forms spoils the uniqueness of the closed form of a given formal power series.

Proposition 7.76 *Two function in $\mathscr{C}^\infty(0)$ are closed forms of the same formal power series if and only if their difference is a closed form of the zero formal power series.*

Proof. If f and g are closed forms of a formal power series $A(X) = \sum_{n=0}^{\infty} a_n X^n$, then for each $n \in \mathbb{N}$ one has

$$(f - g)^{(n)}(0) = f^{(n)}(0) - g^{(n)}(0) = n!(a_n - a_n) = 0$$

and so $f - g$ is a closed form of the zero formal power series.   □

Theorem 7.77 *Every formal power series $A(X)$ admits a closed form. If, moreover, $r = \operatorname{codeg} A(X)$, then every closed form for $A(X)$ is of the type $x \mapsto x^r h(x)$ with h in $\mathscr{C}^\infty(0)$, and $h(0) \neq 0$.*

Proof. Given the formal power series $A(X) = \sum_{k=0}^{\infty} a_k X^k$, by a result of Émile Borel[3] [30, pp. 300–301] there exists a function in $\mathscr{C}^\infty(0)$ such that $f^{(n)}(0) = n!a_n$ for each $n \in \mathbb{N}$: hence one has $f(X) = A(X)$. If $\operatorname{codeg} A(X) = r$, then the derivatives $f^{(k)}(0)$ are zero for $k = 0, \ldots, r - 1$, while $f^{(r)}(0) \neq 0$. Suppose that f is of class $\mathscr{C}^\infty$ on $]-\delta, \delta[$, with $\delta > 0$. Taylor's formula with remainder in integral form (see [2, Theorem 7.6]) then yields

$$f(x) = f(0) + f'(0)x + \cdots + \frac{f^{(r-1)}(0)}{(r-1)!}x^{r-1} + \frac{1}{(r-1)!}\int_0^x (x-t)^{r-1} f^{(r)}(t)\, dt$$

$$= \frac{1}{(r-1)!}\int_0^x (x-t)^{r-1} f^{(r)}(t)\, dt \quad \forall |x| < \delta.$$

Putting $t = xu$ one obtains

$$f(x) = x^r \int_0^1 (1-u)^{r-1} f^{(r)}(xu)\, du.$$

Since f is of class $\mathscr{C}^\infty$, the function $h(x) := \int_0^1 (1-u)^{r-1} f^{(r)}(xu)\, du$ is, by well-known results on derivation under the integral symbol, of class $\mathscr{C}^\infty$ on $]-\delta, \delta[$; furthermore,

$$h(0) = \int_0^1 (1-u)^{r-1} f^{(r)}(0)\, du = -\frac{f^{(r)}(0)}{r}\left[(1-u)^r\right]_{u=0}^{u=1} = \frac{f^{(r)}(0)}{r} \neq 0,$$

---

[3]Émile Borel (1871–1956).

and $f(x) = x^r h(x)$ for each $x \in]-\delta, \delta[$. $\square$

Remark 7.78 We will see in Theorem 8.7 that whenever the formal power series *converges in some non-zero point*, then its *sum function* is one of its closed forms.

It is convenient to recall the definition of the extension by continuity of a function.

Definition 7.79 (*Extension by continuity*) Let f be a continuous function defined on $I \setminus \{x_1, \ldots, x_m\}$, where I is an interval and $x_1, \ldots, x_m \in I$. Assume that the limits $\ell_i := \lim_{x \to x_i} f(x)$, $i = 1, \ldots, m$, exist and are finite. The **extension by continuity** of f to I is the continuous function defined by

$$x \mapsto \begin{cases} f(x) \text{ if } x \in I \setminus \{x_1, \ldots, x_m\} \\ \ell_i \text{ if } x = x_i \ (i = 1, \ldots, m). \end{cases}$$
$\square$

The following result will allow us to make profitable use of the notion of closed form for a formal power series.

Proposition 7.80 (Conservation laws of closed forms) *Let f be a closed form for the formal power series $A(X)$ and g a closed form of the formal power series $B(X)$.*

1. *$f + g$ is a closed form for the formal power series $A(X) + B(X)$:*

$$(f + g)(X) = A(X) + B(X).$$

2. *fg is a closed form for the formal power series $A(X)B(X)$:*

$$(fg)(X) = A(X)B(X).$$

3. *If $B(X)$ is an invertible formal power series then $1/g$ is a closed form for $B^{-1}(X)$:*

$$\frac{1}{g}(X) = B^{-1}(X).$$

4. *If $A(X)/B(X)$ is a formal power series, then the **extension by continuity** of f/g is one of its closed forms:*

$$(f/g)(X) = A(X)/B(X).$$

5. *f' is a closed form for the formal power series $A'(X)$:*

$$f'(X) = A'(X).$$

6. *If $B(X) = A'(X)$, then the primitive of $g(x)$ which equals $[X^0]A(X)$ in 0 is a closed form for $A(X)$.*

7. If the composition $A(B(X))$ is defined, then $f \circ g$ is one of its closed forms:

$$(f \circ g)(X) = A(B(X)).$$

8. If $c \in \mathbb{R}$, then $f(cx)$ is a closed form for $A(cX)$.

Proof. We set

$$A(X) = \sum_{k=0}^{\infty} a_k X^k, \quad B(X) = \sum_{k=0}^{\infty} b_k X^k.$$

By the definition of closed forms one has

$$a_n = \frac{f^{(n)}(0)}{n!}, \quad b_n = \frac{g^{(n)}(0)}{n!} X^n \quad \forall n \in \mathbb{N}.$$

Fix $n \in \mathbb{N}$. We proceed to verify the various points.

1. $\dfrac{(f+g)^{(n)}(0)}{n!} = \dfrac{f^{(n)}(0) + g^{(n)}(0)}{n!} = a_n + b_n = [X^n]\,(A(X)+B(X)).$

2. Using the Leibniz formula for the n-th derivative of a product (see Theorem 3.26, (3.26.b)) one immediately obtains

$$\frac{(fg)^{(n)}(0)}{n!} = \frac{\sum_{k=0}^{n} \binom{n}{k} f^{(k)}(0)g^{(n-k)}(0)}{n!} = \sum_{k=0}^{n} \frac{f^{(k)}(0)g^{(n-k)}(0)}{k!(n-k)!}$$

$$= \sum_{k=0}^{n} \frac{f^{(k)}(0)}{k!} \frac{g^{(n-k)}(0)}{(n-k)!} = \sum_{k=0}^{n} a_k b_{n-k} = [X^n]\,(A(X)B(X)).$$

3. For as much as $g(0) = [X^0]\,B(X) \neq 0$, the function $1/g$ belongs to $\mathscr{C}^{\infty}(0)$. Since $(1/g)g = 1$ by Point 2 one has $(1/g)(X)g(X) = 1$, that is, $(1/g)(X)B(X) = 1$, so that $(1/g)$ is a closed form for $B^{-1}(X)$.

4. By Proposition 7.63 $A(X)/B(X)$ is a formal power series if and only if $s := \operatorname{codeg} A(X) \geq \operatorname{codeg} B(X) =: r$. By Theorem 7.77 one has $f(x) = x^s f_1(x)$ and $g(x) = x^r g_1(x)$ with $f_1, g_1 \in \mathscr{C}^{\infty}(0)$, $f_1(0) \neq 0$ and $g_1(0) \neq 0$. There exists a neighborhood U of zero such that for all $x \in U$ one has $g_1(x) \neq 0$ and hence $g(x) = x^r g_1(x) \neq 0$ for all $x \in U \backslash \{0\}$. Then the function $x \mapsto f(x)/g(x)$ is defined in $U \backslash \{0\}$ and

$$\lim_{x \to 0} \frac{f(x)}{g(x)} = \lim_{x \to 0} x^{s-r} \frac{f_1(x)}{g_1(x)}$$

exists and is finite. It is easy to verify that the extension by continuity h of f/g coincides on U with the function $x \mapsto x^{s-r} \dfrac{f_1(x)}{g_1(x)}$ and it is of class $\mathscr{C}^{\infty}$ in U. Then one has

$$h(X) = X^{s-r}\frac{f_1(X)}{g_1(X)} = \frac{f(X)}{g(X)} = \frac{A(X)}{B(X)}.$$

5. $\dfrac{(f')^{(n)}(0)}{n!} = \dfrac{f^{(n+1)}(0)}{n!} = (n+1)\dfrac{f^{(n+1)}(0)}{(n+1)!} = (n+1)\left[X^{n+1}\right]A(X)$ which coincides with $[X^n]\,A'(X)$ by (7.10.a).

6. One has $B(X) = A'(X) = \displaystyle\sum_{k=1}^{\infty} ka_k X^{k-1}$. Since $g(x)$ is a closed form for $B(X) =$

$A'(X)$, for each $k \geq 1$ the identity $ka_k = \dfrac{g^{(k-1)}(0)}{(k-1)!}$ holds, and from this it follows

that $a_k = \dfrac{g^{(k-1)}(0)}{k!}$. Define $\ell(x) := \displaystyle\int_0^x g(x)\,dx + [X^0]A(X)$; one then has

$$\ell(0) = [X^0]A(X) \text{ and}$$

$$\ell^{(k)}(0) = g^{(k-1)}(0) = k!a_k = k![X^k]A(X),$$

so that ℓ is a closed form for $A(X)$.

7. Since $g(0) = [X^0]B(X) = 0$, clearly one has

$$(f \circ g)^{(0)}(0) = f(g(0)) = f(0) = [X^0]A(X) = [X^0]A(B(X)).$$

We now check that

$$\frac{(f \circ g)^{(n)}(0)}{n!} = [X^n]A(B(X))$$

for all $n \geq 1$. Faà di Bruno's chain rule (5.25.b) yields

$$\frac{(f \circ g)^{(n)}(0)}{n!} = \sum_{k=1}^{n} \frac{f^{(k)}(0)}{k!} \sum_{\substack{(n_1,\ldots,n_k)\,\in\,\mathbb{N}_{\geq 1} \\ n_1 + \cdots + n_k = n}} \frac{g^{(n_1)}(0)}{n_1!} \cdots \frac{g^{(n_k)}(0)}{n_k!}$$

$$= \sum_{k=1}^{n} a_k \sum_{\substack{(n_1,\ldots,n_k)\,\in\,\mathbb{N}_{\geq 1} \\ n_1 + \cdots + n_k = n}} b_{n_1} \cdots b_{n_k}. \qquad (7.80.a)$$

Now, since $b_0 = 0$, for each $k = 1, \ldots, n$ one has precisely

$$[X^n]B^k(X) = \sum_{\substack{(n_1,\ldots,n_k)\,\in\,\mathbb{N} \\ n_1 + \cdots + n_k = n}} b_{n_1} \cdots b_{n_k} = \sum_{\substack{(n_1,\ldots,n_k)\,\in\,\mathbb{N}_{\geq 1} \\ n_1 + \cdots + n_k = n}} b_{n_1} \cdots b_{n_k},$$

and thus from (7.80.a) we find that for each $n \geq 1$

$$\frac{(f \circ g)^{(n)}(0)}{n!} = \sum_{k=1}^{n} a_k [X^n] B^k(X) = \sum_{k=1}^{n} [X^n] \left(a_k B^k(X) \right)$$

$$= [X^n] \left(\sum_{k=0}^{n} a_k B^k(X) \right) = [X^n] \left(\sum_{k=0}^{\infty} a_k B^k(X) \right) = [X^n] A(B(X)).$$

8. Indeed, $A(cX) = \displaystyle\sum_{n=0}^{\infty} c^n a_n X^n$ while, for each n, the n-th derivative of the function $f(cx)$ is given by

$$\frac{d^n f(cx)}{dx^n}(0) = c^n f^{(n)}(0) = c^n a_n n! \,. \qquad \Box$$

☞ *Remark 7.81* Proposition 7.80 is of fundamental importance. In fact, it allows one to pass from the equality between sums, products, composites, and inverses of formal power series to an equality between the corresponding closed forms, and vice versa.

Example 7.82 Let

$$A(X) = 1 - X + X^2 - X^3 + \cdots, \quad \text{and} \quad B(X) = X + \frac{X^2}{2!} + \frac{X^3}{3!} + \cdots.$$

Given that $[X^0] B(X) = 0$, the formal power series $A(B(X))$ is well defined: let us calculate it. On setting

$$A(B(X)) = c_0 + c_1 X + \cdots + c_n X^n + \cdots$$

one has

$$c_n = [X^n] \sum_{k=0}^{n} (-1)^k (B(X))^k = \sum_{k=0}^{n} (-1)^k [X^n] (B(X))^k$$

$$= \sum_{k=0}^{n} (-1)^k [X^n] \left(X + \frac{X^2}{2!} + \cdots \right)^k = \sum_{k=0}^{n} (-1)^k [X^n] \left(X + \frac{X^2}{2!} + \cdots + \frac{X^n}{n!} \right)^k.$$

Proceeding in this way, the calculation of the coefficients c_n appears to be rather complicated. If, however, we use the concept of a closed form, the calculation turns out to be immediate. Indeed, the functions $\dfrac{1}{1+x}$ and $e^x - 1$ are closed forms for $A(X)$ and $B(X)$ respectively, and so

$$A(X) = \frac{1}{1+X} \quad \text{and} \quad B(X) = e^X - 1.$$

By Point 5 of Proposition 7.80, the function $\dfrac{1}{1 + (e^x - 1)} = e^{-x}$ is a closed form for $A(B(X))$ and so

$$A(B(X)) = e^{-X} = 1 - X + \frac{X^2}{2!} + \cdots + (-1)^{n+1}\frac{X^n}{n!} + \cdots$$

from which it follows immediately that $c_n = (-1)^{n+1}/n!$.   □

Here are some examples illustrating the usefulness of the closed forms machinery in applications.

Example 7.83 We reconsider Example 7.29. To determine the number of possible 10-digit codes using the symbols 1, 2 and 3 in which 1 appears at least 4 times but not more than 7 times, while 3 appears an even number of times, we were led to calculate the coefficient of $X^{10}/10!$ in the product of the characteristic EGF's of the sets $E_1 = \{4, 5, 6, 7\}$, $\mathbb{N}$, $E_3 = \{0, 2, 4, 6, \ldots\}$. Now by Example 7.72 and Point 2 of Proposition 7.80 one has

$$\left[\frac{X^{10}}{10!}\right](I_{E_1}^{\mathrm{EGF}}(X) I_{\mathbb{N}}^{\mathrm{EGF}}(X) I_{E_3}^{\mathrm{EGF}}(X)) =$$

$$= \left[\frac{X^{10}}{10!}\right]\left(\frac{X^4}{4!} + \cdots + \frac{X^7}{7!}\right)\left(1 + X + \frac{X^2}{2!} + \cdots\right)\left(1 + \frac{X^2}{2!} + \frac{X^4}{4!} + \cdots\right)$$

$$= \left[\frac{X^{10}}{10!}\right]X^4\left(\frac{1}{4!} + \cdots + \frac{X^3}{7!}\right)e^X \cosh X$$

$$= \left[\frac{X^6}{10!}\right]\left(\frac{1}{4!} + \cdots + \frac{X^3}{7!}\right)\frac{e^{2X} + 1}{2}$$

$$= \frac{1}{2}\left[\frac{X^6}{10!}\right]\left(\frac{1}{4!}\frac{(2X)^6}{6!} + \frac{X}{5!}\frac{(2X)^5}{5!} + \frac{X^2}{6!}\frac{(2X)^4}{4!} + \frac{X^3}{7!}\frac{(2X)^3}{3!}\right)$$

$$= \frac{1}{2}\left[\frac{X^6}{10!}\right]\left(2^6\binom{10}{4}\frac{X^6}{10!} + 2^5\binom{10}{5}\frac{X^6}{10!} + 2^4\binom{10}{6}\frac{X^6}{10!} + 2^3\binom{10}{7}\frac{X^6}{10!}\right)$$

$$= \frac{1}{2}\left(2^6\binom{10}{4} + 2^5\binom{10}{5} + 2^4\binom{10}{6} + 2^3\binom{10}{7}\right) = 12\,912.\qquad□$$

Example 7.84 Let us determine the inverse in $\mathbb{R}[[X]]$ of the formal power series

$$e^X = 1 + X + \frac{X^2}{2!} + \cdots + \frac{X^n}{n!} + \cdots.$$

Given that $\dfrac{1}{e^x} = e^{-x}$, one deduces immediately from Proposition 7.80 (Points 3 and 8), that

$$e^{-X} = 1 - X + \frac{X^2}{2!} + \cdots + (-1)^{n+1}\frac{X^n}{n!} + \cdots$$

is the inverse in $\mathbb{R}[[X]]$ of e^X.   □

Proposition 7.85 *Let f be a closed form for a formal power series $A(X)$.*

1. If f is even then $A(X)$ is even.

2. If f is odd then $A(X)$ is odd.

Proof. By Point 8 of Proposition 7.80 we know that $f(-x)$ is a closed form for $A(-X)$. If f is even one has $f(x) = f(-x)$ for all x and so $A(X) = A(-X)$, that is, $A(X)$ is even; if f is odd one has $-f(x) = f(-x)$ for all x and thus $-A(X) = A(-X)$, that is, $A(X)$ is odd.   □

Example 7.86 Let us determine the coefficients of the formal power series $e^{X^2} \cos X$. Since the function $x \mapsto e^{x^2} \cos x$ is even, by Proposition 7.85 one has

$$[X^n]e^{X^2} \cos X = 0 \text{ for all odd } n.$$

Then $[X^{2n}]e^{X^2} \cos X = \sum_{i=0}^{2n}[X^i]e^{X^2}[X^{2n-i}] \cos X$; since both the functions $x \mapsto e^{x^2}$ and $x \mapsto \cos x$ are even, we have

$$[X^{2n}]e^{X^2} \cos X = \sum_{i=0}^{n}[X^{2i}]e^{X^2}[X^{2n-2i}] \cos X = \sum_{i=0}^{n} \frac{1}{i!}(-1)^{n-i} \frac{1}{(2n-2i)!}. \quad □$$

Definition 7.87 Let $(a_n)_n$ be a sequence.

1. The **sequence of the partial sums** of $(a_n)_n$ is the sequence

$$\left(\sum_{i=0}^{n} a_i \right)_n.$$

2. The **binomial transform** of $(a_n)_n$ is the sequence

$$\left(\sum_{i=0}^{n} \binom{n}{i} a_i \right)_n. \qquad\qquad □$$

The following result allows us to find the OGF of the sequence of partial sums and the EGF of the binomial transform of a sequence without difficulty.

Proposition 7.88 *Let $(a_n)_n$ be a sequence. Then*

$$\mathrm{OGF}\left(\sum_{i=0}^{n} a_i\right)_n = \frac{1}{1-X}\,\mathrm{OGF}(a_n)_n \quad \mathrm{EGF}\left(\sum_{i=0}^{n}\binom{n}{i} a_i\right)_n = e^X\,\mathrm{EGF}(a_n)_n\,.$$

Proof. It suffices to note that the n-th terms of the partial sums and of the binomial transform are respectively the n-th terms of the convolution product and of the binomial convolution product of the given sequence $(a_n)_n$ and the constant sequence $(1)_n$. Since $\dfrac{1}{1-X}$ and e^X are respectively the OGF and the EGF of the latter sequence, the conclusion follows immediately from Proposition 7.21. □

Example 7.89 Let us determine a closed form for the formal power series

$$A(X) = \sum_{n=0}^{\infty} \frac{n(n+1)}{2} X^n \quad \text{and} \quad B(X) = \sum_{n=0}^{\infty} \frac{(1-a)^n + (1+a)^n}{2} \frac{X^n}{n!}$$

where a is any real parameter. Since

$$1 + 2 + \cdots + n = \frac{n(n+1)}{2},$$

$A(X)$ is the OGF of the sequence of partial sums of $(n)_n$. Hence

$$A(X) = \frac{1}{1-X} \sum_{n=0}^{\infty} n X^n = \frac{X}{1-X} \sum_{n=1}^{\infty} n X^{n-1} = \frac{X}{1-X} \left(\sum_{n=0}^{\infty} X^n\right)'$$

$$= \frac{X}{1-X} \left(\frac{1}{1-X}\right)' = \frac{X}{(1-X)^3}.$$

Consider now the formal power series $B(X)$. One has

$$\frac{(1-a)^n + (1+a)^n}{2} = \frac{\displaystyle\sum_{i=0}^{n}\binom{n}{i}(-a)^i + \sum_{i=0}^{n}\binom{n}{i} a^i}{2} = \sum_{i=0}^{n}\binom{n}{i} c_i$$

where $(c_n)_n = (1, 0, a^2, 0, a^4, 0, a^6, 0, \ldots)$. Hence $B(X)$ is the EGF of the binomial transform of $(c_n)_n$. Since $\mathrm{EGF}(1, 0, a^2, 0, a^4, 0, a^6, 0, \ldots) = \cosh(aX)$, one concludes that $B(X) = e^X \cosh(aX)$. □

7.7.3 OGF of the Binomials

The binomials $\binom{n}{k}$ involve two indexes $k, n \in \mathbb{N}$. In this section we study the OGF of the sequences $\binom{n}{k}_n$ when k is a fixed natural number, and $\binom{n}{k}_k$ when n is a fixed natural number.

Proposition 7.90 (OGF of the binomial coefficients $\binom{n}{k}_n$) *Let $k \in \mathbb{N}$. Then*

$$\sum_{n=0}^{\infty} \binom{k+n}{n} X^n = \sum_{n=0}^{\infty} \binom{k+n}{k} X^n = \frac{1}{(1-X)^{k+1}}. \tag{7.90.a}$$

Hence,

$$\mathrm{OGF}\binom{n}{k}_n = \sum_{n=0}^{\infty} \binom{n}{k} X^n = \sum_{n=k}^{\infty} \binom{n}{k} X^n = \frac{X^k}{(1-X)^{k+1}}. \tag{7.90.b}$$

Proof. We prove formula (7.90.a) by induction on k. The assertion holds for $k = 0$: indeed in Example 7.68 we found that

$$\frac{1}{1-X} = 1 + X + X^2 + \cdots = \sum_{n=0}^{\infty} \binom{0+n}{n} X^n.$$

We now suppose that the assertion holds for $k \geq 0$, and prove that it then holds for $k + 1$. By Proposition 7.88 one knows that

$$\frac{1}{(1-X)^{(k+1)+1}} = \frac{1}{1-X} \frac{1}{(1-X)^{k+1}}$$

is the OGF of the partial sums of the sequence $\binom{k+i}{k}_{i \in \mathbb{N}}$. By Point 3 of Proposition 2.56

$$\sum_{i=0}^{n} \binom{k+i}{k} = \sum_{i=0}^{n} \binom{k+i}{i} = \binom{k+n+1}{n} = \binom{k+1+n}{k+1};$$

therefore

$$
\frac{1}{1-X}\frac{1}{(1-X)^{k+1}} = \frac{1}{1-X}\sum_{n=0}^{k}\binom{k+n}{k}X^{n}
$$

$$
= \sum_{n=0}^{\infty}\left(\sum_{i=0}^{n}\binom{k+i}{k}\right)X^{n}
$$

$$
= \sum_{n=0}^{\infty}\binom{k+1+n}{k+1}X^{n}
$$

completing the induction.

As to (7.90.b), on multiplying each term of (7.90.a) by X^{k} one has

$$
\frac{X^{k}}{(1-X)^{k+1}} = \sum_{n=0}^{\infty}\binom{k+n}{k}X^{k+n} = \sum_{n=k}^{\infty}\binom{n}{k}X^{n} = \sum_{n=0}^{\infty}\binom{n}{k}X^{n},
$$

which is the desired result.   □

Example 7.91 We reconsider Example 7.28, the problem of determining in how many ways it is possible to distribute 28 mint gum drops to 6 "kids" if Charlie must receive no more than 5 gum drops (≤ 5) while Al must receive no fewer than 4 gum drops (≥ 4). The solution of the problem was reduced to the calculation of the coefficient of X^{28} in the product of the characteristic OGF's of the sets $E_C = \{0, 1, \ldots, 5\}$, $E_A = \{4, 5, \ldots\}$, $\mathbb{N}$. Remembering that $\dfrac{1-X^{n+1}}{1-X} = 1 + \cdots + X^{n}$ for each $n \in \mathbb{N}$, one has

$$
[X^{28}]I_{E_C}^{OGF}(X)I_{E_A}^{OGF}(X)I_{\mathbb{N}}^{OGF}(X)^{4} =
$$

$$
= [X^{28}](1 + X + X^{2} + \cdots + X^{5})(X^{4} + X^{5} + X^{6} + \cdots)(1 + X + X^{2} + X^{3} + \cdots)^{4}
$$

$$
= [X^{28}]\frac{1-X^{6}}{1-X}X^{4}\frac{1}{1-X}\frac{1}{(1-X)^{4}} = [X^{28}]X^{4}\frac{1-X^{6}}{(1-X)^{6}}
$$

$$
= [X^{24}]\left(\frac{1}{(1-X)^{6}} - \frac{X^{6}}{(1-X)^{6}}\right) = [X^{24}]\frac{1}{(1-X)^{6}} - [X^{18}]\frac{1}{(1-X)^{6}};
$$

by Proposition 7.90 the latter equals

$$
\binom{5+24}{24} - \binom{5+18}{18} = \binom{29}{24} - \binom{23}{18} = 85\,106.
$$ □

In the next example we shall see how to use the relation between the derivative of a formal power series and the derivative of one of its closed forms.

Example 7.92 (*The derivative method*) Let us find a formula for the sum $s_n = 1^2 + 2^2 + \cdots + n^2$ of the first n squares, using the properties of formal power series.

Let $S(X)$ be the OGF of $(s_n)_n$; by Proposition 7.88 one has $S(X) = \dfrac{1}{1 - X} Q(X)$ where $Q(X)$ is the OGF of the sequence of squares of natural numbers. By Point 3 of Proposition 7.80, since $\dfrac{1}{1 - X} = \displaystyle\sum_{n=0}^{\infty} X^n$, the formal power series $\displaystyle\sum_{n=1}^{\infty} nX^{n-1}$ has as closed form the function $x \mapsto \left(\dfrac{1}{1 - x}\right)' = \dfrac{1}{(1 - x)^2}$:

$$\frac{1}{(1 - X)^2} = \left(\frac{1}{1 - X}\right)' = \sum_{n=1}^{\infty} nX^{n-1}.$$

Multiplying both terms by X one has

$$\frac{X}{(1 - X)^2} = \sum_{n=1}^{\infty} nX^n = \sum_{n=0}^{\infty} nX^n.$$

Taking the derivative once again one obtains

$$\left(\frac{X}{(1 - X)^2}\right)' = \sum_{n=1}^{\infty} n^2 X^{n-1}.$$

Finally, multiplying both sides of the last equation by X one has

$$X \left(\frac{X}{(1 - X)^2}\right)' = \frac{X^2 + X}{(1 - X)^3} = \sum_{n=1}^{\infty} n^2 X^n = \sum_{n=0}^{\infty} n^2 X^n = Q(X).$$

Therefore one finds that

$$s_n = [X^n]\, S(X) = [X^n]\, \frac{Q(X)}{1 - X} = [X^n]\, \frac{X^2 + X}{(1 - X)^4}.$$

On applying Proposition 7.90 one obtains

$$s_n = [X^n]\, \frac{X^2 + X}{(1 - X)^4} = [X^n]\, (X^2 + X) \sum_{i=0}^{\infty} \binom{3 + i}{i} X^i;$$

so that $s_0 = 0$, $s_1 = \dbinom{3}{0} = 1$ and, for $n \geq 2$,

$$s_n = \binom{3+n-2}{n-2} + \binom{3+n-1}{n-1}$$

$$= \frac{(n+1)n(n-1)}{6} + \frac{(n+2)(n+1)n}{6} = \frac{n(n+1)(2n+1)}{6}. \qquad \square$$

The OGF of the sequence $\binom{n}{k}_k$ is straightforward when n is a fixed natural number.

Proposition 7.93 (OGF of the binomial coefficients $\binom{n}{k}_k$) *Let $n \in \mathbb{N}$. Then*

$$\mathrm{OGF}\binom{n}{k}_k = (X+1)^n.$$

Proof. It is enough to remark that $(X+1)^n = \sum_{k=0}^{n} \binom{n}{k} X^k$. $\qquad \square$

Let us study a useful extension of the binomial symbol. In Chap. 2 we defined the symbol $\binom{a}{k}$ for $a, k \in \mathbb{N}$; we now allow a to vary over the entire set of real numbers: this will lead us to the binomial series.

Definition 7.94 (*The binomial "a choose k" with $a \in \mathbb{R}$.*) Let $a \in \mathbb{R}$ and $k \in \mathbb{N}$. The **binomial quotient (or symbol)** "a **choose** k" is the number

$$\binom{a}{k} = \begin{cases} \dfrac{a(a-1)\cdots(a-k+1)}{k!} & \text{if } k \geq 1, \\ 1 & \text{if } k = 0. \end{cases} \qquad \square$$

Example 7.95 Given $k \in \mathbb{N}$, let us compute $\binom{1/2}{k}$. If $k = 0$, by definition one has $\binom{1/2}{0} = 1$. Assume $k \geq 1$.

$$\binom{1/2}{k} = \frac{\frac{1}{2}\left(\frac{1}{2}-1\right)\left(\frac{1}{2}-2\right)\cdots\left(\frac{1}{2}-k+1\right)}{k!}$$

$$= \frac{\frac{1}{2}\left(-\frac{1}{2}\right)\left(-\frac{3}{2}\right)\cdots\left(-\frac{2k-3}{2}\right)}{k!}$$

$$= (-1)^{k-1}\frac{\frac{1}{2}\times\frac{1}{2}\times\frac{3}{2}\times\cdots\times\frac{2k-3}{2}}{k!} = (-1)^{k-1}\frac{1\times3\times5\times\cdots\times(2k-3)}{k!2^k}.$$

Multiplying both numerator and denominator by

$$2\times4\times\cdots\times(2k-2) = 2^{k-1}\times(1\times2\times\cdots\times(k-1)) = 2^{k-1}(k-1)!$$

we get

$$\binom{1/2}{k} = (-1)^{k-1}\frac{(2k-2))!}{k!(k-1)!2^k\times2^{k-1}}$$

$$= (-1)^{k-1}\frac{2}{4^k}\frac{1}{k}\binom{2(k-1)}{k-1}$$

$$= (-1)^{k-1}\frac{2}{4^k}\,\mathrm{Cat}_{k-1},$$

where $\mathrm{Cat}_{k-1} = \frac{1}{k}\binom{2(k-1)}{k-1}$ is the $(k-1)$-th Catalan number (see Definition 3.34).

□

Example 7.96 Let $j, k \in \mathbb{N}$. Then

$$\binom{-j}{k} = (-1)^k\binom{k+j-1}{k}.$$

Indeed

$$\binom{-j}{k} = \frac{-j(-j-1)\cdots(-j-(k-1))}{k!} = (-1)^k\frac{j(j+1)\cdots(j+k-1)}{k!}$$

$$= (-1)^k\frac{(k+(j-1))!}{k!(j-1)!} = (-1)^k\binom{k+j-1}{k}.$$

Notice in particular that

$$\binom{-1}{k} = (-1)^k \quad\text{and}\quad \binom{-2}{k} = (-1)^k(k+1).$$

□

Clearly, if $a \in \mathbb{N}$ then $\binom{a}{k} = 0$ if $k > a$. Otherwise, the following estimate at infinity holds.

Proposition 7.97 (Order of magnitude of the sequence $\binom{a}{k}_k$) *Let $a \in \mathbb{R}\backslash\mathbb{N}$. There is a constant*[4] $C_a > 0$ *such that*

$$\left|\binom{a}{k}\right| \sim \frac{C_a}{k^{a+1}} \quad k \to +\infty. \tag{7.97.a}$$

An elementary proof of the above proposition is provided in Sect. 7.7.5. A shorter proof of this result is obtained also in Example 12.75 using the asymptotic Euler-Maclaurin formula.

Remark 7.98 It follows from Proposition 7.97 that $\lim_{k \to +\infty} k^a \binom{a}{k} = 0$ if $a > -1$. Actually the direct proof of this fact is much simpler: indeed if $-1 < a < 0$ we have

$$\left|\binom{a}{k}\right| = \left|\frac{a(a-1)\cdots(a-k+1)}{k!}\right| = \left|\frac{a}{1}\frac{a-1}{2}\cdots\frac{a-k+1}{k}\right| < 1$$

and hence $k^a \left|\binom{a}{k}\right| < \frac{1}{k^{|a|}} \to 0$ as $k \to +\infty$. Otherwise, if $a \geq 0$ and $k > a+1$ we have

$$\left|\binom{a}{k}\right| = \left|\frac{a(a-1)\cdots(a-k+1)}{k!}\right|$$

$$= \frac{a(a-1)\cdots(a-[a])}{k(k-1)\cdots(k-[a])}\left|\frac{(a-[a]-1)\cdots(a-k+1)}{(k-[a]-1)\cdots 2\cdot 1}\right|$$

$$= \frac{a(a-1)\cdots(a-[a])}{k(k-1)\cdots(k-[a])}\left|\frac{(a-[a]-1)}{1}\right|\cdots\left|\frac{a-k+1}{k-[a]-1}\right|$$

$$= \frac{a(a-1)\cdots(a-[a])}{k(k-1)\cdots(k-[a])}\prod_{j=[a]+2}^{k}\frac{|a-j+1|}{j-[a]-1}.$$

Now, for $[a]+2 \leq j \leq k$,

$$\frac{|a-j+1|}{j-[a]-1} = \frac{j-a-1}{j-[a]-1} \leq 1,$$

and thus

---

[4] Actually, $C_a = \dfrac{1}{(|a|-1)!}$ for negative integers and Euler's formula for the Gamma function [38, 12.11] shows that $C_a = \dfrac{1}{\Gamma(-a)}$ for $a \notin \{0, -1, -2, \ldots\}$.

$$k^a \left| \binom{a}{k} \right| \le k^a \frac{a(a-1)\cdots(a-[a])}{k(k-1)\cdots(k-[a])}$$

$$= \frac{k^a}{k^{[a]+1}} \frac{k^{[a]+1}}{k(k-1)\cdots(k-[a])} a(a-1)\cdots(a-[a]).$$

We conclude since $\dfrac{k^{[a]+1}}{k(k-1)\cdots(k-[a])} \to 1$ and $\dfrac{k^a}{k^{[a]+1}} \to 0$ as $k \to +\infty$.

Proposition 7.99 (The binomial series) *Let $a \in \mathbb{R}$. The function $x \mapsto (1+x)^a$ is a closed form of the formal series*

$$\mathrm{OGF}\binom{a}{k}_k := 1 + aX + \binom{a}{2}X^2 + \cdots + \binom{a}{k}X^k + \cdots,$$

that is

$$(1+X)^a = 1 + aX + \binom{a}{2}X^2 + \cdots + \binom{a}{k}X^k + \cdots. \qquad (7.99.\mathrm{a})$$

Proof. For each $k \in \mathbb{N}$, the k-th derivative of the function $x \mapsto (1+x)^a$ is

$$a(a-1)\cdots(a-k+1)(1+x)^{a-k} = k!\binom{a}{k}(1+x)^{a-k},$$

which takes the value $k!\binom{a}{k}$ for $x = 0$.   □

Example 7.100 The explicit expansion of the formal series $(1+X)^{1/2}$ is often useful. The calculation done in Example 7.95 to compute $\binom{1/2}{k}$ and Proposition 7.99 imply

$$(1+X)^{1/2} = 1 + \sum_{k=1}^{\infty} (-1)^{k-1} \frac{2}{4^k} \mathrm{Cat}_{k-1} X^k. \qquad □$$

7.7.4 Second Degree Equations in $\mathbb{R}[[X]]$

When is a formal power series the square of another series? or of a fraction of formal power series?

Lemma 7.101 *A formal power series $A(X)$ is a square in $\mathbb{R}((X))$ if and only if it is a square in $\mathbb{R}[[X]]$. This happens if and only if $m := \operatorname{codeg} A(X)$ is even and $[X^m]A(X)$ belongs to $\mathbb{R}_{>0}$.*

Proof. Let us start proving that if $A(X)$ is a square in $\mathbb{R}((X))$, then it is a square also in $\mathbb{R}[[X]]$. If

$$A(X) = \left(\frac{F(X)}{G(X)}\right)^2,$$

then on putting $G(X) = X^m G_0(X)$ with $G_0(X)$ invertible in $\mathbb{R}[[X]]$, one has

$$A(X) = \left(\frac{F(X)}{X^m G_0(X)}\right)^2 = \frac{F^2(X)(G_0^{-1}(X))^2}{X^{2m}}.$$

Since $A(X)$ is a formal power series and $G_0^{-1}(X)$ is not divisible by X, necessarily X^{2m} must divide $F^2(X)$, so that X^m must divide $F(X)$. If we set $F_1(X) := \dfrac{F(X)}{X^m}$, then

$$A(X) = \left(\frac{F(X)}{G(X)}\right)^2 = (F_1(X)G_0^{-1}(X))^2$$

and so $A(X)$ must be a square in $\mathbb{R}[[X]]$.

Now, suppose that $A(X) = B^2(X)$ for some $B(X) \in \mathbb{R}[[X]]$ and $\operatorname{codeg} B(X) = \ell$. Then $m = \operatorname{codeg} A(X) = \operatorname{codeg} B^2(X) = 2\ell$ is even and $[X^m]A(X) = \left([X^\ell]B(X)\right)^2 > 0$.

Conversely, let $A(X) = X^{2m}C(X)$ with $c_0 := [X^0]C(X) \in \mathbb{R}_{>0}$. Then the function $f_{c_0} : x \mapsto \sqrt{c_0 + x}$ belongs to $\mathscr{C}^\infty(0)$; as usual, we use $f_{c_0}(X)$ to denote the formal Maclaurin series of f_{c_0}. Since $[X^0](C(X) - c_0) = 0$ we may consider the composite formal power series $f_{c_0}(C(X) - c_0)$. Since $f_{c_0}^2(C(X) - c_0) = c_0 + (C(X) - c_0) = C(X)$, one has

$$A(X) = X^{2m}C(X) = X^{2m}f_{c_0}^2(C(X) - c_0) = \left(X^m f_{c_0}(C(X) - c_0)\right)^2. \qquad \square$$

We can now take up the study of second degree equations in $\mathbb{R}[[X]]$:

$$A(X)\mathbb{Y}^2 + B(X)\mathbb{Y} + C(X) = 0.$$

Proposition 7.102 (Equations of second degree) *Let $A(X) \neq 0$, $B(X)$ and $C(X)$ be three formal power series. The equation*

$$A(X)\mathbb{Y}^2 + B(X)\mathbb{Y} + C(X) = 0$$

has a solution in the set of fractions of formal power series $\mathbb{R}((X))$ if and only if the **discriminant**

$$\Delta(X) = B^2(X) - 4A(X)C(X)$$

is a square in $\mathbb{R}[[X]]$. *In this case, on setting* $\Delta(X) = D^2(X)$, *the solutions may be expressed as*

$$\mathbb{Y}_{1,2} = \frac{-B(X) \pm D(X)}{2A(X)} \in \mathbb{R}((X)).$$

Proof. Exactly as it is done in handling ordinary algebraic equations of degree two, we can write that in $\mathbb{R}((X))$ one has

$$A(X)\mathbb{Y}^2 + B(X)\mathbb{Y} + C(X) = A(X)\left(\mathbb{Y}^2 + \frac{B(X)}{A(X)}\mathbb{Y} + \frac{C(X)}{A(X)}\right)$$

$$= A(X)\left(\left(\mathbb{Y} + \frac{B(X)}{2A(X)}\right)^2 - \frac{\Delta(X)}{(2A(X))^2}\right)$$

and so, by Proposition 7.53, $A(X)\mathbb{Y}^2 + B(X)\mathbb{Y} + C(X) = 0$ if and only if

$$\left(\mathbb{Y} + \frac{B(X)}{2A(X)}\right)^2 = \frac{\Delta(X)}{(2A(X))^2} \tag{7.102.a}$$

or, equivalently,

$$\Delta(X) = (2A(X)\mathbb{Y} + B(X))^2.$$

Therefore, in order for the equation to have a solution in $\mathbb{R}((X))$, the formal power series $\Delta(X)$ must be a square in $\mathbb{R}((X))$ and hence, by Lemma 7.101, a square in $\mathbb{R}[[X]]$. Finally, if $\Delta(X) = D^2(X)$ with $D(X) \in \mathbb{R}[[X]]$, then from (7.102.a) one deduces that

$$0 = \left(\mathbb{Y} + \frac{B(X)}{2A(X)}\right)^2 - \frac{D^2(X)}{(2A(X))^2}$$

$$= \left(\mathbb{Y} + \frac{B(X)}{2A(X)} - \frac{D(X)}{2A(X)}\right)\left(\mathbb{Y} + \frac{B(X)}{2A(X)} + \frac{D(X)}{2A(X)}\right)$$

from which it follows that $\mathbb{Y}_{1,2} = \dfrac{-B(X) \pm D(X)}{2A(X)}$. □

Example 7.103 The equation $(1 - X)\mathbb{Y}^2 + (2 + X^2)\mathbb{Y} + (1 - X^3) = 0$ has no solutions in $\mathbb{R}((X))$. Indeed, its discriminant

$$\Delta(X) = (2 + X^2)^2 - 4(1 - X)(1 - X^3) = 4X + 4X^2 + 4X^3 - 3X^4$$

has codegree 1 and so can not be a square in $\mathbb{R}[[X]]$: we conclude by Proposition 7.102. ☐

Example 7.104 Let us solve the equation

$$XY^2 - Y + 1 = 0$$

in $\mathbb{R}((X))$. The discriminant $\Delta(X) = 1 - 4X$ is a square in view of Lemma 7.101. The formal Maclaurin series of the function $x \mapsto (1 - 4x)^{1/2}$, is one of the square roots of $\Delta(X)$. The solutions of the equation in $\mathbb{R}((X))$ are therefore

$$Y_{1,2}(X) = \frac{1 \pm (1 - 4X)^{1/2}}{2X}.$$

In view of Proposition 7.99, one has

$$1 + (1 - 4X)^{1/2} = 2 + \sum_{n=1}^{\infty} \binom{1/2}{n} (-4)^n X^n \quad \text{and}$$

$$1 - (1 - 4X)^{1/2} = -\sum_{n=1}^{\infty} \binom{1/2}{n} (-4)^n X^n.$$

Since $\mathrm{codeg}(1 + (1 - 4X)^{1/2}) = 0 < 1 = \mathrm{codeg}\, X$, by Proposition 7.58 X does not divide $1 + (1 - 4X)^{1/2}$ and so the solution $Y_1(X)$ does not belong to $\mathbb{R}[[X]]$. By contrast, since $\mathrm{codeg}(1 - (1 - 4X)^{1/2}) = 1 = \mathrm{codeg}\, X$, the solution $Y_2(X)$ belongs to $\mathbb{R}[[X]]$:

$$Y_2(X) = -\frac{1}{2} \sum_{n=1}^{\infty} \binom{1/2}{n} (-4)^n X^{n-1}. \qquad ☐$$

7.7.5 A Proof of Proposition 7.97 ☕

We give here an elementary proof of the estimate of the order of magnitude of the sequence $\binom{a}{k}_k$. We consider first the ε−**defocused factorial**, i.e., the product of the ε-translation of the first k integers:

$$\prod_{j=1}^{k} (j - \varepsilon) = (k - \varepsilon)((k - 1) - \varepsilon) \cdots (2 - \varepsilon)(1 - \varepsilon), \qquad 0 \le \varepsilon < 1.$$

Lemma 7.105 (Defocusing the factorial) *Let $0 \le \varepsilon < 1$. There exists $\ell \in]0, 1]$ such that*

$$\prod_{j=1}^{k} (j - \varepsilon) \sim \ell \frac{k!}{k^{\varepsilon}} \quad k \to +\infty, \quad i.e., \quad \lim_{k \to +\infty} k^{\varepsilon} \frac{\prod_{j=1}^{k} (j - \varepsilon)}{k!} = \ell.$$

Proof. We consider the sequence $a_k = k^{\varepsilon} \dfrac{\prod_{j=1}^{k} (j - \varepsilon)}{k!}$, $k \in \mathbb{N}_{\geq 1}$, and we set

$$b_k = \log a_k = \varepsilon \log k + \sum_{j=1}^{k} \log (j - \varepsilon) - \sum_{j=1}^{k} \log j = \varepsilon \log k + \sum_{j=1}^{k} \log \left(1 - \frac{\varepsilon}{j} \right).$$

To prove the thesis it is enough to show that $(b_k)_k$ is an increasing sequence of negative numbers.

- $b_k \leq 0$ for all $k \geq 1$. Indeed, since $\log(1 + x) \leq x$ for all $x > -1$, from (6.44.a) we have

$$b_k \leq \varepsilon \log k - \varepsilon \sum_{j=1}^{k} \frac{1}{j} = \varepsilon(\log k - H_k) \leq 0.$$

- The sequence $(b_k)_{k \geq 1}$ is increasing. Indeed

$$b_{k+1} - b_k = \varepsilon \log(k + 1) - \varepsilon \log k + \log \left(1 - \frac{\varepsilon}{k + 1} \right)$$

$$= \varepsilon \log \left(1 + \frac{1}{k} \right) + \log \left(1 - \frac{\varepsilon}{k + 1} \right).$$

Now, given $k \geq 1$, consider the function

$$\varphi(x) := x \log \left(1 + \frac{1}{k} \right) + \log \left(1 - \frac{x}{k + 1} \right), \quad x \in [0, 1].$$

Clearly $b_{k+1} - b_k = \varphi(\varepsilon)$. Then $\varphi(0) = 0 = \varphi(1)$ and for each $x \in [0, 1]$

$$\varphi'(x) = \log \left(1 + \frac{1}{k} \right) - \frac{1}{k + 1 - x} \quad \text{and} \quad \varphi''(x) = -\frac{1}{(k + 1 - x)^2} < 0.$$

Therefore $\varphi(x) \geq 0$ for each $x \in [0, 1]$. Thus $b_{k+1} - b_k = \varphi(\varepsilon) \geq 0$. □

Proof (of Proposition 7.97). Let $k > |a| + 1$. We have $\left| \dbinom{a}{k} \right| =$ $\dfrac{|a(a - 1) \cdots (a - k + 1)|}{k!}$. Two cases may be considered.

- $a > 0$:

$$\left| \binom{a}{k} \right| = \frac{|a(a-1)\cdots(a-k+1)|}{k!}$$

$$= a(a-1)\cdots(a-[a])\frac{(|a-[a]-1)\cdots(a-k+1)|}{k!}$$

$$= a(a-1)\cdots(a-[a])\frac{\prod\limits_{j=1}^{k-[a]-1}(j-\varepsilon)}{k!}, \quad \varepsilon := a-[a].$$

Since $a \notin \mathbb{N}$ then $\varepsilon < 1$; dividing and multiplying by $\prod\limits_{j=k-[a]}^{k}(j-\varepsilon)$ yields

$$k^{a+1}\left| \binom{a}{k} \right| = a(a-1)\cdots(a-[a])\frac{k^{[a]+1}}{\prod\limits_{j=k-[a]}^{k}(j-\varepsilon)}\; k^{\varepsilon}\frac{\prod\limits_{j=1}^{k}(j-\varepsilon)}{k!}.$$

Since $\lim\limits_{k\to+\infty}\dfrac{k^{[a]+1}}{\prod\limits_{j=k-[a]}^{k}(j-\varepsilon)} = 1$, from Lemma 7.105 we get

$$\lim\limits_{k\to+\infty} k^{a+1}\left| \binom{a}{k} \right| = C_a := a(a-1)\cdots(a-[a])\ell,$$

with $\ell = \lim\limits_{k\to+\infty} k^{\varepsilon}\dfrac{\prod\limits_{j=1}^{k}(j-\varepsilon)}{k!} \in]0, 1]$. Clearly $0 < C_a \le a(a-1)\cdots(a-[a])$.

- $a < 0$:   set $b = -a > 0$ and $\varepsilon := 1 - (b - [b]) \in [0, 1]$.   Then,   since $k > |a| + 1 = b + 1$, one has

$$\left| \binom{a}{k} \right| = \frac{|(-b)(-b-1)\cdots(-b-k+1)|}{k!} = \frac{([b]+1-\varepsilon)\cdots([b]+k-\varepsilon)}{k!}.$$

Since $k^{a+1} = 1/k^{b-1} = 1/k^{[b]-\varepsilon}$, we get

$$k^{a+1}\left|\binom{a}{k}\right| = \frac{\prod\limits_{j=1}^{[b]}(k+j-\varepsilon)}{k^{[b]}}\, k^{\varepsilon}\,\frac{\prod\limits_{j=[b]+1}^{k}(j-\varepsilon)}{k!} \qquad (7.105.\text{a})$$

agreeing that $\prod\limits_{j=1}^{[b]}(k+j-\varepsilon) := 1$ if $[b] = 0$. If $a < 0$ is an integer, then $b \in \mathbb{N}$ and $\varepsilon = 1$; therefore

$$\prod_{j=[b]+1}^{k}(j-\varepsilon) = \prod_{j=b+1}^{k}(j-1) = \frac{(k-1)!}{(b-1)!}$$

and hence for $k \to +\infty$,

$$k^{a+1}\left|\binom{a}{k}\right| = \frac{\prod\limits_{j=1}^{b}(k+j-\varepsilon)}{k^{b}}\,\frac{1}{(b-1)!} \to \frac{1}{(b-1)!} = \frac{1}{(|a|-1)!}.$$

Otherwise, if a is not an integer, then $b \notin \mathbb{N}$ and $\varepsilon < 1$: dividing and multiplying both terms of (7.105.a) by $\prod\limits_{j=1}^{[b]}(j-\varepsilon)$ (we again set this quantity equal to 1 if $[b] = 0$), we get

$$\frac{\prod\limits_{j=1}^{[b]}(k+j-\varepsilon)}{k^{[b]}} k^{\varepsilon}\,\frac{\prod\limits_{j=1}^{k}(j-\varepsilon)}{k!}\,\frac{1}{\prod\limits_{j=1}^{[b]}(j-\varepsilon)}.$$

Since $\lim\limits_{k\to+\infty} \dfrac{\prod\limits_{j=1}^{[b]}(k+j-\varepsilon)}{k^{[b]}} = 1$, from Lemma 7.105 we obtain

$$\lim_{k\to+\infty} k^{a+1}\left|\binom{a}{k}\right| = \frac{1}{\prod\limits_{j=1}^{[b]}(j-\varepsilon)}\lim_{k\to+\infty} k^{\varepsilon}\,\frac{\prod\limits_{j=1}^{k}(j-\varepsilon)}{k!} = \frac{\ell}{\prod\limits_{j=1}^{[b]}(j-\varepsilon)} := C_a.$$

If $[b] = 0$ we have $C_a = \ell \in \,]0,1]$; otherwise

$$0 < C_a = \frac{\ell}{\displaystyle\prod_{j=1}^{[b]}(j - \varepsilon)} \leq \frac{1}{\displaystyle\prod_{j=1}^{[|a|]}(j - \varepsilon)}. \qquad \square$$

7.8 Rational Fractions of Polynomials ☕

In this section it turns out to be particularly useful to conduct our discussion over the field of complex numbers. We recall that if $P(X) = a_0 + a_1 X + \cdots + a_n X^n$, $a_n \neq 0$, is a polynomial with complex coefficients and $\alpha \in \mathbb{C}$, we denote by $P(\alpha)$ the evaluation of $P(X)$ in α:

$$P(\alpha) = a_0 + a_1 \alpha + \cdots + a_n \alpha^n.$$

If $P(\alpha) = 0$, then α is a **root** of the polynomial $P(X)$. By the Fundamental Theorem of Algebra, every polynomial with complex coefficients may be decomposed into a product of polynomials of degree one; precisely one has

$$P(X) = a_n(X - \alpha_1) \cdots (X - \alpha_n)$$

where $\alpha_1, \ldots, \alpha_n$ are the, not necessarily distinct, roots of $P(X)$. We call multiplicity of α_i, $i = 1, \ldots, n$, the number of factors equal to $(X - \alpha_i)$.

7.8.1 The Method of Decomposition into Simple Fractions

Let us consider the subclass of $\mathbb{C}((X))$ of fractions of polynomials.

Definition 7.106 We call **rational fraction of polynomials** the elements of $\mathbb{C}((X))$ of the type $\dfrac{P(X)}{Q(X)}$ with $P(X)$ and $0 \neq Q(X)$ polynomials in $\mathbb{C}[X]$. $\square$

Example 7.107 The fraction of formal power series $\dfrac{1 + X + X^2 + X^3 + \cdots}{X - X^2 + X^3 - X^4 + \cdots}$ coincide with the rational fraction $\dfrac{1 + X}{(1 - X)X}$: indeed

$$(1 + X + X^2 + X^3 + \cdots)(1 - X)X = X = X(1 + X)(X - X^2 + X^3 - X^4 + \cdots).$$

$\square$

Definition 7.108 A rational fraction is said to be:

• **Proper** if it is of the form $\dfrac{P(X)}{Q(X)}$ with deg $P(X) <$ deg $Q(X)$;

- **Simple** if it is of the form $\dfrac{c}{(X - \alpha)^m}$ with $\alpha \in \mathbb{C}, 0 \neq c \in \mathbb{C}$ and $m \in \mathbb{N}$. □

Example 7.109 Consider the rational fraction $\dfrac{1 + X - 3X^2 + 2X^3}{X^2 - X}$. By the Euclidean division of polynomials we get

$$1 + X - 3X^2 + 2X^3 = (X^2 - X)(2X - 1) + 1$$

and hence $\dfrac{1 + X - 3X^2 + 2X^3}{X^2 - X} = 2X - 1 + \dfrac{1}{X^2 - X}$. In the same way, any rational fraction is the sum of a polynomial and a proper rational fraction. □

Hermite's[5] Theorem, for which we give only the statement, asserts that every proper rational fraction of polynomials may be decomposed into a sum of simple ones.

Theorem 7.110 (Hermite decomposition) *Let $P(X)/Q(X)$ be a proper rational fraction. If, for each root α of $Q(X)$, the symbol μ_α denotes the multiplicity of α as a root of $Q(X)$, there exist uniquely determined complex numbers $c_{\alpha,j}$, $j = 1, \ldots, \mu_\alpha$ such that*

$$\frac{P(X)}{Q(X)} = \sum_{\alpha : Q(\alpha) = 0} \sum_{j=1}^{\mu_\alpha} \frac{c_{\alpha,j}}{(X - \alpha)^j}. \tag{7.110.a}$$

Example 7.111 1. Let us decompose the rational fraction $\dfrac{X + 2}{X^2 - 3X + 2}$ into simple fractions. The roots of $X^2 - 3X + 2$ are 1, 2 both with multiplicity 1. Thus there exist a, b such that

$$\frac{a}{X - 1} + \frac{b}{X - 2} = \frac{X + 2}{X^2 - 3X + 2}.$$

The latter equality is equivalent to $a(X - 2) + b(X - 1) = X + 2$, that is, to $(a + b)X - 2a - b = X + 2$ which, on equating coefficients, gives the system

$$a + b = 1, \quad 2a + b = -2.$$

Solving the system one finds that $a = -3, b = 4$ so that

$$\frac{X + 2}{X^2 - 3X + 2} = \frac{-3}{X - 1} + \frac{4}{X - 2}.$$

---

[5]Charles Hermite (1822–1901).

is the desired decomposition.

2. Let us now decompose $\dfrac{1}{X^2 + 1}$ into simple fractions. The roots of $X^2 + 1$ are i and $-i$, both with multiplicity one. Thus there exist complex constants a, b such that

$$\frac{a}{X - i} + \frac{b}{X + i} = \frac{1}{X^2 + 1}.$$

The equality is equivalent to the validity of the equation $a(X + i) + b(X - i) = 1$, which gives rise to the system

$$a + b = 0, \quad ai - bi = 1.$$

Solving the system, one finds that $a = i/2, b = -i/2$ whence

$$\frac{1}{X^2 + 1} = \frac{i/2}{X + i} - \frac{i/2}{X - i}.$$

3. Finally, let us decompose the rational fraction $\dfrac{X}{(X^2 + 1)(X - 1)^2}$ into simple fractions. The roots of the denominator are i, $-i$, both of multiplicity 1, and -1 of multiplicity 2. Hence there exist constants a, b, c, d such that

$$\frac{a}{X - i} + \frac{b}{X + i} + \frac{c}{X - 1} + \frac{d}{(X - 1)^2} = \frac{X}{(X^2 + 1)(X - 1)^2}.$$

This equality determines the following linear system

$$\begin{cases} a + b + c & = 0 \\ -2a - 2b - c + d + i(b - a) & = 0 \\ a + b + c + 2i(a - b) & = 1 \\ d - c + i(b - a) & = 0 \end{cases}$$

which has as its solution $a = -i/4, b = i/4, c = 0, d = 1/2$; therefore

$$\frac{-i/4}{X - i} + \frac{i/4}{X + i} + \frac{1/2}{(X - 1)^2} = \frac{X}{(X^2 + 1)(X - 1)^2}. \qquad \square$$

Illustrating methods for easily deriving such decompositions is not among our goals since they can in fact be found as the output of various CAS'. We mention only the particularly simple case of distinct roots.

Proposition 7.112 *Let $P(X)/Q(X)$ be a proper rational fraction. If $Q(X)$ has all roots of multiplicity 1, and $Q'(X)$ denotes the derivate of $Q(X)$, then one has*

$$\frac{P(X)}{Q(X)} = \sum_{\alpha:Q(\alpha)=0} \frac{P(\alpha)/Q'(\alpha)}{X - \alpha}.$$

Example 7.113 We use the preceding result to decompose $\dfrac{X - 5}{(X - 1)(X - 3)}$ into simple fractions. Here

$$P(X) = X - 5, \qquad Q(X) = (X - 1)(X - 3) = X^2 - 4X + 3$$

and Proposition 7.112 gives

$$\frac{X - 5}{(X - 1)(X - 3)} = \frac{P(1)/Q'(1)}{X - 1} + \frac{P(3)/Q'(3)}{X - 3}.$$

Since $Q'(X) = 2X - 4$ one has

$$\frac{X - 5}{(X - 1)(X - 3)} = \frac{-4/(-2)}{X - 1} + \frac{-2/2}{X - 3} = \frac{2}{X - 1} - \frac{1}{X - 3}. \qquad \square$$

By Proposition 7.63 a rational fraction of polynomials $\dfrac{P(X)}{Q(X)}$ belongs to $\mathbb{C}[X]$ if and only if $\operatorname{codeg} Q(X) \leq \operatorname{codeg} P(X)$; in such a case we can always assume $\operatorname{codeg} Q(X) = 0$, i.e., $Q(0) \neq 0$: indeed, for some $P_1(X), Q_1(X)$ in $\mathbb{R}[[X]]$,

$$\frac{P(X)}{Q(X)} = X^{\operatorname{codeg} P(X) - \operatorname{codeg} Q(X)} \frac{P_1(X)}{Q_1(X)} \quad \text{and } \operatorname{codeg} P_1(X) = 0 = \operatorname{codeg} Q_1(X).$$

We now see how to deduce the coefficients of the corresponding formal power series. First, as we have observed above, the rational fraction of polynomials is a sum of a polynomial and a proper rational fraction. The delicate aspect is the determination of the coefficients of the formal power series corresponding to the proper rational fraction, since one then need only to add the coefficients of the polynomial in the appropriate degrees.

Theorem 7.114 *Let* $A(X) = \displaystyle\sum_{n=0}^{\infty} a_n X^n = \dfrac{P(X)}{Q(X)}$ *be a formal power series with* $Q(0) \neq 0$ *and* $\deg P(X) < \deg Q(X)$. *Then*

$$\forall n \in \mathbb{N} \quad a_n = \sum_{\alpha : Q(\alpha) = 0} \frac{1}{\alpha^n} \left(\sum_{j=1}^{\mu_\alpha} \binom{-j}{n} \frac{(-1)^{n+j} c_{\alpha,j}}{\alpha^j} \right), \qquad (7.114.a)$$

where the $c_{\alpha,j}$ are the coefficients of the decomposition illustrated in Theorem 7.110 and μ_α is the multiplicity of the root α. If, moreover, $Q(X)$ has all its roots of multiplicity 1, then

$$\forall n \in \mathbb{N} \quad a_n = - \sum_{\alpha : Q(\alpha) = 0} \frac{c_{\alpha,1}}{\alpha^{n+1}} = - \sum_{\alpha : Q(\alpha) = 0} \frac{P(\alpha)}{\alpha^{n+1} Q'(\alpha)}. \qquad (7.114.b)$$

Remark 7.115 Note that a CAS allows one to immediately obtain a_n.

🍂 *Proof.* By Theorem 7.110 one has

$$A(X) = \sum_{n=0}^{\infty} a_n X^n = \frac{P(X)}{Q(X)} = \sum_{\alpha : Q(\alpha) = 0} \sum_{j=1}^{\mu_\alpha} \frac{c_{\alpha,j}}{(X - \alpha)^j}.$$

For a given root α (necessarily $\neq 0$) of $Q(X)$ and for $1 \leq j \leq \mu_\alpha$, one has (see Example 7.72)

$$\frac{1}{(X - \alpha)^j} = \frac{(-1)^j}{\alpha^j} \left(1 - \frac{X}{\alpha} \right)^{-j} = \frac{(-1)^j}{\alpha^j} \sum_{n=0}^{\infty} \binom{-j}{n} \frac{(-X)^n}{\alpha^n}$$

$$= \frac{1}{\alpha^j} \sum_{n=0}^{\infty} \binom{-j}{n} (-1)^{j+n} \frac{X^n}{\alpha^n}.$$

Therefore, on setting the coefficients of X^n equal, one obtains

$$a_n = [X^n] A(X) = \sum_{\alpha : Q(\alpha) = 0} [X^n] \sum_{j=1}^{\mu_\alpha} \frac{c_{\alpha,j}}{(X - \alpha)^j}$$

$$= \sum_{\alpha : Q(\alpha) = 0} \sum_{j=1}^{\mu_\alpha} \frac{(-1)^{j+n} c_{\alpha,j}}{\alpha^j} \binom{-j}{n} \frac{1}{\alpha^n},$$

which is what we wished to prove.

In the case for which all roots of $Q(X)$ have multiplicity 1, one easily obtains the simplified formula. Indeed, one has

$$a_n = \sum_{\alpha:Q(\alpha)=0} \frac{(-1)^{1+n}c_{\alpha,1}}{\alpha}\binom{-1}{n}\frac{1}{\alpha^n} = \sum_{\alpha:Q(\alpha)=0} \frac{(-1)^{1+2n}c_{\alpha,1}}{\alpha^{n+1}}$$

$$= -\sum_{\alpha:Q(\alpha)=0} \frac{c_{\alpha,1}}{\alpha^{n+1}} = -\sum_{\alpha:Q(\alpha)=0} \frac{P(\alpha)}{\alpha^{n+1}Q'(\alpha)},$$

since $c_{\alpha,1} = \dfrac{P(\alpha)}{Q'(\alpha)}$ by Proposition 7.112. □

Remark 7.116 Since in practice we will only rarely have to deal with roots of high multiplicity, it is worth the effort to make (7.114.a) explicit in the case of roots of multiplicity less than or equal to 2. Bearing in mind that (see Example 7.96)

$$\forall n \in \mathbb{N} \quad \binom{-1}{n} = (-1)^n, \quad \binom{-2}{n} = (-1)^n(n+1),$$

from (7.114.a) one immediately obtains that when all roots have multiplicity at most two

$$\forall n \in \mathbb{N} \quad a_n = \sum_{\alpha:Q(\alpha)=0} \frac{1}{\alpha^n}\left(-\frac{c_{\alpha,1}}{\alpha^j} + (n+1)\frac{c_{\alpha,2}}{\alpha^2}\right). \tag{7.116.a}$$

☞ In the case of simple roots it is not necessary to remember formula (7.114.b): it is very easy to recover it from the decomposition in simple fractions. Indeed, once one has written

$$\frac{P(X)}{Q(X)} = \frac{c_{1,1}}{X - \alpha_1} + \cdots + \frac{c_{m,1}}{X - \alpha_m}$$

with $\alpha_1, \ldots, \alpha_m$ distinct non-zero roots of $Q(X)$ it suffices to develop each single fraction using the identity

$$(1-X)^{-1} = \frac{1}{1-X} = \sum_{n=0}^{\infty} X^n.$$

In fact, for each $i = 1, \ldots, m$ one has

$$\frac{c_{i,1}}{X - \alpha_i} = -\frac{c_{i,1}}{\alpha_i(1 - X/\alpha_i)}$$

$$= -\frac{c_{i,1}}{\alpha_i}\left(1 - \frac{X}{\alpha_i}\right)^{-1} = -\frac{c_{i,1}}{\alpha_i}\sum_{n=0}^{\infty}\left(\frac{X}{\alpha_i}\right)^n$$

from which one immediately finds that

$$\forall n \in \mathbb{N} \quad [X^n]\frac{P(X)}{Q(X)} = -\sum_{i=1}^{m} \frac{c_{i,1}}{\alpha_i^{n+1}}.$$

Example 7.117 Determine the coefficients of the formal power series $A(X) = \sum_{n=0}^{\infty} a_n X^n$ when $A(X) = \dfrac{X^3 + X^2 - 3X - 1}{(X+1)(X-2)}$.

Solution. First, one has

$$\frac{X^3 + X^2 - 3X - 1}{(X+1)(X-2)} = \frac{X+3}{(X+1)(X-2)} + X + 2,$$

and so we calculate the formal power series $B(X) = \dfrac{X+3}{(X+1)(X-2)}$. We use the formula (7.114.b), after having set $P(X) = X + 3$, $Q(X) = (X+1)(X-2) = X^2 - X - 2$, so that $Q'(X) = 2X - 1$. Since the roots of $Q(X)$ are equal to -1 and 2, one has

$$[X^n]B(X) = -\frac{P(-1)}{(-1)^{n+1}Q'(-1)} - \frac{P(2)}{2^{n+1}Q'(2)} = (-1)^{n+1}\frac{2}{3} - \frac{5}{3}2^{-n-1}.$$

Hence one obtains

$$A(X) = B(X) + X + 2 = \frac{1}{2} + \frac{5X}{4} + \sum_{n=2}^{\infty}\left((-1)^{n+1}\frac{2}{3} - \frac{5}{3}2^{-n+1}\right)X^n. \qquad \square$$

7.8.2 The Recursive Method

Here we illustrate an alternative method for finding the coefficients of a formal power series $A(X) = \dfrac{P(X)}{Q(X)}$ with $Q(0) \neq 0$. This method will be extended and elaborated upon in Sect. 10.4. By setting some coefficients equal to zero if necessary, we may suppose without loss of generality that

$$P(X) = p_0 + p_1 X + \cdots + p_m X^m \quad \text{and} \quad Q(X) = q_0 + q_1 X + \cdots + q_m X^m,$$

where we have set m equal to the larger of the degrees of $P(X)$ and $Q(X)$. Then, since $A(X)Q(X) = P(X)$, we have

$$[X^n]\, A(X)Q(X) = [X^n]\, P(X),$$

that is,

$$\begin{cases} a_0 q_0 & = p_0 \\ a_0 q_1 + a_1 q_0 & = p_1 \\ \cdots \\ a_0 q_m + \cdots + a_m q_0 & = p_m \\ \displaystyle\sum_{k=0}^{n} a_k q_{n-k} & = 0 \qquad \forall n > m. \end{cases} \qquad (7.117.a)$$

Note moreover that since $q_i = 0$ if $i > m$ in reality the latter equation is equivalent to

$$\forall n > m \qquad a_n q_0 + a_{n-1} q_1 + \cdots + a_{n-m} q_m = 0. \qquad (7.117.b)$$

Since $q_0 \neq 0$, Eqs. (7.117.a) and (7.117.b) can be solved one at a time, obtaining the values $a_0, \ldots, a_m, a_{m+1}, a_{m+2}, \ldots$ and so on.

Example 7.118 Consider once again the formal power series of Example 7.117:

$$A(X) = \frac{X+3}{(X+1)(X-2)} = \frac{X+3}{X^2 - X - 2}.$$

On setting $A(X) = \sum_{n=0}^{\infty} a_n X^n$, from the relation $A(X)(-2 - X + X^2) = 3 + X$ we deduce that

$$-2a_0 = 3, \ -2a_1 - a_0 = 1$$

and, for each $n \geq 2$,

$$-2a_n - a_{n-1} + a_{n-2} = 0.$$

Thus one has $a_0 = -3/2$, $a_1 = 1/4$ and, for $n \geq 2$,

$$a_n = \frac{a_{n-2} - a_{n-1}}{2} \qquad (7.118.a)$$

and hence

$$a_2 = \frac{-3/2 - 1/4}{2} = -7/2^3, \ a_3 = \frac{1/4 + 7/2^3}{2} = 9/2^4, \ldots. \qquad \square$$

7.9   Linear Differential Equations ☕

In this section we present another useful application of closed forms for a formal power series.

Proposition 7.119 *Let $A(X) = \sum_n a_n X^n$ be a formal power series and $g_0(x)$, $g_1(x), \ldots, g_n(x), g(x)$ functions in $\mathscr{C}^{\infty}(0)$, with $g_n(0) \neq 0$. If in $\mathbb{R}[[X]]$ one has*

$$\sum_{k=0}^{n} A^{(k)}(X) g_k(X) = g(X),$$

then a solution function f in $\mathscr{C}^{\infty}(0)$ of the Cauchy[6] problem

$$\begin{cases} \displaystyle\sum_{k=0}^{n} y^{(k)} g_k(x) = g(x) \\ y(0) = 0!a_0 = a_0 \\ y'(0) = 1!a_1 = a_1 \\ \cdots\cdots\cdots\cdots\cdots\cdots\cdots \\ y^{(n-1)}(0) = (n-1)!a_{n-1} \end{cases}$$

is a closed form for $A(X)$.

Proof. Since the function $f \in \mathscr{C}^{\infty}(0)$ is a solution of the differential equation under consideration one has $\displaystyle\sum_{k=0}^{n} f^{(k)}(x)g_k(x) = g(x)$. By Proposition 7.80 we have the following equality between formal power series

$$\sum_{k=0}^{n} f^{(k)}(X)g_k(X) = g(X).$$

Consequently, on setting $B(X) = A(X) - f(X)$ one has

$$\sum_{k=0}^{n} B^{(k)}(X)g_k(X) = \sum_{k=0}^{n} \left(A^{(k)}(X) - f^{(k)}(X) \right) g_k(X) = 0. \qquad (7.119.\text{a})$$

We prove that all the coefficients of the series $B(X)$ are equal to 0, and so $B(X) = 0$. The initial conditions on the function f imply that

$$i!\left[X^i\right]B(X) = i!\left[X^i\right]A(X) - i!\left[X^i\right]f(X) = i!a_i - f^{(i)}(0) = 0 \text{ for } 0 \le i \le n-1$$

and so

$$\left[X^0\right]B(X) = \left[X^1\right]B(X) = \cdots = \left[X^{n-1}\right]B(X) = 0.$$

Then we get

$$\left[X^0\right]\left(\sum_{k=0}^{n-1} B^{(k)}(X)g_k(X)\right) = \sum_{k=0}^{n-1} \left[X^0\right](B^{(k)}(X)g_k(X))$$

$$= \sum_{k=0}^{n-1} \left(\left[X^0\right]B^{(k)}(X)\right)\left(\left[X^0\right]g_k(X)\right) = 0.$$

---

[6] Augustin–Louis Cauchy (1789–1857).

It then follows from (7.119.a) that $\left[X^0\right] B^{(n)}(X)g_n(X) = 0$; since $\left[X^0\right] g_n(X) = g_n(0) \neq 0$, one deduces that

$$n!\left[X^n\right] B(X) = \left[X^0\right] B^{(n)}(X) = 0,$$

and thus $[X^n]\, B(X) = 0$. Taking the derivatives of the terms in (7.119.a) one obtains

$$0 = \sum_{k=0}^{n} \left(B^{(k)}(X)g_k(X)\right)' = \sum_{k=0}^{n} \left(B^{(k+1)}(X)g_k(X) + B^{(k)}(X)g_k'(X)\right);$$

since $[X^0]B^{(k)}(X) = 0$ for $0 \le k \le n$ one has

$$0 = [X^0] \sum_{k=0}^{n} \left(B^{(k)}(X)g_k(X)\right)' = [X^0]B^{(n+1)}(X)g_n(X).$$

Given that $\left[X^0\right] g_n(X) = g_n(0) \neq 0$, one has

$$(n+1)![X^{n+1}]B(X) = [X^0]B^{(n+1)}(X) = 0,$$

and so $[X^{n+1}]B(X) = 0$. Proceeding in this way we obtain

$$0 = \left[X^0\right] B^{(m)}(X) = m!\left[X^m\right] B(X) \quad \forall m \in \mathbb{N},$$

and hence $B(X) = 0$. $\square$

Example 7.120 We determine a closed form for a formal power series $A(X)$ satisfying the relation

$$(1 - X)A'(X) = A(X)$$

and the initial condition $\left[X^0\right] A(X) = 1$. The differential equation

$$(1 - x)y' = y, \quad y(0) = 1$$

has as a solution the function $f \in \mathscr{C}^\infty(0)$ defined by

$$f(x) = \frac{1}{1 - x}, \quad x \in]-\infty, 1[.$$

By Proposition 7.119 one has $A(X) = \dfrac{1}{1 - X}$. $\square$

7.10 Infinite Products of Formal Power Series ☕

In some applications it is useful to deal with infinite products of formal series. In this section we study which countable families of formal series allow one to define such a product. Notice that if $B_0(X)$ and $B_1(X)$ are formal power series then

$$(1 + B_0(X))(1 + B_1(X)) = 1 + B_0(X) + B_1(X) + B_0(X)B_1(X).$$

More generally, if $B_0(X), \ldots, B_N(X)$ are formal power series then

$$(1 + B_0(X)) \cdots (1 + B_N(X)) = 1 + \sum_{\substack{0 \leq k \leq N \\ 0 \leq i_0 < i_1 < \cdots < i_k \leq N}} B_{i_0}(X) \cdots B_{i_k}(X).$$

$$(7.120.a)$$

Consider now a countable family of formal power series $\{B_i(X) : i \in \mathbb{N}\}$. In order to extend (7.120.a), we need to be sure that the sum

$$\sum_{\substack{0 \leq k \\ 0 \leq i_0 < i_1 < \cdots < i_k}} B_{i_0}(X) \cdots B_{i_k}(X)$$

is well defined, i.e., the countable family of formal power series

$$\{B_{i_0}(X) \cdots B_{i_k}(X) : 0 \leq i_0 < i_1 < \cdots < i_k, \ k \in \mathbb{N}\}$$

is locally finite.

Lemma 7.121 *Let $\{B_i(X) : i \in \mathbb{N}\}$ be a countable family of formal power series. The family*

$$\{B_{i_0}(X) \cdots B_{i_k}(X) : 0 \leq i_0 < i_1 < \cdots < i_k, \ k \in \mathbb{N}\}$$

is locally finite if and only if $\{B_i(X) : i \in \mathbb{N}\}$ is locally finite.

Proof. Since $\{B_i(X) : i \in \mathbb{N}\} \subseteq \{B_{i_0}(X) \cdots B_{i_k}(X) : 0 \leq i_0 < i_1 < \cdots < i_k, \ k \in \mathbb{N}\}$, clearly if the latter family is locally finite, also the smaller one is locally finite. Conversely, let us now assume that $\{B_i(X) : i \in \mathbb{N}\}$ is locally finite, i.e.,

$$\lim_{i \to \infty} \operatorname{codeg} B_i(X) = +\infty.$$

Fix $n \in \mathbb{N}$; there exists $\mu_n \in \mathbb{N}$ such that

$$\operatorname{codeg} B_i(X) > n \text{ for each } i > \mu_n,$$

or, equivalently, such that

$$\text{codeg } B_i(X) \leq n \ \Rightarrow \ i \leq \mu_n.$$

Given $k \in \mathbb{N}$, since $\text{codeg } B_{i_0}(X) \cdots B_{i_k}(X) = \sum_{j=0}^{k} \text{codeg } B_{i_j}(X)$, if

$$\text{codeg } \left(B_{i_0}(X) \cdots B_{i_k}(X) \right) \leq n$$

then, necessarily, $i_0, \ldots, i_k \leq \mu_n$. Moreover, since $0 \leq i_0 < i_1 < \cdots < i_k$ one has $k \leq i_k \leq \mu_n$. Therefore

$$\{B_{i_0}(X) \cdots B_{i_k}(X) : 0 \leq i_0 < i_1 < \cdots < i_k, \ k \in \mathbb{N}, \ \text{codeg } \left(B_{i_0}(X) \cdots B_{i_k}(X) \right) \leq n\}$$

is a subset of the finite set

$$\{B_{i_0}(X) \cdots B_{i_k}(X) : 0 \leq i_0 < i_1 < \cdots < i_k \leq \mu_n, \ 0 \leq k \leq \mu_n\};$$

hence $\{B_{i_0}(X) \cdots B_{i_k}(X) : 0 \leq i_0 < i_1 < \cdots < i_k, \ k \in \mathbb{N}\}$ is locally finite.   □

The validity of Lemma 7.121 ensures the consistency of the following definition.

Definition 7.122 (*Infinite products of power series*) Let $\{B_i(X) : i \in \mathbb{N}\}$ be a locally finite family of power series. The **product of the infinitely many formal power series** $A_i(X) := 1 + B_i(X)$, $i \in \mathbb{N}$, is defined by

$$\prod_{i=0}^{\infty} A_i(X) = \prod_{i=0}^{\infty} (1 + B_i(X)) := 1 + \sum_{n=0}^{\infty} \left(\sum_{\substack{0 \leq k \\ 0 \leq i_0 < i_1 < \cdots < i_k}} [X^n] B_{i_0}(X) \cdots B_{i_k}(X) \right) X^n \ \square$$

Remark 7.123 If the family $\{B_i(X) : i \in \mathbb{N}\}$ is locally finite, we are able to deal with infinite sums of families of the form $\{0 + B_i(X) : i \in \mathbb{N}\}$ and infinite products of the form $\{1 + B_i(X) : i \in \mathbb{N}\}$: by chance 0 is the neutral element for sums of formal power series, and 1 is the neutral element for product of formal power series in $\mathbb{R}[[X]]$.

We will focus our interest in products of characteristic generating formal series of sets.

Example 7.124 Let $(E_i)_{i \geq 1}$ be a family of subsets of $\mathbb{N}$. The families

$$\{I_{E_i}^{\text{OGF}}(X) - 1 : i \in \mathbb{N}\}, \quad \{I_{E_i}^{\text{EGF}}(X) - 1 : i \in \mathbb{N}\}$$

are locally finite if and only if:

1. 0 belongs to all but a finite number of E_i's;

2. Each $j \geq 1$ belongs to a finite number of E_i's.

Indeed,

$$\operatorname{codeg}(I_{E_i}^{\mathrm{OGF}}(X) - 1) = \operatorname{codeg}(I_{E_i}^{\mathrm{EGF}}(X) - 1) = \begin{cases} \min(E_i \setminus \{0\}) & \text{if } 0 \in E_i \\ 0 & \text{if } 0 \notin E_i \end{cases}$$

is less or equal than $j \in \mathbb{N}$ if and only if

$$0 \in E_i \text{ and } \min(E_i \setminus \{0\}) \leq j, \text{ or } 0 \notin E_i. \qquad (7.124.\mathrm{a})$$

The set of i's satisfying (7.124.a) is finite if and only if $0 \in E_i$ and $\min(E_i \setminus \{0\}) > j$ for all but a finite number of i's, and this occurs if and only if Conditions 1 and 2 both hold. In this case, following Definition 7.122, we are thus allowed to define the products

$$\prod_{i=0}^{\infty} I_{E_i}^{\mathrm{OGF}}(X) \text{ and } \prod_{i=0}^{\infty} I_{E_i}^{\mathrm{EGF}}(X). \qquad \square$$

A remarkable fact about infinite products is that they allow one to extend Theorem 7.27, namely the Basic Principle for occupancy problems in I_n. We shall deal here with collections and sequences with terms not just in a prescribed I_n, but in $\mathbb{N}_{\geq 1}$. Since finite sequences or collections can have non-zero repetitions fur just a finite number of elements of $\mathbb{N}_{\geq 1}$, we shall restrict ourselves to sets $(E_i)_{i \geq 1}$ of multiplicities of $i \geq 1$ that all but a finite number contain 0.

Example 7.125 Let $E_n = \{0, 1, n+1\}$ for all $n \geq 1$. Then

$$(3, 2, 2, 57, 3, 3, 2, 3), \quad [2, 2, 2, 3, 3, 3, 3, 57]$$

are, respectively, a 8-sequence and a 8-collection of $\mathbb{N}_{\geq 1}$ with $0 \in E_1$ repetitions of 1, $3 \in E_2$ repetitions of 2, $4 \in E_3$ repetitions of 3, $1 \in E_{57}$ repetition of 57, $0 \in E_i$ repetitions of i, for every $i \in \mathbb{N} \setminus \{1, 2, 3, 57\}$. $\qquad \square$

Definition 7.126 Let $(E_i)_{i \geq 1}$ be a sequence of subsets of $\mathbb{N}$, with $0 \in E_i$ for all but a finite number of $i \geq 1$, and $k \geq 1$.

- $C(\infty, k; (E_i)_{i \geq 1})$ denotes the number of k-collections of $\mathbb{N}_{\geq 1}$ with $k_i \in E_i$ repetitions of i, for each $i \geq 1$.
- $S(\infty, k; (E_i)_{i \geq 1})$ denotes the number of k-sequences of $\mathbb{N}_{\geq 1}$ with $k_i \in E_i$ repetitions of i, for each $i \geq 1$.

Notice that, for each k-collection or k-sequence with the above occupancies, then $k = \sum_{i \geq 0} k_i$; whence $k_i = 0$ for all but a finite number of i. In particular, assigning a k-collection with $k_i \in E_i$ repetitions of i for each i is equivalent to have what we shall call an **infinite composition** of k, namely a sequence $(k_i)_{i \geq 1}$ with $k_i \in E_i$ for all i and $k = \sum_{i \geq 1} k_i$. $\qquad \square$

Theorem 7.127 (Basic Principle for occupancy problems in $\mathbb{N}_{\geq 1}$) *Let* $\{E_i : i \in \mathbb{N}_{\geq 1}\}$ *be a countable family of subsets of* $\mathbb{N}$, *satisfying the following properties:*

1. $0 \in E_i$ *for all but a finite number of* $i \geq 1$;
2. *Each* $j \in \mathbb{N}_{\geq 1}$ *belongs to a finite number of the sets* E_i's.

Then the products of the characteristic OGF and EGF of the E_i's *are defined, and*

$$C(\infty, k; (E_i)_{i \geq 1}) = \left[X^k\right] \prod_{i=1}^{\infty} I_{E_i}^{\mathrm{OGF}}(X) \quad \forall k \geq 0,$$

$$S(\infty, k; (E_i)_{i \geq 1}) = \left[\frac{X^k}{k!}\right] \prod_{i=1}^{\infty} I_{E_i}^{\mathrm{EGF}}(X) \quad \forall k \geq 0.$$

Proof. It follows from Points 1 and 2 that the family $\{I_{E_i}^{\mathrm{OGF}}(X) - 1 : i \in \mathbb{N}_{\geq 1}\}$ (resp. $\{I_{E_i}^{\mathrm{EGF}}(X) - 1 : i \in \mathbb{N}_{\geq 1}\}$) is locally finite (see Example 7.124).

Fix $k \geq 0$. Our conditions imply the existence of $\mu_k \in \mathbb{N}$ such that E_n does not contain natural numbers up to k other than 0, and $0 \in E_n$, for $n > \mu_k$. Consider a sequence $(k_n)_{n \geq 1}$ in $\mathbb{N}$ such that $\sum_{n=1}^{\infty} k_n = k$ with $k_n \in E_n$ for each $n \geq 1$: since $k_n \leq k$ for each n then $k_n = 0$ for all $n > \mu_k$. It follows that $C(\infty, k; (E_i)_{i \geq 1}) = C(\mu_k, k; (E_1, \ldots, E_{\mu_k}))$; therefore the Basic Principle 7.27 yields

$$C(\infty, k; (E_i)_{i \geq 1}) = \left[X^k\right] \prod_{i=1}^{\mu_k} I_{E_i}^{\mathrm{OGF}}(X).$$

Now, for each $n > \mu_k$ one has $\left[X^j\right] I_{E_n}^{\mathrm{OGF}}(X) = 0$ for each $1 \leq j \leq k$, so that $I_{E_n}^{\mathrm{OGF}}(X) = 1 + X^{k+1} C_n(X)$ for a suitable formal power series $C_n(X)$: we conclude that

$$C(\infty, k; (E_i)_{i \geq 1}) = \left[X^k\right] \prod_{i=1}^{\mu_k} I_{E_i}^{\mathrm{OGF}}(X) = \left[X^k\right] \prod_{i=1}^{\infty} I_{E_i}^{\mathrm{OGF}}(X).$$

Consider a k-sequence of $\mathbb{N}_{\geq 1}$ with $k_n \in E_n$ repetitions of n, for all $n \in \mathbb{N}$. Then, necessarily, $k_n = 0$ for $n > \mu_k$. It follows that $S(\infty, k; (E_i)_{i \geq 1}) = S(\mu_k, k; (E_1, \ldots, E_{\mu_k}))$; therefore the Basic Principle 7.27 yields

$$S(\infty, k; (E_i)_{i \geq 1}) = \left[\frac{X^k}{k!}\right] \prod_{i=1}^{\mu_k} I_{E_i}^{\mathrm{EGF}}(X).$$

Since for each $n > \mu_k$ one has $I_{E_n}^{\mathrm{EGF}}(X) = 1 + X^{k+1} C_n(X)$ for a suitable formal power series $C_n(X)$, we get

$$S(\infty, k; (E_i)_{i \geq 1}) = \left[\frac{X^k}{k!} \right] \prod_{i=1}^{\mu_k} I_{E_i}^{\mathrm{EGF}}(X) = \left[\frac{X^k}{k!} \right] \prod_{i=1}^{\infty} I_{E_i}^{\mathrm{EGF}}(X).$$
□

Example 7.128 Let $\{F_i : i \in \mathbb{N}_{\geq 1}\}$ be a countable family of subsets of $\mathbb{N}_{\geq 1}$, such that $\lim_{i \to \infty} \min F_i = +\infty$. From Theorem 7.127 we get

$$\prod_{i=1}^{\infty} (1 + I_{F_i}^{\mathrm{OGF}}(X)) = \mathrm{OGF}\left(C(\infty, k; (F_i \cup \{0\})_{i \geq 1}) \right)_k.$$
□

We see now what happens if, in Example 7.128, we replace $1 + I_{F_i}^{\mathrm{OGF}}(X)$ with $1 - I_{F_i}^{\mathrm{OGF}}(X)$.

Corollary 7.129 *Let $\{F_i : i \in \mathbb{N}_{\geq 1}\}$ be a countable family of subsets of $\mathbb{N}_{\geq 1}$, such that $\lim_{i \to \infty} \min F_i = +\infty$. Then*

$$\prod_{i=1}^{\infty} \left(1 - I_{F_i}^{\mathrm{OGF}}(X) \right) = \mathrm{OGF}(\mathfrak{c}_k)_k$$

where $\mathfrak{c}_0 = 1$ and, for every $k \geq 1$, $\mathfrak{c}_k$ is the difference of the number of the compositions of k of even length and the number of the compositions of k of odd length, with terms in the sets F_i's.

Proof. Since $\{-I_{F_i}^{\mathrm{OGF}}(X) : i \in \mathbb{N}\}$ is locally finite, the product $\prod_{i=1}^{\infty} \left(1 - I_{F_i}^{\mathrm{OGF}}(X) \right)$ is well defined. For each $k \geq 1$, let $\mu_k \in \mathbb{N}$ such that $\min F_i > k$ for each $i > \mu_k$. Since $[X^k] \left(1 - I_{F_i}^{\mathrm{OGF}}(X) \right) = 0$ for each $i > \mu_k$, we get

$$\mathfrak{c}_k = [X^k] \prod_{i=1}^{\infty} \left(1 - I_{F_i}^{\mathrm{OGF}}(X) \right) = [X^k] \prod_{i=1}^{\mu_k} \left(1 - I_{F_i}^{\mathrm{OGF}}(X) \right).$$

It is an easy computation to verify that

$$\prod_{i=1}^{\mu_k} \left(1 - I_{F_i}^{\mathrm{OGF}}(X) \right) = 1 + \sum_{k=1}^{\infty} \mathfrak{a}_k X^k - \sum_{k=1}^{\infty} \mathfrak{b}_k X^k,$$

where

$$\mathfrak{a}_k := \sum_{\substack{1 \leq 2j \leq \mu_k \\ 1 \leq i_1 < i_2 < \cdots < i_{2j} \leq \mu_k}} [X^k] I_{F_{i_1}}^{\mathrm{OGF}}(X) \cdots I_{F_{i_{2j}}}^{\mathrm{OGF}}(X), \text{ and}$$

$$\mathfrak{b}_k := \sum_{\substack{1 \le 2j+1 \le \mu_k \\ 1 \le i_1 < i_2 < \cdots < i_{2j+1} \le \mu_k}} [X^k]\, I^{\mathrm{OGF}}_{F_{i_1}}(X) \cdots I^{\mathrm{OGF}}_{F_{i_{2j+1}}}(X).$$

Therefore $\mathfrak{c}_0 = [X^0]\prod_{i=1}^{\mu_k}\left(1 - I^{\mathrm{OGF}}_{F_i}(X)\right) = 1$ and, for each $k \ge 1$,

$$\mathfrak{c}_k = [X^k]\prod_{i=1}^{\mu_k}\left(1 - I^{\mathrm{OGF}}_{F_i}(X)\right) = \mathfrak{a}_k - \mathfrak{b}_k.$$

By the Basic Principle 7.27, $\mathfrak{a}_k$ is the number of even compositions of k with terms in the set F_i's and $\mathfrak{b}_k$ is the number of odd compositions of k with terms in the set F_i's.   □

7.10.1 Integer Partitions

An application of infinite products of formal series is the study of the *integer partition problem*, first studied, as is often the case, by Euler. Given a natural number k, we recall that a n-partition of k is a collection $[k_1, \ldots, k_n]$ of numbers in $\mathbb{N}_{\ge 1}$ satisfying $k_1 + \cdots + k_n = k$ (see Definition 1.17) and a partition of k is any n-partition of k, $n \in \mathbb{N}$.

Example 7.130 The partitions of 5 are the following seven collections:

$$[1,1,1,1,1], \quad [1,1,1,2], \quad [1,1,3], \quad [1,4], \quad [1,2,2], \quad [2,3], \quad [5]. \quad □$$

Proposition 7.131 (OGF of integer partitions) *The number $\mathfrak{p}_k$ of partitions of $k \ge 0$ coincides with the number $C(\infty, k; (i\mathbb{N})_{i\ge 1})$ of the infinite compositions $(k_i)_{i\ge 1}$ of k with $k_i \in i\mathbb{N} = \{0, i, 2i, \ldots\}$ for each $i \ge 1$. The ordinary generating formal series of the sequence $(\mathfrak{p}_k)_k$ is*

$$\mathrm{OGF}(\mathfrak{p}_k)_k = \prod_{i=1}^{\infty} I^{\mathrm{OGF}}_{i\mathbb{N}}(X) = \prod_{i=1}^{\infty}(1 + X^i + X^{2i} + X^{3i} + \cdots).$$

Proof. Given a partition $[x_1, \ldots, x_n]$ of k, let us denote by j_i the number of i in $[x_1, \ldots, x_n]$ for each $i \in \mathbb{N}$. Clearly

$$k = \sum_{i=1}^{\infty} ij_i.$$

Set $k_i = i j_i$, we get a bijection between the partitions of k and the infinite compositions $(k_i)_{i \geq 1}$ of k with $k_i \in i\mathbb{N}$ for each $i \geq 1$. Each element in $i\mathbb{N}$ is either 0 or an element greater or equal than i; therefore we can apply Theorem 7.127 obtaining

$$\mathfrak{p}_k = [X^k] \prod_{i=1}^{\infty} I_{i\mathbb{N}}^{\text{OGF}}(X). \qquad \Box$$

The computation of the values of the $\mathfrak{p}_k$'s is not a simple matter; we mention the asymptotic formula due to Hardy and Ramanujan [22], stating that

$$\mathfrak{p}_k \sim \frac{1}{4k\sqrt{3}} \exp\left(\pi\sqrt{\frac{2k}{3}}\right) \qquad k \to +\infty. \qquad (7.131.a)$$

We will be able to say something more in Corollary 7.134, after we acquire some additional tools.

Example 7.132 Since $[X^{<i}] I_{i\mathbb{N}}^{\text{OGF}}(X) = 1$, it follows from Proposition 7.131 that

$$\mathfrak{p}_k = [X^k] \prod_{i=1}^{\infty} (1 + X^i + X^{2i} + X^{3i} + \cdots)$$

$$= [X^k] \prod_{i=1}^{k} \left(\sum_{0 \leq m \leq k/i} X^{mi} \right).$$

For instance, we get

$$\mathfrak{p}_5 = [X^5] (1 + X + X^2 + X^3 + X^4 + X^5)(1 + X^2 + X^4)(1 + X^3)(1 + X^4)(1 + X^5)$$

$$= [X^5] (1 + X + 2X^2 + 3X^3 + 5X^4 + 7X^5 + 7X^6 + \cdots) = 7$$

as we found directly in Example 7.130. Notice that the right-hand term of the asymptotic formula (7.131.a) for $k = 5$ equals $8.9\ldots$, with an error of more than 27 %. $\Box$

7.10.2 Euler's Pentagonal Theorem ☕

In Proposition 7.131 we showed that the ordinary generating formal series $\sum_{k=0}^{\infty} \mathfrak{p}_k X^k$ of the number $\mathfrak{p}_k$ of integer partitions of $k \in \mathbb{N}$ equals

$$\prod_{\ell=1}^{\infty} I_{\ell\mathbb{N}}^{\text{OGF}}(X) = \prod_{\ell=1}^{\infty} (1 + X^{\ell} + X^{2\ell} + X^{3\ell} + \cdots).$$

Since $1 + X + X^2 + X^3 + \cdots = \dfrac{1}{1-X}$, by Point 7 of Proposition 7.80 we get

$$\mathfrak{p}_k = \big[X^k\big]\prod_{\ell=1}^{\infty} I_{\ell\mathbb{N}}^{\mathrm{OGF}}(X) = \big[X^k\big]\prod_{\ell=1}^{\infty}\frac{1}{1-X^\ell}.$$

Euler's celebrated Pentagonal Theorem allows one to obtain an explicit description of the inverse of the latter formal power series

$$\prod_{i=1}^{\infty}(1-X^i) := \sum_{k=0}^{\infty} \Delta_k X^k.$$

Clearly $\Delta_0 = 1$. Henceforth, the following notation will be used.

Notation. For any $k \geq 1$, denote by $\mathscr{R}_k$ the compositions $(i_n, \ldots, i_1)$ of k such that $1 \leq i_n < \cdots < i_1$.

By Corollary 7.129, with $F_i = \{i\}$, $i \in \mathbb{N}_{\geq 1}$, the coefficient Δ_k is the difference between the numbers of the compositions of even and odd *length* in $\mathscr{R}_k$. It is worth noticing that, equivalently, Δ_k equals the difference of distinct partitions of k into an even number of (distinct) terms, with that of the number of distinct partitions of k into an odd number of terms. Euler proved that $\Delta_k = 0$ except when k is a *pentagonal number*, namely of the form $\dfrac{n(3n\pm 1)}{2}$ for some $n \geq 1$; notice that $\dfrac{m(3m-1)}{2} \neq \dfrac{\ell(3\ell+1)}{2}$ for any $\ell, m \in \mathbb{N}_{\geq 1}$: indeed,

$$0 = m(3m-1) - \ell(3\ell+1) = 3(m-\ell)(m+\ell) - (m+\ell) = (m+\ell)[3(m-\ell)-1]$$

has no solutions in $\mathbb{N}_{\geq 1}$.

The adjective *pentagonal* derives from the fact that the sequence

$$(n(3n-1)/2)_{n\geq 1} = \quad 1,\ 5,\ 12,\ 22,\ 35,\ldots$$

can be obtained via the sequence of pentagons of Fig. 7.1.

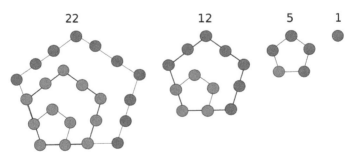

Fig. 7.1 The construction of the first pentagonal numbers 1, 5, 12, 22 (Source: commons wikimedia, https://commons.wikimedia.org/wiki/File:Polygonal_Number_5.gif. Author: Aldoaldoz (Own work), Creative Commons Attribution-Share Alike CC BY-SA 3.0 Unported license)

Theorem 7.133 (Euler's Pentagonal Theorem)

$$\prod_{i=1}^{\infty}(1 - X^i) = 1 + \sum_{n=1}^{\infty}(-1)^n \left(X^{n(3n-1)/2} + X^{n(3n+1)/2} \right).$$

Proof. For each composition $\sigma = (i_n, \ldots, i_1)$ belonging to $\mathscr{R}_k$ we denote by $L(\sigma) = n$ its *length*, by $m(\sigma) := i_n$ its *minimum element*, and we set $s(\sigma)$ to be the number of decreasing consecutive terms from the last term i_1 in the sequence σ, namely

$$s(\sigma) = \begin{cases} n \text{ if } i_2 + 1 = i_1, \ i_3 + 1 = i_2, \ \ldots, \ i_n + 1 = i_{n-1}; \\ \min\{j \in \{1, \ldots, n-1\} : i_{j+1} + 1 < i_j\} \text{ otherwise.} \end{cases}$$

For instance, $(3, 6, 7)$ belongs to $\mathscr{R}_{16}$ and we have $L(3, 6, 7) = 3$, $m(3, 6, 7) = 3$ and $s(3, 6, 7) = 2$. The set $\mathscr{R}_k$ is the disjoint union of the following three sets:

- $S_k := \{\sigma \in \mathscr{R}_k : s(\sigma) < m(\sigma)\} \setminus \{\sigma \in \mathscr{R}_k : s(\sigma) = L(\sigma) = m(\sigma) - 1\}$;
- $T_k := \{\sigma \in \mathscr{R}_k : m(\sigma) \leq s(\sigma)\} \setminus \{\sigma \in \mathscr{R}_k : m(\sigma) = s(\sigma) = L(\sigma)\}$;
- $Z_k = \{\sigma \in \mathscr{R}_k : s(\sigma) = L(\sigma) = m(\sigma) - 1 \text{ or } m(\sigma) = s(\sigma) = L(\sigma)\}$.

For each $\sigma = (i_n, \ldots, i_1) \in S_k$ we define

$$\Phi(\sigma) = \begin{cases} (s(\sigma), i_n, \ldots, i_{s(\sigma)+1}, i_{s(\sigma)} - 1, \ldots, i_1 - 1) & \text{if } s(\sigma) < L(\sigma) = n, \\ (s(\sigma), i_n - 1, \ldots, i_1 - 1) & \text{if } s(\sigma) = L(\sigma) = n. \end{cases}$$

For example, one has that $(3, 6, 7)$ belongs to S_{16} and $\Phi(3, 6, 7) = (2, 3, 5, 6)$. Representing $\sigma = (i_n, \ldots, i_1)$ as a Ferrer's diagram, namely with $n = L(\sigma)$ rows of $i_1 > i_2 > \cdots > i_n$ points, $\Phi(\sigma)$ is obtained moving the rightmost points of the first $s(\sigma)$ lines in a new row at the bottom of the diagram (see Fig. 7.2). We will prove that Φ is a bijection $S_k \to T_k$. Let us start verifying step by step that Φ is an injective map $S_k \to T_k$.

1. $\Phi(S_k) \subseteq \mathscr{R}_k$: given $\sigma = (i_n, \ldots, i_1)$ in S_k, we have $s(\sigma) < m(\sigma) = i_{n-s(\sigma)}$. Let us distinguish the cases $s(\sigma) < L(\sigma)$ and $s(\sigma) = L(\sigma) < m(\sigma) - 1$.

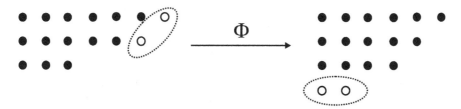

Fig. 7.2 $\Phi(3, 6, 7) = (2, 3, 5, 6)$

a. If $s(\sigma) < L(\sigma) = n$, then $i_{s(\sigma)+1} < i_{s(\sigma)} - 1$; hence

$$s(\sigma) < i_n < \cdots < i_{s(\sigma)+1} < i_{s(\sigma)} - 1 < \cdots < i_1 - 1,$$

and therefore $\Phi(\sigma)$ belongs to $\mathscr{R}_k$.

b. If $s(\sigma) = L(\sigma) = n < m(\sigma) - 1 = i_n - 1$, then $\Phi(\sigma) = (s(\sigma) = n, i_n - 1, \ldots, i_1 - 1)$ belongs to $\mathscr{R}_k$.

2. $\Phi : S_k \to \mathscr{R}_k$ is injective: indeed if $\sigma_1 := (i_n, \ldots, i_1), \sigma_2 := (j_p, \ldots, j_1)$ belong to S_k and $\Phi(\sigma_1) = \Phi(\sigma_2)$ then

$$(s(\sigma_1), i_n, \ldots, i_{s(\sigma_1)+1}, i_{s(\sigma_1)} - 1, \ldots, i_1 - 1) =$$
$$= (s(\sigma_2), j_p, \ldots, j_{s(\sigma_2)+1}, j_{s(\sigma_2)} - 1, \ldots, j_1 - 1),$$

and hence $n = p$ and $\sigma_1 = \sigma_2$.

3. $\Phi(S_k) \subseteq T_k$: for each $\sigma \in S_k$ we have $s(\Phi(\sigma)) \geq s(\sigma) = m(\Phi(\sigma))$. If $s(\Phi(\sigma)) = m(\Phi(\sigma))$, then, since $s(\sigma) = m(\Phi(\sigma))$, both terms are less or equal to $L(\sigma) = L(\Phi(\sigma)) - 1$.

Actually, $\Phi : S_k \to T_k$ is a bijection. Indeed, if $\tau = (j_{n+1}, j_n, \ldots, j_1)$ belongs to T_k, then $j_{n+1} = m(\tau) \leq s(\tau)$, and either $m(\tau) < s(\tau)$ or $m(\tau) = s(\tau) < L(\tau) = n + 1$. Consider

$$\sigma := \begin{cases} (j_n, \ldots, j_{m(\tau)+1}, j_{m(\tau)} + 1, \ldots, j_1 + 1) & \text{if } j_{n+1} = m(\tau) < n, \\ (j_n + 1, \ldots, j_1 + 1) & \text{if } j_{n+1} = m(\tau) = n. \end{cases}$$

Let us check again step by step that σ belongs to S_k (i.e., it belongs to $\mathscr{R}_k$, $s(\sigma) < m(\sigma)$, and $s(\sigma) = L(\sigma)$ implies $s(\sigma) < m(\sigma) - 1$) and $\Phi(\sigma) = \tau$.

4. σ belongs to $\mathscr{R}_k$: indeed

$$j_n + \cdots + j_{m(\tau)+1} + (j_{m(\tau)} + 1) + \cdots + (j_1 + 1) =$$
$$= j_n + \cdots + j_1 + m(\tau) = j_n + \cdots + j_1 + j_{n+1} = k,$$

and $j_n < \cdots < j_{m(\tau)+1} < j_{m(\tau)} < j_{m(\tau)} + 1 < \cdots < j_1 + 1$.

5. $s(\sigma) = m(\tau) < m(\sigma)$: let us distinguish the cases $m(\tau) < n$ and $m(\tau) = n$.

a. $m(\tau) < n$: $j_{m(\tau)+1} < j_{m(\tau)}$ implies $s(\sigma) \leq m(\tau)$; since $m(\tau) \leq s(\tau)$ then $j_{m(\tau)} + 1 = j_{m(\tau)-1}, \ldots, j_2 + 1 = j_1$ and hence $s(\sigma) = m(\tau) = j_{n+1} < j_n = m(\sigma)$.

b. $m(\tau) = n$: then $s(\tau) \geq m(\tau) = n$ implies $s(\sigma) = n = m(\tau) = j_{n+1} < j_n < j_n + 1 = m(\sigma)$.

6. $s(\sigma) = L(\sigma)$ implies $s(\sigma) < m(\sigma) - 1$: assume by contradiction that $s(\sigma) = L(\sigma) = n = m(\sigma) - 1$. Then $\sigma = (n + 1, n + 2, \ldots, n + n = 2n)$. Let us distinguish the cases $m(\tau) < n$ and $m(\tau) = n$.

a. $m(\tau) < n$: we would have

$$(n+1, n+2, \ldots, n+n = 2n) = \sigma = (j_n, \ldots, j_{m(\tau)+1}, j_{m(\tau)} + 1, \ldots, j_1 + 1),$$

and hence $j_{m(\tau)+1} = j_{m(\tau)}$ contradicting $\tau \in \mathscr{R}_k$.

b. $m(\tau) = n$: we would have

$$(n+1, n+2, \ldots, n+n = 2n) = \sigma = (j_n + 1, \ldots, j_1 + 1),$$

and hence $n = m(\tau) = j_{n+1} = j_n$ contradicting $\tau \in \mathscr{R}_k$.

7. $\Phi(\sigma) = \tau$: let us distinguish the cases $m(\tau) < n$ and $m(\tau) = n$.

a. $m(\tau) < n$: we have $\Phi(\sigma) = \Phi(j_n, \ldots, j_{m(\tau)+1}, j_{m(\tau)} + 1, \ldots, j_1 + 1)$; since $m(\tau) = s(\sigma)$, we get

$$\Phi(j_n, \ldots, j_{m(\tau)+1}, j_{m(\tau)} + 1, \ldots, j_1 + 1) =$$

$$= (s(\sigma), j_n, \ldots, j_{s(\sigma)+1}, j_{s(\sigma)} + 1 - 1, \ldots, j_1 + 1 - 1) =$$

$$= (j_{n+1}, \ldots, j_{s(\sigma)+1}, j_{s(\sigma)}, \ldots, j_1) = \tau.$$

b. $m(\tau) = n$: we have $\Phi(\sigma) = \Phi(j_n + 1, \ldots, j_1 + 1)$; since $n = m(\tau) = s(\sigma)$, we get

$$\Phi(j_n + 1, \ldots, j_1 + 1) = (s(\sigma) = m(\tau), j_n, \ldots, j_1) = \tau.$$

It follows that $\Phi : S_k \to T_k$ is one to one and sends n-compositions of k in $(n+1)$-compositions of k, switching the parity of their length. Since Φ maps even (resp. odd) compositions in S_k into odd (resp. even) compositions in T_k, it follows that the difference Δ_k between even and odd compositions in $\mathscr{R}_k$ equals the difference between even and odd compositions in Z_k.

• If $\sigma \in Z_k$ with $s(\sigma) = L(\sigma) = m(\sigma) = n$ then $\sigma = (n, n+1, \ldots, 2n-1)$ and

$$k = n + (n+1) + (n+2) + \cdots + (n + (n-1)) = \frac{n(3n-1)}{2}.$$

In the case where $n = L(\sigma)$ is even (resp. odd), then σ counts $+1$ (resp. -1) in Δ_k: therefore it counts $(-1)^{L(\sigma)} = (-1)^n$ in Δ_k.

• If $\sigma \in Z_k$ with $s(\sigma) = L(\sigma) = m(\sigma) - 1 = n$ then $\sigma = (n+1, \ldots, 2n)$ and

$$k = (n+1) + (n+2) + \cdots + (n+n) = n \times n + \frac{n(n+1)}{2} = \frac{n(3n+1)}{2}.$$

In the case where $n = L(\sigma)$ is even (resp. odd), then σ counts $+1$ (resp. -1) in Δ_k: therefore it counts $(-1)^{L(\sigma)} = (-1)^n$ in Δ_k.

We thus conclude that

$$[X^k]\prod_{i=1}^{\infty}(1 - X^i) = \begin{cases} 1 & \text{if } k = 0, \\ (-1)^n & \text{if } k = \dfrac{n(3n-1)}{2}, \\ (-1)^n & \text{if } k = \dfrac{n(3n+1)}{2}, \\ 0 & \text{otherwise.} \end{cases} \qquad \square$$

We are now ready to describe the number $\mathfrak{p}_k$ of the integer partitions of $k \in \mathbb{N}_{\geq 1}$.

Corollary 7.134 *Set* $\mathfrak{p}_\ell = 0$ *for each* $\ell < 0$ *and* $\mathfrak{p}_0 = 1$, *the numbers* $\mathfrak{p}_k$, $k \in \mathbb{N}_{\geq 1}$, *satisfy the following relation:*

$$\mathfrak{p}_k = \sum_{n=1}^{\infty}(-1)^{n+1}\left(\mathfrak{p}_{k-n(3n-1)/2} + \mathfrak{p}_{k-n(3n+1)/2}\right).$$

Proof. At the beginning of this section we have observed that

$$\sum_{k=0}^{\infty}\mathfrak{p}_k X^k = \prod_{\ell=1}^{\infty}\frac{1}{1 - X^\ell}.$$

Since the formal power series $\displaystyle\prod_{\ell=1}^{\infty}\frac{1}{1 - X^\ell}$ and $\displaystyle\prod_{i=1}^{\infty}(1 - X^i)$ are the inverse one another, by Theorem 7.133 we get

$$1 = \left(\sum_{k=0}^{\infty}\mathfrak{p}_k X^k\right)\left(1 + \sum_{n=1}^{\infty}(-1)^n\left(X^{n(3n-1)/2} + X^{n(3n+1)/2}\right)\right)$$

$$= \left(\sum_{k=0}^{\infty}\mathfrak{p}_k X^k\right)\left(1 - X - X^2 + X^5 + X^7 - X^{12} - X^{15} + X^{22} + X^{26} - \cdots\right).$$

Therefore one gets

$$1 = \mathfrak{p}_0, \quad 0 = \mathfrak{p}_1 - \mathfrak{p}_0, \quad 0 = \mathfrak{p}_2 - \mathfrak{p}_1 - \mathfrak{p}_0, \quad \ldots \quad \text{and for any } k \geq 1$$

$$0 = \mathfrak{p}_k - \mathfrak{p}_{k-1} - \mathfrak{p}_{k-2} + \mathfrak{p}_{k-5} + \mathfrak{p}_{k-7} - \mathfrak{p}_{k-12} - \mathfrak{p}_{k-15} + \cdots$$

$$= \mathfrak{p}_k + \sum_{n=1}^{\infty}(-1)^n\left(\mathfrak{p}_{k-n(3n-1)/2} + \mathfrak{p}_{k-n(3n+1)/2}\right).$$

Hence

$$\mathfrak{p}_0 = 1, \quad \mathfrak{p}_1 = \mathfrak{p}_0, \quad \mathfrak{p}_2 = \mathfrak{p}_1 + \mathfrak{p}_0, \quad \ldots \quad \text{and for any } k \geq 1$$

$$p_k = p_{k-1} + p_{k-2} - p_{k-5} - p_{k-7} + p_{k-12} + p_{k-15} - \cdots$$

$$= \sum_{n=1}^{\infty} (-1)^{n+1} \left(p_{k-n(3n-1)/2} + p_{k-n(3n+1)/2} \right)$$

$$= \sum_{n=1}^{n(3n+1)/2} (-1)^{n+1} \left(p_{k-n(3n-1)/2} + p_{k-n(3n+1)/2} \right). \qquad \square$$

7.11 Problems

Problem 7.1 Let $A(X)$ be a formal power series and $m \in \mathbb{N}$. Prove that for each $n \geq m \in \mathbb{N}$ one has $[X^n] (X^m A(X)) = \left[X^{n-m} \right] A(X)$.

Problem 7.2 Let $A(X)$ be a formal power series. Prove that $A'(X) = 0$ if and only if $A(X)$ is a constant.

Problem 7.3 Let $A(X)$ and $B(X)$ be two formal power series and $m, n \in \mathbb{N}$. Prove that

$$\left([X^m] A(X) \right) \left([X^n] B(X) \right) = [X^n] \left(\left([X^m] A(X) \right) B(X) \right) .$$

Problem 7.4 Let $A(X) = \sum_{k=0}^{\infty} \frac{1}{k!} X^k$ and $B(X) = \sum_{k=1}^{\infty} (-1)^{k-1} X^k$. Determine a closed form for $A(B(X))$ and calculate the coefficient of X^2.

Problem 7.5 Calculate a closed form for the formal power series

$$A(X) = \sum_{n=0}^{\infty} n X^n, \quad B(X) = \sum_{n=0}^{\infty} n(n+1) X^n.$$

Problem 7.6 Let $A(X) = \sum_{n=0}^{\infty} 2^n X^n$; decide whether or not $A(X)$ is invertible in $\mathbb{R}[[X]]$ and, if so, determine its inverse.

Problem 7.7 Prove that the derivative of a formal power series has the same properties (as far as sums and products are concerned) as the derivative of a function: if $A(X)$ and $B(X)$ are two formal power series, then

$$(A(X) + B(X))' = A'(X) + B'(X) \text{ and}$$

$$(A(X) B(X))' = A'(X) B(X) + A(X) B'(X).$$

Problem 7.8 Let $0 < m \in \mathbb{N}$ and $A(X)$ be a formal power series in $\mathbb{R}[[X]]$. Prove that if $A^m(X) = 1$, then $A(X)$ is constant, equal to 1 if m is odd, and to ± 1 if m is even.

Problem 7.9 Calculate a closed form for the formal power series

$$A(X) = \sum_{n=2}^{\infty} (n-1)X^n.$$

Problem 7.10 In view of the upcoming elections, the leader Kingnzi is in the process of choosing candidates. In this first phase it must be decided how many and which positions of the 10 available spots on the ballot should be awarded to various categories of potential candidates; only later will the names of the candidates be selected. The choice will be made respecting the following constraints regarding the four disjoint categories of potential candidates:

1. An even number of party functionaries;
2. An odd number of former council members;
3. At least 5 declaredly traditional conservatives;
4. At most one clearly progressive candidate.

How many possible choices are there in this preliminary phase?

Problem 7.11 How many 6 digit numbers are there having an even number of 7's, an odd number of 9's, and two 5's?

Problem 7.12 Prove that for each $n \geq 0$ one has

$$\binom{-1/2}{n} = \left(-\frac{1}{4}\right)^n \binom{2n}{n}.$$

Problem 7.13 Calculate the number of ways in which it is possible to distribute 25 identical liquorice sticks to Carl, Roberta, Joe, Steve, and Al if Carl wants at least 8 but not more than 12 of them, Roberta no more than 4, and Joe and Al at least 1.

Problem 7.14 A certain pastry shop located in Calle de San Pantalon in Venice displays in its shop windows some "frittelle" with pastry cream (*crema*) filling, some with egg-nog (*zabaione*) filling, and others which are traditional Venetian style "frittelle" with no filling at all. In how many ways can one choose 15 frittelle if one wishes to have an even number ≥ 2 with pastry cream, an odd number ≥ 1 with egg-nog filling, and at least 3 traditional Venetian frittelle? (For the readers' benefit it may be of interest, though of no help in solving the problem, that "frittelle" [singular "frittella"] are a generalization of donuts without holes, but considerably better than their Anglo-Saxon counterpart in the view of many who have tried both.)

Problem 7.15 Calculate the number of ways one can prepare a tray of 12 hors d'oeuvres of five different types with at most four of any given type, and assuming that there must be at least one of each type.

Problem 7.16 In how many ways is it possible to assign 20 mint gum drops and 10 licorice ones to 10 children so that each child receives exactly three gum drops.

Problem 7.17 In a supermarket a lunatic named Pascal throws items at a man, choosing them at random from the supermarket shelves. In the end, before anyone is able to stop him, Pascal manages to throw 11 items. The witnesses present furnish incompatible descriptions of the items thrown at the man; in the end, however, all agree that the items thrown satisfied the following conditions:

1. An even number (≥ 0) of cans of tomato paste;
2. At least 2 bottles of olive oil;
3. Between 4 and 7 cans of beer;
4. A souvenir of the Cathedral of Milan.[7]

Bearing in mind that all possible versions of the story coherent with the conditions given above were actually furnished by witnesses, at least how many people were present in the supermarket besides the man and Pascal?

Problem 7.18 Using Theorem 7.114, compute the coefficients of the formal series

$$\frac{1 + X}{1 - X - X^2}.$$

---

[7]This exercise was inspired by a true story that took place in Milan on December 13, 2009.

Chapter 8
Generating Formal Series and Applications

Abstract We deepen here the insight on formal power series. We temporarily abandon formality and consider the notion of the convergence of a power series; we'll
see in particular how a smart choice of a closed form of a given power series is
useful to recover the sum of the power series. A large part of the chapter is devoted
to determining the generating formal series for some notable sequences, including
sequences of binomial coefficients, harmonic numbers, Stirling and Bell numbers,
Eulerian numbers, as well as sequences of integral powers. One section is devoted to
the Bernoulli numbers: not only do they allow us to express the sum of consecutive
m-th powers of the natural numbers (Faulhaber's formula), but they turn out to be
useful, as we shall see in Chap. 13, in approximating the sum of the consecutive
values on the natural numbers of any given smooth function. Some useful estimates
of the Bernoulli numbers are given via the Riemann zeta function, namely the sum
of the series of the inverses of a given real power of the natural numbers. Finally, a
section is devoted to the applications of formal power series to probabilities.

8.1 Formal Power Series and Their Sum Function

This section contains plenty of insights that will be of interest to the experienced
reader but are by no means essential to the understanding of a large part of the
remainder of the book. It should be accessible by students with an understanding of
the basic notions of convergence of a power series.

In the study of the relations between formal power series and their closed forms,
the notion of *convergence of a series* arises naturally. If $P(X) = \sum_{n=0}^{n} a_n X^n$ is a polynomial it makes sense, for every $x \in \mathbb{R}$ or $\mathbb{C}$, to evaluate the *finite sum* $\sum_{n=0}^{N} a_n x^n$: we
may then consider the *polynomial function* $x \mapsto P(x) := \sum_{n=0}^{N} a_n x^n$ which is clearly a
closed form for $P(X)$ (see Example 7.68). We shall establish here something similar
for a class of formal series more general than polynomials.

© Springer International Publishing Switzerland 2016   275
C. Mariconda and A. Tonolo, *Discrete Calculus*,
UNITEXT - La Matematica per il 3+2 103, DOI 10.1007/978-3-319-03038-8_8

Definition 8.1 We say that a formal power series $A(X) = \sum\limits_{n=0}^{\infty} a_n X^n$ **converges** at $x \in \mathbb{R}$ if the so called **sum of the formal series** $A(X)$ at x, i.e., the limit

$$\sum_{n=0}^{\infty} a_n x^n := \lim_{N\to\infty} \sum_{n=0}^{N} a_n x^n \qquad (8.1.a)$$

exists and is finite. We shall denote the value in (8.1.a) by $A(x)$, hoping no ambiguity occurs. The power series $A(X)$ is said to converge **absolutely** at x if the power series $\sum\limits_{n=0}^{\infty} |a_n| X^n$ converges at $|x|$. $\square$

Remark 8.2 We recall that absolute convergence of a power series at x implies its convergence at x.

We assume that the reader is familiar with the main results concerning the convergence of power series, the essentials of which are recalled here; the details and the proofs of the results formulated can be found in [33].

Proposition 8.3 (Radius and domain of convergence of a power series) *Let $A(X)$ be a formal power series. There exists $R_{A(X)} \in \mathbb{R}_{\geq 0} \cup \{+\infty\}$ such that:*

- *$A(X)$ converges absolutely on $\{x : |x| < R_{A(X)}\}$;*
- *$A(X)$ does not converge on $\{x : |x| > R_{A(X)}\}$.*

*This value $R_{A(X)}$ is called the **radius of convergence** of the formal power series $A(X)$. The **convergence set** of $A(X)$ is the set $D_{A(X)}$ of points at which the formal power series converges; it is an interval since*

$$\{x : |x| < R_{A(X)}\} \subset D_{A(X)} \subset \{x : |x| \leq R_{A(X)}\}.$$

Example 8.4 In order to find the radius of convergence of a power series $A(X) = \sum\limits_{n=0}^{\infty} a_n X^n$ with non-zero coefficients, one often resorts to the quotient method: since, for $x \neq 0$,

$$\left| \frac{a_{n+1} x^{n+1}}{a_n x^n} \right| = |x| \left| \frac{a_{n+1}}{a_n} \right|,$$

it turns out that

$$R_{A(X)} = \lim_{n\to+\infty} \frac{|a_n|}{|a_{n+1}|}$$

whenever the limit exists.

- The radius of convergence of the series

$$e^X = 1 + X + \frac{X^2}{2!} + \cdots + \frac{X^n}{n!} + \cdots$$

is $+\infty$: indeed

$$\lim_{n \to +\infty} \left| \frac{1/n!}{1/(n+1)!} \right| = \lim_{n \to +\infty} (n+1) = +\infty.$$

- The radius of convergence of the series

$$\sum_{n=0}^{\infty} n! X^n = 1 + X + 2X^2 + 3! X^3 + \cdots$$

is 0: indeed

$$\lim_{n \to +\infty} \left| \frac{n!}{(n+1)!} \right| = \lim_{n \to +\infty} \frac{1}{n+1} = 0.$$

- The radius of convergence of

$$\frac{1}{1-X} = \sum_{n=0}^{\infty} X^n = 1 + X + X^2 + X^3 + \cdots$$

equals 1: indeed

$$\lim_{n \to +\infty} \left| \frac{1}{1} \right| = 1.$$

- Let $a \in \mathbb{R} \setminus \{0\}$. The radius of convergence of the series

$$\sum_{n=0}^{\infty} \binom{a}{n} X^n$$

is 1. Indeed

$$\lim_{n \to +\infty} \left| \frac{\binom{a}{n+1}}{\binom{a}{n}} \right| = \lim_{n \to +\infty} \left| \frac{a-n}{n+1} \right| = 1. \qquad \square$$

Definition 8.5 A formal power series $A(X) = \sum_{n=0}^{\infty} a_n X^n$ that converges for at least one point different from 0 is said to be **summable**. The function $x \mapsto A(x)$ defined on the convergence set $D_{A(X)}$ by

$$x \mapsto A(x) := \sum_{n=0}^{\infty} a_n x^n$$

is called the **sum function** of the formal power series $A(X)$.   □

Remark 8.6 If $A(X) = \sum_{n=0}^{\infty} a_n X^n$ is summable, by [33, Theorem 8.1] for each x belonging to the interior of the convergence set $D_{A(X)}$ one has

$$\frac{d}{dx} \left(\sum_{n=0}^{\infty} a_n x^n \right) = \sum_{n=1}^{\infty} n a_n x^{n-1} = A'(x).$$

The sum function of a summable power series is one of its closed forms.

Theorem 8.7 *If $A(X)$ is a summable power series, then the sum function of $A(X)$ is a closed form of $A(X)$.*

Proof. We have
$$A(0) = a_0 = [X^0]A(X);$$

since $\dfrac{d}{dx} \left(\displaystyle\sum_{n=0}^{\infty} a_n x^n \right) = A'(x)$, then

$$A'(0) = a_1 = [X^1]A(X).$$

By applying the above differentiation formula to the power series $A'(X)$ we get

$$A''(0) = 2a_2 = 2[X^2]A(X).$$

We understand (and we invite the reader to prove it by induction) that

$$A^{(m)}(0) = m!a_m = m![X^m]A(X) \qquad \forall m \in \mathbb{N},$$

proving that $x \mapsto A(x)$ is a closed form for $A(X)$.   □

We now consider the following problem: let $A(X)$ be a summable power series, and let f be one of its closed form. Is it true that $f(x) = A(x)$ in a neighborhood of 0? The answer is negative, of course.

Example 8.8 Let $A(X) = 0$ and f be the closed form of 0 defined in Example 7.75. Then, for $x \neq 0$, $f(x) > 0$ whereas $A(x) = 0$.   □

We are naturally led to the notion of an analytic function: we formulate just the essentials of the theory, the details of which are beyond the scope of our textbook: we refer to [34] for a thorough treatise on the subject. Essentially, a function is analytic if its Taylor expansion at every point converges to the function in a neighborhood of that point.

Definition 8.9 (*Analytic functions*) A $\mathscr{C}^\infty$ function defined in an open set U is said to be **analytic** if, for every $x_0 \in U$ there is $r > 0$ such that

$$f(x) = \sum_{n=0}^{\infty} \frac{f^{(n)}(x_0)}{n!}(x - x_0)^n \qquad \forall |x - x_0| < r. \tag{8.9.a}$$

$\square$

Example 8.10 The function f defined in Example 7.75 is not analytic: indeed its derivatives in 0 are all equal to 0, however $f(x) > 0$ if $x \neq 0$, thus (8.9.a) does not hold around 0. $\square$

Remark 8.11 Actually, it is enough, instead of (8.9.a), that for each x_0 in the open set U there exists a sequence $(a_n)_n$, depending on x_0, such that

$$f(x) = \sum_{n=0}^{\infty} a_n (x - x_0)^n \qquad \forall |x - x_0| < r.$$

Indeed the differentiation theorem for power series [33, Theorem 8.1] shows that if the above equality holds, then necessarily $a_n = \dfrac{f^{(n)}(x_0)}{n!}$ for all n.

Example 8.12 The function $x \mapsto e^x$ is analytic. Indeed let $x_0 \in \mathbb{R}$. Since the radius of convergence of e^X is $+\infty$, one has

$$e^x - e^{x_0} = e^{x_0}(e^{x - x_0} - 1) = e^{x_0} \sum_{n=1}^{\infty} \frac{(x - x_0)^n}{n!}$$

whence

$$e^x = \sum_{n=0}^{\infty} \frac{e^{x_0}}{n!}(x - x_0)^n \qquad \forall x \in \mathbb{R}.$$

The conclusion follows from Remark 8.11 or from the fact that $f^{(n)}(x_0) = e^{x_0}$ for all n. $\square$

Example 8.13 Here is a list of useful analytic functions on their domain:

- Polynomials;
- $\log(x)$;
- $\cos x, \sin x$;

- $\cosh x$, $\sinh x$;
- $(1 + x)^a$, for $a \in \mathbb{R}$, $x > -1$.   □

Analytic functions are the best possible closed form for a given summable power series: they allow one to pass from the equality among power series $f(X) = A(X)$ to the pointwise equality $f(x) = A(x)$: a dream for the lazy student that is asked to compute sums of power series. We omit the proof of the following important result, based on some well-known facts concerning analytic functions (such as the Identity Principle) that are beyond the scope of the book.

Theorem 8.14 *Let $A(X)$ be a summable power series.*

1. *The sum function of $A(X)$ is analytic on $] - R_{A(X)}, R_{A(X)}[$.*
2. *Let f be a closed form of $A(X)$, and assume that f is analytic in an open interval I containing 0. Then $f(x) = A(x)$ for each $x \in] - R_{A(X)}, R_{A(X)}[\cap I$.*

Example 8.15 In Example 8.4 we showed that the radius of convergence of the formal power series $\dfrac{1}{1 - X}$ is 1. Since the function $\dfrac{1}{1 - x}$ is analytic on $] - 1, 1[$ it follows from Theorem 8.14 that

$$\frac{1}{1 - x} = 1 + x + \cdots + x^n + \cdots \qquad \forall |x| < 1.$$

Analogously we obtain the following sums of series:

$$\sin x = x - \frac{x^3}{3!} + \cdots + (-1)^n \frac{x^{2n+1}}{(2n+1)!} + \cdots \qquad \forall x \in \mathbb{R},$$

$$\cos x = 1 - \frac{x^2}{2!} + \cdots + (-1)^n \frac{x^{2n}}{(2n)!} + \cdots \qquad \forall x \in \mathbb{R},$$

$$\sinh x = x + \frac{x^3}{3!} + \cdots + \frac{x^{2n+1}}{(2n+1)!} + \cdots \qquad \forall x \in \mathbb{R},$$

$$\cosh x = 1 + \frac{x^2}{2!} + \cdots + \frac{x^{2n}}{(2n)!} + \cdots \qquad \forall x \in \mathbb{R},$$

$$\log(1 + x) = x - \frac{x^2}{2} + \cdots + (-1)^{n+1} \frac{x^n}{n!} + \cdots \qquad \forall |x| < 1,$$

$$(1 + x)^a = \sum_{n=0}^{\infty} \binom{a}{n} x^n \qquad \forall |x| < 1. \quad □$$

In order to apply Theorem 8.14 it is essential to recognize analytic functions at just a glance. The functions listed in Example 8.13 provide, together with the next result, tons of analytic functions.

Proposition 8.16 *1. Sums, products, quotients and composite of analytic functions are analytic on their domains.*
2. Extension by continuity of analytic functions are analytic. More precisely let f be analytic on $I \setminus \{x_1, \ldots, x_m\}$, where I is an open interval and $x_1, \ldots, x_m \in I$. Assume that the limits $\lim_{x \to x_i} f(x)$, $i = 1, \ldots, m$, exist and are finite. Then the extension by continuity of f to I is analytic on I.

Example 8.17 The function $f(x) = \dfrac{e^x}{x-1}$ is a quotient of two analytic functions on $\mathbb{R}$: it follows from Point 1 of Proposition 8.16 that f is analytic on its domain $\mathbb{R} \setminus \{1\}$. Moreover $\lim_{x \to 1} f(x) = 1$: it follows from Point 2 of Proposition 8.16 that the extension by continuity of f on $\mathbb{R}$ defined by

$$x \mapsto \begin{cases} f(x) & \text{if } x \neq 1, \\ 1 & \text{if } x = 1 \end{cases}$$

is analytic in $\mathbb{R}$. □

The reader will not be surprised at this stage to know that given an analytic closed form of a summable power series, one can in principle determine the radius of convergence of the series. We now formulate a result in this direction concerning rational fractions.

Proposition 8.18 *Let $P(X)$ and $Q(X)$ be two polynomials without common factors, with $[X^0]Q(X) \neq 0$. The radius of convergence of $A(X) = \dfrac{P(X)}{Q(X)}$ is the modulus of the root (possibly complex) of the polynomial $Q(X)$ which is closest to the origin, that is,*

$$R_{A(X)} = \min\{|\alpha| : \alpha \in \mathbb{C} \text{ and } Q(\alpha) = 0\}.$$

Remark 8.19 For the advanced reader, let us say that the proof of Proposition 8.18, that we omit, relies on the fact that the function $x \mapsto \dfrac{P(x)}{Q(x)}$ can be seen as a complex valued function on the complex disk of radius $|\alpha|$ centered in the origin; the Cauchy formula for holomorphic functions yields the result, see [34, Theorem 10.16] for other details.

Example 8.20 (The sum function of the OGF of the binomials) Let $k \in \mathbb{N}$; from (7.90.b) a closed form of the OGF of the sequence $\binom{n}{k}_n$ is $\dfrac{X^k}{(1-X)^{k+1}}$, a quotient of two polynomials without common factors. Since 1 is the unique root of $(1-X)^{k+1}$, it follows from Proposition 8.18 that the radius of convergence of this OGF is 1; so that by Theorem 8.14

$$\sum_{n=k}^{\infty} \binom{n}{k} x^n = \frac{x^k}{(1-x)^{k+1}} \qquad \forall |x| < 1. \qquad \square$$

Example 8.21 Let $A(X)$ be a formal power series with $A(X) = \dfrac{X^2 - 1}{(X^2 + 4)(X - 7)}$.
The zeros of $(X^2 + 4)(X - 7)$ are $\pm 2i$ and 7: by Proposition 8.18 the radius of convergence of $A(X)$ is $|2i| = 2$ and so, by Theorem 8.14, one has

$$A(x) = \frac{x^2 - 1}{(x^2 + 4)(x - 7)} \qquad \forall x \in]-2, 2[. \qquad \square$$

Example 8.22 Let $A(X)$ be a formal power series with $A(X) = \dfrac{X^5 - 5}{X^2 + 9}$. The zeros of $X^2 + 9$ are $\pm 3i$: by Proposition 8.18 the radius of convergence of $A(X)$ is $|3i| = 3$ and thus, by Theorem 8.14, one has

$$A(x) = \frac{x^5 - 5}{x^2 + 9} \qquad \forall x \in]-3, 3[. \qquad \square$$

We end this section with a continuity property of the sum function of a power series, that we motivate through the following example.

Example 8.23 We know that the radius of convergence of $\log(1 + x) = \displaystyle\sum_{n=1}^{\infty} (-1)^{n+1} \frac{x^n}{n}$ is 1; therefore

$$\log(1 + x) = x - \frac{x^2}{2} + \cdots + (-1)^{n+1} \frac{x^n}{n} + \cdots \qquad \forall |x| < 1. \qquad (8.23.a)$$

Now by Leibniz test 6.67 the series

$$1 - \frac{1}{2} + \cdots + (-1)^{n+1} \frac{1}{n} + \cdots$$

converges: does the equality in (8.23.a) continue to hold for $x = 1$? The answer is yes, thanks to Abel's theorem that we formulate below. $\square$

Theorem 8.24 (Abel's Theorem) *Let $A(X)$ be a summable formal power series. The sum function of $A(X)$ is continuous on the convergence set $D_{A(X)}$:*

$$A(x) = \lim_{\substack{y \to x \\ y \in D_{A(X)}}} A(y) \qquad \forall x \in D_{A(X)}.$$

We omit the proof of Abel's Theorem, that can be found in [33, Theorem 8.2]. We see now how to apply Abel's Theorem to prove the claim formulated in Example 8.23.

Example 8.25 We come back to Example 8.23 considering

$$A(X) = X - \frac{X^2}{2} + \cdots + (-1)^{n+1}\frac{X^n}{n} + \cdots .$$

We know that $A(x) = \log(1 + x)$ for all $|x| < 1$. Now

$$A(1) = 1 - \frac{1}{2} + \cdots + (-1)^{n+1}\frac{1}{n} + \cdots$$

exists and is finite. By Abel's Theorem 8.24 we get

$$A(1) = \lim_{y \to 1^-} A(y) = \lim_{y \to 1^-} \log(1 + y) = \log 2,$$

due to the continuity of the logarithm function. We thus get the following identity

$$1 - \frac{1}{2} + \cdots + (-1)^{n+1}\frac{1}{n} + \cdots = \log 2. \qquad \square$$

Example 8.26 (Convergence of the binomial series) Let $A(X) = \sum_{k=0}^{\infty} \binom{a}{k} X^k$, $a \in$
$\mathbb{R}$. We know from Example 8.4 that $A(x) = (1 + x)^a$ for all $|x| < 1$. We study now
its convergence at ± 1.

1. The absolute convergence of $A(X)$ at ± 1 is equivalent to the convergence of the
    series

$$\sum_{k=0}^{\infty} \left| \binom{a}{k} \right|.$$

    If $a \in \mathbb{N}$, then $\binom{a}{k}$ does definitively vanish and hence we have trivially the
    convergence. Otherwise, if $a \in \mathbb{R} \setminus \mathbb{N}$, by Proposition 7.97 there exists $C_a > 0$
    such that

$$\left| \binom{a}{k} \right| \sim \frac{C_a}{k^{a+1}} \text{ as } k \to +\infty : \tag{8.26.a}$$

    it follows that the absolute convergence of $A(X)$ at ± 1 occurs if and only if
    $a + 1 > 1$. Putting together the two cases, we have the absolute convergence of
    $A(X)$ at ± 1 if and only if $a \geq 0$.

2. Convergence of $A(X)$ at -1. Notice that $(-1)^k \binom{a}{k}$ has definitively a constant
    sign: the convergence of $A(X)$ at -1 is thus equivalent to its absolute convergence,
    and this occurs if and only if $a \geq 0$.

3. Convergence of $A(X)$ at 1. It follows by (8.26.a) that $\lim\limits_{k \to +\infty} \binom{a}{k} = 0$ just for

   $a > -1$: thus $A(X)$ does not converge at 1 if $a \leq -1$. Notice that $\binom{a}{k} = $

   $\dfrac{a(a-1)\cdots(a-k+1)}{k!}$ has an alternating sign from $k > a+1$: the conver-

   gence of the series is thus equivalent to that of

$$\sum_{k=0}^{\infty}(-1)^{k-1}\left|\binom{a}{k}\right|.$$

Let us remark that, assuming $a > -1$, we have

$$\left|\frac{\binom{a}{k+1}}{\binom{a}{k}}\right| = \frac{|a-k|}{k+1} = \frac{k-a}{k+1} < 1 \quad \forall k > a > -1,$$

so that the sequence $k \mapsto \left|\binom{a}{k}\right|$ is definitively decreasing: Leibniz test 6.67 thus
yields the convergence of the series $A(X)$ at 1 if and only if $a > -1$.

When $A(X)$ converges at ± 1 Abel's Theorem 8.24 tells us what the sum of the series
is. Namely:

- If $a > 0$ then $A(X)$ converges absolutely at -1 and

$$A(-1) = \lim_{y \to (-1)^+} A(y) = \lim_{y \to (-1)^+} (1+y)^a = 0^a = 0.$$

We thus get the following identity

$$\sum_{k=0}^{\infty}(-1)^k \binom{a}{k} = 0 \qquad \forall a > 0,$$

which extends the formula (see Corollary 2.21)

$$\sum_{k=0}^{n}(-1)^k \binom{n}{k} = 0 \qquad \forall n \in \mathbb{N}_{\geq 1}.$$

- If $a > -1$ then $A(X)$ converges at 1 (absolutely if and only if $a \geq 0$) and

$$A(1) = \lim_{y \to 1^-} A(y) = \lim_{y \to 1^-} (1 + y)^a = 2^a.$$

We thus get the following identity

$$\sum_{k=0}^{\infty} \binom{a}{k} = 2^a \qquad \forall a > -1,$$

an extension of the formula (see Corollary 2.20)

$$\sum_{k=0}^{n} \binom{n}{k} = 2^n \qquad \forall n \in \mathbb{N}. \qquad \square$$

Abel's Theorem has an impact also on the situation illustrated in Proposition 8.18.

Corollary 8.27 *Let $P(X)$ and $Q(X)$ be two polynomials, with $[X^0]Q(X) \neq 0$. Let $A(X)$ be the power series $A(X) = \dfrac{P(X)}{Q(X)}$. Assume that $A(X)$ converges at $r \in \mathbb{R}$. Then*

$$A(r) = \lim_{x \to r} \frac{P(x)}{Q(x)}.$$

Proof. It is not restrictive to assume that $P(X)$ and $Q(X)$ have no common factors: erasing common factors modifies neither $A(X)$ nor the limit of $\dfrac{P(x)}{Q(x)}$ at a given point. If $A(X)$ converges at r then its radius of convergence is $R_{A(X)} \geq |r|$. It follows from Theorem 8.14 that $A(x) = \dfrac{P(x)}{Q(x)}$ for all $|x| < R_{A(X)}$. If $|r| < R_{A(X)}$ we immediately get the equality $A(r) = \dfrac{P(r)}{Q(r)}$ from Proposition 8.18. Otherwise, if $r = \pm R_{A(X)}$ then, by Abel's Theorem 8.24 and Theorem 8.14, we get

$$A(r) = \lim_{\substack{x \to r \\ |x| < |r|}} A(x) = \lim_{x \to r} \frac{P(x)}{Q(x)}. \qquad \square$$

Remark 8.28 Notice that, unlike Proposition 8.18, in Corollary 8.27 we do not need that the polynomials be without common factors, quite a useful fact in applications.

8.2   Generating Formal Series for Some Notable Sequences

In this section we calculate some closed forms for ordinary or exponential generating formal series associated to particularly important sequences. For the various sequences discussed here we choose the generating formal series that are most frequently used and easiest to derive.

8.2.1   EGF of the Reciprocals of the Natural Numbers

We have seen in Example 7.74 that $-\log(1 - X)$ is the OGF of the sequence of the inverses of the natural numbers greater or equal than 1. It is also sometimes useful to have a closed form for the EGF of the same sequence. In this regard we first introduce the *integer exponential integral* function.

Definition 8.29 (*Integer exponential integral function* Ein(x)) The **integer exponential integral function** Ein $: \mathbb{R} \to \mathbb{R}$ is defined by setting, for each $x \in \mathbb{R}$,

$$\text{Ein}(x) = \int_0^x \frac{1 - e^{-t}}{t}\, dt. \qquad\qquad \square$$

Proposition 8.30 (EGF of the reciprocals of the natural numbers) *The EGF of the sequence* $\left((-1)^{n-1} \dfrac{1}{n} \right)_{n \geq 1}$ *is*

$$\text{EGF}\left((-1)^{n-1} \frac{1}{n} \right)_{n \geq 1} = \sum_{n=1}^{\infty} (-1)^{n-1} \frac{1}{n} \frac{X^n}{n!} = \text{Ein}(X). \qquad (8.30.\text{a})$$

Thus, the EGF of $\left(\dfrac{1}{n} \right)_{n \geq 1}$ *is*

$$\text{EGF}\left(\frac{1}{n} \right)_{n \geq 1} = \sum_{n=1}^{\infty} \frac{1}{n} \frac{X^n}{n!} = -\text{Ein}(-X).$$

Proof. On setting

$$A(X) := \mathrm{EGF}\left((-1)^{n-1}\frac{1}{n}\right)_{n\geq 1} = \sum_{n=1}^{\infty}(-1)^{n-1}\frac{1}{n}\frac{X^n}{n!},$$

one sees immediately that

$$A'(X) = \sum_{n=1}^{\infty}(-1)^{n-1}\frac{X^{n-1}}{n!} = \frac{1}{X}\sum_{n=1}^{\infty}(-1)^{n-1}\frac{X^n}{n!}$$

$$= -\frac{1}{X}\sum_{n=1}^{\infty}\frac{(-X)^n}{n!} = -\frac{1}{X}(e^{-X}-1) = \frac{1-e^{-X}}{X}.$$

The extension by continuity of $x \mapsto \dfrac{1-e^{-x}}{x}$ is a closed form for the formal power series $A'(X)$; by Point 6 of Proposition 7.80 the function $x \mapsto \mathrm{Ein}(x)$ is then a closed form for $A(X)$. □

In the next example we study the convergence of $\mathrm{EGF}\left(\dfrac{1}{n}\right)_{n\geq 1}$.

☕ *Example 8.31 (The sum function of the EGF of the reciprocals of natural numbers)* It can be shown that the function $\mathrm{Ein}(x)$ is analytic: this is due to the fact that the derivative of $\mathrm{Ein}(x)$, i.e., the extension by continuity of $\dfrac{1-e^{-t}}{t}$, is analytic. The radius of convergence of the power series $\sum_{n=1}^{\infty}\dfrac{1}{n}\dfrac{X^n}{n!}$ is $+\infty$: indeed (see Example 8.4)

$$\lim_{n\to\infty}\frac{\dfrac{1}{nn!}}{\dfrac{1}{(n+1)(n+1)!}} = \lim_{n\to\infty}\frac{(n+1)^2}{n} = +\infty.$$

It follows from Theorem 8.14 that

$$\sum_{n=1}^{\infty}\frac{1}{n}\frac{x^n}{n!} = -\mathrm{Ein}(-x) \qquad \forall x \in \mathbb{R}.$$ □

8.2.2   OGF and EGF of the Harmonic Numbers

The sequence of harmonic numbers $(H_n)_n$ introduced in Chap. 6 is the sequence $H_0 = 0, H_n = \dfrac{1}{1} + \cdots + \dfrac{1}{n}$ for $n \geq 1$.

Proposition 8.32 (OGF of the harmonic numbers) *The OGF of the sequence of harmonic numbers is*

$$\mathrm{OGF}\,(H_n)_n = \sum_{n=0}^{\infty} H_n\, X^n = \frac{\log(1 - X)}{X - 1}. \qquad (8.32.\mathrm{a})$$

Proof. By Example 7.74 one has

$$-\log(1 - X) = \sum_{n=1}^{\infty} \frac{1}{n} X^n;$$

since the harmonic numbers are the partial sums of the sequence $0, 1, 1/2, 1/3, \ldots$ the formula follows immediately from Proposition 7.88. $\qquad \square$

Example 8.33   (The sum function of the OGF of the reciprocals of harmonic numbers) From (8.32.a), $\dfrac{\log(1 - X)}{X - 1}$ is the OGF of the sequence $(H_n)_n$. The function $x \mapsto \dfrac{\log(1 - x)}{x - 1}$ is by Proposition 8.16 analytic on $]-\infty, 1[$. The radius of convergence of the power series $\sum\limits_{n=0}^{\infty} H_n\, X^n$ is 1: indeed, since by Proposition 6.44 the n-th harmonic number H_n is asymptotic to $\log n$ as $n \to +\infty$, one has

$$\lim_{n \to \infty} \frac{H_n}{H_{n+1}} = \lim_{n \to \infty} \frac{\log n}{\log(n + 1)} = 1.$$

It follows from Example 8.4 and Theorem 8.14 that

$$\sum_{n=0}^{\infty} H_n\, x^n = \frac{\log(1 - x)}{x - 1} \qquad \forall |x| < 1. \qquad \square$$

Let us see some other useful OGF of sequences involving harmonic numbers.

Proposition 8.34 *Given $m \in \mathbb{N}$, one has*

$$\sum_{n=0}^{\infty} n(H_n - 1)X^n = -\frac{X}{(X-1)^2}\log(1-X);$$

$$\sum_{n=0}^{\infty} (H_{m+n} - H_m)\binom{m+n}{n} X^n = -\frac{\log(1-X)}{(1-X)^{m+1}}.$$

Proof. Taking the derivative of both sides of (8.32.a) one has

$$\sum_{n=1}^{\infty} n\,H_n\,X^{n-1} = -\frac{1}{(X-1)^2}\log(1-X) + \frac{1}{(X-1)^2}.$$

In view of Proposition 7.90, multiplication by X then gives

$$\sum_{n=1}^{\infty} n\,H_n\,X^n = -\frac{X}{(X-1)^2}\log(1-X) + \frac{X}{(X-1)^2}$$

$$= -\frac{X}{(X-1)^2}\log(1-X) + X\sum_{n=0}^{\infty}\binom{n+1}{n}X^n$$

$$= -\frac{X}{(X-1)^2}\log(1-X) + \sum_{n=1}^{\infty} nX^n,$$

from which it follows that

$$\sum_{n=0}^{\infty} n(H_n - 1)X^n = \sum_{n=1}^{\infty} n(H_n - 1)X^n = -\frac{X}{(X-1)^2}\log(1-X).$$

The second formula may be verified via induction on m. For $m = 0$ it is equivalent to the formula (8.32.a). By the inductive hypothesis we may assume that for a given $m \geq 0$ one has

$$\sum_{n=0}^{\infty} (H_{m+n} - H_m)\binom{m+n}{n} X^n = -\frac{\log(1-X)}{(1-X)^{m+1}}.$$

Taking derivatives, one obtains

$$\sum_{n=1}^{\infty} n\,(H_{m+n} - H_m)\binom{m+n}{n} X^{n-1} = \frac{1}{(1-X)^{m+2}} - (m+1)\frac{\log(1-X)}{(1-X)^{m+2}}.$$

Changing the index of summation and using the expression for $\dfrac{1}{(1-X)^{m+2}}$ obtained in Proposition 7.90 one has

$$\sum_{n=0}^{\infty}(n+1)\,(H_{m+n+1}-H_m)\binom{m+n+1}{n+1}X^n =$$

$$= \sum_{n=0}^{\infty}\binom{m+n+1}{n}X^n - (m+1)\frac{\log(1-X)}{(1-X)^{m+2}}.$$

Hence the formal power series $-\dfrac{\log(1-X)}{(1-X)^{m+2}}$ coincides with

$$\frac{1}{m+1}\sum_{n=0}^{\infty}\left[(n+1)\,(H_{m+n+1}-H_m)\binom{m+n+1}{n+1}-\binom{m+n+1}{n}\right]X^n.$$

It is now easy to verify that

$$(n+1)\,(H_{m+n+1}-H_m)\binom{m+n+1}{n+1}-\binom{m+n+1}{n}=$$

$$= (n+1)\left(H_{m+n+1}-H_{m+1}+\frac{1}{m+1}\right)\frac{m+1}{n+1}\binom{m+n+1}{n}-\binom{m+n+1}{n}=$$

$$= \binom{m+n+1}{n}(H_{m+n+1}-H_{m+1})(m+1).$$

Therefore

$$-\frac{\log(1-X)}{(1-X)^{m+2}} = \sum_{n=0}^{\infty}\binom{m+n+1}{n}(H_{m+n+1}-H_{m+1})X^n,$$

which proves the desired formula for $m+1$. □

In Proposition 8.32 we determined the OGF of the sequence of harmonic numbers. Let us determine now its EGF.

Proposition 8.35 (EGF of the harmonic numbers) *The EGF of the sequence of harmonic numbers is*

$$\mathrm{EGF}\,(H_n)_n = \sum_{n=1}^{\infty}H_n\frac{X^n}{n!} = e^X\,\mathrm{Ein}(X). \tag{8.35.a}$$

Proof. It follows from Proposition 6.46 that the sequence $(H_n)_{n\geq 1}$ is the binomial transform of the sequence $\left((-1)^{n-1}\dfrac{1}{n}\right)_{n\geq 1}$: the result now follows from Propositions 7.88 and 8.30.   □

Next we see how the EGF of the harmonic numbers gives a new proof of the binomial identity of Problem 6.9.

Example 8.36 Let us prove that for each integer $m \geq 0$ one has

$$\sum_{k=0}^{m}(-1)^k\frac{1}{(k+1)^2}\binom{m}{k}=\frac{1}{m+1}H_{m+1}\,.$$

One recognizes that the sum under consideration is the m-th term of the binomial transform of the sequence $\left((-1)^n\dfrac{1}{(n+1)^2}\right)_n$. On setting

$$A(X)=\sum_{n=0}^{\infty}(-1)^n\frac{1}{(n+1)^2}\frac{X^n}{n!}\,,$$

then Proposition 7.88 gives

$$\mathrm{EGF}\left(\sum_{k=0}^{n}(-1)^k\frac{1}{(k+1)^2}\binom{n}{k}\right)_n=e^X A(X). \tag{8.36.a}$$

Clearly $A(X)$ is the derivative of $B(X)=\displaystyle\sum_{n=1}^{\infty}(-1)^{n-1}\frac{1}{n^2}\frac{X^n}{n!}$. Now

$$B'(X)=\sum_{n=1}^{\infty}(-1)^{n-1}\frac{1}{n}\frac{X^{n-1}}{n!}$$

$$=\frac{1}{X}\sum_{n=1}^{\infty}(-1)^{n-1}\frac{1}{n}\frac{X^n}{n!}=\frac{1}{X}\,\mathrm{Ein}\,X=\frac{1}{X}e^{-X}\,\mathrm{EGF}\,(H_n)_n\,,$$

by Proposition 8.35. One then has

$$\mathrm{EGF}\left(\sum_{k=0}^{n}(-1)^k\frac{1}{(k+1)^2}\binom{n}{k}\right)_n=e^X A(X)=e^X B'(X)=\frac{1}{X}\,\mathrm{EGF}\,(H_n)_n$$

$$=\sum_{n=1}^{\infty}H_n\frac{X^{n-1}}{n!}=\left(\sum_{n=1}^{\infty}\frac{H_n}{n}\frac{X^n}{n!}\right)'=\left(\mathrm{EGF}\left(\frac{H_n}{n}\right)_n\right)',$$

which in view of Point 6 of Proposition 7.21 equals EGF $\left(\dfrac{H_{n+1}}{n+1}\right)_n$. □

Example 8.37 (*The sum function of the EGF of the reciprocals of harmonic numbers*)
From (8.35.a), it follows that $e^X \operatorname{Ein}(X)$ is a closed form of the EGF of the sequence $(H_n)_n$. The function $e^x \operatorname{Ein}(x)$ is a product of two analytic functions, and hence it is analytic on $\mathbb{R}$. The radius of convergence of $\displaystyle\sum_{n=0}^{\infty} H_n \dfrac{X^n}{n!}$ is $+\infty$: indeed, since H_n is asymptotic to $\log n$ as $n \to +\infty$ (see Proposition 6.44),

$$\lim_{n\to\infty} \frac{H_n / n!}{H_{n+1}/(n+1)!} = \lim_{n\to\infty} (n+1)\frac{\log n}{\log(n+1)} = +\infty.$$

It follows from Theorem 8.14 that

$$\sum_{n=0}^{\infty} H_n \frac{x^n}{n!} = e^x \operatorname{Ein}(x) \qquad \forall x \in \mathbb{R}.$$

□

8.2.3 Generating Formal Series for the Stirling and Bell Numbers

We recall that the Stirling number of the first kind $\begin{bmatrix} n \\ k \end{bmatrix}$ (see Definition 5.43) is the number of k-partitions into cycles of I_n $(k, n \in \mathbb{N}_{\geq 1})$. For fixed $n \geq 1$, the closed form of the OGF of the sequence $\begin{bmatrix} n \\ k \end{bmatrix}_k$ is obtained by simply reformulating Corollary 5.54.

Proposition 8.38 (OGF of the Stirling numbers of the first kind) *For fixed $n \geq 1$ one has*

$$\operatorname{OGF}\begin{bmatrix} n \\ k \end{bmatrix}_k = \sum_{k=0}^{\infty} \begin{bmatrix} n \\ k \end{bmatrix} X^k = \sum_{k=0}^{n} \begin{bmatrix} n \\ k \end{bmatrix} X^k = X(X+1)(X+2)\cdots(X+(n-1)).$$

In Sect. 5.1.1 we introduced the Stirling numbers of the second kind. The Stirling number $\begin{Bmatrix} n \\ k \end{Bmatrix}$ represents the number of k-partitions of I_n. For fixed k, we now determine a closed form for the EGF of the sequence $\begin{Bmatrix} n \\ k \end{Bmatrix}_n$ of Stirling numbers of the second kind.

Proposition 8.39 (EGF for the Stirling numbers of the second kind) *For fixed $k \geq 1$ one has*

$$\text{EGF}\left\{ {n \atop k} \right\}_n = \sum_{n=0}^{\infty} \left\{ {n \atop k} \right\} \frac{X^n}{n!} = \sum_{n=k}^{\infty} \left\{ {n \atop k} \right\} \frac{X^n}{n!} = \frac{(e^X - 1)^k}{k!}.$$

Proof. For $k \geq 1$ one has (see Theorem 5.6)

$$\left\{ {n \atop k} \right\} = \frac{1}{k!} \sum_{i=0}^{k} (-1)^i \binom{k}{i} (k-i)^n.$$

Since

$$(e^X - 1)^k = \sum_{i=0}^{k} \binom{k}{i} (-1)^i e^{(k-i)X} = \sum_{i=0}^{k} \binom{k}{i} (-1)^i \left(\sum_{n=0}^{\infty} (k-i)^n \frac{X^n}{n!} \right)$$

$$= \sum_{n=0}^{\infty} \left(\sum_{i=0}^{k} \binom{k}{i} (-1)^i (k-i)^n \right) \frac{X^n}{n!}$$

one finds that

$$\sum_{n=0}^{\infty} \left\{ {n \atop k} \right\} \frac{X^n}{n!} = \frac{1}{k!} \sum_{n=0}^{\infty} \left(\sum_{i=0}^{k} (-1)^i \binom{k}{i} (k-i)^n \right) \frac{X^n}{n!} = \frac{(e^X - 1)^k}{k!}. \qquad \square$$

Finally, we recall that the Bell number $\mathfrak{B}_n$ ($n \in \mathbb{N}$), introduced in Sect. 5.1.1, is the number of partitions of I_n: we now determine the exponential generating formal series of the sequence $(\mathfrak{B}_n)_n$.

Proposition 8.40 (EGF of the Bell numbers) *The exponential generating formal series of the Bell numbers is*

$$\mathfrak{B}(X) := \sum_{n=0}^{\infty} \mathfrak{B}_n \frac{X^n}{n!} = e^{(e^X - 1)}.$$

Proof. We give two different proofs: the first uses the fact that $\mathfrak{B}_n = \sum_{k=0}^{n} \left\{ {n \atop k} \right\}$ and applies the formula for the EGF of the Stirling numbers of the second kind (see Proposition 8.39); the second, on the other hand, uses the recursive formula for the

Bell numbers (see Proposition 5.16) and Proposition 7.119 to determine a closed form
for a formal power series that satisfies a differential equation.

1. (First proof) Fix $m \in \mathbb{N}$. We prove that

$$\left[X^m \right] \mathcal{B}(X) = \left[X^m \right] e^{(e^X - 1)}. \tag{8.40.a}$$

One has

$$\left[X^m \right] \mathcal{B}(X) = \frac{\mathcal{B}_m}{m!} = \sum_{k=0}^{m} \begin{Bmatrix} m \\ k \end{Bmatrix} \frac{1}{m!}$$

and by Proposition 8.39

$$\left[X^m \right] e^{(e^X - 1)} = \left[X^m \right] \sum_{k=0}^{\infty} \frac{(e^X - 1)^k}{k!} = \left[X^m \right] \sum_{k=0}^{\infty} \sum_{n=0}^{\infty} \begin{Bmatrix} n \\ k \end{Bmatrix} \frac{X^n}{n!}$$

$$= \sum_{k=0}^{\infty} \begin{Bmatrix} m \\ k \end{Bmatrix} \frac{1}{m!} = \sum_{k=0}^{m} \begin{Bmatrix} m \\ k \end{Bmatrix} \frac{1}{m!}.$$

2. (Second proof) By Proposition 5.16 one has

$$\mathcal{B}_{n+1} = \sum_{i=0}^{n} \binom{n}{i} \mathcal{B}_i .$$

Therefore $(\mathcal{B}_{n+1})_n$ is the binomial transform of the sequence $(\mathcal{B}_n)_n$. Thus by
Proposition 7.88 one has

$$\sum_{n=0}^{\infty} \mathcal{B}_{n+1} \frac{X^n}{n!} = e^X \sum_{n=0}^{\infty} \mathcal{B}_n \frac{X^n}{n!}.$$

On the other side, by Point 6 of Proposition 7.21, the EGF of the sequence $(\mathcal{B}_{n+1})_n$
also coincides with the derivative of the EGF for the sequence $(\mathcal{B}_n)_n$ and therefore
one has

$$\mathcal{B}'(X) = e^X \mathcal{B}(X).$$

The function $e^{e^x - 1}$ is a solution to the Cauchy problem

$$\begin{cases} y'(x) = e^x y(x), \\ y(0) = \mathcal{B}_0 = 1; \end{cases}$$

hence in view of Proposition 7.119 one has $\mathcal{B}(X) = e^{(e^X - 1)}$. $\qquad\qquad \square$

8.2.4 OGF of Integer Powers and Eulerian Numbers

In this section we calculate the OGF of the integer power sequences; more precisely, for fixed n in $\mathbb{N}_{\geq 1}$, we determine the OGF of the sequence

$$0^n, \ 1^n, \ 2^n, \ 3^n, \ 4^n, \ldots, \ k^n, \ldots .$$

In order to express the closed form of that OGF in a concise manner, it is convenient to introduce Eulerian polynomials first. We recall that the Eulerian number $\left\langle\begin{matrix} n \\ k \end{matrix}\right\rangle$ counts the number of permutations $(a_1, \ldots, a_n)$ of $(1, \ldots, n)$ with exactly k strict excesses, i.e., with k indexes i such that $a_i > i$.

Definition 8.41 Let $n \in \mathbb{N}_{\geq 1}$. The n-**th Eulerian polynomial** $A_n(X)$ is the polynomial of degree $n - 1$ defined by

$$A_n(X) := \mathrm{OGF}\left\langle\begin{matrix} n \\ k \end{matrix}\right\rangle_k = 1 + \left\langle\begin{matrix} n \\ 1 \end{matrix}\right\rangle X + \cdots + \left\langle\begin{matrix} n \\ n-1 \end{matrix}\right\rangle X^{n-1} \quad (n \geq 1). \qquad \square$$

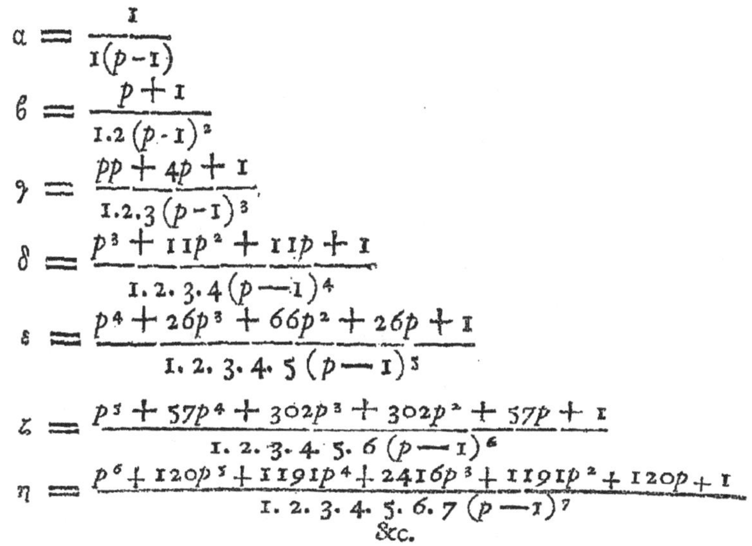

$$\alpha = \frac{1}{1(p-1)}$$

$$\beta = \frac{p+1}{1.2\,(p-1)^2}$$

$$\gamma = \frac{pp+4p+1}{1.2.3\,(p-1)^3},$$

$$\delta = \frac{p^3+11p^2+11p+1}{1.2.3.4\,(p-1)^4}$$

$$\varepsilon = \frac{p^4+26p^3+66p^2+26p+1}{1.2.3.4.5\,(p-1)^5}$$

$$\zeta = \frac{p^5+57p^4+302p^3+302p^2+57p+1}{1.2.3.4.5.6\,(p-1)^6}$$

$$\eta = \frac{p^6+120p^5+1191p^4+2416p^3+1191p^2+120p+1}{1.2.3.4.5.6.7\,(p-1)^7}$$
$$\&c.$$

Fig. 8.1 The first few Eulerian polynomials in Euler's original text "*Institutiones calculi differentialis*" of 1755 (in reality, with the present terminology, what we have here is actually a list of the quotients $\dfrac{A_n(p)}{n!(p-1)^n}$ for $n = 1, \ldots, 7$)

Example 8.42 The first few Eulerian polynomials (see also Fig. 8.1) are

$A_1(X) = 1$, $A_2(X) = 1 + X$, $A_3(X) = 1 + 4X + X^2$, $A_4(X) = 1 + 11X + 11X^2 + X^3$,
$A_5(X) = 1 + 26X + 66X^2 + 26X^3 + X^4$,
$A_6(X) = 1 + 57X + 302X^2 + 302X^3 + 57X^4 + X^5$,
$A_7(X) = 1 + 120X + 1191X^2 + 2416X^3 + 1191X^4 + 120X^5 + X^6$.   □

The OGF of the powers $(k^n)_k$ is easily expressed in terms of the Eulerian polynomials; the proof is an application of Worpitzky's formula (Proposition 5.84).

Proposition 8.43 (OGF of the powers $(k^n)_k$) *For a fixed integer $n \geq 1$ one has*

$$\text{OGF}(k^n)_k = \sum_{k=0}^{\infty} k^n X^k = \frac{X A_n(X)}{(1 - X)^{n+1}} = \frac{X}{(1 - X)^{n+1}} \sum_{k=0}^{n-1} \left\langle {n \atop k} \right\rangle X^k. \quad (8.43.a)$$

Proof. Given $k \in \mathbb{N}$, Worpitzky's formula (see Proposition 5.84) yields

$$k^n = \sum_{i=0}^{n-1} \left\langle {n \atop i} \right\rangle \binom{k + i}{n}.$$

Then one obtains

$$\sum_{k=0}^{\infty} k^n X^k = \sum_{k=0}^{\infty} \left(\sum_{i=0}^{n-1} \left\langle {n \atop i} \right\rangle \binom{k + i}{n} \right) X^k;$$

using the definition of sum of formal power series, one immediately obtains the
interchange of the order of summation

$$\sum_{k=0}^{\infty} k^n X^k = \sum_{i=0}^{n-1} \left\langle {n \atop i} \right\rangle \left(\sum_{k=0}^{\infty} \binom{k + i}{n} X^k \right).$$

Now, in view of (7.90.b), for each given $i \in \{0, \ldots, n - 1\}$ one has

$$\sum_{k=0}^{\infty} \binom{k + i}{n} X^k = \frac{1}{X^i} \sum_{k=0}^{\infty} \binom{k + i}{n} X^{k+i} = \frac{1}{X^i} \sum_{j=i}^{\infty} \binom{j}{n} X^j$$

$$= \frac{1}{X^i} \sum_{j=n}^{\infty} \binom{j}{n} X^j \quad \text{(it is } i + 1 \leq n \text{ and } \binom{j}{n} = 0 \text{ if } j < n)$$

$$= \frac{1}{X^i} \frac{X^n}{(1 - X)^{n+1}} = \frac{X^{n-i}}{(1 - X)^{n+1}},$$

thanks to (7.90.b). Therefore

$$\sum_{k=0}^{\infty} k^n X^k = \sum_{i=0}^{n-1} \left\langle {n \atop i} \right\rangle \frac{X^{n-i}}{(1-X)^{n+1}}.$$

Changing the index to $j = n - 1 - i$ in the last formal power series we obtain

$$\sum_{k=0}^{\infty} k^n X^k = \sum_{j=0}^{n-1} \left\langle {n \atop n-1-j} \right\rangle \frac{X^{j+1}}{(1-X)^{n+1}}.$$

The conclusion now follows from the fact that $\left\langle {n \atop n-1-j} \right\rangle = \left\langle {n \atop j} \right\rangle$ (see Proposition 5.73). $\qquad\square$

Example 8.44 Applying Proposition 8.43 for $n = 3$ we obtain

$$\sum_{k=0}^{\infty} k^3 X^k = \frac{X A_3(X)}{(1-X)^4} = \frac{X + 4X^2 + X^3}{(1-X)^4}. \qquad\square$$

Note that in order to prove Proposition 8.43 we have made no use of the explicit expression for the Eulerian numbers obtained in Proposition 5.75. We now see how Proposition 8.43 allows us to recover an explicit expression for the Eulerian numbers in a more expeditious fashion than the considerably longer and tedious inductive proof carried out in Proposition 5.75.

Corollary 8.45 *Let $n \in \mathbb{N}_{\geq 1}$. The explicit formula for the Eulerian numbers is*

$$\left\langle {n \atop k} \right\rangle = \begin{cases} \displaystyle\sum_{i=0}^{k} (-1)^i \binom{n+1}{i} (k+1-i)^n & \text{if } 0 \leq k < n, \\ 0 & \text{otherwise.} \end{cases}$$

Proof. By Proposition 8.43 one has

$$A_n(X) = (1-X)^{n+1} \sum_{k=1}^{\infty} k^n X^{k-1}.$$

Now

$$(1-X)^{n+1} = \sum_{i=0}^{n+1} (-1)^i \binom{n+1}{i} X^i;$$

and so it follows that

$$A_n(X) = \sum_{i=0}^{n+1} (-1)^i \binom{n+1}{i} \left(\sum_{k=1}^{\infty} k^n X^{k+i-1} \right).$$

Fix i in $\{0, \dots, n+1\}$ and make the change of index $j = k + i - 1$ in the summation between the two parentheses of the foregoing formula:

$$A_n(X) = \sum_{i=0}^{n+1} (-1)^i \binom{n+1}{i} \left(\sum_{j=i}^{\infty} (j+1-i)^n X^j \right).$$

Changing the order of summation, since $0 \le i \le j$ and $i \le n+1$ one has that for each given j, the index i varies between 0 and $\min\{j, n+1\}$, whence

$$A_n(X) = \sum_{j=0}^{\infty} \left(\sum_{i=0}^{\min\{j, n+1\}} (-1)^i \binom{n+1}{i} (j+1-i)^n \right) X^j,$$

from which it follows that $\forall k \in \{0, \dots, n-1\}$,

$$\left\langle {n \atop k} \right\rangle = [X^k] A_n(X) = \sum_{i=0}^{k} (-1)^i \binom{n+1}{i} (k+1-i)^n. \qquad \square$$

8.2.5 OGF of Catalan Numbers

In this section we calculate the OGF of the Catalan numbers (see Sect. 3.3). We recall that the Catalan numbers are

$$\mathrm{Cat}_n = \frac{1}{n+1} \binom{2n}{n}, \quad n \in \mathbb{N}.$$

Proposition 8.46 *The extension by continuity of* $x \mapsto \dfrac{1 - (1-4x)^{1/2}}{2x}$ *is a closed form for the OGF of the Catalan numbers, that is, one has*

$$\mathrm{OGF}(\mathrm{Cat}_n)_n = \sum_{n=0}^{\infty} \mathrm{Cat}_n X^n = \frac{1 - (1-4X)^{1/2}}{2X}.$$

Proof. Let us explicitly calculate the coefficients of $\dfrac{1 - (1-4X)^{1/2}}{2X}$. One has

$$1 - (1 - 4X)^{1/2} = -\sum_{n=1}^{\infty} \binom{1/2}{n} (-4)^n X^n.$$

We have seen in Example 7.95 that

$$\binom{1/2}{n} = (-1)^{n-1} \frac{2}{4^n} \operatorname{Cat}_{n-1}.$$

Consequently,

$$\frac{1 - (1 - 4X)^{1/2}}{2X} = -\sum_{n=1}^{\infty} (-1)^{n-1} \frac{1}{4^n} \operatorname{Cat}_{n-1} (-4)^n X^{n-1}$$

$$= \sum_{n=1}^{\infty} \operatorname{Cat}_{n-1} X^{n-1} = \sum_{n=0}^{\infty} \operatorname{Cat}_n X^n. \qquad \square$$

8.3 Bernoulli Numbers and Their EGF

Bernoulli[1] numbers appear often in many formulas: they serve, for example, in expressing the sum of the m-th powers of integers which we shall calculate in Example 8.52; we will again find them in the general Euler-Maclaurin formula of Chap. 13. The inquisitive reader will find motivation for them in Proposition 13.1.

8.3.1 Bernoulli Numbers

The Bernoulli numbers are recursively defined as follows:

Definition 8.47 The **Bernoulli numbers** B_m, $m \in \mathbb{N}$, are defined by the relations $B_0 = 1$ and, for $m \geq 1$,

$$\sum_{j=0}^{m} \binom{m+1}{j} B_j = 0. \qquad (8.47.a)$$

Therefore, once $B_0 = 1, B_1, \ldots, B_{m-1}$ are known one can find B_m by using the formula

$$B_m = -\frac{1}{m+1} \sum_{j=0}^{m-1} \binom{m+1}{j} B_j. \qquad \square$$

---

[1] Jacob Bernoulli (1654–1705).

Example 8.48 We write the first Bernoulli numbers. Applying the definition one finds $B_0 = 1$; then

$$B_1 = -\frac{1}{2} B_0 = -\frac{1}{2}, \quad B_2 = -\frac{1}{3}(B_0 + 3B_1) = \frac{1}{6}, \quad \ldots$$

and so on, until one obtains

$$B_0 = 1, \quad B_1 = -\frac{1}{2}, \quad B_2 = \frac{1}{6}, \quad B_3 = 0, \quad B_4 = -\frac{1}{30}, \quad B_5 = 0, \quad B_6 = \frac{1}{42},$$

$$B_7 = 0, \quad B_8 = -\frac{1}{30}, \quad B_9 = 0, \quad B_{10} = \frac{5}{66}, \quad B_{11} = 0, \quad B_{12} = -\frac{691}{2730}. \qquad \Box$$

Example 8.49 For each $m \geq 2$ one has

$$\sum_{j=0}^{m} \binom{m}{j} B_j = B_m. \tag{8.49.a}$$

Indeed, by definition, since $m - 1 \geq 1$ one has

$$\sum_{j=0}^{m} \binom{m}{j} B_j = \sum_{j=0}^{m-1} \binom{m}{j} B_j + \binom{m}{m} B_m = 0 + B_m = B_m.$$

☞ This formula is equivalent to the definition of the Bernoulli numbers. It has the advantage of being easy to remember by way of the formula

$$B_m = (1 + B)_m,$$

where $(1 + B)_m$ is expanded like $(1 + B)^m$, but replacing each occurrence of a B^j with the Bernoulli number B_j. $\qquad \Box$

8.3.2 The EGF of the Bernoulli Numbers

Let us calculate the EGF of the Bernoulli numbers:

Proposition 8.50 (EGF of the Bernoulli numbers) *One has*

$$\mathrm{EGF}(\mathrm{B}_n)_n = \sum_{n=0}^{\infty} \mathrm{B}_n \frac{X^n}{n!} = \frac{X}{e^X - 1}. \tag{8.50.a}$$

It follows that

$$\mathrm{B}_0 + \sum_{n=2}^{\infty} \mathrm{B}_n \frac{X^n}{n!} = \frac{X}{2} \coth \frac{X}{2}, \tag{8.50.b}$$

where $\coth(x)$ *is the hyperbolic cotangent* $\coth(x) := \dfrac{\cosh(x)}{\sinh(x)}.$

Proof. We set

$$A(X) := \sum_{n=0}^{\infty} \mathrm{B}_n \frac{X^n}{n!} \left(e^X - 1\right) = \left(\sum_{n=0}^{\infty} \mathrm{B}_n \frac{X^n}{n!}\right)\left(\sum_{n=1}^{\infty} \frac{X^n}{n!}\right)$$

$$= X \left(\sum_{n=0}^{\infty} \mathrm{B}_n \frac{X^n}{n!}\right)\left(\sum_{n=0}^{\infty} \frac{X^n}{(n+1)!}\right).$$

One has $[X^0]A(X) = 0$, $[X]A(X) = 1$, and for $n \geq 2$

$$\left[\frac{X^n}{n!}\right] A(X) = n! \left(\sum_{i=0}^{n-1} \frac{1}{i!} \mathrm{B}_i \frac{1}{(n-i)!}\right) = \sum_{i=0}^{n-1} \binom{n}{i} \mathrm{B}_i = 0$$

and so $A(X) = X$, from which one obtains (8.50.a). By adding the term $\dfrac{X}{2} = -\mathrm{B}_1 X$ to both terms of (8.50.a) one gets

$$\mathrm{B}_0 + \sum_{n=2}^{\infty} \mathrm{B}_n \frac{X^n}{n!} = \frac{X}{e^X - 1} + \frac{X}{2} = \frac{X}{2}\left(1 + \frac{2}{e^X - 1}\right) = \frac{X}{2} \frac{e^X + 1}{e^X - 1}$$

$$= \frac{X}{2} \frac{e^{X/2}(e^{X/2} + e^{-X/2})}{e^{X/2}(e^{X/2} - e^{-X/2})} = \frac{X}{2} \frac{e^{X/2} + e^{-X/2}}{e^{X/2} - e^{-X/2}}. \quad \square$$

An immediate consequence of Proposition 8.50 is the vanishing of the Bernoulli numbers of odd index, from 3 onward.

Corollary 8.51 *The Bernoulli numbers* B_m *with* $m \geq 3$ *odd are zero.*

Proof. It follows by (8.50.b) that $B_0 + \sum_{n=2}^{\infty} B_n \dfrac{X^n}{n!}$ is even. By Proposition 7.45 one has $B_n = 0$ for all odd n greater than or equal to 3. □

Proposition 8.50 yields the calculation of the sum of the m-th powers of the natural numbers ($m \in \mathbb{N}$) known as Faulhaber's[2] formula; later we shall give another derivation of the same result using the Euler-Maclaurin formula (Chap. 12).

Example 8.52 (Faulhaber's formula) For each $m \in \mathbb{N}$ and $n \in \mathbb{N}_{\geq 2}$ we use $S_m(n)$ to denote the sum of the m-th powers of $0, 1, \ldots, (n-1)$, that is,

$$S_m(n) = 0^m + 1^m + \cdots + (n-1)^m,$$

where we set $0^0 = 1$. Note that $S_0(n) = 0^0 + 1^0 + \cdots + (n-1)^0 = n$. We verify the formula

$$S_m(n) = 0^m + 1^m + \cdots + (n-1)^m = \frac{1}{m+1} \sum_{i=0}^{m} \binom{m+1}{i} B_i\, n^{m+1-i}.$$

$$(8.52.a)$$

Observe that, using the notation introduced in Example 8.49, the foregoing formula may easily be remembered by writing

$$S_m(n) = 0^m + 1^m + \cdots + (n-1)^m = \frac{(B+n)_{m+1} - B_{m+1}}{m+1},$$

where $(B+n)_{m+1}$ is expanded like $(B+n)^{m+1}$, but replacing each occurrence of B^j with the Bernoulli number B_j.

For fixed n, we determine the EGF of the sequence $(S_m(n))_m$. Here is the trick: recall that

$$e^{kX} = 1 + kX + k^2 \frac{X^2}{2!} + \cdots + k^m \frac{X^m}{m!} + \cdots.$$

$$(8.52.b)$$

Summing this formula for $k = 0, \ldots, n-1$ one obtains

$$1 + e^X + e^{2X} + \cdots + e^{(n-1)X} = S_0(n) + S_1(n)X + S_2(n)\frac{X^2}{2!} + \cdots + S_m(n)\frac{X^m}{m!} + \cdots.$$

Now

[2]Johann Faulhaber (1580–1635).

$$1 + e^X + e^{2X} + \cdots + e^{(n-1)X} = 1 + e^X + (e^X)^2 + \cdots + (e^X)^{n-1} = \frac{e^{nX} - 1}{e^X - 1}$$

and therefore

$$S_0(n) + S_1(n)X + S_2(n)\frac{X^2}{2!} + \cdots = \frac{e^{nX} - 1}{e^X - 1}.$$

Obviously one has

$$\frac{e^{nX} - 1}{e^X - 1} = \frac{e^{nX} - 1}{X} \frac{X}{e^X - 1}$$

and moreover

$$\frac{e^{nX} - 1}{X} = \frac{1}{X}\left(\sum_{k=1}^{\infty} n^k \frac{X^k}{k!}\right) = \sum_{k=0}^{\infty} n^{k+1} \frac{X^k}{(k+1)!},$$

while, by Proposition 8.50, one has

$$\frac{X}{e^X - 1} = \sum_{k=0}^{\infty} B_k \frac{X^k}{k!}.$$

It follows that for each $m \in \mathbb{N}$

$$S_m(n) = \left[\frac{X^m}{m!}\right] \frac{e^{nX} - 1}{X} \frac{X}{e^X - 1} = \left[\frac{X^m}{m!}\right] \sum_{k=0}^{\infty} n^{k+1} \frac{X^k}{(k+1)!} \sum_{k=0}^{\infty} B_k \frac{X^k}{k!}$$

$$= m! \sum_{i=0}^{m} \frac{B_i}{i!} \frac{n^{m-i+1}}{(m-i+1)!} = \frac{1}{m+1} \sum_{i=0}^{m} \binom{m+1}{i} B_i\, n^{m+1-i},$$

proving (8.52.a).

It must be said that Faulhaber [15] computed the formula for $m \in \{1, \ldots, 23\}$ without knowing the Bernoulli numbers that allowed Bernoulli [7] to prove the general formula about 70 years later. We may now calculate the sum $S_m(n)$, involving n terms, by way of a sum with $m + 1$ terms; of course, the greater n is compared to m, the more useful this result becomes. Thus, for example, since $B_0 = 1, B_1 = -1/2, B_2 = 1/6$ one immediately obtains the sum $S_1(n)$ of the integers from 1 to n and the sum $S_2(n)$ of the squares of integers between 1 and n:

$$S_1(n) = \frac{1}{2}\left(\binom{2}{0} B_0\, n^2 + \binom{2}{1} B_1\, n^1\right) = \frac{1}{2}n^2 - \frac{1}{2}n,$$

$$S_2(n) = \frac{1}{3}\left(\binom{3}{0} B_0\, n^3 + \binom{3}{1} B_1\, n^2 + \binom{3}{2} B_2\, n^1\right) = \frac{1}{3}n^3 - \frac{1}{2}n^2 + \frac{1}{6}n.$$

Bernoulli amazed his contemporaries by calculating "in seven and a half minutes" the sum

$$S_{10}(1\,000) := \sum_{k=0}^{999} k^{10} :$$

indeed, in view of Example 8.52 one has

$$S_{10}(1\,000) = \frac{1}{11} \sum_{i=0}^{10} \binom{11}{i} B_i (1\,000)^{11-i}.$$

To calculate the sum it suffices to have available the first 11 Bernoulli numbers (half of which are zero): the sum turns out to be

91 409 924 241 424 243 424 241 924 242 500.   □

8.3.3   An Estimate of the Bernoulli Numbers

The Riemann[3] zeta function appears in many fields in Mathematics and is still a big source of open problems, in particular the celebrated Riemann hypothesis concerning the zeros of the (complex) Riemann zeta function. Let us recall its definition.

Definition 8.53   (*The Riemann ζ function*) The (real) Riemann ζ function is defined in $]1, +\infty[$ by

$$\zeta(s) = \sum_{k=1}^{\infty} \frac{1}{k^s}. \tag{8.53.a}$$

□

Remark 8.54   It is of course well known that the series (8.53.a) converges for $s > 1$. This is, for instance, a consequence of the integral test (see, e.g., Theorem 12.108 or Corollary 12.78) since the integral $\int_1^{+\infty} \frac{1}{x^s}\,dx$ is finite if and only if $s > 1$.

The Bernoulli numbers of even index can be expressed in terms of the Riemann ζ function: the following equality will be proved in Corollary 13.9.

$$B_n = (-1)^{\frac{n}{2}-1} 2 \frac{n!}{(2\pi)^n} \zeta(n) \qquad \forall n \geq 2,\ n \text{ even}. \tag{8.54.a}$$

---

[3] Bernhard Riemann (1826–1866).

A rough estimate of the values of the Riemann ζ function yields immediately a subsequent estimate of the Bernoulli numbers of even index (remind that those of odd index are zero, except B_1).

Proposition 8.55 (Estimate of Bernoulli numbers) *One has*

$$1 \leq \zeta(s) \leq \frac{s}{s-1} \qquad \forall s > 1, \tag{8.55.a}$$

$$\forall n \in \mathbb{N}, \quad n \geq 2,\ n \text{ even}, \quad 2\frac{n!}{(2\pi)^n} \leq |B_n| \leq 4\frac{n!}{(2\pi)^n}. \tag{8.55.b}$$

Proof. Regarding (8.55.a), the estimate from below follows from the fact that, obviously,

$$\zeta(s) = \sum_{k=1}^{\infty} \frac{1}{k^s} \geq \frac{1}{1^s} = 1.$$

The upper estimate follows immediately from (6.11.a) by passing to the limit for $n \to +\infty$ in

$$1 + \frac{1}{2^s} + \cdots + \frac{1}{n^s} \leq \frac{s}{s-1}.$$

Finally, from (8.54.a) and (8.55.a), for every even $n \geq 2$ we have

$$2\frac{n!}{(2\pi)^n} \leq |B_n| \leq 2\frac{n!}{(2\pi)^n} \frac{n}{n-1} \leq 4\frac{n!}{(2\pi)^n},$$

proving (8.55.b). $\square$

The radius of convergence of the EGF of the Bernoulli numbers series can be easily deduced from Proposition 8.55.

Example 8.56 (The sum function of the EGF of the Bernoulli numbers) The radius of convergence of the series $B(X) := \sum_{n=0}^{\infty} \frac{B_n}{n!} X^n$ is 2π. Indeed, the estimate in (8.55.b) and the comparison principle implies that the series $B(X)$ is absolutely convergent at $x \in \mathbb{R}$ as soon as the series

$$\sum_{n=0}^{\infty} \frac{X^n}{(2\pi)^n} \tag{8.56.a}$$

converges absolutely at x. It follows that the radius of convergence of $B(X)$ is the same of that of (8.56.a), which is actually equal to 2π. Since x and $e^x - 1$ are analytic functions, and $\lim_{x \to 0} \dfrac{x}{e^x - 1} = 1$, then from Proposition 8.16 the extension by continuity of $\dfrac{x}{e^x - 1}$ is analytic in $\mathbb{R}$. Theorem 8.14 yields the equality

$$\sum_{n=0}^{\infty} \frac{B_n}{n!} x^n = \begin{cases} \dfrac{x}{e^x - 1} & \text{if } 0 < |x| < 2\pi, \\[2mm] 1 \text{ if } x = 0. \end{cases} \tag{8.56.b}$$

As an application, we suggest in Problem 8.4 to prove that

$$\lim_{m \to +\infty} \sum_{0 \le k < m} \left(\frac{k}{m}\right)^m = \frac{1}{e - 1}. \qquad \square$$

The list of the first Bernoulli numbers (see Example 8.48) may lead one to believe that they remain bounded, and perhaps tend to 0 at infinity, but this is not the case: the even indexed Bernoulli numbers diverge quite fast to infinity, with alternating signs.

Corollary 8.57 *The sequence* $(B_{2m})_m$ *has alternating signs; moreover*

$$\lim_{m \to +\infty} |B_{2m}| = +\infty.$$

Proof. The fact that the given sequence has alternate signs follows from (8.54.a): $B_{2m} = (-1)^{m-1} \dfrac{2(2m)!}{(2\pi)^{2m}} \zeta(2m)$ has the sign of $(-1)^{m-1}$. From the lower estimate (8.55.b), we get

$$\frac{2(2m)!}{(2\pi)^{2m}} \le |B_{2m}| \qquad \forall m \ge 1.$$

Now, the sequence $\left(x_m := \dfrac{(2m)!}{(2\pi)^{2m}}\right)_m$ diverges to $+\infty$: indeed

$$\lim_{m \to +\infty} \frac{x_{m+1}}{x_m} = \lim_{m \to +\infty} \frac{(2m+2)(2m+1)}{(2\pi)^2} = +\infty.$$

The conclusion is straightforward. $\qquad \square$

8.3.4 Explicit Form for the Bernoulli Numbers ☕

In view of Proposition 8.50 and Point 4 of Proposition 7.80, a closed form for the
EGF of the Bernoulli numbers is given by the extension by continuity of the function
$x \mapsto \dfrac{x}{e^x - 1}$ to $\mathbb{R}$. The formula for the successive derivatives allows us to give an
explicit description of the Bernoulli numbers.

Lemma 8.58 *Let f be the extension by continuity of $x \mapsto \dfrac{x}{e^x - 1}$. For each $n \in \mathbb{N}$
one has*

$$f^{(n)}(0) = \sum_{k=1}^{n}(-1)^k \binom{n+1}{k+1} \frac{k!n!}{(n+k)!} \begin{Bmatrix} n+k \\ k \end{Bmatrix}. \tag{8.58.a}$$

Proof. Clearly $f = 1/g$ where g is the function defined by

$$g(x) := \begin{cases} \dfrac{e^x - 1}{x} & \text{if } x \neq 0 \\ 1 & \text{if } x = 0. \end{cases}$$

☞ The formula (5.28.a) for the n-th derivative of functions of the type $1/g$ here yields

$$f^{(n)}(0) = \sum_{k=1}^{n}(-1)^k \binom{n+1}{k+1} g^{-k-1}(0) \left(g^k\right)^{(n)}(0). \tag{8.58.b}$$

For each $k \in \mathbb{N}_{\geq 1}$ one then has

$$g^k(x) = \begin{cases} \dfrac{(e^x - 1)^k}{x^k} & \text{if } x \neq 0 \\ 1 & \text{if } x = 0. \end{cases}$$

To determine the derivatives, recall that the function $(e^x - 1)^k$ appears in the for-
mula furnishing a closed form for the EGF of the sequence $\left\{\begin{matrix} n \\ k \end{matrix}\right\}_n$: more precisely
Proposition 8.39 states that

$$\sum_{n=k}^{\infty} \begin{Bmatrix} n \\ k \end{Bmatrix} \frac{X^n}{n!} = \frac{(e^X - 1)^k}{k!},$$

from which

$$(e^X - 1)^k = k! \sum_{n=k}^{\infty} \begin{Bmatrix} n \\ k \end{Bmatrix} \frac{X^n}{n!}.$$

Reasoning as above (to obtain the expansion of g) and making the change of variable $m = n - k$, one obtains

$$g^k(X) = k! \sum_{n=k}^{\infty} \begin{Bmatrix} n \\ k \end{Bmatrix} \frac{X^{n-k}}{n!} = k! \sum_{m=0}^{\infty} \begin{Bmatrix} m+k \\ k \end{Bmatrix} \frac{X^m}{(m+k)!}.$$

Then by definition of a closed form one gets

$$\frac{(g^k)^{(n)}(0)}{n!} = [X^n] g^k(X) = \frac{k!}{(n+k)!} \begin{Bmatrix} n+k \\ k \end{Bmatrix} \quad \forall n \in \mathbb{N},$$

from which it follows that

$$(g^k)^{(n)}(0) = \frac{k!n!}{(n+k)!} \begin{Bmatrix} n+k \\ k \end{Bmatrix} \quad \forall n \in \mathbb{N}.$$

Bearing in mind that $g(0) = 1$, relation (8.58.b) then yields

$$f^{(n)}(0) = \sum_{k=1}^{n} (-1)^k \binom{n+1}{k+1} \frac{k!n!}{(n+k)!} \begin{Bmatrix} n+k \\ k \end{Bmatrix}. \qquad \square$$

Corollary 8.59 (Explicit form for the Bernoulli numbers) *For each $n \in \mathbb{N}_{\geq 1}$ one has*

$$B_n = n! \sum_{k=1}^{n} (-1)^k \binom{n+1}{k+1} \frac{k!}{(n+k)!} \begin{Bmatrix} n+k \\ k \end{Bmatrix}$$

$$= n! \sum_{k=1}^{n} \binom{n+1}{k+1} \frac{1}{(n+k)!} \sum_{j=1}^{k} (-1)^j \binom{k}{j} j^{n+k}. \tag{8.59.a}$$

Proof. Let $n \geq 1$. By Proposition 8.50, and Point 4 of Proposition 7.80, the function

$$f(x) := \begin{cases} \dfrac{x}{e^x - 1} & \text{if } x \neq 0, \\[2mm] 1 & \text{if } x = 0 \end{cases}$$

is a closed form for the EGF of the Bernoulli numbers, namely, one has

$$f^{(n)}(0) = B_n.$$

Since by Lemma 8.58

$$f^{(n)}(0) = n! \sum_{k=1}^{n} (-1)^k \binom{n+1}{k+1} \frac{k!}{(n+k)!} \begin{Bmatrix} n+k \\ k \end{Bmatrix},$$

one obtains the first equality in (8.59.a). The second follows from the explicit formula for the Stirling numbers of the second kind stated in Theorem 5.6: since

$$\begin{Bmatrix} n+k \\ k \end{Bmatrix} = \frac{1}{k!} \sum_{i=0}^{k} (-1)^i \binom{k}{i} (k-i)^{n+k} \quad \forall k \in \mathbb{N},$$

the change of variable $j = k - i$ in the summation yields

$$\begin{Bmatrix} n+k \\ k \end{Bmatrix} = \frac{1}{k!} \sum_{j=0}^{k} (-1)^{k-j} \binom{k}{j} j^{n+k} \quad \forall k \in \mathbb{N},$$

and then one immediately obtains the second equality. $\square$

8.4   The Probability Generating Formal Series ☕

The material in this section assumes some familiarity with the theory of discrete random variables. Given a sample space Ω, a discrete random variable $Y : \Omega \to \mathbb{N}$ and a probability function $P : \mathscr{P}(\Omega) \to \mathbb{R}$, we use $(p_n)_n$ to denote the sequence of values assumed by the discrete density of Y, that is,

$$\forall n \in \mathbb{N} \qquad p_n := P(Y = n).$$

To the sequence $(p_n)_n$ we associate the formal power series

$$P_Y(X) := \sum_{n=0}^{\infty} p_n X^n = \sum_{n=0}^{\infty} P(Y = n) X^n,$$

called the **probability generating formal series** of Y.

Example 8.60 (The problem of the two dice) Can one create two 6-sided dice such that when one throws with both dice and sums their values the probability of any sum (from 2 to 12) is the same? Clearly at least one of the two dice should not be fair: indeed if one considers the usual uniform probability on each outcome (i, j) of the two dice (say the red first, the green second), with i, j in $\{1, \ldots, 6\}$, then the probability that the sum of the two dice equals 5 is 4/36, whereas the probability that their sum equals 6 is 5/36. It turns that, actually, having the sums with equal probability is an impossible task. We give a proof of this statement by means of probability generating formal series. Let Y_1 and Y_2 the random variables denoting the score shown on the top face of the two dice. Clearly

$$P_{Y_i}(X) = \sum_{j=1}^{6} P(Y_i = j)X^j = X \sum_{j=0}^{5} P(Y_i = j + 1)X^j, \quad i = 1, 2,$$

are X times a polynomial of fifth degree. Assuming that the throws of the two dice are independent, for each $k = 2, \ldots, 12$ the probability that the sum of the two dice equals k is given by

$$p_k := P(Y_1 + Y_2 = k) = \sum_{i=1}^{6} P(Y_1 = i)P(Y_2 = k - i),$$

so that, from Proposition 7.21,

$$P_{Y_1+Y_2}(X) = \sum_{k=2}^{12} p_k X^k = P_{Y_1}(X)P_{Y_2}(X). \tag{8.60.a}$$

Assuming that the outcomes of the experiment are equally probable, then necessarily $p_2 = \cdots = p_{12} = \dfrac{1}{11}$, so that

$$P_{Y_1+Y_2}(X) = \frac{1}{11}(X^2 + \cdots + X^{12}) = \frac{X^2}{11}(1 + X + \cdots + X^{10}) = \frac{X^2}{11}\frac{1 - X^{11}}{1 - X}.$$

It follows from (8.60.a) that

$$X^2(1 - X^{11}) = 11(1 - X)P_{Y_1}(X)P_{Y_2}(X)$$

$$= 11(1 - X)X^2 \sum_{j=0}^{5} P(Y_1 = j + 1)X^j \sum_{j=0}^{5} P(Y_2 = j + 1)X^j,$$

and hence

$$1 - X^{11} = 11(1 - X) \sum_{j=0}^{5} P(Y_1 = j + 1)X^j \sum_{j=0}^{5} P(Y_2 = j + 1)X^j.$$

This equality of polynomials is not possible! Indeed the roots of $1 - X^{11}$ are the eleventh roots of 1, given by

$$\cos \frac{2k\pi}{11} + i \cos \frac{2k\pi}{11} \quad (k = 0, \ldots, 10).$$

Now all of these roots, except 1, are not reals. It follows that the roots of the fifth degree polynomials $\sum_{j=0}^{5} P(Y_i = j + 1)X^j$, $i = 1, 2$, are all complex, contradicting the fact that any polynomial of odd degree has at least one real root. $\square$

Example 8.61 *(Uniform discrete variable)* Let Y be a uniform discrete variable with parameter $n \in \mathbb{N}_{\geq 1}$: here one is dealing with a variable that assumes each of the values $0, 1, 2, \ldots, n - 1$ with equal probability $1/n$. One has

$$P_Y(X) = P(Y = 0) + P(Y = 1)X + \cdots + P(Y = n - 1)X^{n-1}$$
$$= \frac{1}{n}\left(1 + X + \cdots + X^{n-1}\right) = \frac{1}{n}\frac{1 - X^n}{1 - X}.$$
$\square$

Example 8.62 *(Bernoulli variable)* Let Y be a Bernoulli variable with parameter $p \in [0, 1]$. It assumes the values $0, 1$ respectively with probabilities $q := 1 - p$ and p. One has
$$P_Y(X) = P(Y = 0) + P(Y = 1)X = q + pX.$$
$\square$

Example 8.63 *(Binomial variable)* Let Y be a binomial variable of parameter (N, p), with $N \in \mathbb{N}_{\geq 1}$ and $p \in [0, 1]$. It assumes the values $0, 1, \ldots, N$. On setting $q = 1 - p$, the value i is assumed with probability $\binom{N}{i} p^i q^{N-i}$. One has

$$P_Y(X) = \sum_{i=0}^{N} P(Y = i)X^i = \sum_{i=0}^{N} \binom{N}{i} p^i q^{N-i} X^i = (q + pX)^N.$$
$\square$

Example 8.64 *(Geometric variable)* Let Y be a geometric variable of parameter $p \in [0, 1]$. It assumes the non-zero natural numbers as values. On setting $q = 1 - p$, the value i is assumed with probability $q^{i-1} p$. One has

$$P_Y(X) = \sum_{i=1}^{\infty} P(Y = i)X^i = \sum_{i=1}^{\infty} pq^{i-1} X^i = pX \sum_{i=0}^{\infty} q^i X^i = \frac{pX}{1 - qX}.$$
$\square$

Example 8.65 *(Poisson variable)* Let Y be a Poisson[4] variable of parameter $\lambda > 0$. It assumes as values all the positive integers, and the value i is assumed with probability $\frac{\lambda^i}{i!}e^{-\lambda}$. One has

$$P_Y(X) = \sum_{i=0}^{\infty} P(Y = i)X^i = \sum_{i=0}^{\infty} \frac{\lambda^i}{i!}e^{-\lambda} X^i = e^{-\lambda} \sum_{i=0}^{\infty} \frac{\lambda^i}{i!} X^i = e^{\lambda(X-1)}.$$
$\square$

---

[4]Siméon Denis Poisson (1781–1840).

8.4.1 Expected Value and Variance for Random Variables

Let us recall the definitions of the expected value and variance of a random variable with non-negative integer values.

Definition 8.66 Let Y be a random variable with natural integer values.

1. If the series $\displaystyle\sum_{i=1}^{\infty} i P(Y = i)$ converges, we say that the **expected value** of Y is the number

$$E[Y] := \sum_{i=1}^{\infty} i P(Y = i) = \sum_{i=1}^{\infty} i p_i.$$

2. If Y admits expected value $\mu := E[Y]$ and the series $\displaystyle\sum_{i=0}^{\infty} i^2 p_i$ converges, then the **variance** of Y is the number

$$\mathrm{Var}(Y) := E[Y^2] - E[Y]^2 = \sum_{i=1}^{\infty} i^2 p_i - \mu^2. \qquad \Box$$

The generating formal series of probability facilitates the calculation of the expected value and the variance of a discrete random variable.

Proposition 8.67 *Let Y be a random variable with non-negative integer values, and let $P_Y(X)$ be its probability generating formal series.*

1. *If Y admits its expected value, then $P_Y'(X)$ converges in 1 and one has*

$$E[Y] = P_Y'(1).$$

2. *If Y has finite variance, then $P_Y'(X)$ and $P_Y''(X)$ converge in 1 and one has*

$$\mathrm{Var}(Y) = P_Y''(1) + P_Y'(1) - \left(P_Y'(1)\right)^2.$$

Proof. Note that $P_Y'(X) = \sum\limits_{n=1}^{\infty} n p_n X^{n-1}$ and $P_Y''(X) = \sum\limits_{n=2}^{\infty} n(n-1) p_n X^{n-2}$. Thus, Point 1 follows immediately from the definition. As to Point 2, one has

$$P_Y''(1) = \sum_{n=2}^{\infty} n(n-1) p_n = \sum_{n=2}^{\infty} n^2 p_n - \sum_{n=2}^{\infty} n p_n$$

$$= \sum_{n=1}^{\infty} n^2 p_n - \sum_{n=1}^{\infty} n p_n = \sum_{n=1}^{\infty} n^2 p_n - \mu$$

and so

$$\mathrm{Var}(Y) = \sum_{n=1}^{\infty} n^2 p_n - \mu^2 = \left(\sum_{n=1}^{\infty} n^2 p_n - \mu \right) + \mu - \mu^2$$

$$= P_Y''(1) + P_Y'(1) - \left(P_Y'(1) \right)^2. \qquad \square$$

We calculate now the expected values and variances for the random variables given in the preceding examples.

Example 8.68 (Uniform discrete variable) Let Y be a uniform discrete variable with parameter $n \in \mathbb{N}_{\geq 1}$ (see Example 8.61) and $P_Y(X)$ its probability generating formal series. One then has

$$P_Y(X) = \frac{1}{n}(1 + X + \cdots + X^{n-1}),$$

$$P_Y'(X) = \frac{1}{n}\left(1 + 2X + \cdots + (n-1)X^{n-2}\right),$$

$$P_Y''(X) = \frac{1}{n}\left(2 + (2 \times 3)X + \cdots + (n-2)(n-1)X^{n-3}\right),$$

from which, in view of what has been seen in Sect. 6.1.1 and Example 6.30,

$$P_Y'(1) = \frac{1}{n}(1 + 2 + \cdots + (n-1)) = \frac{1}{n}\frac{n(n-1)}{2} = \frac{n-1}{2},$$

$$P_Y''(1) = \frac{1}{n}\left(2 + (2 \times 3) + \cdots + (n-2)(n-1)\right)$$

$$= \frac{1}{n}\sum_{k=2}^{n-1} k^2 = \frac{1}{3}(n-1)(n-2).$$

Therefore

$$E[Y] = P'_Y(1) = \frac{n-1}{2} \quad \text{and} \quad \text{Var } Y = P''_Y(1) + P'_Y(1) - P'_Y(1)^2 = \frac{n^2-1}{12}. \quad \square$$

Example 8.69 (Bernoulli variable) Let Y be a Bernoulli variable with parameter p and $P_Y(X)$ its probability generating formal series (see Example 8.62). One has

$$P_Y(X) = (1-p) + pX, \quad P'_Y(X) = p, \quad P''_Y(X) = 0,$$

from which it follows that

$$E[Y] = P'_Y(1) = p, \quad \text{Var}(Y) = P''_Y(1) + P'_Y(1) - P'_Y(1)^2 = p - p^2 = p(1-p).$$
$$\square$$

Example 8.70 (Binomial variable) Let Y be a binomial variable with parameter (N, p) and let $P_Y(X)$ be its probability generating formal series (see Example 8.63). Setting $q = 1 - p$, one has $P_Y(X) = (pX + q)^N$. Therefore one finds

$$P'_Y(X) = Np(pX+q)^{N-1}, \quad P''_Y(X) = N(N-1)p^2(pX+q)^{N-2},$$

so that
$$E[Y] = P'_Y(1) = Np(p+q) = Np \quad \text{and}$$

$$\text{Var}(Y) = P''_Y(1) + P'_Y(1) - P'_Y(1)^2 = N(N-1)p^2 + Np - N^2p^2 = Npq. \quad \square$$

Example 8.71 (Geometric variable) Let Y be a geometric variable of parameter p and let $P_Y(X)$ be its probability generating formal series (see Example 8.64). On setting $q = 1 - p$, one has

$$P_Y(X) = \frac{pX}{1-qX}, \quad P'_Y(X) = \frac{p}{(1-qX)^2}, \quad P''_Y(X) = \frac{2pq}{(1-qX)^3}.$$

Therefore,
$$E[Y] = P'_Y(1) = \frac{p}{(1-q)^2} = \frac{1}{p},$$

$$\text{Var}(Y) = P''_Y(1) + P'_Y(1) - P'_Y(1)^2 = \frac{2pq}{p^3} + \frac{1}{p} - \frac{1}{p^2} = \frac{1-p}{p^2}. \quad \square$$

Example 8.72 (Poisson variable) Let Y be a Poisson variable of parameter λ and let $P_Y(X)$ be its probability generating formal series (see Example 8.65). In this case one has

$$P_Y(X) = e^{\lambda(X-1)}, \quad P_Y'(X) = \lambda e^{\lambda(X-1)}, \quad P_Y''(X) = \lambda^2 e^{\lambda(X-1)},$$

so that

$$E[Y] = P_Y'(1) = \lambda \quad \text{and} \quad \text{Var}(Y) = P_Y''(1) + P_Y'(1) - P_Y'(1)^2 = \lambda. \qquad \square$$

8.4.2 Functions of Independent Random Variables

Let us see how to use the results we have developed in the previous sections to carry out some calculations with discrete random variables with values in $\mathbb{N}$.

Definition 8.73 Given a sample space Ω, a family of n discrete random variables

$$Y_i : \Omega \to \mathbb{N}, \quad i = 1, \ldots, n$$

is **independent** if for each n-sequence $(a_1, \ldots, a_n)$ of natural numbers one has

$$P(Y_1 = a_1, \ldots, Y_n = a_n) = P(Y_1 = a_1) \times \cdots \times P(Y_n = a_n). \qquad \square$$

Given the probability generating formal series $P_Y(X)$ and $P_Z(X)$ of two independent discrete random variables it is easy to obtain the probability generating formal series of their sum $Y + Z$:

Proposition 8.74 *The probability generating formal series $P_{Y+Z}(X)$ of the sum of two independent discrete random variables Y, Z with values in $\mathbb{N}$ is the product of the generating formal series $P_Y(X)$ and $P_Z(X)$.*

Proof. Since Y and Z are independent one has

$$P(Y + Z = n) = P\left(\bigcup_{k=0}^{n} \{Y = k\} \cap \{Z = n - k\}\right) = \sum_{k=0}^{n} P(Y = k)P(Z = n - k).$$

Then, by Definition 7.5, we have

$$P_{Y+Z}(X) = \sum_{n=0}^{\infty} P(Y + Z = n)X^n$$

$$= \sum_{n=0}^{\infty} \left(\sum_{k=0}^{n} P(Y = k)P(Z = n - k)\right) X^n = P_Y(X)P_Z(X). \qquad \square$$

Corollary 8.75 *Given a family of n independent discrete random variables*

$$Y_i : \Omega \to \mathbb{N}, \quad i = 1, \ldots, n$$

then $P_{Y_1 + \cdots + Y_n}(X) = P_{Y_1}(X) \cdots P_{Y_n}(X)$.

Proof. Observe that if three random variables Y_1, Y_2, Y_3 are independent, then $Y_1 + Y_2$ and Y_3 are also independent. Therefore

$$P_{Y_1+Y_2+Y_3}(X) = P_{Y_1+Y_2}(X)P_{Y_3}(X) = P_{Y_1}(X)P_{Y_2}(X)P_{Y_3}(X).$$

The general case follows easily.   □

Example 8.76 Let us compute the probability generating formal series of the sum of $N \geq 1$ independent Bernoulli variables Y_1, ..., Y_N of parameter p. Since $P_{Y_i}(X) = q + pX$, with $q := 1 - p$, we have

$$P_{Y_1+\cdots+Y_N}(X) = (q + pX)^N = \sum_{n=0}^{N} \binom{N}{n} p^n q^{N-n} X^n.$$

Therefore $Y_1 + \cdots + Y_N$ is a binomial variable of parameters (N, p).   □

Also the composite of two probability generating formal series has an interesting probabilistic interpretation.

Proposition 8.77 *Let $T : \Omega \to \mathbb{N}$ be a discrete random variable, and $Y_1, \ldots, Y_n, \ldots$ independent and identically distributed discrete random variables with values in $\mathbb{N}$. Assume either:*

1. *T assumes a finite number of values;*
2. *$P(Y_1 = 0) = 0$.*

Then, denoted by Y_T the random variable $\Omega \to \mathbb{N}$, $\omega \mapsto Y_{T(\omega)}(\omega)$, one has

$$P_{Y_1+\cdots+Y_T}(X) = P_T(P_{Y_1}(X)).$$

Remark 8.78 Observe that under assumptions 1 or 2 in Proposition 8.77 the composite of formal series $P_T(P_{Y_1}(X))$ is well defined (see Definition 7.41).

Proof (of Proposition 8.77). By definition one has

$$P_{Y_1+\cdots+Y_T}(X) = \sum_{n=0}^{\infty} P(Y_1 + \cdots + Y_T = n)X^n.$$

Since $P(Y_1 + \cdots + Y_T = n) = \sum_{k=0}^{\infty} P(Y_1 + \cdots + Y_T = n|T = k)P(T = k)$, then one obtains

$$P_{Y_1+\cdots+Y_T}(X) = \sum_{n=0}^{\infty} P(Y_1 + \cdots + Y_T = n)X^n$$

$$= \sum_{n=0}^{\infty} \left(\sum_{k=0}^{\infty} P(Y_1 + \cdots + Y_T = n|T = k)P(T = k) \right) X^n.$$

Under our assumptions, then $\sum_{k=0}^{\infty} P(Y_1 + \cdots + Y_T = n | T = k) P(T = k)$ has a finite number of summands: indeed, if Condition 1 holds, then $P(T = k)$ is different from zero for a finite number of $k \in \mathbb{N}$, while, if Condition 2 holds, then $P(Y_1 = 0) = \cdots = P(Y_T = 0) = 0$ and hence $P(Y_1 + \cdots + Y_T = n | T = k)$ equals zero for $k > n$. Therefore, the sums commute and one gets

$$
\begin{aligned}
P_{Y_1 + \cdots + Y_T}(X) &= \sum_{k=0}^{\infty} P(T = k) \sum_{n=0}^{\infty} P(Y_1 + \cdots + Y_T = n | T = k) X^n \\
&= \sum_{k=0}^{\infty} P(T = k) \sum_{n=0}^{\infty} P(Y_1 + \cdots + Y_k = n) X^n,
\end{aligned}
$$

which, by Corollary 8.75, coincides with

$$
\sum_{k=0}^{\infty} P(T = k) (P_{Y_1}(X))^k = P_T(P_{Y_1}(X)). \qquad \square
$$

Example 8.79 A die whose faces are numbered from 1 to 6 is tossed until the outcome is 1. We want to determine the expected value of the discrete random variable which computes the sum of all the outcomes. Denote by Y_i, $i \in \mathbb{N}$, the discrete random variable that assumes each of the values $1, 2, \ldots, 6$ with equal probability $1/6$ and by T the geometric variable of parameter $1/6$. Similarly to Example 8.61, one gets

$$
P_{Y_i}(X) = \frac{X + \cdots + X^6}{6} = \frac{1}{6} X \frac{1 - X^6}{1 - X}.
$$

Let us compute the probability generating formal series of $Y_1 + \cdots + Y_T$. By Proposition 8.77 we have

$$
P_{Y_1 + \cdots + Y_T}(X) = P_T(P_{Y_1}(X)).
$$

Thus from Examples 8.61, and 8.64 we get

$$
\begin{aligned}
P_{Y_1 + \cdots + Y_T}(X) &= \frac{\frac{1}{6} P_{Y_1}(X)}{1 - \frac{5}{6} P_{Y_1}(X)} = \frac{\dfrac{X(1 - X^6)}{36(1 - X)}}{1 - \dfrac{5X(1 - X^6)}{36(1 - X)}} \\
&= \frac{X(1 - X^6)}{5X^7 - 41X + 36}.
\end{aligned}
$$

Testing the limits of our patience (or perhaps, more sensibly, using a CAS), we get

$$\left(P'_{Y_1+\cdots+Y_T}(X)\right) = \frac{-6X^6 + 1 - X^6}{5X^7 - 41X + 36} - \frac{\left(1 - X^6\right)\left(35X^6 - 41\right)X}{\left(5X^7 - 41X + 36\right)^2}$$

$$= \frac{36\left(6X^5 + 5X^4 + 4X^3 + 3X^2 + 2X + 1\right)}{\left(5X^6 + 5X^5 + 5X^4 + 5X^3 + 5X^2 + 5X - 36\right)^2}.$$

Then by Proposition 8.67 we obtain

$$E[Y_1 + \cdots + Y_T] = \frac{36(6 + 5 + 4 + 3 + 2 + 1)}{(5 + 5 + 5 + 5 + 5 + 5 - 36)^2} = 21. \qquad \square$$

8.5 Problems

Problem 8.1 Let $f(x)$ be a closed form for the derivative $A'(X)$ of a formal power series $A(X)$ and let $A(0) = 0$. Then $g(x) = \int_0^x f(t)\,dt$ is a closed form for $A(X)$.

Problem 8.2 Calculate the following sums:

$$\sum_{k=1}^{149} k^5; \quad \sum_{k=1}^{299} k^7; \quad \sum_{k=1}^{9999} k^3.$$

Problem 8.3 Use generating formal series to give an alternative proof of the formula (2.57.a):

$$\sum_{k \in \mathbb{Z}} \binom{r}{m+k}\binom{s}{n+k} = \binom{r+s}{r-m+n} \qquad \forall m, n, r, s \in \mathbb{N}.$$

Problem 8.4 1. Prove, from (8.52.a), that

$$\frac{1}{m^m} \sum_{1 \le k < m} k^m = \frac{m}{m+1} + \sum_{i=1}^{m} \frac{B_i}{i!}\left(1 + O\left(\frac{1}{m}\right)\right) \quad m \to +\infty.$$

2. Deduce that $\displaystyle \lim_{m \to +\infty} \sum_{0 \le k < m} \left(\frac{k}{m}\right)^m = \frac{1}{e - 1}.$

Chapter 9
Recurrence Relations

Abstract In the upcoming chapter we introduce recurrence relations. These are *equations* that define *in recursive fashion*, via suitable functions, the terms appearing in a real or complex sequence. The first section deals with some well-known examples that show how these relations may arise in real life, e.g., the Lucas Tower game problem or the death or life Titus Flavius Josephus problem. We then devote a large part of the chapter to discrete dynamical systems, namely recurrences of the form $x_{n+1} = f(x_n)$ where f is a real valued function: in this context the sequence that solves the recurrence, starting from an initial datum, is called the orbit of the initial point. We thoroughly study the case where f is monotonic, and the periodic orbits. The last part of the chapter is devoted to the celebrated *Sarkovskii theorem*, stating that the existence of a periodic orbit of minimum period 3 implies the existence of a periodic orbit of arbitrary minimum period: we thus give to the reader the taste of a chaotic dynamical system, although that notion is not explicitly developed in this book.

9.1 Basic Definitions and Models

In this section we present to our readers the basic notions underlying recurrence relations, as well as several examples of such relations.

9.1.1 First Definitions

A recurrence relation is a countable family of equations that define sequences in a recursive fashion. The sequences thus arising are called *solutions of the recurrence*, depending on one or more of the *initial data*: every term that follows the initial data in such sequences is defined as a function of the preceding terms.

© Springer International Publishing Switzerland 2016

C. Mariconda and A. Tonolo, *Discrete Calculus*,
UNITEXT - La Matematica per il 3+2 103, DOI 10.1007/978-3-319-03038-8_9

Example 9.1   1. The system of equations with real coefficients in the infinite collection of unknowns $x_0, x_1, \ldots, x_n, \ldots$

$$\begin{cases} x_1 = 3x_0 \\ x_2 = 3x_1 \\ \ldots\ldots\ldots \\ x_{n+1} = 3x_n \\ \ldots\ldots\ldots \end{cases}$$

which may be indicated more concisely by $x_{n+1} = 3x_n, n \geq 0$, is a *recurrence relation*. The sequence $(3^n)_{n \geq 0}$ is a solution of the given recurrence with *initial datum* $x_0 = 1$. It is easy to convince oneself that in general, for every real number $c \in \mathbb{R}$, the sequence $(c3^n)_{n \geq 0}$ is the unique solution of the given *recurrence* having *initial datum* $x_0 = c$.

2. With suitable care it is easy to verify that the real sequence

$$x_0 = 1, \ x_1 = 1, \ x_2 = 2, \ x_3 = 2, \ x_4 = 4, \ \ldots, \ x_7 = 4, \ \ldots, \ x_{2^m} = 2^m, \ \ldots$$

is the solution of the *recurrence relation* with real coefficients

$$x_n = \begin{cases} 2x_{n/2} & \text{if } n \geq 2 \text{ is even,} \\ x_{n-1} & \text{if } n \text{ is odd,} \end{cases}$$

and *initial datum* $x_0 = 1$.

3. The real sequence

$$x_0 = 2, x_1 = 1, x_2 = 2^{1/2}, x_3 = 1, \ldots, x_{2m-1} = 1, x_{2m} = 2^{1/2^m}, \ldots$$

is the solution of the *recurrence relation* with real coefficients

$$x_n = \sqrt{x_{n-2}}, \quad n \geq 2,$$

and having *initial data* $x_0 = 2, x_1 = 1$.   □

The question that now naturally arises is that of defining general recurrence relations. We seek to expound in a rigorous fashion what we have just inferred from the preceding examples.

Definition 9.2 A recurrence relation   in the unknowns $x_i, i \in \mathbb{N}$, is a family of equations

$$x_n = f_n(x_0, \ldots, x_{n-1}), \quad n \geq r,$$

where $r \in \mathbb{N}_{\geq 1}$, and $(f_n)_{n \geq r}$ are functions

$$f_n : D_n \to \mathbb{R}, \ D_n \subseteq \mathbb{R}^n, \text{ or } f_n : D_n \to \mathbb{C}, \ D_n \subseteq \mathbb{C}^n.$$

Depending on the case encountered, we will speak either of **real recurrences** or
complex recurrences. The unknowns $x_0, \ldots, x_{r-1}$ are called **free**. Their number r
is the **order** of the relation.   □

By replacing n with $n + r$, the recurrence relation of order r

$$x_n = f_n(x_0, \ldots, x_{n-1}), \quad n \geq r,$$

can also be rewritten as

$$x_{n+r} = f_{n+r}(x_0, \ldots, x_{n+r-1}), \quad n \geq 0.$$

Definition 9.3 A sequence $(a_n)_n$ is a **solution** to the recurrence relation of order r

$$x_n = f_n(x_0, \ldots, x_{n-1}), \quad n \geq r, \tag{9.3.a}$$

with $f_n : D_n \to \mathbb{R}, D_n \in \mathbb{R}^n$, if

$$(a_0, \ldots, a_{n-1}) \in D_n, \qquad a_n = f_n(a_0, a_1, \ldots, a_{n-1}) \quad \forall n \geq r.$$

The sequence $(a_0, \ldots, a_{r-1})$ of values assigned to the r free unknowns is called the
r-sequence of **initial data** or of the **initial conditions** of the solution. We define
the **real (resp. complex) general solution** of the recurrence to be the family of all
the solutions with elements belonging to $\mathbb{R}$ (respectively, to $\mathbb{C}$).   □

Example 9.4 Consider the first order recurrence relation defined by

$$x_n = \frac{1}{x_{n-1} - 1}, \quad n \geq 1.$$

The 1-sequence (2) is NOT the sequence of initial data of a solution; indeed 2
belongs to the domain of $f_0(x) = \dfrac{1}{x - 1}$ but $(2, f_0(2)) = (2, 1)$ does not belong to
the domain of $f_1(x_0, x_1) = \dfrac{1}{x_1 - 1}$. On the other hand, the 1-sequence (3) is indeed
the sequence of initial data of the solution (sequence)

$$(a_n)_n := (3, 1/2, -2, -1/3, -3/4, -4/7, -7/11, \ldots).$$

Note that for $n \geq 2$ one has $a_n < 0$ and so $a_{n+1} = \dfrac{1}{a_n - 1} < 0$ is different from 1. □

Example 9.5 In many occasions a recurrence relation of order r involves just the last r items and is of the form

$$x_n = g_n(x_{n-r}, \ldots, x_{n-1}), \quad n \geq r,$$

where $(g_n)_{n \geq r}$ are functions defined on a subset E_n of $\mathbb{R}^r$ or $\mathbb{C}^r$. The latter is indeed a recurrence relation: it is enough to set $f_n(x_0, \ldots, x_{n-1}) := g_n(x_{n-r}, \ldots, x_{n-1})$ for $(x_0, \ldots, x_{n-1}) \in D_n := \mathbb{R}^{n-r} \times E_n$ (or $\mathbb{C}^{n-r} \times E_n$) in order to fulfill the requirements of Definition 9.2. □

9.1.2 Some Models of Recurrence Relations

We now give a series of examples illustrating how to reduce the solution of a problem to that of finding the solutions of an appropriate recurrence relation.

Example 9.6 A child decides to climb a staircase with $n \geq 1$ steps in such a way that with each pace he clears one or two of the steps of the staircase (see Fig. 9.1). Find a recurrence relation that serves to calculate the number of different possible ways of climbing the staircase.

Solution. We use the unknown variable x_n to denote the number of ways in which the child can climb the staircase of $n \geq 1$ steps. It is easy to observe that $x_1 = 1$ and $x_2 = 2$ (two paces each of length one, or one pace of length two staircase steps) Now let $n \geq 3$: if with the first pace the child moves only one staircase step, there are clearly x_{n-1} possible ways to climb those remaining. If instead with the first pace two staircase steps are climbed, there are then x_{n-2} ways to climb the remaining steps of the staircase. Thus one has the recurrence relation

$$x_n = x_{n-1} + x_{n-2}, \quad n \geq 3,$$

in the unknown variables $x_1, x_2, \ldots$ with initial data $x_1 = 1, x_2 = 2$. Thus one meets a recurrence relation of order 2: the free unknowns (variables) are x_1 and x_2. □

Remark 9.7 The recurrence relation

Fig. 9.1 The 5 ways of climbing a staircase with 4 steps

$$x_n = x_{n-1} + x_{n-2}, \quad n \geq 2, \tag{9.7.a}$$

in the unknowns $x_0, x_1, \ldots$ is the **Fibonacci**[1] **recurrence**. The numbers generated by the Fibonacci recurrence with initial data $x_0 = x_1 = 1$ are called the **Fibonacci numbers**: the first few Fibonacci numbers are $1, 1, 2, 3, 5, 8, 13, 21, 34, 55, 89$. The solution to (9.7.a) with initial data $x_0 = 2, x_1 = 1$ is called the Lucas[2] sequence; the first few Lucas numbers are $2, 1, 3, 4, 7, 11, 18, 29, 47, 76, 123$. A bit further on in our discussion (in Example 10.19) we shall find an explicit description of the n-th Fibonacci and Lucas numbers.

Example 9.8 A bank gives a yearly payment of 8 % interest on the money deposited there. Find a recurrence relation for the total sum of money one succeeds in accumulating after n years if one follows an investment strategy of the type:

1. Invest 1000 euro and leaves them in the bank for n years;
2. Invest 100 euro at the beginning of the first year, and then 100 euro at the end of each year for n years.

Solution. We use the unknowns x_n and y_n, $n \geq 0$, to indicate the total sum accumulated after n years by following investment strategies 1 and 2 respectively. One has

$$x_n = 1.08\, x_{n-1} \quad \text{and} \quad y_n = 1.08\, y_{n-1} + 100, \quad n \geq 1,$$

with initial data $x_0 = 1000$ and $y_0 = 100$. It is obviously possible, though boring, to recursively deduce the terms of the solutions of the two recurrence relations by repeatedly applying the functions that define them. We shall find an explicit description of the n-th terms of such solutions in Example 10.37 which will appear in the next chapter. □

Example 9.9 We now use recurrence relations to calculate the number of n-sequences without repetition of $I_n = \{1, \ldots, n\}$. We employ the unknown variable x_n to indicate the number of n-sequences without repetition of I_n, for every $n \geq 0$. If $n \geq 1$, one must have $x_n = n x_{n-1}$: indeed, having chosen the position of 1 among the n possible, we have x_{n-1} possibilities for placing the other $n - 1$ elements. Since the empty sequence is the unique 0-sequence of I_n, one must have $x_0 = 1$; therefore one has

$$x_n = n x_{n-1} = n(n - 1)x_{n-2} = \cdots = n(n - 1)(n - 2)\cdots 2 \times 1 = n!, \quad n \geq 0.$$
□

Example 9.10 (*Number of regions of a plane divided by n lines*) Suppose that we have designed $n \geq 0$ lines in general position in a given plane: by "general position" we mean that no pair of the lines designed are parallel, and moreover that no three of our lines meet in just a single point. Into how many regions does the plane turn out to be divided?

---

[1] Leonardo Pisano called Leonardo Fibonacci (1170–1240).
[2] François Édouard Anatole Lucas (1842–1891).

Solution. We use the unknown variable x_n to indicate the number of regions in which the given plane is divided by n lines in general position. Let us examine the situation for small values of n: if no line is present in the given plane then the plane is clearly divided into a single region, if there is one line present, then there are two regions, while the presence of exactly two lines produces a division into four regions, and the presence of exactly three lines divides the plane into seven regions.

Now let $n \geq 2$. Tracing the n-th line, we will intersect our preceding $n - 1$ lines. In reaching each new point of intersection, we divide an already existing region into two parts; and, finally, leaving the last point of intersection we divide another pre-existing region into two parts. Thus by the tracing the n-th line the number of regions of decomposition of the plane increases by n units: thus the recurrence relation that describes the problem is

$$x_n = x_{n-1} + n, \quad n \geq 2,$$

with initial datum $x_0 = 1$. It is rather easy to use recursion in order to obtain the first few terms of a solution for this recurrence relation:

$$1, 2, 4, 7, 11, 16, 22, 29, 37, 46, \ldots.$$

In Example 10.36 we shall find an explicit description for the n-th term of such a solution.   □

Example 9.11 (*Lucas Tower*) The Lucas (or Hanoi or Brahma) Tower game consists of n rings of various size and three poles; at the start of the game the rings are all positioned on pole A, the most leftward of the three poles, and the rings appear in decreasing order, that is, starting from the largest ring at the bottom of the pole up to the smallest ring at the highest level used on pole A (see Fig. 9.2). By shifting the rings among the three poles one wishes to reconstruct the same tower as that specified by the initial condition of rings on pole A to the most rightward pole C: however, every time that a ring is shifted onto a pole it must be smaller than all the rings already present on that pole. In other words, at each step in the game there must be a decreasing pile of rings (or no ring at all) on each pole.

The problem of finding the minimal number of steps in order to conclude the game was proposed in Europe by the mathematician Lucas, under the pseudonym of N. Claus de Siam, actually an anagram of Lucas d'Amiens, his birth town. There are many legends about this game, all of them concern a temple (either in Hanoi, or in Kashi Vishwanath, …) which contains a large room with three time-worn posts

Fig. 9.2 Lucas Tower

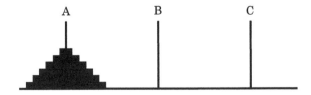

in it surrounded by 64 golden disks. Priests or monks, acting out the command of an ancient prophecy, have been moving these disks since that time. According to the legend, when the last move of the puzzle will be completed, the world will end.

We set out to find a recurrence relation that describes the minimal number of moves necessary to shift the n rings from pole A to pole C. We use the unknown x_n to denote the number of moves necessary to complete the Lucas Tower game involving n rings. For $n = 1$ a single move is obviously sufficient. Now let $n \geq 2$. The fundamental observation to be made is that if the n rings are on A and we wish to move them to C, we must first "liberate" the largest ring on A by shifting all of A's other rings on to pole B: at that point the largest ring of A can be moved to pole C with one further move. To shift the $n - 1$ smallest rings from A to B is equivalent to ending the Lucas Tower game based on $n - 1$ rings. Thus, this configuration may be reached in $x_{n-1} + 1$ moves. We must then shift the tower with $n - 1$ rings from B onto C: this requires x_{n-1} additional moves. Thus one obtains the recurrence relation involving the unknowns x_i, $i \geq 1$,

$$x_n = 2x_{n-1} + 1, \quad n \geq 2,$$

with initial datum $x_1 = 1$. It is easy to verify that $x_n = 2^n - 1, n \geq 1$, is the solution for this recurrence relation. Therefore, if the above legends were true, and if the priests were able to move disks at a rate of one per second, using the smallest number of moves, it would take them $2^{64} - 1$ s or roughly 585 billion years, i.e., about 127 times the current age of the sun [26]. □

Example 9.12 We use the unknown variable x_n to denote the number of n-sequences of I_3 that do not contain the subsequence $(1, 2, 3)$. If $n \leq 2$, no n-sequence of I_3 contains the subsequence $(1, 2, 3)$: thus, one has

$$x_0 = 3^0 = 1, \ x_1 = 3^1 = 3, \ x_2 = 3^2 = 9.$$

Let $n \geq 3$. If the first figure is 2 or 3, then the remaining figures form an arbitrary $(n - 1)$-sequence of I_3 which does not contain the subsequence $(1, 2, 3)$: in both cases we have x_{n-1} possibilities. If instead the first digit is 1, the remaining digits form an $(n - 1)$-sequence of I_3 that does not contain the subsequence $(1, 2, 3)$ and that does not begin with the subsequence $(2, 3)$. The $(n - 1)$-sequences of I_3 that do not contain the subsequence $(1, 2, 3)$ are x_{n-1} in number; among these, the ones that begin with the subsequence $(2, 3)$, being formed by the 2-sequence $(2, 3)$ followed by an $(n - 3)$-sequence that does not contain the subsequence $(1, 2, 3)$, are x_{n-3} in number. Thus the n-sequences that begin with 1 and do not contain the subsequence $(1, 2, 3)$ are in number $x_{n-1} - x_{n-3}$. Hence, one has

$$x_n = x_{n-1} + x_{n-1} + (x_{n-1} - x_{n-3}) = 3x_{n-1} - x_{n-3}, \quad n \geq 3,$$

with sequence of initial data $(1, 3, 9)$. Repeatedly applying the functions of the recurrence relation we find that

$$x_3 = 3 \times 9 - 1 = 26, \quad x_4 = 3 \times 26 - 3 = 75, \quad x_5 = 3 \times 75 - 26 = 199,$$

and so on. In the Example 10.21 we shall find an explicit description of the n-th term of this solution. □

Example 9.13 We use the unknown variable x_n to denote the number of n-sequences of I_4 that contain the subsequence $(1, 2, 3, 4)$. One has $x_0 = x_1 = x_2 = x_3 = 0$ in view of the fact that if $n < 4$ no n-sequence of I_4 contains the subsequence $(1, 2, 3, 4)$. Now let $n \geq 4$. The n-sequences that contain the subsequence $(1, 2, 3, 4)$ are of two types: those that contain it among their first $n - 1$ terms, and those that end with the subsequence $(1, 2, 3, 4)$, but do not contain it in an earlier position. There are $4x_{n-1}$ sequences of the first type: indeed, an $(n - 1)$-sequence that contains $(1, 2, 3, 4)$ must be completed by one of the 4 elements of I_4 to form a n-sequence. The sequences of the second type number the same as the $(n - 4)$-sequences of I_4 which do not contain $(1, 2, 3, 4)$ and thus are in number $4^{n-4} - x_{n-4}$: therefore, one has

$$x_n = 4x_{n-1} - x_{n-4} + 4^{n-4}, \quad n \geq 4,$$

with initial data sequence $(0, 0, 0, 0)$. Repeatedly applying the functions of the recurrence relation one finds that

$$x_4 = 4^0 + 4 \times 0 - 0 = 1, \, x_5 = 4^1 + 4 \times 1 - 0 = 8, \, x_6 = 4^2 + 4 \times 8 - 0 = 48,$$

and so on. We shall find an explicit description of the n-th term of the solution in Example 10.38. □

Example 9.14 (The Titus Flavius Josephus problem) The Josephus problem or the Josephus permutation is a problem connected with an autobiographical episode recounted by the historian Titus Flavius Josephus,[3] and which we now translate into mathematical terms. The problem presents n people arranged in a circle. Having chosen an initial person and direction of rotation, one moves multiple times along the circle eliminating every second person that one meets in the chosen direction until there no longer remains a single person. Given $n \geq 1$, one asks for the determination of the initial position of the remaining person if the circle is initially formed by n people. For example, if $n = 10$ the order of elimination of the people located in positions $1, 2, \ldots, 10$ is

$$2, 4, 6, 8, 10, 3, 7, 1, 9,$$

---

[3] Titus Flavius Iosephus (37–100).

and so it is the person in position 5 who will be the last remaining. We use the unknown variable x_n to denote the initial position of the last person remaining on a circle of n people with $n \geq 1$. Obviously $x_1 = 1$; then, for $n \geq 2$, we distinguish two cases:

- $n = 2m$ is even, with $m \geq 1$: on executing the first circuit the people in positions $2, 4, \ldots, 2m$ are all eliminated. At this point the new circle is formed by the m people who were located in positions $1, 3, \ldots, 2m - 1$. In the new circle the person who first occupied position $2i - 1$, now occupies position i, where $i = 1, \ldots, m$. The last remaining person is the one who occupies position x_m of the new circle, or, in other words, the one who occupied position $2x_m - 1 = 2x_{n/2} - 1$ in the initial circle: therefore, $x_n = 2x_{n/2} - 1$.
- $n = 2m + 1$ is odd, with $m \geq 1$: on making a first circuit the people who occupied positions $2, 4, \ldots, 2m$ are eliminated; on making a second pass one eliminates also who was in position 1. At this point there remain the m people who occupied the positions $3, \ldots, 2m + 1$. In the new circle the person who began by occupying position $2i + 1$ of the original circle, now occupies position i, where $i = 1, \ldots, m$. Hence the last person to remain is the one who occupies position x_m of the new circle, namely the person who occupied position $2x_m + 1 = 2x_{(n-1)/2} + 1$ and thus is $x_n = 2x_{(n-1)/2} - 1$.

Thus we obtain the following recurrence relation for the unknowns x_i, $i \geq 1$,

$$\forall n \geq 2 \quad x_n = \begin{cases} 2x_{n/2} - 1 & \text{if } n \geq 2 \text{ is even;} \\ 2x_{(n-1)/2} + 1 & \text{if } n \geq 3 \text{ is odd.} \end{cases}$$

On repeatedly applying the functions giving the recurrence relation, starting from the initial datum $x_1 = 1$, we find that

$$x_2 = 2x_1 - 1 = 1, \ x_3 = 2x_1 + 1 = 3, \ x_4 = 2x_2 - 1 = 1, \ x_5 = 2x_2 + 1 = 3;$$

for example, continuing up to $n = 16$ we find that:

n	1	2	3	4	5	6	7	8	9	10	11	12	13	14	15	16
x_n	1	1	3	1	3	5	7	1	3	5	7	9	11	13	15	1

It is possible to describe the terms of the solution with an explicit formula. Indeed, one has

$$\forall m \in \mathbb{N} \quad x_n = 1 \text{ if } n = 2^m, \ x_n = 2\ell + 1 \text{ if } n = 2^m + \ell, \ 1 \leq \ell < 2^m.$$

We leave as an exercise for the reader a proof of the fact that this is effectively the solution of the recurrence relation which we found above (Problem 9.9). □

9.2   Discrete Dynamical Systems

In mathematical physics and in the theory of systems, the concept of a dynamical system arises from the requirement of constructing a mathematical model capable of describing the temporal evolution of a system. The discrete dynamical systems comprise a subclass of the class of recurrence relations.

9.2.1   Terminology and Notation

A notable class of recurrence relations is constituted by the *dynamical discrete systems*, namely, by the recurrence relations of the type

$$x_{n+1} = f(x_n), \quad n \in \mathbb{N},$$

in which the function $f : D \to D \subseteq \mathbb{R}$ that permits one to obtain x_{n+1} starting from x_n remains the same for every $n \geq 0$. Here we limit ourselves to a brief illustration of some notable facts regarding dynamical systems. Those readers interested in greater detail in this regard are invited to consult [12, 11].

In the context of dynamical systems, the solution of the recursion relation

$$x_{n+1} = f(x_n), \quad n \in \mathbb{N},$$

with initial datum x_0 is called the **orbit** of x_0 with respect to f. Notice that $x_n = f^n(x_0)$ where, here and henceforth in this chapter, f^n denotes the n times composition $\underbrace{f \circ f \circ \cdots \circ f}_{n \text{ times}}$ of f with itself (and not the n-th power of f).

The graph of f often allows one to understand how the orbit of a point x_0 evolves through the use of a procedure called *graphical analysis*.

Graphical Analysis. We use Δ to denote the diagonal $\{(x, x) : x \in \mathbb{R}\}$ of the plane. The vertical line through (x_0, x_0) meets the graph of f at $(x_0, f(x_0))$. The horizontal line through $(x_0, f(x_0))$ meets Δ in $(f(x_0), f(x_0))$. Proceeding in this way, a vertical line that goes from the diagonal to the graph followed by a horizontal line that goes from the graph to the diagonal specifies the points $(f^n(x_0), f^n(x_0))$ on Δ, with the variation of $n = 1, 2, \ldots$. The Fig. 9.3 describes the orbit of $x_0 = 0.1$ with respect to the function $f : \mathbb{R} \to \mathbb{R}$ defined by setting $f(x) = 2x - x^2$.

Definition 9.15   Let $f : D \to D$ be a function, and let $a_0 \in D$.

- The point a_0 is a **fixed point** for f if $f(a_0) = a_0$. In this case the constant sequence $(a_n := a_0)_n$ is a solution of the dynamical system

$$x_{n+1} = f(x_n), \quad n \in \mathbb{N}.$$

Fig. 9.3 Graphical analysis of the orbit of $x_0 = 0.1$ for $x_{n+1} = 2x_n - x_n^2$

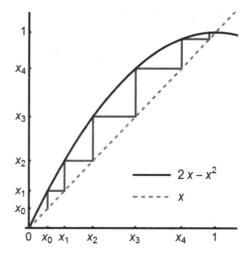

- The point a_0 is said to be **periodic** if there exists ℓ, called a **period** of a_0, such that $a_\ell = f^\ell(a_0) = a_0$: we also say that a_0 is ℓ-**periodic**. The **minimum period** a_0 is the minimum among the natural numbers $m > 0$ such that $a_m = f^m(a_0) = a_0$. It is not difficult (Problem 9.17) to verify that the periods of a_0 are all multiples of the minimum period. If a_0 is ℓ-periodic, then its orbit is the sequence

$$\overline{(a_0, f(a_0), \ldots, f^{\ell-1}(a_0))} := (a_0, f(a_0), \ldots, f^{\ell-1}(a_0), a_0, f(a_0), \ldots, f^{\ell-1}(a_0), a_0, \ldots).$$

In this case, every point a_k of the orbit of a_0 satisfies $f^\ell(a_k) = a_k$: we thus also say that the **orbit of a_0 is periodic of period ℓ**. □

Remark 9.16 It is easy to convince oneself that, in the graphical analysis, the fixed points are the coordinates of the intersections of the graph of f with the diagonal: the fixed points of f are the solutions to $f(x) = x$. Analogously, the points of period ℓ are the coordinates of the intersections of the graph of f^ℓ with the diagonal.

Example 9.17 Let $f(x) = x^2 - \dfrac{11}{10}$. We wish to find the periodic points of f: we thus have to solve the second degree equation

$$x^2 - x - \frac{11}{10} = 0, \tag{9.17.a}$$

whose solutions are

$$\alpha_1 = \frac{5 - 3\sqrt{15}}{10}, \qquad \alpha_2 = \frac{5 + 3\sqrt{15}}{10}. \tag{9.17.b}$$

We now look for the points of period 2, namely the solutions to $f^2(x) = x$. Now

$$f^2(x) - x = f\left(x^2 - \frac{11}{10}\right) - x = \left(x^2 - \frac{11}{10}\right)^2 - \frac{11}{10} - x = x^4 - \frac{22}{10}x^2 - x - \frac{11}{100}$$

is a polynomial function of degree 4. We can however easily find its roots: indeed, the solutions of $f(x) = x$ satisfy $f^2(x) - x = 0$, so that the polynomial $X^2 - X - \frac{11}{10}$ divides $X^4 - \frac{22}{10}X^2 - X - \frac{11}{100}$. The division yields

$$X^4 - \frac{22}{10}X^2 - X - \frac{11}{100} = \left(X^2 - X - \frac{11}{10}\right)\left(X^2 + X - \frac{1}{10}\right).$$

Therefore, the roots of $X^4 - \frac{22}{10}X^2 - X - \frac{11}{100}$ are α_1, α_2 found in (9.17.b), and the roots β_1, β_2 of $X^2 + X - \frac{1}{10}$ given by

$$\beta_1 = \frac{-5 - \sqrt{35}}{10}, \qquad \beta_2 = \frac{-5 + \sqrt{35}}{10},$$

which correspond to the 2-periodic points of f that are not fixed (Fig. 9.4).   □

Example 9.18 The point $x_0 = 0$ is a periodic point of minimum period 2 for the dynamical system $x_{n+1} = \frac{\pi}{2}\cos(x_n)$, $n \in \mathbb{N}$. Indeed, one has $x_1 = \frac{\pi}{2}$, $x_2 = \frac{\pi}{2}\cos(x_1) = \frac{\pi}{2}\cos\frac{\pi}{2} = 0$. The orbit of $x_0 = 0$ is the sequence $\left(0, \frac{\pi}{2}\right) := \left(0, \frac{\pi}{2}, 0, \frac{\pi}{2}, 0, \frac{\pi}{2}, \ldots\right)$; its graphical analysis is described in Fig. 9.5.   □

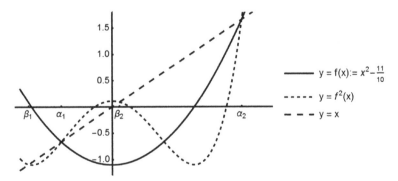

Fig. 9.4 The fixed points α_1, α_2 and the 2-periodic non fixed points β_1, β_2 of the function $x^2 - \frac{11}{10}$

Fig. 9.5 Graphical analysis of the orbit of $x_0 = 0$ for $x_{n+1} = \dfrac{\pi}{2}\cos x_n$

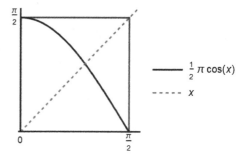

—— $\frac{1}{2}\pi\cos(x)$

---- x

Example 9.19 We look for the orbits of minimum period 3 of the dynamical system

$$x_{n+1} = f(x_n), \qquad n \in \mathbb{N}, \qquad f(x) := -\frac{3}{2}x^2 + \frac{5}{2}x + 1.$$

We know that the points of period 3 correspond to the intersections of the graph of f^3 with the diagonal $y = x$. It seems from Figs. 9.6 and 9.7 that $x_0 := 0$ is not a fixed point of f and is of period 3, thus of minimum period 3 since the periods of a point are all multiples of the minimum period. This graphical conjecture turns out to be true: indeed one has

$$x_1 = -\frac{3}{2}x_0^2 + \frac{5}{2}x_0 + 1 = 1, \quad x_2 = -\frac{3}{2}x_1^2 + \frac{5}{2}x_1 + 1 = 2, \quad x_3 = -\frac{3}{2}x_2^2 + \frac{5}{2}x_2 + 1 = 0.$$

As we shall see in Theorem 9.41, since the function $f(x) = \dfrac{3}{2}x^2 + \dfrac{5}{2}x + 1$ is continuous, the existence of periodic points of minimum period 3 implies the existence of periodic points having minimum period an *arbitrary* natural number. □

One of the main issues of discrete dynamical systems is that of understanding the behavior of its solutions. In the next result we see that, when the orbit of a point with

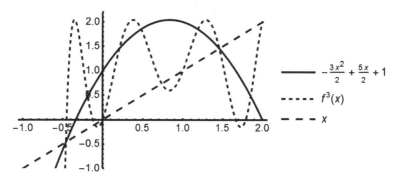

—— $-\frac{3x^2}{2} + \frac{5x}{2} + 1$

····· $f^3(x)$

-- - x

Fig. 9.6 Fixed and 3-periodic points of $-\dfrac{3}{2}x^2 + \dfrac{5}{2}x + 1$

Fig. 9.7 Graphical analysis
of the orbit of $x_0 = 0$ for
$x_{n+1} = -\dfrac{3}{2}x_n^2 + \dfrac{5}{2}x_n + 1$

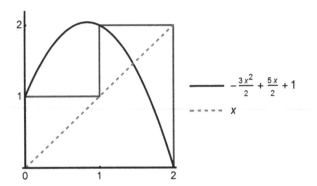

$$\rule{2em}{0.4pt}\quad -\tfrac{3x^2}{2} + \tfrac{5x}{2} + 1$$

$$\text{-----}\quad x$$

respect to a *continuous* function f is a convergent sequence, its limit is then a fixed point.

Proposition 9.20 *Let I be a real interval and let $f : I \to I$ be a* continuous *function. Let $a_0 \in I$ and suppose that $\lim\limits_{n \to +\infty} f^n(a_0) = \ell \in I$. Then ℓ is a fixed point of f.*

Proof. Indeed, one has

$$\ell = \lim_{n \to +\infty} f^{n+1}(a_0) = \lim_{n \to +\infty} f(f^n(a_0)) = f\left(\lim_{n \to +\infty} f^n(a_0)\right) = f(\ell). \qquad \square$$

We now see an example of a discrete dynamical system whose orbits always converge.

Example 9.21 (Approximation of square roots) Here we give a recurrence that has been known since Babylonian times, and which is useful for approximating the square root of a given number. Let p be a strictly positive real number. In an attempt to approximate its square root, we construct the following recurrence relation.

We begin by giving an arbitrary estimate $a_0 > 0$ for $\sqrt{p}$. If $a_0 = \sqrt{p}$ we have finished, and otherwise

$$a_0 < \sqrt{p} \text{ or else } a_0 > \sqrt{p},$$

or equivalently (on multiplying both terms of the inequality by $\sqrt{p}$) one has either

$$a_0 < \sqrt{p} < \frac{p}{a_0} \text{ or else } a_0 > \sqrt{p} > \frac{p}{a_0}.$$

Hence we consider, always in our effort to approximate $\sqrt{p}$, the midpoint of the points a_0 and $\dfrac{p}{a_0}$: thus we set

$$a_1 = \frac{1}{2}\left(a_0 + \frac{p}{a_0}\right).$$

Proceeding in this way we are induced to define the recurrence relation

$$x_{n+1} = f(x_n), \quad n \geq 0, \qquad f(x) = \frac{1}{2}\left(x + \frac{p}{x}\right), \tag{9.21.a}$$

with initial condition $x_0 = a_0$. Clearly we are considering a discrete dynamical system. It is not difficult to show that whatever the value of $a_0 \neq 0$ may be, the orbit of a_0 converges to some $\ell > 0$. By what has been seen in Proposition 9.20 ℓ is a fixed point; since

$$\ell = \frac{1}{2}\left(\ell + \frac{p}{\ell}\right),$$

one has $\ell^2 = p$ and so $\ell = \sqrt{p}$. For example, in searching for $\sqrt{5}$ we begin the preceding recurrence by setting $x_0 = 1$. One thus obtains

$$x_1 = 3, \quad x_2 = \frac{7}{3}, \quad x_3 = \frac{47}{21}, \quad x_4 = \frac{2207}{987} \approx 2.23607;$$

note that in reality one has $\sqrt{5} \approx 2.2360679$. The graphical analysis of the sequence is illustrated in Fig. 9.8. □

In general the orbits of apparently very similar dynamical systems can actually behave in extremely different fashions.

Example 9.22 Consider the dynamical systems

$$x_{n+1} = f_a(x_n), \quad n \in \mathbb{N},$$

obtained from polynomial functions of the type $f_a(x) = ax(1-x)$ with $a > 0$. We can graphically "represent" the orbits of a value $x_0 = a_0$ by indicating on the horizontal axis the integers n and, in correspondence with them, on the vertical axis the values a_n that constitute the terms of the solution sequence for the dynamical system with initial datum $x_0 = a_0$. The graphs that follow were obtained by taking $x_0 = 0.2$

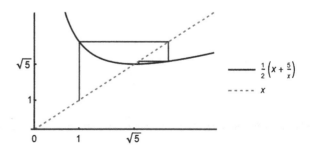

Fig. 9.8 Graphical analysis of $x_{n+1} = \dfrac{1}{2}\left(x_n + \dfrac{5}{x_n}\right)$, $x_0 = 1$

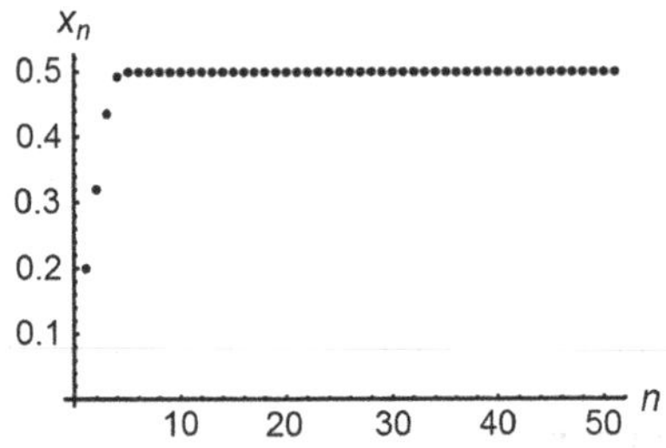

Fig. 9.9 The system
$x_{n+1} = 2x_n(1 - x_n)$ with
$x_0 = 0.2$

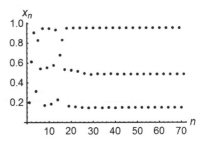

Fig. 9.10 The system
$x_{n+1} = 3.839x_n(1 - x_n)$
with $x_0 = 0.2$

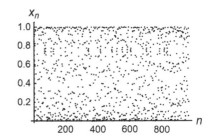

Fig. 9.11 The system
$x_{n+1} = 4x_n(1 - x_n)$ with
$x_0 = 0.2$

as initial point; one can see explicitly that the various behaviors of the dynamical systems turn out to be quite diverse as one varies the parameter a.

- In the case $a = 2$ the values reported in Fig. 9.9 suggest that the orbit converges to 0.5; effectively the orbit does indeed converge to 0.5 which is therefore a fixed point.
- In the case $a = 3.839$ the values reported in Fig. 9.10 suggest that the orbit tends to be 3-periodic. One proves that actually there is a 3-periodic point close to 0.149.
- In the case $a = 4$ the values reported in Fig. 9.11 reveal a much more chaotic situation.

The situation described above does not depend on the given value of x_0, but remains substantially the same if in the place of 0.2, one chooses an arbitrary value for x_0 from the open interval $]0, 1[$.   □

In the next example we will deal with Newton's[4] celebrated method of tangents for determining the solutions of an equation of the type $f(x) = 0$, where f is a given function.

Example 9.23 (Newton's method of the tangents) Let $f : \mathbb{R} \to \mathbb{R}$ be of class $\mathscr{C}^1$ and fix $a_0 \in \mathbb{R}$. The equation of the tangent line to the graph of f in the point $(a_0, f(a_0))$ is

$$y = f'(a_0)(x - a_0) + f(a_0).$$

Therefore, if $f'(a_0) \neq 0$, the tangent line intersects the x-axis in the point

$$a_1 = a_0 - \frac{f(a_0)}{f'(a_0)}.$$

Next, we consider the equation of the tangent line to the graph of f in the point $(a_1, f(a_1))$:

$$y = f'(a_1)(x - a_1) + f(a_1).$$

If $f'(a_1) \neq 0$, the tangent line intersects the x-axis in the point

$$a_2 = a_1 - \frac{f(a_1)}{f'(a_1)}.$$

Iterating the preceding procedure, we can construct, as in Fig. 9.12, a sequence $(a_n)_n$, assuming that the derivative of f does not vanish in the points of the sequence thus obtained. The sequence so constructed provides a solution of the recurrence relation

$$x_{n+1} = x_n - \frac{f(x_n)}{f'(x_n)}, \quad n \geq 0, \tag{9.23.a}$$

with initial datum $x_0 = a_0$. The recurrence (9.23.a) is clearly the discrete dynamical system associated to the function $x \mapsto x - \dfrac{f(x)}{f'(x)}$ defined on the points x where $f'(x) \neq 0$.

If a solution $(a_n)_n$ of the recurrence relation (9.23.a), or, in the language of dynamical systems, the orbit of a_0, converges to a finite number ℓ with $f'(\ell) \neq 0$, then on passing to the limit in (9.23.a) one obtains

$$\ell = \ell - \frac{f(\ell)}{f'(\ell)},$$

from which it follows that $f(\ell) = 0$: the sequence $(a_n)_n$ thus converges to a zero of f. For example, if $f(x) = x^3 + 2x - 1$, one obtains the recurrence relation

---

[4]Sir Isaac Newton (1642–1727).

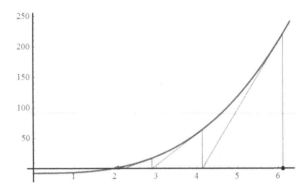

Fig. 9.12 Newton's method of tangents (adapted from "Learning Newton's Method" from the Wolfram Demonstrations Project http://demonstrations.wolfram.com/LearningNewtonsMethod. Contributed by: Angela Sharp, Chad Pierson, and Joshua Fritz. Creative Commons AttributionNonCommercial-ShareAlike CC BY-NC-SA 3.0 Unported License)

$$x_{n+1} = x_n - \frac{x_n^3 + 2x_n - 1}{3x_n^2 + 2}, \quad n \ge 1;$$

starting from $x_0 = 1$ one gets

$$x_1 \approx 0.6, \quad x_2 \approx 0.46493, \quad x_3 \approx 0.453467, \quad x_4 \approx 0.45339765$$

and $f(0.45339765) \approx -3.96799 \times 10^{-9}$ is already very close to zero.

If $f(x) = x^2 - p$ with $p > 0$, one obtains the recurrence relation

$$x_{n+1} = x_n - \frac{x_n^2 - p}{2x_n} = \frac{x_n^2 + p}{2x_n} = \frac{1}{2}\left(x_n + \frac{p}{x_n}\right).$$

This is the recurrence relation already studied in Example 9.21 in order to determine the square root of p.

It is not always the case that the solutions of the recurrence relation

$$x_{n+1} = x_n - \frac{f(x_n)}{f'(x_n)}, \quad n \ge 0,$$

are convergent sequences: for example, if $f(x) = x^2 + 1$, then no solution of

$$x_{n+1} = x_n - \frac{x_n^2 + 1}{2x_n}, \quad n \ge 0,$$

is a convergent sequence, in view of the fact that any possible limit would be a zero of $x^2 + 1$. In Fig. 9.13 one sees the corresponding graphical analysis and the graph of the values of the orbit of $x_0 = 0.2$.

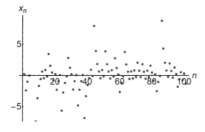

Fig. 9.13 Values of the orbit of $x_0 = 0.2$ for (9.23.a) with $f(x) = x^2 + 1$

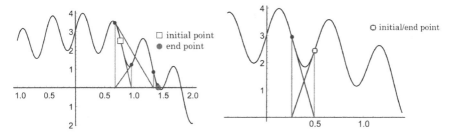

Fig. 9.14 Newton's method of tangents for $f(x) = -x^2 + \sin(12x) + 4$ with two different initial data

Also functions f that have real zeroes can give rise to more variable behaviors: in Fig. 9.14 the case $f(x) = -x^2 + \sin(12x) + 3$ is discussed with two different sets of initial data: in the first case the sequence converges to a zero of the function, in the second case the sequence is a periodic orbit and so does not converge.   □

9.2.2 Dynamical Systems Generated by Monotonic Functions

It is quite common to encounter dynamical systems of the form $x_{n+1} = f(x_n)$, $n \in \mathbb{N}$, where f is monotonic. We first introduce some notation.

- Let $(x_n)_n$ be a sequence. We write $x_n \uparrow \ell \in [a, b]$ (resp. $x_n \downarrow \ell$) if the sequence $(x_n)_n$ is increasing (resp. decreasing) and tends to ℓ as $n \to +\infty$, where $\ell \in \mathbb{R} \cup \{-\infty\} \cup \{+\infty\}$.
- If $S \subset \mathbb{R} \cup \{\pm\infty\}$ we set $M = \max S$ if M is the maximum of S, i.e., if $M \in S$ and $M \geq s$ for all $s \in S$, with the agreement that $M = +\infty$ if $+\infty \in S$. Analogously we set $m = \min S$ if m is the minimum of S, i.e., if $m \in S$ and $m \leq s$ for all $s \in S$, with the agreement that $m = -\infty$ if $-\infty \in S$.

We recall the important fact that monotonic sequences do always admit a limit, finite or infinite [33]. We first consider the easiest case where f is increasing ($f(x) \leq f(y)$ whenever $x \leq y$).

Proposition 9.24 *Let I be an interval, with $a := \inf I$ and $b := \sup I$, possibly not finite (and even not in I). Assume that $f : I \to I$ is continuous and **increasing**. Consider the discrete dynamical system*

$$x_0 \in I, \qquad x_{n+1} = f(x_n) \qquad \forall n \in \mathbb{N}.$$

The following cases may arise:

1. *If $f(x_0) \geq x_0$ then $x_n \uparrow \beta := \min \{\{x \in I : f(x) = x, x \geq x_0\} \cup \{b\}\}$;*
2. *If $f(x_0) \leq x_0$ then $x_n \downarrow \alpha := \max \{\{x \in I : f(x) = x, x \leq x_0\} \cup \{a\}\}$.*

Proof. We prove only Point 1; Point 2 follows similarly. Since $f(x_0) \geq x_0$ one has $f^2(x_0) \geq f(x_0)$ and, by induction, $f^{n+1}(x_0) \geq f^n(x_0)$ for every n. Thus the sequence $(x_n)_n$ is increasing; therefore there exists, either finite or infinite, $\ell = \lim_{n \to +\infty} x_n$. Suppose that there is a fixed point x of f which is larger than x_0: then by the hypothesis of monotonicity one has

$$x_0 \leq x, \; x_1 = f(x_0) \leq f(x) = x, \; x_2 = f^2(x_0) \leq f(x) = x, \ldots, x_n = f^n(x_0) \leq x, \; n \in \mathbb{N}$$

and consequently, $\ell \leq x$: by Proposition 9.20 ℓ is a fixed point of f, and it is necessarily the minimum among those that are larger than x_0. If instead f does not have fixed points in the interval I which are larger than x_0, then the sequence tends necessarily to b. $\qquad\qquad\qquad\qquad\qquad\qquad\qquad\qquad\qquad\qquad\qquad\qquad\qquad\qquad \square$

☞ *Remark 9.25 (How to find the limit: a recipe when f is increasing)* Following Proposition 9.24, here is how to proceed, when $f : I \to I$ is continuous and **increasing**, in order to find the limit point ℓ of the sequence defined by $x_{n+1} = f(x_n)$:

1. Compare $f(x_0)$ with x_0;
2. ℓ is the nearest fixed point of f from x_0 on the same side of $f(x_0)$, whenever it exists; otherwise ℓ is the endpoint of I on the same side of $f(x_0)$.

Example 9.26 Discuss the existence and the value of the limit of the sequence

$$x_{n+1} = \sqrt{x_n + 1}, \quad x_0 = 1,$$

whose first terms are depicted in Fig. 9.15. The function $f(x) = \sqrt{x + 1}$ is increasing on the open interval $]-1, +\infty[$; its fixed points are those $x > -1$ such that $x = \sqrt{x + 1}$ from which it follows that $x^2 - x - 1 = 0$, which yields $x = \dfrac{1 \pm \sqrt{5}}{2}$. Now $f(1) = \sqrt{2} > 1$, and the unique fixed point of f greater than 1 is $\beta := \dfrac{1 + \sqrt{5}}{2}$. It follows from Proposition 9.24 that $x_n \uparrow \beta$. In some texts this fact also is described as follows:

Fig. 9.15 Graphical analysis
of the sequence
$x_{n+1} = \sqrt{1 + x_n}$, $x_0 = 1$

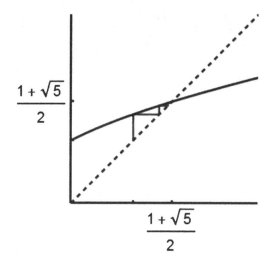

$$\sqrt{1 + \sqrt{1 + \sqrt{1 + \sqrt{1 + \cdots}}}} = \frac{1 + \sqrt{5}}{2}.$$ □

We consider now the case where the function f is decreasing; here the argument is based on the fact that $f^2 = f \circ f$ is increasing, so one is allowed to apply Proposition 9.24 to f^2.

Proposition 9.27 *Let I be an interval, with $a := \inf I$ and $b := \sup I$, possibly not finite (and even not in I). Assume that $f : I \to I$ is continuous and **decreasing**. Consider the discrete dynamical system*

$$x_0 \in I, \qquad x_{n+1} = f(x_n) \quad \forall n \in \mathbb{N}. \tag{9.27.a}$$

The following cases may arise:

1. *If $f^2(x_0) \geq x_0$ then*

$$x_{2n} \uparrow \min\left\{\{x \in I : f^2(x) = x, \ x \geq x_0\} \cup \{b\}\right\},$$

$$x_{2n+1} \downarrow \max\left\{\{x \in I : f^2(x) = x, \ x \leq x_1\} \cup \{a\}\right\};$$

2. *If $f^2(x_0) \leq x_0$ then*

$$x_{2n} \downarrow \max\left\{\{x \in I : f^2(x) = x, \ x \leq x_0\} \cup \{a\}\right\},$$

$$x_{2n+1} \uparrow \min\left\{\{x \in I : f^2(x) = x, \ x \geq x_1\} \cup \{b\}\right\}.$$

Fig. 9.16 The convergence
to the 2-periodic orbit of β

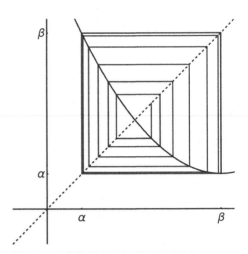

*Moreover, $\alpha := \lim\limits_{n \to +\infty} x_{2n} \in I$ if and only if $\beta := \lim\limits_{n \to +\infty} x_{2n+1} \in I$, in which case $f(\alpha) = \beta$, $f(\beta) = \alpha$: α and β are periodic of period 2. If $\alpha = \beta$ then $(x_n)_n$ converges to α, otherwise we say that the orbit of x_0 **converges to the 2-periodic orbit** of α (Fig. 9.16).*

Proof. The first part of the claim follows from the fact that f^2 is increasing: indeed if $y_1 \le y_2$ belongs to I, we have $f(y_1) \ge f(y_2)$ and hence

$$f^2(y_1) = f(f(y_1)) \le f(f(y_2)) = f^2(y_2).$$

Therefore applying Proposition 9.24 to the discrete dynamical system

$$y_0 := x_0 \in I, \quad y_{n+1} = f^2(y_n) \quad \forall n \in \mathbb{N}$$

we get $y_n = x_{2n}$ for each n and thus:

- If $f^2(x_0) \ge x_0$, then the sequence $(x_{2n})_n$ is increasing and converges to

$$\min\left\{\{x \in I : f^2(x) = x, \ x \ge x_0\} \cup \{b\}\right\};$$

- if $f^2(x_0) \le x_0$ then $(x_{2n})_n$ is decreasing and converges to

$$\max\left\{\{x \in I : f^2(x) = x, \ x \le x_0\} \cup \{a\}\right\}.$$

Analogously, applying Proposition 9.24 to the discrete dynamical system

$$y_0 := x_1 \in I, \quad y_{n+1} = f^2(y_n) \quad \forall n \in \mathbb{N}$$

we get that $y_n = x_{2n+1}$ for each n and thus:

- If $f^2(x_1) \geq x_1$, then the sequence $(x_{2n+1})_n$ is increasing and converges to

$$\min\{\{x \in I : f^2(x) = x,\ x \geq x_1\} \cup \{b\}\};$$

- If $f^2(x_1) \leq x_1$, then the sequence $(x_{2n+1})_n$ is decreasing and converges to

$$\max\{\{x \in I : f^2(x) = x,\ x \geq x_1\} \cup \{a\}\}.$$

Now observe that, since f is decreasing, we have

$$f^2(x_0) \geq x_0 \text{ if and only if } f^2(x_1) = f(f^2(x_0)) \leq f(x_0) = x_1 :$$

we thus get in one fell swoop both Points 1 and 2.

Finally, assume that $\alpha = \lim_{n \to +\infty} x_{2n} \in I$. Then $\beta = \lim_{n \to +\infty} f(x_{2n}) = f(\alpha) \in I$, thanks to the continuity of f, and analogously $f(\beta) = \alpha$. $\qquad \square$

Proposition 9.27 yields a simple characterization of the convergence of the sequence defined by (9.27.a).

Corollary 9.28 *Let I be an interval, with $a := \inf I$ and $b := \sup I$, possibly not finite (and even not in I). Assume that $f : I \to I$ is continuous and **decreasing** and x_0 belongs to I and consider the sequence $(x_n)_n$ defined by $x_{n+1} := f(x_n)$, $n \in \mathbb{N}$. The sequence $(x_n)_n$ has a limit if and only if, denoted by I_{x_0,x_1} the closed interval whose endpoints are x_0 and x_1, the following conditions are satisfied:*

1. *x_2 and a fixed point ℓ of f belong to I_{x_0,x_1};*
2. *I_{x_0,x_1} does not contain any 2-periodic point of f different from ℓ.*

In this case $\lim_{n \to +\infty} x_n = \ell \in \mathbb{R}$.

Proof. Assume that the sequence has a limit ℓ. If $f^2(x_0) \geq x_0$, from Proposition 9.27 we have

$$x_0 \leq x_2 \leq \cdots \leq \ell \leq \cdots \leq x_3 \leq x_1;$$

thus x_2 and ℓ belongs to I_{x_0,x_1} and ℓ is a fixed point of f (Proposition 9.20). If ℓ_1 is any 2-periodic point of f in I_{x_0,x_1} then, by Proposition 9.27, for any n we have

$$x_{2n} \leq \ell_1 \leq x_{2n+1}.$$

Passing to the limit as $n \to +\infty$ we get $\ell_1 = \ell$. The case where $f^2(x_0) \leq x_0$ can be treated similarly and yields the same conclusion.

Conversely, assume that Conditions 1 and 2 are satisfied. Suppose, for instance, that $x_1 \leq x_2 = f^2(x_0) \leq x_0$. Since by hypothesis the set $\{x \in I_{x_0,x_1} : f^2(x) = x\}$ of 2-periodic points in I_{x_0,x_1} coincides with $\{\ell\}$, by Point 2 of Proposition 9.27 both $(x_{2n})_n$ and $(x_{2n+1})_n$ converge to ℓ. The case where $x_0 \leq x_1$ can be treated similarly. $\qquad \square$

☞ *Remark 9.29* (*A recipe when f is decreasing*) Following Proposition 9.27, here is
how to proceed, when $f : I \to I$ is continuous and **decreasing**, in order to study
the orbits of the system $x_{n+1} = f(x_n)$.

1. Compare x_0 with $f^2(x_0)$.
2. Find the fixed points of f^2 from x_0 on the same side of $f^2(x_0)$:

   a. If these points do not exist then the limit of $(x_{2n})_n$ is the endpoint of
      I on the same side of $f^2(x_0)$, and the limit of $(x_{2n+1})_n$ is the other
      endpoint of I;
   b. Otherwise $(x_{2n})_n$ converges to the minimum point α among of the
      above fixed points, and $(x_{2n+1})_n$ converges to $f(\alpha)$. If $f(\alpha) = \alpha$ then
      $(x_n)_n$ converges to α, otherwise the orbit of x_0 converges to the 2-
      periodic orbit of α.

Remark 9.30 (*A necessary condition*) Assume that $f : I \to I$ is continuous and
decreasing. It follows from Corollary 9.28 that a **necessary condition** for the convergence of the sequence $(x_n)_n$ is that x_2 belongs to the closed interval I_{x_0,x_1} whose
endpoints are x_0, x_1 (see Problem 9.15 for a generalization). We weren't able to find
a reference for the necessary and sufficient condition formulated here in the same
corollary.

Example 9.31 Discuss the behavior of the sequence, the existence and possibly the
value of the limit of the sequence

$$x_{n+1} = 1 + \frac{1}{x_n}, \quad x_0 = 1,$$

whose first iterations are depicted in Fig. 9.17. The function $f :]0, +\infty[\to]0, +\infty[$
defined by $f(x) = 1 + (1/x)$ is decreasing. Now $x_1 = f(1) = 2$ and $x_2 = f^2(0) =$

Fig. 9.17 Graphical analysis
of the sequence
$x_{n+1} = 1 + (1/x_n)$, $x_0 = 1$

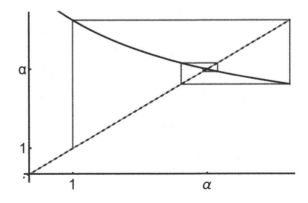

$f(2) = 3/2 \in [x_0, x_1] = [1, 2]$. It follows from Proposition 9.27 that the sequence $(x_{2n})_n$ is increasing, whereas $(x_{2n+1})_n$ is decreasing. We study the behavior of the sequence, by following the procedure of Remark 9.29.

- We look for the 2-periodic points of f in $[1, 2]$: $f^2(x) = x$ if and only if $\dfrac{2x + 1}{x + 1} = x$, i.e., $x^2 - x - 1 = 0$ so that $x = \dfrac{1 \pm \sqrt{5}}{2}$. Thus the unique 2-periodic point of f in $[1, 2]$ is $\alpha := \dfrac{1 + \sqrt{5}}{2}$.

- $f(\alpha) = \dfrac{\alpha + 1}{\alpha} = \alpha$: it follows that the sequence $(x_n)_n$ converges to α.

It is now clear what is meant by the frequently used written statement

$$1 + \cfrac{1}{1 + \cfrac{1}{1 + \cfrac{1}{1 + \cdots}}} = \frac{1 + \sqrt{5}}{2}.$$ $\square$

Example 9.32 Discuss the behavior, the existence and possibly the value of the limit of the sequence

$$x_{n+1} = (1 - x_n)^2, \quad x_0 = \frac{1}{2}.$$

The function $f(x) = (1 - x)^2$ is not monotonic on $\mathbb{R}$. However it is easy to see, inductively, that $x_n \in [0, 1]$ for every n and f is decreasing on $[0, 1]$. So that actually $(x_n)_n$ satisfies

$$x_{n+1} = f(x_n) \quad n \in \mathbb{N},$$

with $f : [0, 1] \to [0, 1]$. Now $x_0 = \dfrac{1}{2}, x_1 = \dfrac{1}{4}$ and $x_2 = \dfrac{9}{16} \notin [x_1, x_0]$: the necessary condition formulated in Corollary 9.28 shows that the sequence does not converge. Since $x_0 < x_2$, it follows from Proposition 9.27 that the sequence $(x_{2n})_n$ is increasing and that $(x_{2n+1})_n$ is decreasing. For sure both converge, since the points of both the monotonic sequences are all in $[0, 1]$: in order to find their limits we proceed as suggested in Remark 9.29:

- We look for the 2-periodic points of f in the interval $\left[\dfrac{1}{2}, 1\right]$: $f^2(x) = x$ as long as $(1 - (1 - x)^2)^2 = x$, i.e., if and only if $x^4 - 4x^3 + 4x^2 - x = 0$. In order to find the solutions to this fourth degree equation we use the fact that the solutions to $f(x) = x$, namely the roots of $x^2 - 3x + 1 = 0$, solve the given equation. It follows that $x^2 - 3x + 1$ divides $x^4 - 4x^3 + 4x^2 - x$: the polynomial division yields

$$x^4 - 4x^3 + 4x^2 - x = (x^2 - x)(x^2 - 3x + 1).$$

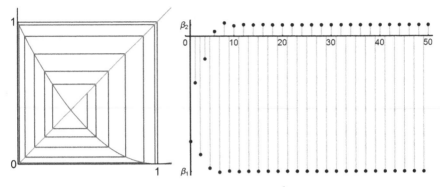

Fig. 9.18 Graphical analysis and graph of the sequence $x_0 = \dfrac{1}{2}$, $x_{n+1} = (1 - x_n)^2$

Therefore the 2-periodic points of f^2 are the zeroes 0, 1 of $x^2 - x$ and the zeroes $\dfrac{3 \pm \sqrt{5}}{2}$ of $x^2 - 3x + 1$; only $\alpha := 1$, among these numbers, belongs to the interval $\left[\dfrac{1}{2}, 1\right]$.

- Since $f(\alpha) = f(1) = 0$ then $x_{2n} \uparrow 1$, whereas $x_{2n+1} \downarrow 0$, as it is confirmed by the graphical analysis in Fig. 9.18. The orbit of $\dfrac{1}{2}$ thus converges to the 2-periodic orbit of 1. □

9.2.3 Periodic Orbits: Sarkovskii Theorem

We now introduce an ordering of the natural numbers different from that usually employed, but which will be particularly useful in the study of the existence of orbits with an assigned period in a discrete dynamical system. This ordering takes its name from Oleksandr Mykolaiovych Sarkovskii,[5] the mathematician who first came up with this ingenious notion.

Recall that every natural number different from 0 is the product of a power of 2 and an odd integer.

Definition 9.33 (*Sarkovskii Order*) The Sarkovskii order $\triangleright$ on $\mathbb{N}_{\geq 1}$, from the greatest to the smallest, is as follows:

- First are the odd numbers greater than 3 in increasing order:

$$3 \triangleright 5 \triangleright 7 \triangleright \cdots ;$$

---

[5]Oleksandr Mykolaiovych Sarkovskii (1936-).

- Then are 2 times the above numbers, in the same order:

$$2 \times 3 \triangleright 2 \times 5 \triangleright 2 \times 7 \triangleright \cdots ;$$

- Then are 2^2 times the numbers of Point 1, in the same order:

$$2^2 \times 3 \triangleright 2^2 \times 5 \triangleright 2^2 \times 7 \triangleright \cdots ;$$

- · · · · · · · · · · · · · · · ·
- Then are 2^n times the numbers of Point 1, in the same order:

$$2^n \times 3 \triangleright 2^n \times 5 \triangleright 2^n \times 7 \triangleright \cdots \quad (n \geq 1);$$

- · · · · · · · · · · · · · · · ·
- Finally, one lists the powers of 2 in decreasing order:

$$\cdots \triangleright 2^3 \triangleright 2^2 \triangleright 2 \triangleright 1.$$

Formulated in more concise fashion, one has

$$3 \triangleright 5 \triangleright 7 \triangleright \cdots \triangleright 2 \times 3 \triangleright 2 \times 5 \triangleright 2 \times 7 \triangleright \cdots$$
$$\cdots \triangleright 2^2 \times 3 \triangleright 2^2 \times 5 \triangleright 2^2 \times 7 \triangleright \cdots \triangleright 2^3 \triangleright 2^2 \triangleright 2 \triangleright 1. \qquad \square$$

Remark 9.34 Notice that 1 is the minimum of $\mathbb{N}_{\geq 1}$ both in the usual and in the Sarkovskii order; however $(\mathbb{N}_{\geq 1}, \triangleright)$ has a maximum, namely the number 3, whereas a maximum of $\mathbb{N}_{\geq 1}$ does not exist for the usual order.

Theorem 9.35 (Sarkovskii) *Let I be an interval of $\mathbb{R}$ and let $f : I \to I$ be continuous. Suppose that the dynamical system*

$$x_{n+1} = f(x_n), \quad n \in \mathbb{N},$$

admits a periodic point of minimum period $\ell \geq 1$. Then, for each k with $\ell \triangleright k$, f has a periodic point of minimum period k.

Example 9.36 Suppose that the function $f : I \to I \subseteq \mathbb{R}$ admits a point of minimum period $56 = 2^3 \times 7$. Now:

- $2^3 \times 7 \triangleright 2^3 \times m$ for every $m \geq 9$ odd;
- $2^3 \times 7 \triangleright 2^p \times m$ for every $p \geq 4$ and m odd;
- $2^3 \times 7 \triangleright 2^p$ for every $p \geq 0$.

Therefore f has points of minimum period $2^3 \times m$ with $m \geq 9$ odd, points of minimum period $2^p \times m$ for every $p \geq 4$ and m odd and points of minimum period 2^p for every $p \geq 0$. $\qquad \square$

Remark 9.37 • The conclusion of Theorem 9.35 does not remain true in general
if the domain and codomain of f is not an interval of $\mathbb{R}$. For example, the map
$F : \mathbb{R}\backslash\{0\} \to \mathbb{R}\backslash\{0\}$, $x \mapsto -\dfrac{1}{x}$ has 2-periodic points (any $r \in \mathbb{R}\backslash\{0\}$), but does
not have fixed points.
- The statement of Theorem 9.35 is optimal: if $\ell \rhd k$ ($k \neq \ell$) there exists a function
  f continuous on an interval that has a point of minimum period k, and that is
  without points of minimum period ℓ; Problem 9.21 discusses the case in which
  $k = 5$ and $\ell = 3$.
- Under the hypotheses of Theorem 9.35:

  - If f has no fixed points then f has no periodic points;
  - If the set of periodic points of f is finite, then their minimum periods are powers
    of 2.

We limit ourselves to proving only some particular cases of Theorem 9.35, direct-
ing the reader to [11] for a complete proof of the theorem. The simplest case is the
passage from an arbitrary period to the period 1 (that is, to the existence of a fixed
point).

Example 9.38 (Periodic Point ⇒ existence of a fixed point) If the continuous map
$f : I \to I$ has a point of minimum period $k > 1$, then Theorem 9.35 guarantees the
existence of a fixed point: we give a direct proof of this fact. Let $w_1, \dots, w_k$ be all
the distinct values of a periodic orbit with minimum period k listed from the smallest
to the largest values, that is, with $w_1 < w_2 < \cdots < w_k$, as in Fig. 9.19. Necessarily

$$f(w_1) \in \{w_2, \dots, w_k\}, \quad f(w_k) \in \{w_1, \dots, w_{k-1}\}.$$

Therefore $f(w_1) > w_1$ and $f(w_k) < w_k$: by the Intermediate Zero Theorem,
$f(x) - x$ has a zero ξ in I, and hence it is $f(\xi) = \xi$.   □

A particularly relevant case of Theorem 9.35 regards the passage from period 3,
the maximum of the natural numbers according to the Sarkovskii ordering, to every
other period. We state this result and we prove it, given its importance.
We first prove the following well-known properties of continuous functions.

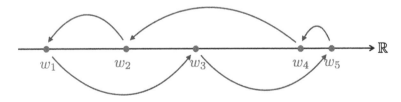

Fig. 9.19 The orbit $\{w_1, w_2, w_3, w_4, w_5\}$ of a point of minimum period $k = 5$

Lemma 9.39 *Let* $f : [a, b] \to \mathbb{R}$ *be a continuous map.*

1. If $\gamma \leq \delta \in f([a, b])$, *then there exist* $\alpha \leq \beta \in [a, b]$ *such that* $f([\alpha, \beta]) = [\gamma, \delta]$.
2. If $a, b \in f([a, b])$, *then* f *has at least one fixed point in* $[a, b]$.

Proof. 1. Let $c, d \in [a, b]$ be such that $f(c) = \gamma$ and $f(d) = \delta$. Suppose that $c \leq d$; the case $c > d$ can be proved in the same way. Let

$$\alpha := \max\{x \in [c, d] : f(x) = \gamma\};$$

this maximum exists since the set under consideration is closed and bounded. For the same reason there exists

$$\beta := \min\{x \in [\alpha, d] : f(x) = \delta\}.$$

By the Intermediate Value Theorem, the image under a continuous function of an interval contains all the values included between the images of the endpoints of the interval; thus, $[\gamma, \delta] \subseteq f([c, d]) \subseteq f([\alpha, \beta])$. If there were an $x \in [\alpha, \beta]$ with $f(x) < \gamma$, since $f(\beta) = \delta$, again by the Intermediate Value Theorem, there would exist $\alpha' > x \geq \alpha$ such that $f(\alpha') = \gamma$, which would contradict the maximality of α. Analogously, one proves that $f(x) \leq \delta$ for each $x \in [\alpha, \beta]$ and hence $f([\alpha, \beta]) = [\gamma, \delta]$, as depicted in Fig. 9.20.
2. Let c, d in $[a, b]$ be such that $f(c) = a$ and $f(d) = b$. If $c = a$ or $d = b$ we are done. Otherwise

$$a = f(c) < c \quad \text{and} \quad b = f(d) > d.$$

The continuous function $f(x) - x$ assumes values of opposite sign in the points c and d, and so it has a zero $\xi \in [c, d] \subseteq [a, b]$: clearly ξ is a fixed point for f. $\quad\square$

Remark 9.40 Point 2 of Lemma 9.39 ensures the existence of a fixed point once f is continuous and $f([a, b]) \supseteq [a, b]$. This is somewhat similar to the more celebrated result of the existence of a fixed point once $f([a, b]) \subseteq [a, b]$ (Problem 9.10).

Fig. 9.20 $f([\alpha, \beta]) = [\gamma, \delta]$

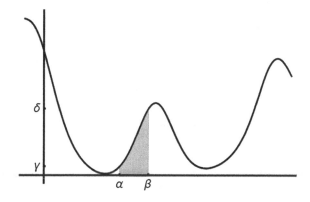

Theorem 9.41 (The period 3) *Let I be an interval of $\mathbb{R}$ and let $f : I \to I$ be continuous. Suppose that the dynamical system*

$$x_{n+1} = f(x_n), \quad n \in \mathbb{N},$$

admits a periodic point of minimum period 3. Then f has points of arbitrary minimum period.

Proof. Let a in I be a point of minimum period 3. We set

$$b = f(a), \qquad c = f(b).$$

Necessarily a, b, c are distinct, and one has $f(c) = a$. It is not restrictive to suppose that $a = \min\{a, b, c\}$. Two cases are then possible: either $a < b < c$ or $a < c < b$. Here we will suppose that

$$a < b < c;$$

the proof in the other case is carried out in analogous fashion, and we leave it as an exercise for the reader. The initial setting is illustrated in Fig. 9.21. The Intermediate Value Theorem applied to $I_0 := [a, b]$ and $I_1 := [b, c]$ yields

$$f(I_0) \supseteq I_1, \qquad f(I_1) \supseteq [a, c] \supseteq I_0, \qquad f(I_1) \supseteq [a, c] \supseteq I_1.$$

Let $n \in \mathbb{N}_{\geq 1}$: we prove the existence of a point of minimum period n for f.

- $n = 1$: since $f(I_1) \supseteq I_1$, by Point 2 of Lemma 9.39 one deduces the existence of a fixed point for f in I_1.
- $n = 2$: since $f(I_1) \supseteq I_0$, by Point 1 of Lemma 9.39 there exists a closed interval $B_1 \subseteq I_1$ such that $f(B_1) = I_0$. The situation is depicted in Fig. 9.22.

Fig. 9.21 $f(a) = b$, $f(b) = c$, $f(c) = a$

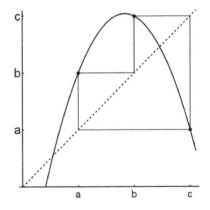

Fig. 9.22 The case $n = 2$

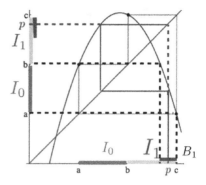

As a consequence one has

$$f^2(B_1) = f(f(B_1)) = f(I_0) \supseteq I_1 \supseteq B_1 :$$

applying Point 2 of Lemma 9.39 to the continuous function $f^2 = f \circ f$ one obtains the existence of $p \in B_1$ such that $f^2(p) = p$. If it were the case that $f(p) = p$ then

$$p \in B_1 \cap f(B_1) = B_1 \cap I_0 \subseteq I_1 \cap I_0 = \{b\},$$

from which it would follow that $p = b$, which, however, is absurd because b is not of period 2. Hence, as a consequence p has minimum period equal to 2.

• Now let $n > 3$. Since $f(I_1) \supseteq I_1$, by Point 1 of Lemma 9.39 there exists a closed interval $A_1 \subseteq I_1$ such that $f(A_1) = I_1$. But then, since $f(A_1) = I_1 \supseteq A_1$, for the same reason there exists a closed interval $A_2 \subseteq A_1$ such that $f(A_2) = A_1$ (Fig. 9.23).

Proceeding inductively one obtains a sequence of closed intervals

$$A_{n-2} \subseteq A_{n-3} \subseteq \cdots \subseteq A_1 \subseteq I_1$$

Fig. 9.23 The case $n > 3$: the closed intervals A_1 and A_2

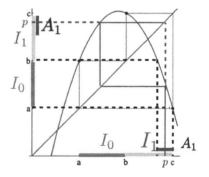

Fig. 9.24 The construction
of the sets A_k

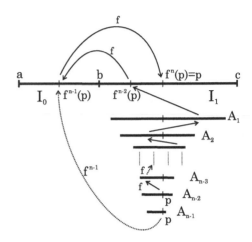

satisfying

$$f(A_{n-2}) = A_{n-3}, \; f(A_{n-3}) = A_{n-4}, \ldots, f(A_2) = A_1, \; f(A_1) = I_1.$$

One notes then that

$$f^{n-2}(A_{n-2}) = f^{n-3}(A_{n-3}) = \cdots = f(A_1) = I_1, \qquad (9.41.\text{a})$$

so that $f^{n-1}(A_{n-2}) = f(I_1) \supseteq I_0$: by Point 1 of Lemma 9.39 applied to f^{n-1} there
exists a closed interval A_{n-1} with the property that

$$A_{n-1} \subseteq A_{n-2}, \qquad f^{n-1}(A_{n-1}) = I_0.$$

From this it follows that $f^n(A_{n-1}) = f(I_0) \supseteq I_1 \supseteq A_{n-1}$: applying Point 2 of
Lemma 9.39 to f^n one deduces the existence of $p \in A_{n-1}$ such that $f^n(p) = p$,
namely the existence of a point having period n: we invite the reader to give a look
at Fig. 9.24 before going on.
There remains to prove that p has *minimum* period n. It will be useful to observe
that, since

$$p \in A_{n-1} \subseteq A_{n-2} \subseteq \cdots \subseteq A_1 \subseteq I_1,$$

from relation (9.41.a) one deduces that

$$p, \, f(p), \, f^2(p), \ldots, f^{n-2}(p) \in I_1; \qquad (9.41.\text{b})$$

while

$$f^{n-1}(p) \in f^{n-1}(A_{n-1}) = I_0. \qquad (9.41.\text{c})$$

We note that $p \notin \{a, b, c\}$: if instead that were the case, then from (9.41.b) it would follow that $f^k(p) \in I_1$ for $k = 0, 1, 2$ and from that one would have $\{a, b, c\} \subseteq I_1 = [b, c]$, which is absurd.

Assume that $f^k(p) = p$ for some $1 \le k \le n - 1$. By applying $f^{(n-1)-k}$ to both sides of the equality we get $f^{n-1}(p) = f^i(p)$, with $i = (n-1) - k \le n - 2$. Now $f^i(p) \in I_1$ whereas $f^{n-1}(p) \in I_0$ implying that $f^{n-1}(p) \in I_0 \cap I_1 = \{b\}$ and $p = f^n(p) = f(f^{n-1}(p)) = c$, a contradiction. Thus the minimum period of p is n. $\qquad\qquad\qquad\qquad\qquad\qquad\qquad\qquad\qquad\qquad\qquad\qquad\qquad\qquad$ $\square$

9.3 Problems

Problem 9.1 Find a recurrence relation for the number of possible distributions of n distinct objects on 5 shelves of a closet. What is the initial condition?

Problem 9.2 Find a recurrence relation for the number of sequences of automobiles of 3 different types—Audi, Fiat and Mercedes—in a row of n parking spaces, bearing in mind that an Audi or a Mercedes occupies two parking spaces while a Fiat occupies only one parking space.

Problem 9.3 Suppose that during every month from the second month on, every pair of rabbits generates a new couple (a male and a female) of rabbits. Find the recurrence relation that describes the number of couples of rabbits month after month (assuming that all the couples survive). If initially there is only a single couple of newly born rabbits, what is the number of pairs of rabbits after 5 months?

Problem 9.4 Fix $k \in \mathbb{N}_{\ge 1}$. Find the recursive relation for the number of regions of the plane created by n lines if all the following conditions are satisfied:

1. Exactly k of these lines are parallel;
2. Each of the $n - k$ non-parallel lines among themselves intersect all the other lines;
3. Three distinct lines never have a point in common.

Find the number of such regions if $n = 9, k = 3$.

Problem 9.5 Find a recursive relation for the accumulation of the money deposited in a bank account after n years if the interest is a rate of 6 % and in each year 50 euro are added to the bank account.

Problem 9.6 Find a recurrence relation to count the number of binary n-sequences with at least a pair of consecutive 0 digits.

Problem 9.7 Find a recurrence relation to count the number of Gilbreath permutations (see Definition 2.48) of a deck of n cards. [Hint: Think at who could be the last card in a Gilbreath permutation.]

Problem 9.8 If 500 euros are invested in a fund that gives 8 % annual interest, find a formula for calculating the quantity of money accumulated after n years.

Problem 9.9 Prove that the sequence $(a_n)_{n\geq 1}$ defined by setting

$$a_n = \begin{cases} 1 & \text{if } n = 2^m, \\ 2\ell + 1 & \text{if } n = 2^m + \ell, 1 \leq \ell < 2^m \end{cases}$$

is a solution of the recurrence relation

$$x_n = \begin{cases} 2x_{n/2} - 1 & \text{if } n \geq 2 \text{ is even}, \\ 2x_{(n-1)/2} + 1 & \text{if } n \geq 3 \text{ is odd} \end{cases}$$

with initial datum $x_1 = 1$.

Problem 9.10 Let $f : [a, b] \to [a, b]$ be continuous. Prove that f has at least a fixed point.

Problem 9.11 Discuss the existence and the value of the limit of the sequence

$$x_{n+1} = \sin x_n + x_n, \quad x_0 = 7\pi/2,$$

whose first iterations are shown in Fig. 9.25.

Problem 9.12 Let $(a_n)_n$ be the sequence defined by

$$a_0 = \sqrt{2}, \qquad a_{n+1} = \sqrt{2 + a_n}.$$

Fig. 9.25 Graphical analysis of the sequence $x_{n+1} = x_n + \sin x_n$, $x_0 = 7\pi/2$

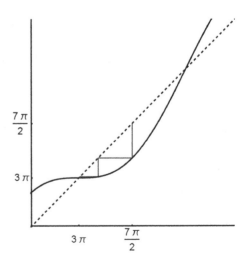

1. Prove that $(a_n)_n$ is bounded above by 2.
2. Study the existence of the limit of the sequence.

Problem 9.13 1. Prove that $f(x) = \dfrac{x+1}{x+2}$ is monotonic on $]-2, +\infty[$ and that
$f([0, +\infty[) \subseteq [0, +\infty[$.
2. Study the existence of a limit for the sequence

$$x_0 = 0, \qquad x_{n+1} = \frac{x_n + 1}{x_n + 2}.$$

Problem 9.14 Discuss, depending on the value of $\lambda > 0$, the existence of a limit for the sequence defined by

$$x_0 = \lambda, \qquad x_{n+1} = \frac{1}{8} x_n^2 + \frac{1}{8}.$$

Problem 9.15 Let I be an interval of $\mathbb{R}$, and $f : I \to I$ be continuous and decreasing. Let $(a_n)_n$ be a solution of

$$x_0 \in I, \qquad x_{n+1} = f(x_n) \quad \forall n \in \mathbb{N}.$$

Assume that the sequence $(a_n)_n$ converges to $\ell \in \mathbb{R}$. Show that, for any k even, m odd and N larger than both, the point a_N is between a_k and a_m.

Problem 9.16 Let $[a, b]$ be a closed and bounded interval of $\mathbb{R}$, and $f : [a, b] \to [a, b]$ be continuous and decreasing. Let $(b_n)_n$ be the solution to

$$x_0 \in I, \qquad x_{n+1} = f(x_n) \quad \forall n \in \mathbb{N}.$$

Assume that f has no points of minimum period 2. Show that the sequence $(b_n)_n$ does converge.

Problem 9.17 Let $f : I \to I$ be a function from an interval I into itself. Suppose that p is periodic of period $N \geq 1$. Prove that then the minimum period of p divides N.

Problem 9.18 Let $f : \mathbb{R} \to \mathbb{R}$ be continuous, with a periodic point of minimum period 20. Prove that f has a point of minimum period 48.

Problem 9.19 Let $f : \mathbb{R} \to \mathbb{R}$ be continuous, with a periodic point of minimum period 40. Does this necessarily imply that f has a point of minimum period 30?

Problem 9.20 Let $f : [a, b] \to \mathbb{R}$ be strictly decreasing (as in Fig. 9.26). Show that f has at most a single fixed point.

Fig. 9.26 A strictly
decreasing function has at
most a unique fixed point

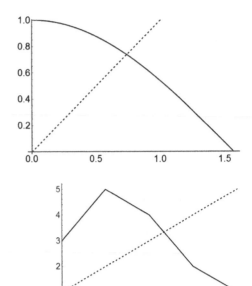

Fig. 9.27 The function f
defined in Problem 9.21

Problem 9.21 In this exercise we give an example of a function that has a point of
minimum period 5 but does not have points of minimum period 3.

Let $f : [1, 5] \rightarrow [1, 5]$ be such that

$$f(1) = 3, \ f(3) = 4, \ f(4) = 2, \ f(2) = 5, \ f(5) = 1,$$

and be affine on each interval $[1, 2]$, $[2, 3]$, $[3, 4]$, $[4, 5]$ (see Fig. 9.27).

1. Prove that f has a point of minimum period 5.
2. Determine $f^3([1, 2])$, $f^3([2, 3])$, $f^3([4, 5])$: deduce that f does not have points
    of minimum period 3 in the intervals $[1, 2]$, $[2, 3]$, $[4, 5]$.
3. Show that f^3 is strictly decreasing on the interval $[3, 4]$, and deduce that f^3 has
    a unique fixed point in $[3, 4]$, and that the uniquely determined point is indeed a
    fixed point of f.
4. Deduce that f does not have points of minimum period 3.

Chapter 10
Linear Recurrence Relations

Abstract In the following chapter we address the techniques for the resolution of some celebrated recurrence relations. We will discuss in detail the linear recurrences with constant coefficients. Our emphasis goes to the application of the theory: the proofs, though elementary, are relegated to the end of the chapter. We proceed step by step in showing first how to solve just homogeneous recurrences, then how to find a particular solution in some special cases and only finally how to obtain all the solutions to the original problem. We also consider linear recurrences with variable coefficients and the divide and conquer recurrences: here we focus on the order of magnitude of the solutions, a fact which has an impact in the analysis of algorithms. There are about 40 examples and 50 classified problems.

10.1 Linear Recurrences with Constant Coefficients

In this section we will try to present the main results on the resolution of linear recurrence relations with *constant coefficients* and their applicability by presenting several examples. We leave the more technical proofs for Sect. 10.5, as they are non-essential in the first reading.

Definition 10.1 A **linear recurrence relation of order** r **with constant coefficients** is a recurrence of the type

$$c_0 x_n + c_1 x_{n-1} + \cdots + c_r x_{n-r} = h_n, \quad n \geq r, \qquad (R)$$

where $c_0, c_1, \ldots, c_r$ are real or complex constants, with c_0 and c_r both different from zero (this will always be assumed in the chapter, even if not stated explicitly), and $(h_n)_{n \geq r}$ is a sequence of real or complex numbers called the **sequence of non-homogeneous terms** of the recurrence. The recurrence is called **homogeneous** if the sequence of non-homogeneous terms is the null sequence, **non-homogeneous** if $h_n \neq 0$ for some n.

© Springer International Publishing Switzerland 2016

C. Mariconda and A. Tonolo, *Discrete Calculus*,

UNITEXT - La Matematica per il 3+2 103, DOI 10.1007/978-3-319-03038-8_10

The recurrence relation

$$c_0 x_n + c_1 x_{n-1} + \cdots + c_r x_{n-r} = 0, \quad n \geq r, \qquad (R_o)$$

is called the **associated homogeneous recurrence**, or the **homogeneous part** of the recurrence (R).   ☐

As we have already noted, the recurrence

$$c_0 x_n + c_1 x_{n-1} + \cdots + c_r x_{n-r} = h_n, \quad n \geq r,$$

can be written equivalently as

$$c_0 x_{n+r} + c_1 x_{n+(r-1)} + \cdots + c_r x_n = h_{n+r}, \quad n \geq 0.$$

It makes no difference which formulation one uses.

Remark 10.2 Each r-sequence of values assigned to the r free unknowns of the linear recurrence relation

$$c_0 x_n + c_1 x_{n-1} + \cdots + c_r x_{n-r} = h_n, \quad n \geq r,$$

uniquely determines a solution.

When solving a linear recurrence relation, the following principle is of fundamental importance.

Proposition 10.3 (Superposition Principle)

1. Let $(u_n)_n$, $(v_n)_n$ be respectively solutions of the linear recurrence relations

$$c_0 x_n + c_1 x_{n-1} + \cdots + c_r x_{n-r} = h_n, \quad n \geq r, \quad and$$

$$c_0 x_n + c_1 x_{n-1} + \cdots + c_r x_{n-r} = k_n, \quad n \geq r,$$

with equal homogeneous part and sequences of non homogeneous terms $(h_n)_n$ and $(k_n)_n$. For any pair of constants A and B, the sequence $(Au_n + Bv_n)_n$ is a solution of the recurrence relation

$$c_0 x_n + c_1 x_{n-1} + \cdots + c_r x_{n-r} = Ah_n + Bk_n, \quad n \geq r.$$

☞ *2. The general solution of the linear recurrence relation*

$$c_0 x_n + c_1 x_{n-1} + \cdots + c_r x_{n-r} = h_n \quad n \geq r, \qquad (R)$$

is obtained by adding a particular solution to the general solution of the associated homogeneous recurrence.

Proof. 1. One has easily

$$c_0(Au_n + Bv_n) + c_1(Au_{n-1} + Bv_{n-1}) + \cdots + c_r(Au_{n-r} + Bv_{n-r}) =$$
$$= A(c_0 u_n + c_1 u_{n-1} + \cdots + c_r u_{n-r}) + B(c_0 v_n + c_1 v_{n-1} + \cdots + c_r v_{n-r})$$
$$= Ah_n + Bk_n.$$

2. Let $(u_n)_n$ be a particular solution of (R). By the previous point we know that $(v_n)_n = (u_n)_n + (v_n - u_n)_n$ is a solution of (R) if and only if $(v_n - u_n)_n$ is a solution of the associated homogeneous recurrence. Therefore each solution of (R) is obtained by adding a solution of the associated homogeneous recurrence to $(u_n)_n$. $\square$

10.1.1 Homogeneous Linear Recurrence Relation with Constant Coefficients

The null sequence is a solution of any homogeneous linear recurrence relation. The structure of the general solution of a homogeneous recurrence relation corresponds to the structure of the general solution of a system of homogeneous linear equations.

Proposition 10.4 *Consider the homogeneous linear recurrence relation of order r,*

$$c_0 x_n + c_1 x_{n-1} + \cdots + c_r x_{n-r} = 0, \quad n \geq r \quad (c_0 c_r \neq 0). \qquad (R_o)$$

1. *Any linear combination of solutions of (R_o) is again a solution of (R_o).*
2. *There exist r solutions of (R_o) such that any other solution of (R_o) can be expressed uniquely as their linear combination.*

Proof. 1. This follows immediately by the Superposition Principle 10.3.
2. For every $i \in \{0, \ldots, r-1\}$ let $(u_n^i)_n$ be the solution of (R_o) with r-sequence of initial data equal to 0 at places $j \neq i$, equal to 1 at places i, that is

$$u_j^i = 0 \text{ if } j \neq i, \quad u_i^i = 1 \quad j \in \{0, \ldots, r-1\}.$$

Consider now any solution $(a_n)_n$ of (R_o); the linear combination

$$a_0(u_n^0)_n + a_1(u_n^1)_n + \cdots + a_{r-1}(u_n^{r-1})_n$$

is a solution of (R_o) with sequence of initial data $(a_0, \ldots, a_{r-1})$. Since by Remark 9.4 the sequence of initial data determines the solution of a recurrence relation, one has

$$(a_n)_n = a_0(u_n^0)_n + a_1(u_n^1)_n + \cdots + a_{r-1}(u_n^{r-1})_n. \qquad \Box$$

☞   The reader who is familiar with the concept of a *vector space* can easily deduce from Point 2 of Proposition 10.4 that the general solution of a homogeneous linear recurrence relation of order r is a vector space of dimension r: the family of solutions considered in the proof is one of the *bases* of this vector space.

Definition 10.5 We define **characteristic polynomial** of the linear recurrence relation with constant coefficients of order r

$$c_0 x_n + c_1 x_{n-1} + \cdots + c_r x_{n-r} = h_n, \quad n \geq r \ (c_0 c_r \neq 0),$$

to be the polynomial of degree r

$$P_{\text{char}}(X) := c_0 X^r + c_1 X^{r-1} + \cdots + c_r. \qquad \Box$$

☞   *Remark 10.6* A way to recover easily the characteristic polynomial of the recurrence relation

$$c_0 x_n + c_1 x_{n-1} + \cdots + c_r x_{n-r} = h_n, \quad n \geq r,$$

is that to replace formally in the left-hand side of the above equality each x_i with X^i, and to divide the resulting polynomial by X^{n-r} (the minimum exponent of X).

Remark 10.7 Obviously, a given polynomial $c_0 X^r + c_1 X^{r-1} + \cdots + c_r$ with $c_0 c_r \neq 0$, is the characteristic polynomial of any linear recurrence relation of the form (R), and is the characteristic polynomial of just one linear homogeneous recurrence relation, namely (R_o).

Each polynomial of degree r has exactly r complex roots counted with their multiplicity. We see now that the sequence of the natural powers of a given root of the characteristic polynomial of a linear recurrence relation is a solution to the corresponding homogeneous relation.

Proposition 10.8 *Let $\lambda \in \mathbb{C}$. The sequence $(\lambda^n)_n$ of the powers of λ is a solution of the homogeneous linear recurrence relation*

$$c_0 x_n + c_1 x_{n-1} + \cdots + c_r x_{n-r} = 0, \quad n \geq r \ (c_0 c_r \neq 0), \qquad (R_o)$$

if and only if λ is a root of its characteristic polynomial.

Proof. Since $c_r \neq 0$, the roots of the characteristic polynomial must necessarily be non-zero. Substituting the values of the sequence $(\lambda^n)_n$ in the recurrence, one has

$$c_0 \lambda^n + c_1 \lambda^{n-1} + \cdots + c_r \lambda^{n-r} = 0$$

and, dividing by $\lambda^{n-r} \neq 0$,

$$c_0 \lambda^r + c_1 \lambda^{r-1} + \cdots + c_r = 0.$$

Therefore the sequence $(\lambda^n)_n$ is a solution of (R_o) if and only if λ is a root of the polynomial $c_0 X^r + c_1 X^{r-1} + \cdots + c_r$. $\qquad\square$

In general, it is not easy to find the roots of a polynomial of degree greater than two, though one can always use a suitable CAS for the purpose. The following simple criterion, however, demonstrates how to find the rational roots of a polynomial with integer coefficients.

Proposition 10.9 (The rational roots of a polynomial with integer coefficients) *Let $P(X) = c_0 X^r + c_1 X^{r-1} + \cdots + c_{r-1} X + c_r$ be a polynomial with integer coefficients $c_0, \ldots, c_r \in \mathbb{Z}$, with $c_0 \neq 0$. If the fraction $\dfrac{a}{b}$, with a, b integers with no common factors, is a root of $P(X)$, then a divides c_r and b divides c_0. In particular, if $c_0 = \pm 1$ the rational roots of the polynomial $P(X)$ are integers dividing c_r.*

Proof. Since $c_0 \left(\dfrac{a}{b}\right)^r + c_1 \left(\dfrac{a}{b}\right)^{r-1} + \cdots + c_{r-1} \left(\dfrac{a}{b}\right) + c_r = 0$, multiplying by b^r one obtains
$$c_0 a^r + c_1 a^{r-1} b + \cdots + c_{r-1} a b^{r-1} + c_r b^r = 0.$$

As a divides $c_0 a^r + c_1 a^{r-1} b + \cdots + c_{r-1} a b^{r-1}$, then it has to divide $c_r b^r$ too, and hence, not having a and b common factors, a divides c_r; analogously b divides $c_0 a^r$ and hence it divides c_0. $\qquad\square$

Remark 10.10 (*Roots of a polynomial with real coefficients*) It is well known that each polynomial $P(X) = c_0 X^r + c_1 X^{r-1} + \cdots + c_r$ factors in a product of linear polynomial with complex coefficients:

$$P(X) = c_0 X^r + c_1 X^{r-1} + \cdots + c_r = (X - z_1)^{\mu_1} \cdots (X - z_k)^{\mu_k}.$$

Since the conjugate of a product of two polynomials is equal to the product of the two conjugates, if $P(X)$ has real coefficients one has

$$P(X) = \overline{P(X)} = (X - \overline{z_1})^{\mu_1} \cdots (X - \overline{z_k})^{\mu_k}.$$

Thus, if $z_j = a_j + i b_j$ is a root of $P(X)$, also $\overline{z_j} = a_j - i b_j$ is a root of $P(X)$. Therefore the non real roots of a polynomial with real coefficients are pairs of complex conjugate roots with the same multiplicity.

☞ *Remark 10.11* (*Trigonometric form of a complex number*) We recall that the complex numbers of modulus 1 can be written as $\cos\alpha + i\sin\alpha =: e^{i\alpha}$ for an appropriate $0 \le \alpha < 2\pi$. Therefore any complex number can be written in its *trigonometric form* as $\rho(\cos\alpha + i\sin\alpha) = \rho e^{i\alpha}$ with $\rho \in \mathbb{R}_{\ge 0}$ and $0 \le \alpha < 2\pi$. For example, consider the complex number $2 + i2\sqrt{3}$; one has

$$2 + i2\sqrt{3} = |2 + i2\sqrt{3}|\frac{2 + i2\sqrt{3}}{|2 + i2\sqrt{3}|}$$

$$= 4\frac{2 + i2\sqrt{3}}{4} = 4\left(\frac{1}{2} + i\frac{\sqrt{3}}{2}\right) = 4\left(\cos\frac{\pi}{3} + i\sin\frac{\pi}{3}\right).$$

Proposition 10.8 gives a way to find *some* solutions to a homogeneous recurrence relation once one knows the roots of its characteristic polynomial. Actually it turns out that, if λ is a root of multiplicity $\mu > 1$ of the characteristic polynomial of (R_o) then

$$(\lambda^n)_n, (n\lambda^n)_n, \ldots, (n^{\mu-1}\lambda^n)_n$$

are solutions to (R_o), and actually, they generate the space of solutions to (R_o). The proof of the next Theorem 10.12, somewhat technical, is postponed in Sect. 10.5.

Theorem 10.12 (Basis-solutions) *Consider the homogeneous linear recurrence relation of order r*

$$c_0 x_n + c_1 x_{n-1} + \cdots + c_r x_{n-r} = 0, \quad n \ge r \quad (c_0 c_r \ne 0). \qquad (R_o)$$

1. *Let $\lambda_1,\ldots,\lambda_m$ be the distinct complex roots of the characteristic polynomial and $\mu_1, \ldots, \mu_m$ their multiplicity. The general complex solution of the recurrence (R_o) is given by the linear combinations with complex coefficients of the $r = \mu_1 + \cdots + \mu_m$ sequences*

$$(\lambda_j^n)_n, \ldots, (n^{\mu_j-1}\lambda_j^n)_n \quad j = 1, \ldots, m$$

called basis-solutions of the recurrence (R_o).

2. *Suppose that the coefficients $c_0,\ldots,c_r$ are real numbers. Let*

   • *$\rho_1(\cos\alpha_1 \pm i\sin\alpha_1), \ldots, \rho_h(\cos\alpha_h \pm i\sin\alpha_h)$ the pairs of the non real complex conjugate roots of the characteristic polynomial and $\mu_1, \ldots, \mu_h$ their multiplicity.*

   • *$\lambda_1, \ldots, \lambda_\ell$ be the real roots of the characteristic polynomial and $\mu'_1, \ldots, \mu'_\ell$ their multiplicity.*

Then, the general real solution of the recurrence (R_o) is the set of all linear combinations with real coefficients of the $r = 2\mu_1 + \cdots + 2\mu_h + \mu_1' + \cdots + \mu_\ell'$ sequences

$$(\rho_j^n \cos(n\alpha_j))_n, \ldots, (n^{\mu_j-1}\rho_j^n \cos(n\alpha_j))_n \quad j = 1, \ldots, h,$$

$$(\rho_j^n \sin(n\alpha_j))_n, \ldots, (n^{\mu_j-1}\rho_j^n \sin(n\alpha_j))_n \quad j = 1, \ldots, h,$$

$$(\lambda_j^n)_n, \ldots, (n^{\mu_j'-1}\lambda_j^n)_n \quad j = 1, \ldots, \ell,$$

*called **real basis-solutions** of the recurrence (R_o).*

Remark 10.13 (Structure of the solutions to (R_o)) By referring to Point 1 of Theorem 10.12, denoted by $\lambda_1, \ldots, \lambda_m$ the distinct complex roots of the characteristic polynomial and by $\mu_1, \ldots, \mu_m$ their multiplicity, $(a_n)_n$ is a solution of (R_o) if and only if there are polynomials $P_j(X)$ of degree strictly less than μ_j $(j = 1, \ldots, m)$ satisfying

$$a_n = P_1(n)\lambda_1^n + \cdots + P_m(n)\lambda_m^n \quad \forall n \in \mathbb{N}. \tag{10.13.a}$$

The general solution of a homogeneous linear recurrence of order r depends on r parameters $A_1, \ldots, A_r$. Such parameters are bijectively determined by the assignation of the sequence of initial data.

Example 10.14 The second-order recurrence relation

$$x_n = 5x_{n-1} - 6x_{n-2}, \quad n \geq 2,$$

has characteristic polynomial $X^2 - 5X + 6$ whose roots are $\lambda_1 = 2$ and $\lambda_2 = 3$. The sequences $(2^n)_n$ and $(3^n)_n$ are the basis-solutions of the recurrence. The general real (resp. complex) solution of the recurrence is then $x_n = A_1 2^n + A_2 3^n$, $n \geq 0$, with A_1 and A_2 varying among the real (resp. complex) numbers. □

Example 10.15 Determine the general solution of the recurrence relation

$$x_n = 3x_{n-1} - 4x_{n-3}, \quad n \geq 3,$$

and next its solution with initial data $x_0 = 4$, $x_1 = 1$ and $x_2 = 15$. The characteristic polynomial of the recurrence is

$$X^3 - 3X^2 + 4 = (X - 2)^2(X + 1).$$

The basis-solutions of the recurrence are $(2^n)_n$, $(n2^n)_n$ and $((-1)^n)_n$: thus, the general solution is

$$x_n = A_1 2^n + A_2 n 2^n + A_3 (-1)^n, \quad n \in \mathbb{N},$$

with the variation of A_1, A_2 and A_3 in $\mathbb{R}$. Imposing the initial data, we obtain the linear system

$$\begin{cases} A_1 + A_3 & = 4 \\ 2A_1 + 2A_2 - A_3 = 1 \\ 4A_1 + 8A_2 + A_3 = 15, \end{cases}$$

which has the unique solution $A_1 = 1$, $A_2 = 1$ and $A_3 = 3$. Thus, the required solution is $x_n = 2^n (1 + n) + 3(-1)^n$, $n \geq 0$.   □

Example 10.16 Let us compute the solution of the recurrence

$$x_n = -x_{n-2}, \quad n \geq 2,$$

with initial data $x_0 = 0$, $x_1 = 1$. Clearly, in this case, without using the above results, one can easily get

$$x_0 = 0, x_1 = 1, x_2 = -x_0 = 0, x_3 = -x_1 = -1, x_4 = -x_2 = 0$$

and hence deduce the solution

$$x_n = \begin{cases} 0 & \text{if } n \text{ is even,} \\ 1 & \text{if } n = 1 + 4k, \ k \in \mathbb{N}, \\ -1 & \text{if } n = 3 + 4k, \ k \in \mathbb{N}. \end{cases} \quad (10.16.a)$$

Let us solve now the same recurrence relation using Theorem 10.12. The characteristic polynomial $X^2 + 1$ has $\pm i$ as roots; since $\pm i = \cos \dfrac{\pi}{2} \pm i \sin \dfrac{\pi}{2}$, the sequences $\left(\cos \dfrac{n\pi}{2} \right)_n$ and $\left(\sin \dfrac{n\pi}{2} \right)_n$ are the real basis-solutions of the recurrence. Therefore the real general solution is

$$x_n = A_1 \cos \frac{n\pi}{2} + A_2 \sin \frac{n\pi}{2}, \quad n \geq 0,$$

with the variation of A_1, A_2 in $\mathbb{R}$. By imposing the initial data one has

$$0 = x_0 = A_1 \cos 0 + A_2 \sin 0 = A_1 \quad \text{and} \quad 1 = x_1 = A_1 \cos \frac{\pi}{2} + A_2 \sin \frac{\pi}{2} = A_2.$$

Therefore the solution we were looking for is $x_n = \sin \dfrac{n\pi}{2}$, $n \geq 0$, as we found in (10.16.a).   □

Example 10.17 The second-order homogeneous recurrence

$$x_n = 2x_{n-1} - 2x_{n-2}, \quad n \geq 2,$$

has characteristic polynomial $X^2 - 2X + 2$ whose roots are $\lambda_1 = 1 - i$ and $\lambda_2 = 1 + i$. The sequences $((1 - i)^n)_n$ and $((1 + i)^n)_n$ are the basis-solutions of the recurrence. The general complex solution of the recurrence is

$$x_n = A_1(1 - i)^n + A_2(1 + i)^n, \quad n \geq 0,$$

with the variation of A_1 and A_2 among the complex numbers. Let us look for the general real solution. One has

$$\lambda_1 = 1 - i = \sqrt{2}\left(\frac{\sqrt{2}}{2} - \frac{\sqrt{2}}{2}i\right) = \sqrt{2}\left(\cos\left(\frac{\pi}{4}\right) - i\sin\left(\frac{\pi}{4}\right)\right) \text{ and}$$

$$\lambda_2 = 1 + i = \overline{\lambda_1} = \sqrt{2}\left(\cos\left(\frac{\pi}{4}\right) - i\sin\left(\frac{\pi}{4}\right)\right).$$

Then the sequences $\left(2^{n/2}\cos\left(\frac{n\pi}{4}\right)\right)_n$ and $\left(2^{n/2}\sin\left(\frac{n\pi}{4}\right)\right)_n$ are the real basis-solutions of the recurrence. Therefore, the general real solution of the recurrence is

$$x_n = A_1 2^{n/2}\cos\left(\frac{n\pi}{4}\right) + A_2 2^{n/2}\sin\left(\frac{n\pi}{4}\right), \quad n \geq 0,$$

with the variation of A_1 and A_2 among the real numbers. □

Example 10.18 Let us compute the general real solution of the recurrence relation

$$x_{n+4} + 8x_{n+3} + 48x_{n+2} + 128x_{n+1} + 256x_n = 0, \quad n \geq 0.$$

Its characteristic polynomial is

$$(X^2 + 4X + 16)^2 = (X + 2 + 2i\sqrt{3})^2(X + 2 - 2i\sqrt{3})^2.$$

Since $-2 \mp 2i\sqrt{3} = 4\left(\cos\frac{2\pi}{3} \mp i\sin\frac{2\pi}{3}\right)$, the sequences

$$\left(4^n\cos\left(\frac{2\pi n}{3}\right)\right)_n, \left(4^n\sin\left(\frac{2\pi n}{3}\right)\right)_n, \left(n4^n\cos\left(\frac{2\pi n}{3}\right)\right)_n, \left(n4^n\sin\left(\frac{2\pi n}{3}\right)\right)_n$$

are its real basis-solutions and thus the general real solution is

$$x_n = A_1 4^n\cos\left(\frac{2\pi n}{3}\right) + A_2 4^n\sin\left(\frac{2\pi n}{3}\right) + A_3 n4^n\cos\left(\frac{2\pi n}{3}\right) + A_4 n4^n\sin\left(\frac{2\pi n}{3}\right),$$

$n \in \mathbb{N}$, with the variation of A_1, A_2, A_3, A_4 in $\mathbb{R}$. □

Example 10.19 Determine the general real solution of the Fibonacci recurrence relation

$$x_n = x_{n-1} + x_{n-2}, \quad n \geq 2.$$

The characteristic polynomial $X^2 - X - 1$ of the recurrence has roots

$$\lambda_1 = \frac{1 + \sqrt{5}}{2}, \qquad \lambda_2 = \frac{1 - \sqrt{5}}{2};$$

the general real solution is therefore

$$x_n = A_1 \lambda_1^n + A_2 \lambda_2^n, \quad n \in \mathbb{N},$$

with the variation of A_1, A_2 in $\mathbb{R}$. Imposing the initial data ($x_0 = 0, x_1 = 1$), one obtains the famous Fibonacci numbers $(F_n)_n$. One has

$$A_1 = \frac{1}{\sqrt{5}}, \quad A_2 = -\frac{1}{\sqrt{5}};$$

substituting these values in the general solution we obtain

$$F_n = \frac{1}{\sqrt{5}} \left[\left(\frac{1 + \sqrt{5}}{2} \right)^n - \left(\frac{1 - \sqrt{5}}{2} \right)^n \right], \quad n \in \mathbb{N}.$$

Notice that, since $\left| \frac{1}{\sqrt{5}} \left(\frac{1 - \sqrt{5}}{2} \right)^n \right| < \left| \frac{1}{\sqrt{5}} (1)^n \right| < 0.5$ and the required solution assumes values in $\mathbb{N}$, one has that each F_n is the closest integer number to $\frac{1}{\sqrt{5}} \left(\frac{1 + \sqrt{5}}{2} \right)^n$.

Imposing the initial data ($x_0 = 2, x_1 = 1$), one has $A_1 = A_2 = 1$. Substituting these values in the general solution one obtains the famous Lucas numbers $(L_n)_n$:

$$L_n = \left(\frac{1 + \sqrt{5}}{2} \right)^n + \left(\frac{1 - \sqrt{5}}{2} \right)^n, \quad n \in \mathbb{N}.$$

It turns easily out that L_n is the closest integer to $\left(\frac{1 + \sqrt{5}}{2} \right)^n$.   □

Example 10.20 In the Example 7.52 we have seen that the inverse of the formal series $A(X) = 3 - X - X^2$ is $B(X) = \sum_{n=0}^{\infty} b_n X^n$, where the coefficients $(b_n)_n$ satisfy the recurrence relation:

$$3b_n = b_{n-1} + b_{n-2} \quad n \geq 2,$$

with initial data $b_0 = 1/3$, $b_1 = 1/9$. The characteristic polynomial $3X^2 - X - 1$ of the recurrence has roots $\lambda_1 = \dfrac{1 + \sqrt{13}}{6}$ and $\lambda_2 = \dfrac{1 - \sqrt{13}}{6}$: thus, the general real solution is

$$b_n = A_1 \frac{\left(1 + \sqrt{13}\right)^n}{6^n} + A_2 \frac{\left(1 - \sqrt{13}\right)^n}{6^n},$$

with the variation of $A, B \in \mathbb{R}$. The initial data impose

$$A_1 + A_2 = 1/3, \quad A_1 - A_2 = \frac{1}{3\sqrt{13}},$$

from which one obtains

$$A_1 = \frac{1 + \sqrt{13}}{6\sqrt{13}}, \quad A_2 = \frac{\sqrt{13} - 1}{6\sqrt{13}}.$$

Substituting these values in the general solution we obtain the required solution

$$b_n = \frac{1}{\sqrt{13}} \left[\left(\frac{1 + \sqrt{13}}{6}\right)^{n+1} - \left(\frac{1 - \sqrt{13}}{6}\right)^{n+1} \right], \quad n \in \mathbb{N}. \qquad \square$$

It is not always possible to determine in an explicit way the roots of the characteristic polynomial of a recurrence relation. Nevertheless, one can compute approximated roots, which can give an idea of how the solution evolves.

Example 10.21 In this example we deal with approximate solutions to a recurrence relation. It follows from Theorem 10.12 that each term of a solution to a recurrence relation is a continuous function of the roots of its characteristic polynomial: thus approximate roots of the characteristic polynomial provide approximate solutions of the recurrence. We wish here to determine the general real solution of the recurrence relation

$$x_n = 3x_{n-1} - x_{n-3}, \quad n \geq 3,$$

and next the solution with initial data $x_0 = 1$, $x_1 = 3$ and $x_2 = 9$. The characteristic polynomial of the recurrence is $X^3 - 3X^2 + 1$; by Proposition 10.9, the possible rational roots would be integers dividing 1. Since neither 1 nor -1 are roots, the polynomial has no rational roots. Using an adequate CAS we have computed the approximate values of the roots:

$$\lambda_1 \approx -0.532089, \quad \lambda_2 \approx 0.652704, \quad \lambda_3 \approx 2.879385.$$

The approximate basis-solution of the recurrence are

$$((-0.532089)^n)_n, \quad ((0.652704)^n)_n, \quad ((2.879385)^n)_n.$$

By Theorem 10.12, the approximate general solution is

$$x_n = A_1(-0.532089)^n + A_2(0.652704)^n + A_3(2.879385)^n, \quad n \in \mathbb{N},$$

with the variation of A_1, A_2 and A_3 among the real numbers. Imposing the initial data, one has $A_1 \approx 0.070046$, $A_2 \approx -0.161485$ and $A_3 \approx 1.091439$. We are now able to calculate the approximate values of the sequence that solves the recurrence relation: for each $n \geq 0$ one has

$$x_n \approx 0.070046 \times (-0.532089)^n - 0.161485 \times (0.652704)^n + 1.091439 \times (2.879385)^n.$$

Taking into account that the solution is a sequence of integer numbers, one can guess the exact values. For example

$$x_3 \approx 0.070046 \times (-0.532089)^3 - 0.161485 \times (0.652704)^3 +$$

$$+1.091439 \times (2.879385)^3 \approx 26.0000006,$$

from which we deduce $x_3 = 26$ (as an immediate analysis of the recurrence confirms). Analogously one obtains $x_7 = 1\,791$.   □

We end this section with an obvious, though useful, remark: an immediate consequence of Theorem 10.12 is that a sort of converse is also true: any sequence $(a_n)_n$ of the form (10.13.a) is a solution to a suitable linear recurrence relation.

Corollary 10.22 *Let $P_1(X), \ldots, P_m(X)$ be m polynomials in $\mathbb{C}[X]$ and $\lambda_1, \ldots, \lambda_m$ be non zero complex numbers. The sequence*

$$(P_1(n)\lambda_1^n + \cdots + P_m(n)\lambda_m^n)_n$$

is a solution to a suitable linear homogeneous recurrence.

Proof. Fix natural numbers $n_1 > \deg P_1(X), \ldots, n_m > \deg P_m(X)$. By Theorem 10.12 and Remark 10.13, the solutions to the homogeneous recurrence relation whose characteristic polynomial is

$$(X - \lambda_1)^{n_1} \cdots (X - \lambda_m)^{n_m}$$

are precisely all the sequences of the form

$$(Q_1(n)\lambda_1^n + \cdots + Q_m(n)\lambda_m^n)_n,$$

for any choice of polynomials $Q_1(X), \ldots, Q_m(X)$ with $\deg Q_i(X) < n_i$, $i = 1, \ldots, m$. □

Remark 10.23 In the particular case of a polynomial $Q(X)$, Corollary 10.22 shows that for each $m > \deg Q(X)$, $(Q(n))_n$ solves the homogeneous recurrence whose characteristic polynomial is

$$(X - 1)^m = \sum_{i=0}^{m} (-1)^i \binom{m}{i} X^{m-i}.$$

Thus $\sum_{i=0}^{m} (-1)^i \binom{m}{i} Q(m - i) = 0$, or equivalently $\sum_{i=0}^{m} (-1)^i \binom{m}{i} Q(i) = 0$ for all polynomial $Q(X)$ with $\deg Q(X) < m$, which is what we obtained with different methods in Corollary 4.11.

Example 10.24 The sequence $((n^2 - 7n + 1)3^n - 5n)_n$ is, for instance, a solution to the linear homogeneous recurrence whose characteristic polynomial is

$$(X - 3)^3 (X - 1)^2 = X^5 - 11X^4 + 46X^3 - 90X^2 + 81X - 27,$$

i.e., of
$$x_n - 11x_{n-1} + 46x_{n-2} - 90x_{n-3} + 81x_{n-4} - 27x_{n-5} = 0.$$ □

10.1.2 Particular Solutions to a Linear Recurrence Relation

By Proposition 10.3, the general solution of the order r linear recurrence relation

$$c_0 x_n + c_1 x_{n-1} + \cdots + c_r x_{n-r} = h_n, \quad n \geq r,$$

is obtained by adding a particular solution to the general solution of the associated homogeneous recurrence. Imposing the initial data, one obtains the values of the r parameters on which the general solution depends and thus determines univocally the solution.

Example 10.25 As we have seen in Example 9.11, the minimum number x_n of steps to end the Lucas Tower game with n rings satisfies the recurrence

$$x_n = 2x_{n-1} + 1, \quad n \geq 2,$$

with initial data $x_1 = 1$. The characteristic polynomial is $X - 2$ and therefore the basis-solution of the associated homogeneous recurrence is the sequence $(2^n)_n$. Since the constant sequence $(-1)_n$ is a particular solution of $x_n = 2x_{n-1} + 1$, the general real solution of the recurrence is the family of sequences $(A2^n - 1)_n$ with the variation of $A \in \mathbb{R}$. Next, imposing the initial data $x_1 = 1$ we get $A = 1$, and thus the desired solution is

$$x_n = 2^n - 1, \quad n \geq 1.$$

$\square$

It is not always as straightforward as in the preceding example to find a particular solution of a linear recurrence relation just by following one's nose. At the end of this section (see Proposition 10.41) we will give a general formula to find a particular solution of a recurrence relation; nevertheless it has the caveat of being difficult to apply.

Let us see now a simple recipe to find a particular solution in case the non homogeneous term is the product of a polynomial $Q(n)$ in n by an exponential q^n, for some constant q. We return to the proof of the following proposition in Sect. 10.5.

Proposition 10.26 (Particular solution with non homogeneous term $Q(n)q^n$) *Given a constant q and a polynomial $Q(X)$, let us consider the recurrence relation of order r with constant coefficients*

$$c_0 x_n + c_1 x_{n-1} + \cdots + c_r x_{n-r} = Q(n)q^n, \quad n \geq r \quad (c_0 c_r \neq 0). \qquad (R)$$

The above recurrence relation have a particular solution of the type

$$\left(n^\mu \widetilde{Q}(n)q^n\right)_n,$$

where:

- $\widetilde{Q}(X)$ *is a polynomial of degree less or equal than the degree of $Q(X)$;*
- μ *is the multiplicity of q as root of the characteristic polynomial ($\mu = 0$ if q is not a root).*

☞ *Remark 10.27* 1. To find $\widetilde{Q}(X)$ one sets $\widetilde{Q}(X) = \alpha_k X^k + \cdots + \alpha_1 X + \alpha_0$; then one looks for coefficients $\alpha_0, \ldots, \alpha_k$ such that the sequence

$$\left((\alpha_k X^k + \cdots + \alpha_1 X + \alpha_0)\, n^\mu q^n\right)_n$$

is a solution of the recurrence.

2. By the Superposition Principle 10.3, Proposition 10.26 guarantees to find particular solutions also when the non homogeneous term is a sum of terms of the type $Q(n)q^n$: indeed, one can simply add the solutions which correspond to each of the terms of the sum.

Remark 10.28 The conclusion of Proposition 10.26 holds true also if one considers a recurrence of the form

$$c_0 x_{n+r} + \cdots + c_r x_n = Q(n)q^n, \quad n \geq 0.$$

Indeed the latter can be rewritten in the equivalent form

$$c_0 x_n + \cdots + c_r x_{n-r} = Q(n-r)q^{n-r} = \hat{Q}(n)q^n, \quad n \geq r,$$

where $\hat{Q}(X) := q^{-r}Q(X - r)$ is a polynomial of the same degree of $Q(X)$.

Example 10.29 Compute a particular solution of each of these linear recurrences:

1. $x_n = 4x_{n-1} - 4x_{n-2} + 5(3^n), \quad n \geq 2$;
2. $x_n = 4x_{n-1} - 4x_{n-2} + 7(2^n), \quad n \geq 2$;
3. $x_{n+2} = 4x_{n+1} - 4x_n + 7 + 3n, \quad n \geq 0$.

Solution. The characteristic polynomial of all the three recurrences is

$$X^2 - 4X + 4 = (X - 2)^2.$$

1. Since 3 is not a root of the characteristic polynomial and 5 is a polynomial of degree 0, one looks for a particular solution of the form $(\alpha 3^n)_n$, where α is a polynomial of degree 0, i.e., a constant. Substituting this in the recurrence we obtain

$$\alpha 3^n = 4\alpha 3^{n-1} - 4\alpha 3^{n-2} + 5(3^n).$$

Dividing by 3^n we have

$$\alpha = 4\frac{\alpha}{3} - 4\frac{\alpha}{9} + 5,$$

and hence $\alpha = 45$. Thus, a particular solution is the sequence $(45 \times 3^n)_n$.

2. Since 2 is a root of multiplicity 2 of the characteristic polynomial and 7 is a polynomial of degree 0, we look for a particular solution of the form $(\alpha n^2 2^n)_n$. Substituting this in the recurrence we obtain

$$\alpha n^2 2^n = 4\alpha(n-1)^2 2^{n-1} - 4\alpha(n-2)^2 2^{n-2} + 7(2^n).$$

Dividing by 2^n we have

$$\alpha n^2 = 2\alpha(n-1)^2 - \alpha(n-2)^2 + 7,$$

and hence $\alpha = 7/2$. Thus, a particular solution is the sequence $(7n^2 2^{n-1})_n$.

3. Since $7 + 3n = (7 + 3n)1^n$ and 1 is not a root of the characteristic polynomial, we look for a particular solution of the form $(\alpha_0 + \alpha_1 n)_n$. Substituting this in the recurrence we get

$$\alpha_0 + \alpha_1(n+2) = 4(\alpha_0 + \alpha_1(n+1)) - 4(\alpha_0 + \alpha_1 n) + 7 + 3n,$$

or equivalently

$$\alpha_0 - 2\alpha_1 + \alpha_1 n = 7 + 3n,$$

and hence $\alpha_1 = 3$ and $\alpha_0 = 7 + 2\alpha_1 = 13$. Thus, a particular solution is the sequence $(13 + 3n)_n$.   □

Proposition 10.26 also allows to determine particular solutions of recurrences with real coefficients whose non-homogeneous term is of the form $Q(n)\rho^n \cos(n\gamma)$ or $Q(n)\rho^n \sin(n\gamma)$.

Remark 10.30 Suppose we have a linear recurrence relation with *real coefficients* of one of the following types

$$c_0 x_n + c_1 x_{n-1} + \cdots + c_r x_{n-r} = Q(n)\rho^n \cos(n\gamma), \quad n \ge r, \qquad (10.30.\mathrm{a})$$

$$c_0 y_n + c_1 y_{n-1} + \cdots + c_r y_{n-r} = Q(n)\rho^n \sin(n\gamma), \quad n \ge r, \qquad (10.30.\mathrm{b})$$

with $c_0, \ldots, c_r \in \mathbb{R}$ ($c_0 c_r \ne 0$), $\gamma, \rho \in \mathbb{R}$ and $Q(X)$ is a polynomial with real coefficients. Then one obtains a particular real solution of (10.30.a) (resp. (10.30.b)) considering the real (resp. imaginary) part of a particular solution of the recurrence

$$c_0 z_n + c_1 z_{n-1} + \cdots + c_r z_{n-r} = Q(n)q^n, \; n \ge r, \; q := \rho(\cos\gamma + i\sin\gamma).$$
$$(10.30.\mathrm{c})$$

By Proposition 10.26, the recurrence (10.30.c) has a particular solution of the form $z_n = n^\mu \tilde{Q}(n)q^n$ where μ is the multiplicity of q as root of the characteristic polynomial and $\tilde{Q}(X)$ has the same degree of $Q(X)$.

Example 10.31 Let us find a particular solution of the recurrence

$$x_{n+1} = 2x_n + 3^n \cos(5n). \qquad (10.31.\mathrm{a})$$

Remark 10.30 suggests to take the real part of the solution $(z_n)_n$ to

$$z_{n+1} = 2z_n + 3^n(\cos(5n) + i\sin(5n)) = 2z_n + \left(3e^{5i}\right)^n. \qquad (10.31.\mathrm{b})$$

The basis-solution of the homogeneous recurrence associated with (10.31.b) is $(2^n)_n$. By Proposition 10.26, the recurrence (10.31.b) has a particular solution of the type

$$\left(\alpha \left(3e^{5i}\right)^n\right)_n = \left(\alpha 3^n \left(\cos(5n) + i\sin(5n)\right)\right)_n.$$

To compute the parameter α, we substitute in the recurrence (10.31.b):

$$\alpha \left(3e^{5i}\right)^{n+1} = 2\alpha \left(3e^{5i}\right)^n + \left(3e^{5i}\right)^n.$$

Dividing by $\left(3e^{5i}\right)^n$, one obtains

$$3\alpha e^{5i} = 2\alpha + 1$$

and hence

$$\alpha = \frac{1}{3e^{5i} - 2} = \frac{3\cos 5 - 2 - 3i\sin 5}{(3\cos 5 - 2)^2 + 9\sin^2 5} = \frac{3\cos 5 - 2 - 3i\sin 5}{13 - 12\cos 5}.$$

Thus, a particular solution of the recurrence (10.31.b) is

$$\frac{1}{3e^{5i} - 2}\left(3e^{5i}\right)^n = \frac{3\cos 5 - 2 - 3i\sin 5}{13 - 12\cos 5}3^n(\cos 5 + i\sin 5)^n, \quad n \in \mathbb{N}.$$

Its real part,

$$\frac{3\cos 5 - 2}{13 - 12\cos 5}3^n\cos(5n) + \frac{3\sin 5}{13 - 12\cos 5}3^n\sin(5n), \quad n \in \mathbb{N},$$

is therefore a particular solution to (10.31.a). □

Example 10.32 We want to determine a particular solution of the recurrence

$$x_{n+2} = 2x_{n+1} - 2x_n + (\sqrt{2})^n \sin\left(n\frac{\pi}{4}\right), \quad n \geq 0.$$

Observe that the non homogeneous term is the imaginary part of q^n, where

$$q = \sqrt{2}\left(\cos\left(\frac{\pi}{4}\right) + i\sin\left(\frac{\pi}{4}\right)\right) = 1 + i.$$

The characteristic polynomial $X^2 - 2X + 2 = (X - 1)^2 + 1$ has the two complex conjugate roots $1 \pm i$, so that q is a root of multiplicity 1 of the characteristic polynomial. By Proposition 10.26 the complex recurrence

$$z_{n+2} = 2z_{n+1} - 2z_n + q^n, \quad n \geq 0,$$

has a particular solution of the form $(\alpha n q^n)_n$ where α is a suitable constant. Substituting in the recurrence, one has

$$\alpha(n + 2)q^{n+2} = 2\alpha(n + 1)q^{n+1} - 2\alpha n q^n + q^n.$$

Dividing by q^n, one obtains

$$\alpha(n+2)q^2 = 2\alpha(n+1)q - 2\alpha n + 1,$$

or equivalently

$$\alpha n(q^2 - 2q + 2) + 2\alpha q^2 - 2\alpha q - 1 = 0.$$

Since q is a root of the characteristic polynomial $X^2 - 2X + 2$, one has

$$\alpha = \frac{1}{2(q^2 - q)} = \frac{1}{4i - 2(1+i)} = \frac{1}{-2+2i} = -\frac{1+i}{4} :$$

therefore a particular solution of the recurrence

$$z_{n+2} = 2z_{n+1} - 2z_n + q^n \quad n \geq 0$$

is given by

$$-\frac{1+i}{4}nq^n = -\frac{1+i}{4}n(\sqrt{2})^n \left(\cos\left(\frac{n\pi}{4}\right) + i \sin\left(\frac{n\pi}{4}\right) \right)$$

$$= \frac{n(\sqrt{2})^n}{4} \left(-\cos\left(\frac{n\pi}{4}\right) + \sin\left(\frac{n\pi}{4}\right) - i\left(\cos\left(\frac{n\pi}{4}\right) + \sin\left(\frac{n\pi}{4}\right) \right) \right),$$

with $n \in \mathbb{N}$. Now, a particular solution of the initial recurrence is the imaginary part:

$$-\frac{n(\sqrt{2})^n}{4} \left(\cos\left(\frac{n\pi}{4}\right) + \sin\left(\frac{n\pi}{4}\right) \right), \quad n \in \mathbb{N}.$$

Observe that

$$\cos\left(\frac{n\pi}{4}\right) + \sin\left(\frac{n\pi}{4}\right) = \sqrt{2}\left(\cos\left(\frac{n\pi}{4}\right)\frac{1}{\sqrt{2}} + \sin\left(\frac{n\pi}{4}\right)\frac{1}{\sqrt{2}} \right)$$

$$= \sqrt{2}\left(\cos\left(\frac{n\pi}{4}\right)\sin\frac{\pi}{4} + \sin\left(\frac{n\pi}{4}\right)\cos\frac{\pi}{4} \right)$$

$$= \sqrt{2}\sin\left(\frac{(n+1)\pi}{4}\right).$$

Thus, we can then write the particular solution of the initial recurrence in the following more compact form

$$-\frac{n}{4}(\sqrt{2})^{n+1}\sin\left(\frac{(n+1)\pi}{4}\right), \quad n \in \mathbb{N}. \qquad \square$$

10.1.3 General Solution to a Linear Recurrence

We know now how to find a particular solution to a linear recurrence. Unless we are very lucky, its initial terms do not match with the given initial conditions: we need to find the general solution of the recurrence, which will depend on as many parameters as is the order of the recurrence, and finally choose them in the convenient way. Let us recap, for the convenience of the reader, the strategy we have developed for determining the solutions of linear recurrence relation.

Remark 10.33 (*Method for resolving a linear recurrence relation*) Given a linear recurrence of order r with constant coefficients, one proceeds with the following plan:

1. Determine the general solution of the associated homogeneous recurrence: this is a family of sequences, depending on r parameters, obtained by considering the linear combinations of the r basis-solutions of the associated homogeneous recurrence;
2. Determine a particular solution of the recurrence;
3. The general solution of the recurrence is the sum of a particular solution and the general solution of the associated homogeneous recurrence; one obtains a family of sequences which depend on the r parameters of Point 1.

   Next, if initial data are given:

4. Obtain the values of the r parameters, forcing the general solution to satisfy the r initial data.

Example 10.34 In Example 10.31 it turned out that a particular solution to the recurrence

$$x_{n+1} = 2x_n + 3^n \cos(5n), \quad n \in \mathbb{N},$$

is

$$\frac{3\cos 5 - 2}{13 - 12\cos 5} 3^n \cos(5n) + \frac{3\sin 5}{13 - 12\cos 5} 3^n \sin(5n), \quad n \in \mathbb{N}.$$

Let us find the general solution. The basis-solution of the homogeneous recurrence relation $x_{n+1} = 2x_n$, $n \in \mathbb{N}$, is $(2^n)_n$. Thus, the solution of the given recurrence is

$$x_n = A_1 2^n + \frac{3\cos 5 - 2}{13 - 12\cos 5} 3^n \cos(5n) + \frac{3\sin 5}{13 - 12\cos 5} 3^n \sin(5n), \quad n \in \mathbb{N},$$

with A_1 varying in $\mathbb{R}$. ☐

Example 10.35 In Example 10.32 we found that

$$-\frac{n}{4}(\sqrt{2})^{n+1} \sin\left(\frac{(n+1)\pi}{4}\right), \quad n \in \mathbb{N}, \tag{10.35.a}$$

is a particular solution of the recurrence

$$x_{n+2} = 2x_{n+1} - 2x_n + (\sqrt{2})^n \sin\left(n\frac{\pi}{4}\right).$$

Let us find the solution of the above recurrence, with initial data

$$x_0 = 0, \ x_1 = 1.$$

First, we settle the set of solutions of the homogeneous recurrence

$$x_{n+2} = 2x_{n+1} - 2x_n, \quad n \geq 0.$$

The characteristic polynomial $X^2 - 2X + 2 = (X - 1)^2 + 1$ has the two complex conjugate roots $1 \pm i$. We write the roots in their trigonometric form:

$$1 \pm i = \sqrt{2}\left(\frac{1}{\sqrt{2}} \pm i\frac{1}{\sqrt{2}}\right) = \sqrt{2}\left(\cos\left(\frac{\pi}{4}\right) \pm i \sin\left(\frac{\pi}{4}\right)\right).$$

By Theorem 10.12, the space of real solutions of the associated homogeneous recurrence is generated by the basis-solutions

$$\left((\sqrt{2})^n \cos\left(\frac{n\pi}{4}\right)\right)_n \quad \text{and} \quad \left((\sqrt{2})^n \sin\left(\frac{n\pi}{4}\right)\right)_n.$$

Therefore, taking (10.35.a) into account, the general real solution of the recurrence is

$$x_n = A(\sqrt{2})^n \cos\left(\frac{n\pi}{4}\right) + B(\sqrt{2})^n \sin\left(\frac{n\pi}{4}\right) - \frac{n}{4}(\sqrt{2})^{n+1} \sin\left(\frac{(n+1)\pi}{4}\right),$$

with the variation of A, B in $\mathbb{R}$. In order that $x_0 = 0$, it must be $A = 0$; moreover $x_1 = 1$ only if

$$B\sqrt{2} \sin\left(\frac{\pi}{4}\right) - \frac{1}{4}(\sqrt{2})^2 \sin\left(\frac{2\pi}{4}\right) = 1,$$

or equivalently $B - \frac{1}{2} = 1$, i.e., $B = 3/2$. Thus

$$x_n = \frac{3}{2}(\sqrt{2})^n \sin\left(\frac{n\pi}{4}\right) - \frac{n}{4}(\sqrt{2})^{n+1} \sin\left(\frac{(n+1)\pi}{4}\right)$$

is the desired solution.   □

Example 10.36 In the Example 9.10 we found that the sequence whose n-th term is
the number of regions of the plane created by n lines in general position, is a solution
of the recurrence relation

$$x_n = x_{n-1} + n, \quad n \geq 1,$$

with initial datum $x_0 = 1$. To find a solution we follow the pattern described in
Remark 10.33.

1. The characteristic polynomial is $X - 1$; thus, the sequence $(1^n)_n = (1)_n$ is the
   basis-solution of the associated homogeneous recurrence. Its general solution is
   $(A)_n$, with variation of the constant A.
2. Let us now compute a particular solution. Observe that the non homogeneous term
   of the recurrence is $n = n \cdot 1^n$ and 1 is a root of the characteristic polynomial.
   Therefore there exists a particular solution of the form

$$n^1 (\alpha_0 + \alpha_1 n) 1^n = n(\alpha_0 + \alpha_1 n) = \alpha_0 n + \alpha_1 n^2.$$

   Let us compute the parameters α_0 and α_1:

$$\alpha_0 n + \alpha_1 n^2 = \alpha_0 (n - 1) + \alpha_1 (n - 1)^2 + n,$$

$$n^2 (\alpha_1 - \alpha_1) + n(\alpha_0 - \alpha_0 + 2\alpha_1 - 1) + \alpha_0 - \alpha_1 = 0,$$

   and hence $\alpha_1 = 1/2 = \alpha_0$.
3. Thus, the general solution of the recurrence relation is the sequence

$$x_n = A + \frac{1}{2}(n + n^2), n \geq 0,$$

   with variation of the constant A.
4. At this stage, imposing the initial datum $x_0 = 1$ yields

$$A + \frac{1}{2}(0 + 0) = A = 1;$$

   therefore the desired solution is $x_n = 1 + \frac{1}{2}(n + n^2)$.   □

Example 10.37 The recurrence relations

$$x_n = 1.08x_{n-1} \quad \text{and} \quad y_n = 1.08y_{n-1} + 100 \quad n \geq 1,$$

respectively with initial datum $x_0 = 1\,000$ and $y_0 = 100$ describe the total sum of
money one will succeed in accumulating after n years if one follows the two different

investment strategies considered in Example 9.8. The characteristic polynomial is in both the cases $X - 1.08$; thus the general solution of the homogeneous part is the family of sequences $(A(1.08)^n)_n$, with the variation of the constant A. In the first recurrence, imposing the initial datum one obtains

$$x_0 = A(1.08)^0 = 1\,000$$

and hence the solution is the sequence $(1\,000 \cdot (1.08)^n)_n$. To solve the second recurrence, we have to find a particular solution. The non homogeneous part is $100 = 100 \cdot 1^n$; since 1 is not a root of the characteristic polynomial, the recurrence has a particular solution of the form $(\alpha)_n$ where α is a real constant. To compute α, we substitute in the recurrence, obtaining

$$\alpha = 1.08\alpha + 100$$

and hence $\alpha = -100/0.08$. Thus, the general solution of the second recurrence is the family of sequences

$$\left(A(1.08)^n - \frac{100}{0.08} \right)_n ,$$

with the variation of A in $\mathbb{R}$. Imposing the initial datum

$$y_0 = A - \frac{100}{0.08} = 100$$

we obtain $A = \dfrac{108}{0.08}$ and hence the solution

$$x_n = \frac{1}{0.08}(108 \cdot (1.08)^n - 100), \quad n \in \mathbb{N}.$$

We have obtained explicitly the money we would have accumulated in n years following the two investment strategies. $\square$

Example 10.38 (*Approximate solutions*) Let us solve the recurrence relation

$$x_n = 4x_{n-1} - x_{n-4} + 4^{n-4}, \quad n \geq 4,$$

obtained in the Example 9.13 to compute the number of n-sequences of I_4 containing the subsequence $(1, 2, 3, 4)$. The characteristic polynomial $X^4 - 4X^3 + 1$ has no rational roots; using an appropriate CAS we have computed the approximate values of the roots:

$$\lambda_1 \approx 0.669632, \quad \lambda_2 \approx 3.984188$$

and the conjugate complex roots

$$\lambda_3 \approx -0.32691 - 0.51764i, \quad \lambda_4 \approx -0.32691 + 0.51764i.$$

Since $|-0.32691 - 0.51764i| \approx 0.612226$, one has

$$-0.32691 - 0.51764i = 0.612226 \left(\frac{-0.32691 - 0.51764i}{0.612226} \right)$$

$$\approx 0.612226(\cos 2 + i \sin 2).$$

Thus, by Theorem 10.12, the family of sequences

$$(A_1(0.669632)^n + A_2(3.984188)^n +$$
$$+ A_3(0.612226)^n \cos(2n) + A_4(0.612226)^n \sin(2n))_n$$
$$(10.38.\text{a})$$

with the variation of A_1, A_2, A_3 and A_4 among the real numbers, is an approximation of the general solution of the homogeneous recurrence. Let us now compute a particular solution. Since the non homogeneous term is $\dfrac{4^n}{4^4}$, we look for a particular solution of the form $\alpha 4^n$. Substituting in the recurrence we get

$$\alpha 4^n = 4\alpha 4^{n-1} - \alpha 4^{n-4} + 4^{n-4};$$

hence $\alpha = 1$. Therefore the family of sequences

$$(A_1(0.669632)^n + A_2(3.984188)^n +$$

$$+ A_3(0.612226)^n \cos(2n) + A_4(0.612226)^n \sin(2n) + 4^n)_n,$$

with the variation of the constants A_1, A_2, A_3 and A_4 in $\mathbb{R}$, is the (estimated) general solution of the recurrence. Imposing the initial data we obtain a linear system; using a CAS we get

$$A_1 = 0.0718, \quad A_2 = -1.012, \quad A_3 = -0.0598, \quad A_4 = -0.0685.$$

Thus, the desired solution is

$$x_n = 0.0718(0.669632)^n - 1.012(3.984188)^n +$$

$$- 0.0598(0.612226)^n \cos(2n) - 0.0685(0.612226)^n \sin(2n) + 4^n, \quad n \in \mathbb{N}.$$

Taking into account that the solution is a sequence of integer values, one can guess the exact values of the solution. For example

$$x_4 = 1, x_5 = 8, x_6 = 48, x_7 = 257, \dots .$$

Observe that x_4, x_5 and x_6 are the correct values, while the exact value for x_7 is 256.   □

Example 10.39 We solve the recurrence relation $z_n = \dfrac{z_{n-1} + i}{2}$ with initial datum $z_0 = 1$. The characteristic polynomial is $X - 1/2$; then, the general solution of the homogeneous part of the recurrence is $A(1/2^n)$ with the variation of $A \in \mathbb{C}$. Since the non homogeneous term is a constant and 1 is not a root of the characteristic polynomial, the recurrence admits a particular solution of the form $(\alpha)_n$ for a suitable constant α. To compute the value of α, let us force the sequence $(\alpha)_n$ to be a solution: $\alpha = \dfrac{\alpha + i}{2}$, hence $\alpha = i$. Therefore, the general solution of the recurrence is $z_n = A(1/2^n) + i$. Imposing the initial datum $z_0 = 1$, one obtains $A = 1 - i$; thus the wished solution is $z_n = \dfrac{1 - i}{2^n} + i$. The geometric representation of the sequence in the complex plane is depicted in Fig. 10.1. Geometrically, z_n is the midpoint between i and z_{n-1}, starting from $z_0 = 1$.   □

Example 10.40 We wish to compute the sum of the squares of the first n positive integers. Setting $x_n = 0^2 + \dots + n^2$, one has $x_n = x_{n-1} + n^2$ for all $n \geq 1$, with initial datum $x_0 = 0$. The characteristic polynomial of this recurrence is $X - 1$. Therefore, the family of sequences $(A)_n$, with variation of the constant A, is the general solution of the homogeneous part. The non homogeneous term is $n^2 = n^2 \times 1^n$; since 1 is a root of the characteristic polynomial of multiplicity 1, a particular solution of the recurrence has the form $(n(\alpha_0 + \alpha_1 n + \alpha_2 n^2))_n$ for suitable values of

Fig. 10.1 Geometric interpretation of the recurrence $z_n = \dfrac{z_{n-1} + i}{2}$

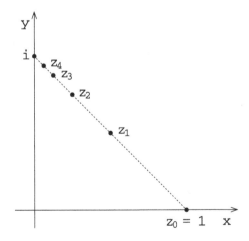

the parameters α_0, α_1, α_2. To compute such parameters, let us force the sequence to be a solution:

$$n(\alpha_0 + \alpha_1 n + \alpha_2 n^2) = (n-1)(\alpha_0 + \alpha_1(n-1) + \alpha_2(n-1)^2) + n^2.$$

Comparing the coefficients of n^0, n^1, n^2, n^3 we get: $\alpha_0 = 1/6$, $\alpha_1 = 1/2$, $\alpha_2 = 1/3$. Therefore the general solution is $x_n = A + \frac{1}{6}n + \frac{1}{2}n^2 + \frac{1}{3}n^3$. The initial datum $x_0 = 0$ implies $A = 0$; thus, the desired solution is $x_n = \frac{1}{6}n + \frac{1}{2}n^2 + \frac{1}{3}n^3$.   □

We have thus seen some techniques that allow us to find a particular solution of a linear recurrence relation when the non-homogeneous term belongs to a particular class of sequences. For completeness we insert a formula valid for any non homo-geneous term. However, satisfaction with the generality of the result is countered by the difficult application of the same; we will postpone the proof to Sect. 10.5.

Proposition 10.41 (An explicit particular solution) *Consider the recurrence relation of order r*

$$c_0 x_n + c_1 x_{n-1} + \cdots + c_r x_{n-r} = h_n, \quad n \geq r \quad (c_0 c_r \neq 0). \tag{10.41.a}$$

Let $(w_n)_n$ be the solution of the associated homogeneous recurrence with initial data $x_0 = 0$, ..., $x_{r-2} = 0$, $x_{r-1} = 1$. Then the sequence $(a_n)_n$ defined by

$$a_n = \begin{cases} 0 & \text{if } n \leq r-1, \\ \dfrac{h_r}{c_0} & \text{if } n = r, \\ \dfrac{1}{c_0} \displaystyle\sum_{k=1}^{n} w_{n-k} h_{k+r-1} = \dfrac{1}{c_0}\left(h_n + \displaystyle\sum_{k=1}^{n-r} w_{n-k} h_{k+r-1}\right) & \text{if } n > r, \end{cases}$$

is a particular solution of the recurrence relation (10.41.a).

Example 10.42 We wish to determine a particular solution of the recurrence relation of order 1

$$x_n - x_{n-1} = n \quad n \geq 1.$$

The characteristic polynomial is $X - 1$; therefore the family of constant sequences $(A)_n$ with $A \in \mathbb{R}$ is the general solution of the homogeneous part. Let us try to use Proposition 10.41 to find a particular solution of the recurrence. The constant sequence $(1)_n$ is the solution of the homogeneous part with initial datum $x_0 = 1$. Therefore the sequence $(a_n)_n$ defined by

$$a_n = \begin{cases} 0 & \text{if } n = 0, \\ 1 & \text{if } n = 1, \\ \dfrac{1}{1} \displaystyle\sum_{k=1}^{n} 1(k + 1 - 1) = \dfrac{n(n+1)}{2} & \text{if } n > 1, \end{cases}$$

actually equal to $\dfrac{n(n+1)}{2}$ for all n, is a particular solution. The verification is immediate: $\dfrac{n(n+1)}{2} - \dfrac{(n-1)n}{2} = n$ for each $n \geq 1$. $\square$

10.2   Linear Recurrences with Variable Coefficients

A linear recurrence with variable coefficients of order $r \geq 1$ is a recurrence of the form

$$x_n + c_{1,n} x_{n-1} + \cdots + c_{r,n} x_{n-r} = h_n, \quad n \geq r,$$

where $(c_{1,n})_{n \geq r}, \ldots, (c_{r,n})_{n \geq r}$ and $(h_n)_{n \geq r}$ are sequences, with $c_{r,n} \neq 0$ for all $n \geq r$. If $h_n = 0$ for each $n \geq r$, the above recurrence is said to be homogeneous. Clearly the recurrence has a unique solution once the sequence of the initial data is given. Due to the linearity of the recurrence, the set of solutions keeps the same structure as that of the linear recurrences with constant coefficients: a particular solution plus the space of solutions of the associated homogeneous recurrence.

We limit ourselves to studying the case of linear recurrences with variable coefficients of order 1.

Proposition 10.43 (Linear recurrence of order 1) *Let us consider the linear recurrence of order 1 with variable coefficients*

$$x_n = c_n x_{n-1} + h_n, \quad n \geq 1,$$

where $(c_n)_{n \geq 1}$ and $(h_n)_{n \geq 1}$ are fixed sequences with $c_n \neq 0$ for all $n \geq 1$.

1. The sequence $(u_n)_n$ defined setting

$$u_n = \begin{cases} 1 & \text{if } n = 0, \\ \displaystyle\prod_{j=1}^{n} c_j = c_1 \cdots c_n & \text{if } n \geq 1, \end{cases}$$

is a basis-solution of the associated homogeneous recurrence.

2. *The sequence $(b_n)_n$ defined by*

$$b_n = \begin{cases} 0 & \text{if } n = 0, \\ u_n \sum_{j=1}^{n} \dfrac{h_j}{u_j} & \text{if } n \geq 1, \end{cases}$$

is a particular solution of the recurrence.

3. *The solution with initial datum $x_0 = a_0$ is the sequence $(a_n)_n$ whose n-th term is*

$$a_n = u_n a_0 + b_n = \begin{cases} a_0 & \text{if } n = 0, \\ u_n a_0 + u_n \sum_{j=1}^{n} \dfrac{h_j}{u_j} = c_1 c_2 \cdots c_n \left(a_0 + \sum_{j=1}^{n} \dfrac{h_j}{c_1 \cdots c_j} \right) & \text{if } n \geq 1. \end{cases}$$

Proof. 1. If $x_0 = 1 = u_0$, one has

$$x_1 = c_1 \cdot 1 = c_1 = u_1, x_2 = c_2 c_1 = u_2, \ldots, x_n = c_n c_{n-1} \cdots c_1 = u_n.$$

Thus, the sequence $(Au_n)_n$ is the solution with initial datum $x_0 = A$.

2. It is easy to verify that $b_1 = h_1 = c_1 \cdot 0 + h_1 = c_1 \cdot b_0 + h_1$ and for each $n \geq 1$

$$b_n = u_n \sum_{j=1}^{n} \frac{h_j}{u_j} = u_n \left(\sum_{j=1}^{n-1} \frac{h_j}{u_j} + \frac{h_n}{u_n} \right) = c_n u_{n-1} \sum_{j=1}^{n-1} \frac{h_j}{u_j} + h_n = b_{n-1} + h_n.$$

3. By the Superposition Principle 10.3, the general solution of the recurrence is the family of the sequences $(Au_n + b_n)_n$ with the variation of the constant A; imposing the initial datum $x_0 = a_0$, taking into account that $u_0 = 1$ and $b_0 = 0$, one finds $A = a_0$. $\qquad \square$

Example 10.44 Let us determine the solution of

$$x_n = n x_{n-1} + n!, \quad n \geq 1,$$

with initial datum $x_0 = 5$. Following Proposition 10.43, one has that

$$u_n = n \times (n-1) \times \cdots \times 2 \times 1 = n!$$

is a basis-solution,

$$b_n = n! \sum_{j=1}^{n} \frac{j!}{J!} = n \times n!$$

is a particular solution and hence the desired solution is

$$
x_n = \begin{cases}
5 & \text{if } n = 0, \\
5n! + n! \displaystyle\sum_{j=1}^{n} \frac{j!}{j!} = (5+n)n! & \text{if } n \geq 1.
\end{cases}
$$

Notice that, actually, $x_n = (5+n)n!$ for all $n \geq 0$.   □

Example 10.45 (*Quicksort*) The *quicksort* is one of the most important algorithms for ordering data in a computer. Given a list of $n > 0$ distinct numbers, let us denote by x_n the average number of steps made to put them in ascending order. The algorithm starts by choosing randomly a number p, called the *pivot*. We compare then each of the other $n - 1$ numbers in the list with the pivot p, putting on his left elements strictly less p, and on his right elements strictly greater than p: this first phase is composed of $n - 1$ steps. At this point we are left to sort the $i \in \{0, \ldots, n - 1\}$ numbers to the left of the pivot (x_i steps) and the $n - 1 - i$ numbers to the right of the pivot (x_{n-1-i} steps); on average, we will have to make $\dfrac{1}{n} \displaystyle\sum_{i=0}^{n-1} (x_i + x_{n-1-i})$ steps to end this second phase. We obtain for x_n the recurrence relation

$$
x_n = n - 1 + \frac{1}{n} \sum_{i=0}^{n-1} (x_i + x_{n-1-i}) = n - 1 + \frac{2}{n} \sum_{i=0}^{n-1} x_i, \quad n \geq 1.
$$

Multiplying by n both sides of the equality one obtains

$$
n x_n = n(n - 1) + 2 \sum_{i=0}^{n-1} x_i, \quad n \geq 1.
$$

Then, for each $n \geq 2$ one has

$$
n x_n - (n - 1) x_{n-1} = n(n - 1) - (n - 1)(n - 2) + 2 \left(\sum_{i=0}^{n-1} x_i - \sum_{i=0}^{n-2} x_i \right)
$$

$$
= 2(n - 1) + 2 x_{n-1};
$$

therefore we obtain

$$
x_n = \frac{n + 1}{n} x_{n-1} + \frac{2n - 2}{n}, \quad n \geq 2.
$$

Bearing in mind that a set of zero numbers or of one number need 0 steps to be sorted, the above recurrence is also true for $n = 1$ with the initial condition $x_0 = 0$. By Proposition 10.43, set $u_0 = 1$ and

$$u_n = \prod_{j=1}^{n} \frac{j+1}{j} = \frac{(n+1)!}{n!} = n+1,$$

it turns out that the desired solution is the sequence $(a_n)_n$ defined by

$$a_n = \begin{cases} 0 & \text{if } n = 0, \\ (n+1) \sum_{j=1}^{n} \frac{2j-2}{j(j+1)} & \text{otherwise.} \end{cases} \qquad (10.45.a)$$

Now, denoted by H_n the harmonic numbers (see Definition 6.40), for each $n \geq 1$, one has

$$\sum_{j=1}^{n} \frac{2j-2}{j(j+1)} = 2 \sum_{j=1}^{n} \frac{j-1}{j(j+1)} = 2 \sum_{j=1}^{n} \left(\frac{2}{j+1} - \frac{1}{j} \right)$$

$$= 2(2(H_{n+1} - 1) - H_n) = 2(H_{n+1} + H_{n+1} - H_n - 2)$$

$$= 2 \left(H_{n+1} + \frac{1}{n+1} - 2 \right) = 2 \left(H_{n+1} + \frac{1 - 2(n+1)}{n+1} \right)$$

$$= 2 H_n - \frac{4n}{n+1}.$$

It follows from (10.45.a) that

$$a_n = 2(n+1) H_n - 4n \quad \forall n \in \mathbb{N}.$$

Note in particular that, since $H_n \sim \log n$ by Proposition 6.44, the number of steps to reorder a set of n elements is asymptotic to $2n \log n$ for $n \to \infty$. We will solve the recurrence of the quicksort also in the Example 10.69, using the alternative method of the generating formal series. $\qquad \square$

10.3 Divide and Conquer Recurrences

In this section we introduce a special class of recurrence relations that appear frequently in the analysis of recursive algorithms, especially in computer science. These recurrences aim to calculate the number of steps of an algorithm constructed with an approach known by the name *divide and conquer*. This approach consists in breaking a problem of *size n* into two (resp. $b \in \mathbb{N}$, $b \geq 2$) problems of *size n/2* (resp. n/b).

10.3.1   *Examples and definition*

Example 10.46 We wish to calculate the number of operations required to determine the minimum and maximum of a set S of $n = 2^k$ ($k \geq 1$) distinct integers. If $n = 2$ one comparison is sufficient to solve the problem. If $n \geq 2^2$, suppose we have found the minimum m_1 and the maximum M_1 of the first half of S (formed by the first $n/2$ numbers) and the minimum m_2 and the maximum M_2 of the second half of S (formed by the last $n/2$ numbers). It is sufficient at this point to make 2 comparisons: the minimum of S will be the smallest between m_1 and m_2, the maximum of S will be the greatest between M_1 and M_2.

Denoting by y_n the number of comparisons needed to solve the problem using the method described above with $|S| = n = 2^k$, one obtains

$$y_n = 2y_{n/2} + 2, \quad n = 2^k, \ k \geq 2,$$

with $y_2 = 1$.

Strictly speaking we have not obtained a recurrence relation according to the definition; it in fact involves only indices that are powers of two. But it certainly looks like a lot! For each $n = 2^k$, set $x_k = y_n = y_{2^k}$; the previous recurrence is transformed into the linear recurrence relation

$$x_k = 2x_{k-1} + 2, \quad k \geq 2,$$

with initial datum $x_1 = 1$. The sequence $(2^k)_k$ is a basis-solution of the associated homogeneous recurrence. A particular solution instead is of the type $(\alpha)_k$; substituting in the recurrence one has

$$\alpha = 2\alpha + 2$$

and hence $\alpha = -2$. Thus, the general solution of the recurrence is

$$x_k = A2^k - 2;$$

imposing the initial datum $x_1 = 1$ one obtains $A2^1 - 2 = 1$, and hence $A = \dfrac{3}{2}$. Recalling that $x_k = y_{2^k}$, one obtains

$$y_n = y_{2^k} = x_k = \frac{3}{2}2^k - 2 = \frac{3}{2}n - 2, \quad n = 2^k, k \geq 1. \qquad \square$$

Example 10.47 Generally, one need n^2 multiplications of numbers with a figure to make the product of two numbers with n digits. We use the approach of *divide and conquer* to develop an alternative algorithm. Suppose as in the previous example that n is a power of 2. If m is a number of n digits, denote by m_1 its first $n/2$ digits and by m_2 its last $n/2$ digits; then we have $m = m_1 10^{n/2} + m_2$. Let also ℓ be an n-digit number; similarly, one has $\ell = \ell_1 10^{n/2} + \ell_2$. Consequently we have

$$m\ell = (m_1\ell_1)10^n + (m_1\ell_2 + m_2\ell_1)10^{n/2} + m_2\ell_2.$$

Since

$$m_1\ell_2 + m_2\ell_1 = (m_1 + m_2)(\ell_1 + \ell_2) - m_1\ell_1 - m_2\ell_2,$$

it is sufficient to include just the 3 products of numbers with $n/2$ digits

$$(m_1 + m_2)(\ell_1 + \ell_2), \quad m_1\ell_1, \quad m_2\ell_2,$$

to obtain xy (actually $m_1 + m_2$ and $\ell_1 + \ell_2$ could have $(n/2 + 1)$ figures, but this small difference does not alter the order of magnitude of the solution). If y_n is the number of products of single digits required to multiply two n-digit numbers with the procedure described above, we have

$$y_n = 3y_{n/2}, \quad n = 2^k, \ k \geq 1,$$

with $y_1 = 1$. Setting $n = 2^k$, and $x_k = y_{2^k}$ one obtains the recurrence

$$x_k = 3x_{k-1}, \quad k \geq 1,$$

with initial datum $x_0 = 1$. This is a homogeneous recurrence that, by Theorem 10.12, has the general solution

$$x_k = A3^k,$$

with the variation of the constant A; the initial datum implies $A = 1$. Taking into account that $x_k = y_{2^k}$, one has

$$y_n = y_{2^k} = x_k = 3^k = 3^{\log_2 n}, \quad n = 2^k, \ k \geq 0;$$

since $\log_n(3^{\log_2 n}) = (\log_2 n)(\log_n 3) = \log_2 3$, we obtain $y_n = n^{\log_2 3}$. Observe that $\log_2 3 \approx 1.58496 \leq 1.6$; thus $y_n \leq n^{1.6}$. The described algorithm for the product of numbers therefore requires less than the n^2 steps of the traditional method.   □

Example 10.48 (*Binary Search*) Imagine having to find *Matilde* in a list of n names alphabetically ordered. A sequential search involves browsing all the names, and from time to time comparing them with *Matilde*: it requires at worst n comparisons and on average $\dfrac{1 + \cdots + n}{n} = \dfrac{n + 1}{2}$ comparisons. An alternative method, assuming that n is a power of two, consists in dividing the list into two and figuring out in which of the two parties lies *Matilde*. To do this, just compare *Matilde* with the first name of the second part; depending on whether this name follows or precedes *Matilde* in alphabetical order, the name we are looking for will wither be in the first or in the second part. In all, we have made two steps: first, we divided the list, then we have established what part contains *Matilde*. We continue in the same manner in the identified part, dividing it into two new parts and identifying in which part is the name *Matilde*. Denoting by y_n the maximum number of steps that may be needed to find *Matilde*, we have

$$y_n = y_{n/2} + 2, \quad n = 2^k, \ k \geq 1,$$

with initial datum $y_1 = 1$. Setting $n = 2^k$ and $x_k = y_{2^k}$, we get the recurrence relation

$$x_k = x_{k-1} + 2, \quad k \geq 1,$$

with initial datum $x_0 = 1$. The sequence $(1^k)_k = (1)_k$ is a basis-solution of the associated homogeneous recurrence. Since the non homogeneous term is $2 \cdot 1^n$ and 1 is a root of the characteristic polynomial, there exists a particular solution of the form $(k\alpha)_k$. Substituting in the recurrence, one obtains $k\alpha = (k-1)\alpha + 2$, and hence $\alpha = 2$. Thus, the general solution of the recurrence is

$$x_k = A + 2k,$$

with the variation of the constant A. Imposing the initial datum $1 = x_0$, one obtains

$$x_k = 1 + 2k.$$

Since $x_k = y_{2^k}$, we have

$$y_n = y_{2^k} = x_k = 1 + 2k = 1 + 2\log_2 n, \quad n = 2^k, \ k \geq 0.$$

Observe that $\dfrac{2\log_2 n + 1}{(n+1)/2}$ tends to 0 as n tends to infinity: the method of binary search is therefore more advantageous than that of the sequential search. $\qquad\square$

We formalize the type of recurrences we have seen in the previous examples.

Definition 10.49 A **"divide and conquer" recurrence** of order r is a family of equations of the form

$$c_0 y_n + c_1 y_{n/b} + c_2 y_{n/b^2} + \cdots + c_r y_{n/b^r} = h_n, \quad n = n_0 b^k, \ k \geq r, \quad (10.49.a)$$

where $b \in \mathbb{N}_{\geq 2}$, $n_0 \in \mathbb{N}_{\geq 1}$, $r \in \mathbb{N}$, $c_0 c_r \neq 0$ and $(h_n)_n$ is a sequence.

We say that $(a_n)_{n \in \{n_0 b^k : k \in \mathbb{N}\}}$ (most often we write $(a_n)_n$, with $n = n_0 b^k$) is a solution of (10.49.a) if $(a_{n_0 b^k})_{k \in \mathbb{N}}$ solves the **linear recurrence of order r induced by** (10.49.a) defined by

$$c_0 x_k + c_1 x_{k-1} + \cdots + c_r x_{k-r} = h_{n_0 b^k}, \quad k \geq r. \tag{10.49.b}$$

The r-sequence of the initial data $(a_{n_0}, a_{n_0 b}, \ldots, a_{n_0 b^{r-1}})$ of (10.49.b) is called the sequence of the initial data of (10.49.a). $\qquad \square$

☞ *Remark 10.50* In order to solve the recurrence (10.49.a) with initial conditions $(y_{n_0}, \ldots, y_{n_0 b^{r-1}})$ it is enough to set $x_k = y_{n_0 b^k}$ ($k \in \mathbb{N}$) and find the solution $(a_k)_k$ of (10.49.b) with initial conditions $(y_{n_0}, \ldots, y_{n_0 b^{r-1}})$. The solution of (10.49.a) is $y_n = a_k$, where $n = n_0 b^k$ for some $k \in \mathbb{N}$.

Example 10.51 Consider the recurrence relation

$$y_n - 5 y_{n/2} + 6 y_{n/4} = n, \quad n = 2^k, \ k \geq 2.$$

Determine the general solution and solve it with initial data $y_1 = 2$ and $y_2 = 1$.

Solution. Setting $n = 2^k$ and $x_k = y_{2^k}$, the previous recurrence becomes the linear recurrence of order 2

$$x_k - 5 x_{k-1} + 6 x_{k-2} = 2^k, \quad k \geq 2,$$

with initial conditions $x_0 = 2$, and $x_1 = 1$. The roots of the characteristic polynomial $X^2 - 5X + 6$ are 2 and 3. Therefore, the general solution of the induced homogeneous recurrence is $x_k = A 2^k + B 3^k$ with the variation of the constants A and B. Since 2 is a root of the characteristic polynomial, we look for a particular solution of the form $x_k = \alpha k 2^k$. Substituting in the recurrence $x_k - 5 x_{k-1} + 6 x_{k-2} = 2^k$, one deduces

$$\alpha k 2^k = 5\alpha(k-1)2^{k-1} - 6\alpha(k-2)2^{k-2} + 2^k;$$

hence $4\alpha k = 10\alpha(k-1) - 6\alpha(k-2) + 4$, or equivalently $2\alpha + 4 = 0$, i.e., $\alpha = -2$. Thus, the general solution is $x_k = A 2^k + B 3^k - 2^{k+1} k$ with variation of the constants A and B. The initial data impose that $2 = x_0 = A + B$ and $1 = x_1 = 2A + 3B - 4$: hence $A = 1$ and $B = 1$. Thus, the solution of the linear recurrence induced by the "divide and conquer" one is

$$x_k = 2^k + 3^k - 2^{k+1} k = (1 - 2k)2^k + 3^k.$$

Recalling that $x_k = y_{2^k}$, one obtains

$$y_n = y_{2^k} = x_k = (1 - 2k)2^k + 3^k = (1 - 2\log_2 n)n + 3^{\log_2 n}$$
$$= (1 - 2\log_2 n)n + n^{\log_2 3}, \quad n = 2^k, \ k \geq 0. \qquad \square$$

10.3.2   Order of Magnitude of the Solutions

Often, rather than the explicit solution of a "divide and conquer" recurrence, we are more interested in the order of magnitude of the solutions.

Recall that if $(a_n)_n$ and $(b_n)_n$ are two sequences, we say that $a_n = O(b_n)$ (we say that a_n is "big O" of b_n) as $n \to +\infty$ if there exists a constant $M > 0$ such that

$$|a_n| \le M|b_n| \quad \forall n \in \mathbb{N}.$$

In particular, if b_n is definitively non-zero and $\lim\limits_{n\to\infty} \dfrac{|a_n|}{|b_n|}$ exists and is finite, then $a_n = O(b_n)$ as $n \to +\infty$. For instance:

- $n = O(n^2)$ as $n \to +\infty$, since $\lim\limits_{n\to+\infty} \dfrac{n}{n^2} = 0$;
- $n \log^{100} n = O(n^{11/10})$ as $n \to +\infty$, since $\lim\limits_{n\to+\infty} \dfrac{n \log^{100} n}{n^{11/10}} = 0$;
- $n \sin(n^2 + \log(n^2 + 5)) = O(n)$ as $n \to \infty$ since $|n \sin(n^2 + \log(n^2 + 5))| \le n$ for every $n \in \mathbb{N}$.

We refer to Sect. 14.1 for more details on this concept. We introduce a further notation, more suitable for the solutions to the divide and conquer recurrences.

Definition 10.52   Given $b \in \mathbb{N}_{\ge 2}$, let $(a_n)_n$ and $(\beta_n)_n$ be two sequences. We write

$$a_n = O(\beta_n) \text{ for } n = b^k, \ k \to +\infty,$$

if $a_{b^k} = O(\beta_{b^k})$ as $k \to +\infty$.   □

Using the results on the linear recurrence relations with constant coefficients, one immediately obtains the following estimate.

Theorem 10.53   *Let $\lambda > 0$, $c \in \mathbb{R} \setminus \{0\}$, $\rho \ge 0$ and $b \in \mathbb{N}_{\ge 2}$.*

1. Let $(a_n)_n$, $n = b^k$, be a solution to the divide and conquer recurrence

$$y_n = \lambda y_{n/b} + cn^\rho, \quad n = b^k, \ k \ge 1.$$

Then

$$a_n = \begin{cases} O(n^\rho) & \text{if } \lambda < b^\rho, \\ O(n^\rho \log n) & \text{if } \lambda = b^\rho, \quad \text{for } n = b^k, \ k \to +\infty. \\ O(n^{\log_b \lambda}) & \text{if } \lambda > b^\rho, \end{cases}$$

2. *Let $(a_n)_n$, $n = b^k$, be a solution to the divide and conquer recurrence*

$$y_n = \lambda y_{n/b} + c \log_b n \quad n = b^k, k \geq 1.$$

Then

$$a_n = \begin{cases} O(\log n) & \text{if } \lambda < 1, \\ O(\log^2 n) & \text{if } \lambda = 1, \quad \text{for } n = b^k, \, k \to +\infty. \\ O(n^{\log_b \lambda}) & \text{if } \lambda > 1, \end{cases}$$

Proof. 1. Setting $x_k = y_{b^k}$, $k \in \mathbb{N}$, one obtains the linear recurrence of order 1

$$x_k = \lambda x_{k-1} + c b^{k\rho} \quad k \geq 1.$$

The family of sequences $(A\lambda^k)_k$, with variation of A in $\mathbb{R}$, is the general solution of the induced homogeneous recurrence. The non homogeneous term is $c(b^\rho)^k$. To find a particular solution of the recurrence we distinguish two cases:

- $\lambda \neq b^\rho$: the linear recurrence has a particular solution of the type $(\alpha(b^\rho)^k)_k$ for some $\alpha \in \mathbb{R}$; in this case the general solution of the recurrence is of the type $(A\lambda^k + \alpha b^{\rho k})_k$, with variation of $A \in \mathbb{R}$ and for a suitable $\alpha \in \mathbb{R}$;
- $\lambda = b^\rho$: the linear recurrence has a particular solution of the type $(\alpha k(b^\rho)^k)_k$ for some $\alpha \in \mathbb{R}$; in this case the general solution of the recurrence is of the type $(A\lambda^k + \alpha k b^{\rho k})_k$, with variation of $A \in \mathbb{R}$ and for a suitable $\alpha \in \mathbb{R}$.

Setting $n = b^k$, or equivalently $k = \log_b n$, one has $\lambda^k = \lambda^{\log_b n} = n^{\log_b \lambda}$ and $b^{\rho k} = n^\rho$; therefore the general solution of the recurrence is:

- $\lambda \neq b^\rho$: $y_n = An^{\log_b \lambda} + \alpha n^\rho$, for suitable $A, \alpha \in \mathbb{R}$, and $n = b^k, k \geq 0$. The conclusion follows from the fact that if $\lambda < b^\rho$, then $\rho > \log_b \lambda$ and $n^{\log_b \lambda} = O(n^\rho)$; while, if $\lambda > b^\rho$, then $\rho < \log_b \lambda$ and $n^\rho = O(n^{\log_b \lambda})$;
- $\lambda = b^\rho$: $y_n = An^\rho + \alpha(\log_b n)n^\rho = An^\rho + \alpha(\log_b n)n^\rho$ for suitable $A, \alpha \in \mathbb{R}$, and $n = b^k, k \geq 0$. The conclusion follows from

$$An^\rho + \alpha(\log_b n)n^\rho = O(n^\rho \log n), \ n \to +\infty.$$

2. Setting $x_k = y_{b^k}$, $k \in \mathbb{N}$, one obtains the linear recurrence of order 1

$$x_k = \lambda x_{k-1} + ck, \quad k \geq 1.$$

The family of sequences $(A\lambda^k)_k$, with the variation of A in $\mathbb{R}$, is the general solution of the induced homogeneous recurrence. The non homogeneous term is ck. To find a particular solution of the recurrence we distinguish two cases:

- $\lambda \neq 1$: the linear recurrence has a particular solution of the type $(\alpha_1 k + \alpha_2)_k$ for some $\alpha_1, \alpha_2 \in \mathbb{R}$; in this case the general solution of the recurrence is of the type $(A\lambda^k + \alpha_1 k + \alpha_2)_k$, with variation of $A \in \mathbb{R}$ and for suitable $\alpha_1, \alpha_2 \in \mathbb{R}$;
- $\lambda = 1$: the linear recurrence has a particular solution of the type $k(\alpha_1 k + \alpha_2)$ for some $\alpha_1, \alpha_2 \in \mathbb{R}$; in this case the general solution of the recurrence is of the type $(A + \alpha_1 k^2 + \alpha_2 k)_k$, with variation of $A \in \mathbb{R}$ and for suitable $\alpha_1, \alpha_2 \in \mathbb{R}$.

Setting $n = b^k$, or equivalently $k = \log_b n$, one has $\lambda^k = \lambda^{\log_b n} = n^{\log_b \lambda}$; therefore the general solution of the recurrence is:

- $\lambda \neq 1$: $y_n = An^{\log_b \lambda} + \alpha_1 \log_b n + \alpha_2$ for suitable $A, \alpha_1, \alpha_2 \in \mathbb{R}$, and $n = b^k, k \geq 0$. The conclusion follows from the fact that if $\lambda < 1$, then $n^{\log_b \lambda} = O(1) \subseteq O(\log n)$; while, if $\lambda > 1$, then $\log_b n = O(n^{\log_b \lambda})$.
- $\lambda = 1$: $y_n = A + \alpha_1 \log_b^2 n + \alpha_2 \log_b n = O(\log^2 n)$ for $A, \alpha_1, \alpha_2 \in \mathbb{R}$, and $n = b^k, k \geq 0$. $\square$

Example 10.54 We wish to determine the order of magnitude of the solutions of the "divide and conquer" recurrence

$$y_n = 9y_{n/3} + 5, \quad n = 3^k, \quad k \geq 1,$$

with initial datum $y_1 = 2$. Applying Point 1 of Theorem 10.53 with $\lambda = 9$, $b = 3, \rho = 0, c = 5$, since $\lambda = 9 > 3^0 = 1 = b^\rho$, we immediately obtain the estimate $y_n = O(n^{\log_3 9}) = O(n^2)$.

Alternatively, one could solve the recurrence. Setting $x_k = y_{3^k}$, one has

$$x_k = 9x^{k-1} + 5, \quad k \geq 1.$$

The characteristic polynomial of this linear recurrence is $X - 9$; then the family of sequences $(A9^k)_k$, with the variation of the constant A, is the general solution of the homogeneous part. The recurrence has a constant solution $(\alpha)_n$; substituting in the recurrence, one obtains $\alpha = 9\alpha + 5$, and hence $\alpha = -5/8$. Thus, the general solution of the induced linear recurrence is

$$x_k = A9^k - \frac{5}{8}, \quad k \geq 0.$$

Setting $n = 3^k$, one has $k = \log_3 n$; then

$$y_n = y_{3^k} = x_k = A9^k - \frac{5}{8}$$
$$= A9^{\log_3 n} - \frac{5}{8} = An^{\log_3 9} - \frac{5}{8} = An^2 - \frac{5}{8}, \quad n = 3^k, k \geq 0,$$

showing again that $y_n = O(n^2)$ for $n = 3^k, k \to +\infty$.

Imposing the initial condition $y_1 = 2$, or equivalently $x_0 = 2$, to the general solution, one has

$$2 = x_0 = A - \frac{5}{8}, \text{ i.e., } A = \frac{21}{8};$$

thus, we obtain the solution

$$y_n = y_{3^k} = x_k = \frac{21}{8}9^k - \frac{5}{8} = \frac{21}{8}n^2 - \frac{5}{8}, \quad n = 3^k, k \geq 0. \qquad \square$$

Example 10.55 Let us determine the order of magnitude of the solutions of the "divide and conquer" recurrence

$$y_n = \frac{1}{4}y_{n/2} + n \quad n = 2^k, k \geq 1.$$

By applying Point 1 of Theorem 10.53 with $\lambda = \frac{1}{4}, b = 2, \rho = 1, c = 1$, since $\lambda = \frac{1}{4} < 2^1 = b^\rho$, we immediately get that $y_n = O(n)$, $n = 2^k, k \to +\infty$. Alternatively one could solve the recurrence. Setting $x_k = y_{2^k}$ one has

$$x_k = \frac{1}{4}x_{k-1} + 2^k, \quad k \geq 1.$$

This is a linear recurrence whose characteristic polynomial is $X - 1/4$: the family of sequences $\left(\dfrac{A}{4^k}\right)_k$, with the variation of the constant A, is the general solution of the homogeneous part. A particular solution is of the form $(\alpha 2^k)_k$; substituting this in the recurrence we have

$$\alpha 2^k = \frac{1}{4}\alpha 2^{k-1} + 2^k$$

and hence $\alpha = \frac{8}{7}$. Therefore, the general solution of the linear recurrence is

$$x_k = A\left(\frac{1}{4}\right)^k + \frac{8}{7}2^k.$$

Setting $n = 2^k$, one has $k = \log_2 n$; thus

$$y_n = y_{2^k} = x_k = A4^{-k} + \frac{8}{7}2^k$$

$$= A4^{-\log_2 n} + \frac{8}{7}n = An^{-\log_2 4} + \frac{8}{7}n = \frac{A}{n^2} + \frac{8}{7}n = O(n), \quad n = 2^k, k \to +\infty.$$

$\square$

Example 10.56 Let us determine the order of magnitude of the solutions of the "divide and conquer" recurrence

$$y_n = 5y_{n/5} + 4\log_5 n, \quad n = 5^k, k \geq 1,$$

and its solution with initial datum $y_1 = 1$. By applying Point 2 of Theorem 10.53 with $\lambda = 5, b = 5, c = 4$, since $\lambda = 5 > 1$, we immediately obtain the estimate $y_n = O(n^{\log_5 5}) = O(n), n = 5^k, k \rightarrow +\infty$. Alternatively, one could solve the recurrence. Setting $x_k = y_{5^k}$ we have

$$x_k = 5x_{k-1} + 4k, \quad k \geq 1.$$

It is a linear recurrence whose characteristic polynomial is $X - 5$: the family of sequences $(A5^k)_k$, with variation of the constant A, is the general solution of the homogeneous part. A particular solution is of the form $(\alpha_1 k + \alpha_2)_k$. Substituting this in the recurrence one has

$$\alpha_1 k + \alpha_2 = 5(\alpha_1(k-1) + \alpha_2) + 4k,$$

or equivalently $k(\alpha_1 - 5\alpha_1 - 4) = 0$ and $\alpha_2 + 5\alpha_1 - 5\alpha_2 = 0$; hence $\alpha_1 = -1$ and $\alpha_2 = -5/4$. Therefore, the general solution of the linear recurrence is

$$x_k = A5^k - k - \frac{5}{4}.$$

Setting $n = 5^k$, we have $k = \log_5 n$; thus

$$y_n = y_{5^k} = x_k = A5^k - k - \frac{5}{4}$$

$$= A5^{\log_5 n} - \log_5 n - \frac{5}{4} = An - \log_5 n - \frac{5}{4} = O(n), \quad n = 5^k, k \rightarrow +\infty.$$

Imposing to the general solution the initial condition $y_1 = 1$, or equivalently $x_0 = 1$, one obtains $A = 9/4$; thus, one gets the solution

$$y_n = y_{5^k} = x_k = \frac{9}{4}5^k - k - \frac{5}{4} = \frac{9}{4}n - \log_5 n - \frac{5}{4}, \quad n = 5^k, k \geq 0. \qquad \square$$

Example 10.57 Let us determine the order of magnitude of the solutions of the "divide and conquer" recurrence

$$y_n = y_{n/5} + 4\log_5 n, \quad n = 5^k, k \geq 1,$$

and its solution with initial datum $y_1 = 1$. By applying Point 2 of Theorem 10.53 with $\lambda = 1, b = 5, c = 4$, since $\lambda = 1$ we immediately get that $y_n = O(\log^2 n), n = 5^k, k \rightarrow +\infty$. Alternatively, one could solve the recurrence. Setting $x_k = y_{5^k}$ one has

$$x_k = x_{k-1} + 4k, \quad k \geq 1.$$

It is a linear recurrence whose characteristic polynomial is $X - 1$: the family of sequences $(A1^k)_k = (A)_k$, with variation of the constant A, is the general solution of the homogeneous part. Since 1 is a root of the characteristic polynomial, a particular solution is of the form $(k(\alpha_1 k + \alpha_2))_k$. Substituting this in the recurrence one has

$$\alpha_1 k^2 + \alpha_2 k = \alpha_1 (k - 1)^2 + \alpha_2 (k - 1) + 4k,$$

or equivalently $k(\alpha_2 + 2\alpha_1 - \alpha_2 - 4) = 0$ and $\alpha_1 = \alpha_2$; hence $\alpha_1 = \alpha_2 = 2$. Therefore, the general solution of the linear recurrence is

$$x_k = A + 2k^2 + 2k.$$

Setting $n = 5^k$ we have $k = \log_5 n$; thus

$$y_n = y_{5^k} = x_k = A + 2k^2 + 2k = A + 2\log_5^2 n + 2\log_5 n = O(\log^2 n), \quad n = 5^k, \ k \to +\infty.$$

Imposing to the general solution the initial condition $y_1 = 1$, or equivalently $x_0 = 1$, one obtains $A = 1$; thus, one obtains the solution

$$y_n = y_{5^k} = x_k = 1 + 2k^2 + 2k = 1 + 2\log_5^2 n + 2\log_5 n, \quad n = 5^k, \ k \geq 0. \qquad \square$$

10.4 Recurrences and Generating Formal Series

In this section we will see how to use formal series to determine the solutions of a linear recurrence relation.

10.4.1 Linear Recurrences with Constant Coefficients and Their OGF

Let us consider the linear recurrence of order $r \geq 1$

$$c_0 x_n + c_1 x_{n-1} + \cdots + c_r x_{n-r} = h_n, \quad n \geq r, \tag{R}$$

where $c_0, c_1, \ldots, c_r$ are constants with $c_0 c_r \neq 0$, and $(h_n)_{n \geq r}$ is a given sequence.

The characteristic polynomial of (R) is $P_{\text{char}}(X) = c_0 X^r + c_1 X^{r-1} + \ldots + c_r$. The OGF of a solution of the given recurrence admits a closed form that is easily obtained from the data of the recurrence itself. The solution is then found by using Theorem 7.114 or a CAS.

Proposition 10.58 *The formal series* $A(X) = \sum_{n=0}^{\infty} a_n X^n$ *is the OGF of a solution of the linear recurrence* (R) *if and only if there exists a polynomial* $S(X)$ *with* $\deg S(X) < r$ *satisfying*

$$A(X) = \frac{S(X) + \sum_{n=r}^{\infty} h_n X^n}{c_0 + c_1 X + \cdots + c_r X^r}. \tag{10.58.a}$$

In that case $S(X)$ *is uniquely determined by the initial data and the coefficients* $c_0, \ldots, c_{r-1}$:

$$S(X) = [X^{\leq r-1}] \left(c_0 + c_1 X + \cdots + c_{r-1} X^{r-1} \right) A(X)$$

$$= [X^{\leq r-1}] \left(\sum_{n=0}^{r-1} c_n X^n \right) \left(\sum_{n=0}^{r-1} a_n X^n \right). \tag{10.58.b}$$

Proof. Consider the recurrence (R). Let $A(X) = \sum_{n=0}^{\infty} a_n X^n$ be any formal series. For each $n \geq r$, the coefficient of the term of degree n in the product $A(X)(c_0 + c_1 X + \cdots + c_r X^r)$ is

$$[X^n] A(X)(c_0 + c_1 X + \cdots + c_r X^r) = c_0 a_n + c_1 a_{n-1} + \cdots + c_r a_{n-r}.$$

Therefore

$$[X^n] A(X)(c_0 + c_1 X + \cdots + c_r X^r) = h_n, \quad \forall n \geq r$$

if and only if

$$c_0 a_n + c_1 a_{n-1} + \cdots + c_r a_{n-r} = h_n, \quad \forall n \geq r,$$

or equivalently, if and only if the sequence $(a_n)_n$ is a solution of (R). In this case we have

$$A(X)(c_0 + c_1 X + \cdots + c_r X^r) =$$

$$= [X^{\leq r-1}] \left(c_0 + c_1 X + \cdots + c_r X^r \right) A(X) + [X^{\geq r}] \left(c_0 + c_1 X + \cdots + c_r X^r \right) A(X)$$

$$= [X^{\leq r-1}] \left(\sum_{n=0}^{r-1} c_n X^n \right) \left(\sum_{n=0}^{r-1} a_n X^n \right) + \sum_{n=r}^{\infty} h_n X^n,$$

from which we obtain (10.58.a) and (10.58.b).   □

We will see in Sect. 10.5 that Proposition 10.58 is very useful in proving most of the example left open in Sect. 10.1.

Remark 10.59 About Proposition 10.58, observe that:

1. The order r of the recurrence relation is the degree of the polynomial in the denominator of the formula (10.58.a);
2. The polynomial $S(X)$ is formed from the terms of degree strictly less than r of the formal series that appears in the numerator of (10.58.a) and its coefficients depend only on $c_0, \ldots, c_{r-1}$ and $a_0, \ldots, a_{r-1}$;
3. The non homogeneous terms h_n, $n \geq r$, of the recurrence relation are the coefficients of the terms of degree greater or equal than r of the formal series in the numerator of (10.58.a);
☞ 4. The polynomial $Q(X) := c_0 + c_1 X + \cdots + c_r X^r$ appearing in the denominator of (10.58.a) is *not* the characteristic polynomial $P_{\text{char}}(X) = c_0 X^r + c_1 X^{r-1} + \cdots + c_r$ of the linear recurrence (R). The roots of $P_{\text{char}}(X)$ and $Q(X)$ have a strict link: indeed λ is a root of $P_{\text{char}}(X)$ (necessarily $\neq 0$, since $c_r \neq 0$) if and only if

$$c_0 \lambda^r + c_1 \lambda^{r-1} + \cdots + c_r = 0.$$

Multiplying by $1/\lambda^r$, this equality holds if and only if one has

$$c_0 + c_1 \frac{1}{\lambda} + \cdots + c_r \left(\frac{1}{\lambda}\right)^r = 0,$$

or equivalently, if and only if $1/\lambda$ is a root of $Q(X)$; in this case the roots λ and $1/\lambda$ have the same multiplicity.

Proposition 10.58 becomes particularly simple in the case of homogeneous recurrence relations.

Corollary 10.60 *The formal series* $A(X) = \sum_{n=0}^{\infty} a_n X^n$ *is the OGF of a solution of the* homogeneous *linear recurrence*

$$c_0 x_n + c_1 x_{n-1} + \cdots + c_r x_{n-r} = 0, \quad n \geq r,$$

if and only if

$$A(X) = \frac{S(X)}{c_0 + c_1 X + \cdots + c_r X^r}, \quad \deg S(X) < r.$$

10.4.2 Applications

Solving Recurrence Relations

A first use of Proposition 10.58 is that of solving linear recurrence relations.

Example 10.61 We want to determine the ordinary generating formal series $A(X)$ of the solution of the recurrence relation

$$x_n = x_{n-1} + 1, \quad n \geq 1,$$

with initial datum $x_0 = 0$. It follows by Proposition 10.58 that

$$A(X) = \frac{S(X) + \displaystyle\sum_{k=1}^{\infty} X^k}{1 - X},$$

with $S(X) = [X^0](1)(a_0) = a_0 = 0$. Now, we have

$$\sum_{k=1}^{\infty} X^k = X \sum_{k=0}^{\infty} X^k = \frac{X}{1 - X},$$

and hence

$$A(X) = \frac{X}{(1 - X)^2}.$$

From formula (7.90.b), we obtain

$$\frac{X}{(1 - X)^2} = \sum_{n=0}^{\infty} \binom{n}{1} X^n = \sum_{n=0}^{\infty} n X^n$$

and thus the sequence $(n)_n$ is the desired solution of the recurrence.   □

Example 10.62 Let us determine the solution of the Fibonacci recurrence using the generating formal series. The Fibonacci numbers are the solution of the homogeneous recurrence

$$x_n = x_{n-1} + x_{n-2}, \quad n \geq 2,$$

with initial data $x_0 = 0$, $x_1 = 1$. Let $A(X)$ be the OGF of Fibonacci numbers. By Corollary 10.60, we have that

$$A(X) = \frac{S(X)}{1 - X - X^2},$$

with $S(X) = [X^{\leq 1}](1 - X)(0 + X) = X$; hence $A(X) = \dfrac{X}{1 - X - X^2}$. The roots of $1 - X - X^2$ are

$$\lambda_1 = \frac{-1 + \sqrt{5}}{2}, \quad \lambda_2 = -\frac{1 + \sqrt{5}}{2}.$$

By formula (7.114.b) with $P(X) = X$ and $Q(X) = 1 - X - X^2$, we get that for each $n \in \mathbb{N}$

$$\left[X^n\right]A(X) = a_n = -\sum_{\alpha:Q(\alpha)=0} \frac{P(\alpha)}{\alpha^{n+1} Q'(\alpha)}$$

$$= -\frac{(-1 + \sqrt{5})/2}{((-1 + \sqrt{5})/2)^{n+1}(-1 - 2(-1 + \sqrt{5})/2)} -$$

$$-\frac{(-1 - \sqrt{5})/2}{((-1 - \sqrt{5})/2)^{n+1}(-1 - 2(-1 - \sqrt{5})/2)}$$

$$= -\frac{1}{\sqrt{5}}\frac{(1 - \sqrt{5})/2}{((-1 + \sqrt{5})/2)^{n+1}} + \frac{1}{\sqrt{5}}\frac{(1 + \sqrt{5})/2}{((-1 - \sqrt{5})/2)^{n+1}}.$$

Since $\dfrac{1}{(-1 + \sqrt{5})/2} = \dfrac{1 + \sqrt{5}}{2}$ and $\dfrac{1}{(-1 - \sqrt{5})/2} = \dfrac{1 - \sqrt{5}}{2}$, one has

$$[X^n]A(X) = \frac{1}{\sqrt{5}}\left(\frac{\sqrt{5} + 1}{2}\right)^n - \frac{1}{\sqrt{5}}\left(\frac{1 - \sqrt{5}}{2}\right)^n,$$

as we had already obtained in Example 10.19. □

Example 10.63 Let us solve the homogeneous linear recurrence

$$x_n = 2x_{n-1} + x_{n-2} - 2x_{n-3}, \quad n \geq 3,$$

with initial data $x_0 = 0$, $x_1 = x_2 = 1$. By Proposition 10.9, the possible rational roots of the characteristic polynomial $X^3 - 2X^2 - X + 2$ are integer numbers (the coefficient of X^3 is 1) dividing 2: thus we have to look at the divisors $\{\pm 1, \pm 2\}$ of 2. Evaluating the polynomial at these values one finds that the roots are $1, -1, 2$: the general real solution of the recurrence is the family of sequences

$$(A_1 + A_2(-1)^n + A_3 2^n)_n,$$

with the variation of A_1, A_2 and A_3 in $\mathbb{R}$. Imposing the initial data we obtain the linear system

$$\begin{cases} A_1 + A_2 + A_3 = 0 \\ A_1 - A_2 + 2A_3 = 1 \\ A_1 + A_2 + 4A_3 = 1. \end{cases}$$

We obtain easily $A_3 = 1/3$ and hence $A_1 = 0$ and $A_2 = -1/3$: the desired solution is the sequence

$$x_n = \frac{1}{3}\left(2^n - (-1)^n\right).$$

We could also solve the recurrence using generating formal series. By Corollary 10.60, the OGF of the solution we are looking for is the formal series

$$A(X) = \frac{S(X)}{1 - 2X - X^2 + 2X^3},$$

with

$$S(X) = [X^{\le 2}](1 - 2X - X^2)(0 + X + X^2) = X - X^2;$$

therefore

$$A(X) = \frac{X - X^2}{1 - 2X - X^2 + 2X^3}.$$

It is easily seen, using Proposition 10.9 or the fact that the roots of the denominator are the inverse of the roots of the characteristic polynomial, that

$$1 - 2X - X^2 + 2X^3 = (1 - X)(1 + X)(1 - 2X)$$

and hence

$$A(X) = \frac{X}{(1 + X)(1 - 2X)} = \frac{X}{1 - X - 2X^2}.$$

By formula (7.114.b) with $P(X) = X$ and $Q(X) = 1 - X - 2X^2$, we find that for each $n \in \mathbb{N}$

$$\left[X^n\right] A(X) = -\sum_{\alpha: Q(\alpha)=0} \frac{P(\alpha)}{\alpha^{n+1} Q'(\alpha)}$$

$$= -\left(\frac{-1}{(-1)^{n+1}3} + \frac{1/2}{(1/2)^{n+1}(-3/2)}\right) = \frac{1}{3}\left(2^n - (-1)^n\right). \qquad \square$$

Example 10.64 Consider the recurrence of Example 10.63, changing only the initial data:

$$x_n = 2x_{n-1} + x_{n-2} - 2x_{n-3}, \quad n \ge 3,$$

with initial data $x_0 = x_1 = x_2 = 1$. By Proposition 10.58, the OGF of the solution is

$$A(X) = \frac{S(X)}{1 - 2X - X^2 + 2X^3},$$

where

$$S(X) = [X^{\le 2}](1 - 2X - X^2)(1 + X + X^2)$$
$$= 1 - X - 2X^2.$$

Therefore, we have

$$A(X) = \frac{1 - X - 2X^2}{1 - 2X - X^2 + 2X^3} = \frac{(1 + X)(1 - 2X)}{(1 - X)(1 + X)(1 - 2X)} = \frac{1}{1 - X} = \sum_{n=0}^{\infty} X^n.$$

Thus the constant sequence $(1)_n$ is the solution of the recurrence relation. Observe that while in Example 10.63 the solution of the recurrence is a sequence that grows exponentially, here, changing only the initial data, the solution is a constant sequence.   □

Finding the Coefficients of a Rational Power Series

Another use of Proposition 10.58 is that of finding the coefficients of a formal power series equal to a rational fraction of polynomials.

Remark 10.65 Assume that

$$\sum_{n=0}^{\infty} a_n X^n := A(X) = \frac{B(X)}{c_0 + c_1 X + \cdots + c_r X^r}, \qquad (10.65.a)$$

where $B(X) := \sum_{n=0}^{\infty} b_n X^n \in \mathbb{R}[[X]]$ and $c_0, \ldots, c_r$ with $c_0 c_r \neq 0$ are given. It follows from Proposition 10.58 that:

- $(a_n)_n$ solves the linear recurrence

$$c_0 x_n + \cdots + c_r x_{n-r} = b_n, \quad \forall n \geq r ;$$

- The first r values $a_0, \ldots, a_{r-1}$ can be recovered uniquely from the equality

$$A(X)(c_0 + c_1 X + \cdots + c_r X^r) = B(X),$$

due to the fact that $c_0 + c_1 X + \cdots + c_r X^r$ is invertible.

The above holds in particular when $A(X)$ is expressed as a quotient of polynomials: indeed in this case it follows from Proposition 7.63 that $A(X)$ can be written as in (10.65.a), for a suitable polynomial $B(X)$.

Example 10.66 Let us consider the formal series $A(X)$ of Example 7.117:

$$A(X) = \frac{3 + X}{(X + 1)(X - 2)} = \frac{X + 3}{-2 - X + X^2}.$$

Let us calculate explicitly the coefficients of the powers of X in $A(X)$. By Corollary 10.60, the formal series $A(X) := a_0 + a_1 X + a_2 X^2 + \cdots$ is the OGF of the solution of the linear recurrence $-2x_n - x_{n-1} + x_{n-2} = 0, n \geq 2$, with sequence of initial data (a_0, a_1). Since

$$S(X) := 3 + X = [X^{\leq 1}](-2 - X)(a_0 + a_1 X)$$
$$= -2a_0 - (a_0 + 2a_1)X,$$

we get the initial data $a_0 = -3/2$ and $a_1 = 1/4$.

The characteristic polynomial of the recurrence is

$$-2X^2 - X + 1 = -2(X + 1)(X - 1/2);$$

therefore the family of sequences

$$(A_1(-1)^n + A_2(1/2)^n)_n$$

is the general solution of the recurrence. Imposing the initial data we get

$$A_1 + A_2 = -\frac{3}{2}, \quad -A_1 + \frac{1}{2}A_2 = \frac{1}{4}$$

and hence $A_1 = -2/3$, $A_2 = -5/6$. Therefore the coefficients of the formal series $A(X)$ are

$$[X^n]A(X) = -\frac{2}{3}(-1)^n - \frac{5}{6}\left(\frac{1}{2}\right)^n = \frac{2}{3}(-1)^{n+1} - \frac{5}{3}\left(\frac{1}{2}\right)^{n+1},$$

confirming what was found in Example 7.117.   □

Example 10.67 We want to determine the coefficients of $A(X) = \dfrac{X^2}{2 + 3X}$. By Proposition 10.58, the coefficients of $A(X) := a_0 + a_1 X + \cdots + a_n X^n + \cdots$ satisfy a recurrence relation of order 1 (i.e., the degree of the polynomial in the denominator $2 + 3X$):

$$2x_n + 3x_{n-1} = h_n, \quad n \geq 1,$$

with initial datum a_0. Since $S(X) = 2a_0$ is formed by terms of degree strictly less than 1 of the formal series appearing in the numerator of $A(X) = \dfrac{X^2}{2 + 3X}$, one has

$$2a_0 = [X^{<1}](X^2) = 0,$$

and hence $a_0 = 0$. The non homogeneous terms $h_n, n \geq 1$, of the recurrence relation are the coefficients of the terms of degree greater or equal than 1 of the formal series in the numerator of $A(X) = \dfrac{X^2}{2 + 3X}$: hence $h_1 = 0$, $h_2 = 1$ and $h_n = 0$ for each $n \geq 3$. Thus, the desired recurrence relation is

$$2x_n + 3x_{n-1} = h_n = \begin{cases} 0 \text{ if } n \neq 2, \\ 1 \text{ if } n = 2, \end{cases} \quad \forall n \geq 1,$$

with initial datum $x_0 = a_0 = 0$; therefore $x_1 = 0$, $x_2 = 1/2$ and

$$x_n = -\frac{3}{2}x_{n-1}, \quad n \geq 3.$$

The general solution of this recurrence is $x_n = A(-3/2)^n$; imposing the initial datum $x_2 = 1/2$, one obtains $A = 2/9$. Thus, the desired formal series is

$$A(X) = \frac{1}{2}X^2 - \frac{3}{4}X^3 + \cdots = \sum_{n=2}^{\infty} \frac{2}{9}\left(\frac{-3}{2}\right)^n X^n. \qquad \square$$

Example 10.68 (*Fibonacci convolution*) We have seen in Example 10.62 that the OGF of the Fibonacci sequence is the formal series

$$A(X) = \frac{X}{1 - X - X^2}.$$

The coefficients $(a_n)_n$ of $A(X)$ satisfy the homogeneous recurrence

$$x_n = x_{n-1} + x_{n-2}, \quad n \geq 2,$$

with initial data $x_0 = 0$, $x_1 = 1$. We now calculate the sequence $(b_n)_n$ obtained by the convolution product of the Fibonacci sequence with itself. This sequence occurs in the analysis of some important algorithms. By Proposition 7.21, the ordinary generating formal series $B(X)$ of $(b_n)_n$ is

$$B(X) = \frac{X^2}{(1 - X - X^2)^2}.$$

By Corollary 10.60, the sequence of coefficients $(b_n)_n$ satisfies a linear homogeneous recurrence relation. Since $(1 - X - X^2)^2 = 1 - 2X - X^2 + 2X^3 + X^4$, this recurrence is

$$x_n - 2x_{n-1} - x_{n-2} + 2x_{n-3} + x_{n-4} = 0, \quad n \geq 4.$$

Again by Corollary 10.60, we have

$$
\begin{aligned}
X^2 &= [X^{\leq 3}](1 - 2X - X^2 + 2X^3)(b_0 + b_1 X + b_2 X^2 + b_3 X^3) \\
&= b_0 + (-2b_0 + b_1)X + (-b_0 - 2b_1 + b_2)X^2 + (2b_0 - b_1 - 2b_2 + b_3)X^3,
\end{aligned}
$$

from which we get the initial data

$$x_0 = b_0 = 0, \ x_1 = b_1 = 0, \ x_2 = b_2 = 1, \ x_3 = b_3 = 2.$$

The characteristic polynomial of the recurrence is

$$X^4 - 2X^3 - X^2 + 2X + 1 = (X^2 - X - 1)^2;$$

its root are $\dfrac{1 \pm \sqrt{5}}{2}$ both with multiplicity 2. Therefore, the general solution of the recurrence is

$$(A_1 + A_2 n)\left(\frac{1 - \sqrt{5}}{2}\right)^n + (A_3 + A_4 n)\left(\frac{1 + \sqrt{5}}{2}\right)^n,$$

with the variation of A_1, A_2, A_3, A_4 in $\mathbb{R}$. Imposing the initial data we get the linear system

$$\begin{cases} A_1 + A_3 &= 0 \\ (A_1 + A_2)\left(\dfrac{1 - \sqrt{5}}{2}\right) + (A_4 - A_1)\left(\dfrac{1 + \sqrt{5}}{2}\right) &= 0 \\ (A_1 + 2A_2)\left(\dfrac{1 - \sqrt{5}}{2}\right)^2 + (2A_4 - A_1)\left(\dfrac{1 + \sqrt{5}}{2}\right)^2 &= 1 \\ (A_1 + 3A_2)\left(\dfrac{1 - \sqrt{5}}{2}\right)^3 + (3A_4 - A_1)\left(\dfrac{1 + \sqrt{5}}{2}\right)^3 &= 2. \end{cases}$$

Using a suitable CAS for solving linear systems we get $A_1 = \sqrt{5}/25$, $A_2 = 1/5$, $A_3 = -\sqrt{5}/25$ and $A_4 = 1/5$. Thus, the solution we are looking for is

$$[X^n]B(X) = b_n = \left(\frac{\sqrt{5}}{25} + \frac{1}{5}n\right)\left(\frac{1 - \sqrt{5}}{2}\right)^n + \left(-\frac{\sqrt{5}}{25} + \frac{1}{5}n\right)\left(\frac{1 + \sqrt{5}}{2}\right)^n. \qquad \square$$

10.4.3   The Quicksort OGF ☕

The method of generating formal series shown in Proposition 10.58 also applies (with a bit of luck) to other types of recurrence, such as those which are non-constant with linear coefficients. We shall just treat the example of the quicksort algorithm.

Example 10.69 We solve, as an example of using the generating formal series, the quicksort recurrence, which was introduced and solved by a stratagem in Example 10.45. The number a_n of steps needed to sort a list of $n \geq 1$ distinct names is the n-th term of the solution of the linear recurrence relation

$$n x_n = 2 \sum_{0 \leq i < n} x_i + n(n - 1), \quad n > 0,$$

with initial datum $x_0 = 0$. Let us consider the OGF of the sequence $(a_n)_n$:

$$A(X) = \sum_{n=0}^{\infty} a_n X^n.$$

The terms of the sequence $(a_n)_n$ satisfy by hypothesis

$$na_n = 2 \sum_{0 \le i < n} a_i + n(n-1), \quad n \ge 1.$$

Multiplying both sides by X^n and formally summing we get

$$\sum_{n=1}^{\infty} na_n X^n = 2 \sum_{n=1}^{\infty} \left(\sum_{0 \le i < n} a_i \right) X^n + \sum_{n=1}^{\infty} n(n-1) X^n.$$

Let's now evaluate each term that appears in the above formula.

Since $A'(X) = \sum_{n=0}^{\infty} na_n X^{n-1}$, the left side of the equality is equal to $XA'(X)$. Next, by Proposition 7.88 we have

$$\sum_{n=1}^{\infty} \left(\sum_{0 \le i < n} a_i \right) X^n = X \frac{1}{1-X} A(X).$$

Finally, thanks to the formula (7.90.b), we obtain

$$\sum_{n=1}^{\infty} n(n-1) X^n = 2 \sum_{n=2}^{\infty} \binom{n}{2} X^n = 2 \frac{X^2}{(1-X)^3}.$$

Therefore

$$A'(X) = 2 \frac{1}{1-X} A(X) + \frac{2X}{(1-X)^3}.$$

Let us solve the Cauchy problem

$$\begin{cases} y'(x) = 2 \dfrac{1}{1-x} y(x) + \dfrac{2x}{(1-x)^3}, \\ y(0) = 0, \end{cases}$$

(the choice of $y(0) = 0$ is due to the fact that $a_0 = 0$). Multiplying both sides of the differential equation by $(x-1)^2 = (1-x)^2$ we get

$$(x-1)^2 y'(x) + 2(x-1)y(x) = \frac{2x}{1-x},$$

or equivalently

$$((x-1)^2 y(x))' = \frac{2x}{1-x}.$$

Integrating, taking in account that $y(0) = 0$, we get that for $x < 1$

$$y(x) = -2\frac{x}{(x-1)^2} - \frac{2}{(1-x)^2}\log(1-x).$$

By Proposition 7.119 we have

$$A(X) = -2\frac{X}{(X-1)^2} - \frac{2}{(1-X)^2}\log(1-X).$$

Now, thanks to the formula (7.90.b), for each $n \in \mathbb{N}$ we obtain

$$[X^n]\frac{X}{(X-1)^2} = n.$$

By Propositions 7.88, 8.32, and Example 6.50, we get

$$[X^n]\frac{\log(1-X)}{(1-X)^2} = [X^n]\frac{1}{1-X}\frac{\log(1-X)}{1-X} = -\sum_{k=0}^{n}H_k = n+1-(n+1)H_{n+1}$$

and therefore

$$a_n = [X^n]A(X) = -2n - 2(n+1-(n+1)H_{n+1})$$

$$= 2(n+1)H_{n+1} - 4n - 2 = 2(n+1)H_n - 4n,$$

as we have found in Example 10.45.　　　　　　　　　　　　　　　□

10.5　Proofs ☕

In this section we give, for completeness' sake, the proofs of Theorem 10.12, of Proposition 10.26 and of Proposition 10.41; all these results have been widely used in the examples of the previous sections. The reading will certainly allow for a better understanding of the theory; at the same time its omission will not in any way affect the study of the book.

10.5.1　Proof of Theorem 10.12

In this section we use the difference and shift operators Δ and θ introduced in Chap. 6. We show first that the difference operator Δ somewhat preserves the sequences of the forms $(P(n)\lambda^n)_n$, when $P(X)$ is a polynomial.

Lemma 10.70 *Let $P(X) \neq 0$ be a polynomial and $\lambda \in \mathbb{C} \setminus \{0\}$. Then there exists a polynomial $Q(X)$ such that $\Delta(P(n)\lambda^n) = Q(n)\lambda^n$, where:*

- If $\lambda \neq 1$ then $\deg Q(X) = \deg P(X)$ and $Q(X) \neq 0$;
- If $\lambda = 1$ then $\deg Q(X) = \deg P(X) - 1 \geq 0$ if $\deg P(X) \geq 1$, $Q(X) = 0$ otherwise.

Proof. By Proposition 6.13 one has

$$\Delta(P(n)\lambda^n) = \lambda^n \Delta P(n) + \theta P(n)\Delta\lambda^n$$
$$= (\Delta P(n) + (\lambda - 1)P(n+1))\lambda^n.$$

If $P(X) = c$ is constant then

$$\Delta(P(n)\lambda^n) = \Delta(c\lambda^n) = c(\lambda - 1)\lambda^n = 0 \Leftrightarrow \lambda = 1.$$

We thus henceforth assume that $\deg P(X) \geq 1$. We have $\Delta P(n) = \widetilde{P}(n)$ where $\widetilde{P}(X) = P(X+1) - P(X)$ and $\theta P(n) = P(n+1)$. It it easy to see that

$$\deg(P(X+1) - P(X)) = \deg P(X) - 1, \quad \deg P(X+1) = \deg P(X).$$

Now

$$\deg(\widetilde{P}(X) + (\lambda - 1)P(X+1)) = \begin{cases} \deg P(X+1) = \deg P(X) & \text{if } \lambda \neq 1; \\ \deg \widetilde{P}(X) = \deg P(X) - 1 & \text{if } \lambda = 1. \end{cases} \quad \square$$

The following result, showing the linear independence of the basis-solutions to (R_o), has an interest in itself.

Proposition 10.71 (Linear independence of the basis-solutions) *For every $m \in \mathbb{N}_{\geq 1}$, and non-zero distinct complex numbers $\lambda_1, \ldots, \lambda_m$, the sequences*

$$(n^{s_1}\lambda_1^n)_n, \ldots, (n^{s_m}\lambda_m^n)_n, \quad s_1, \ldots, s_m \in \mathbb{N}$$

are linearly independent, i.e., whenever

$$P_1(n)\lambda_1^n + \cdots + P_m(n)\lambda_m^n = 0 \quad \forall n \in \mathbb{N},$$

for some polynomials $P_1(X), \ldots, P_m(X)$, then necessarily $P_1(X) = \cdots = P_m(X) = 0$.

Proof. We use the technique used in [9] for the independence of the quasi polynomials functions $t^s e^{\lambda t}$. Assume by contradiction that the claim is not true: let $m \in \mathbb{N}_{\geq 1}$ the minimal integer for which a counterexample exists. Then we can find non-zero complex distinct numbers $\lambda_1, \ldots, \lambda_m$ and, by the minimality of m, non-zero polynomials $P_1(X), \ldots, P_m(X)$ satisfying

$$P_1(n)\lambda_1^n + \cdots + P_m(n)\lambda_m^n = 0 \quad \forall n \in \mathbb{N}. \tag{10.71.a}$$

Certainly $m > 1$, because the equality $(P_1(n)\lambda_1^n)_n = (0)_n$ implies $P_1(n) = 0$ for each $n \in \mathbb{N}$ and hence $P_1(X) = 0$. Set $\eta_i = \lambda_i/\lambda_m$ for $i = 1, \dots, m$, the formula (10.71.a) is equivalent to

$$-P_m(n) = P_1(n)\eta_1^n + \cdots + P_{m-1}(n)\eta_{m-1}^n \quad \forall n \in \mathbb{N}.$$

Let d be the degree of $P_m(X)$. By applying the operator Δ^{d+1} to both members of the latter equality we get

$$\Delta^{d+1}(-P_m(n)) = \Delta^{d+1}(P_1(n)\eta_1^n) + \cdots + \Delta^{d+1}(P_{m-1}(n)\eta_{m-1}^n) \quad \forall n \in \mathbb{N}.$$

Since $\eta_i \neq 1$ for $i = 1, \dots, m - 1$, Lemma 10.70 shows that $\Delta^d(-P_m(n))$ is a constant so that $\Delta^{d+1}(-P_m(n)) = 0$ and there are non-zero polynomials $Q_1(X), \dots, Q_{m-1}(X)$ satisfying

$$\Delta^{d+1}(P_i(n)\eta_i^n) = Q_i(n)\eta_i^n, \quad n \in \mathbb{N}, \ i = 1, \dots, m - 1,$$

whence

$$Q_1(n)\eta_1^n + \cdots + Q_{m-1}(n)\eta_{m-1}^n = 0 \quad \forall n \in \mathbb{N},$$

where $Q_i(X)$ are non-zero polynomials, thus violating the minimality of m, a contradiction. $\square$

Proof (of Theorem 10.12). 1. We first show that the elements of the basis-solutions are indeed solutions of the recurrence (R_o). Let $f : \mathbb{N} \to \mathbb{N}$. Notice that $(f(n))_n$ is a solution to (R_o) if and only if

$$(c_0\theta^r + \cdots + c_r)f(n) = 0 \quad \forall n \in \mathbb{N}.$$

Moreover,

$$c_0\theta^r + \cdots + c_r = c_0(\theta - \lambda_1)^{\mu_1} \cdots (\theta - \lambda_m)^{\mu_m}. \tag{10.71.b}$$

Let $P(X)$ be a polynomial. For $\lambda \in \mathbb{C}$ we have

$$(\theta - \lambda)(P(n)\lambda^n) = P(n + 1)\lambda^{n+1} - P(n)\lambda^{n+1} = \lambda^{n+1}\Delta P(n),$$

and $\Delta P(n) = Q(n)$ where, from Lemma 10.70, $Q(X)$ is a polynomial of degree $\deg P(X) - 1$ if $\deg P(X) \geq 1$, or $Q(X) = 0$ if $P(X)$ is a constant. Therefore, if $\mu \in \mathbb{N}_{\geq 1}$, inductively we have

$$(\theta - \lambda)^\mu(P(n)\lambda^n) = \lambda^{n+\mu}\Delta^\mu P(n),$$

where $\Delta^\mu P(n)$ is a polynomial in n of degree $\deg P(X) - \mu$ if $\deg P(X) \geq \mu$ or $\Delta^\mu P(n) = 0$ if $\deg P(X) < \mu$. In particular, for $0 \leq s < \mu$ in $\mathbb{N}$ we have $\deg X^s = s < \mu$ so that

$$(\theta - \lambda)^\mu (n^s \lambda^n) = \lambda^{n+\mu} \Delta^\mu (n^s) = 0.$$

It follows from (10.71.b) that $(n^s \lambda^n)_n$ is a solution to (R_o) whenever λ is a root of the characteristic polynomial of (R_o) of multiplicity μ, and $s < \mu$. The conclusion follows since the space of solutions to (R_o) has dimension r (Proposition 10.4) and, by Proposition 10.71, the r solutions

$$(\lambda_j^n)_n, \ldots, (n^{\mu_j - 1} \lambda_j^n)_n \qquad j = 1, \ldots, m,$$

are independent.

2. By Point 1, the sequences

$$n^s \rho_j^n (\cos \alpha_j \pm i \sin \alpha_j)^n \qquad 0 \leq s < \mu_j \qquad j = 1, \ldots, h,$$

$$n^s \lambda_j^n, \qquad 0 \leq s < \mu_j' \qquad j = 1, \ldots, \ell,$$

generate the set of complex solutions of (R_o). Now we have

$$(\cos \alpha_j \pm i \sin \alpha_j)^n = \cos(n\alpha_j) \pm i \sin(n\alpha_j).$$

Therefore every basis-solution of the type $n^k \rho_j^n (\cos \alpha_j \pm i \sin \alpha_j)^n$ is the sum of $n^k \rho_j^n \cos(n\alpha_j)$ and of $\pm i\, n^k \rho_j^n \sin(n\alpha_j)$; at the same time

$$n^k \rho_j^n \cos(n\alpha_j) = \frac{1}{2} n^k \rho_j^n (\cos(n\alpha_j) + i \sin(n\alpha_j)) + \frac{1}{2} n^k \rho_j^n (\cos(n\alpha_j) - i \sin(n\alpha_j)) \text{ and}$$

$$n^k \rho_j^n \sin(n\alpha_j) = \frac{-i}{2} n^k \rho_j^n (\cos(n\alpha_j) + i \sin(n\alpha_j)) + \frac{i}{2} n^k \rho_j^n (\cos(n\alpha_j) - i \sin(n\alpha_j))$$

are linear combinations of the basis-solutions $n^k \rho_j^n (\cos \alpha_j \pm i \sin \alpha_j)^n$. It follows that the sequences

$$(n^s \rho_j^n \cos(n\alpha_j))_n, \quad 0 \leq s < \mu_j, \quad 1 \leq j \leq h,$$

$$(n^s \rho_j^n \sin(n\alpha_j))_n, \quad 0 \leq s < \mu_j, \quad 1 \leq j \leq h,$$

$$(n^s \lambda_j^n)_n, \quad 0 \leq s < \mu_j', \quad 1 \leq j \leq \ell,$$

are r generators of the space of the complex solutions of (R_o). Since these are all real sequences, each real solution is obtained via their combination with real coefficients. $\qquad\qquad\square$

10.5.2 Proof of Proposition 10.26

The idea of the proof is to show that any solution of (R) is actually a solution to a suitable homogeneous linear recurrence with constant coefficients. We will need the following preliminary result.

Lemma 10.72 (Composition of linear recurrences) *Let $(b_n)_n$ be a solution of the linear homogeneous recurrence*

$$\sum_{k=0}^{r_1} c_{1,k} x_{n-k} = 0, \quad n \geq r_1 \quad (c_{1,0} c_{1,r_1} \neq 0), \tag{10.72.a}$$

and let $(a_n)_n$ be a solution to the linear recurrence

$$\sum_{\ell=0}^{r_2} c_{2,\ell} x_{n-\ell} = b_n, \quad n \geq r_2 \quad (c_{2,0} c_{2,r_2} \neq 0). \tag{10.72.b}$$

Then $(a_n)_n$ solves the linear homogeneous recurrence whose characteristic polynomial is the product of the characteristic polynomials of the recurrences (10.72.a) and (10.72.b).

Proof. Since

$$\sum_{k=0}^{r_1} c_{1,k} b_{n-k} = 0, \quad \forall n \geq r_1$$

and, for each $k \in \{0, \ldots, r_1\}$,

$$b_{n-k} = \sum_{\ell=0}^{r_2} c_{2,\ell} a_{n-k-\ell}, \quad \forall n \geq k + r_2,$$

then

$$\sum_{k=0}^{r_1} c_{1,k} \left(\sum_{\ell=0}^{r_2} c_{2,\ell} a_{n-k-\ell} \right) = 0, \quad \forall n \geq r_1 + r_2,$$

or equivalently,

$$\sum_{m=0}^{r_1+r_2} \left(\sum_{k+\ell=m} c_{1,k} c_{2,\ell} \right) a_{n-m} = 0, \quad \forall n \geq r_1 + r_2.$$

Therefore $(a_n)_n$ solves the linear homogeneous recurrence

$$\sum_{m=0}^{r_1+r_2} c_m x_{n-m} = 0, \quad n \geq r_1 + r_2,$$

where $c_m := \sum_{k+\ell=m} c_{1,k} c_{2,\ell}$. The characteristic polynomial of the latter recurrence is

$$\sum_{m=0}^{r_1+r_2} c_m X^{r_1+r_2-m} = \sum_{m=0}^{r_1+r_2} \left(\sum_{k+\ell=m} c_{1,k} c_{2,\ell} X^{r_1+r_2-k-\ell} \right)$$

$$= \left(\sum_{k=0}^{r_1} c_{1,k} X^{r_1-k} \right) \left(\sum_{\ell=0}^{r_2} c_{2,\ell} X^{r_2-\ell} \right),$$

proving the claim. ☐

Proof (of Proposition 10.26). Let $d := \deg Q(X)$; it follows from Corollary 10.22 that $(Q(n)q^n)_n$ is a solution to the homogenous recurrence whose characteristic polynomial is $(X-q)^{d+1}$. Let $(a_n)_n$ be a solution of (R): by Lemma 10.72 $(a_n)_n$ is a solution to the linear homogeneous recurrence whose characteristic polynomial is $P(X)(X-q)^{d+1}$, where $P(X)$ is the characteristic polynomial of (R). Two cases may occur, in each of them we invoke Theorem 10.12 and Remark 10.13.

- If q is not a root of $P(X)$, there are polynomials $P_1(X), \ldots, P_m(X)$ and $\widetilde{Q}(X)$ with $\deg P_i(X) < \mu_i$ for $i = 1, \ldots, m$, and $\deg \widetilde{Q}(X) \leq d$ such that

$$a_n = P_1(n)\lambda_1^n + \cdots + P_m(n)\lambda_m^n + \widetilde{Q}(n)q^n \quad \forall n \in \mathbb{N}.$$

Since $P_1(n)\lambda_1^n + \cdots + P_m(n)\lambda_m^n$ is a solution to the associated homogeneous relation (R_o), the Superposition Principle 10.3 implies that $(\widetilde{Q}(n)q^n)_n$ is a solution of (R).

- If q is a root of $P(X)$, say $q = \lambda_m$, then q is a root of multiplicity $\mu_m + d + 1$ of $P(X)(X-q)^{d+1}$: there are polynomials $P_1(X), \ldots, P_{m-1}(X)$ and $Q_1(X)$ with $\deg P_i(X) < \mu_i$ $(i = 1, \ldots, m-1)$, $\deg Q_1(X) \leq \mu_m + d$ such that

$$a_n = P_1(n)\lambda_1^n + \cdots + P_{m-1}(n)\lambda_{m-1}^n + Q_1(n)q^n \quad \forall n \in \mathbb{N}.$$

We may write

$$Q_1(X) = P_m(X) + X^{\mu_m} \widetilde{Q}(X),$$

with $\deg P_m(X) < \mu_m$, $\deg \widetilde{Q}(X) \leq d$, so that

$$a_n = P_1(n)\lambda_1^n + \cdots + P_m(n)\lambda_m^n + n^{\mu_m} \widetilde{Q}(n)q^n \quad \forall n \in \mathbb{N}.$$

Since $P_1(n)\lambda_1^n + \cdots + P_m(n)\lambda_m^n$ is a solution to (R_o), the Superposition Principle 10.3 implies that $(n^{\mu_m} \widetilde{Q}(n)q^n)_n$ is a solution of (R). ☐

10.5.3   Proof of Proposition 10.41

Proof (of Proposition 10.41). First we observe that since $w_0 = \cdots = w_{r-2} = 0$ and $w_{r-1} = 1$, we have that for each $n \geq 1$

$$\frac{1}{c_0}\sum_{k=1}^{n} w_{n-k}h_{k+r-1} = \frac{1}{c_0}\left(\sum_{k=1}^{n-r} w_{n-k}h_{k+r-1} + w_{r-1}h_n + w_{r-2}h_{n+1} + \cdots + w_0 h_{n+r-1}\right)$$

$$= \frac{1}{c_0}\left(\sum_{k=1}^{n-r} w_{n-k}h_{k+r-1} + h_n\right).$$

Let $W(X)$ and $H(X)$ be the OGF of the sequences $(w_n)_n$ and $(h_{n+r-1})_{n\geq 1}$. By Proposition 10.58, we have

$$W(X) = \frac{c_0 X^{r-1}}{c_0 + c_1 X + \cdots + c_r X^r}.$$

Therefore

$$\frac{1}{c_0}W(X)H(X) = \frac{X^{r-1}H(X)}{c_0 + c_1 X + \cdots + c_r X^r}.$$

Since

$$X^{r-1}H(X) = X^{r-1}\sum_{k=1}^{\infty} h_{k+r-1}X^k = \sum_{k=1}^{\infty} h_{k+r-1}X^{k+r-1} = \sum_{n=r}^{\infty} h_n X^n,$$

then it turns out that

$$\frac{1}{c_0}W(X)H(X) = \frac{\displaystyle\sum_{n=r}^{\infty} h_n X^n}{c_0 + c_1 X + \cdots + c_r X^r}.$$

Again by Proposition 10.58, one sees that $\frac{1}{c_0}W(X)H(X)$ is the OGF of a particular solution of the recurrence

$$c_0 x_n + c_1 x_{n-1} + \cdots + c_r x_{n-r} = h_n, \quad n \geq r.$$

The conclusion follows from the fact that the sequence of the coefficients of $W(X)H(X)$ is the convolution product of the sequences of the coefficients of $W(X)$ and $H(X)$.   □

10.6 Problems

Homogeneous recurrences

Problem 10.1 Solve the following recurrence relations:

(a) $x_n = 3x_{n-1} + 4x_{n-2}$, $n \geq 2$, with initial data $x_0 = x_1 = 1$;
(b) $x_n = x_{n-2}$, $n \geq 2$, with initial data $x_0 = x_1 = 1$;
(c) $x_n = 2x_{n-1} - x_{n-2}$, $n \geq 2$, with initial data $x_0 = x_1 = 2$;
(d) $x_n = 3x_{n-1} - 3x_{n-2} + x_{n-3}$, $n \geq 3$, with initial data $x_0 = x_1 = 1$, $x_2 = 2$.

Problem 10.2 Find and solve a recurrence relation to compute the number of possible ways of filling a row of n places in a parking using blue cars, red cars and trucks, taking into account that the trucks take up two spaces, whilst the cars will occupy one.

Problem 10.3 A multinational pharmaceutical company decides to double the increase in the price of its flagship product every year. Find and solve the recurrence relation for the price p_n of the product in the year n, supposing that $p_0 = 1$, $p_1 = 4$.

Problem 10.4 Assume the recurrence relation $x_n = c_1 x_{n-1} + c_2 x_{n-2}$ $(n \geq 2)$ has general solution $x_n = A\,13^n + B\,26^n$ with $A, B \in \mathbb{R}$: determine the constants c_1, c_2.

Problem 10.5 Find the real and complex solutions to the recurrence

$$x_{n+2} - 6x_{n+1} + 9x_n = 0, \quad n \geq 0.$$

Problem 10.6 Compute the general real solution of the recurrence relation

$$x_{n+2} + 4x_{n+1} + 16x_n = 0, \quad n \geq 0.$$

Problem 10.7 Compute the general real solution of the recurrence relation

$$x_n = x_{n-1} + 8x_{n-2} - 12x_{n-3}, \quad n \geq 3.$$

Problem 10.8 Determine the real solutions of the recurrence

$$x_n - 9x_{n-2} = 0, \quad n \geq 2,$$

with initial data:

1. $x_0 = 6$, $x_1 = 12$;
2. $x_3 = 324$, $x_4 = 486$;
3. $x_0 = 6$, $x_2 = 54$;
4. $x_0 = 6$, $x_2 = 10$.

Problem 10.9 Solve the recurrence relation

$$x_n = 3x_{n-1} + 4x_{n-2} - 12x_{n-3}, \quad n \geq 3,$$

with initial data $x_0 = 2$, $x_1 = 5$, $x_2 = 13$.

Problem 10.10 Solve the homogeneous linear recurrence of order 3

$$x_{n+3} = 6x_{n+2} - 12x_{n+1} + 8x_n, \quad n \geq 0,$$

with initial data $x_0 = 1$, $x_1 = 0$, $x_2 = 4$.

Particular solutions

Problem 10.11 Determine a particular solution of the recurrence $x_n = cx_{n-1} + h_n$ with $c \in \mathbb{R}$ and:

(a) $h_n = 1$;
(b) $h_n = n$;
(c) $h_n = n^2$;
(d) $h_n = q^n$ $(q \in \mathbb{R})$.

Problem 10.12 Assume the characteristic polynomial of a given recurrence relation is $(X - 1)^2(X - 2)(X - 3)^2$; determine the general solution of the induced homogeneous recurrence. Determine the type of a particular solution of the given recurrence if its non homogeneous part h_n is defined by one of the following:

(a) $h_n = 4n^3 + 5n$;
(b) $h_n = 4^n$;
(c) $h_n = 3^n$.

General solutions

Problem 10.13 Let $(h_n)_n$ be a sequence and c a constant. Determine the solution of $x_n = cx_{n-1} + h_n$ $(n \geq 1)$ with initial datum $x_0 = 1$.

Problem 10.14 Solve the following recurrence relations:

(a) $x_n = x_{n-1} + 3(n - 1)$, $x_0 = 1$;
(b) $x_n = x_{n-1} + n(n - 1)$, $x_0 = 3$;
(c) $x_n = x_{n-1} + 3n^2$, $x_0 = 10$.

Problem 10.15 Determine $(x_n)_{n \in \mathbb{N}}$ knowing that $x_0 = 3$ and $\dfrac{x_n + x_{n-1}}{2} = 2n + 5$ for each $n \geq 1$.

Problem 10.16 Find the general solution of the recurrence

$$x_n = 5x_{n-1} - 6x_{n-2} + 6(4^n).$$

Problem 10.17 Find the general solution of the recurrence

$$x_n = 2x_{n-1} + 2^n + n.$$

Problem 10.18 Solve the following recurrence relations:

(a) $x_n = 3x_{n-1} - 2$, $x_0 = 0$;
(b) $x_n = 2x_{n-1} + (-1)^n$, $x_0 = 2$;
(c) $x_n = 2x_{n-1} + n$, $x_0 = 1$;
(d) $x_n = 2x_{n-1} + 2n^2$, $x_0 = 3$.

Problem 10.19 Solve the recurrence relation $x_n = 3x_{n-1} - 2x_{n-2} + 3$, with initial data $x_0 = x_1 = 1$.

Problem 10.20 Find and solve a recurrence relation for the profit of a company if the growth rate of the profit in the n-th year is 10×2^n euros more than the growth rate of the previous year and in the first year the profit is of 20 euros, while in the second year the profit is 1 020 euros.

Problem 10.21 Find the general solution of the recurrence relation

$$x_n - 5x_{n-1} + 6x_{n-2} = 2 + 3n.$$

Problem 10.22 Solve the following recurrence relation with initial datum $y_0 = 1$:

$$y_n^2 = 2y_{n-1}^2 + 1 \quad n \geq 1.$$

(Hint: set $x_n = y_n^2$).

Problem 10.23 Determine the general solution of the recurrence

$$x_n = 4x_{n-1} - 4x_{n-2} + 2^n, \quad n \geq 2.$$

Problem 10.24 Solve the recurrence $x_n = x_{n-1} + 12n^2$, with initial datum $x_0 = 5$.

Problem 10.25 Determine the general solution of the recurrence

$$x_n = 3x_{n-1} - 4n + 3 \times 2^n \quad n \geq 1.$$

Find the solution with initial datum $x_1 = 8$.

Problem 10.26 Find the real solution of the recurrence

$$x_n = -x_{n-2} + 1, \quad x_0 = 0, \ x_1 = 0.$$

Problem 10.27 Determine the solution of the following recurrences:

1. $x_{n+1} = 5x_n + 2^n \cos\left(n\frac{\pi}{3}\right),$ $x_0 = 0;$
2. $y_{n+1} = 5y_n + 2^n \sin\left(n\frac{\pi}{3}\right),$ $y_0 = 0.$

Problem 10.28 Determine the general solution of the recurrence:

$$x_{n+1} = x_n + (1+n)2^n \sin\left(n\frac{\pi}{3}\right), \quad x_0 = 1.$$

Problem 10.29 Determine the solution of the recurrence

$$x_{n+2} - 4\sqrt{3}x_{n+1} + 16x_n = 32 \times 4^n \cos\left(n\frac{\pi}{6}\right),$$

with initial data $x_0 = 0$, $x_1 = 0$.

Divide and conquer

Problem 10.30 Determine the solution of the following "divide and conquer" recurrences:

(a) $y_n = 2y_{n/2} + 5$, $n = 2^k$, $k \geq 1$ with initial datum $y_1 = 1$;
(b) $y_n = 2y_{n/4} + n$, $n = 4^k$, $k \geq 1$ with initial datum $y_1 = 3$;
(c) $y_n = 2y_{n/2} + 2n$, $n = 2^k$, $k \geq 1$ with initial datum $y_1 = 5$;
(d) $y_n = y_{n/3} + 4$, $n = 3^k$, $k \geq 1$ with initial datum $y_1 = 7$.

Problem 10.31 Determine the solution of the following "divide and conquer" recurrences:

(a) $y_n = y_{n/3} + 2$, $n = 4 \times 3^k$, $k \geq 1$ with initial datum $y_4 = 5$;
(b) $y_n = 2y_{n/3} + 2$, $n = 3^k$, $k \geq 1$ with initial datum $y_1 = 1$;
(c) $y_n = y_{n/3} + 2n$, $n = 3^k$, $k \geq 1$ with initial datum $y_1 = 5$;
(d) $y_n = 2y_{n/3} + 2n$, $n = 2 \times 3^k$, $k \geq 1$ with initial datum $y_2 = -1$.

Problem 10.32 Describe an approach of type "divide and conquer" to determine the maximum among the elements of a set of n numbers. Write a recurrence relation for the number of necessary comparisons and solve it.

Problem 10.33 Describe an approach of type "divide and conquer" to determine the first and second largest among the elements of a set of n numbers. Write a recurrence relation for the number of necessary comparisons and solve it.

Problem 10.34 Determine an estimate for the order of magnitude (i.e., $y_n = O(\cdots)$) of the solution $(y_n)_n$ of the "divide and conquer" recurrence

$$y_n = 3y_{n/2} + 4n^2 \quad n = 2^k, \quad k \geq 1,$$

with initial datum $y_1 = 5$ in two different ways: (1) Use Theorem 10.53 (2) Solve the recurrence.

Problem 10.35 Determine an estimate for the order of magnitude (i.e., $y_n = O(\cdots)$) of the solution $(y_n)_n$ of the "divide and conquer" recurrence

$$y_n = 2y_{n/2} + 2, \quad n = 2^k, \quad k \geq 1,$$

with initial datum $y_1 = 2$ in two different ways: (1) Use Theorem 10.53 (2) Solve the recurrence.

Problem 10.36 Determine an estimate for the order of magnitude (i.e., $y_n = O(\cdots)$) of the solution $(y_n)_n$ of the "divide and conquer" recurrence

$$y_n = 2y_{n/2} + 2\log_2 n, \quad n = 2^k, \ k \geq 1,$$

with initial datum $y_1 = 1$ in two different ways: (1) Use Theorem 10.53 (2) Solve the recurrence.

Problem 10.37 Determine an estimate for the order of magnitude (i.e., $y_n = O(\cdots)$) of the solution $(y_n)_n$ of the "divide and conquer" recurrence

$$y_n = \frac{1}{2}y_{n/4} + 3\log_4 n, \quad n = 4^k, \ k \geq 1,$$

with initial datum $y_1 = 0$ in two different ways: (1) Use Theorem 10.53 (2) Solve the recurrence.

Problem 10.38 Determine an estimate for the order of magnitude (i.e., $y_n = O(\cdots)$) of the solution $(y_n)_n$ of the "divide and conquer" recurrence

$$y_n = y_{n/2} + 2\log_2 n, \quad n = 2^k, \ k \geq 1,$$

with initial datum $y_1 = 1$ in two different ways: (1) Use Theorem 10.53 (2) Solve the recurrence.

Recurrences and generating formal series

Problem 10.39 Solve the following recurrence relation using the generating formal series:

$$x_n = 2x_{n-1} + 1, \quad n \geq 1,$$

with initial datum $x_0 = 1$.

Problem 10.40 Solve the following recurrence relation using the generating formal series:

$$x_n = 2x_{n-1} - x_{n-2} + 2x_{n-3}, \quad n \geq 3,$$

with initial data $x_0 = 1, x_1 = 0, x_2 = -1$.

Problem 10.41 Solve the following recurrence relations using the generating formal series:

(a) $x_n = -x_{n-1} + 6x_{n-2} \quad n > 1 \quad x_0 = 0, x_1 = 1$;
(b) $x_n = 3x_{n-1} - 4x_{n-2} \quad n > 1 \quad x_0 = 0, x_1 = 1$.

Problem 10.42 Solve the following recurrence relation using the generating formal series:

$$(n + 1)x_{n+1} = (n + 100)x_n, \quad n > 0, \quad x_0 = 1.$$

Problem 10.43 Solve the following recurrence relation using the generating formal series:

$$x_n = 11x_{n-2} - 6x_{n-3}, \quad n > 2, \quad x_0 = 0, x_1 = x_2 = 1.$$

Problem 10.44 Solve the following recurrence relation using the generating formal series:

$$x_n = 3x_{n-1} - 3x_{n-2} + x_{n-3}, \quad n > 2, \quad x_0 = x_1 = 0, x_2 = 1.$$

Problem 10.45 Solve the following recurrence relations using the generating formal series:

(a) $x_n = 5x_{n-1} - 8x_{n-2} + 4x_{n-3}, \quad n > 2, \quad x_0 = 1, x_1 = 2, x_2 = 4$;
(b) $x_n = 2x_{n-2} - x_{n-4}, \quad n > 4, \quad x_0 = x_1 = 0, x_2 = x_3 = 1$.

Problem 10.46 Let $m \in \mathbb{N}_{\geq 1}$. Solve the following recurrence relation using the generating formal series:

$$\sum_{k=0}^{m} \binom{m}{k} x_{n-k} = 0, \quad n \geq m, \quad x_0 = \cdots = x_{m-2} = 0, x_{m-1} = 1.$$

Problem 10.47 Solve the following recurrence relation using the generating formal series:

$$x_{n+6} - 21x_{n+5} + 175x_{n+4} - 735x_{n+3} + 1624x_{n+2} - 1764x_{n+1} + 720x_n = 0, \quad n \geq 0,$$

with initial data

$$x_0 = 0, \ x_1 = 1, \ x_2 = 0, \ x_3 = 1, \ x_4 = 0, \ x_5 = 1.$$

Problem 10.48 Let $t \in \mathbb{R}, t \neq 2$. Solve the following recurrence relations using the generating formal series:

$$x_n = n + 1 + \frac{t}{n} \sum_{k=1}^{n} x_{k-1}, \qquad n \geq 1, \quad x_0 = 0,$$

with $t = 2 - \varepsilon$ and then with $t = 2 + \varepsilon$, for a sufficiently small positive constant ε.

Chapter 11
Symbolic Calculus

Abstract The notion of *combinatorial class* provides a deep method in order to solve a huge class of combinatorial problems: indeed we shall see a very efficient technique for calculating, by way of generating formal series, the cardinality of various sets. In particular we will concentrate on the number of sequences containing a given *pattern*, the number of *triangulations* of a convex polygon, and the number of *rooted plane trees*.

11.1 Combinatorial Classes

The objects of the symbolic calculus are sets where the elements are divided into finite subsets, according to their *size*.

Definition 11.1 A **combinatorial class** is a couple $(\mathcal{U}, \nu)$ formed by a set $\mathcal{U}$ and a function $\nu : \mathcal{U} \to \mathbb{N}$, called **valuation** or **size**, such that

$$\nu^{-1}(n) := \{x \in \mathcal{U} : \nu(u) = n\} \text{ is finite for every } n \geq 0.$$

The finite subset $\nu^{-1}(n)$ is called the n-**fibre** of $\mathcal{U}$ and its elements are said to be of size n. □

Remark 11.2 From the definition it follows immediately that if $(\mathcal{U}, \nu)$ is a combinatorial class then $\mathcal{U}$ is finite or countable: indeed, $\mathcal{U} = \bigcup_{n \in \mathbb{N}} \nu^{-1}(n)$ and the fibres $\nu^{-1}(n)$ are finite. Furthermore, one notes that any finite or countable set always admits a valuation (it suffices, in fact, to list the elements and assign them their position as value); in general there can be various valuations on the same set. The choice of one valuation over another depends on the combinatorial problem under consideration.

Example 11.3 1. Given $n \in \mathbb{N}_{\geq 1}$, consider the set of sequences of I_n. The function ν that associates to every sequence its length (the empty sequence has length 0) is a valuation that makes the set of sequences of I_n into a combinatorial class. The function ν' that associates to every sequence the number of 1's appearing in it does not make the set of sequences of I_n into a combinatorial class: indeed, for all $m \in \mathbb{N}$ there are infinite sequences of I_n with m entries equal to 1 and hence the fibres are not finite.

© Springer International Publishing Switzerland 2016

C. Mariconda and A. Tonolo, *Discrete Calculus*,

UNITEXT - La Matematica per il 3+2 103, DOI 10.1007/978-3-319-03038-8_11

2. A trivial valuation on $\mathbb{N}$ that makes the set of natural numbers into a combinatorial class is the identity function; a different valuation which makes $\mathbb{N}$ into a combinatorial class may be obtained, for example, by associating to every n the number $\nu(n)$ of its digits.

□

☞   Once the valuation on a combinatorial class $(\mathscr{U}, \nu)$ is specified, we often indicate both the set as well as the combinatorial class with the same symbol $\mathscr{U}$.

Definition 11.4 (*The class* $\mathbb{1}$) We speak of a **neutral combinatorial class** $\mathbb{1}$ as being a class consisting of a set with a single element of size 0.   □

To every combinatorial class we can associate a sequence of natural numbers:

Definition 11.5 We define the **sequence of the fibres** of a combinatorial class $(\mathscr{U}, \nu)$ to be the sequence $(u_n)_n$ defined by setting

$$u_n := |\nu^{-1}(n)| \qquad \forall n \in \mathbb{N}. \qquad\qquad □$$

The sequence of the fibres describes, size by size, the *width* of the combinatorial class.

Example 11.6 Given $n \in \mathbb{N}$, we consider the combinatorial class formed by the set of sequences of I_n and by the valuation function that associates to every sequence its length. For every $k \in \mathbb{N}$, the number of elements of size k is $S((n, k)) = n^k$; therefore $(n^k)_k$ is the sequence of fibres associated to this combinatorial class.   □

Definition 11.7 Let $(\mathscr{U}, \nu)$ be a combinatorial class, and let $\mathscr{V}$ be a subset of $\mathscr{U}$. The restriction $\nu_{|\mathscr{V}}$ of ν to $\mathscr{V}$ is called the **induced valuation** and yields $(\mathscr{V}, \nu_{|\mathscr{V}})$ a combinatorial class, called the **combinatorial subclass** of $\mathscr{U}$.   □

In what follows, except where explicitly stated, the subsets of a combinatorial class will be considered as combinatorial subclasses.

11.2 Operations Between Combinatorial Classes

We introduce some operations between combinatorial classes.

Definition 11.8 Let $(\mathscr{U}_i, \nu_i)$, $1 \le i \le n$, $n \in \mathbb{N}_{\ge 1}$ and $(\mathscr{U}, \nu)$ be combinatorial classes.

1. The **product** of the classes $\mathscr{U}_i$ is the couple

$$\prod_{i=1}^{n} \mathscr{U}_i = (\mathscr{U}_1 \times \cdots \times \mathscr{U}_n, \nu_{\Pi_i \mathscr{U}_i}),$$

formed by the Cartesian product $\mathscr{U}_1 \times \cdots \times \mathscr{U}_n$ and the valuation

$$\nu_{\Pi_i \mathscr{U}_i} : \mathscr{U}_1 \times \cdots \times \mathscr{U}_n \to \mathbb{N}, \quad (x_1, \ldots, x_n) \mapsto \nu_1(x_1) + \cdots + \nu_n(x_n)$$

In particular one sets

$$\mathscr{U}^n := \begin{cases} \prod_{i=1}^{n} \mathscr{U} & \text{if } n \geq 1, \\ \mathbb{1} & \text{if } n = 0, \end{cases}$$

where $\mathbb{1}$ represents the neutral combinatorial class. The elements of $\mathscr{U}^n$ with $n \in \mathbb{N}$, are said to be n-sequences of $\mathscr{U}$. The element in $\mathscr{U}^0 = \mathbb{1}$ is said to be the **empty sequence** of $\mathscr{U}$ and is indicated with ().

2. Suppose that $\mathscr{U}_i \cap \mathscr{U}_j = \emptyset$ for every $i \neq j$. Then the **direct sum** of the classes $\mathscr{U}_i$ is the couple

$$\bigoplus_{i=1}^{n} \mathscr{U}_i := (\mathscr{U}_1 \cup \cdots \cup \mathscr{U}_n, \nu_{\bigoplus_i \mathscr{U}_i}),$$

formed by the set $\mathscr{U}_1 \cup \cdots \cup \mathscr{U}_n$ and the valuation $\nu_{\bigoplus_i \mathscr{U}_i} : \mathscr{U}_1 \cup \cdots \cup \mathscr{U}_n \to \mathbb{N}$ defined by putting

$$\nu_{\bigoplus \mathscr{U}_i}(x) := \nu_k(x) \text{ if } x \in \mathscr{U}_k.$$

3. If $\mathscr{U}$ does not have elements of size 0, the class of the **sequences** of $\mathscr{U}$ is the couple

$$\text{SEQ } \mathscr{U} := \left(\bigcup_{k=0}^{\infty} \mathscr{U}^k, \nu_{\text{SEQ } \mathscr{U}} \right),$$

formed by the set of all the finite sequences of $\mathscr{U}$ and by the valuation $\nu_{\text{SEQ}\,\mathscr{U}}$ defined by setting

$$\begin{cases} \nu_{\text{SEQ}\,\mathscr{U}}\,(()) & = 0, \\ \nu_{\text{SEQ}\,\mathscr{U}}\,(x_1, \ldots, x_k) & = \nu(x_1) + \cdots + \nu(x_k)\ \forall k \geq 1. \end{cases}$$

4. If $\mathscr{U}$ does not have elements of size 0, and $\Omega \subseteq \mathbb{N}$ we use $\text{SEQ}_\Omega\,\mathscr{U}$ to indicate the subclass of $\text{SEQ}\,\mathscr{U}$ defined by

$$\text{SEQ}_\Omega\,\mathscr{U} = \bigcup_{k \in \Omega} \mathscr{U}^k,$$

formed by the sequences of $\mathscr{U}$, with length varying in Ω. If $m \in \mathbb{N}$ we shall write $\text{SEQ}_{\{\geq m\}}\,\mathscr{U}$ (resp. $\text{SEQ}_{\{\leq m\}}\,\mathscr{U}$) in place of $\text{SEQ}_{\{m,m+1,\ldots\}}\,\mathscr{U}$ (resp. $\text{SEQ}_{\{0,1,\ldots,m\}}\,\mathscr{U}$).   □

Remark 11.9 If $\mathscr{U}$ is a combinatorial class without elements of size 0, then $\mathscr{U}^k$ is a subclass of $\text{SEQ}\,\mathscr{U}$ for each $k \in \mathbb{N}$: indeed the valuation $\nu_{\mathscr{U}^k}$ coincides with the restriction of $\nu_{\text{SEQ}\,\mathscr{U}}$ to $\mathscr{U}^k$.

Example 11.10 Let $\mathscr{U}$ be a combinatorial class without elements of size 0. The class $\text{SEQ}_{\{0,2\}}\,\mathscr{U}$ is formed by $\mathbb{1} \oplus \mathscr{U}^2$, that is, by the *empty* sequence and by the 2-sequences of elements of $\mathscr{U}$. One notes that, given $x_1, x_2 \in \mathscr{U}$, the elements (x_1, x_2) and () belong to $\mathbb{1} \oplus \mathscr{U}^2$, while the triple $((), x_1, x_2)$ is not an element of $\mathbb{1} \oplus \mathscr{U}^2$, but it belongs to $\mathbb{1} \times \mathscr{U}^2$.   □

The operations of product, direct sum, and sequence introduced above result in combinatorial classes.

Proposition 11.11 *1. Let $\mathscr{U}_i, i = 1, \ldots, n$, be combinatorial classes. Then, $\prod_i \mathscr{U}_i$ and $\bigoplus_i \mathscr{U}_i$ have finite fibres, and so are combinatorial classes.*
  *2. Let $\mathscr{U}$ be a combinatorial class **without elements of null size**. Then $\text{SEQ}\,\mathscr{U}$ has finite fibres, and so is a combinatorial class. In particular, for every $\Omega \subseteq \mathbb{N}$, the set $\text{SEQ}_\Omega\,\mathscr{U}$ endowed with the valuation induced by $\nu_{\text{SEQ}\,\mathscr{U}}$ is a combinatorial subclass of $\text{SEQ}\,\mathscr{U}$.*

Proof 1. Let us verify that $\nu_{\prod_i \mathscr{U}_i}^{-1}(m)$ and $\nu_{\bigoplus_i \mathscr{U}_i}^{-1}(m)$ are finite for each $m \in \mathbb{N}$. Given $(x_1, \ldots, x_n) \in \mathscr{U}_1 \times \cdots \times \mathscr{U}_n$, if $\nu_{\prod_i \mathscr{U}_i}(x_1, \ldots, x_n) = m$, then certainly each u_i has size $\leq m$ in $\mathscr{U}_i, i = 1, \ldots, n$. Since there are only a finite number of elements of size $\leq m$ in every $\mathscr{U}_i$, one concludes that $\nu_{\prod_i \mathscr{U}_i}^{-1}(m)$ is finite.
The set $\nu_{\bigoplus_i \mathscr{U}_i}^{-1}(m)$ of the elements of size m in $\bigoplus_i \mathscr{U}_i$ is the union of the sets of elements of size m in the various $\mathscr{U}_i$. Since the sets of elements of size m in the various $\mathscr{U}_i$ are finite, $\nu_{\bigoplus_i \mathscr{U}_i}^{-1}(m)$ is also finite.

2. Let us verify that $\nu_{\text{SEQ}\,\mathscr{U}}^{-1}(m)$ is finite for each $m \in \mathbb{N}$. If $m = 0$, then the unique element of size 0 in SEQ $\mathscr{U}$ is the empty sequence () belonging to $\mathbb{1}$. Let $m \geq 1$. Not having $\mathscr{U}$ elements of size 0, if a k-sequence $(x_1, \ldots, x_k)$ of $\mathscr{U}$ $(k \geq 1)$ belongs to $\nu_{\text{SEQ}\,\mathscr{U}}^{-1}(m)$, then one has

$$m = \nu_{\text{SEQ}\,\mathscr{U}}(x_1, \ldots, x_k) = \nu_{\mathscr{U}}(x_1) + \cdots + \nu_{\mathscr{U}}(x_k) \geq k.$$

Therefore, the set $\nu_{\text{SEQ}\,\mathscr{U}}^{-1}(m)$ of the elements of size m is the union of the subsets of elements of size m in $\mathscr{U}, \ldots, \mathscr{U}^m$. Since each of these subsets is finite, it follows that $\nu_{\text{SEQ}\,\mathscr{U}}^{-1}(m)$ is also finite. $\qquad\square$

Remark 11.12 It is essential for SEQ $\mathscr{U}$ to be a combinational class that $\mathscr{U}$ has no elements of size 0. If there were an element $\overline{x} \in \mathscr{U}$ of size 0, for every $m \in \mathbb{N}_{\geq 1}$ one would have

$$\nu_{\text{SEQ}\,\mathscr{U}}(\underbrace{\overline{x}, \ldots, \overline{x}}_{m}) = \underbrace{\nu_{\mathscr{U}}(\overline{x}) + \cdots + \nu_{\mathscr{U}}(\overline{x})}_{m} = m\nu_{\mathscr{U}}(\overline{x}) = 0,$$

and there would be infinitely many elements of size 0.

Different sets can have substantially the same valuation structure; they are different, but have the *same form*.

Definition 11.13 Two combinatorial classes $\mathscr{U}$ and $\mathscr{W}$ are **isomorphic** (we will write $\mathscr{U} \cong \mathscr{W}$), if there exists a bijective application $\phi : \mathscr{U} \to \mathscr{W}$ which preserves the fibres, that is, such that

$$\forall x \in \mathscr{U} \quad \nu_{\mathscr{W}}(\phi(x)) = \nu_{\mathscr{U}}(x).$$ $\qquad\square$

A combinatorial class is determined, up to isomorphisms, when one knows the cardinality of its n-fibres, for every $n \in \mathbb{N}$.

Proposition 11.14 (Characterization of the isomorphisms) *Let $\mathscr{U}$ and $\mathscr{W}$ be two combinatorial classes. The following assertions are then equivalent:*

1. $\mathscr{U}$ and $\mathscr{W}$ are isomorphic classes;
2. The fibres of $\mathscr{U}$ and $\mathscr{W}$ have the same cardinality.

Proof A function $\phi : \mathscr{U} \to \mathscr{W}$ induces an isomorphism between $\mathscr{U}$ and $\mathscr{W}$ if and only if for each $n \in \mathbb{N}$ it induces a bijection between the n-fibres $\nu_{\mathscr{U}}^{-1}(n)$ and $\nu_{\mathscr{W}}^{-1}(n)$. Therefore $\mathscr{U}$ and $\mathscr{W}$ are isomorphic if and only if for every n one has $|\nu_{\mathscr{U}}^{-1}(n)| = |\nu_{\mathscr{W}}^{-1}(n)|$. $\qquad\square$

Thanks to the proposition just proven, every sequence $(a_n)_n$ of natural numbers uniquely determines, up to isomorphisms, the combinatorial class of which it is the sequence of the fibres (see Definition 11.5), that is, in which the n-fibre has a_n elements.

Example 11.15 (Role of the neutral class in the product) Let $\mathcal{U}$ be a combinatorial class and let $\mathbb{1}$ be the neutral class. Then $\mathcal{U} \times \mathbb{1} \cong \mathbb{1} \times \mathcal{U} \cong \mathcal{U}$. Indeed, if we indicate the unique element of $\mathbb{1}$ with x, then the bijective applications

$$\mathcal{U} \times \mathbb{1} \longrightarrow \mathbb{1} \times \mathcal{U} \longrightarrow \mathcal{U}$$
$$(u,x) \longmapsto (x,u) \longmapsto u$$

are isomorphisms of combinatorial classes:

$$\nu_{\mathcal{U} \times \mathbb{1}}(u, x) = \nu_{\mathcal{U}}(u) + 0 = \nu_{\mathcal{U}}(u) = 0 + \nu_{\mathcal{U}}(u) = \nu_{\mathbb{1} \times \mathcal{U}}(x, u).$$

The name "neutral class" for $\mathbb{1}$ derives from the fact that it plays the role of a neutral element in the product of combinatorial classes. $\qquad\square$

Example 11.16 A finite set Γ endowed with the constant function equal to 1 as valuation is a combinatorial class. Then the combinatorial class SEQ Γ of the sequences of Γ is endowed with the valuation which measures the length of the sequences. Indeed, one has $\nu_{\text{SEQ }\Gamma}() = 0$ and

$$\nu_{\text{SEQ }\Gamma}(x_1, \ldots, x_k) = \nu_{\Gamma}(x_1) + \cdots + \nu_{\Gamma}(x_k) = \underbrace{1 + \cdots + 1}_{k} = k, \qquad \forall k \geq 1.$$

It is easy to verify the following isomorphism:

$$\text{SEQ }\Gamma \times \Gamma \cong \text{SEQ}_{\{\geq 1\}} \Gamma.$$

Indeed, the bijective application

$$\text{SEQ}\Gamma \times \Gamma \longrightarrow \text{SEQ}_{\{\geq 1\}} \Gamma$$
$$((a_1,..,a_k),b) \longmapsto (a_1,..,a_k,b)$$
$$((),b) \longmapsto (b)$$

induces an isomorphism of combinatorial classes:

$$\nu_{\text{SEQ }\Gamma \times \Gamma}((a_1, \ldots, a_k), b) = k + 1 = \nu_{\text{SEQ}_{\{\geq 1\}} \Gamma}(a_1, \ldots, a_k, b),$$

$$\nu_{\text{SEQ }\Gamma \times \Gamma}((), b) = 0 + 1 = \nu_{\text{SEQ}_{\{\geq 1\}} \Gamma}(b). \qquad\square$$

11.3 OGF of Combinatorial Classes

The generating formal series studied in Chaps. 7, and 8 allow one to calculate the cardinality of the subsets of a given size of a combinatorial class in an efficient fashion.

Definition 11.17 We define the **OGF of a combinatorial class** $\mathcal{U}$ to be the ordinary generating formal series of its sequence of the fibres:

$$\mathcal{U}(X) := \mathrm{OGF}(u_n)_n = \sum_{n=0}^{\infty} u_n X^n, \qquad u_n := |v_{\mathcal{U}}^{-1}(n)|. \qquad \square$$

Lemma 11.18 *Two combinatorial classes $\mathcal{U}$ and $\mathcal{W}$ are isomorphic if and only if their ordinary generating formal series $\mathcal{U}(X)$ and $\mathcal{W}(X)$ coincide.*

Proof By Proposition 11.14 two classes are isomorphic if and only if they have the same sequence of fibres, and that occurs if and only if they have the same OGF. $\square$

The operations of product, direct sum, and sequences of combinatorial classes correspond to appropriate operations between the OGF of the associated sequences of the fibres.

Proposition 11.19 (OGF and operations on the combinatorial classes) *Let $\mathcal{U}_i$, $i = 1, \dots, m$, and $\mathcal{U}$ be combinatorial classes.*

1. *The sequence of the fibres of $\prod_{i=1}^{m} \mathcal{U}_i$ is the convolution product of the sequences of the fibres of the $\mathcal{U}_i$. Therefore,*

$$\left(\prod_{i=1}^{m} \mathcal{U}_i\right)(X) = \prod_{i=1}^{m} \mathcal{U}_i(X).$$

2. *If $\mathcal{U}_i \cap \mathcal{U}_j = \emptyset$ for $i \neq j$, the sequence of the fibres of $\bigoplus_{i=1}^{m} \mathcal{U}_i$ is the sum of the sequences of the fibres of the $\mathcal{U}_i$. Therefore,*

$$\left(\bigoplus_{i=1}^{m} \mathcal{U}_i\right)(X) = \mathcal{U}_1(X) + \cdots + \mathcal{U}_m(X).$$

3. *If $\mathcal{U}$ does not have elements of size 0, then the OGF of SEQ $\mathcal{U}$ is*

$$(SEQ\, \mathscr{U})\, (X) = 1 + \mathscr{U}(X) + \mathscr{U}^2(X) + \cdots = \frac{1}{1 - \mathscr{U}(X)}.$$

4. *If $\mathscr{U}$ does not have elements of size 0, and $\Omega \subseteq \mathbb{N}$, then the OGF of $SEQ_\Omega\, \mathscr{U}$ is*

$$(SEQ_\Omega\, \mathscr{U})\, (X) = \sum_{k \in \Omega} \mathscr{U}^k(X).$$

Proof We verify 1 and 2 for $m = 2$; the general case may then be obtained with a simple inductive argument.

1. If $(x, y) \in \mathscr{U}_1 \times \mathscr{U}_2$, then $v_{\mathscr{U}_1 \times \mathscr{U}_2}(x, y) = n$ if and only if $v_{\mathscr{U}_1}(x) = k$ and $v_{\mathscr{U}_2}(y) = n - k$ for some $0 \le k \le n$. Therefore one has

$$v_{\mathscr{U}_1 \times \mathscr{U}_2}^{-1}(n) = \bigcup_{k=0}^{n} v_{\mathscr{U}_1}^{-1}(k) \times v_{\mathscr{U}_2}^{-1}(n - k).$$

Since the sets of the preceding union are two by two disjoint, one has

$$|v_{\mathscr{U}_1 \times \mathscr{U}_2}^{-1}(n)| = \sum_{k=0}^{n} |v_{\mathscr{U}_1}^{-1}(k)|\, |v_{\mathscr{U}_2}^{-1}(n - k)|.$$

Consequently, the sequence of the fibres of $\mathscr{U}_1 \times \mathscr{U}_2$ is the convolution product of the sequences of the fibres of $\mathscr{U}_1$ and of $\mathscr{U}_2$, and hence, by Point 3 of Proposition 7.21, the OGF of $\mathscr{U}_1 \times \mathscr{U}_2$ is the product of the ordinary generating formal series of $\mathscr{U}_1$ and of $\mathscr{U}_2$.

2. For every $n \in \mathbb{N}$ one has

$$v_{\mathscr{U}_1 \oplus \mathscr{U}_2}^{-1}(n) = v_{\mathscr{U}_1}^{-1}(n) \cup v_{\mathscr{U}_2}^{-1}(n).$$

Since $\mathscr{U}_1 \cap \mathscr{U}_2 = \emptyset$, one deduces that $|v_{\mathscr{U}_1 \oplus \mathscr{U}_2}^{-1}(n)| = |v_{\mathscr{U}_1}^{-1}(n)| + |v_{\mathscr{U}_2}^{-1}(n)|$. Consequently, the sequence of the fibres of $\mathscr{U}_1 \oplus \mathscr{U}_2$ is the sum of the sequences of the fibres of $\mathscr{U}_1$ and of $\mathscr{U}_2$ and hence, by Point 1 of Proposition 7.21, the OGF of $\mathscr{U}_1 \oplus \mathscr{U}_2$ is the sum of the ordinary generating formal series of $\mathscr{U}_1$ and of $\mathscr{U}_2$.

3. Since $\mathscr{U}^k$ is the product of $k \ge 1$ copies of $\mathscr{U}$, one immediately obtains that the ordinary generating formal series of $\mathscr{U}^k$ is $\mathscr{U}(X)^k$; moreover, $\mathbb{1}(X) = 1 = \mathscr{U}(X)^0$ given that $\mathbb{1}$ is constituted by a unique element of size zero. Since there are no elements of size 0 in $\mathscr{U}$, one has that $[X^0]\mathscr{U}(X) = 0$ and therefore, by Proposition 7.33, the family of formal power series $\{\mathscr{U}(X)^k : k \in \mathbb{N}\}$ is locally finite and every $\mathscr{U}(X)^k$ has only terms of degree $\ge k$. So, for every $k > m \ge 0$ one has

$$|\nu^{-1}_{\mathcal{U}^k}(m)| = [X^m]\mathcal{U}(X)^k = 0;$$

and therefore, for every $m \geq 0$

$$|\nu^{-1}_{\mathrm{SEQ}\,\mathcal{U}}(m)| = \sum_{k=0}^{\infty}|\nu^{-1}_{\mathcal{U}^k}(m)| = \sum_{k=0}^{m}|\nu^{-1}_{\mathcal{U}^k}(m)| = \sum_{k=0}^{m}[X^m]\mathcal{U}(X)^k = [X^m]\sum_{k=0}^{\infty}\mathcal{U}(X)^k.$$

Thus one concludes that $\displaystyle\sum_{k=0}^{\infty}\mathcal{U}(X)^k = \frac{1}{1 - \mathcal{U}(X)}$ is the OGF of SEQ $\mathcal{U}$.

4. It follows immediately from what was observed in Point 3 that

$$|\nu^{-1}_{\mathrm{SEQ}_\Omega\,\mathcal{U}}(m)| = \sum_{k\in\Omega}|\nu^{-1}_{\mathcal{U}^k}(m)| = [X^m]\sum_{k\in\Omega}\mathcal{U}(X)^k.$$

Thus, one concludes that $\displaystyle\sum_{k\in\Omega}\mathcal{U}(X)^k$ is the OGF of SEQ$_\Omega\,\mathcal{U}$.   □

Example 11.20 Let us try to recalculate, in order to undertake a bit of practice with the ordinary generating formal series of combinatorial classes, the number of binary sequences of length n by way of the symbolic calculus. The combinatorial class $\{0, 1\}$ with constant valuation equal to 1 has as its OGF the formal power series $2X$: indeed, the 1-fibre has two elements, while there are no elements with size $\neq 1$. The binary n-sequences are exactly the elements of the n-fibre of SEQ$\{0, 1\}$. Since by Proposition 11.19 one has

$$\mathrm{SEQ}\{0, 1\}(X) = \frac{1}{1 - 2X} = \sum_{k=0}^{\infty}(2X)^k,$$

the number of binary n-sequences is $[X^n]\,\mathrm{SEQ}\{0, 1\}(X) = 2^n$ for every $n \in \mathbb{N}$.   □

Example 11.21 Given $n \in \mathbb{N}_{\geq 1}$, let us use symbolic calculus to determine the number of binary n-sequences that have exactly $k \geq 0$ elements equal to 1. We consider the combinatorial class $\{0, 1\}$ with valuation ν_1 that counts the number of ones, that is, defined by setting $\nu_1(0) = 0$ and $\nu_1(1) = 1$. We must determine the cardinality of the k-fibre in the product of combinatorial classes $\underbrace{\{0, 1\} \times \cdots \times \{0, 1\}}_{n \text{ times}}$: every element of size k is in fact a binary n-sequence with k elements equal to 1. The OGF associated to the combinatorial class $\{0, 1\}$ is $1 + X$: there is, in fact, 1 element with zero ones and 1 element with a one. By Proposition 11.19 one has

$$\left(\prod_{i=1}^{n}\{0, 1\}\right)(X) = (1 + X)^n = \sum_{i=0}^{n}\binom{n}{i}X^i;$$

thus, the number of binary sequences of length n that have exactly k ones is

$$[X^k](1+X)^n = \begin{cases} \binom{n}{k} & \text{if } k \leq n, \\ 0 & \text{otherwise,} \end{cases}$$

as we could have deduced easily via our combinatorial calculus.   □

11.4 Patterns in Strings

In this section we determine the number of sequences of a given length that contain or do not contain a determinate *pattern*, that is, a prescribed subsequence. We begin with some examples.

Example 11.22 (Binary sequences without two consecutive zeroes) Let us determine the number of binary n-sequences in which two consecutive zeroes do not appear (one such subsequence of only consecutive zeroes is called a *strip* or *run* of zeroes). We use $\text{SEQ}^{\neg 00}\{0,1\}$ to indicate the combinatorial class formed by the binary sequences in which there do not appear two consecutive zeroes and whose valuation function measures the length of the sequences.

The sequences in $\text{SEQ}^{\neg 00}\{0,1\}$ can be the *empty* one, that formed by only 0, or else they can begin with 1 or with (0, 1) followed by a sequence (possibly *empty*) that does not contain two consecutive zeroes. In other words, one has

$$\text{SEQ}^{\neg 00}\{0,1\} \cong \{()\} \bigoplus \{(0)\} \bigoplus \left(\{(1),(0,1)\} \times \text{SEQ}^{\neg 00}\{0,1\}\right).$$

Since the OGF's of the subclasses $\{()\}$, $\{(0)\}$, and $\{(1),(0,1)\}$ are 1, X, and $X+X^2$ respectively, by Proposition 11.19 one has that

$$\text{SEQ}^{\neg 00}\{0,1\}(X) = 1 + X + (X+X^2)\,\text{SEQ}^{\neg 00}\{0,1\}(X),$$

from which it follows that

$$\text{SEQ}^{\neg 00}\{0,1\}(X) = \frac{1+X}{1-X-X^2}.$$

The coefficient of X^n in the formal power series $\text{SEQ}^{\neg 00}\{0,1\}(X)$ is the number of the binary n-sequences in which two consecutive zeroes do not appear. As we have seen in Sect. 7.8 we can determine the coefficients $(s_n)_n$ of $\text{SEQ}^{\neg 00}\{0,1\}(X) := \sum_{n=0}^{\infty} s_n X^n$ with a recursive method. From the relation found above one has

$$\mathrm{SEQ}^{\neg 00}\{0, 1\}(X)(1 - X - X^2) = 1 + X.$$

Therefore,

$$\forall n \geq 2 \quad 0 = \left[X^n\right] \mathrm{SEQ}^{\neg 00}\{0, 1\}(X)(1 - X - X^2) = s_n - s_{n-1} - s_{n-2},$$

from which one obtains the relation

$$s_n = s_{n-1} + s_{n-2}, \quad \forall n \geq 2.$$

Bearing in mind that $s_0 = 1$ and $s_1 = 2$ one immediately deduces the first coefficients and finds that

$$s_2 = 1 + 2 = 3, \ s_3 = 3 + 2 = 5, \ s_4 = 5 + 3 = 8, \ldots.$$

The recurrence relation can be easily solved using the techniques developed in Chap. 10 determining the general formula for s_n:

$$s_n = \frac{5 + 3\sqrt{5}}{10}\left(\frac{1 + \sqrt{5}}{2}\right)^n + \frac{5 - 3\sqrt{5}}{10}\left(\frac{1 - \sqrt{5}}{2}\right)^n.$$

As an alternative, we could have used Theorem 7.114 which allows one to find (in principle) the coefficients of every rational fraction of polynomials.   □

Let us take inspiration from Example 11.22 in order to determine the number of binary sequences of a given length that do not contain a strip formed by a prescribed number of zeroes.

Example 11.23 (*Sequences without strips of zeroes*) Let $k, n \in \mathbb{N}$; let us determine the number of binary n-sequences in which there do not appear $k + 1$ consecutive zeroes. Let us use $\mathrm{SEQ}^{\neg 0^{k+1}}\{0, 1\}$ to indicate the combinatorial class formed by the set of binary sequences in which there do not appear $k + 1$ consecutive zeroes and with the valuation function that measures the length of the sequences. The sequences in $\mathrm{SEQ}^{\neg 0^{k+1}}\{0, 1\}$ can be of the following types:

(a) The *empty* one, or that formed by a sequence of i zeroes with $i = 1, \ldots, k$;
(b) A sequence of type (a) followed by a 1 and by a sequence (possibly *empty*) that does not have $k + 1$ consecutive zeroes.

In other words, one has

$$\mathrm{SEQ}^{\neg 0^{k+1}}\{0, 1\} \cong \{(), (0), (0, 0), \ldots, \underbrace{(0, \ldots, 0)}_{k}\} \bigoplus$$

$$\bigoplus \left(\{(1), (0, 1), \ldots, \underbrace{(0, \ldots, 0}_{k}, 1)\} \times \mathrm{SEQ}^{\neg 0^{k+1}}\{0, 1\}\right).$$

Since the OGF's of the classes

$$\{(), (0), (0, 0), \ldots, \underbrace{(0, \ldots, 0)}_{k}\}, \text{ and } \{(1), (0, 1), \ldots, \underbrace{(0, \ldots, 0, 1)}_{k}\}$$

are $1 + X + \cdots + X^k$ and $X + \cdots + X^{k+1}$, respectively, by Proposition 11.19 one has

$$\text{SEQ}^{\neg 0^{k+1}}\{0, 1\}(X) = (1 + X + X^2 + \cdots + X^k) + (X + \cdots + X^{k+1}) \text{ SEQ}^{\neg 0^{k+1}}\{0, 1\}(X),$$

from which it follows that

$$\text{SEQ}^{\neg 0^{k+1}}\{0, 1\}(X) = \frac{1 + X + \cdots + X^k}{1 - X - X^2 - \cdots - X^{k+1}}. \qquad (11.23.\text{a})$$

Given that

$$1 + X + \cdots + X^k = \frac{1 - X^{k+1}}{1 - X}$$

and

$$1 - X - \cdots - X^{k+1} = 1 - X(1 + \cdots + X^k) = 1 - X\left(\frac{1 - X^{k+1}}{1 - X}\right)$$

$$= \frac{1 - 2X + X^{k+2}}{1 - X},$$

one obtains $\text{SEQ}^{\neg 0^{k+1}}\{0, 1\}(X) = \dfrac{1 - X^{k+1}}{1 - 2X + X^{k+2}}$.

Using a CAS one finds the initial terms of $\text{SEQ}^{\neg 0^{k+1}}\{0, 1\}(X)$ in the cases $k = 0, 1, 2$:

$$\frac{1 - X}{1 - 2X + X^2} = 1 + X + X^2 + X^3 + X^4 + X^5 + X^6 + X^7 + \cdots;$$

$$\frac{1 - X^2}{1 - 2X + X^3} = 1 + 2X + 3X^2 + 5X^3 + 8X^4 + 13X^5 + 21X^6 + 34X^7 + \cdots;$$

$$\frac{1 - X^3}{1 - 2X + X^4} = 1 + 2X + 4X^2 + 7X^3 + 13X^4 + 24X^5 + 44X^6 + 81X^7 + \cdots.$$

For example, the binary 7-sequences without three consecutive zeroes number 81 in total.   □

11.4.1 Sequences that Do Not Contain a Given Pattern

In Example 11.23 we calculated the number of binary strings of a given length which do not contain a prescribed number of consecutive zeroes. Now let us consider the more general problem of counting the number of sequences of a given length which do not contain a certain string of pre-assigned characters.

In all this section $\Gamma \neq \emptyset$ is a finite set endowed with the constant function equal to 1 as valuation, which makes it a combinatorial class.

Definition 11.24 We use the word **pattern** of Γ for a given element of SEQ Γ. In this situation, we agree to indicate with the symbol $a_0 \ldots a_n$ the sequence $(a_0, \ldots, a_n)$ of elements of Γ. Then, having set $a := a_0 \ldots a_n$ and $b := b_0 \ldots b_m$, we will denote with ab the sequence $ab := a_0 \ldots a_n b_0 \ldots b_m$. □

The number of sequences that do not contain a given pattern will change depending on the pattern under consideration. For example, there are five binary 3-sequences that do not contain the pattern 00, while there are only four that do not contain the pattern 01:

$$111, \ 011, \ 101, \ 110, \ 011 \text{ do not contain the pattern } 00,$$

$$111, \ 110, \ 100, \ 000 \text{ do not contain the pattern } 01.$$

Hence it is necessary to distinguish between the various patterns.

Definition 11.25 Let $k \in \mathbb{N}$ and $p = p_0 \ldots p_k$ be a pattern of Γ of length $\ell(p) := k + 1$. The **autocorrelation polynomial** of the pattern p is the polynomial of degree less than or equal to k defined by

$$C_p(X) = 1 + c_1 X + \cdots + c_k X^k,$$

where, for every $i = 1, \ldots, k$, $c_i = 1$ if the *tail* of the last $k + 1 - i$ elements of p (from p_i onwards) coincides with the first $k + 1 - i$ elements from the beginning of p, $c_i = 0$ otherwise; that is to say

$$c_i = \begin{cases} 1 & \text{if } p_i \ldots p_k = p_0 \ldots p_{k-i}, \\ 0 & \text{otherwise,} \end{cases} \quad \forall i \in \{1, \ldots, k\}. \qquad \square$$

Example 11.26 The pattern 00 has autocorrelation polynomial 1+X, while the pattern 01 has autocorrelation polynomial 1. The autocorrelation polynomial of a strip p formed by $k + 1$ consecutive zeroes is

$$C_p(X) = 1 + X + \cdots + X^k. \qquad \square$$

Remark 11.27 The autocorrelation polynomial of a pattern $p_0 \ldots p_k$ of Γ does not depend on Γ, but only on the symbols that make up the pattern. One notes that if $p_i \neq p_0$ then necessarily $c_i = 0$ given that one certainly has $p_i \ldots p_{k-1} \neq p_0 \ldots p_{(k-1)-i}$.

Remark 11.28 In order to avoid making errors with the indices, we suggest to proceed in the exercises as follows: write the pattern $p_0 \ldots p_k$ on a line; on a lower line, write 1 under p_0, and 0 under any $p_i \neq p_0$, and then complete the sequence of the coefficients c_i in correspondence to the other symbols p_j's. The autocorrelation polynomial is obtained by taking as coefficients of the growing powers of X those that appear in the second line. Let us consider, for example, the pattern

$$p = p_0 p_1 p_2 p_3 p_4 p_5 p_6 := \# @ * \# \# @ * .$$

1. We write 1 under $p_0 = \#$:

pattern p	#	@	*	#	#	@	*
coefficient c	1						
power of X	1	X	X^2	X^3	X^4	X^5	X^6

2. For every $i \geq 1$, if $p_i \neq p_0$ we write $c_i = 0$:

pattern p	#	@	*	#	#	@	*
coefficient c	1	0	0			0	0
power of X	1	X	X^2	X^3	X^4	X^5	X^6

3. Then to compute c_i, $i = 3, 4$, we start by considering the block $p_i \ldots p_6$ from p_i onward, and then we compare it with the block of equal length starting from p_0; if the two coincide, we write $c_i = 1$, while otherwise we write $c_i = 0$:

pattern p	#	@	*	#	#	@	*
coefficient c	1	0	0	0	1	0	0
power of X	1	X	X^2	X^3	X^4	X^5	X^6

The autocorrelation polynomial is $1 + X^4$.

Example 11.29 Determine the autocorrelation polynomial of the pattern

$$p = z@axbz@ax.$$

Proceeding as indicated above, one has

pattern p	z	$@$	a	x	b	z	$@$	a	x
coefficient c	1	0	0	0	0	1	0	0	0
power of X	1	X	X^2	X^3	X^4	X^5	X^6	X^7	X^8

Indeed, $c_1 = c_2 = c_3 = c_4 = c_6 = c_7 = c_8 = 0$, given that $p_i \neq p_0$ for $i = 1, 2, 3, 4, 6, 7$ and $c_5 = 1$ since the tail $z@ax$ from $p_5 = z$ coincides with the beginning of the same size. The autocorrelation polynomial of p is therefore $C_p(X) = 1 + X^5$. □

In general, the more symbols are present in the pattern the more rare is that there are tails coinciding with beginnings: in such a case it is rather common to encounter, via autocorrelation, polynomials reduced to the constant 1.

Example 11.30 Let p be a pattern formed by distinct symbols. Then $C_p(X) = 1$. Indeed, it is $p_i \neq p_0$ for each $i \geq 1$. □

The autocorrelation polynomial can be equal to 1 whilst not being made up of distinct symbols.

Example 11.31 Determine the autocorrelation polynomial of the pattern

$$p = MATHEMATICS.$$

Proceeding as indicated above, one has

pattern p	M	A	T	H	E	M	A	T	I	C	S
coefficient c	1	0	0	0	0	0	0	0	0	0	0
power of X	1	X	X^2	X^3	X^4	X^5	X^6	X^7	X^8	X^9	X^{10}

Indeed, $c_1 = c_2 = c_3 = c_4 = c_6 = c_7 = c_8 = c_9 = c_{10} = 0$, given that $p_i \neq p_0$ for $i = 1, 2, 3, 4, 6, 7, 8, 9, 10$ and $c_5 = 0$ since $MATICS \neq MATHEM$. The autocorrelation polynomial of $p = MATHEMATICS$ is therefore $C_p(X) = 1$. □

Example 11.32 Determine the autocorrelation polynomial of 101001010. Proceeding as indicated above one has

pattern p	1	0	1	0	0	1	0	1	0
coefficient c	1	0	0	0	0	1	0	1	0
power of X	1	X	X^2	X^3	X^4	X^5	X^6	X^7	X^8

Indeed, $c_1 = c_3 = c_4 = c_6 = 0$ given that $p_i \neq p_0$ for $i = 1, 3, 4, 6$; then $c_2 = 0$ since the tail 1001010 is different from the beginning 1010010, and $c_5 = c_7 = 1$, the corresponding tails being equal to the beginnings. The autocorrelation polynomial of 101001010 is therefore $1 + X^5 + X^7$.   □

Example 11.33 Determine the autocorrelation polynomial of

$$TOBEORNOTTOBE.$$

Proceeding as above we obtain the table

T	O	B	E	O	R	N	O	T	T	O	B	E
1	0	0	0	0	0	0	0	0	1	0	0	0
1	X	X^2	X^3	X^4	X^5	X^6	X^7	X^8	X^9	X^{10}	X^{11}	X^{12}

The autocorrelation polynomial of the pattern is thus given by $C(X) = 1 + X^9$.   □

We here introduce the class of sequences that do not contain a given pattern p.

Definition 11.34 (*The class* SEQ$^{\neg p}$ Γ) Given a pattern $p = p_0 \ldots p_k$, $k \in \mathbb{N}$, let us consider the following combinatorial subclasses of SEQ Γ:

- The set SEQ$^{\neg p}$ Γ of sequences in SEQ Γ that *do not contain* p as a subsequence;
- The set SEQ$^{\cdot p}$ Γ of the sequences in SEQ Γ which terminate with the pattern p and which do not contain p in any other position.   □

☞   We want to analyze the class SEQ$^{\neg p}$ Γ; the set SEQ$^{\cdot p}$ Γ will merely be a useful tool in studying the formal series SEQ$^{\neg p}$ $\Gamma(X)$ associated to the combinatorial class SEQ$^{\neg p}$ Γ.

Example 11.35 The first $k + 2$ coefficients of SEQ$^{\neg p}$ $\Gamma(X)$ are immediate. Indeed, there not being sequences of length less than $k + 1$ which contain p, one has

$$[X^n] \, \text{SEQ}^{\neg p} \, \Gamma(X) = \begin{cases} |\Gamma|^n & \text{if } 0 \leq n \leq k, \\ |\Gamma|^n - 1 & \text{if } n = k + 1. \end{cases}$$ □

Lemma 11.36 *Let* $p = p_0 \ldots p_k$ *be a pattern of length* $k + 1$ *of* Γ. *If* $C_p(X) = 1 + c_1 X + \cdots + c_k X^k$, *one has the following isomorphisms of combinatorial classes:*

$$\mathrm{SEQ}^{\neg p}\,\varGamma \bigoplus \mathrm{SEQ}^{\cdot p}\,\varGamma \cong \mathbb{1} \bigoplus \left(\mathrm{SEQ}^{\neg p}\,\varGamma \times \varGamma\right), \tag{11.36.a}$$

$$\mathrm{SEQ}^{\neg p}\,\varGamma \times \{p\} \cong \mathrm{SEQ}^{\cdot p}\,\varGamma \,\oplus \left(\bigoplus_{\{1 \le i \le k : c_i = 1\}} \mathrm{SEQ}^{\cdot p}\,\varGamma \times \{p_{k-i+1}\ldots p_k\}\right). \tag{11.36.b}$$

Proof We prove that (11.36.a) holds. First one observes that certainly

$$\mathrm{SEQ}^{\neg p}\,\varGamma \cap \mathrm{SEQ}^{\cdot p}\,\varGamma = \emptyset = \mathbb{1} \cap \left(\mathrm{SEQ}^{\neg p}\,\varGamma \times \varGamma\right);$$

therefore the direct sums of the combinatorial classes considered in (11.36.a) are well defined. There remains to verify that

$$\mathrm{SEQ}^{\neg p}\,\varGamma \cup \mathrm{SEQ}^{\cdot p}\,\varGamma = \mathbb{1} \cup \left(\mathrm{SEQ}^{\neg p}\,\varGamma \times \varGamma\right).$$

Observe first that one can identify the elements of $\mathrm{SEQ}^{\neg p}\,\varGamma \times \varGamma$ with the sequences $(a_1, \ldots, a_{n-1}, a_n)$ of $\varGamma$ for which $(a_1, \ldots, a_{n-1})$ belong to $\mathrm{SEQ}^{\neg p}\,\varGamma$. Then let us prove separately the two inclusions.

- $\mathrm{SEQ}^{\neg p}\,\varGamma \cup \mathrm{SEQ}^{\cdot p}\,\varGamma \subseteq \mathbb{1} \cup \left(\mathrm{SEQ}^{\neg p}\,\varGamma \times \varGamma\right)$: a sequence in $\mathrm{SEQ}^{\neg p}\,\varGamma$ either is the *empty* sequence () that belongs to $\mathbb{1}$, or else has length greater than or equal to 1 and is in that case an element of $\mathrm{SEQ}^{\neg p}\,\varGamma \times \varGamma$. A sequence in $\mathrm{SEQ}^{\cdot p}\,\varGamma$ has length greater than or equal to 1: in every case it belongs to $\mathrm{SEQ}^{\neg p}\,\varGamma \times \varGamma$ given that it contains the pattern p only in its final part.
- $\mathbb{1} \cup \left(\mathrm{SEQ}^{\neg p}\,\varGamma \times \varGamma\right) \subseteq \mathrm{SEQ}^{\neg p}\,\varGamma \cup \mathrm{SEQ}^{\cdot p}\,\varGamma$: certainly $() \in \mathrm{SEQ}^{\neg p}\,\varGamma$; finally, a sequence of $\mathrm{SEQ}^{\neg p}\,\varGamma \times \varGamma$ either does not contain p, and therefore belongs to $\mathrm{SEQ}^{\neg p}\,\varGamma$, or else contains p only up to its final part, and so belongs to $\mathrm{SEQ}^{\cdot p}\,\varGamma$.

Now in order to prove (11.36.b), we verify that

$$\mathrm{SEQ}^{\neg p}\,\varGamma \times \{p\} = \mathrm{SEQ}^{\cdot p}\,\varGamma \,\cup \left(\bigcup_{\{1 \le i \le k : c_i = 1\}} \mathrm{SEQ}^{\cdot p}\,\varGamma \times \{p_{k-i+1}\ldots p_k\}\right).$$

- $\mathrm{SEQ}^{\neg p}\,\varGamma \times \{p\} \subseteq \mathrm{SEQ}^{\cdot p}\,\varGamma \cup \left(\bigcup_{\{1 \le i \le k : c_i = 1\}} \mathrm{SEQ}^{\cdot p}\,\varGamma \times \{p_{k-i+1}\ldots p_{k-1}\}\right)$: a sequence in $\mathrm{SEQ}^{\neg p}\,\varGamma \times \{p\}$ either belongs to $\mathrm{SEQ}^{\cdot p}\,\varGamma$, or else is of the type

$$q := app_{k-i+1}\ldots p_k,$$

with $p_0 \ldots p_{k-i} = p_i \ldots p_k$, i.e., $c_i = 1$, for some $i \in \{1, \ldots, k\}$, for a suitable sequence a of elements of $\varGamma$. In the last case the initial part ap of q is in $\mathrm{SEQ}^{\cdot p}\,\varGamma$, so that $q \in \mathrm{SEQ}^{\cdot p}\,\varGamma \times \{p_{k-i+1}\ldots p_{k-1}\}$ with $c_i = 1$.

- $\text{SEQ}^{\cdot p}\,\Gamma \cup \big(\displaystyle\bigcup_{\{1\le i\le k:c_i=1\}} \text{SEQ}^{\cdot p}\,\Gamma \times \{p_{k-i+1}\dots p_{k-1}\}\big) \subseteq \text{SEQ}^{\neg p}\,\Gamma \times \{p\}$: having

chosen $i \in \{1,\dots,k\}$ such that $c_i = 1$, a sequence q in $\text{SEQ}^{\cdot p}\,\Gamma \times \{p_{k-i+1}\dots p_k\}$ is necessarily of the type $q = app_{k-i+1}\dots p_k$ with $ap_0\dots p_{k-1}$ in $\text{SEQ}^{\neg p}\,\Gamma$. Since $c_i = 1$, one has $p_i\dots p_k = p_0\dots p_{k-i}$, and so, being $i-1 \le k-1$, the sequence $q = ap_0\dots p_{i-1}p$ belongs to $\text{SEQ}^{\neg p}\,\Gamma \times \{p\}$.   □

We can now give an explicit description of the OGF of $\text{SEQ}^{\neg p}\,\Gamma$, as a function of the autocorrelation polynomial.

Theorem 11.37 *Let p be a pattern of Γ of length $\ell(p) \ge 1$. Then the OGF of the combinatorial class $\text{SEQ}^{\neg p}\,\Gamma$ is*

$$\text{SEQ}^{\neg p}\,\Gamma(X) = \frac{C_p(X)}{X^{\ell(p)} + (1 - |\Gamma|X)C_p(X)},$$

where $C_p(X)$ is the autocorrelation polynomial of p.

Proof By Lemma 11.36 and Proposition 11.19, set $m := |\Gamma|$ and $\ell(p) = k+1$, one has

$$\text{SEQ}^{\neg p}\,\Gamma(X) + \text{SEQ}^{\cdot p}\,\Gamma(X) = 1 + \text{SEQ}^{\neg p}\,\Gamma(X)\cdot mX \quad \text{and}$$

$$\text{SEQ}^{\neg p}\,\Gamma(X)\cdot X^{k+1} = \text{SEQ}^{\cdot p}\,\Gamma(X) + \sum_{\{1\le i\le k:c_i=1\}} X^i\text{SEQ}^{\cdot p}\,\Gamma(X)$$

$$= C_p(X)\text{SEQ}^{\cdot p}\,\Gamma(X).$$

Multiplying both members of the first equation by $C_p(X)$ we obtain

$$C_p(X)\text{SEQ}^{\neg p}\,\Gamma(X) + C_p(X)\text{SEQ}^{\cdot p}\,\Gamma(X) = C_p(X) + mXC_p(X)\text{SEQ}^{\neg p}\,\Gamma(X).$$

Substituting $C_p(X)\text{SEQ}^{\cdot p}\,\Gamma(X)$ with $\text{SEQ}^{\neg p}\,\Gamma(X)X^{k+1}$ one gets

$$C_p(X)\text{SEQ}^{\neg p}\,\Gamma(X) + \text{SEQ}^{\neg p}\,\Gamma(X)X^{k+1} = C_p(X) + mXC_p(X)\text{SEQ}^{\neg p}\,\Gamma(X),$$

and easily concludes.   □

Remark 11.38 One notes that, differing from the autocorrelation polynomial, the formal series $\text{SEQ}^{\neg p}\,\Gamma(X)$ depends explicitly on the cardinality $|\Gamma|$ of the set Γ under consideration.

It follows from Theorem 11.37 that the coefficients of $\text{SEQ}^{\neg p}\,\Gamma(X)$ satisfy a linear homogeneous recurrence relation.

Corollary 11.39 *Let p be a pattern of Γ of length $\ell(p)$, and c_i, $i = 1,\dots, \ell(p)-1$, be the coefficient of X^i in the autocorrelation polynomial of p. Set for convenience $c_0 = c_{\ell(p)} := 1$, the sequence $(a_n)_n$ of the fibres of $\text{SEQ}^{\neg p}\,\Gamma$ satisfies the recurrence*

$$a_n = \sum_{i=1}^{\ell(p)} \left(|\Gamma| c_{i-1} - c_i \right) a_{n-i} \quad \forall n \geq \ell(p),$$

with initial data $a_i = |\Gamma|^i$, $i = 0, \ldots, \ell(p) - 1$.

Proof It follows from Theorem 11.37 that $\mathrm{SEQ}^{\neg p} \Gamma(X) = \sum_{n=0}^{\infty} a_n X^n$ is a proper rational fraction whose denominator is

$$X^{\ell(p)} + (1 - |\Gamma| X) C_p(X)) = 1 + \sum_{i=1}^{\ell(p)} \left(c_i - |\Gamma| c_{i-1} \right) X^i.$$

We deduce from Corollary 10.60 that $(a_n)_n$ satisfies

$$a_n + \sum_{i=1}^{\ell(p)} \left(c_i - |\Gamma| c_{i-1} \right) a_{n-i} = 0 \quad \forall n \geq \ell(p).$$

Since all sequences of length strictly less than $\ell(p)$ do not contain the pattern p, clearly $a_i = |\Gamma|^i$ for $i = 0, \ldots, \ell(p) - 1$. □

Example 11.40 In the context of binary strings, if p is a strip formed by $\ell \geq 1$ consecutive zeroes, let $\mathrm{SEQ}^{\neg p} \{0, 1\}(X) := a_0 + a_1 X + a_2 X^2 + \cdots$. Here

$$|\Gamma| = |\{0, 1\}| = 2, \quad C_p(X) = 1 + X + \cdots + X^{\ell-1}.$$

Corollary 11.39 shows that $(a_n)_n$ satisfies the **generalized Fibonacci relation**

$$a_n = a_{n-1} + \cdots + a_{n-\ell}, \quad \forall n \geq \ell,$$

with initial data $a_0 = 1$, $a_1 = 2$, $a_2 = 2^2$, $\ldots$, $a_{\ell-1} = 2^{\ell-1}$. We can then find all the other coefficients a_n recursively. Let us calculate, for example, a_ℓ, $a_{\ell+1}$ and $a_{\ell+2}$:

$$a_\ell = a_0 + \cdots + a_{\ell-1} = 1 + 2 + 2^2 + \cdots + 2^{\ell-1} = 2^\ell - 1,$$
$$a_{\ell+1} = a_1 + \cdots + a_\ell = 2 + 2^2 + \cdots + 2^{\ell-1} + (2^\ell - 1)$$
$$= 2(1 + 2 + \cdots + 2^{\ell-1}) - 1 = 2(2^\ell - 1) - 1 = 2^{\ell+1} - 3,$$
$$a_{\ell+2} = a_2 + \cdots + a_{\ell+1} = 2^2 + \cdots + 2^{\ell-1} + (2^\ell - 1) + (2^{\ell+1} - 3)$$
$$= 2^2(1 + 2 + \cdots + 2^{\ell-1}) - 4 = 2^2(2^\ell - 1) - 4 = 2^{\ell+2} - 8. \quad □$$

Example 11.41 Consider the pattern $p = aba$ in the sets $\Gamma_1 := \{a, b\}$ and $\Gamma_2 := \{a, b, c\}$. In both sets the autocorrelation polynomial is equal to $C_p(X) = 1 + X^2$. However, from Theorem 11.37 one has

$$\text{SEQ}^{-p} \, \Gamma_1(X) = \frac{X^2 + 1}{X^3 + (1 - 2X)(X^2 + 1)},$$

while

$$\text{SEQ}^{-p} \, \Gamma_2(X) = \frac{X^2 + 1}{X^3 + (1 - 3X)(X^2 + 1)}.$$

Using a CAS one gets

$$\text{SEQ}^{-p} \, \Gamma_1(X) = 1 + 2X + 4X^2 + 7X^3 + 12X^4 + 21X^5 + 37X^6 + 65X^7 + \cdots,$$

$$\text{SEQ}^{-p} \, \Gamma_2(X) = 1 + 3X + 9X^2 + 26X^3 + 75X^4 + 217X^5 + 628X^6 + 1817X^7 + \cdots.$$

Alternatively, by Corollary 11.39 the coefficients of $\text{SEQ}^{-p} \, \Gamma_1(X)$ satisfy the recurrence relation

$$a_n = 2a_{n-1} - a_{n-2} + a_{n-3}, \quad \forall n \geq 3$$

with initial data $a_0 = 1$, $a_1 = 2$, $a_2 = 4$, whereas the coefficients of $\text{SEQ}^{-p} \, \Gamma_2(X)$ satisfy the recurrence relation

$$a_n = 3a_{n-1} - a_{n-2} + 2a_{n-3}, \quad \forall n \geq 3$$

with initial data $a_0 = 1$, $a_1 = 3$, $a_2 = 9$.   □

Example 11.42 Let us compute the number of n-sequences of the Dutch alphabet Γ of 26 letters which do not contain the name *RADEMAKER*. The autocorrelation polynomial of $p = RADEMAKER$ is $C_p(X) = 1 + X^8$. Corollary 11.39 shows that the coefficients $(a_n)_n$ of $\text{SEQ}^{-p} \, \Gamma$ satisfy

$$a_n = 26a_{n-1} - a_{n-8} + 25a_{n-9}, \quad \forall n \geq 9.$$

Given that

$$a_i = 26^i \text{ if } i = 0, 1, \ldots, 8,$$

one recovers successively from the previous relation the values of $a_9, a_{10}, \ldots$. For high values of n, it is more convenient the use of a CAS: for example, one gets that the number of 20-sequences of the Dutch alphabet that do not contain *RADEMAKER*, i.e., the coefficient of X^{20} in $\text{SEQ}^{-p} \, \Gamma(X)$ is

$$19\,928\,148\,895\,165\,365\,018\,496\,418\,424.$$

The probability that a word of 20 letters written by chance does not contain *RADEMAKER* is therefore equal to

$$\frac{19\,928\,148\,895\,165\,365\,018\,496\,418\,424}{26^{20}} \approx 0.999999999998. \qquad \square$$

11.4.2 *The Monkey's Theorem* 🖐

We conclude this section with an application that uses the notion of the *expected value of a random variable*. Here we give a formula for the average number of random keystrokes on a keyboard with m keys necessary to make a given word appear.

🖐 **Theorem 11.43** *Let p be a pattern of a non-empty set Γ of length $\ell(p)$, whose autocorrelation polynomial is $C_p(X)$. The expected length of a randomly selected sequence of Γ in which appears the pattern p is $|\Gamma|^{\ell(p)} C_p(1/|\Gamma|)$.*

Proof Let $|\Gamma| = m$ and $\ell(p) = k + 1 \geq 1$. Consider the random variable Z which counts the number of attempts so that, choosing at random $(k + 1)$-sequences of Γ, one obtains the pattern p. The function Z is a geometric variable of parameter $1/m^{k+1}$ and so is $E[Z] = m^{k+1}$. Let Y be the length of a randomly selected sequence of Γ in which appears the pattern p. It is now easy to see that $Y \leq (k + 1)Z$ and therefore $E[Y] \leq (k + 1)m^{k+1}$ is finite. The expected value of Y is given (see [31, Chap. IV]) by

$$E[Y] = \sum_{n=0}^{\infty} nP(Y = n) = \sum_{n=0}^{\infty} P(Y > n).$$

Now one has that $Y > n$ if and only if after having selected the first n symbols, the pattern p does not appear in the obtained sequence. Therefore, the probability that $Y > n$ is given by the quotient between the number of n-sequences of Γ in which the pattern p does not appear and the total number of n-sequences of Γ:

$$P(Y > n) = \frac{[X^n]\mathrm{SEQ}^{\neg p}\,\Gamma(X)}{m^n} = \left([X^n]\mathrm{SEQ}^{\neg p}\,\Gamma(X)\right)\left(\frac{1}{m}\right)^n.$$

Therefore,

$$E[Y] = \sum_{n=0}^{\infty} P(Y > n) = \sum_{n=0}^{\infty} \left([X^n]\mathrm{SEQ}^{\neg p}\,\Gamma(X)\right)\left(\frac{1}{m}\right)^n = \mathrm{SEQ}^{\neg p}\,\Gamma(1/m).$$

Since $E[Y]$ is finite, the formal power series $\mathrm{SEQ}^{-p}\,\Gamma(X)$ converges at $1/m$; moreover, by Theorem 11.37, one has

$$\mathrm{SEQ}^{-p}\,\Gamma(X) = \frac{C_p(X)}{X^{k+1} + (1 - mX)C_p(X)}.$$

It follows from Corollary 8.27 that

$$
\begin{aligned}
E[Y] &= \lim_{x \to 1/m} \frac{C_p(x)}{x^{k+1} + (1 - mx)C_p(x)} \\
&= \frac{C_p(1/m)}{(1/m)^{k+1} + (1 - m(1/m))C_p(1/m)} = \frac{C_p(1/m)}{(1/m)^{k+1}} = m^{k+1}C_p(1/m). \quad \square
\end{aligned}
$$

Remark 11.44 It follows easily from Theorem 11.43 that, given the alphabet Γ and the length $k + 1$ of the pattern, one maximises the expected length of a randomly selected sequence of Γ in which appears the pattern choosing to repeat $k + 1$ times the same symbol: in such a way indeed the autocorrelation polynomial is equal to $1 + X + \cdots + X^k$ and the expected length is

$$|\Gamma|^{k+1}\left(1 + \frac{1}{|\Gamma|} + \cdots + \frac{1}{|\Gamma|^k}\right).$$

Conversely, one minimises the expected length in particular choosing the last k symbols of the pattern different from the first one: in such a case surely the autocorrelation polynomial is equal to 1 and the expected length is $|\Gamma|^{k+1}$.

Example 11.45 How many binary characters must one write on average in order to obtain the sequence 1111? How many for the sequence 1000?
Solution. The polynomial of autocorrelation of 1111 is $C(X) = 1 + X + X^2 + X^3$; Theorem 11.43 furnishes an average value of $2^4C(1/2) = 30$ characters. In the second case, for 1000 one has that $C(X) = 1$ from which an average value of $2^4 = 16$ characters are needed in order to obtain the pattern 1000. $\square$

Example 11.46 A monkey bashes the keys of a keyboard with 32 characters at random. What is the average number of keys that must be hit in order for the monkey pictured in Fig. 11.1, a fan of Shakespeare's Hamlet, to write the famous phrase TO BE OR NOT TO BE?
Solution. The pattern in this case is the sequence $p = TOBEORNOTTOBE$, and is composed by 13 letters. The autocorrelation polynomial of p is $C_p(X) = 1 + X^9$ (see Example 11.33). By Theorem 11.43 one has that the average number searched for has value $32^{13}(1 + 1/32^9) = 32^{13} + 32^4 \approx 368 \times 10^{17}$: even if the monkey strikes 3 keys per second, the time required to type our pattern is equal to approximately 20 times the age of the universe! $\square$

Fig. 11.1 Chimp does Hamlet. A monkey who wasn't hanging about waiting for his cousins to pass him the manuscript hot off the typewriter (https://www.flickr.com/photos/33122834@N06/3601626998. Author: Rhys Davenport. Creative Commons Attribution CC BY 4.0 International License)

11.4.3 The Overlapping Word Paradox

In this section we consider a generalization of the coin flipping game Penney Ante, invented in 1969 by Walter Penney. Here, given a finite set $\Gamma \neq \emptyset$, two players select distinct patterns in SEQ Γ of length $k + 1$; then they toss a $|\Gamma|$-sided dice, whose faces are labelled with the elements of Γ, until the last $k + 1$ results match one player's pattern. At first glance it looks like the two players have the same chance, since the two patterns have the same probability to be selected. It is not difficult to realize that this is not: imagine to select the sequence HHH of 3 Heads, whereas your opponent selects THH (Tail, Head, Head). Now in a run of coin tosses, either the pattern HHH appears at the beginning of the sequence (with probability 1/8) or, whenever HH appears, it is preceded by T: the odds in favor of pattern HHH, i.e., the ratio of the probability that HHH appears before THH to the probability that THH appears before HHH, are 1/7! We will compute here the odds in favor of one of two patterns, in the general case.

Definition 11.47 Let $\Gamma \neq \emptyset$ be a finite set, $k \in \mathbb{N}$ and $p = p_0 \ldots p_k, q = q_0 \ldots q_k$ be two patterns of the same length $\ell(p) = \ell(q) = k + 1$ in SEQ Γ. The **correlation polynomial** of (p, q) is the polynomial of degree less than or equal to k defined by

$$C_{(p,q)}(X) = c_0 + c_1 X + \cdots + c_k X^k,$$

where, for every $i = 0, \ldots, k$, we set $c_i = 1$ if the tail of the last $k + 1 - i$ elements of p (from p_i onwards) coincides with the first $k + 1 - i$ elements from the beginning of q, and $c_i = 0$ otherwise; that is to say

$$c_i = \begin{cases} 1 & \text{if } p_i \ldots p_k = q_0 \ldots q_{k-i}, \\ 0 & \text{otherwise}, \end{cases} \quad \forall i \in \{0, \ldots, k\}. \qquad \square$$

Remark 11.48 It is easy to verify the following useful properties of the correlation polynomial of a pair (p, q) of patterns of length $k + 1$:

1. $C_{(p,p)}(X)$ coincides with the autocorrelation polynomial $C_p(X)$ of the pattern p;

2. $[X^0]C_{(p,q)}(X) = \begin{cases} 0 & \text{if } p \neq q, \\ 1 & \text{if } p = q. \end{cases}$

Given the correlation polynomial $C_{(p,q)}(X) = \sum_{i=0}^{k} c_i X^i$, let $c_{i_1}, \ldots, c_{i_j}$ with $i_1 < \cdots < i_j$, the coefficients equal to 1. Let us denote by $\mathscr{C}_{(p,q)}$ the combinatorial class of binary sequences

$$\mathscr{C}_{(p,q)} := \{(c_0, \ldots, c_{i_1-1}), \ldots, (c_0, \ldots, c_{i_j-1})\}$$

endowed with the valuation which measures the length of the sequences (if, e.g., $i_1 = 0$, then the corresponding binary sequence is the empty one ()). Then $C_{(p,q)}(X)$ is the OGF of the combinatorial class $\mathscr{C}_{(p,q)}$. In particular observe that:

- $\mathbb{1} = \{()\} \subseteq \mathscr{C}_{(p,q)}$ if and only if $c_0 = 1$, i.e., if and only if $p = q$;
- $C_{(p,q)}(X) = 0$ if and only if $\mathscr{C}_{(p,q)} = \emptyset$.

Example 11.49 Let us consider, for example, the patterns

$$p = p_0 p_1 p_2 p_3 p_4 p_5 p_6 := \#@ * \#\#@\# \quad \text{and}$$
$$q = q_0 q_1 q_2 q_3 q_4 q_5 q_6 := \#@\# * @@ * .$$

1. For every $i \geq 0$, if $p_i \neq q_0$ we write $c_i = 0$:

pattern p	#	@	*	#	#	@	#
pattern q	#	@	#	*	@	@	*
coefficient c		0	0			0	
power of X	1	X	X^2	X^3	X^4	X^5	X^6

2. Then to compute c_i, $i = 0, 3, 4, 6$, we start by considering the block $p_i \ldots p_6$ from p_i onward, and then we compare it with the block of equal length starting from q_0; if the two coincide, we write $c_i = 1$, while otherwise we write $c_i = 0$:

pattern p	#	@	*	#	#	@	#
pattern q	#	@	#	*	@	@	*
coefficient c	0	0	0	0	1	0	1
power of X	1	X	X^2	X^3	X^4	X^5	X^6

The correlation polynomial of (p, q) is $C_{(p,q)}(X) = X^4 + X^6$ and

$$\mathscr{C}_{(p,q)} = \{(0, 0, 0, 0), (0, 0, 0, 0, 1, 0)\}.$$

Analogously we have:

pattern q	#	@	#	*	@	@	*
pattern p	#	@	*	#	#	@	#
coefficient c	0	0	0	0	0	0	0
power of X	1	X	X^2	X^3	X^4	X^5	X^6

The correlation polynomial of (q, p) is $C_{(q,p)}(X) = 0$ and

$$\mathscr{C}_{(q,p)} = \emptyset.$$

pattern p	#	@	*	#	#	@	#
pattern p	#	@	*	#	#	@	#
coefficient c	1	0	0	0	0	0	1
power of X	1	X	X^2	X^3	X^4	X^5	X^6

The correlation polynomial of (p, p) is $C_{(p,p)}(X) = C_p(X) = 1 + X^6$ and

$$\mathscr{C}_{(p,p)} = \{(), (1, 0, 0, 0, 0, 0)\}.$$

pattern q	#	@	#	*	@	@	*
pattern q	#	@	#	*	@	@	*
coefficient c	1	0	0	0	0	0	0
power of X	1	X	X^2	X^3	X^4	X^5	X^6

The correlation polynomial of (q, q) is $C_{(q,q)}(X) = C_q(X) = 1$ and

$$\mathscr{C}_{(q,q)} = \{()\}.$$ ∎

Following [27] we give the following definitions:

Definition 11.50 Let p and q be two patterns of the same length. A sequence $r \in \text{SEQ}\,\Gamma$ is a:

- p-**victory over** q if it does not contain q and p appears only at the end as suffix of r, i.e., following Definition 11.34, r belongs to $\text{SEQ}^{\cdot p}\,\Gamma \cap \text{SEQ}^{\neg q}\,\Gamma$;
- p-**previctory over** q if rp is a p-victory;
- **no-victory over** p **and** q if it contains neither p nor q.

We denote by $\mathscr{S}_{(p,q)} \subseteq \text{SEQ}\,\Gamma$ the set of all p-previctories over q, by $\mathscr{S}_{(q,p)} \subseteq \text{SEQ}\,\Gamma$ the set of all q-previctories over p, and by $\mathscr{N}_{(p,q)} \subseteq \text{SEQ}\,\Gamma$ the set of all no-victories. ∎

The combinatorial class $\mathscr{N}_{(p,q)}$ can be seen in two different ways.

Theorem 11.51 *The following isomorphisms of combinatorial classes hold:*

$$\mathscr{N}_{(p,q)} \cong \left(\mathscr{S}_{(p,q)} \times \mathscr{C}_{(p,p)}\right) \bigoplus \left(\mathscr{S}_{(q,p)} \times \mathscr{C}_{(q,p)}\right),$$

$$\mathscr{N}_{(p,q)} \cong \left(\mathscr{S}_{(q,p)} \times \mathscr{C}_{(q,q)}\right) \bigoplus \left(\mathscr{S}_{(p,q)} \times \mathscr{C}_{(p,q)}\right).$$

Proof Let n be an element of $\mathscr{N}_{(p,q)}$. The sequence np has certainly an initial part which is a p-victory over q or a q-victory over p.

If np has an initial part which is a p-victory over q, then $n = n^1 n^2$ and $p = p^1 p^2$ with $n^2 p^1 = p$ first occurrence of p in the sequence np. Since $n^1 p$ is a p-victory, then n^1 is a p-previctory over q. Since $n^2 p^1 = p^1 p^2 = p$, then p has a tail and a beginning equal to p^1: therefore if as tail $p^1 = p_i \dots p_k$, then the i-th coefficient c_i in $C_p(X)$ is equal to 1 and hence $(c_0, \dots, c_{i-1})$ belongs to $\mathscr{C}_{(p,p)}$. In such a case we associate to $n = n^1 n^2$ the equal-length sequence $n^1 c_0, \dots, c_{i-1} \in \mathscr{S}_{(p,q)} \times \mathscr{C}_{(p,p)}$. Observe that in this case $c_0 = 1$.

If np has an initial part which is a q-victory over p, then we have $n = n^1 n^2$ and $p = p^1 p^2$ with $n^2 p^1 = q$ first occurrence of q in the sequence np. Since $n^1 q = n^1 n^2 p^1$ is a q-victory, then n^1 is a q-previctory. Since $n^2 p^1 = q$, then q has a tail and p a beginning equal to p^1: therefore if $p^1 = q_i \dots q_k$, then the i-th coefficient c_i in $C_{(q,p)}(X)$ is equal to 1 and hence $(c_0, \dots, c_{i-1})$ belongs to $\mathscr{C}_{(q,p)}$. In such a case we associate to $n = n^1 n^2$ the equal-length sequence $n^1 c_0 \dots c_{i-1} \in \mathscr{S}_{(q,p)} \times \mathscr{C}_{(q,p)}$. Observe that in this case $c_0 = 0$. It is easy to verify that in such a way we have defined an isomorphism of combinatorial classes

$$\mathscr{N}_{(p,q)} \xrightarrow{\cong} \left(\mathscr{S}_{(p,q)} \times \mathscr{C}_{(p,p)}\right) \bigoplus \left(\mathscr{S}_{(q,p)} \times \mathscr{C}_{(q,p)}\right).$$

Analogously, distinguishing if the sequence nq has an initial part which is a p-victory over q or a q-victory over p, one proves the existence of the other isomorphism of combinatorial classes. □

We are now ready to prove the key result of this section.

Corollary 11.52 *In* $\mathbb{R}[[X]]$ *the following equality holds*

$$\mathscr{S}_{(p,q)}(X)\left(C_p(X) - C_{(p,q)}(X)\right) = \mathscr{S}_{(q,p)}(X)\left(C_q(X) - C_{(q,p)}(X)\right).$$

Proof From Proposition 11.19 and Theorem 11.51, passing to the corresponding OGF, one has

$$\mathscr{N}_{(p,q)}(X) = \mathscr{S}_{(p,q)}(X)C_p(X) + \mathscr{S}_{(q,p)}(X)C_{(q,p)}(X)$$
$$= \mathscr{S}_{(q,p)}(X)C_q(X) + \mathscr{S}_{(p,q)}(X)C_{(p,q)}(X);$$

then we get

$$\mathscr{S}_{(p,q)}(X)\left(C_p(X) - C_{(p,q)}(X)\right) = \mathscr{S}_{(q,p)}(X)\left(C_q(X) - C_{(q,p)}(X)\right). \qquad □$$

John Conway[1] was the first to discover the following elegant formula, though his proof was never published. Martin Gardner (see [17]) wrote about it:

> [...] I have no idea why it works. It just cranks out the answer as if by magic, like so many of Conway's other algorithms.

Theorem 11.53 (Conway equation) *Let* $p \neq q$ *be two patterns of the same length. The odds in favor of the appearing of q before p are equal to*

$$\frac{C_p(1/|\Gamma|) - C_{(p,q)}(1/|\Gamma|)}{C_q(1/|\Gamma|) - C_{(q,p)}(1/|\Gamma|)}.$$

Proof We follows the proof given in [27] in the case $|\Gamma| = 2$. Any p-victory over q is a sequence $v = xp$ with $x \in \mathscr{S}_{(p,q)}$, the set of all p-previctories over q. For each $n \in \mathbb{N}$, the pattern p appears before q for the first time in a sequence of length n with probability equal to the quotient of the number of the p-victories over q of length n, by the number of all possible n-sequences of Γ. The number of p-victories over q of length n is equal to the number of p-previctories over q of length $n-\ell(p) = n-(k+1)$, i.e., the cardinality of the $(n-k-1)$-fibre of the combinatorial class $\mathscr{S}_{(p,q)}$. Therefore, denoted by $(s_n)_n$ the sequence of the fibres of $\mathscr{S}_{(p,q)}$, the probability that p appears before q is

$$\frac{s_0}{|\Gamma|^{k+1}} + \frac{s_1}{|\Gamma|^{k+2}} + \cdots = \frac{1}{|\Gamma|^{k+1}} \sum_{i=0}^{\infty} s_i \frac{1}{|\Gamma|^i} = \frac{1}{|\Gamma|^{k+1}} \mathscr{S}_{(p,q)}\left(\frac{1}{|\Gamma|}\right) \neq 0.$$

---

[1] John Conway (1937-).

Analogously the probability that q appears before p is $\dfrac{1}{|\Gamma|^{k+1}} \mathscr{S}_{(q,p)}\left(\dfrac{1}{|\Gamma|}\right) \neq 0$.

Therefore the odds in favor of q are

$$\frac{\frac{1}{|\Gamma|^{k+1}} \mathscr{S}_{(q,p)}\left(\frac{1}{|\Gamma|}\right)}{\frac{1}{|\Gamma|^{k+1}} \mathscr{S}_{(p,q)}\left(\frac{1}{|\Gamma|}\right)} = \frac{\mathscr{S}_{(q,p)}\left(\frac{1}{|\Gamma|}\right)}{\mathscr{S}_{(p,q)}\left(\frac{1}{|\Gamma|}\right)}.$$

Since $|\Gamma|$ does not divide the coefficient of the leading term of $C_q(X) - C_{(q,p)}(X)$ (the coefficients of this non-zero polynomial are only -1, 0, and 1), the rational number $1/|\Gamma|$ is not a root of $C_q(X) - C_{(q,p)}(X)$. The conclusion then follows from Corollary 11.52.   □

Remark 11.54 If the pattern p is declared in advance, how can the pattern q be chosen in order to maximize its odds? It is proved in [21] that if $p = p_0 p_1 \ldots p_k$, then choosing $q := q_0 p_0 \ldots p_{k-1}$ for some appropriate $q_0 \in \Gamma$ is the best strategy to maximize the odds of q. More (see [16, Theorem 4.1]), the optimal leading symbol q_0 is any minimizer of

$$W_p(y) := \sum_{x \in \Gamma} C_{(p_0 \ldots p_{k-1} x, y p_0 \ldots p_{k-1})}\left(\frac{1}{|\Gamma|}\right),$$

as y varies in Γ.

Example 11.55 Consider again the patterns $p = \#@ * \#\#@\#$ and $q = \#@\# * @@*$ with $\Gamma = \{\#, @, *\}$. Since

$$C_q(X) = 1, \ C_{(q,p)}(X) = 0, \ C_p(X) = 1 + X^6, \ C_{(p,q)}(X) = X^4 + X^6,$$

the odds of q are equal to

$$\begin{aligned}
\frac{C_p(1/|\Gamma|) - C_{(p,q)}(1/|\Gamma|)}{C_q(1/|\Gamma|) - C_{(q,p)}(1/|\Gamma|)} &= 1 + \left(\frac{1}{3^6}\right) - \left(\frac{1}{3^4}\right) - \left(\frac{1}{3^6}\right) \\
&= 1 - \left(\frac{1}{3^4}\right) = \frac{3^4 - 1}{3^4} = \frac{80}{81} \approx 0.9876 < 1.
\end{aligned}$$

Therefore pattern p has an edge over pattern q. Following Remark 11.54, an optimal choice for the pattern q is to choose, instead of $\#@\# * @@*$, the pattern $q_0 \#@ * \#\#@$ with $q_0 \in \Gamma = \{\#, @, *\}$ which minimizes

$$W_p(y) := \sum_{x \in \{\#, @, *\}} C_{(\#@*\#\#@x, y\#@*\#\#@)}\left(\frac{1}{3}\right), \qquad y \in \Gamma.$$

One obtains that the element of Γ which minimizes this sum is $q_0 = *$. Therefore the pattern $q = *\#@ * \#\#@$ is an optimal choice: one gets that the odds of q are

$$\frac{C_p(1/|\Gamma|) - C_{(p,q)}(1/|\Gamma|)}{C_q(1/|\Gamma|) - C_{(q,p)}(1/|\Gamma|)} = \frac{1 + \frac{1}{3^6} - 0}{1 - (\frac{1}{3} + \frac{1}{3^6})} = \frac{146}{97} \approx 1.5 > 1. \qquad \Box$$

Example 11.56 (A bar trick for three-bit sequences) We consider here a three-bit sequence, say of Heads and Tails. Assuming that the first player chooses the sequence $p = p_0 p_1 p_2$, with $p_0, p_1, p_2 \in \{H, T\}$, a bar trick suggests that the second player, in order to optimize her/his odds, should play $q = \overline{p_1} p_0 p_1$, where we set $\overline{H} = T$ and $\overline{T} = H$. Let us show the validity of this strategy. Remark 11.54 suggests to choose a pattern of the form $q_0 p_0 p_1$, where q_0 minimizes the function

$$W(y) = \sum_{x \in \{H,T\}} C_{(p_0 p_1 x, y p_0 p_1)}(1/2), \qquad y \in \{H, T\}.$$

Since $\Gamma = \{H, T\} = \{p_1, \overline{p_1}\}$, just two cases may occur:

1. $p_1 = p_0$;

$$W(\overline{p_1}) = W(\overline{p_0}) = C_{(p_0 p_0 p_0, \overline{p_0} p_0 p_0)}(1/2) + C_{(p_0 p_0 \overline{p_0}, \overline{p_0} p_0 p_0)}(1/2)$$
$$= 0 + \left(\frac{1}{2^2}\right) = \frac{1}{2^2};$$

$$W(p_1) = W(p_0) = C_{(p_0 p_0 p_0, p_0 p_0 p_0)}(1/2) + C_{(p_0 p_0 \overline{p_0}, p_0 p_0 p_0)}(1/2)$$
$$= \left(1 + \frac{1}{2} + \frac{1}{2^2}\right) + 0 = 1 + \frac{1}{2} + \frac{1}{2^2};$$

2. $p_1 = \overline{p_0}$:

$$W(\overline{p_1}) = W(p_0) = C_{(p_0 \overline{p_0} p_0, p_0 p_0 \overline{p_0})}(1/2) + C_{(p_0 \overline{p_0}\, \overline{p_0}, p_0 p_0 \overline{p_0})}(1/2)$$
$$= \left(\frac{1}{2^2}\right) + 0 = \frac{1}{2^2};$$

$$W(p_1) = W(\overline{p_0}) = C_{(p_0 \overline{p_0} p_0, \overline{p_0} p_0 \overline{p_0})}(1/2) + C_{(p_0 \overline{p_0}\, \overline{p_0}, \overline{p_0} p_0 \overline{p_0})}(1/2)$$
$$= \left(\frac{1}{2}\right) + \left(\frac{1}{2^2}\right) = \frac{1}{2} + \frac{1}{2^2}.$$

In both situations it turns out that $W(\overline{p_1}) < W(p_1)$: it follows from Remark 11.54 that $q = \overline{p_1} p_0 p_1$ is indeed an optimal choice for the second player. Notice that the odds in favor of the second player depend, in any case, on the pattern p. For instance:

- If $p = $ HHH then $q = $ THH is optimal for the second player and the odds in her/his favor are

$$\frac{C_p(1/2) - C_{(p,q)}(1/2)}{C_q(1/2) - C_{(q,p)}(1/2)} = \frac{1 + \frac{1}{2} + \frac{1}{2^2}}{1 - \frac{1}{2} - \frac{1}{2^2}} = 7;$$

- If $p = $ HTH then $q = $ HHT is optimal for the second player and the odds in her/his favor are

$$\frac{C_p(1/2) - C_{(p,q)}(1/2)}{C_q(1/2) - C_{(q,p)}(1/2)} = \frac{1}{1 - \frac{1}{2}} = 2.$$

By considering the 8 possible patterns p, one sees that the odds in favor of the second player may just take the values 2, 3, 7 whenever q is optimal with respect to p.   □

Example 11.57 In Example 11.46 we left a monkey bashing the keys of a keyboard with 32 characters at random, waiting for the appearance of the pattern TOBE-ORNOTTOBE. We want to choose, in an optimal way, a pattern of the same length which has a better chance to appear. Let Γ be the set of 32 characters of the keyboard. Following Remark 11.54, we can choose a pattern q_0TOBEORNOTTOB with $q_0 \in \Gamma$ minimizing

$$W(y) := \sum_{x \in \Gamma} C_{(\text{TOBEORNOTTOB}x, y\text{TOBEORNOTTOB})} \left(\frac{1}{32}\right), \quad y \in \Gamma. \qquad (11.57.\text{a})$$

Let us compute the correlation polynomials. Comparing the patterns leaving as unknown x and y one easily gets

T	O	B	E	O	R	N	O	T	T	O	B	x
y	T	O	B	E	O	R	N	O	T	T	O	B
0	0	0	0	0	0	0	0	$c_8(x,y)$	0	0	$c_{11}(x,y)$	$c_{12}(x,y)$
1	X	X^2	X^3	X^4	X^5	X^6	X^7	X^8	X^9	X^{10}	X^{11}	X^{12}

where $c_8(x, y) = 1$ if and only if $x = $ E and $y = $ T, $c_{11}(x, y) = 1$ if and only if $x = $ T and $y = $ B, $c_{12} = 1$ if and only if $x = y$. Therefore to mimimize the sum (11.57.a) it is sufficient to choose $y \notin \{$B, T$\}$: for such a choice of y we get

$$C_{(\text{TOBEORNOTTOB}x, y\text{TOBEORNOTTOB})}(X) = \begin{cases} 0 & \text{if } x \neq y, \\ X^{12} & \text{if } x = y. \end{cases}$$

Therefore the minimal value of W is

$$W(y) = C_{(\text{TOBEORNOTTOB}y, y\text{TOBEORNOTTOB})} \left(\frac{1}{32}\right) = \left(\frac{1}{32}\right)^{12} \quad \forall y \in \Gamma \setminus \{\text{B, T}\}.$$

For instance, let us choose $y = $ E; if

$$p := \text{TOBEORNOTTOBE}, \quad q := \text{ETOBEORNOTTOB},$$

then $C_p(X) = 1 + X^9$, $C_q(X) = 1$, $C_{(p,q)}(X) = X^{12}$, and $C_{(q,p)}(X) = X + X^{10}$. Therefore the odds in favor of q are

$$\frac{C_p(1/|\Gamma|) - C_{(p,q)}(1/|\Gamma|)}{C_q(1/|\Gamma|) - C_{(q,p)}(1/|\Gamma|)} = \frac{1 + \frac{1}{32^9} - \frac{1}{32^{12}}}{1 - \left(\frac{1}{32} + \frac{1}{32^{10}}\right)} = \frac{32}{31} > 1. \qquad \square$$

11.5 Triangulations of a Convex Polygon ☕

Historically, the application that follows here initiated the birth of symbolic combinatorics.

Definition 11.58 (*Triangulations*) For every $n \in \mathbb{N}$, let P_{n+2} be a given convex polygon with $n + 2$ consecutive vertices $(1, \ldots, n + 2)$. We call:

1. **Diagonal** of P_{n+2} any segment $]i, j[$ (excluding the extremes) that conjoins two non-consecutive vertices i, j;
2. **Triangulation** of P_{n+2} any decomposition of P_{n+2} in triangles (i, j, k), with distinct vertices i, j, k, by way of diagonals which do not intersect two by two. $\square$

The polygon P_2 with two vertices, that is, a segment, admits only the *empty triangulation*.

Remark 11.59 It is easy to verify that for each $n \in \mathbb{N}$ one has that in P_{n+2}:

1. Every triangulation is formed by n triangles;
2. If $n \geq 3$, given a triangulation of the polygon, every side of the border of the polygon lies in a single triangle of the triangulation.

Figure 11.2 depicts the five triangulations of P_5. The problem of calculating the number of triangulations of a given convex polygon dates back Euler, who in 1751 spoke of it in a letter to his colleague Goldbach,[2] referring to a solution, without, however, furnishing the proof.

There are various ways of resolving the problem; here we will use the method of symbolic calculus. The following result furnishes a further interpretation of the Catalan numbers.

---

[2]Christian Goldbach (1690–1764).

Fig. 11.2 The case $n = 3$: the triangulations of P_5 are 5

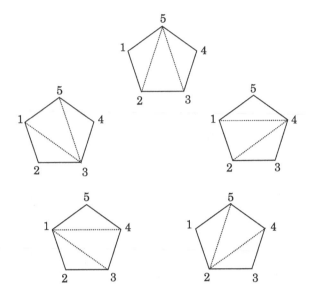

Proposition 11.60 *For every $n \in \mathbb{N}$, the number of triangulations of P_{n+2} is equal to the n-th Catalan number* $\mathrm{Cat}_n = \dfrac{1}{n+1}\dbinom{2n}{n}$.

Proof The idea of the proof consists in proving that every triangulation of P_{n+2} consists of the union of a triangle that contains the side $(1, 2)$ and the triangulations of the two convex polygons obtained by "removing" the aforementioned triangle from the initial polygon.

We introduce the combinatorial class $\mathscr{T}$ of the triangulations of all the polygons with valuation equal to the number of triangles. For each $n \in \mathbb{N}$, the n-fibre of $\mathscr{T}$ is the set of all triangulations of P_{n+2}; the 0-fibre of $\mathscr{T}$ contains only the empty triangulation $t_\varnothing$. Given a triangulation $t \in \mathscr{T}$ of P_{n+2}, the triangle that contains the side of vertices 1, 2 is uniquely determined: we indicate by h_t, $3 \le h_t \le n + 2$, the third vertex of that triangle. The convex polygon P_{n+2}, deprived of the side of vertices 1, 2, is the union of two convex polygons $Q_1(t)$ and $Q_2(t)$ of consecutive vertices respectively $(1, h_t, h_t + 1, \ldots, n + 2)$ and $(2, 3, \ldots, h_t)$: one notes that effectively the two polygons both have at least 2 vertices. The triangles of t with vertices at $\{1, h_t, h_t + 1, \ldots, n+2\}$ form a triangulation of $Q_1(t)$ (possibly *empty* if $h_t = n + 2$); in similar fashion the triangles of t with vertices at $\{2, 3, \ldots, h_t\}$ form a triangulation of $Q_2(t)$ (possible *empty* if $h_t = 3$). Renaming in ordered fashion the vertices of $Q_1(t)$ from 1 to $k(t) := (n + 2) - h_t + 2 = n + 4 - h_t$ and the vertices of $Q_2(t)$ from 1 to $m(t) := h_t - 1$, the triangulations on $Q_1(t)$ and $Q_2(t)$ induce a triangulation $t_1(t)$ on $P_{k(t)}$ and a triangulation $t_2(t)$ on $P_{m(t)}$.

Fig. 11.3 The action of the map Φ on a triangulation

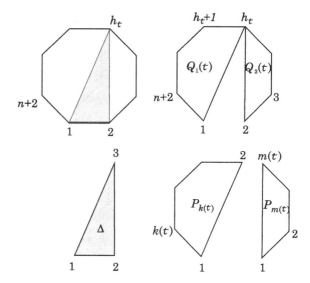

Having indicated with Δ the unique triangulation of P_3, we define the map Φ : $\mathcal{T} \setminus \{t_\emptyset\} \to \{\Delta\} \times \mathcal{T}^2$ which to $t \in \mathcal{T}$ associates the triple

$$\Phi(t) := (\Delta, t_1(t), t_2(t)).$$

The action of Φ on a triangulation is depicted in Fig. 11.3. Let us prove now that the map Φ is an isomorphism of combinatorial classes.

- One sees immediately that Φ conserves the valuations, given that the number of triangles of t is equal to one plus the number of triangles of $t_1(t)$, plus the number of triangles of $t_2(t)$.
- The map Φ is injective: if s and t are two triangulations of P_{n+2} that have the same image, then $t_2(t)$ and $t_2(s)$ are the same triangulation of $P_{m(t)} = P_{m(s)}$. From

$$h_t - 1 = m(t) = m(s) = h_s - 1$$

it follows that $h_t = h_s$ and so s and t have the same triangle with side $(1, 2)$. Since s and t must then induce the same triangulations on $P_{k(t)} = P_{k(s)}$ and $P_{m(t)} = P_{m(s)}$, they necessarily coincide.
- The map Φ is surjective: let t_1, t_2 be two triangulations of the polygons P_k and P_m respectively with $k \geq 2$ consecutive vertices $(1, \ldots, k)$ and with $m \geq 2$ consecutive vertices $(1, \ldots, m)$. We invite the reader to follow the proof by looking at Fig. 11.4. One substitutes the names of the vertices in t_1 replacing in ordered fashion $(1, \ldots, k)$ with $(1, m+1, m+2, \ldots, m+k-1)$: in this fashion one obtains a triangulation t_1' of a convex polygon of consecutive vertices $(1, m+1, m+2, \ldots, m+k-1)$. One substitutes the names of the vertices in t_2 by replacing in ordered fashion $(1, 2, \ldots, m)$ with $(2, \ldots, m+1)$: thus one obtains

Fig. 11.4 The surjectivity of the map Φ

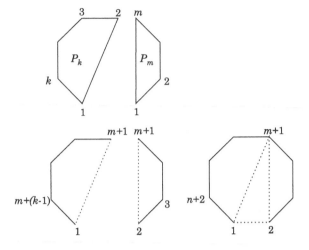

a triangulation t_2' of a convex polygon with consecutive vertices $(2, \ldots, m + 1)$. Having set $n + 2 = m + k - 1$, the triangulation t of P_{n+2} formed by the triangles in t_1', by the triangles in t_2' and by the triangle of vertices $(1, 2, m + 1)$ has as its own image (t_1, t_2, Δ).

Passing to the corresponding OGF, thanks to the isomorphism Φ just established between $\mathscr{T} \setminus \{t_\emptyset\}$ and $\{\Delta\} \times \mathscr{T}^2$, one obtains the equality

$$\mathscr{T}(X) - 1 = X \mathscr{T}(X)^2;$$

therefore, $\mathscr{T}(X)$ is a solution in $\mathbb{R}[[X]]$ of the equation

$$X\mathbb{Y}^2 - \mathbb{Y} + 1 = 0.$$

As has been seen in Example 7.104, the unique solution in $\mathbb{R}[[X]]$ of that equation is

$$\mathbb{Y}_2(X) = \frac{1 - \sqrt{1 - 4X}}{2X},$$

and one has $[X^n]\mathbb{Y}_2(X) = \mathrm{Cat}_n$ for every n. Therefore, the number of triangulations of the convex polygon P_{n+2} is equal to Cat_n. $\square$

11.6 Rooted Trees ☕

A **graph** consists of a finite set of **vertices** and one of **edges** that connect some couples of vertices. If there is at most one edge connecting any two vertices, the graph is said to be **simple**. Such graphs are generally represented with points (the

Fig. 11.5 The vertices d, g have degree 1; a, e, f have degree 2; b, c have degree 3; h has degree 4. The sum of the degrees is 18: twice the number of edges

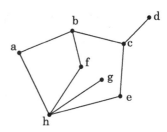

vertices) connected by arcs or lines (the edges). The **degree** of a vertex is the number of distinct edges of which it is an extreme. Since every edge has two extremes, it is easy to verify that the sum of the degrees of the vertices is equal to twice the number of edges (see Fig. 11.5).

Rooted plane trees are a class of simple graphs that appear often in the study of numerous algorithms; they are defined recursively.

Definition 11.61 A **rooted plane tree** is a simple graph consisting of a vertex, called the **root**, possibly linked by way of edges to the roots of a finite sequence of rooted plane trees, called the **sequence of the maximal sub-trees**. The tree consisting of the root alone is said to be **trivial**. The vertices of degree ≤ 1 are said to be **external**, while those of degree > 1 are said to be **internal**.   □

Remark 11.62 By definition, a rooted plane tree has at least one vertex, namely its root. If the rooted tree is not trivial, all its external vertices have degree equal to 1. The root can be both an internal vertex or an external vertex. Two rooted plane trees are *equal* if they are both trivial or if both their sequences of maximal subtrees coincide.

An important subfamily of the rooted plane trees is that of the *binary trees*:

Definition 11.63 A **binary tree** is a plane tree consisting of a vertex called **root** possibly connected by two edges to the roots of two binary trees (Fig. 11.6).   □

Remark 11.64 The sequence of maximal subtrees of a non trivial binary tree have length 2. It is easy to show by induction on the number of vertices that a binary tree has an odd number of vertices: indeed, the trivial tree has one vertex; if the binary tree is not trivial, its two maximal subtrees have, by the induction hypothesis, odd numbers n_1, n_2 of vertices and so the binary tree from which we have started has $(n_1 + n_2) + 1$ vertices.

Fig. 11.6 Two binary trees with root, indicated by ω. The external vertices are indicated with *white squares*

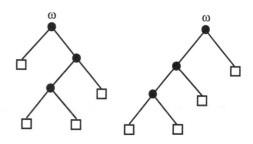

Fig. 11.7 All the possible
rooted plane trees that have 4
vertices. The root is
indicated with a *white circle*

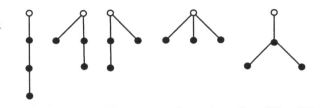

How many distinct rooted plane trees with a given number of vertices are there? And what about binary trees? It is easy to see that there is only a single rooted plane tree having one or two vertices, and that there are two with three vertices. There is, instead, a unique binary tree with one or three vertices and none with two vertices. The rooted trees with four vertices are listed in Fig. 11.7. None of these is a binary tree. The symbolic calculus allows us to answer the question in general. Here, too, the Catalan numbers intervene once again.

Proposition 11.65 *Let $n \in \mathbb{N}_{\geq 1}$; there are:*

1. Cat_{n-1} *rooted plane trees having n vertices;*
2. $\mathrm{Cat}_{\frac{n-1}{2}}$ *binary trees with an odd number n of vertices.*

Proof 1. Let $\mathscr{R}$ be the combinatorial class formed by the rooted plane trees having valuation that counts the number of vertices for every tree. We indicate with ω the tree formed by the single root. Every rooted plane tree consists of the root and the sequence (possibly empty) of its maximal subtrees. For every A in $\mathscr{R}$, we use $(A_1, \dots, A_{m_A})$ to denote the sequence of its maximal subtrees, possibly reduced to the empty sequence if $A = \omega$. On putting

$$A \mapsto \Phi(A) := \begin{cases} (\omega, (A_1, \dots, A_{m_A})) \text{ if } A \neq \omega, \\ (\omega, ()) \text{ if } A = \omega, \end{cases}$$

we define a bijective function

$$\Phi : \mathscr{R} \to \{\omega\} \times \mathrm{SEQ}\,\mathscr{R}$$

which induces an isomorphism of combinatorial classes.

Passing to the corresponding OGF, by Point 3 of Proposition 11.19 one obtains

$$\mathscr{R}(X) = X \, (\mathrm{SEQ}\,\mathscr{R}(X)) = X \frac{1}{1 - \mathscr{R}(X)},$$

which is to say,

$$\mathscr{R}(X)(1 - \mathscr{R}(X)) = X.$$

Therefore, $\mathscr{R}(X)$ is a solution in $\mathbb{R}[[X]]$ of the equation $\mathbb{Y}^2 - \mathbb{Y} + X = 0$. By Proposition 7.102, the solutions of that equation are

$$\mathbb{Y}_1(X) = \frac{1 + (1 - 4X)^{1/2}}{2}, \quad \mathbb{Y}_2(X) = \frac{1 - (1 - 4X)^{1/2}}{2}.$$

Let us develop $(1 - 4X)^{1/2}$ using Proposition 7.99:

$$\mathbb{Y}_1(X) = \frac{1 + (1 - 4X)^{1/2}}{2} = \frac{1}{2}\left(1 + \sum_{k=0}^{\infty} \binom{1/2}{k} (-4)^k X^k\right),$$

which, in particular, gives us

$$[X^0]\,\mathbb{Y}_1(X) = 1.$$

Now there are no trees in $\mathscr{R}$ having 0 vertices, so that

$$\mathscr{R}(X) = \mathbb{Y}_2(X) = \frac{1 - (1 - 4X)^{1/2}}{2}.$$

It follows that the number of rooted plane trees having n vertices is

$$[X^n]\mathscr{R}(X) = [X^n]\frac{1 - (1 - 4X)^{1/2}}{2}.$$

We recall that by Proposition 8.46 one has

$$\frac{1 - (1 - 4X)^{1/2}}{2X} = \sum_{n=0}^{\infty} \mathrm{Cat}_n\, X^n,$$

so that

$$\frac{1 - (1 - 4X)^{1/2}}{2} = \sum_{n=0}^{\infty} \mathrm{Cat}_n\, X^{n+1} = \sum_{n=1}^{\infty} \mathrm{Cat}_{n-1}\, X^n.$$

Consequently, the number of rooted plane trees having $n \geq 1$ vertices is

$$[X^n]\mathscr{R}(X) = [X^n]\frac{1 - (1 - 4X)^{1/2}}{2} = \mathrm{Cat}_{n-1}.$$

2. Let us indicate with $\mathscr{B}$ the combinatorial subclass of $\mathscr{R}$ formed by the binary trees. The restriction to the binary trees of the correspondence Φ considered in the previous point produces

$$A \mapsto \Phi(A) := \begin{cases} (\omega, (A_1, A_2)) & \text{if } A \neq \omega, \\ (\omega, ()) & \text{if } A = \omega. \end{cases}$$

It induces the following isomorphism of combinatorial classes:

$$\Phi : \mathscr{B} \to \{\omega\} \times \left(\mathbb{1} \bigoplus \mathscr{B}^2\right).$$

Passing to the corresponding OGF's, one obtains

$$\mathscr{B}(X) = X(1 + \mathscr{B}(X)^2).$$

It follows that $\mathscr{B}(X)$ is a solution in $\mathbb{R}[[X]]$ of the equation $X\mathbb{Y}^2 - \mathbb{Y} + X = 0$. By Proposition 7.102, the solutions of this equation in $\mathbb{R}((X))$ are

$$\mathbb{Y}_1(X) = \frac{1 + (1 - 4X^2)^{1/2}}{2X}, \quad \mathbb{Y}_2(X) = \frac{1 - (1 - 4X^2)^{1/2}}{2X}.$$

Using the expansion of $(1 + X)^{1/2}$ found in Example 7.100 one sees immediately that

$$\left[X^0\right]\left(1 + (1 - 4X^2)^{1/2}\right) = 2,$$

and hence X does not divide $1 + (1 - 4X^2)^{1/2}$: it follows that $\mathbb{Y}_1(X)$ is not a formal power series. Consequently one must have $\mathscr{B}(X) = \mathbb{Y}_2(X)$. We know from Proposition 8.46 that

$$\frac{1 - (1 - 4X)^{1/2}}{2X} = \sum_{m=0}^{\infty} \mathrm{Cat}_m X^m.$$

By "replacing" X^2 for X one finds immediately that

$$\frac{1 - (1 - 4X^2)^{1/2}}{2X^2} = \sum_{m=0}^{\infty} \mathrm{Cat}_m X^{2m},$$

from which it follows that

$$\mathscr{B}(X) = \frac{1 - (1 - 4X^2)^{1/2}}{2X} = X\frac{1 - (1 - 4X^2)^{1/2}}{2X^2} = \sum_{m=0}^{\infty} \mathrm{Cat}_m X^{2m+1}.$$

In particular, we reobtain that there are no binary trees with an even number of vertices, and that for every $m \in \mathbb{N}$ the number of binary trees with $n = 2m + 1$ vertices is equal to

$$\left[X^{2m+1}\right]\mathbb{Y}_2(X) = \mathrm{Cat}_m = \mathrm{Cat}_{\frac{n-1}{2}}. \qquad \square$$

Remark 11.66 It follows from Proposition 11.65 and from Proposition 3.37 that the number of plane trees with $n + 1$ vertices coincides with the number of $2n$-sequences of Dyck. One establishes an isomorphism between these two classes of objects by associating to every rooted plane tree formed by $n+1$ vertices, the $2n$-sequence which

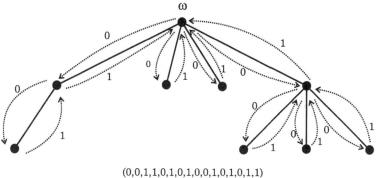

(0,0,1,1,0,1,0,1,0,0,1,0,1,0,1,1)

Fig. 11.8 The correspondence between a plane tree with root ω having 9 vertices and the 16-sequence of Dyck $(0, 0, 1, 1, 0, 1, 0, 1, 0, 0, 1, 0, 1, 0, 1, 1)$

is derived as follows, as in Fig. 11.8: one runs over the edges of the tree starting from the root "keeping the tree at one's left", running over every edge first in one direction and then in the opposite direction when coming back; attributing sequentially the value 0 each time that one moves over a edge the first time, and 1 the second time, one obtains a $2n$-sequence of Dyck. We leave the verification of these statements to the careful reader (Problem 11.14).

One can deal with the same problem using various combinatorial classes: here we shall recover the number of binary trees using an alternative combinatorial class to that in the proof of Proposition 11.65.

Example 11.67 It is not too difficult to prove that in a binary tree the number of internal vertices determines the total number of vertices: more precisely, a binary tree with n internal vertices has exactly $2n + 1$ vertices (Problem 11.15). To calculate the number of binary trees with a given number of vertices is therefore equivalent to finding the number of binary trees with a given number of internal vertices. Let us indicate by $\mathscr{B}^{int}$ the combinatorial class formed by the binary trees, with valuation ν^{int} equal to the number of internal vertices.

For each A in $\mathscr{B}^{int}$, let us denote by (A_1, A_2) the sequence (possibly empty) of the maximal subtrees of A. The trivial binary tree ω formed by the single root does not have internal vertices; a non-trivial binary tree has an internal vertex (its root) of degree 2, and the other internal vertices have degree 3. These last vertices are internal vertices also of the maximal subtrees. Indicated with Λ the binary tree with three vertices (a root connected to two vertices), it is easy to verify that the position

$$A \mapsto \begin{cases} \omega & \text{if } A = \omega, \\ (\Lambda, (A_1, A_2)) & \text{otherwise,} \end{cases}$$

defines an isomorphism of combinatorial classes

$$\mathscr{B}^{int} \cong \{\omega\} \bigoplus (\{\Lambda\} \times \mathscr{B}^{int} \times \mathscr{B}^{int}).$$

By Proposition 11.19 one recovers

$$\mathscr{B}^{int}(X) = 1 + X\mathscr{B}^{int}(X)^2.$$

Therefore $\mathscr{B}^{int}(X)$ is a solution in $\mathbb{R}[[X]]$ of the equation $X\mathbb{Y}^2 - \mathbb{Y} + 1 = 0$. By what has been seen in Example 7.104, such an equation has only one solution in $\mathbb{R}[[X]]$ and so necessarily

$$\mathscr{B}^{int}(X) = \frac{1 - \sqrt{1 - 4X}}{2X} \quad \text{and} \quad [X^m]\mathscr{B}^{int}(X) = \mathrm{Cat}_m, \; \forall m \in \mathbb{N}.$$

Thus there are Cat_m binary trees of ν^{int}-size $m \geq 0$, or with $2m + 1$ vertices. In particular, having set $n = 2m + 1$, there are $\mathrm{Cat}_{\frac{n-1}{2}}$ binary trees with n vertices, which is in agreement with what we have found in Proposition 11.65.

Alternatively, one could have proceeded by considering the class of the binary trees $\mathscr{B}^{ext}$ with valuation equal to the number of the *external* vertices: in this case $[X^m]\,\mathscr{B}^{ext}(X)$ is the number of binary trees that have m external vertices, or $2m - 1$ vertices. Proceeding as above one finds that $\mathscr{B}^{ext}(X)$ is a root in $\mathbb{R}[[X]]$ of the equation $\mathbb{Y}^2 - \mathbb{Y} + X = 0$: we leave the simple verification to the reader. □

Example 11.68 Let us consider the subclass $\mathscr{P}$ of the class $\mathscr{R}$ of the rooted plane trees containing the trivial plane tree with the single root ω and the rooted plane trees having a number ≥ 2 of maximal subtrees, all belonging to $\mathscr{P}$ (Fig. 11.9).

1. Prove that the class $\mathscr{P}$ is isomorphic to a suitable subclass of $\{\omega\} \times$ SEQ $\mathscr{P}$.
2. Deduce that the OGF of $\mathscr{P}$ satisfies an equation of second degree with coefficients in $\mathbb{R}[[X]]$.
3. Determine the OGF of $\mathscr{P}$.
4. Find the number of trees of $\mathscr{P}$ with 10 vertices.

Fig. 11.9 A tree of the class $\mathscr{P}$; the root is indicated by ω

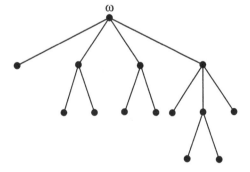

Solution. 1. Given that every tree A of $\mathscr{P}$ consists of a unique root ω, or of the root connected to a sequence $(A_1, A_2, \ldots, A_m)$ with $m \geq 2$ maximal subtrees, the correspondence Φ established in the proof of Proposition 11.65 allows one to obtain the following isomorphism of combinatorial classes

$$\mathscr{P} \cong \{\omega\} \times \left(\mathbb{1} \bigoplus \mathrm{SEQ}_{n \geq 2} \mathscr{P}\right).$$

2. Passing to the corresponding OGF's, one obtains that

$$\begin{aligned}
\mathscr{P}(X) &= X \left(1 + \mathscr{P}(X)^2 + \mathscr{P}(X)^3 + \cdots\right) \\
&= X + X \mathscr{P}(X)^2 \left(1 + \mathscr{P}(X) + \mathscr{P}(X)^2 + \cdots\right) \\
&= X + X \frac{\mathscr{P}(X)^2}{1 - \mathscr{P}(X)}.
\end{aligned}$$

It follows that $\mathscr{P}(X)$ is a solution in $\mathbb{R}[[X]]$ of the equation

$$(X + 1)\mathbb{Y}^2 - (1 + X)\mathbb{Y} + X = 0.$$

The solutions in $\mathbb{R}((X))$, are

$$\mathbb{Y}_1(X) = \frac{1 + X + \sqrt{-3X^2 - 2X + 1}}{2(1 + X)}, \quad \mathbb{Y}_2(X) = \frac{1 + X - \sqrt{-3X^2 - 2X + 1}}{2(1 + X)}.$$

Since the formal power series $2(1 + X)$ is invertible, both the solutions belong to $\mathbb{R}[[X]]$. Now the first term of the expansion of

$$\frac{\sqrt{-3X^2 - 2X + 1}}{1 + X} = \sqrt{-3X^2 - 2X + 1}(1 - X + X^2 + \cdots)$$

is 1 (the calculation is carried out in Example 7.65), and therefore

$$\left[X^0\right] \mathbb{Y}_1(X) = \frac{1}{2} + \frac{1}{2} = 1.$$

If one had $\mathscr{P}(X) = \mathbb{Y}_1(X)$ one would then have $\left[X^0\right] \mathscr{P}(X) = 1$, but that is absurd (the number of trees of $\mathscr{P}$ with zero vertices is equal to 0, not to 1!). Consequently,

$$\mathscr{P}(X) = \mathbb{Y}_2(X) = \frac{1 + X - \sqrt{-3X^2 - 2X + 1}}{2(1 + X)}.$$

3. We set $\mathscr{P}(X) = \sum_{n=0}^{\infty} p_n X^n$. From the relation

$$(X+1)\mathscr{P}(X)^2 - (1+X)\mathscr{P}(X) + X = 0,$$

we obtain

$$[X^n]\left((X+1)\mathscr{P}(X)^2 - (1+X)\mathscr{P}(X) + X\right) = 0 \quad \forall n \in \mathbb{N}. \qquad (11.68.a)$$

Now, for a given $n \in \mathbb{N}_{\geq 1}$ one has

$$[X^n]\mathscr{P}(X)^2 = \sum_{k=0}^{n} p_k p_{n-k},$$

$$[X^n](1+X)\mathscr{P}(X)^2 = \sum_{k=0}^{n} p_k p_{n-k} + \sum_{k=0}^{n-1} p_k p_{n-1-k},$$

$$[X^n](1+X)\mathscr{P}(X) = p_n + p_{n-1}.$$

The relation (11.68.a) then gives

$$\sum_{k=0}^{n} p_k p_{n-k} + \sum_{k=0}^{n-1} p_k p_{n-1-k} - p_n - p_{n-1} = \begin{cases} -1 & \text{if } n = 1, \\ 0 & \forall n \geq 2. \end{cases}$$

Bearing in mind that $p_0 = 0$ one obtains $p_1 = 1$, $p_2 = 0$ and

$$p_n = \sum_{k=1}^{n-1} p_k p_{n-k} + \sum_{k=1}^{n-2} p_k p_{n-1-k} - p_{n-1} \quad \forall n \geq 3.$$

Thus we have

$$p_3 = (2p_1 p_2) + p_1^2 - p_2 = 1,$$
$$p_4 = (2p_1 p_3 + p_2^2) + (2p_1 p_2) - p_3 = 2 + 0 - 1 = 1,$$
$$p_5 = (2p_1 p_4 + 2p_2 p_3) + (2p_1 p_3 + p_2^2) - p_4 = 2 + 2 - 1 = 3.$$

Proceeding patiently, or making use of a CAS one obtains the first 11 terms of the expansion of $\mathscr{P}(X)$:

$$[X^{\leq 10}]\mathscr{P}(X) = X + X^3 + X^4 + 3X^5 + 6X^6 + 15X^7 + 36X^8 + 91X^9 + 232X^{10}.$$

In particular in the class $\mathscr{P}$ there are 232 trees with 10 vertices. $\square$

11.7 Problems

Problem 11.1 Consider the combinatorial class $\text{SEQ}_{\{\geq 1\}}\{0, 1\}$ of the non-*empty* binary sequences with valuation function that associates to each sequence its length. Every such sequence can be either a sequence of length 1 ((0) or (1)), or else a longer sequence beginning with 0 or with 1. Using this fact, prove the following isomorphism of combinatorial classes

$$\text{SEQ}_{\{\geq 1\}}\{0, 1\} \cong \{(0)\} \bigoplus \{(1)\} \bigoplus \left(\{0, 1\} \times \text{SEQ}_{\{\geq 1\}}\{0, 1\}\right).$$

Problem 11.2 Consider the combinatorial class $\text{SEQ}\{0, 1\}$ of the binary sequences with the valuation function that associates to every sequence its length. Prove the following isomorphism of combinatorial classes:

$$\text{SEQ}\{0, 1\} \cong \{()\} \bigoplus \left(\{0, 1\} \times \text{SEQ}\{0, 1\}\right).$$

Problem 11.3 Determine the autocorrelation polynomial of the pattern $p = (0, 1)$ and then the OGF of the combinatorial class $\text{SEQ}^{\neg p}\{0, 1\}$ of the binary sequences which do not contain p. In particular, deduce the number of binary 4-sequences that do not contain p.

Problem 11.4 Find the autocorrelation polynomial of the pattern $p = (1, 0, 1, 0)$ and then the OGF of the combinatorial class of the binary sequences that do not contain p. In particular deduce the number of binary 6-sequences which do not contain p.

Problem 11.5 Find the autocorrelation polynomial of the pattern $p = (0, 0, 0, 0, 0)$ and thus the OGF of the combinatorial class of the binary sequences that do not contain p. In particular, deduce the number of binary 7-sequences which do not contain p.

Problem 11.6 What is the number of binary sequences of length 8 which do not contain a strip of 3 consecutive zeroes?

Problem 11.7 Determine the number of binary 5-sequences that do not contain the pattern 010.

Problem 11.8 Determine the autocorrelation polynomial of the pattern $p = (1, 1)$ of $\Gamma = \{1, 2, 3\}$ and thus the OGF of the combinatorial class of the sequences that do not contain p. In particular, deduce the number of 4-sequences of Γ which do not contain p.

Problem 11.9 What is the probability that a word of 6 letters written by chance from an alphabet Γ of 26 letters contains the Indian name $p := TATA$? And with the alphabet $\Gamma' = \{A, T\}$?

Problem 11.10 Determine the number of 6-sequences of $\Gamma = \{a, b, c, d, e\}$ which do not contain the pattern $p = abc$.

Problem 11.11 Determine the number of 4-sequences of $\Gamma = \{0, 1, 2, 3\}$ that contain the pattern $p = 11$.

Problem 11.12 We toss a coin until when one of the following two patterns appears:

$$p := HTHTTHH, \quad q := THTTHTH.$$

1. Compute the odds in favor of p.
2. Change the pattern p in order to maximize the odds in its favor.

Problem 11.13 Prove that a rooted plane tree with n vertices has $n - 1$ edges.

Problem 11.14 Let $n \geq 1$. Prove that the correspondence between rooted plane trees with $n + 1$ vertices and the $2n$-sequences of $\{0, 1\}$ established in Remark 11.66 is a bijection between rooted plane trees and the $2n$-sequences of Dyck.

Problem 11.15 Prove that, in a binary tree, the number of external vertices is exactly one plus the number of internal vertices.

Problem 11.16 A ternary tree is a plane tree constituted by a vertex called its root possibly connected by way of three edges to the roots of three ternary trees, as the one depicted in Fig. 11.10. We use $\mathcal{T}_3$ to indicate the combinatorial class formed by the ternary trees, with the valuation equal to the number of vertices.

1. Prove that the class $\mathcal{T}_3$ is isomorphic to a suitable subclass of $\{\omega\} \times$ SEQ $\mathcal{T}_3$.
2. Deduce that the OGF of $\mathcal{T}$ satisfies an equation of third degree with coefficients in $\mathbb{R}[[X]]$.
3. Verify the correctness of the result on the number of ternary trees that have 7 vertices.

Problem 11.17 A unary–binary tree is a plane tree consisting of a vertex, called its root, possibly connected by 0, 1 or 2 edges to the roots of unary–binary trees (Fig. 11.11). Let $\mathcal{U}$ indicate the combinatorial class formed by the unary–binary trees, with the valuation equal to the number of vertices.

Fig. 11.10 A ternary tree;
the root is indicated with ω

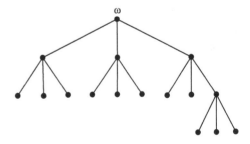

Fig. 11.11 A unary–binary
tree; the root is indicated
with ω

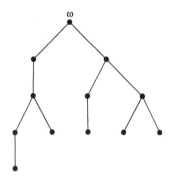

1. Prove that the class $\mathcal{U}$ is isomorphic to a suitable subclass of $\{\omega\} \times \mathrm{SEQ}\, \mathcal{U}$.
2. Deduce that the ordinary generating formal series $\mathcal{U}(X)$ of $\mathcal{U}$ satisfies an equation of second degree with coefficients in $\mathbb{R}[[X]]$.
3. Determine the OGF of $\mathcal{U}$.

Chapter 12
The Euler–Maclaurin Formulas
of Order 1 and 2

The Euler–MacLaurin summation formula is one of the most
remarkable formulas of mathematics.

G.C. Rota [32]

Euler's summation formula and its relation to Bernoulli
numbers and polynomials provides a treasure trove of interesting
enrichment material suitable for elementary calculus courses.

T.M. Apostol [3]

Abstract Let $a < b$ be two integers and $f : [a, b] \to \mathbb{R}$ a function. In Chap. 6 we
saw how to calculate the sum $\displaystyle\sum_{a \leq k < b} f(k)$ through the notion of a discrete primitive. In
this chapter we instead study the problem of estimating such a sum. It is well known
that when f is a monotonic function one can make such an approximation by way of
the integral $\displaystyle\int_a^b f(t)\, dt$: in this case the modulus of the difference between the sum
and the integral of f is at most equal to $|f(b) - f(a)|$. The integral formula may be
extended for non-monotonic functions provided that they are regular, and is referred
to as the *Euler–Maclaurin formula*. The Euler–Maclaurin formula represents both a
method for calculating the sum $\displaystyle\sum_{a \leq k < b} f(k)$ by means of the integral of f, and also a
way of calculating the integral $\displaystyle\int_a^b f(x)\, dx$ once one knows the sum $\displaystyle\sum_{a \leq k < b} f(k)$. These
two different points of view gave Euler and Maclaurin their respective motivations
for establishing the formula bearing their names, which was subsequently treated
in detail [28] by Poisson. In this chapter we deal only with the Euler–Maclaurin
formulas of order 1 and 2 for functions that are, respectively, of class $\mathscr{C}^1$ or $\mathscr{C}^2$:
these cases are the source of most of the subsequent applications in the book. We
treat the Euler–Maclaurin formula in full generality for functions of class $\mathscr{C}^m$ in the
next chapter. In the last section we recover Euler–Maclaurin type formulas assuming
just the monotonicity (no smoothness) of the function: the methods employed there
are very different from those concerning regular functions and arise from the proof of
the celebrated integral test for the convergence of series with monotonic terms. Last
but not least, in many cases we make use of a CAS to perform hard computations

© Springer International Publishing Switzerland 2016
465
C. Mariconda and A. Tonolo, *Discrete Calculus*,
UNITEXT - La Matematica per il 3+2 103, DOI 10.1007/978-3-319-03038-8_12

(like difficult integrals, sums, sums of series). Some remarks about the notation used throughout the following chapters: remainders in formulas will be typically denoted by a letter R; we will replace the R by ε if the remainder may tend to 0 by passing to the limit in some parameter, i.e., when the formula truly provides an approximation.

12.1   Decimal Representations, Basic Integral Estimates

This chapter deals with estimates of various kinds. Most of the time, however, we will truncate decimal numbers and give an upper bound for integrals (that may be difficult to compute). We now spend some time clarifying these procedures.

12.1.1   Decimal Representation and Approximations of a Real Number

We will often formulate the results of examples by means of decimal numbers. We begin with a universally well-known notation. We consider only positive numbers: for negative numbers one applies the same arguments to the opposite ones.

Definition 12.1 (*n-digit numbers*) Let $x_0 \in \mathbb{N}$ and $x_1, \ldots, x_n \in \{0, \ldots, 9\}$. We set

$$x_0 . x_1 \ldots x_n := x_0 + \frac{x_1}{10} + \cdots + \frac{x_n}{10^n}.$$

Such a number is called a n-**digit number**.   □

Remark 12.2 Notice that $x_0 . x_1 \ldots x_n$ defined above is a rational number.

Example 12.3 For instance, $3.14 = 3 + \dfrac{3}{10} + \dfrac{4}{10^2}$.   □

Example 12.4 The sequence $(0.\underbrace{9 \ldots 9}_{n \text{ times}})_n$ converges to 1 as $n \to +\infty$. Indeed

$$0.\underbrace{9 \ldots 9}_{n \text{ times}} = \frac{9}{10}\left(1 + \frac{1}{10} + \cdots + \frac{1}{10^{n-1}}\right) = \frac{9}{10}\frac{1 - \dfrac{1}{10^n}}{1 - \dfrac{1}{10}} \to 1$$

as $n \to +\infty$.   □

Proposition 12.5 *Let $x \in \mathbb{R}_{\geq 0}$. There exist $x_0 \in \mathbb{N}$ and a (unique) sequence $(x_n)_{n \geq 1}$ in $\{0, \ldots, 9\}$ such that*

$$x_0 . x_1 \ldots x_n \leq x < x_0 . x_1 \ldots x_n + \frac{1}{10^n} \quad \forall n \in \mathbb{N}. \tag{12.5.a}$$

*Thus $(x_0 . x_1 \ldots x_n)_n$ converges to x. We write $x = x_0 . x_1 \ldots x_n \ldots$ and say that the latter is the **decimal representation** of x.*

Proof. Let $x_0 = [x]$, the floor of x. Then $x_0 \le x < x_0 + 1$. Let $x_1 = [10(x - x_0)]$: then $x_1 \le 10(x - x_0) < x_1 + 1$ so that

$$x_0 . x_1 \le x < x_0 . x_1 + \frac{1}{10}.$$

Assume that for $n \ge 1$ we find $x_1, \ldots, x_n$ satisfying

$$x_0 . x_1 \ldots x_m \le x < x_0 . x_1 \ldots x_m + \frac{1}{10^m} \qquad \forall m = 1, \ldots, n.$$

Set $x_{n+1} := \left[10^{n+1} \left(x - x_0 - \frac{x_1}{10} - \cdots - \frac{x_n}{10^n} \right) \right]$. Then

$$x_0 . x_1 \ldots x_{n+1} \le x < x_0 . x_1 \ldots x_{n+1} + \frac{1}{10^{n+1}}.$$

Inductively, we obtain the desired conclusion. □

☞ *Remark 12.6* Small dots makes the difference: in general

$$x = x_0 . x_1 \ldots x_n \ldots \ne x_0 . x_1 \ldots x_n.$$

For instance $\pi = 3 . 14 \ldots$ is not equal to the rational number 3.14.

Remark 12.7 The decimal representation of a real number never ends with an infinite sequence of 9. Indeed if, say, $x = x_0 . x_1 \ldots x_n 999 \ldots$ for some $n \ge 1$ then we would have

$$x_0 . x_1 \ldots x_n \le x = x_0 . x_1 \ldots x_n + \sum_{k=n+1}^{\infty} \frac{9}{10^k} = x_0 . x_1 \ldots x_n + \frac{1}{10^n},$$

thus violating the strict inequality in (12.5.a). Any sequence $(x_n)_n$ of natural numbers not definitively equal to 9 is the unique decimal representation of the positive real number

$$x := \lim_{n \to \infty} \sum_{i=0}^{n} \frac{x_n}{10^n}.$$

The n-digit numbers can be characterized as follows.

Proposition 12.8 *A real number x is a n-digit number if and only if $10^n x \in \mathbb{N}$.*

Proof. If $x = x_0 . x_1 \ldots x_n$ with $x_0 \in \mathbb{N}$ and $x_1, \ldots, x_n \in \{0, \ldots, 9\}$ then

$$10^n x = 10^n x_0 + 10^{n-1} x_1 + \cdots + 10 x_{n-1} + x_n \in \mathbb{N}.$$

Conversely, assume that $10^n x \in \mathbb{N}$ and let $x_0 . x_1 \ldots x_n \ldots$ be the decimal representation of x. Then, by (12.5.a), we have

$$10^n x = \left[10^n x\right] = 10^n (x_0 . x_1 \ldots x_n),$$

whence $x = x_0 . x_1 \ldots x_n$. $\qquad\qquad\qquad\qquad\qquad\qquad\qquad\qquad\qquad$ □

Example 12.9 The sum of two n-digit numbers is an n-digit number. Indeed if $x = x_0 . x_1 \ldots x_n$ and $y = y_0 . y_1 \ldots y_n$ then

$$10^n (x + y) = 10^n (x_0 + y_0) + 10^{n-1}(x_1 + y_1) + \cdots + (x_n + y_n) \in \mathbb{N}. \qquad □$$

The conclusion follows from Preposition 12.8

Definition 12.10 (*Decimals*) Let $x \in \mathbb{R}_{\geq 0}$ and $x_0 . x_1 \ldots x_n \ldots$ be its decimal representation. The number x_0 is the **integer part** or **floor** of x, the numbers $x_1, \ldots, x_n, \ldots$ are the **decimals** of x and x_n is the n-th **decimal** of x, or the decimal of x in the n-th place ($n \geq 1$). The **truncation** of $x_0 . x_1 \ldots x_n \ldots$ to the n-th decimal is $x_0 . x_1 \ldots x_n$. $\qquad\qquad\qquad\qquad\qquad$ □

Two numbers with the same integer parts whose first $n \geq 0$ decimals are equal differ by at most 10^{-n}.

Proposition 12.11 *Let $x = x_0 . x_1 \ldots x_n \ldots$ and $y = y_0 . y_1 \ldots y_n \ldots$ be two decimal numbers together with their decimal representation. Assume that*

$$x_0 = y_0, \quad x_1 = y_1, \quad \ldots, \quad x_n = y_n.$$

Then $|x - y| < 10^{-n}$.

Proof. Since $x - y = \displaystyle\sum_{k=n+1}^{\infty} \frac{x_k - y_k}{10^k}$ then

$$|x - y| = \left| \sum_{k=n+1}^{\infty} \frac{x_k - y_k}{10^k} \right| \leq \sum_{k=n+1}^{\infty} \frac{|x_k - y_k|}{10^k}$$

$$\leq \sum_{k=n+1}^{\infty} \frac{9}{10^k} = \frac{9}{10^{n+1}} \frac{10}{9} = \frac{1}{10^n}.$$

Notice that, in the above inequality, the equality cannot hold, otherwise

$$x_k - y_k = 9 \quad \forall k > n \text{ or } y_k - x_k = 9 \quad \forall k > n,$$

and this occurs if and only if the decimal representation of x or y has a queue ending with a sequence of 9, contradicting Remark 12.7: thus $|x - y| < 10^{-n}$. $\qquad$ □

Remark 12.12 Notice that the converse of Proposition 12.11 does not hold. For instance $1.01 - 0.99 = 0.02 < \dfrac{1}{10}$ but the two numbers have different corresponding decimals.

Everybody is supposed to know that the rounding of 1.45799 to the 3rd decimal is 1.458 and that this number should be the decimal number with 3 digits that is nearest to 1.45799. We now justify and clarify these notions.

Lemma 12.13 *Let $x \in \mathbb{R}_{\geq 0}$, and let $x_0 . x_1 \ldots x_n \ldots$ its decimal representation. The n-digit number nearest to x is either $x_0 . x_1 \ldots x_n$ or $x_0 . x_1 \ldots x_n + \dfrac{1}{10^n}$.*

Proof. It follows from (12.5.a) that $10^n (x_0 . x_1 \ldots x_n) = [10^n x]$. Let y be a n-digit number. Then, by Proposition 12.8, $10^n y$ is an integer. It follows that either $10^n y \leq 10^n (x_0 . x_1 \ldots x_n)$ or $10^n y \geq 10^n (x_0 . x_1 \ldots x_n) + 1$: in any case $y \leq x_0 . x_1 \ldots x_n$ or $y \geq x_0 . x_1 \ldots x_n + \dfrac{1}{10^n}$. $\qquad\square$

Definition 12.14 (*Rounding decimals*) Let $x \in \mathbb{R}_{\geq 0}$ and $x_0 . x_1 \ldots x_n \ldots$ its decimal representation. The **rounding of** x to the n-th digit is the n-digit number y closest to x. Thus, from Lemma 12.13:

- $y = x_0 . x_1 \ldots x_n$ if $x_{n+1} < 5$;
- $y = x_0 . x_1 \ldots x_n + \dfrac{1}{10^n}$ if $x_{n+1} > 5$.

We set $y = x_0 . x_1 \ldots x_n + \dfrac{1}{10^n}$ if $x_{n+1} = 5$. $\qquad\square$

Here is the trick in order to get the explicit form of the rounding of a decimal number.

Proposition 12.15 *Let $x \in \mathbb{R}_{\geq 0}$ and $x_0 . x_1 \ldots x_n \ldots$ its decimal representation. The rounding of x to the n-th digit is the truncation of $x_0 . x_1 \ldots x_n x_{n+1} + \dfrac{5}{10^{n+1}}$ to the n-th digit.*

Proof. Indeed, if $x_{n+1} < 5$ then $x_{n+1} + 5 \leq 9$ and thus

$$x_0 . x_1 \ldots x_n x_{n+1} + \frac{5}{10^{n+1}} = x_0 . x_1 \ldots x_n (x_{n+1} + 5),$$

so that its truncation $x_0 . x_1 \ldots x_n$ coincides with the rounding of x to the n-th digit. If $x_{n+1} \geq 5$ then

$$x_0 . x_1 \ldots x_n + \frac{5}{10^{n+1}} \leq x_0 . x_1 \ldots x_n x_{n+1} < x_0 . x_1 \ldots x_n + \frac{1}{10^n}$$

so that, adding $5/10^{n+1}$ to the terms of the above inequalities, we get

$$x_0 . x_1 \ldots x_n + \frac{1}{10^n} \leq x_0 . x_1 \ldots x_n x_{n+1} + \frac{5}{10^{n+1}} < \left(x_0 . x_1 \ldots x_n + \frac{1}{10^n} \right) + \frac{5}{10^{n+1}},$$

proving that $x_0 . x_1 \ldots x_n + \dfrac{1}{10^n}$, which is the truncation of $x_0 . x_1 \ldots x_n x_{n+1} + \dfrac{5}{10^{n+1}}$ to the n-th digit, is the rounding of x to the n-th digit. $\qquad\square$

Example 12.16 For instance, consider $\pi = 3.1415926535897932385 \ldots$. It follows
from Definition 12.14 that:

1. The rounding of π to the second digit is 3.14;
2. The rounding of π to the fourth digit 3.1416;
3. The rounding of π to the fifth digit 3.14159;
4. The rounding of π to the twelfth digit is 3.141592653590.   □

Remark 12.17 As it is common in many books we will also use the symbol $\approx$
as a substitute for the word "approximates": namely if $x, y \in \mathbb{R}_{\geq 0}$ we write $x \approx y$
with an **error less than** $\varepsilon > 0$ whenever $|x - y| \leq \varepsilon$. Also, the **relative error** in
approximating $x \neq 0$ with y is the quotient $|x - y|/x$.

Example 12.18 We write $6.9876 \ldots \approx 6.95$ with an error less than $4/100$: indeed

$$|6.9876\ldots - 6.95| \leq 6.9877 - 6.95 = 0.04.$$

The relative error in approximating $6.9876\ldots$ with 6.95 is

$$\frac{|6.9876\ldots - 6.95|}{6.9876\ldots} < \frac{0.04}{6.9876} < 0.6\,\%.$$ □

Remark 12.19 A word on negative numbers: if $x, y \leq 0$ we will write $x \approx y$ whenever $-x \approx -y$. For instance $-6.346 \approx -6.35$ since $6.346 \approx 6.35$.

12.1.2 Integrals and Their Estimates

We will often deal with integrals and their estimates. We now set the notation and
the basic necessary facts.

Definition 12.20 *(The sup-norm)* Let $f : I \to \mathbb{R}$ be a bounded function on a set I.
The sup-norm of f on I is

$$\|f\|_{L^\infty(I)} = \sup\{|f(x)| : x \in I\}. \tag{12.20.a}$$

We will simply write $\|f\|_\infty$ instead of $\|f\|_{L^\infty(I)}$ if there is no risk of ambiguity.   □

Example 12.21 If $f : [a, b] \to \mathbb{R}$ is continuous then by Weierstrass[1] Theorem the
supremum in (12.20.a) is actually a maximum, i.e.,

$$\|f\|_{L^\infty([a,b])} = \max\{|f(x)| : x \in [a, b]\}.$$ □

The integral norm of a function on an interval may be estimated easily by means
of its sup-norm.

---

[1] Karl Theodor Wilhelm Weierstrass (1815–1897).

Proposition 12.22 *Let $f : [a, b] \to \mathbb{R}$ be an integrable function. Then*

$$\int_a^b |f(x)| \, dx \leq (b - a) \|f\|_{L^\infty([a,b])}. \tag{12.22.a}$$

Proof. Indeed for all x in $[a, b]$ we have $|f(x)| \leq \|f\|_\infty$: it is enough to integrate both terms of the inequality to get the conclusion. □

☞ *Remark 12.23* The inequality (12.22.a) is strict, unless $|f|$ is not constant. Actually the right-hand side of (12.22.a) can be much bigger than the left-hand side term, even when f is regular.

In the following example we see that the quotient $\dfrac{(b - a) \|f\|_{L^\infty([a,b])}}{\int_a^b |f(x)| \, dx}$ may be as big as one wishes.

Example 12.24 Let g be a $\mathscr{C}^\infty$ function on $\mathbb{R}$, equal to 0 out of $[-1, 1]$, with $\int_{-1}^1 g(x) \, dx = 1$, $g(0) = 1$ and $0 \leq g \leq 1$. Let $n \in \mathbb{N}_{\geq 1}$ and $f_n(x) := ng(nx)$. Then

$$\int_{-1}^1 |f_n(x)| \, dx = \int_{-1}^1 n|g(nx)| \, dx = \int_{-n}^n g(x) \, dx = \int_{-1}^1 g(x) \, dx = 1,$$

whereas $\|f_n\|_\infty = n$. □

There is no need for integral estimates in (12.22.a) if f is the derivative of a monotonic function g.

Lemma 12.25 *Let g: $[a, b]$ be of class $\mathscr{C}^1$ and monotonic. Then*

$$\int_a^b \left|g'(x)\right| \, dx = |g(b) - g(a)|.$$

Proof. Indeed either $|g'| = g'$ or $|g'| = -g'$ depending on the fact that g is, respectively, increasing or decreasing. In the first case we have

$$\int_a^b \left|g'(x)\right| \, dx = \int_a^b g'(x) \, dx = g(b) - g(a) = |g(b) - g(a)|,$$

whereas, in the second,

$$\int_a^b |g'(x)| \, dx = -\int_a^b g'(x) \, dx = -g(b) + g(a) = |g(b) - g(a)| \,. \qquad \square$$

12.2   The Euler–Maclaurin Formulas

In this chapter we will consider, if not explicitly stated otherwise, functions defined on intervals of the form $[a, b]$ or $[a, +\infty[$, where a, b are *natural numbers*. If $f : [a, b] \to \mathbb{R}$ is a function of class $\mathscr{C}^m$, that is, differentiable up to order m with its m-th derivative continuous on the interval $[a, b]$, then we use f', f'' and $f^{(m)}$ respectively to denote the first, second, and m-th derivatives of f.

In what follows we will make use of the *Bernoulli polynomials*.

Definition 12.26   The **Bernoulli polynomials** of degrees 0, 1 and 2 are respectively

$$B_0(X) := 1, \quad B_1(X) := X - \frac{1}{2}, \quad B_2(X) := X^2 - X + \frac{1}{6}.$$

We use the notation $x \mapsto B_0(x)$, $x \mapsto B_1(x)$ and $x \mapsto B_2(x)$ to indicate the corresponding polynomial functions. $\qquad \square$

We set

$$\left[f \right]_a^b := f(b) - f(a).$$

Theorem 12.27   (First- and second- order Euler–Maclaurin formulas) *Let* $a, b \in \mathbb{N}$ *and* $f : [a, b] \to \mathbb{R}$ *be a function.*

- **First-order Euler–Maclaurin formula.** *If* f *is of class* $\mathscr{C}^1$ *on* $[a, b]$ *one has*

$$\sum_{a \le k < b} f(k) = \int_a^b f(x) \, dx - \frac{1}{2} \left[f \right]_a^b + R_1,$$

$$R_1 = \int_a^b B_1(x - [x]) f'(x) \, dx, \quad |R_1| \le \frac{1}{2} \int_a^b |f'(x)| \, dx. \qquad (12.27.a)$$

• **Second-order Euler–Maclaurin formula.** *If f is of class $\mathscr{C}^2$ on $[a, b]$ one has*

$$\sum_{a \le k < b} f(k) = \int_a^b f(x)\, dx - \frac{1}{2} [f]_a^b + \frac{1}{12} [f']_a^b + R_2,$$

$$R_2 = -\frac{1}{2} \int_a^b B_2(x - [x]) f''(x)\, dx, \quad |R_2| \le \frac{1}{12} \int_a^b |f''(x)|\, dx. \qquad (12.27.\text{b})$$

Proof. Let $g : [0, 1] \to \mathbb{R}$ be of class $\mathscr{C}^1$. Give that $B_1'(x) = 1$, integration by parts gives

$$\int_0^1 g(x)\, dx = \int_0^1 g(x)\, B_1'(x)\, dx = \big[g(x)\, B_1(x)\big]_0^1 - \int_0^1 g'(x)\, B_1(x)\, dx,$$

and since $B_1(0) = -1/2$ and $B_1(1) = 1/2$ one obtains

$$\int_0^1 g(x)\, dx = \int_0^1 g(x)\, B_1'(x)\, dx = \frac{1}{2}(g(1) + g(0)) - \int_0^1 g'(x)\, B_1(x)\, dx.$$

It follows that

$$\frac{g(1) + g(0)}{2} = \int_0^1 g(x)\, dx + \int_0^1 g'(x)\, B_1(x)\, dx. \qquad (12.27.\text{c})$$

On applying (12.27.c) to $g(x) = f(x + k)$, one finds that, for $k = a, \dots, b - 1$,

$$\frac{f(k) + f(k + 1)}{2} = \int_k^{k+1} f(x)\, dx + \int_k^{k+1} f'(x) B_1(x - k)\, dx.$$

If $x \in [k, k + 1[$, one has $x - k = x - [x]$ where $[x]$ represents the *integral part* or *floor* of x. Therefore, summing both the terms of the preceding equation over k as it varies from a to $b - 1$ one obtains

$$\frac{f(a)}{2} + \sum_{a+1 \le k < b} f(k) + \frac{f(b)}{2} = \int_a^b f(x)\, dx + \int_a^b f'(x)\, B_1(x - [x])\, dx,$$

from which, on adding $f(a)/2$ and subtracting $f(b)/2$ to both terms of the inequality one obtains (12.27.a). Since $|B_1(x)| \le 1/2$ for $x \in [0, 1]$, one gets the desired estimate of R_1.

Now let g be a function of class $\mathscr{C}^2$ on $[0, 1]$. Given that $B_2'(x) = 2\,B_1(x)$, integration by parts yields

$$\int_0^1 g'(x)\,B_1(x)\,dx = \int_0^1 g'(x)\frac{B_2'(x)}{2}\,dx = \left[g'(x)\frac{B_2(x)}{2} \right]_0^1 - \int_0^1 g''(x)\frac{B_2(x)}{2}\,dx$$

$$= \frac{1}{12}\left[g'(x) \right]_0^1 - \int_0^1 g''(x)\frac{B_2(x)}{2}\,dx,$$

since $B_2(0) = B_2(1) = 1/6$. By substituting what was found above into (12.27.c) for the term $\int_0^1 g'(x)\,B_1(x)\,dx$ one obtains

$$\frac{g(1) + g(0)}{2} = \int_0^1 g(x)\,dx + \frac{1}{12}\left[g'(x) \right]_0^1 - \int_0^1 g''(x)\frac{B_2(x)}{2}\,dx. \qquad (12.27.d)$$

If f is of class $\mathscr{C}^2$ on $[a, b]$, applying the formula (12.27.d) to $g(x) = f(x + k)$ one obtains that, for $k = a, \ldots, b - 1$,

$$\frac{f(k) + f(k + 1)}{2} = \int_k^{k+1} f(x)\,dx + \frac{1}{12}\left[f'(x) \right]_k^{k+1} - \int_k^{k+1} f''(x)\frac{B_2(x - k)}{2}\,dx.$$

Summing for k as it varies from a to $b - 1$ one has

$$\frac{f(a)}{2} + \sum_{a+1\leq k<b} f(k) + \frac{f(b)}{2} = \int_a^b f(x)\,dx + \frac{1}{12}\left[f' \right]_a^b - \frac{1}{2}\int_a^b B_2(x - [x])f''(x)\,dx,$$

from which, on adding $f(a)/2$ and subtracting $f(b)/2$ to both terms of the equation one obtains relation (12.27.b). Since $|B_2(x)| \leq 1/6$ for $x \in [0, 1]$, one thus gets the desired estimate for R_2.   □

The terms in (12.27.a) and (12.27.b), apart the remainders, deserve to be named after Euler and Maclaurin.

Definition 12.28 (*The first- and second- order Euler–Maclaurin expansions*)

• Let $f : [a, b] \to \mathbb{R}$ be an integrable function, with $a, b \in \mathbb{N}$. The term

$$EM_1(f; [a, b]) := \int_a^b f(t)\,dt - \frac{1}{2}\left[f \right]_a^b$$

is called the **first-order Euler–Maclaurin expansion** of f on $[a, b]$.

- Let f be of class $\mathscr{C}^1$ on $[a, b]$. The term

$$\text{EM}_2(f; [a, b]) := \int_a^b f(x)\, dx - \frac{1}{2}\,[f\,]_a^b + \frac{1}{12}\,[f'\,]_a^b$$

is called the **second-order Euler–Maclaurin expansion** of f on $[a, b]$. □

Most of the time we will deal with monotonic functions; in this case the estimates of the remainders in the Euler–Maclaurin formulas are particularly simple.

Remark 12.29 (*Estimates of the remainder under the hypothesis of monotonicity*) If f is *monotonic* then, from (12.27.a) together with Lemma 12.25, one immediately obtains

$$|R_1| \le \frac{1}{2}\left|[f\,]_a^b\right| = \frac{1}{2}|f(b) - f(a)|.$$

Analogously, if f' is *monotonic* then from (12.27.b) one has

$$|R_2| \le \frac{1}{12}\left|[f'\,]_a^b\right| = \frac{1}{12}|f'(b) - f'(a)|. \tag{12.29.a}$$

☞ Note that in both cases, the remainder R_1 or R_2 turns out to be at its maximum (worst) equal to the last term of the Euler–Maclaurin expansion.

12.2.1 Examples of First-Order Expansion

In the following examples we compute the first-order Euler–Maclaurin expansions of some functions.

Example 12.30 In Sect. 6.1.1 we computed the sum

$$1 + 2 + \cdots + (n - 1) = \frac{n(n - 1)}{2},$$

but we ask that one forgets this result for the moment. The application of (12.27.a), with $f(x) = x$, yields

$$\sum_{1 \le k < n} k = \int_1^n x \, dx - \frac{1}{2} [f]_1^n + R_1$$

$$= \frac{1}{2}(n^2 - 1) - \frac{1}{2}(n - 1) + R_1 = \frac{n(n-1)}{2} + R_1,$$

with a remainder R_1 such that $|R_1| \le \dfrac{1}{2} \displaystyle\int_1^n 1 \, dx = \dfrac{1}{2}(n-1)$; actually we know that $R_1 = 0$.   □

Example 12.31 Let us give an estimate of the sum of the first 999 inverses of the square roots of the natural numbers

$$\sum_{1 \le k < 1000} \frac{1}{\sqrt{k}}.$$

It follows from (12.27.a), with $f(x) = 1/\sqrt{x}$, that

$$\sum_{1 \le k < 1000} \frac{1}{\sqrt{k}} = \int_1^{1000} \frac{1}{\sqrt{x}} \, dx - \frac{1}{2} [f]_1^{1000} + R_1$$

$$= 2(\sqrt{1000} - 1) - \frac{1}{2}(1/\sqrt{1000} - 1) + R_1$$

$$= 2\sqrt{1000} - 3/2 - 1/(2\sqrt{1000}) + R_1 \approx 61.73 + R_1$$

and

$$|R_1| \le \frac{1}{2} \left| [f]_1^{1000} \right| < \frac{f(1)}{2} = \frac{1}{2}.$$

Actually the given sum equals[2] 61.76938...; we obtained one correct digit, and a relative error of $(61.76938 - 61.73)/61.76938 \approx 0.06 \%$.   □

Example 12.32 We wish to estimate the value of $H_{100} = \displaystyle\sum_{1 \le k \le 100} \frac{1}{k}$, the sum of the first 100 inverses of the natural numbers. The first-order Euler–Maclaurin expansion of $f(x) = 1/x$ on $[1, 101]$ equals

$$\text{EM}_1 \left(\frac{1}{x}; [1, 101] \right) = \left[\log x - \frac{1}{2x} \right]_1^{101} \approx 5.1102.$$

Theorem 12.27 ensures that the difference R_1 between H_{100} and the current expansion satisfies

$$|R_1| \le \frac{1}{2}(f(1) - f(101)) \approx 0.495.$$

---

[2]This can be seen with a CAS.

Actually it turns out that $H_{100} = 5.17737\ldots$: we obtained one correct digit in our estimate. □

Example 12.33 We compare the value of the sum $\displaystyle\sum_{1 \leq k < 100} k^{5/3} = 79\,717.06142\ldots$ with the first-order Euler–Maclaurin expansion, equal to

$$\mathrm{EM}_1(x^{5/3}; [1, 100]) = \left[\frac{3}{8}x^{8/3} - \frac{1}{2}x^{5/3}\right]_1^{100} = 79\,714.20\ldots .$$

We know from Theorem 12.27 that, a priori, the difference R_1 between the given sum and the above expansion satisfies the estimate

$$|R_1| \leq \frac{1}{2}(f(100) - f(1)) \approx 1\,077.$$

Actually here $R_1 \approx 2.86$: much smaller in norm than the upper bound above. □

Example 12.34 *(Estimate of H_n)* Let $f(x) = \dfrac{1}{x}$, $x \geq 1$ and $n \in \mathbb{N}_{\geq 1}$. The application of (12.27.a) yields

$$\sum_{1 \leq k < n} \frac{1}{k} = \int_1^n \frac{1}{x}\,dx - \frac{1}{2}\left[\frac{1}{x}\right]_1^n + R_1(n)$$

$$= \log n - \frac{1}{2n} + \frac{1}{2} + R_1(n),$$

with $|R_1(n)| \leq \left|\dfrac{1}{2}\left[\dfrac{1}{x}\right]_1^n\right| = \dfrac{1}{2} - \dfrac{1}{2n}$. Thus

$$H_n = \sum_{1 \leq k < n} \frac{1}{k} + \frac{1}{n} = \log n + R_1'(n), \quad |R_1'(n)| < 1.$$

We shall see in Example 12.64 that $R_1'(n)$ tends to a constant, as $n \to +\infty$. □

In the next example we see how the first-order Euler–Maclaurin formula, as simple as it is, yields already a simplified version of the celebrated Stirling formula.

Example 12.35 *(A first step towards Stirling's formula)* We wish to study the behavior of $n!$ as $n \to +\infty$. Notice that the logarithm of $n!$ is a sum of terms, namely

$$\log n! = \log 1 + \log 2 + \cdots + \log n.$$

We are thus in a situation where we are able to apply the first-order Euler–Maclaurin formula. Let $f(x) = \log x$, $x \geq 1$. By (12.27.a) we have

$$\log 1 + \cdots + \log(n-1) = \int_1^n \log x \, dx - \frac{1}{2}\left[\log x\right]_1^n + R_1(n)$$

$$= \left[x \log x - x\right]_1^n - \frac{1}{2}\log n + R_1(n)$$

$$= \left(n - \frac{1}{2}\right)\log n - n + 1 + R_1(n),$$

with $|R_1(n)| \leq \left|\dfrac{1}{2}\left[\log x\right]_1^n\right| = \dfrac{\log n}{2}$. It follows that

$$\log n! = \left(n + \frac{1}{2}\right)\log n - n + 1 + R_1(n), \quad |R_1(n)| \leq \frac{\log n}{2}.$$

The composition with the exponential function yields

$$n! = e^{\log n!} = n^n \sqrt{n}\, e^{-n} g(n), \quad |g(n)| = e^{1+R_1(n)} \leq e\sqrt{n}. \tag{12.35.a}$$

□

☞   *Remark 12.36* The previous examples may lead one to think that the Euler–Maclaurin expansions are a good approximation of the sum $\displaystyle\sum_{a \leq k < b} f(k)$: actually the remainders in (12.27.a) and (12.27.b) may be quite large in general, as the next example shows.

Example 12.37 Let $b > 0$ and set $S := \displaystyle\sum_{0 \leq k < b} e^k$. Of course we know that

$$\sum_{0 \leq k < b} e^k = 1 + \cdots + e^{b-1} = \frac{e^b - 1}{e - 1}.$$

The first-order Euler–Maclaurin expansion $\mathrm{EM}_1(e^x; [0, b])$ is

$$\int_0^b e^x \, dx - \frac{e^b - 1}{2} = \frac{e^b - 1}{2} = \frac{e - 1}{2} S.$$

The relative error in approximating S is thus

$$\frac{\left|S - \dfrac{e-1}{2}S\right|}{S} = \frac{3 - e}{2} \approx 14\%.$$

□

12.2.2 Examples of Second-Order Expansion

In the following examples we compute the second-order Euler–Maclaurin expansions of some functions.

Example 12.38 In Example 12.30 we estimated the sum $\sum\limits_{1\le k<n} k$ using the first-order Euler–Maclaurin expansion. Given that the second derivative of $f(x) = x$ is zero, the second-order expansion permits one to determine the exact value of the sum. Indeed, from (12.27.b) one obtains

$$\sum_{1\le k<n} k = \int_1^n x\,dx - \frac{1}{2}\left[f\right]_1^n + \frac{1}{12}\left[f'\right]_1^n + 0$$

$$= \frac{1}{2}(n^2 - 1) - \frac{1}{2}(n - 1) = \frac{n^2 - n}{2}. \qquad \square$$

Example 12.39 Let us reconsider Example 12.31, where we used the first-order Euler–Maclaurin expansion applied to $f(x) = \dfrac{1}{\sqrt{x}}$ in order to estimate the sum $\sum\limits_{1\le k<1000} \dfrac{1}{\sqrt{k}}$, finding a value of 61.73. The second-order Euler–Maclaurin expansion $\mathrm{EM}_2\left(\dfrac{1}{\sqrt{x}}; [1, 1\,000]\right)$ yields

$$\int_1^{1000} \frac{1}{\sqrt{x}}\,dx - \frac{1}{2}\left[x^{-1/2}\right]_1^{1000} + \frac{1}{12}\left[-\frac{1}{2}x^{-3/2}\right]_1^{1000} =$$

$$= \left[2\sqrt{x} - \frac{1}{2}x^{-1/2} - \frac{1}{24}x^{-3/2}\right]_1^{1000} = 61.7714\ldots.$$

Given that $f'(x)$ is monotonic, by Remark 12.29 this expansion allows one to approximate the given sum up to an error R_2 satisfying

$$|R_2| \le \frac{1}{12}\left[f'\right]_1^{1000} = 0.11\ldots.$$

In reality the sum is equal to about 61.76938... and the relative error here is $(61.7714 - 61.76938)/61.76938 \approx 0.003\,\%$: much better than the 0.06 % found in Example 12.31. $\qquad \square$

Example 12.40 In Example 12.33 we found that $\mathrm{EM}_1(x^{5/3}; [1, 100]) = 79\,714.20\ldots$, whereas $\sum\limits_{1\le k<100} k^{5/3} = 79\,717.06142\ldots$, with an a priori bound of the remainder R_1 given by $|R_1| \le 1\,077$. The second-order Euler–Maclaurin expansion of $x^{5/3}$ on $[1, 100]$ is

$$EM_2(x^{5/3}; [1, 100]) = \int_1^{100} x^{5/3}\, dx - \frac{1}{2}\left[x^{5/3}\right]_1^{100} + \frac{1}{12}\left[\frac{3}{8}x^{8/3}\right]_1^{100} = 79\,717.0608\ldots ,$$

with a remainder R_2 satisfying

$$|R_2| \le \left|\frac{1}{12}\left[\frac{3}{8}x^{8/3}\right]_1^{100}\right| \le 2.9\ldots .$$

We see here how in this case the second-order expansion, despite the above estimate of the remainder R_2, surprisingly gives a quite accurate approximation of
the value of the original sum: actually the two terms coincide up to the first two
decimals.   □

Example 12.41 In Example 12.37 we saw that the relative error of the sum $S :=$
$\sum_{0 \le k < b} e^k$ $(b > 0)$ with respect to the first-order Euler–Maclaurin expansion of $f(x) =$
e^x is about 14%. Things go better with the second-order expansion of f on $[0, b]$,
given by

$$\int_0^b e^x\, dx - \frac{e^b - 1}{2} + \frac{1}{12}(e^b - 1) = (e^b - 1)\left(1 - \frac{1}{2} + \frac{1}{12}\right)$$

$$= \frac{7}{12}(e^b - 1) = \frac{7(e - 1)}{12}S \approx 1.0023S.$$

Thus the relative error is

$$\frac{\left|\frac{7(e - 1)}{12}S - S\right|}{|S|} = \frac{7e - 19}{12} \approx 0.2\%.$$ □

Example 12.42 In Example 12.32 we furnished an estimate of H_{100} through use
of the first-order Euler–Maclaurin expansion of the function $f(x) = 1/x$ $(x > 0)$,
obtaining an approximate value equal to 5.1102 whereas the value of H_{100} is equal
to $5.18737\ldots$: we thus got a relative error of $\dfrac{5.18737 - 5.1102}{5.18737} \approx 1.49\%$. By
Remark 12.29 and the monotonicity of $f'(x)$, we know that the difference R_2 between
the value of the sum H_{100} under consideration and the second-order expansion of f
on $[1, 101]$ a priori satisfies the inequality

$$|R_2| \le \frac{1}{12}\left|\left[-\frac{1}{x^2}\right]_1^{101}\right| = 0.0833\ldots .$$

Actually, the second-order expansion is given by

$$EM_2\left(\frac{1}{x}; [1, 101]\right) = \left[\log x - \frac{1}{2x} - \frac{1}{12x^2}\right]_1^{101} = 5.1935\ldots .$$

The relative error here is therefore $\dfrac{5.18737 - 5.1935}{5.18737} \approx 0.1\,\%.$ □

Example 12.43 (A further step towards Stirling's formula) In Example 12.35 the first-order Euler–Maclaurin formula applied to $f(x) = \log x$, $x \geq 1$ gave (12.35.a), a simplified version of Stirling's formula, namely

$$n! = e^{\log n!} = n^n \sqrt{n}\, e^{-n} g(n), \quad |g(n)| \leq e\sqrt{n}.$$

The application of (12.27.b) yields

$$\log 1 + \cdots + \log(n-1) = \int_1^n \log x\, dx - \frac{1}{2}\left[\log x\right]_1^n + \frac{1}{12}\left[\frac{1}{x}\right]_1^n + R_2(n)$$

$$= [x\log x - x]_1^n - \frac{1}{2}\log n + \frac{1}{12n} - \frac{1}{12} + R_2(n)$$

$$= \left(n - \frac{1}{2}\right)\log n - n + \frac{11}{12} + \frac{1}{12n} + R_2(n),$$

with $|R_2(n)| \leq \left|\dfrac{1}{12}\left[\dfrac{1}{x}\right]_1^n\right| = \dfrac{1}{12}\left(1 - \dfrac{1}{n}\right).$ Thus

$$\log n! = \log 1 + \cdots + \log(n-1) + \log n = \left(n + \frac{1}{2}\right)\log n - n + R_2'(n), \quad |R_2'(n)| \leq 1,$$

so that

$$n! = e^{\log n!} = n^n \sqrt{n}e^{-n}g(n), \quad |g(n)| \leq e, \tag{12.43.a}$$

giving us more details on the behavior of the sequence $g(n)$. We will see in Proposition 12.71 that $g(n)$ has a limit as n tends to infinity, and in Corollary 12.73 that it actually equals $\sqrt{2\pi}$.   □

Remark 12.44 The explicit differences R_1 and R_2 found in Theorem 12.27 are linked to the first- and second- order derivatives of the function f; if these happen to be "large", the expansions turn out to be of little significance for the purpose of the approximation. For example, consider the sum $S := \sum\limits_{10 \leq k < 20} ke^{k^2} \approx 1.14 \times 10^{158}$: on setting $f(x) = xe^{x^2}$, the first-order Euler–Maclaurin expansion of f gives

$$\text{EM}_1\left(xe^{x^2}; [10, 20]\right) = \int_{10}^{20} xe^{x^2}\, dx - \frac{1}{2}\left[xe^{x^2}\right]_{10}^{20}$$

$$= \frac{1}{2}\left(9e^{100} - 19e^{400}\right) \approx -4.96 \times 10^{174},$$

a negative number! The second-order Euler–Maclaurin expansion of f yields

$$EM_2(xe^{x^2}; [10, 20]) = \int_1^{10} xe^{x^2}\, dx - \frac{1}{2}\left[xe^{x^2} \right]_1^{10} + \frac{1}{12}\left[(2x^2 + 1)e^{x^2} \right]_1^{10}$$

$$= \frac{1}{4}e^{100}\left(229e^{300} - 49 \right) \approx 2.98 \times 10^{175}.$$

Both the results are very far away from the correct value of S.

12.3   First- and Second- Order Euler–Maclaurin Approximation Formulas

We have seen that the Euler–Maclaurin expansion of a function f on an interval $[a, b]$ does not allow one, in general, to approximate the sum $\displaystyle\sum_{a \leq k < b} f(k)$. However, it does allow, for integers $a \leq n \leq N$, to *approximate* $\displaystyle\sum_{a \leq k < N} f(k)$ in terms of the briefer sum

$$\sum_{a \leq k < n} f(k).$$

Corollary 12.45 (First- and second- order approximations) *Let* $f : [a, +\infty[\rightarrow \mathbb{R}$ *and* $n, N \in \mathbb{N}$ *with* $a \leq n \leq N$.

1. *If* $f \in \mathscr{C}^1$, *then it satisfies the following **first-order approximation***

$$\sum_{a \leq k < N} f(k) = \sum_{a \leq k < n} f(k) + \int_n^N f(x)\, dx - \frac{1}{2}\left[f \right]_n^N + \varepsilon_1(n, N),$$

$$|\varepsilon_1(n, N)| \leq \frac{1}{2}\int_n^N |f'(x)|\, dx \leq \frac{1}{2}\int_n^{+\infty} |f'(x)|\, dx. \qquad (12.45.\text{a})$$

2. *If* $f \in \mathscr{C}^2$, *then it satisfies the following **second-order approximation***

$$\sum_{a \leq k < N} f(k) = \sum_{a \leq k < n} f(k) + \int_n^N f(x)\, dx - \frac{1}{2}\left[f \right]_n^N + \frac{1}{12}\left[f' \right]_n^N + \varepsilon_2(n, N),$$

$$|\varepsilon_2(n, N)| \leq \frac{1}{12}\int_n^N |f''(x)|\, dx \leq \frac{1}{12}\int_n^{+\infty} |f''(x)|\, dx. \qquad (12.45.\text{b})$$

Proof. It is sufficient to apply (12.27.a) and (12.27.b) to the sum

$$\sum_{n \le k < N} f(k) = \sum_{a \le k < N} f(k) - \sum_{a \le k < n} f(k).$$ □

Here we introduce a notation that will be used quite frequently in this context.

Definition 12.46 (*The notation $g(\infty)$*) Let $g : [a, +\infty[\to \mathbb{R}$ be a function. Its limit at $+\infty$, if it exists (finite or infinite), will be denoted by $g(\infty)$. □

Remark 12.47 (*The remainders under monotonicity assumptions*) If f is *monotonic* then (12.45.a) offers, thanks to Lemma 12.25,

$$|\varepsilon_1(n, N)| \le \frac{1}{2} |f(n) - f(N)| \le \frac{1}{2} |f(n) - f(\infty)| .$$

Analogously if f' is monotonic, from (12.45.b) one obtains

$$|\varepsilon_2(n, N)| \le \frac{1}{12} |f'(n) - f'(N)| \le \frac{1}{12} |f'(n) - f'(\infty)| .$$

☞ *Remark 12.48* If f is $\mathscr{C}^2$, the Corollary 12.45 proposes us two distinct approximations for the sum $\displaystyle\sum_{a \le k < N} f(k)$; clearly the (12.45.b) is convenient, in terms of precision, with respect to (12.45.a) if

$$\frac{1}{12} \int_n^N |f''(x)| \, dx \le \frac{1}{2} \int_n^N |f'(x)| \, dx.$$

12.3.1 Examples of First-Order Approximation

Example 12.49 We wish to approximate the sum

$$\sum_{1 \le k < 1000} \frac{1}{\sqrt{k}}$$

within an error of 10^{-1}. Let $f(x) = \dfrac{1}{\sqrt{x}}$, for $x \ge 1$. Since f is monotonic and $f(\infty) = 0$, it follows from (12.45.a) that, for all $n \le 1000$, we may approximate the given sum with

$$\sum_{1 \le k < n} \frac{1}{\sqrt{k}} + \int_n^{1000} \frac{1}{\sqrt{x}} \, dx - \frac{1}{2} \left(\frac{1}{\sqrt{1000}} - \frac{1}{\sqrt{n}} \right)$$

and an error less than $\dfrac{1}{2\sqrt{n}}$. Thus it is enough that n satisfies

$$\frac{1}{2\sqrt{n}} \leq 10^{-1}$$

or, equivalently, $n \geq 25$. For $n = 25$ we get the following approximated value:

$$\sum_{1 \leq k < 1000} \frac{1}{\sqrt{k}} \approx \sum_{1 \leq k < 25} \frac{1}{\sqrt{k}} + \int_{25}^{1000} \frac{1}{\sqrt{x}}\,dx - \frac{1}{2}\left(\frac{1}{\sqrt{1000}} - \frac{1}{\sqrt{25}}\right) = 61.7691\ldots,$$

with an error of at most 10^{-1}. Actually $\displaystyle\sum_{1 \leq k < 1000} \frac{1}{\sqrt{k}} \approx 61.7694\ldots$: two digits are correct.   □

Example 12.50 By means of a CAS we check easily that

$$\sum_{1 \leq k < 100} \frac{1}{k} = 5.177377518\ldots\,.$$

For $n = 100$ and $N \geq n$, (12.45.a) and the monotonicity of $x \mapsto \dfrac{1}{x}$ yield

$$\sum_{1 \leq k < N} \frac{1}{k} = \sum_{1 \leq k < 100} \frac{1}{k} + \int_{100}^{N} \frac{1}{x}\,dx - \frac{1}{2}\left(\frac{1}{N} - \frac{1}{100}\right) + \varepsilon_1(100, N)$$

$$= 5.17737\ldots + \log N - \log 100 - \frac{1}{2N} + \frac{1}{200} + \varepsilon_1(100, N)$$

$$= 0.57720\ldots + \log N - \frac{1}{2N} + \varepsilon_1(100, N),$$

with $|\varepsilon_1(100, N)| \leq \dfrac{1}{200}$. For instance, with $N = 1000^{1000}$ we get

$$\sum_{1 \leq k < 1000^{1000}} \frac{1}{k} \approx 0.57720\ldots + 1000\log 1000 - \frac{1}{2000} = 6908.33198\ldots,$$

with an error at most equal to $1/200$. Notice that, actually, it turns out that

$$\sum_{1 \leq k < 1000^{1000}} \frac{1}{k} = 6908.33249\ldots\,.$$ □

Example 12.51 Consider the sum $\displaystyle\sum_{1 \leq k < 100} \frac{1}{k^3}$. In order to approximate it with an error within 10^{-3} it is enough to apply (12.45.a) with $N = 100$ and $n \leq 100$ in such a way that

$$\frac{1}{2n^3} \leq 10^{-3},$$

i.e., $n \geq 8$. With the choice of $n = 8$ we obtain

$$\sum_{1 \leq k < 100} \frac{1}{k^3} \approx \sum_{1 \leq k < 8} \frac{1}{k^3} + \int_8^{100} \frac{1}{x^3}\, dx - \frac{1}{2}\left[\frac{1}{x^3}\right]_8^{100}$$

$$= \frac{9822481}{8232000} - \left[\frac{1}{2x^2}\right]_8^{100} - \frac{1}{2}\left[\frac{1}{x^3}\right]_8^{100}$$

$$= \frac{19788833693}{16464000000} = 1.20194\ldots,$$

with a precision of 10^{-3}. Now using a CAS we get $\displaystyle\sum_{1 \leq k < 100} \frac{1}{k^3} = 1.20200\ldots$: our approximation is thus precise up to $0.00007 < 10^{-4}$. □

12.3.2 Examples of Second-Order Approximation

Example 12.52 In Example 12.49 we saw how to approximate the sum $\displaystyle\sum_{1 \leq k < 1000} \frac{1}{k}$ with an error within 10^{-1} by means of (12.45.a); there we had to compute the first 24 terms of the sum. Now, the function $f(x) = \dfrac{1}{\sqrt{x}}$ is of class $\mathscr{C}^\infty$; moreover f' is monotonic and $f'(\infty) = 0$. Then (12.45.b) allows us to approximate the given sum with an error of at most 10^{-1} with the formula

$$\sum_{1 \leq k < n} f(k) + \int_n^{1000} f(x)\, dx - \frac{1}{2}\left[f\right]_n^{1000} + \frac{1}{12}\left[f'\right]_n^{1000}$$

if just $\dfrac{1}{12}\left|f'(n) - f'(1000)\right| \leq 10^{-1}$. For this purpose it is sufficient that

$$\frac{1}{12}|f'(x) - f'(\infty)| = \frac{1}{24}n^{-3/2} \leq 10^{-1},$$

and this happens for $n \geq (5/12)^{2/3} \approx 0.56$. At this point we might try a better approximation, say with an error of at most 10^{-2}. In this case it would be enough that $\dfrac{1}{24}n^{-3/2} \leq 10^{-2}$, or equivalently $n \geq (25/6)^{2/3} \approx 2.6$: with the choice of $n = 3$ we get, with an error of at most 10^{-2},

$$\sum_{1\le k<1000}\frac{1}{\sqrt{k}} \approx \sum_{1\le k<3}\frac{1}{\sqrt{k}} + \int_3^{1000}\frac{1}{\sqrt{x}}\,dx - \frac{1}{2}\left[\frac{1}{\sqrt{x}}\right]_3^{1000} + \frac{1}{12}\left[-\frac{1}{2}\frac{1}{x^{3/2}}\right]_3^{1000}$$
$$= 61.76943\ldots.$$

It can be shown that the sum equals $61.76938\ldots$: we obtained the first 3 digits of $\sum_{1\le k<1000}\frac{1}{k}$ with just the first two terms of the sum!   □

Example 12.53 We wish to compute

$$S := \sum_{1\le k<1000^{1000}}\frac{1}{k}$$

with an error within 10^{-3}. The function $f(x) = \frac{1}{x}$ is of class $\mathscr{C}^\infty$, both f and f' are monotonic on $[1, +\infty[$, and tend to 0 at infinity. Equation (12.45.a) gives, for all $n \le 1000^{1000}$ an approximation of S with an error

$$|\varepsilon_1(n, 1000^{1000})| \le \frac{1}{2n},$$

and $|\varepsilon_1(n, 1000^{1000})| \le 10^{-3}$ whenever $n \ge 500$. Instead, (12.45.b), gives an approximation of S with an error

$$|\varepsilon_2(n, 1000^{1000})| \le \frac{1}{12n^2}.$$

In particular $|\varepsilon_2(n, 1000^{1000})| \le 10^{-3}$ as soon as $n \ge \sqrt{\frac{1000}{12}} \approx 9.1$. With $n = 10$ we thus get

$$\sum_{1\le k<1000^{1000}}\frac{1}{k} \approx \sum_{1\le k<10}\frac{1}{k} + \int_{10}^{1000^{1000}}\frac{1}{x}\,dx - \frac{1}{2}\left[\frac{1}{x}\right]_{10}^{1000^{1000}} + \frac{1}{12}\left[-\frac{1}{x^2}\right]_{10}^{1000^{1000}}$$
$$= 6908.332495476\ldots,$$

with an error of at most 10^{-3}. Actually, $\sum_{1\le k<1000^{1000}}\frac{1}{k} = 6908.332494647\ldots$: to obtain 5 correct digits we just needed to add the first ten terms of the sum. Notice that, in order to approximate S with an error within 10^{-2} by means of (12.45.b) it is enough to choose n in such a way that $\frac{1}{12n^2} \le 10^{-2}$, i.e., $n = 3$.   □

12.4   A Glimpse at Infinity: The Euler Constant and the Asymptotic Formulas of Euler–Maclaurin

In this section we want to study the asymptotic behavior of the difference

$$\sum_{k=a}^{n} f(k) - \int_{a}^{n} f(x)\, dx,$$

to get a better estimate of $\sum_{k=a}^{n} f(k)$. We need first to introduce the reader to some basic facts about improper integrals and asymptotic comparison.

12.4.1   Generalized Integrals and Summable Functions

We recall here when and how is possible to integrate functions up to $+\infty$; we refer to [33] for more insights on the subject. We will need to integrate functions up to $+\infty$.

Definition 12.54 (*Improper integrals and summable functions*) A function $g : [a, +\infty[\to \mathbb{R}$ is said to be **integrable in a generalized sense** on $[a, +\infty[$ if g is locally integrable, i.e., integrable on every bounded interval, and if the limit

$$\lim_{b \to +\infty} \int_{a}^{b} g(x)\, dx := \int_{a}^{+\infty} g(x)\, dx$$

exists and is finite. In this case the limit is called the **generalized integral** or **improper integral** of f in $[a, +\infty[$. In this case we also say that the integral $\int_{a}^{+\infty} g(x)\, dx$ is convergent; whereas if the above limit exists but is not finite we sometimes say that the integral $\int_{a}^{+\infty} g(x)\, dx$ **diverges at infinity**. The function g is said to be **summable** if $|g|$ is integrable in a generalized sense: in this case we will write that $g \in L^{1}([a, +\infty[)$.   □

The following comparison test holds.

Proposition 12.55 (Comparison test) *Let f, g be two locally integrable functions on $[a, +\infty[$, with $|f| \le g$. If $g \in L^{1}([a, +\infty[)$ then f is integrable in a generalized sense.*

Remark 12.56 It follows from Proposition 12.55 that a summable function is integrable in a generalized sense. The converse, however, is not true: we quote, as an example, the function $f(x) = \dfrac{\sin x}{x}$ on $[1, +\infty]$.

For a smooth function there are some useful links between the existence of a finite limit at $+\infty$ and the summability of its derivative. Remember that a property holds *definitively* if it holds on an interval $[b, +\infty[$, for some $b > 0$.

Lemma 12.57 *Let $g \in \mathscr{C}^1([a, +\infty[)$.*

1. *If g' is integrable in a generalized sense on $[a, +\infty[$ then the limit $g(\infty)$ exists and is finite;*
2. *If g is definitively monotonic, and $g(\infty) \in \mathbb{R}$ then $g' \in L^1([a, +\infty[)$.*

Proof. 1. For all $b \geq a$ we may write

$$g(b) = g(a) + \int_a^b g'(x)\,dx;$$

we obtain the desired conclusion by passing to the limit as $b \to +\infty$.

2. Assume that g is monotonic on $[b, +\infty[$, for some $b \geq a$. For every $c \geq b$, thanks to Lemma 12.25, we have

$$\int_b^c |g'(x)|\,dx = |g(c) - g(b)| \leq |g(\infty) - g(b)|,$$

so that the integral

$$\int_b^{+\infty} |g'(x)|\,dx = \lim_{c \to +\infty} \int_b^c |g'(x)|\,dx$$

is finite, due to the fact that the limit of a bounded monotonic function does always exist and is finite. □

12.4.2 The Euler–Maclaurin Asymptotic Formulas

Let us investigate, for a given locally integrable function $f : [a, +\infty[\to \mathbb{R}$ and $n \geq a$, the asymptotic behavior of the difference

$$\sum_{a \leq k < n} f(k) - \int_a^n f(x)\,dx$$

for $n \to +\infty$.

Definition 12.58 (*Euler constant of a function*) Let $f : [a, +\infty[\to \mathbb{R}$ be a locally integrable function, with $a \in \mathbb{N}$. For every $n \in \mathbb{N}$ we denote by γ_n^f the difference between $\sum_{a \leq k < n} f(k)$ and the integral $\int_a^n f(x)\,dx$:

$$\gamma_n^f := \sum_{a \le k < n} f(k) - \int_a^n f(x)\, dx \qquad \forall n \in \mathbb{N}_{\ge a}.$$

Whenever the sequence $(\gamma_n^f)_n$ converges, its limit, denoted by γ^f, is called the **Euler constant** of f:

$$\gamma^f := \lim_{n \to +\infty} \gamma_n^f.$$

☐

☞ *Remark 12.59* Just to be clear: the Euler constant is defined only when the above limit exists and is finite.

Let us briefly recall the asymptotic comparisons "big O" among sequences; further details, that we will not use at the moment, can be found in Sect. 14.1.

Definition 12.60 (*Big O relation*) Let $(a_n)_n$ and $(b_n)_n$ be two sequences. We write $a_n = O(b_n)$, and say "a_n is big o of b_n", as $n \to +\infty$, if there is $M \ge 0$ satisfying $|a_n| \le M|b_n|$ for all n in $\mathbb{N}$. ☐

12.4.3 The First-Order Asymptotic Formula

By Theorem 12.27, if $f : [a, +\infty[\to \mathbb{R}$ is a locally integrable function ($a \in \mathbb{N}$), then

$$\gamma_n^f = \sum_{a \le k < n} f(k) - \int_a^n f(x)\, dx = -\frac{1}{2}\big[f \big]_a^n + R_1(n).$$

If f is a $\mathscr{C}^1$ function and moreover f' is summable, or if f is unbounded and monotonic then the sequence γ_n^f converges.

Theorem 12.61 (First-order criterion for the existence of the Euler constant and Euler–Maclaurin asymptotic formula) *Let $f : [a, +\infty[\to \mathbb{R}$ be a function of class $\mathscr{C}^1$.*

*1. If $f' \in L^1([a, +\infty[)$ the Euler constant γ^f exists and the following **first-order asymptotic formula** holds:*

$$\sum_{a \le k < n} f(k) = \int_a^n f(x)\, dx + \gamma^f + O\left(\int_n^{+\infty} |f'(x)|\, dx \right) \qquad n \to +\infty. \quad (12.61.a)$$

*Moreover, for $n \in \mathbb{N}_{\ge a}$ the following **first-order estimate** of γ^f holds:*

$$\gamma^f = \gamma_n^f - \frac{1}{2} [f]_n^\infty + \varepsilon_1(n), \quad |\varepsilon_1(n)| \leq \frac{1}{2} \int_n^{+\infty} |f'(x)| \, dx. \qquad (12.61.b)$$

*1'. If f is **bounded** and **definitively monotonic**, then $f' \in L^1([a, +\infty[)$, and (12.61.a) may be rewritten as*

$$\sum_{a \leq k < n} f(k) = \int_a^n f(x) \, dx + \gamma^f + O(f(n) - f(\infty)) \quad n \to +\infty. \qquad (12.61.c)$$

Moreover in the first-order estimate (12.61.b) of γ^f we have:

$$|\varepsilon_1(n)| \leq \frac{|f(n) - f(\infty)|}{2}. \qquad (12.61.d)$$

*2. Assume now that f is **unbounded** and **definitively monotonic**. The following **first-order asymptotic formula** holds:*

$$\sum_{a \leq k < n} f(k) = \int_a^n f(x) \, dx + O(f(n)) \quad n \to +\infty. \qquad (12.61.e)$$

Remark 12.62 It follows from Theorem 12.61 that:

☞   • The Euler constant of f exists whenever $f' \in L^1([a, +\infty[)$;
     • A first-order asymptotic Euler–Maclaurin formula holds if $f' \in L^1([a, +\infty[)$ or if f is definitively monotonic.

The proof of Theorem 12.61 is based on the explicit integral form of the remainder R_1 in the Euler–Maclaurin formula.

Proof (of Theorem 12.61). Let $n \geq a$. It follows from the Euler–Maclaurin formula (12.27.a) that

$$\sum_{a \le k < n} f(k) = \int_a^n f(x)\, dx - \frac{1}{2}\left[f\right]_a^n + \int_a^n B_1(x - [x])f'(x)\, dx. \qquad (12.62.a)$$

1. If $f' \in L^1([a, +\infty[)$ we have

$$\left[f\right]_a^n = \int_a^n f'(x)\, dx = \int_a^{+\infty} f'(x)\, dx - \int_n^{+\infty} f'(x)\, dx$$

$$= \int_a^{+\infty} f'(x)\, dx + O\left(\int_n^{+\infty} |f'(x)|\, dx\right) \quad n \to +\infty.$$

Since B_1 is bounded on $[0, 1]$, then $B_1(x - [x])f'(x) \in L^1([a, +\infty[)$: following the lines of the first part of this proof we obtain

$$\int_a^n B_1(x - [x])f'(x)\, dx = \int_a^{+\infty} B_1(x - [x])f'(x)\, dx - \int_n^{+\infty} B_1(x - [x])f'(x)\, dx$$

$$= \int_a^{+\infty} B_1(x - [x])f'(x)\, dx + O\left(\int_n^{+\infty} |f'(x)|\, dx\right) \quad n \to +\infty.$$

We deduce from (12.62.a) that

$$\sum_{a \le k < n} f(k) = \int_a^n f(x)\, dx + C_1^f + O\left(\int_n^{+\infty} |f'(x)|\, dx\right) \quad n \to +\infty, \qquad (12.62.b)$$

where we set

$$C_1^f = -\frac{1}{2}\int_a^{+\infty} f'(x)\, dx + \int_a^{+\infty} B_1(x - [x])f'(x)\, dx.$$

It follows from (12.62.b) that

$$C_1^f = \lim_{n \to +\infty}\left(\sum_{a \le k < n} f(k) - \int_a^n f(x)\, dx\right),$$

whence $\gamma^f = C_1^f$. Next, by formula (12.62.a) we have

$$\gamma_n^f = -\frac{1}{2}[f]_a^n + \int_a^{+\infty} B_1(x - [x])f'(x)\, dx$$

$$= \frac{1}{2}\int_n^{+\infty} f'(x)\, dx - \frac{1}{2}\int_a^{+\infty} f'(x)\, dx +$$

$$+ \int_a^{+\infty} B_1(x - [x])f'(x)\, dx - \int_n^{+\infty} B_1(x - [x])f'(x)\, dx$$

$$= \gamma^f + \frac{1}{2}[f]_n^{+\infty} - \int_n^{+\infty} B_1(x - [x])f'(x)\, dx.$$

Since $|B_1(x)| \le 1/2$ in $[0, 1]$, one gets

$$\varepsilon_1(n) := \int_n^{+\infty} B_1(x - [x])f'(x)\, dx \le \frac{1}{2}\int_n^{+\infty} |f'(x)|\, dx.$$

1'. Assume now that f is monotonic on $[b, +\infty[$, for some $b \ge a$, and bounded. Then $f(\infty) \in \mathbb{R}$ and thus, from Point 2 of Lemma 12.57, $f' \in L^1([a, +\infty[)$. At this stage (12.61.c) and (12.61.d) follow directly from (12.61.a), (12.61.b), and Lemma 12.25.
2. Assume now that f is monotonic on $[b, +\infty[$, with $b \ge a$, and that $|f(\infty)| = +\infty$. It follows from Lemma 12.25 that for $n \ge b$ we have

$$\int_b^n |f'(x)|\, dt = |f(n) - f(b)| \le |f(n)| + |f(b)|. \tag{12.62.c}$$

By (12.27.a) $R_1(n) = \int_a^n B_1(x - [x])f'(x)\, dx$ satisfies

$$|R_1(n)| \le \frac{1}{2}\int_a^n |f'(x)|\, dx \le \frac{1}{2}\int_a^b |f'(x)|\, dx + \frac{1}{2}\int_b^n |f'(x)|\, dx$$

$$\le c + \frac{1}{2}|f(n)|,$$

where $c = \frac{1}{2}\left(\int_a^b |f'(x)|\, dx + |f(b)|\right)$ is a constant that depends just on f, in any case not on n. Of course $|f(n)|$, and every constant, are $O(f(n))$ as $n \to +\infty$: (12.62.a) thus yields

$$\sum_{a \le k < n} f(k) = \int_a^n f(x)\, dx + O(f(n)) \qquad n \to +\infty. \qquad \square$$

Remark 12.63 If $f' \in L^1([a, +\infty[)$ then Theorem 12.61 yields the existence of

$$\gamma^f = \lim_{n \to +\infty} \left(\sum_{a \le k < n} f(k) - \int_a^n f(x)\, dx\right).$$

Of course, if both the series $\sum_{k=a}^{\infty} f(k)$ and the integral $\int_a^{+\infty} f(x)\,dx$ converge, then

$$\gamma^f = \sum_{k=a}^{\infty} f(k) - \int_a^{+\infty} f(x)\,dx.$$

It may happen however that neither the series nor the integral converge, as it is described in the following Example 12.64, with $f(x) = \dfrac{1}{x}, x \geq 1$.

The next example illustrates the difference between (12.27.a) and the asymptotic formulas (12.61.a)–(12.61.c).

Example 12.64 (The Euler–Mascheroni constant γ, part I) We saw in Example 12.34 that

$$H_n = \log n + R_1'(n), \qquad |R_1'(n)| \leq \frac{1}{2}, \tag{12.64.a}$$

from which we deduce the following asymptotic estimate for H_n:

$$H_n \sim \log n \qquad n \to +\infty.$$

The above estimate, though important, was already deduced out by means of the elementary methods of the discrete calculus in Proposition 6.44. However we cannot deduce from (12.64.a) any information on the behavior of the sequence $(R_1'(n))_n$ other than the fact that it is bounded by $1/2$. If, instead, we apply Theorem 12.61 to the function $f(x) = \dfrac{1}{x}, x \geq 1$, we find something more. The Euler constant of the function $x \mapsto 1/x$, $x \geq 1$, is denoted simply by γ and called **Euler–Mascheroni**[3] constant. By applying (12.61.c) we get

$$\sum_{1 \leq k < n} \frac{1}{k} = \int_1^n \frac{1}{x}\,dx + \gamma + O\left(\frac{1}{n}\right) \qquad n \to +\infty,$$

which, in terms of H_n, may be rewritten as

$$H_n = \log n + \gamma + O\left(\frac{1}{n}\right) \qquad n \to +\infty. \tag{12.64.b}$$

We now wish to approximate the Euler–Mascheroni constant with an error at most equal to 10^{-2}: by means of (12.61.b) it is enough that $\varepsilon_1(n) \leq \dfrac{1}{2n} \leq 10^{-2}$, or equivalently $n \geq 50$. For $n = 50$, from (12.61.b) we have

---

[3]Lorenzo Mascheroni (1750–1800).

$$\gamma \approx \gamma_{50} + \frac{1}{2}\frac{1}{50},$$

with an error within 10^{-2}. Now, using a CAS, we get

$$\gamma_{50} = \sum_{1 \le k < 50} \frac{1}{k} - \int_1^{50} \frac{1}{t}\, dt = \frac{1388125668713913502663l}{3099044504245996706400} - \log 50 = 0.56718\dots,$$

whence

$$\gamma \approx 0.57718\dots,$$

with an error within 10^{-2}. Actually $\gamma = 0.57721\dots$, we thus got the first 3 digits of γ. This Euler–Mascheroni constant appears in many formulas[4] and is still a source of mathematical challenges: it is not known for instance, at the date of publication of this book, whether γ is rational nor whether γ is an algebraic number.   □

12.4.4 The Second-Order Asymptotic Formula

The second-order asymptotic formula for a function f holds whenever f is of class at least $\mathscr{C}^2$ and if either its second derivative f'' is summable or its derivative f' is definitively monotonic. The Euler constant of f exists under the more severe conditions that f'' is summable and the limit $f(\infty)$ is finite. We postpone the proof of Theorem 12.65 to that of the general case of order $m \ge 1$ in Theorem 13.45.

Theorem 12.65 (Second-order criterion for the existence of the Euler constant and Euler–Maclaurin asymptotic formula) *Let $f : [a, +\infty[\to \mathbb{R}$ be of class $\mathscr{C}^2$, $a \in \mathbb{N}$.*

1. If $f'' \in L^1([a, +\infty[)$ there exists a constant C_2^f satisfying

$$\sum_{a \le k < n} f(k) = \int_a^n f(x)\, dx + C_2^f - \frac{f(n)}{2} + O\left(\int_n^{+\infty} |f''(x)|\, dx\right) \quad n \to +\infty.$$

$$(12.65.a)$$

One has $f'(\infty) \in \mathbb{R}$; assuming further that $f(\infty) \in \mathbb{R}$, the Euler constant γ^f of f is defined and it is given by $\gamma^f = C_2^f - \dfrac{f(\infty)}{2}$. Moreover, for $n \in \mathbb{N}_{\ge a}$ we have the following second-order estimate of γ^f:

---

[4]http://en.wikipedia.org/wiki/Euler-Mascheroni_constant.

$$\gamma^f = \gamma^f_n - \frac{1}{2}\left[f\right]^\infty_n + \frac{1}{12}\left[f'\right]^\infty_n + \varepsilon_2(n), \quad |\varepsilon_2(n)| \le \frac{1}{12}\int^{+\infty}_n |f''(x)|\, dx.$$

(12.65.b)

*1'. If f' is **bounded** and **definitively monotonic** then $f'' \in L^1([a, +\infty[)$, and (12.65.a) may be rewritten as*

$$\sum_{a \le k < n} f(k) = \int^n_a f(x)\, dx + C^f_2 - \frac{f(n)}{2} + O\left(f'(n) - f'(\infty)\right) \quad n \to +\infty.$$

(12.65.c)

Moreover in the second-order estimate (12.65.b) of γ^f we have

$$|\varepsilon_2(n)| \le \frac{|f'(n) - f'(\infty)|}{12}.$$

(12.65.d)

*2. If f' is **unbounded** and **definitively monotonic** then*

$$\sum_{a \le k < n} f(k) = \int^n_a f(x)\, dx - \frac{f(n)}{2} + O\left(f'(n)\right) \quad n \to +\infty.$$

(12.65.e)

Remark 12.66 Notice that, if f' is *monotonic* then, thanks to Lemma 12.25, the estimate in (12.65.b) may be rewritten as

$$|\varepsilon_2(n)| \le \frac{|f'(n) - f'(\infty)|}{12}.$$

☞ *Remark 12.67* In the first-order asymptotic formula (12.61.a) the summability of the first derivative ensures both the validity of the formula and the existence of the Euler constant. By way of contrast, the second-order conditions that ensure the existence of the Euler constant are more severe than those that ensure the validity of the asymptotic formula. In general, the summability of f'' is not enough for the existence of the Euler constant: we refer the skeptical reader to Example 12.69.

Example 12.68 (The Euler–Mascheroni constant γ, part II) In Example 12.64 we found that

$$\sum_{1 \le k < n} \frac{1}{k} = \int^n_1 \frac{1}{x}\, dx + \gamma + O\left(\frac{1}{n}\right) \quad n \to +\infty.$$

Since $f(\infty) = 0$ and f' is monotonic, in (12.65.c) we have $C_2^f = \gamma^f = \gamma$ and thus

$$H_n = \sum_{1 \le k < n} \frac{1}{k} + \frac{1}{n} = \log n + \gamma + \frac{1}{2n} + O\left(\frac{1}{n^2}\right) \qquad n \to +\infty. \qquad (12.68.a)$$

In Example 12.64 we got an estimate of γ with an error below 10^{-2} by means of (12.61.b): actually this required some work, namely the computation of $\gamma_{50} = 0.5671823329\ldots$, the sum of the first 50 terms of the original sum. Now, for each n, the modulus of the remainder $\varepsilon_2(n)$ in (12.65.b) is bounded above by $\frac{1}{12}|f'(n)| = \frac{1}{12n^2}$: therefore $|\varepsilon_2(n)| \le 10^{-2}$ if just $n \ge \frac{5}{\sqrt{3}} \approx 2.89$. For $n = 3$ the application of (12.65.b) yields, with an error within 10^{-2}:

$$\gamma \approx \gamma_3 - \frac{1}{2}\left[\frac{1}{x}\right]_3^\infty + \frac{1}{12}\left[-\frac{1}{x^2}\right]_3^\infty$$
$$= \left(1 + \frac{1}{2}\right) - \log 3 + \frac{1}{6} + \frac{1}{12}\frac{1}{3^2} = \frac{181}{108} - \log(3) = 0.577314\ldots.$$

Notice that $\gamma = 0.577215664901\ldots$: we obtained 3 digits of γ, without hard computations. We can also benefit from our computation of γ_{50}, that was carried out in Example 12.64. For $n = 50$ the application of (12.65.b) yields

$$\gamma \approx \gamma_{50} - \frac{1}{2}\left[\frac{1}{x}\right]_{50}^\infty + \frac{1}{12}\left[-\frac{1}{x^2}\right]_{50}^\infty$$
$$= 0.5671823329\ldots + \frac{1}{2}\frac{1}{50} + \frac{1}{12}\frac{1}{50^2} = 0.57721566623\ldots$$

The error here is at most equal to $\varepsilon_2(50) \le \frac{1}{12 \times 50^2} \le \frac{10^{-4}}{3}$; actually, by comparison with the correct value of γ we now correctly obtained its first 8 digits.   □

In the next example we see that it may happen that a function f whose second derivative is summable might not admit the existence of the Euler constant even though the constant C_2^f in (12.65.a) is non-zero. We recall that the Riemann zeta function is defined by

$$\zeta(s) = \sum_{n=1}^{\infty} \frac{1}{n^s}, \qquad s > 1.$$

Example 12.69 (Second-order asymptotic formula and non existence of the Euler constant) Let $f(x) = \sqrt{x}, x > 0$. By (12.61.e) we get the asymptotic estimate

$$\sum_{1 \le k < n} \sqrt{k} = \int_1^n \sqrt{x}\,dx + O(\sqrt{n}) = \frac{2}{3}n^{3/2} + O(\sqrt{n}) \quad n \to +\infty. \qquad (12.69.\mathrm{a})$$

Since $f'(x) = \frac{1}{2}x^{-1/2}$ is decreasing and tends to 0 at infinity, (12.65.c) yields the existence of $C_2^f \in \mathbb{R}$ satisfying

$$\sum_{1 \le k < n} \sqrt{k} = \int_1^n \sqrt{x}\,dx + C_2^f - \frac{\sqrt{n}}{2} + O(1/\sqrt{n})$$

$$\qquad\qquad (12.69.\mathrm{b})$$

$$= \frac{2}{3}n^{3/2} - \frac{2}{3} + C_2^f - \frac{\sqrt{n}}{2} + O(1/\sqrt{n}) \qquad n \to +\infty.$$

This constitutes a refinement of (12.69.a). The Euler constant of f does not exist here since, from (12.69.b),

$$\gamma_n^f = \sum_{1 \le k < n} \sqrt{k} - \int_1^n \sqrt{x}\,dx$$

is asymptotic to $-\dfrac{\sqrt{n}}{2}$ and thus tends to $-\infty$ as $n \to +\infty$. Notice that, since $1/\sqrt{n}$ tends to 0 at infinity, the constant C_2^f is given by

$$C_2^f = \lim_{n \to +\infty} \left(\sum_{1 \le k < n} \sqrt{k} - \frac{2}{3}n^{3/2} + \frac{\sqrt{n}}{2} \right) + \frac{2}{3}.$$

It can be shown that[5] $C_2^f = -\dfrac{\zeta(3/2)}{4\pi} + \dfrac{2}{3}.$ \hfill $\square$

The next example, a generalization of Example 12.69, yields some asymptotic estimates for the sum $\displaystyle\sum_{1 \le k < n} k^\alpha$.

Example 12.70 Let $\alpha \in \mathbb{R} \setminus \{-1, 0\}$, and consider the function $f_\alpha(x) = x^\alpha$ $(x \ge 1)$. We look for an asymptotic estimate of $\displaystyle\sum_{1 \le k < n} k^\alpha$ and the existence of the Euler constant of f_α. The case $\alpha = 0$ is trivial since $f_0(x) = 1$, $\forall x \ge 1$, and hence $\gamma_n^{f_0} = \displaystyle\sum_{1 \le k < n} 1 = \displaystyle\int_1^n 1\,dx = 0$ for each $n \in \mathbb{N}$. The case $\alpha = -1$ was considered in Examples 12.64 and 12.68. We are generalizing here the case $\alpha = 1/2$ considered in Example 12.69. Both f_α and f_α' are monotonic and either converge to 0 or diverge to infinity for $x \to$

---

[5]We obtained this with a CAS; the proof of the result is due to Ramanujuan [29], [6, Chap. VII, Corollary 7], decades before the advent of computers.

$+\infty$: Theorems 12.61 and 12.65 yield an asymptotic estimate of the sum $\displaystyle\sum_{1\le k<n} k^\alpha$
with a remainder that is $O(n^\alpha)$ or $O(n^{\alpha-1})$ as $n \to +\infty$. Since $\alpha n^{\alpha-1} = O(n^\alpha)$ as $n \to +\infty$ it is clear that the first-order formulas can be derived from the second-order ones (12.65.c)–(12.65.e) (see Problem 12.21). It is convenient to consider few cases, depending on the fact that $f_\alpha(x) = x^\alpha, f'_\alpha(x) = \alpha x^{\alpha-1}$ converge or diverge at infinity.

- $\alpha < 0$: $f_\alpha(\infty) = f'_\alpha(\infty) = 0$. From (12.65.a) and (12.65.b) we deduce the existence of the Euler constant γ^{f_α} of f_α, and

$$\sum_{1\le k<n} k^\alpha = \frac{1}{\alpha+1}(n^{\alpha+1} - 1) + \gamma^{f_\alpha} + O(n^{\alpha-1}),$$

  from which we get

$$\gamma^{f_\alpha} = \lim_{n\to+\infty}\left(\sum_{1\le k<n} k^\alpha - \frac{n^{\alpha+1} - 1}{\alpha+1}\right).$$

  – If $\alpha < -1$ then

$$\gamma^{f_\alpha} = \lim_{n\to+\infty}\left(\sum_{1\le k<n} k^\alpha - \frac{n^{\alpha+1} - 1}{\alpha+1}\right) = \zeta(-\alpha) + \frac{1}{\alpha+1}.$$

  – If $-1 < \alpha < 0$ then both $\displaystyle\sum_{1\le k<n} k^\alpha$ and $\dfrac{n^{\alpha+1} - 1}{\alpha+1}$ diverge to $+\infty$, their difference however converges to γ^{f_α}: we are in a situation similar to that of Example 12.69, and the explicit computation of γ^{f_α} is usually very difficult.

- $\alpha \in]0, 1[$: $f_\alpha(\infty) = +\infty$ and $f'_\alpha(\infty) = 0$: from (12.65.c) we deduce

$$\sum_{1\le k<n} k^\alpha = \frac{n^{\alpha+1} - 1}{\alpha+1} + C_2^f - \frac{n^\alpha}{2} + O(n^{\alpha-1}),$$

  and the remainder $O(n^{\alpha-1})$ tends to 0 at infinity.

- $\alpha > 1$: $|f_\alpha(\infty)| = |f'_\alpha(\infty)| = +\infty$: from (12.65.e) we have

$$\sum_{1\le k<n} k^\alpha = \frac{n^{\alpha+1} - 1}{\alpha+1} - \frac{n^\alpha}{2} + O(n^{\alpha-1})$$

$$= \frac{1}{\alpha+1}n^{\alpha+1} - \frac{n^\alpha}{2} + O(n^{\alpha-1}) \qquad n \to +\infty.$$

Finally, we notice that, if $\alpha \in \mathbb{N}$, the sum $\sum\limits_{1 \le k < n} k^\alpha$ is a polynomial of degree $\alpha + 1$ in n: this case was considered in Example 8.52. □

12.4.5   Estimates of the Factorials and the Binomials

An important consequence of the second-order asymptotic formula is a better estimate of the factorial than the one found in Example 12.43.

Proposition 12.71   (Stirling's formula, up to a multiplicative constant) *There exists a real constant C such that*

$$\log(n!) = \sum_{k=1}^{n} \log k = n \log n - n + \frac{\log n}{2} + C + O(1/n) \qquad n \to +\infty,$$

$$(12.71.a)$$

from which one obtains the formula

$$n! \sim e^C n^n \sqrt{n}\, e^{-n} \qquad n \to +\infty, \qquad (12.71.b)$$

i.e., $\lim\limits_{n \to +\infty} \dfrac{n!}{n^n \sqrt{n}\, e^{-n}} = e^C.$

Proof. Let $f(x) = \log x$, $x \ge 1$; then $f'(x) = 1/x$ is decreasing and tends to 0 at infinity. Through the application of the second-order asymptotic formula (12.65.c) we get, for some $C_2^f \in \mathbb{R}$,

$$\sum_{k=1}^{n} \log k = \sum_{1 \le k < n} \log k + \log n$$

$$= \int_1^n \log x \, dx + \frac{\log n}{2} + C_2^f + O(1/n) \quad n \to +\infty.$$

Integration by parts yields

$$\int_1^n \log x \, dx = n \log n - n + 1,$$

from which it follows that

$$\sum_{k=1}^{n} \log k = n \log n - n + \frac{\log n}{2} + C + O(1/n) \qquad n \to +\infty,$$

with $C = C_2^f + 1$. Given that $n! = \exp(\log n!) = \exp\left\{\sum_{k=1}^{n} \log k\right\}$, one finds that

$$n! = \exp\left\{n \log n - n + \frac{\log n}{2} + C + O(1/n)\right\}$$

$$= e^C \frac{n^{n+\frac{1}{2}}}{e^n} e^{O(1/n)} \sim e^C \frac{n^{n+\frac{1}{2}}}{e^n} \qquad n \to +\infty,$$

given that a function that is $O(1/n)$ tends to 0 as $n \to +\infty$. □

We need Wallis[6] formula in order to determine C in (12.71.b).

Proposition 12.72 (Wallis formula)

$$\lim_{n \to +\infty} \frac{2^2 \cdot 4^2 \cdots (2n)^2}{1 \cdot 3^2 \cdots (2n-1)^2} \frac{1}{2n+1} = \frac{\pi}{2}. \qquad (12.72.a)$$

Proof. We consider the sequence

$$a_n := \int_0^{\pi/2} \sin^n x \, dx \qquad \forall n \in \mathbb{N}.$$

For all $n \geq 2$ integration by parts gives

$$a_n = \left[-(\sin x)^{n-1} \cos x\right]_0^{\pi/2} + \int_0^{\pi/2} (n-1)(\sin x)^{n-2} \cos^2 x \, dx$$

$$= \int_0^{\pi/2} (n-1)(\sin x)^{n-2} \cos^2 x \, dx = \int_0^{\pi/2} (n-1)(\sin x)^{n-2}(1 - \sin^2 x) \, dx$$

$$= (n-1)(a_{n-2} - a_n),$$

so that

$$a_n = \frac{n-1}{n} a_{n-2} \qquad n \geq 2. \qquad (12.72.b)$$

---

[6]John Wallis (1616–1703).

Notice that

$$\lim_{n \to +\infty} \frac{a_{2n}}{a_{2n+1}} = 1. \tag{12.72.c}$$

Indeed $(a_n)_n$ is decreasing (due to the fact that $(\sin^n x)_n$ is decreasing for all $0 \le x \le \pi/2$), so that from (12.72.b)

$$1 \le \frac{a_{2n}}{a_{2n+1}} \le \frac{a_{2n-1}}{a_{2n+1}} \le \frac{2n+1}{2n} \to 1 \text{ as } n \to +\infty.$$

Since $a_0 = \dfrac{\pi}{2}$, we deduce from (12.72.b) that

$$a_{2n} = \frac{\pi}{2} \times \frac{1}{2} \times \frac{3}{4} \times \cdots \times \frac{2n-1}{2n} \qquad \forall n \ge 1;$$

since $a_1 = 1$, again from (12.72.b) we get

$$a_{2n+1} = 1 \times \frac{2}{3} \times \frac{4}{5} \times \cdots \times \frac{2n}{2n+1} \qquad \forall n \ge 1.$$

Therefore we obtain

$$\frac{a_{2n}}{a_{2n+1}} = \frac{\pi}{2} \times \frac{1 \cdot 3^2 \cdots (2n-1)^2}{2^2 \cdot 4^2 \cdots (2n)^2} (2n+1) \qquad \forall n \ge 1;$$

the conclusion follows then directly from (12.72.c). □

We are now able to determine C in Proposition 12.71.

Corollary 12.73 (Stirling's formula) *The constant C in (12.71.b) equals* $\log \sqrt{2\pi}$, *and thus*

$$\log(n!) = \sum_{k=1}^{n} \log k = n \log n - n + \frac{\log n}{2} + \frac{1}{2} \log 2\pi + O(1/n) \qquad n \to +\infty,$$
$$\tag{12.73.a}$$

$$n! \sim \sqrt{2\pi} n^n \sqrt{n} \, e^{-n} \qquad n \to +\infty. \tag{12.73.b}$$

Proof. It turns out easily that Wallis formula is equivalent to

$$\frac{(2^n n!)^2}{(2n)!} = \frac{2n(2n-2)\cdots 2}{(2n-1)\cdots 3\cdot 1} \sim \sqrt{\frac{(2n+1)\pi}{2}} \sim \sqrt{\pi n} \qquad n \to +\infty. \qquad (12.73.\text{c})$$

Thus, knowing that $n! \sim e^C n^n \sqrt{n} e^{-n}$ as $n \to +\infty$ we obtain

$$(2n)! \sim e^C (2n)^{2n} \sqrt{2n} e^{-2n} = \sqrt{2} e^C 2^{2n} n^{2n} \sqrt{n} e^{-2n} \quad n \to +\infty,$$

whereas

$$(2^n n!)^2 \sim (2^n e^C n^n \sqrt{n} e^{-n})^2 = e^{2C} 2^{2n} n^{2n} n e^{-2n} \quad n \to +\infty.$$

It follows that

$$\frac{(2^n n!)^2}{(2n)!} \sim \frac{e^{2C} 2^{2n} n^{2n} n e^{-2n}}{\sqrt{2} e^C 2^{2n} n^{2n} \sqrt{n} e^{-2n}} = \frac{e^C}{\sqrt{2}} \sqrt{n} \quad n \to +\infty.$$

Comparing with (12.73.c), we conclude that $\dfrac{e^C}{\sqrt{2}} = \sqrt{\pi}$, so that $e^C = \sqrt{2\pi}$.   □

Remark 12.74 The application of the first-order asymptotic Euler–Maclaurin formula is not enough to get (12.71.a). Indeed, (12.61.e) applied to the function $f(x) = \log x, x \geq 1$ yields

$$\sum_{1 \leq k < n} \log k + \log n = \int_1^n \log x \, dx + O(\log n) \qquad n \to +\infty,$$

whence

$$\log n! = n \log n - n + O(\log n) \qquad n \to +\infty.$$

By composing both members of the inequality with the exponential function we obtain

$$n! = n^n e^{-n} h(n),$$

where $h(n) = e^{O(\log n)}$ as $n \to +\infty$. Proposition 12.71 states, more precisely, that $h(n)$ is asymptotic to $e^C \sqrt{n}$ as $n \to +\infty$.

In Proposition 7.97 we showed that, for $a \in \mathbb{R} \setminus \mathbb{N}$, $\left| \binom{a}{n} \right|$ is asymptotic to a constant times $\dfrac{1}{n^{a+1}}$ as $n \to +\infty$: the proof was based on elementary argument upon the estimate of the defocused factorial $(n - \varepsilon)(n - 1 - \varepsilon) \cdots (1 - \varepsilon)$. The asymptotic Euler–Maclaurin formula provides a direct shorter proof of the result.

Example 12.75 Let $a \in \mathbb{R} \setminus \mathbb{N}$. We prove that (7.97.a) holds, i.e., there is $C_a > 0$ such that

$$\left| \binom{a}{n} \right| \sim \frac{C_a}{n^{a+1}} \quad n \to +\infty.$$

Since $\left| \binom{-1}{n} \right| = 1$, the claim is true for $a = -1$: we henceforth assume $a \neq -1$. Let $n > [a] + 2$. Then

$$x_n := n^{a+1} \left| \binom{a}{n} \right| = n^{a+1} \frac{|a(a-1) \cdots (a-n+1)|}{n!}$$

$$= n^{a+1} \frac{|(a+1)-1|}{1} \frac{|(a+1)-2|}{2} \cdots \frac{|(a+1)-n|}{n}.$$

Thus

$$\log x_n = (a+1) \log n + \sum_{k=1}^{n} \log \frac{|(a+1)-k|}{k}$$

$$= (a+1) \log n + C_1 + \sum_{[a]+2 \le k < n} \log \left(1 - \frac{a+1}{k}\right) + \log \left(1 - \frac{a+1}{n}\right),$$

$$(12.75.a)$$

with $C_1 = \sum_{k=1}^{[a]+1} \log \frac{(a+1)-k}{k}$.

Let $f(x) := \log \left(1 - \frac{a+1}{x}\right)$; then the derivative $f'(x) = \frac{a+1}{x(x-(a+1))}$ is monotonic and tends to 0 at infinity. Therefore (12.65.c) yields

$$\sum_{[a]+2 \le k < n} f(k) = \int_{[a]+2}^{n} f(x) \, dx - \frac{1}{2} f(n) + C_2^f + O\left(f'(n)\right) \qquad (12.75.b)$$

as $n \to +\infty$. Now $f'(n) = \frac{a+1}{n(n-(a+1))} \sim \frac{a+1}{n^2}$ as $n \to +\infty$, and

$$\int f(x) \, dx = \int (\log(x-(a+1)) - \log x) \, dx$$

$$= (x-(a+1)) \log(x-(a+1)) - (x-(a+1)) - (x \log x - x) + \mathbb{R}$$

$$= x \log \left(1 - \frac{a+1}{x}\right) - (a+1) \log(x-(a+1)) + \mathbb{R}.$$

It then follows from (12.75.b) that

$$\sum_{[a]+2\leq k<n} \log\left(1-\frac{a+1}{k}\right) = n\log\left(1-\frac{a+1}{n}\right) - (a+1)\log(n-(a+1))+$$

$$-\frac{1}{2}\log\left(1-\frac{a+1}{n}\right) + C_3 + O\left(\frac{1}{n^2}\right) \quad n\to+\infty,$$

for a suitable constant C_3. By (12.75.a) we finally obtain

$$\log x_n = (a+1)\log\left(\frac{n}{n-(a+1)}\right) + n\log\left(1-\frac{a+1}{n}\right) + C_4 + \varepsilon(n)$$

$$= \log\left(1-\frac{a+1}{n}\right)^n + C_4 + \varepsilon'(n),$$

for some constant C_4 and $\varepsilon(n)$, $\varepsilon'(n)$ converging to 0 as $n\to+\infty$. Passing to the limit we obtain $\lim_{n\to+\infty} \log x_n = C_4 - (a+1)$, whence $\lim_{n\to+\infty} x_n = C_a$ $:= e^{C_4-(a+1)}$. $\qquad\square$

12.5   A True Step to Infinity: The Integral Test ☕

The celebrated integral test for the convergence of a series states that if a function $f : [a, +\infty[\to \mathbb{R}$ is bounded and monotonic the series

$$\sum_{k=a}^{\infty} f(k) \tag{12.75.c}$$

converges if, and only if, the generalized integral

$$\int_a^{+\infty} f(x)\,dx \tag{12.75.d}$$

is finite. The precise statement and proof of the result can be found in Corollary 12.108. It is a rather frequent error among students to sustain the validity of the result even without assuming the monotonicity of the function. In the following two examples we see that in reality such belief is completely lacking in justification.

Example 12.76 (The integral converges, the series does not) Let $f : [1, +\infty[\to \mathbb{R}$ be the continuous function that, for each $n \in \mathbb{N}_{\geq 1}$, equals $\frac{1}{n}$ at n, 0 at $n+\frac{1}{2n}$ and $(n+1)-\frac{1}{2(n+1)}$, and is affine elsewhere (Fig. 12.1). Then:

Fig. 12.1 $\displaystyle\sum_{k=1}^{\infty} f(k)$ diverges

but $\displaystyle\int_{1}^{+\infty} f(x)\,dx < +\infty$

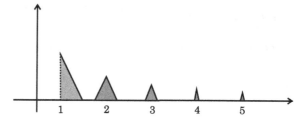

Fig. 12.2 $\displaystyle f \geq 0,\ \sum_{k=1}^{\infty} f(k)$

converges but

$\displaystyle\int_{1}^{+\infty} f(x)\,dx = +\infty$

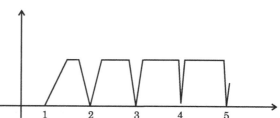

1. $f \geq 0$ is continuous and $\displaystyle\lim_{x\to+\infty} f(x) = 0$;

2. $\displaystyle\sum_{k=1}^{\infty} f(k) = \sum_{k=1}^{\infty} \frac{1}{k} = +\infty, \quad \int_{1}^{+\infty} f(x)\,dx = \frac{1}{2}\left(\frac{1}{2} + \sum_{k=2}^{+\infty} \frac{1}{k^2}\right) < +\infty.$ ☐

Example 12.77 (The series converges, the integral does not) Let $f : [1, +\infty[\to \mathbb{R}$ be the continuous function whose values are, for each $n \in \mathbb{N}_{\geq 1}$, 0 at n, 1 at $n + \dfrac{1}{2n}$ and $(n + 1) - \dfrac{1}{2(n + 1)}$, affine elsewhere (Fig. 12.2). Then:

1. $f \geq 0$ is continuous;

2. $\displaystyle\sum_{k=1}^{\infty} f(k) = \sum_{k=1}^{\infty} 0 = 0, \quad \int_{1}^{+\infty} f(x)\,dx = +\infty.$ ☐

At the same time, there are conditions on a function, others than monotonicity, that ensure the validity of the integral test, i.e., that relate the convergence of the series (12.75.c) to that of the integral (12.75.d): they are the very same that ensure the existence of the Euler constant.

Corollary 12.78 (First-order integral test and approximation of the sum of a series) *Let $f : [a, +\infty[\to \mathbb{R}$ be of class $\mathscr{C}^1$, $a \in \mathbb{N}$, and assume that*

• $f' \in L^1([a, +\infty[).$

Then:

1. *The series $\displaystyle\sum_{k=a}^{\infty} f(k)$ and the generalized integral $\displaystyle\int_{a}^{+\infty} f(x)\,dx$ have the same behavior (both convergent, divergent, or do not exist);*

2. *If the series* $\sum\limits_{k=a}^{\infty} f(k)$ *converges, then for all* $n \in \mathbb{N}_{\geq a}$, *the following **first-order** approximation holds:*

$$\sum_{k=a}^{\infty} f(k) = \sum_{a \leq k < n} f(k) + \int_{n}^{+\infty} f(x)\, dx - \frac{1}{2} [f]_{n}^{\infty} + \varepsilon_1(n),$$

$$|\varepsilon_1(n)| \leq \frac{1}{2} \int_{n}^{+\infty} |f'(x)|\, dx. \qquad (12.78.a)$$

Proof. 1. There exists, from Theorem 12.61, the Euler constant $\gamma^f \in \mathbb{R}$ satisfying

$$\sum_{a \leq k < n} f(k) = \int_{a}^{n} f(x)\, dx + \gamma^f + \varepsilon_n, \quad \forall n \in \mathbb{N},$$

with ε_n converging to 0 as $n \to +\infty$; therefore the series $\sum\limits_{k=a}^{\infty} f(k)$ and the limit $\lim\limits_{\substack{n \to +\infty \\ n \in \mathbb{N}}} \int_{a}^{n} f(x)\, dx$ have the same behavior. Lemma 12.57 ensures the existence of $f(\infty) \in \mathbb{R}$: the behavior of $\lim\limits_{\substack{n \to +\infty \\ n \in \mathbb{N}}} \int_{a}^{n} f(x)\, dx$ then coincides with that of $\int_{a}^{+\infty} f(x)\, dx$ (see Problem 12.15).

2. Since there exists the Euler constant of f, for all $n \geq a$ (12.61.b) may be rewritten as

$$\gamma^f = \sum_{k=a}^{\infty} f(k) - \int_{a}^{+\infty} f(x)\, dx = \sum_{a \leq k < n} f(k) - \int_{a}^{n} f(x)\, dx - \frac{1}{2} [f]_{n}^{\infty} + \varepsilon_1(n).$$

By adding $\int_{a}^{+\infty} f(x)\, dx$ to both terms of the last equality, we deduce that

$$\sum_{k=a}^{\infty} f(k) = \sum_{a \leq k < n} f(k) + \int_{n}^{\infty} f(x)\, dx - \frac{1}{2} [f]_{n}^{\infty} + \varepsilon_1(n).$$

The estimate of $\varepsilon_1(n)$ was established in Theorem 12.61.   □

Remark 12.79 It is worth noticing that, if f is *monotonic*, then in (12.78.a) we obtain

$$|\varepsilon_1(n)| \leq \frac{1}{2}|f(n)|.$$

Indeed Lemma 12.25 implies that $|\varepsilon_1(n)| \leq \frac{1}{2}|f(n) - f(\infty)|$, and the convergence of the series $\sum_{k=a}^{\infty} f(k)$ implies that $f(\infty) = 0$.

Remark 12.80 In view of Lemma 12.57, a smooth (e.g., $\mathscr{C}^1$) monotonic function f satisfies the assumption of Corollary 12.78 if, and only if, $f(\infty)$ is finite. As we pointed out, the smoothness assumption can be dropped in this case: the proof of the result is of course different and involves an Euler–Maclaurin first-order type formula, we refer to Corollary 12.108 for the details. At the same time, if f is smooth, the only existence of a finite limit at ∞ is not, in general, sufficient for the validity of Corollary 12.78: an example to illustrate this situation may be built by smoothing the function defined in Example 12.76.

Let us now see some applications of Corollary 12.78.

Example 12.81 (*The value $\zeta(2)$*) We wish to approximate $\zeta(2) = \sum_{k=1}^{\infty} \frac{1}{k^2}$ with an error within 10^{-2}: by Corollary 12.78, with $f(x) = \frac{1}{x^2}$, $x \geq 1$, it is enough that in (12.78.a) the remainder $\varepsilon_1(n)$ satisfies $|\varepsilon_1(n)| \leq 10^{-2}$; this occurs as soon as if $\frac{1}{2}|f(n) - f(\infty)| = \frac{1}{2n^2} \leq 10^{-2}$, i.e., for $n \geq 8$. With the choice of $n = 10$ we get, with an error within 10^{-2},

$$\sum_{k=1}^{\infty} \frac{1}{k^2} \approx \sum_{1 \leq k < 10} \frac{1}{k^2} + \int_{10}^{+\infty} \frac{1}{t^2}\,dt - \frac{1}{2}\left[\frac{1}{t^2}\right]_{10}^{\infty} = \frac{9778141}{6350400} + \left[-\frac{1}{t}\right]_{10}^{\infty} + \frac{1}{200}$$

$$= \frac{10444933}{6350400} = 1.64476\ldots.$$

It seems that this approximation led Euler to conjecture that $\zeta(2) = \frac{\pi^2}{6} = 1.64493\ldots$; notice that he was not aware of the estimate of the remainder, which is due to Poisson years later. Note that we obtained the 3 digits of $\zeta(2)$ by means of just the first nine terms of the series. □

Example 12.82 (*Apéry's constant $\zeta(3)$*) We wish to approximate $\zeta(3) = \sum_{k=1}^{\infty} \frac{1}{k^3}$ with an error within 10^{-2}: by Corollary 12.78, with $f(x) = \frac{1}{x^3}$, $x \geq 1$ it is enough that, in (12.78.a), $|\varepsilon_1(n)| \leq 10^{-2}$; this occurs if $\frac{1}{2}|f(n) - f(\infty)| = \frac{1}{2n^3} \leq 10^{-2}$, i.e., for (just!) $n \geq 4$. With $n = 4$ we obtain, with an error at most equal to 10^{-2},

$$\sum_{k=1}^{\infty} \frac{1}{k^3} \approx \sum_{1 \leq k < 4} \frac{1}{k^3} + \int_4^{+\infty} \frac{1}{t^3}\, dt - \frac{1}{2}\left[\frac{1}{t^3}\right]_4^{\infty} = \frac{251}{216} + \left[-\frac{1}{2t^2}\right]_4^{\infty} + \frac{1}{2 \times 4^3}$$

$$= \frac{4151}{3456} = 1.20109\ldots.$$

The value of $\zeta(3)$ is, approximately, $1.20205\ldots$: two digits are correct in our approximation. Unlike the case of $\zeta(2) = \frac{\pi^2}{6}$, the explicit form of $\zeta(3)$ is still unknown; its value is also called *Apéry's*[7] *constant* in honour of R. Apéry who proved in [1] the irrationality of this number. It is worth noticing that its inverse is the limit, as n goes to infinity, of the probability that three natural numbers less than or equal to n be coprime.[8]   □

In the next example we apply Corollary 12.78 to a function that is not monotonic, so the integral test for monotonic function (Corollary 12.108) does not apply.

Example 12.83 We study the convergence of the series $\displaystyle\sum_{k=1}^{\infty} \frac{\sin(\sqrt{k})}{k}$. It is easy to see that the generalized integral

$$\int_1^{+\infty} \frac{\sin(\sqrt{x})}{x}\, dx = 2 \int_1^{+\infty} \frac{\sin u}{u}\, du$$

is finite, since an integration by parts gives

$$\int_1^{+\infty} \frac{\sin u}{u}\, du = \cos(1) - \int_1^{+\infty} \frac{\cos u}{u^2}\, du,$$

and $\frac{\cos u}{u^2}$ is summable, due to the fact that its absolute value is bounded above by $\frac{1}{u^2} \in L^1([1, +\infty[)$. Set $f(x) = \frac{\sin(\sqrt{x})}{x}$. Then we have:

- $f(\infty) = 0$;
- $f'(x) = \frac{\cos(\sqrt{x})}{2x^{3/2}} - \frac{\sin(\sqrt{x})}{x^2}$ is summable since $|f'(x)| \leq \frac{1}{2x^{3/2}} + \frac{1}{x^2} \in L^1([1, +\infty[)$.

---

[7]Roger Apéry (1916–1994).
[8]http://en.wikipedia.org/wiki/Ap%C3%A9ry%27s_constant.

The application of the first point of Corollary 12.78 allows to conclude that the series
$$\sum_{k=1}^{\infty} \frac{\sin(\sqrt{k})}{k} \text{ converges.} \qquad \Box$$

We formulate here the second-order integral test and approximation for the sum of a convergent series; its proof is carried out in greater generality in Corollary 13.49. Here, again, the assumptions that ensure the validity of the test are those that guarantee the existence of the Euler constant.

Corollary 12.84 (Second-order integral test and approximation of the sum of a series) *Let $f : [a, +\infty[\to \mathbb{R}$ be a function of class $\mathscr{C}^2$, $a \in \mathbb{N}$, such that:*

- $f(\infty)$ *is finite;*
- $f'' \in L^1([a, +\infty[).$

Then:

1. *The series $\sum_{k=a}^{\infty} f(k)$ and the generalized integral $\int_a^{+\infty} f(x)\, dx$ have the same behavior (both convergent, divergent, or do not exist);*

2. *If the series $\sum_{k=a}^{\infty} f(k)$ converges, then for all $n \in \mathbb{N}_{\geq a}$, the following **second-order approximation** holds:*

$$\sum_{k=a}^{\infty} f(k) = \sum_{a \leq k < n} f(k) + \int_n^{+\infty} f(x)\, dx - \frac{1}{2}\left[f\right]_n^{\infty} + \frac{1}{12}\left[f'\right]_n^{\infty} + \varepsilon_2(n),$$

$$|\varepsilon_2(n)| \leq \frac{1}{12}\int_n^{+\infty} |f''(x)|\, dx. \quad (12.84.a)$$

Remark 12.85 Assume that a function f is of class $\mathscr{C}^2$. The conditions that ensure the validity of the first-order and the second-order integral tests are of different nature: it is convenient, due to the simplicity of its assumption on the first derivatives, to use the first-order test if f' is summable. If not, one has to check the validity of the more complex assumptions of Corollary 12.84.

We come back to some examples considered at a first-order level.

Example 12.86 (*The value $\zeta(2)$*) In Example 12.81 we found an approximated value $1.64476\ldots$ of $\zeta(2)$ by means of (12.78.a) with $n = 10$ and $f(x) = \dfrac{1}{x^2}$, $x \geq 1$. Since $f(\infty) = 0$, and f' is monotonic and tends to zero at infinity, by Lemma 12.57 $f'' \in L^1([a, +\infty[)$. One can then apply (12.84.a) to approximate $\zeta(2)$ within 10^{-4}

if $|\varepsilon_2(n)| \leq 10^{-4}$, or equivalently, $\dfrac{1}{12}\dfrac{2}{n^3} \leq 10^{-4}$, i.e., for $n \geq \dfrac{10^{4/3}}{6^{1/3}} \approx 11.86$. For $n = 12$ we obtain, with an error of at most 10^{-4},

$$\zeta(2) \approx \sum_{1 \leq k < 12} \frac{1}{k^2} + \int_{12}^{+\infty} \frac{1}{x^2}\, dx - \frac{1}{2}\left[\frac{1}{x^2}\right]_{12}^{\infty} + \frac{1}{12}\left[-\frac{2}{x^3}\right]_{12}^{\infty}$$

$$= \frac{168528641}{102453120} = 1.6449342,$$

which actually shares 6 digits with $\zeta(2) = \dfrac{\pi^2}{6} = 1.6449340\ldots$.   □

Example 12.87 (*Apéry's constant* $\zeta(3)$) In Example 12.82 we obtained an approximated value of Apéry's constant $\zeta(3)$ with an error of at most 10^{-2} by means of (12.78.a), with $f(x) = \dfrac{1}{x^3}, x \geq 1$ and $n = 4$. We wish here to push forward the approximation and get the value of $\zeta(3)$ with an error of at most 10^{-6}. Since $f'(x) = -\dfrac{3}{x^4}$ is monotonic and $f'(\infty) = 0 \in \mathbb{R}$, by Lemma 12.57 we get $f'' \in L^1([a, +\infty[)$, and hence we can apply Corollary 12.84. From (12.84.a) it is enough to find n in such a way that $|\varepsilon_2(n)| \leq 10^{-6}$, and this occurs if $\dfrac{1}{12}\dfrac{3}{n^6} \leq 10^{-6}$, i.e., for $n \geq \left(\dfrac{10^6}{4}\right)^{1/6} \approx 7.94$. With $n = 8$ we obtain, with an error below 10^{-6},

$$\zeta(3) \approx \sum_{1 \leq k < 8} \frac{1}{k^3} + \int_{8}^{+\infty} \frac{1}{x^3}\, dx - \frac{1}{2}\left[\frac{1}{x^3}\right]_{8}^{\infty} + \frac{1}{12}\left[-\frac{3}{x^4}\right]_{8}^{\infty}$$

$$= \frac{2533205761}{2107392000} \approx 1.202057216\ldots,$$

which actually differs from $\zeta(3) = 1.2020569031\ldots$ for at most 4×10^{-7}.   □

12.6 Estimates of the Integral of a Function and the Trapezoidal Method

So far, we have considered the Euler–Maclaurin formula from the Eulerian point of view, namely as a way to approximate finite sums. We consider here the dual approach, followed initially by Maclaurin, to view these formulas as a way to approximate integrals.

12.6.1 First- and Second- Order Euler–Maclaurin Approximations of an Integral

Let $g : [\alpha, \beta] \to \mathbb{R}$ be a function with $\alpha < \beta \in \mathbb{R}$ (and not necessarily integers!). For any $n > 0$ we consider the subdivision $\alpha = x_0 < x_1 < \cdots < x_n = \beta$ of $[\alpha, \beta]$ into n intervals of length $\dfrac{\beta - \alpha}{n}$, defined by

$$x_k = \alpha + h_n k \quad k = 0, \dots, n - 1, \quad h_n := \frac{\beta - \alpha}{n}.$$

Definition 12.88 Let $g \in \mathscr{C}([\alpha, \beta])$, with $\alpha < \beta \in \mathbb{R}$, and $n \in \mathbb{N}_{\geq 1}$. We denote by T_n^g the sum of the areas of the trapezes with vertices

$$(x_k, 0), (x_k, g(x_k)), (x_{k+1}, g(x_{k+1})), (x_{k+1}, 0) \quad k = 0, \dots, n - 1,$$

represented in Fig. 12.3, and given by

$$
\begin{aligned}
T_n^g :&= h_n \frac{g(x_0) + g(x_1)}{2} + h_n \frac{g(x_1) + g(x_2)}{2} + \cdots + h_n \frac{g(x_{n-1}) + g(x_n)}{2} \\
&= \frac{h_n}{2} \left(g(x_0) + 2g(x_1) + \cdots + 2g(x_{n-1}) + g(x_n) \right) = h_n \left(\sum_{0 \leq k < n} g(x_k) + \frac{1}{2} [g]_\alpha^\beta \right). \quad \square
\end{aligned}
$$

The last equality in the above definition shows, incidentally, how T_n^g is related to the Euler–Maclaurin expansions:

Fig. 12.3 The sum T_n^g of the areas of the trapezes

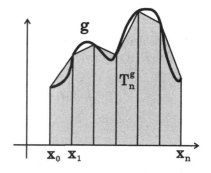

$$T_n^g = h_n \left(\sum_{0 \le k < n} f(k) + \frac{1}{2} [f]_0^n \right) \qquad f(t) := g(\alpha + th_n). \qquad (12.88.\text{a})$$

It is not surprising, from (12.88.a), that one may use the Euler–Maclaurin formulas to deduce an approximation of the integral of g.

Theorem 12.89 (First- and second- order Euler–Maclaurin approximation of integrals) *Let $\alpha < \beta \in \mathbb{R}$.*

1. If $g \in \mathscr{C}^1([\alpha, \beta])$ then

$$\int_\alpha^\beta g(x)\, dx = T_n^g + \varepsilon_1(n), \quad |\varepsilon_1(n)| \le \frac{\beta - \alpha}{2n} \int_\alpha^\beta |g'(x)|\, dx \le \frac{(\beta - \alpha)^2}{2n} \|g'\|_\infty.$$

$$(12.89.\text{a})$$

2. If $g \in \mathscr{C}^2([\alpha, \beta])$ then

$$\int_\alpha^\beta g(x)\, dx = T_n^g - \frac{(\beta - \alpha)^2}{12n^2} [g']_\alpha^\beta + \varepsilon_2(n),$$

$$|\varepsilon_2(n)| \le \frac{(\beta - \alpha)^2}{12n^2} \int_\alpha^\beta |g''(x)|\, dx \le \frac{(\beta - \alpha)^3}{12n^2} \|g''\|_\infty. \qquad (12.89.\text{b})$$

Proof. Set $f(t) = g(\alpha + th_n)$, $t \in [0, n]$. The change of variables $x = \alpha + th_n$ gives

$$\int_\alpha^\beta g(x)\, dx = h_n \int_0^n f(t)\, dt.$$

1. Assume that g is of class $\mathscr{C}^1$. The first-order Euler–Maclaurin formula (12.27.a), applied to f on $[0, n]$ yields

$$\int_0^n f(t)\, dt = \sum_{0 \le k < n} f(k) + \frac{1}{2} [f]_0^n - R_1(n), \qquad (12.89.\text{c})$$

with

$$|R_1(n)| \le \frac{1}{2} \int_0^n |f'(t)|\, dt.$$

Now

$$[f]_0^n = [g]_\alpha^\beta, \quad \int_0^n |f'(t)|\, dt = \int_\alpha^\beta |g'(x)|\, dx.$$

By rewriting (12.89.c) in terms of g we get

$$\frac{1}{h_n} \int_\alpha^\beta g(x)\, dx = \left(\sum_{0 \le k < n} g(x_k) + \frac{1}{2} [g]_\alpha^\beta \right) - R_1(n), \qquad (12.89.d)$$

with

$$|R_1(n)| \le \frac{1}{2} \int_\alpha^\beta |g'(x)|\, dx.$$

If we multiply both members of (12.89.d) by h_n we obtain

$$\int_\alpha^\beta g(x)\, dx = T_n^g - h_n R_1(n).$$

Set $\varepsilon_1(n) = -h_n R_1(n)$; the following estimate holds:

$$|\varepsilon_1(n)| \le h_n |R_1(n)| \le \frac{h_n}{2} \int_\alpha^\beta |g'(x)|\, dx,$$

showing the validity of (12.89.a).

2. Let g be of class $\mathscr{C}^2$. The second-order Euler–Maclaurin (12.27.b) applied to f on $[0, n]$ yields

$$\int_0^n f(t)\, dt = \sum_{0 \le k < n} f(k) + \frac{1}{2} [f]_0^n - \frac{1}{12} [f']_0^n - R_2(n), \qquad (12.89.e)$$

with

$$|R_2(n)| \le \frac{1}{12} \int_0^n |f''(t)|\, dt.$$

Now

$$[f']_0^n = h_n [g']_\alpha^\beta, \quad \int_0^n |f''(t)|\, dt = h_n \int_\alpha^\beta |g''(x)|\, dx.$$

By rewriting (12.89.e) in terms of g we get

$$\frac{1}{h_n} \int_\alpha^\beta g(x)\, dx = \left(\sum_{0 \le k < n} g(x_k) + \frac{1}{2} [g]_\alpha^\beta \right) - \frac{h_n}{12} [g']_\alpha^\beta - R_2(n), \qquad (12.89.f)$$

with

$$|R_2(n)| \leq \frac{h_n}{12} \int_\alpha^\beta |g''(x)|\, dx.$$

If we multiply both members of (12.89.f) by h_n we get

$$\int_\alpha^\beta g(x)\, dx = T_n^g - \frac{h_n^2}{12}\left[g'\right]_\alpha^\beta - h_n R_2(n).$$

Set $\varepsilon_2(n) = -h_n R_2(n)$; the following estimate holds:

$$|\varepsilon_2(n)| = |h_n R_2(n)| \leq \frac{h_n^2}{12} \int_\alpha^\beta |g''(x)|\, dx,$$

showing the validity of (12.89.b).   □

Remark 12.90 Here, the second-order approximation formula (12.89.b) does always give a better approximation of the integral of a function with respect the first-order one (12.89.a), at least for high values of n: indeed

$$\frac{(\beta-\alpha)^2}{12n^2} \int_\alpha^\beta |g''(x)|\, dx \leq \frac{\beta-\alpha}{2n} \int_\alpha^\beta |g'(x)|\, dx,$$

for sufficiently large values of n.

The reader will not be surprised to see that the remainders in (12.89.a) and (12.89.b) have a much simpler upper estimate under certain monotonicity assumptions.

Remark 12.91 (*Estimates of the remainders under monotonicity assumptions*) If g is *monotonic*, in (12.89.a) we have

$$|\varepsilon_1(n)| \leq \frac{\beta-\alpha}{2n} |g(\beta) - g(\alpha)|; \qquad (12.91.a)$$

if g' is *monotonic* then in (12.89.b) we have

$$|\varepsilon_2(n)| \leq \frac{(\beta-\alpha)^2}{12n^2} |g'(\beta) - g'(\alpha)|. \qquad (12.91.b)$$

Example 12.92 Let $g(x) = x^2$ on $[0, 1]$; we compare here the integral $I := \int_0^1 x^2\, dx$ $= \frac{1}{3}$ with its first- and second- order approximations. Subdividing the interval into

10 intervals of width $1/10$ the first-order approximation (12.89.a) of I is given by

$$T_{10}^g = \frac{1}{10} \sum_{0 \le k < 10} g\left(\frac{k}{10}\right) + \frac{1}{10} \frac{g(1) - g(0)}{2} = \frac{1}{10} \sum_{0 \le k < 10} \frac{k^2}{10^2} + \frac{1}{20} = \frac{67}{200} = 0.335;$$

whereas the second-order approximation of (12.89.b) is given by

$$T_{10}^g - \frac{1}{1200} \left[g'\right]_0^1 = 0.335 - \frac{1}{600} = 1/3,$$

and thus equals the integral I. □

Example 12.93 We wish to approximate here the integral $I := \int_0^1 e^{x^2}\, dx$ with an error less than 10^{-2}, motivated by the fact that e^{x^2} is not elementarily integrable. Since g is monotonic, (12.89.a) together with (12.91.a) yield

$$\left| \int_0^1 g(x)\, dx - T_n^g \right| \le \frac{1}{2n} |g(1) - g(0)| = \frac{e - 1}{2n}.$$

Thus, T_n^g approximates I within 10^{-2} if $\dfrac{e - 1}{2n} \le \dfrac{1}{100} \Leftrightarrow n \ge 50(e - 1) \approx 85.9$. For $n = 86$ we find $T_{86}^g = 1.46271\ldots$. Now, g is actually of class $\mathscr{C}^2$: the application of (12.89.b) and (12.91.b) with $n = 86$ yield

$$I \approx T_{86}^g - \frac{1}{12 \times 86^2} \left[g'\right]_0^1 = T_{86}^g - \frac{2e}{12 \times 86^2} = 1.462651745\ldots,$$

with an error at most equal to

$$|\varepsilon_2(n)| \le \frac{1}{12 \times 86^2} |g'(1) - g'(0)| = \frac{e}{6 \times 86^2} = \frac{e}{88\,752} = 0.00003\ldots < 10^{-4}.$$

Notice that, by means of (12.89.b), in order to get an approximation with an error at most equal to 10^{-2} it is enough that $\dfrac{e}{6 \times n^2} \le 10^{-2}$, and this occurs *just* for $n \ge 6.73$! For $n = 8$, from (12.89.b) we get

$$I \approx T_8^g - \frac{1}{12 \times 8^2} \left[g'\right]_0^1 = T_8^g - \frac{e}{384} = 1.46263\ldots.$$

It can be shown, for instance with the help of a CAS, that

$$\int_0^1 e^{x^2}\, dx = 1.462651746\ldots;$$

the approximation $I \approx 1.462651745\ldots$ is thus correct up 8 digits. □

Example 12.94 The function $g(x) = x \sin\left(\dfrac{1}{x}\right)$ is not elementarily integrable. Let us approximate the value of the integral

$$I := \int_1^{10} x \sin\left(\frac{1}{x}\right) dx,$$

with an error at most equal to 10^{-1}. Since $\|g'\|_{L^\infty(1,10)} \le 1 + 1 \times 1 = 2$, it follows from (12.89.a) that

$$\left| \int_1^{10} g(x)\, dx - T_n^g \right| \le \frac{9^2}{n},$$

and $\dfrac{9^2}{n} \le 10^{-1}$ if $n \ge 810$: it would take quite a lot of time to compute T_{810}^g with a pocket calculator! With a CAS we get $T_{810}^g = 8.85273\ldots$. If, instead of (12.89.a) we make use of (12.89.b), we get an approximation of I within 10^{-1} if

$$|\varepsilon_2(n)| \le \frac{9^3}{12n^2} \|g''\|_\infty \le 10^{-1}.$$

Now

$$|g''(x)| = \left| \frac{\sin\left(\frac{1}{x}\right)}{x^3} \right| \le \frac{1}{x^3},$$

so that $\|g''\|_{L^\infty(1,10)} \le 1$. Therefore $|\varepsilon_2(n)| \le 10^{-1}$ if $\dfrac{9^3}{12n^2} \le 10^{-1}$, or equivalently, *just* $n \ge \sqrt{5 \times 9^3/6} \approx 17.43$. With $n = 18$ we get

$$
\begin{aligned}
I &\approx T_{18}^g - \frac{9^2}{12 \times 18^2} \left[g' \right]_1^{10} \\
&= T_{18}^g - \frac{1}{48}\left(-\sin(1) + \sin\left(\frac{1}{10}\right) + \cos(1) - \frac{1}{10}\cos\left(\frac{1}{10}\right) \right) \\
&= 8.85297\ldots .
\end{aligned}
$$

Again, by means of a suitable CAS, we find

$$\int_1^{10} x \sin\left(\frac{1}{x}\right) dx = 8.8527357087\ldots .$$

With the first-order approximation we got 4 digits of I, whereas just 3 with the second-order approximation. However, in the first case we needed a subdivision of $[1, 10]$ into 810 intervals, whereas in the second case 18 intervals were enough. Notice that, with $n = 810$, (12.89.b) yields

$$I \approx T_{810}^g - \frac{9^2}{12 \times 810^2} \left[g' \right]_1^{10} =$$

$$= T_{810}^g - \frac{1}{97\,200} \left(-\sin(1) + \sin\left(\frac{1}{10}\right) + \cos(1) - \frac{1}{10}\cos\left(\frac{1}{10}\right) \right)$$

$$= 8.8527357088\ldots:$$

we achieve precision up to 9 digits.   □

12.6.2 The Reduced Euler–Maclaurin Formula ☕

We now study the effect, on the Euler–Maclaurin formulas, of getting rid of the last term of the first- and second- order expansions. This will lead us to the famous rectangle and trapezoidal methods for the approximation of integrals.

Remark 12.95 More precisely, assume that $f \in \mathscr{C}^1([a, b])$. We may add the term $-\frac{1}{2}\left[f\right]_a^b$ of the first-order Euler–Maclaurin expansion of f to the remainder R_1 in (12.27.a), and obtain

$$\sum_{a \le k < b} f(k) = \int_a^b f(x)\,dx + R_0, \quad R_0 := -\frac{1}{2}\left[f\right]_a^b + R_1. \qquad (12.95.a)$$

Analogously, if $f \in \mathscr{C}^2([a, b])$ we may add the last term $\frac{1}{12}\left[f'\right]_a^b$ of the second-order expansion of f to the remainder R_2 in (12.27.b) and write

$$\sum_{a \le k < b} f(k) = \int_a^b f(x)\,dx - \frac{1}{2}\left[f\right]_a^b + R_1, \quad R_1 = \frac{1}{12}\left[f'\right]_a^b + R_2. \qquad (12.95.b)$$

Notice that R_1 here is exactly the remainder that appears in (12.27.a), with a different representation that makes use of the fact that f is of class $\mathscr{C}^2$.

A rough estimate of R_0 and of R_1, written as above, may be carried out without difficulty as it is illustrated in the following example.

Example 12.96 (*A naive estimate of R_0 and R_1 in formulas* (12.95.a) *and* (12.95.b)) Let $f : [a, b] \to \mathbb{R}$ be a function, with $a, b \in \mathbb{N}$.

1. If f is of class $\mathscr{C}^1$ on $[a, b]$ then

$$\sum_{a \le k < b} f(k) = \int_a^b f(x)\,dx + R_0, \quad |R_0| \le \int_a^b |f'(x)|\,dx. \qquad (12.96.a)$$

Indeed, from (12.95.a) we have

$$R_0 = -\frac{1}{2}\left[f\right]_a^b + R_1 = -\frac{1}{2}\left[f\right]_a^b + \int_a^b B_1(x - [x])f'(x)\,dx.$$

The integral estimate of R_0 follows directly from the fact that

$$\left|\frac{1}{2}\left[f\right]_a^b\right| = \frac{1}{2}\left|\int_a^b f'(x)\,dx\right| \leq \frac{1}{2}\int_a^b |f'(x)|\,dx$$

and for the same upper estimate of $|R_1|$ established in Theorem 12.27.

2. If f is of class $\mathscr{C}^2$ on $[a, b]$ then

$$\sum_{a \leq k < b} f(k) = \int_a^b f(x)\,dx - \frac{1}{2}\left[f\right]_a^b + R_1, \quad |R_1| \leq \frac{1}{6}\int_a^b |f''(x)|\,dx. \quad (12.96.b)$$

Indeed from (12.95.b) we have $R_1 = \frac{1}{12}\left[f'\right]_a^b + R_2$; as above the estimate follows directly from the integral upper estimate of R_2 established in Theorem 12.27. $\square$

It is not often easy to compute integrals: it may be convenient sometimes to replace the estimates in terms of integrals (12.27.a) and (12.27.b), or in (12.96.a) and (12.96.b), in terms of the sup-norms of the functions that are involved by means of Proposition 12.22. For instance, in (12.96.a) and (12.96.b), one gets

$$|R_0| \leq (b-a)\|f'\|_\infty, \qquad |R_1| \leq \frac{b-a}{6}\|f'\|_\infty. \quad (12.96.c)$$

It is quite surprising, though not trivial, that the above upper bounds can be halved. In order to show it, we need an improved version of the well-known (first) Mean Value Integral Theorem.

Lemma 12.97 (The *2nd* Mean Value Theorem for integrals) *Let $f, \varphi : [a, b] \to \mathbb{R}$ be continuous, with φ of constant sign (i.e., $\varphi \geq 0$ or $\varphi \leq 0$ on $[a, b]$). There is $\xi \in [a, b]$ satisfying*

$$\int_a^b f(x)\varphi(x)\,dx = f(\xi)\int_a^b \varphi(x)\,dx.$$

Proof. It is not restrictive to assume that $\varphi \geq 0$ on $[a, b]$: if not one applies the forthcoming proof to $-\varphi$. The claim is also obvious if φ vanishes identically, so we may assume that $\int_a^b \varphi(x)\,dx > 0$. Let m and M be, respectively, the minimum and the maximum of f on $[a, b]$. Then

$$\int_a^b m\varphi(x)\,dx \le \int_a^b f(x)\varphi(x)\,dx \le \int_a^b M\varphi(x)\,dx$$

and hence

$$m \le \frac{\displaystyle\int_a^b f(x)\varphi(x)\,dx}{\displaystyle\int_a^b \varphi(x)\,dx} \le M.$$

The Intermediate Value Theorem implies that there is $\xi \in [a, b]$ such that

$$f(\xi) = \frac{\displaystyle\int_a^b \varphi(x)f(x)\,dx}{\displaystyle\int_a^b \varphi(x)\,dx},$$

thus proving the claim. $\qquad\square$

Proposition 12.98 (The first- and second- order reduced Euler–Maclaurin formulas) *Let $f : [a, b] \to \mathbb{R}$ be a function, with $a, b \in \mathbb{N}$.*

1. If $f \in \mathscr{C}^1([a, b])$ then

$$\sum_{a \le k < b} f(k) = \int_a^b f(x)\,dx + R_0, \quad |R_0| \le \frac{b-a}{2}\|f'\|_\infty. \qquad (12.98.a)$$

2. If $f \in \mathscr{C}^2([a, b])$ then

$$\sum_{a \le k < b} f(k) = \int_a^b f(x)\,dx - \frac{1}{2}\,[f]_a^b + R_1, \quad |R_1| \le \frac{b-a}{12}\|f''\|_\infty. \qquad (12.98.b)$$

Proof. From (12.27.a) we have

$$R_0 = -\frac{1}{2}\,[f]_a^b + R_1 = -\frac{1}{2}\,[f]_a^b + \int_a^b B_1(x - [x])f'(x)\,dx.$$

If we write that

$$[f]_a^b = \int_a^b f'(x)\,dx$$

we get

$$R_0 = \int_a^b \left(B_1(x - [x]) - \frac{1}{2} \right) f'(x)\, dx.$$

Here comes the key argument that enables us to halve the estimates in (12.96.c). Recall that $B_1(x) = x - \frac{1}{2}$, implying that $B_1(x) - \frac{1}{2} \le 0$ on $[0, 1]$. Lemma 12.97 yields the existence of $\xi \in [a, b]$ satisfying

$$R_0 = \int_a^b \left(B_1(x - [x]) - \frac{1}{2} \right) f'(x)\, dx = f'(\xi) \int_a^b \left(B_1(x - [x]) - \frac{1}{2} \right) dx.$$

Since $\int_0^1 B(x)\, dx = 0$, we get

$$\int_a^b \left(B_1(x - [x]) - \frac{1}{2} \right) dx = - \int_a^b \frac{1}{2} = -\frac{1}{2}(b - a).$$

Thus

$$|R_0| = \frac{(b - a)}{2} |f'(\xi)| \le \frac{b - a}{2} \|f'\|_\infty.$$

2. In (12.27.b) we have

$$R_1 = \frac{1}{12} [f']_a^b - \frac{1}{2} \int_a^b B_2(x - [x]) f''(x)\, dx$$

$$= \frac{1}{2} \int_a^b \left(\frac{1}{6} - B_2(x - [x]) \right) f''(x)\, dx.$$

Now $B_2(x) - \frac{1}{6} = x(x - 1) \le 0$ on $[0, 1]$ and

$$\int_0^1 B_2(x)\, dx = \int_0^1 \left(x^2 - x + \frac{1}{6} \right) dx = 0.$$

The proof goes now on as in Point 1. In any case the curious reader can follow all the details in the proof of the general case of any order, formulated in Proposition 13.37.   □

Remark 12.99 • If one does not care to the estimate of the remainder term, formula (12.98.b) *is* the Euler–Maclaurin formula of order 1 established in (12.27.a): the new fact concerns the upper estimate of R_1 that is here carried now in terms of f''.
• The upper estimate of the remainder R_1 in (12.98.b) equals the estimate in terms of the sup-norm of f'' that one could get, in (12.27.b):

$$|R_2| \le \frac{1}{12} \int_a^b |f''(x)|\, dx \le \frac{b-a}{12} \|f''\|_\infty. \qquad (12.99.\text{a})$$

This, however, *does not mean* that the estimate of $\sum\limits_{a \le k < b} f(k)$ given in (12.98.b) is
as good as the one given in (12.27.b): indeed the second inequality in (12.99.a) is
strict, in general (see Remark 12.23). For instance, in Example 12.30 with $f(x) = x$,
the sum

$$\sum_{0 \le k < 10} f(k) = 1 + \cdots + 9 = \frac{9 \times 10}{2} = 45$$

coincides with the Euler–Maclaurin expansion of order 1, whereas the reduced
expansion of order 1 in (12.98.a) equals

$$\int_0^{10} f(x)\, dx = 50.$$

12.6.3 The Rectangle and the Trapezoidal Method ☕

In Sect. 12.6.1 we presented the approximation of the integral $\int_\alpha^\beta g(x)\, dx$ of a function g, by means of the sum T_n^g of the trapezes built upon the graph of g. The rectangle
method, obtained in a similar way for a function of class $\mathscr{C}^1$, gives an approximation
of the integral in terms of the sum D_n^g of the areas of the rectangles of side $[x_k, x_{k+1}]$
and height $g(x_k)$, as k varies in $\{0, \ldots, n-1\}$.

Definition 12.100 For any $n \ge 1$ we define

$$x_k = \alpha + h_n k \qquad k = 0, \ldots, n-1, \qquad h_n := \frac{\beta - \alpha}{n},$$

$$D_n^g := h_n \sum_{0 \le k < n} g(x_k). \qquad \qquad \square$$

The following result gives the classical approximation of a definite integral by
means of rectangles and trapezes.

Theorem 12.101 (The rectangle and the trapezoidal methods)

1. Let $g \in \mathscr{C}^1([\alpha, \beta])$. Then

$$\int_\alpha^\beta g(x)\,dx = D_n^g + \varepsilon_0(n), \quad |\varepsilon_0(n)| \le \frac{(\beta - \alpha)^2}{2n} \|g'\|_\infty. \tag{12.101.a}$$

2. Let $g \in \mathscr{C}^2([\alpha, \beta])$. Then

$$\int_\alpha^\beta g(x)\,dx = T_n^g + \varepsilon_1(n), \quad |\varepsilon_1(n)| \le \frac{(\beta - \alpha)^3}{12n^2} \|g''\|_\infty. \tag{12.101.b}$$

Remark 12.102 If one ignores the estimates in the remainder terms, the equality in (12.101.b) is exactly the approximation formula (12.89.a). The difference between the two formulas relies on the estimate of the remainder $\varepsilon_1(n)$. The advantage of (12.101.b) with respect to (12.89.a) is due to the fact that the further regularity of f ensures in (12.101.b) that $\varepsilon_1(n) = O\left(\dfrac{1}{n^2}\right)$ as $n \to +\infty$, whereas in (12.89.a) we are just supposed to know that $\varepsilon_1(n) = O\left(\dfrac{1}{n}\right)$ as $n \to +\infty$. At the same time, there is a price to pay if one adopts (12.101.b) instead of (12.89.b). Indeed the upper bound of the estimate of $\varepsilon_1(n)$ in (12.101.b) is bigger than that of $\varepsilon_2(n)$ in (12.89.b):

$$\frac{1}{12} \frac{(\beta - \alpha)^2}{n^2} \int_\alpha^\beta |g''(x)|\,dx \le \frac{1}{12} \frac{(\beta - \alpha)^3}{n^2} \|g''\|_\infty,$$

and the inequality is usually strict, as we realized in Remark 12.23. The advantage of the rectangle and trapezoidal methods, with respect to the Euler–Maclaurin approximations of the integral in Theorem 12.89, is to deal with some simpler expressions without being significantly less accurate. We may approximate the integral $\int_a^b g(x)\,dx$ by means of the following formulas, in an *increasing* order of precision:

1. The approximation formula (12.89.a);
2. The trapezoidal method formula (12.101.b);
3. The approximation formula (12.89.b).

The proof of Theorem 12.101 is similar to that of Theorem 12.89; the only difference is to use the reduced Euler–Maclaurin formulas (12.98.a) and (12.98.b) instead of the original ones (12.27.a) and (12.27.b).

Proof. [of Theorem 12.101] Let $f(t) = g(\alpha + th_n)$, $t \in [0, n]$. The change of variables $x = \alpha + th_n$ leads to

$$\int_\alpha^\beta g(x)\, dx = h_n \int_0^n f(t)\, dt.$$

1. Assume that g is of class $\mathscr{C}^1$. The reduced first-order Euler–Maclaurin formula (12.98.a) applied to f on $[0, n]$ yields

$$\int_0^n f(t)\, dt = \sum_{0 \leq k < n} f(k) - R_0, \qquad (12.102.\text{a})$$

with $|R_0| \leq \dfrac{n}{2}\|f'\|_\infty$. Since

$$\|f'\|_\infty = h_n \|g'\|_\infty,$$

if we rewrite (12.102.a) in terms of g we obtain

$$\frac{1}{h_n}\int_\alpha^\beta g(x)\, dx = \sum_{0 \leq k < n} g(x_k) - R_0, \qquad (12.102.\text{b})$$

with

$$|R_0| \leq \frac{nh_n}{2}\|g'\|_\infty = \frac{\beta - \alpha}{2}\|g'\|_\infty.$$

Identity (12.101.a) follows by multiplying both members of (12.102.b) by h_n.
2. Assume that g is of class $\mathscr{C}^2$. The second-order reduced Euler–Maclaurin formula (12.98.b) applied to f on $[0, n]$ yields

$$\int_0^n f(t)\, dt = \sum_{0 \leq k < n} f(k) + \frac{1}{2}\left[f\right]_0^n - R_1, \qquad (12.102.\text{c})$$

with $|R_1| \leq \dfrac{n}{12}\|f''\|_\infty$. Now $\|f''\|_\infty = h_n^2 \|g''\|_\infty$: if we rewrite (12.102.c) in terms of g we get

$$\frac{1}{h_n}\int_\alpha^\beta g(x)\, dx = \left(\sum_{0 \leq k < n} g(x_k) + \frac{1}{2}\left[g\right]_\alpha^\beta\right) - R_1, \qquad (12.102.\text{d})$$

with

$$|R_1| \leq \frac{n}{12}h_n^2\|g''\|_\infty = \frac{(\beta - \alpha)^2}{12n}\|g''\|_\infty.$$

Identity (12.101.b) follows by multiplying both members of (12.102.d) by h_n. $\square$

Remark 12.103 As a consequence of (12.101.a) we obtain the well-known interpretation of the integral as a limit of the Riemann sums [33, Theorem 6.7]:

$$\int_\alpha^\beta g(x)\,dx = \lim_{n\to+\infty} \sum_{0\le k<n} \frac{\beta-\alpha}{n} g\left(\alpha + k\frac{\beta-\alpha}{n}\right). \tag{12.103.a}$$

12.7   Formulas for Non-Smooth Monotonic Functions

As surprising as it might be, every first-order Euler–Maclaurin type formula established for $\mathscr{C}^1$ functions may be formulated for functions that are just monotonic, without assuming any kind of differentiability. The only difference is, of course, in the remainder term, which has an explicit formula in terms of the derivatives in the smooth case, and is merely estimated here. The Euler–Maclaurin type formula for monotonic functions in particular implies the celebrated integral test for the convergence of a series that inspired the one that we formulated for smooth functions in Corollary 12.78.

12.7.1   Euler–Maclaurin Type Formula

The following formula looks like the first-order Euler–Maclaurin formula; it is obtained immediately by proceeding just as in the proof of the integral test for series with decreasing terms [33]. We recall that a monotonic function is integrable on every closed and bounded interval [33, Theorem 6.9].

Theorem 12.104 *Let a, b in $\mathbb{N}$ and $f : [a, b] \to \mathbb{R}$ be a monotonic function. Then*

$$\sum_{a\le k<b} f(k) = \int_a^b f(x)\,dx - \frac{1}{2}\left[f\right]_a^b + R_1, \quad |R_1| \le \frac{1}{2}\left|\left[f\right]_a^b\right|. \tag{12.104.a}$$

Proof. Suppose that f is decreasing. On every interval $[k, k+1]$ $(k \in \mathbb{N})$ contained in $[a, b]$ one has (see Fig. 12.4)

$$f(k) \ge f(t) \ge f(k+1) \quad \forall t \in [k, k+1],$$

Fig. 12.4 The idea used in the proof of Theorem 12.104

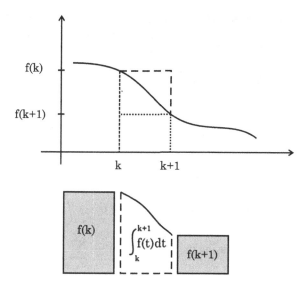

from which it follows that

$$f(k) = \int_k^{k+1} f(k)\,dt \geq \int_k^{k+1} f(t)\,dt \geq \int_k^{k+1} f(k+1)\,dt = f(k+1).$$

Summing the terms of the foregoing inequalities, as k varies between a and $b-1$, one obtains

$$\sum_{a \leq k < b} f(k) \geq \int_a^b f(t)\,dt \geq \sum_{a \leq k < b} f(k) + f(b) - f(a).$$

Subtracting the term $\frac{1}{2}(f(b) - f(a))$ from the various members of the preceding inequalities one finds

$$\sum_{a \leq k < b} f(k) - \frac{1}{2}(f(b) - f(a)) \geq \int_a^b f(t)\,dt - \frac{1}{2}(f(b) - f(a))$$

$$\geq \sum_{a \leq k < b} f(k) + \frac{1}{2}(f(b) - f(a)),$$

and hence

$$|R_1| = \left| \sum_{a \leq k < b} f(k) - \int_a^b f(t)\,dt + \frac{1}{2}(f(b) - f(a)) \right| \leq \frac{1}{2} \left| [f]_a^b \right|.$$

If f is an increasing function it suffices to apply the obtained result to $-f$. $\qquad\square$

Remark 12.105 Theorem 12.104 may be paraphrased by affirming that if f is *monotonic*, the Euler–Maclaurin expansion of order 1 of f is an estimate of the sum $\sum_{a \leq k < b} f(k)$, up to a remainder whose modulus is less than the last term of that expansion. In particular, from (12.104.a), we deduce the following estimate

$$\left| \sum_{a \leq k < b} f(k) - \int_a^b f(t)\, dt \right| \leq |f(b) - f(a)|. \tag{12.105.a}$$

More precisely, from the proof of Theorem 12.104 it turns out that:

- If f is decreasing then $0 \leq \sum_{a \leq k < b} f(k) - \int_a^b f(t)\, dt \leq f(a) - f(b)$;

- If f is increasing then $f(a) - f(b) \leq \sum_{a \leq k < b} f(k) - \int_a^b f(t)\, dt \leq 0$.

12.7.2   The Approximation Formula for Finite Sums

Similarly to what we achieved for regular functions in Corollary 12.45 we can deduce an approximation formula of the sums $\sum_{a \leq k < N} f(k)$ in terms of $\sum_{a \leq k < n} f(k)$, whenever $n \leq N$.

Corollary 12.106 (The approximation formula for finite sums) *Let $f : [a, +\infty[\to \mathbb{R}$ be monotonic, $a \in \mathbb{N}$. For every $N \geq n$ the following **approximation formula** holds:*

$$\sum_{a \leq k < N} f(k) = \sum_{a \leq k < n} f(k) + \int_n^N f(x)\, dx - \frac{1}{2}\left[f\right]_n^N + \varepsilon_1(n, N),$$

$$|\varepsilon_1(n, N)| \leq \frac{1}{2}|f(n) - f(N)| \leq \frac{1}{2}|f(n) - f(\infty)|. \tag{12.106.a}$$

Proof. It is enough to remark that

$$\sum_{a \leq k < N} f(k) - \sum_{a \leq k < n} f(k) = \sum_{n \leq k < N} f(k)$$

and to apply (12.104.a) with $a = n$ and $b = N$.   □

12.7.3 The Integral Test for the Convergence of a Series with Bounded Monotonic Terms

We assume here that, given $a \in \mathbb{N}$, the monotonic function $f : [a, +\infty[\to \mathbb{R}$ is also *bounded*: the function f being monotonic, it is equivalent to suppose that $f(\infty) \in \mathbb{R}$. We see from (12.105.a) that

$$\gamma_n^f := \sum_{a \leq k < n} f(k) - \int_a^n f(x)\, dx \qquad \forall n \in \mathbb{N}_{\geq a}$$

remains bounded. We prove here that, similarly to the case of a smooth function as stated in Theorem 12.61, the sequence γ_n^f tends to a finite value as n goes to infinity.

Theorem 12.107 (The Euler constant) *Let $f : [a, +\infty[\to \mathbb{R}$ be monotonic and bounded, $a \in \mathbb{N}$. The Euler constant γ^f of f is defined and the following **estimate** of γ^f holds:*

$$\gamma^f = \gamma_n^f - \frac{1}{2} \big[f \big]_n^\infty + \varepsilon_1(n), \quad |\varepsilon_1(n)| \leq \frac{1}{2} |f(n) - f(\infty)| \quad \forall n \in \mathbb{N}_{\geq a}.$$
$$(12.107.a)$$

Proof. Assume that f is decreasing. For all $n \geq a$ we have

$$\gamma_{n+1}^f - \gamma_n^f = f(n) - \int_n^{n+1} f(x)\, dx \geq 0,$$

thus proving the monotonicity of $(\gamma_n^f)_n$. From (12.104.a) we have

$$\gamma_n^f = -\frac{1}{2} \big[f \big]_a^n + R_1,$$

so that

$$|\gamma_n^f| \leq \frac{1}{2} |f(n) - f(a)| + \frac{1}{2} |f(n) - f(a)| = |f(n) - f(a)|,$$

which is bounded, due to the convergence of $(f(n))_n$. It follows that the $\lim_{n \to +\infty} \gamma_n^f$ exists in $\mathbb{R}$ and hence γ^f is defined.

Let us prove now (12.107.a). Applying (12.104.a) between n and $N \geq n$ we get:

$$\gamma_N^f - \gamma_n^f = \sum_{n \le k < N} f(k) - \int_n^N f(x)\,dx$$

$$= -\frac{1}{2}\left[f\right]_n^N + \varepsilon_1(n,N), \quad |\varepsilon_1(n,N)| \le \frac{1}{2}\left|\left[f\right]_n^N\right|.$$

By passing to the limit for $N \to +\infty$ we obtain

$$\left|\gamma^f - \left(\gamma_n^f - \frac{1}{2}\left[f\right]_n^\infty\right)\right| \le \frac{1}{2}\left|f(n) - f(\infty)\right|,$$

proving (12.107.a). Finally, if f is increasing, it is enough to apply the above results to the decreasing function $-f$.   □

An immediate consequence of Theorem 12.107 is the well-known integral test for the convergence of the series $\sum_{k=a}^{\infty} f(k)$, as f is bounded and monotonic.

Corollary 12.108 (The integral test) *Let* $f : [a, +\infty[\to \mathbb{R}$ *be monotonic,* $a \in \mathbb{N}$. *Then the series* $\sum_{k=a}^{\infty} f(k)$ *and the generalized integral* $\int_a^{+\infty} f(x)\,dx$ *have the same behavior: both are either convergent or divergent.*

Proof. If $f(\infty)$ is $\pm\infty$ or not 0, then both the series and the integral do diverge. Henceforth we assume now that $f(\infty) = 0$. We know from Theorem 12.107 that there exists $\gamma^f \in \mathbb{R}$ such that

$$\gamma^f = \lim_{n \to \infty}\left(\sum_{a \le k < n} f(k) - \int_a^n f(x)\,dx\right).$$

Thus $\sum_{k=a}^{\infty} f(k)$ and the limit $\lim_{\substack{n \to +\infty \\ n \in \mathbb{N}}} \int_a^n f(x)\,dx$ have the same behavior. Since $f(\infty) = 0$ belongs to $\mathbb{R}$, the value of $\lim_{\substack{n \to +\infty \\ n \in \mathbb{N}}} \int_a^n f(x)\,dx$ coincides with that of $\int_a^{+\infty} f(x)\,dx$ (Problem 12.15): the conclusion follows.   □

12.7.4   The Approximation of the Sum of a Convergent Series

We now see how one may approximate the sum of a convergent series with monotonic terms: the result was obtained in Corollary 12.78 for functions of class $\mathscr{C}^1$.

Corollary 12.109 (Approximation of the sum of a series) *Let $a \in \mathbb{N}$ and $f : [a, +\infty[\rightarrow \mathbb{R}$ be a monotonic function such that the series $\sum\limits_{k=a}^{\infty} f(k)$ converges. The following* **approximation** *holds:*

$$\sum_{k=a}^{\infty} f(k) = \sum_{a \leq k < n} f(k) + \int_n^{+\infty} f(x)\,dx - \frac{1}{2}\left[f\,\right]_n^{\infty} + \varepsilon_1(n),$$

$$\hspace{8cm} (12.109.a)$$

$$|\varepsilon_1(n)| \leq \frac{1}{2}|f(n)| \qquad \forall n \in \mathbb{N}_{\geq a}.$$

Proof. The convergence of the series $\sum\limits_{k=a}^{\infty} f(k)$ implies that $f(\infty) = 0$. It follows from (12.106.a) that for every $N \geq n$ we have

$$\sum_{a \leq k < N} f(k) = \sum_{a \leq k < n} f(k) + \int_n^N f(x)\,dx - \frac{1}{2}\left[f\,\right]_n^N + \varepsilon_1(n, N), \qquad (12.109.b)$$

with $|\varepsilon_1(n, N)| \leq \frac{1}{2}|f(n) - f(\infty)| = \frac{1}{2}|f(n)|$. It follows from Corollary 12.108 that f is integrable in a generalized sense on $[a, +\infty[$. Passing to the limit for $N \rightarrow +\infty$ in (12.109.b) we deduce that $\varepsilon_1(n) := \lim\limits_{N \rightarrow +\infty} \varepsilon_1(n, N)$ is finite, and the validity of (12.109.a). $\qquad\square$

12.7.5 Asymptotic Formulas

We are now in the position to study the asymptotic behavior of the difference

$$\gamma_n^f := \sum_{a \leq k < n} f(k) - \int_a^n f(x)\,dx$$

as $n \rightarrow +\infty$, for a monotonic function $f : [a, +\infty[\rightarrow \mathbb{R}$. We get nothing more than a reformulation of some results of the previous sections, and are analogous to the claim of Theorem 12.61, that was established for functions of class $\mathscr{C}^1$.

Theorem 12.110 (Asymptotic formulas) *Let $f : [a, +\infty[\rightarrow \mathbb{R}$ be a monotonic function, $a \in \mathbb{N}$.*

1. If $f(\infty) = \pm\infty$, then for every $n \in \mathbb{N}_{\geq a}$ we have

$$\sum_{a \leq k < n} f(k) = \int_a^n f(x)\, dx + O\left(f(n)\right) \quad n \to +\infty. \tag{12.110.a}$$

2. If $f(\infty) \in \mathbb{R}$ then for every $n \in \mathbb{N}_{\geq a}$ we have

$$\sum_{a \leq k < n} f(k) = \gamma^f + \int_a^n f(x)\, dx + \varepsilon_1'(n),$$

$$|\varepsilon_1'(n)| \leq |f(n) - f(\infty)| = O\left(f(n) - f(\infty)\right) \quad n \to +\infty. \tag{12.110.b}$$

Proof. 1. It follows from Theorem 12.104 that, for every $n \in \mathbb{N}_{\geq a}$ we have

$$\sum_{a \leq k < n} f(k) = \int_a^n f(x)\, dx - \frac{1}{2}(f(n) - f(a)) + R_1(n),$$

with $|R_1(n)| \leq \frac{1}{2}|f(n) - f(a)|$. Since $\lim_{n \to +\infty} f(n) = \pm\infty$, then

$$f(n) - f(a) = O(f(n)) \quad n \to +\infty,$$

whence $-(f(n) - f(a))/2 + R_1(n) = O(f(n))$ for $n \to +\infty$: the conclusion follows.
2. From (12.107.a) we obtain

$$\gamma_n^f = \gamma^f + \varepsilon_1'(n),$$

where $\varepsilon_1'(n) := \frac{1}{2}\left[f\right]_n^\infty - \varepsilon_1(n)$ and the following estimate holds

$$|\varepsilon_1'(n)| = \left|\frac{1}{2}\left[f\right]_n^\infty - \varepsilon_1(n)\right| \leq |f(n) - f(\infty)|.$$

The conclusion follows. $\square$

Remark 12.111 Theorem 12.104, and all the subsequent results of this section, can be easily extended to functions f that are of *bounded variation*: indeed any such a function is the difference of two increasing functions. Namely, (12.104.a) does still hold with

$$|R_1| \leq \frac{1}{2}\mathrm{pV}(f, [a, b]),$$

where $pV(f, [a, b])$ is the pointwise variation of f, given by

$$pV(f, [a, b]) = \sup \left\{ \sum_{i=0}^{n} |f(x_{i+1}) - f(x_i)| : a := x_0 < x_1 < \cdots < x_n < x_{n+1} := b \right\}.$$

The interested reader can find the details in [10].

12.8 Problems

Problem 12.1 Let $x_0 \in \mathbb{N}$ and $(x_n)_{n \geq 1}$ be a sequence in $\{0, \ldots, 9\}$. Show that the sequence $x_0 . x_1 \ldots x_n$ converges.

Problem 12.2 Let $x_0 \in \mathbb{N}$ and $x_1, \ldots, x_n \in \{0, \ldots, 9\}$ and $n \in \mathbb{N}_{\geq 1}$. Prove that

$$x_0 . x_1 \ldots x_n + \frac{1}{10^n} = \begin{cases} x_0 . x_1 \ldots x_{n-1}(x_n + 1) \text{ if } x_n < 9, \\ x_0 + 1 \text{ if } x_n = x_{n-1} = \cdots = x_1 = 9, \\ x_0 . x_1 \ldots x_{i-1}(x_i + 1) \text{ if } x_i \neq 9, \ x_{i+1} = \cdots = x_n = 9. \end{cases}$$

Problem 12.3 Using the Euler–Maclaurin expansion of first-order for a suitable function, estimate the following sums and the error made:

1. $\sum_{1 \leq k < 10} \log k$;

2. $\sum_{1 \leq k < 10} k^{-2}$;

3. $\sum_{0 \leq k < 10} e^{-k}$ (compare this also with the exact sum);

4. $\sum_{1 \leq k < 10} \sqrt{k}$;

5. $\sum_{0 \leq k < 10} k^{3/2}$.

Problem 12.4 Find the minimum value of n such that the first-order Euler–Maclaurin approximation formula yields an approximation of the sum $\sum_{1 \leq k < 10000} \frac{1}{k^4}$ by means of $\sum_{1 \leq k < n} \frac{1}{k^4}$ with an error at most equal to 10^{-1}; for such a value compute the approximated value of the given sum.

Problem 12.5 Using the Euler–Maclaurin expansion of second-order estimate the sums and the errors made in the sums of Problem 12.3.

Problem 12.6 Find the minimum value of n such that the second-order Euler–Maclaurin approximation formula yields an approximation of the sum $\displaystyle\sum_{k=1}^{+\infty}\frac{1}{k^4}$ by means of $\displaystyle\sum_{1\le k<n}\frac{1}{k^4}$ with an error at most equal to 10^{-2}; for such a value compute the approximated value of the given sum.

Problem 12.7 Prove the existence of the Euler constant for $f(x) := \dfrac{1}{\sqrt{x}}$, $x \ge 1$; give its first-order approximation with an error less than 10^{-1}.

Problem 12.8 Approximate the integral

$$\int_1^8 xe^{-x^{0.8}}\,dx$$

with an error of at most 10^{-2} by means of:

1. The integral approximation of order 2 (12.89.b);
2. The trapezoidal method.

Compare the results with the numerical approximation given by a CAS: which of the two results gives the best approximation?

Problem 12.9 Prove that

$$\left(n + \frac{1}{2n}\right)(n^2 + 1) = n^3 + \frac{3n}{2} + o(1) \quad n \to +\infty.$$

Problem 12.10 Show that

$$1 + \frac{2}{n} + O\left(\frac{1}{n^2}\right) = \left(1 + \frac{2}{n}\right)\left(1 + O\left(\frac{1}{n^2}\right)\right) \quad n \to +\infty.$$

Problem 12.11 Let a_n and b_n be two sequences, and suppose furthermore that $b_n \ne 0$ for every n. Prove that $a_n = O(b_n)$ if and only if there exists a $C > 0$ such that $|a_n| \le C|b_n|$ for every n.

Problem 12.12 In each row, which of the two statements implies the other, for $n \to +\infty$?

$$1.a)\ a_n = n^2 + O(n\log n) \quad 1.b)\ a_n = n^2 + 2n\log n + O(n).$$

$$2.a)\ a_n = n^2 + o(n^2) \quad 2.b)\ a_n = n^2 + 2n\log n - 6n + O(n).$$

Problem 12.13 Let $\alpha, \beta \in \mathbb{R}$, $a > 0$ and $a_n \sim \alpha n^a$ for $n \to +\infty$ and $b_n \sim \beta n^a$ for $n \to +\infty$, with $\alpha + \beta \ne 0$. Prove that $a_n + b_n \sim (\alpha + \beta)n^a$ for $n \to +\infty$.

Problem 12.14 Let a_n be a sequence such that $\lim\limits_{n\to+\infty} a_n = \ell \in \mathbb{R}$. Then $a_n = \ell + O(1)$ for $n \to +\infty$.

Problem 12.15 Let $f : [a, +\infty[\to \mathbb{R}$ be locally integrable, and assume that the limit $\ell := \lim\limits_{x\to+\infty} f(x)$ exists in $\mathbb{R} \cup \{\pm\infty\}$. Prove that the limit $\lim\limits_{\substack{n\to+\infty \\ n\in\mathbb{N}}} \int_a^n f(t)\,dt$ and the generalized integral $\int_a^{+\infty} f(t)\,dt$ have the same behavior: both exist and are equal or both do not exist.

Problem 12.16 The function $\dfrac{\sin x}{x^2}$ is summable on $[1, +\infty[$, due to the fact that $\left|\dfrac{\sin x}{x^2}\right| \le 1/x^2 \in L^1([1, +\infty[)$. Set $a := \int_1^{+\infty} \dfrac{\sin t}{t^2}\,dt$; it can be shown that $a \approx 0.504067$. Let $f(x) := \int_1^x \dfrac{\sin t - a}{t^2}\,dt$, for $x > 1$. Study the convergence of the series $\sum\limits_{k=1}^{\infty} f(k)$:

1. Prove that f is not monotonic, $f(\infty) = 0$, and $f' \in L^1([1, +\infty[)$;
2. Deduce that $\sum\limits_{k=1}^{\infty} f(k)$ and $\lim\limits_{n\to+\infty} \int_1^n f(x)\,dx$ have the same behavior;
3. Prove that $\lim\limits_{n\to+\infty} \int_1^n f(x)\,dx = +\infty$.

Problem 12.17 Prove relation (12.65.c) in the case $f'(\infty) \in \{0, \pm\infty\}$.

Problem 12.18 Furnish a first-order asymptotic estimate for:

1. $\sum\limits_{1\le k<n} \log k$;
2. $\sum\limits_{1\le k<n} k^{-2}$;
3. $\sum\limits_{0\le k<n} e^{-k}$;
4. $\sum\limits_{0\le k<n} \sqrt{k}$;
5. $\sum\limits_{0\le k<n} k^{3/2}$.

Problem 12.19 Using the asymptotic versions of the Euler–Maclaurin formula (Theorem 12.61), provide approximations for $\sum\limits_{1\le k<n} k \log k$.

Problem 12.20 Let $p > 0$. Show that

$$\sum_{1 \leq k < n} \frac{\log^p(k)}{k} = \frac{1}{p+1} \log^{p+1}(n) + O\left(\frac{\log^p(n)}{n}\right) \quad n \to +\infty.$$

Problem 12.21 Assume that $f : [a, +\infty[\to \mathbb{R}$ is of class $\mathscr{C}^1$, definitively monotonic and tends to 0 at infinity, and moreover that $f'(n) = O(f(n))$ as $n \to +\infty$. If (12.65.c) holds true, deduce the validity of (12.61.c).

Chapter 13
The Euler–Maclaurin Formula of Arbitrary Order

Abstract In this chapter we extend the results formulated in Chap. 12 to order m, i.e., to functions of class $\mathscr{C}^m$, for any $m \in \mathbb{N}_{\geq 1}$. Namely, we formulate the Euler–Maclaurin formula, its asymptotic version, its applications to the approximation of finite sums and of sums of convergent series in great generality. The reader will find here not only the general statements of the formulas of order m, but also their detailed proofs, some of which were just sketched out or even skipped entirely in the previous sections. Among the applications, we discover the Hermite formula for the approximations of integrals: it is a refinement of the trapezoidal method which is even more accurate than Simpson's method.

☞   Throughout this chapter we assume, if not explicitly stated otherwise, that the functions are defined on an *integer interval*, i.e., an interval whose extreme points are natural numbers.

13.1   Bernoulli Polynomials

The Bernoulli polynomial functions enter naturally into the general Euler–Maclaurin formulas. We introduce them by way of one of their characteristic properties which also allows us to motivate the definition of Bernoulli numbers already presented in Definition 8.47. First we recall that the Bernoulli numbers are defined recursively by the formula

$$B_0 = 1, B_m = -\frac{1}{m+1} \sum_{j=0}^{m-1} \binom{m+1}{j} B_j \;\; \forall m \geq 1.$$

The first few Bernoulli numbers are

$$B_0 = 1, \; B_1 = -\frac{1}{2}, \; B_2 = \frac{1}{6}, \; B_3 = 0, \; B_4 = -\frac{1}{30}, \; B_5 = 0, \; B_6 = \frac{1}{42}.$$

© Springer International Publishing Switzerland 2016
C. Mariconda and A. Tonolo, *Discrete Calculus*,
UNITEXT - La Matematica per il 3+2 103, DOI 10.1007/978-3-319-03038-8_13

13.1.1 Recursive Definition of the Bernoulli Polynomials

Let us introduce the Bernoulli polynomials: they will be the key tool in studying the remainder term in the Euler–Maclaurin formulas.

Proposition 13.1 *There exists a unique sequence of polynomials* $(B_m(X))_{m \in \mathbb{N}}$ *such that:*

1. $B_0'(X) = 0$ *and* $B_m'(X) = m\,B_{m-1}(X)$ *for every integer* $m \geq 1$;
2. $B_0(0) = 1$ *and* $B_m(0) = B_m(1)$ *for every integer* $m \geq 2$.

The sequence $(B_m(X))_{m \in \mathbb{N}}$ *is called the sequence of* **Bernoulli polynomials.** *For each* $m \in \mathbb{N}$, *one has*

$$B_m(X) = \sum_{k=0}^{m} \binom{m}{k} B_k\, X^{m-k}.$$

In particular $B_m(0) = B_m$.

Proof. Conditions 1 allows for a recursive determination of the polynomials $B_m(X)$. up to an additive constant γ_m. Indeed:

- From $B_0'(X) = 0$, it follows that $B_0(X) = \gamma_0 \in \mathbb{R}$;
- From $B_1'(X) = B_0(X) = \gamma_0$ it follows that $B_1(X) = \gamma_0 X + \gamma_1$;
- From $B_2'(X) = 2B_1(X) = 2\gamma_0 X + 2\gamma_1$ it follows that

$$B_2(X) = \gamma_0 X^2 + 2\gamma_1 X + \gamma_2 = \binom{2}{0}\gamma_0 X^2 + \binom{2}{1}\gamma_1 X + \binom{2}{2}\gamma_2;$$

- From $B_3'(X) = 3\,B_2(X) = 3\gamma_0 X^2 + 6\gamma_1 X + 3\gamma_2$ it follows that

$$B_3(X) = \gamma_0 X^3 + 3\gamma_1 X^2 + 3\gamma_2 X + \gamma_3 = \binom{3}{0}\gamma_0 X^3 + \binom{3}{1}\gamma_1 X^2 + \binom{3}{2}\gamma_2 X + \binom{3}{3}\gamma_3.$$

It is now easy to prove by induction that

$$B_m(X) = \binom{m}{0}\gamma_0 X^m + \binom{m}{1}\gamma_1 X^{m-1} + \cdots + \binom{m}{m-1}\gamma_{m-1} X + \binom{m}{m}\gamma_m.$$

Thanks to Condition 2 one obtains that $\gamma_0 = 1 = B_0$ and for $m \geq 1$

$$0 = B_{m+1}(1) - B_{m+1}(0)$$
$$= \binom{m+1}{0}\gamma_0 + \binom{m+1}{1}\gamma_1 + \cdots + \binom{m+1}{m-1}\gamma_{m-1} + \binom{m+1}{m}\gamma_m,$$

or equivalently

$$\gamma_m = -\frac{1}{m+1}\left(\binom{m+1}{0}\gamma_0 + \binom{m+1}{1}\gamma_1 + \cdots + \binom{m+1}{m-1}\gamma_{m-1}\right).$$

This is in fact the recursive relation that defines the Bernoulli numbers (see Sect. 8.3); thus one has $\gamma_m = B_m$ for every $m \geq 0$. □

Remark 13.2 Note that Condition 2 of Proposition 13.1 does not hold if $m = 1$: indeed $B_1(0) = -1/2$ while $B_1(1) = 1/2$. This difference between the case $m = 1$ and the cases $m > 1$ will have a key role in the proof of the Euler–Maclaurin formula (Theorem 13.15).

It is useful to list the first instances of the Bernoulli polynomials.

Example 13.3 (The first few Bernoulli polynomials)

$B_0(X) = 1,\ B_1(X) = -\dfrac{1}{2} + X$	$B_5(X) = -\dfrac{X}{6} + \dfrac{5X^3}{3} - \dfrac{5X^4}{2} + X^5$
$B_2(X) = \dfrac{1}{6} - X + X^2$	$B_6(X) = \dfrac{1}{42} - \dfrac{X^2}{2} + \dfrac{5X^4}{2} - 3X^5 + X^6$
$B_3(X) = \dfrac{X}{2} - \dfrac{3X^2}{2} + X^3$	$B_7(X) = \dfrac{X}{6} - \dfrac{7X^3}{6} + \dfrac{7X^5}{2} - \dfrac{7X^6}{2} + X^7$
$B_4(X) = -\dfrac{1}{30} + X^2 - 2X^3 + X^4$	$B_8(X) = -\dfrac{1}{30} + \dfrac{2X^2}{3} - \dfrac{7X^4}{3} + \dfrac{14X^6}{3} - 4X^7 + X^8$

□

13.1.2 Qualitative Behavior of the Bernoulli Polynomial Functions on [0, 1] ☕

We study here the qualitative behavior of the Bernoulli polynomial functions on the interval $[0, 1]$; the graphs of the first Bernoulli polynomial functions on $[0, 1]$ are depicted in Fig. 13.1.

Proposition 13.4 *Let $m \in \mathbb{N}_{\geq 1}$.*

1. Assume that m is even. Then:

- *The graph of $B_m(x)$, for $x \in [0, 1]$, is symmetric with respect to the line $x = 1/2$;*

Fig. 13.1 Graphs of the
Bernoulli polynomial
functions $B_1(x)$, $B_2(x)$,
$B_3(x)$, $B_4(x)$ and $B_5(x)$ on
$[0, 1]$

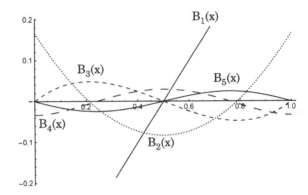

- $B_m(x)$ *has a unique zero and is monotonic (in the opposite directions) on both* $[0, 1/2]$ *and* $[1/2, 1]$.

2. *Assume that m is odd. Then:*

- *The graph of* $B_m(x)$, *for* $x \in [0, 1]$, *is symmetric with respect to the point* $(1/2, 0)$;
- $B_m(x)$ *vanishes only for* $x = 1/2$ *on* $]0, 1[$;
- *If* $m \geq 3$ *then* $B_m(0) = B_m(1) = 0$, *and there exists* $\alpha \in]0, 1/2[$ *such that* $B_m(x)$ *is monotonic on* $[\alpha, 1 - \alpha]$, *and monotonic (in the opposite directions) on* $[0, \alpha]$ *and* $[1 - \alpha, 1]$.

Proof. We prove the two points by induction on m. The statement is true for $m = 1$ in view of the fact that $B_1(x) = x - \dfrac{1}{2}$. Suppose now that both the statements are true up to a given $m \geq 1$.

If $m + 1$ is even, then, by the induction hypothesis, the polynomial function $B_m(x)$ does not change sign on $[0, 1/2]$ and on $[1/2, 1]$; furthermore by the symmetry with respect to the point $(1/2, 0)$, $B_m(x)$ has opposite signs on the two intervals. Since $B'_{m+1}(x) = (m + 1) B_m(x)$ (see Condition 1 in Proposition 13.1), one has that $B_{m+1}(x)$ is monotonic (in the opposite directions) on both $[0, 1/2]$ and on $[1/2, 1]$. Moreover, by the symmetry of the graph of $B_m(x)$ with respect to $(1/2, 0)$, it is $B_m(1 - x) = -B_m(x)$ and hence

$$\frac{d}{dx}(B_{m+1}(1 - x) - B_{m+1}(x)) = -(m + 1) B_m(1 - x) - (m + 1) B_m(x) = 0.$$

Therefore $B_{m+1}(1 - x) - B_{m+1}(x)$ is constant; since $B_{m+1}(1 - 1/2) = B_{m+1}(1/2)$, one gets $B_{m+1}(1 - x) - B_{m+1}(x) = 0$ for every $x \in [0, 1]$, i.e., $B_{m+1}(x)$ is symmetric with respect to the axis $x = 1/2$. In view of Proposition 13.1, one has

$$\int_0^1 B_{m+1}(x)\, dx = \frac{1}{m + 2} \left[B_{m+2}(x) \right]_0^1 = \frac{1}{m + 2} (B_{m+2} - B_{m+2}) = 0;$$

therefore $B_{m+1}(x)$ can not have a constant sign on $[0, 1]$ and hence, in view of its symmetry with respect to the axis $x = 1/2$, $B_{m+1}(x)$ can not have a constant sign on $[0, 1/2]$. Thus, by the monotonicity of $B_{m+1}(x)$ there exists a unique internal point of the interval $[0, 1/2]$ at which B_{m+1} vanishes.

If $m + 1$ is odd, then $B_m(x)$ changes its sign exactly once on $[0, 1/2]$: let α be the unique zero of $B_m(x)$ on that interval. Since $B'_{m+1}(x) = (m + 1) B_m(x)$, one has that $B_{m+1}(x)$ is monotonic in the opposite directions on both $[0, \alpha]$ and $[\alpha, 1/2]$. Moreover, by the symmetry of the graph of $B_m(x)$ with respect to the axis $x = 1/2$, it is $B_m(1 - x) = B_m(x)$ and hence

$$\frac{d}{dx}(B_{m+1}(1 - x) + B_{m+1}(x)) = -(m + 1) B_m(1 - x) + (m + 1) B_m(x) = 0.$$

Therefore $B_{m+1}(1 - x) + B_{m+1}(x)$ is constant. By Proposition 13.1 and Corollary 8.51 one has $B_{m+1}(1) = B_{m+1}(0) = B_{m+1} = 0$. Then $B_{m+1}(1 - x) + B_{m+1}(x) = 0$ for every $x \in [0, 1]$, i.e., $B_{m+1}(x)$ is symmetric with respect to the point $(1/2, 0)$. In particular, for $x = 1/2$ we get $B_{m+1}(1/2) = 0$. Since $B_{m+1}(0) = B_{m+1}(1) = 0$, then $B_{m+1}(x)$ does not vanish on both $]0, 1/2[$ and $]1/2, 1[$: indeed, it vanishes at 0, $1/2$ and 1, and it is monotonic on $[0, \alpha]$, $[\alpha, 1/2]$, $[1/2, 1 - \alpha]$ and $[1 - \alpha, 1]$. □

Remark 13.5 The above Proposition 13.4 shows that the behavior of the Bernoulli polynomial functions $B_m(x)$ on $[0, 1]$ changes drastically depending on the parity of m. Actually, one can obtain a more subtle classification depending on the divisibility of m by 4: the interested reader may find the details in Problem 13.1.

In the following example we take a look at the behavior of the Bernoulli polynomial functions with odd index, investigating the sup-norm of $B_3(x)$, $B_5(x)$ and $B_7(x)$ in $[0, 1]$.

Example 13.6 By Proposition 13.4, the function $|B_3(x)|$ on $[0, 1]$ reaches its maximum value in a point $a \in]0, 1/2[$ and, by symmetry, in $1 - a$. The function

$$B'_3(x) = 3 B_2(x) = 3\left(x^2 - x + \frac{1}{6}\right)$$

vanishes at $x = \dfrac{3 \pm \sqrt{3}}{6}$; then

$$\| B_3(x) \|_{L^\infty([0,1])} = \left| B_3\left(\frac{3 \pm \sqrt{3}}{6}\right) \right| = \frac{1}{12\sqrt{3}} < \frac{1}{21}.$$

Analogously one can prove

$$\| B_5(x)\|_{L^\infty([0,1])} = \frac{1}{900}\left(5\sqrt{2}+\sqrt{15}\right)\sqrt{15-2\sqrt{30}} < 0.02446 < \frac{1}{40},$$

$$\| B_7(x)\|_{L^\infty([0,1])} < 0.02607 < \frac{1}{38}.$$

The sup-norms $\| B_3(x)\|_{L^\infty([0,1])}$, $\| B_5(x)\|_{L^\infty([0,1])}$, $\| B_7(x)\|_{L^\infty([0,1])}$ are irrational numbers: we shall see now that the sup-norm of the even indexed Bernoulli polynomial functions $B_m(x)$ on $[0, 1]$ is a rational number, precisely the absolute value $|B_m|$ of the m-th Bernoulli number.	□

13.1.3   The Sup-Norm of the Bernoulli Polynomials on $[0, 1]$

The estimate of the sup-norm of the Bernoulli polynomials is a key tool to understand the magnitude of the remainder term in the Euler–Maclaurin formulas. We find very convenient here to consider the Fourier expansions of the Bernoulli polynomials: the reader may skip the details with no risks and jump directly to the claim of Corollary 13.9.

Let us recall the following basic facts about Fourier series [33].

- The **Fourier[1] series** of a locally integrable *periodic* function $f : \mathbb{R} \to \mathbb{R}$ of period 1 is the series

$$\frac{a_0}{2} + \sum_{k=1}^{+\infty} a_k \cos(2\pi kx) + \sum_{k=1}^{+\infty} b_k \sin(2\pi kx),$$

where the **Fourier coefficients** of f are given by

$$a_k = 2\int_0^1 f(x)\cos(2\pi kx)\,dx \quad (k \geq 0),$$

$$b_k = 2\int_0^1 f(x)\sin(2\pi kx)\,dx \quad (k \geq 1).$$

Notice that, since f is periodic of period 1, $a_k = 0$ if f is odd, and $b_k = 0$ if f is even.
- (**Fourier pointwise convergence**). The Fourier series of f converges to $f(x)$ at every point x where f is continuous and has left and right derivatives [33, Theorem 8.14].

Lemma 13.7 (Fourier expansion of the Bernoulli polynomials) *The following identities hold:*

---
[1] Jean Baptiste Joseph Fourier (1768–1830).

$$B_1(x) = -\frac{1}{\pi} \sum_{k=1}^{\infty} \frac{\sin(2\pi kx)}{k} \quad \forall x \in]0, 1[; \tag{13.7.a}$$

$$m \geq 2 \text{ even: } B_m(x) = 2(-1)^{\frac{m}{2}+1} \frac{m!}{(2\pi)^m} \sum_{k=1}^{\infty} \frac{\cos(2\pi kx)}{k^m} \forall x \in [0, 1]; \tag{13.7.b}$$

$$m \geq 2 \text{ odd: } B_m(x) = 2(-1)^{\frac{m+1}{2}} \frac{m!}{(2\pi)^m} \sum_{k=1}^{\infty} \frac{\sin(2\pi kx)}{k^m} \quad \forall x \in [0, 1]. \tag{13.7.c}$$

Proof. For every $m \geq 1$ let $B_m^p(x)$ be the extension by periodicity to $\mathbb{R}$ of the restriction of $B_m(x)$ to $[0, 1[$. Since $B_m^p(x) = B_m(x)$ on $[0, 1[$, in the computation of the Fourier coefficients of $B_m^p(x)$ we can substitute $B_m^p(x)$ with $B_m(x)$. By Proposition 13.4 the function $B_m^p(x)$ is even if m is even, and odd if m is odd. In particular, the Fourier series of $B_1^p(x)$ is of the form $\sum_{k=1}^{\infty} b_k^1 \sin(2\pi kx)$, where

$$b_k^1 = 2 \int_0^1 B_1^p(x) \sin(2\pi kx)\, dx = 2 \int_0^1 B_1(x) \sin(2\pi kx)\, dx$$
$$= 2 \int_0^1 \left(x - \frac{1}{2}\right) \sin(2\pi kx)\, dx = -\frac{1}{k\pi}.$$

Since $B_1^p(x)$ is piecewise $\mathscr{C}^1$ and continuous on $]0, 1[$, Fourier pointwise convergence theorem yields (13.7.a). For each $m \geq 2$ one has

$$a_0^m = 2 \int_0^1 B_m(x)\, dx = 2 \int_0^1 \frac{1}{m+1} B_{m+1}'(x)\, dx = \frac{2}{m+1}[B_{m+1}(x)]_0^1 = 0;$$

therefore the Fourier series of B_m^p is of the form

$$\sum_{k=1}^{\infty} a_k^m \cos(2\pi kx) \text{ if } m \text{ even}, \quad \sum_{k=1}^{\infty} b_k^m \sin(2\pi kx) \text{ if } m \text{ odd}.$$

Integration by parts, for $m \geq 3$ odd, yields

$$b_k^m = 2 \int_0^1 B_m^p(t) \sin(2\pi k t)\, dt = 2 \int_0^1 B_m(t) \sin(2\pi k t)\, dt$$

$$= 2 \left[-B_m(t) \frac{1}{2\pi k} \cos(2\pi k t) \right]_0^1 + \frac{1}{\pi k} \int_0^1 B_m'(t) \cos(2\pi k t)\, dt$$

$$= \frac{m}{\pi k} \int_0^1 B_{m-1}(t) \cos(2\pi k t)\, dt = \frac{m}{2\pi k} a_k^{m-1}.$$

Similarly, for $m \geq 2$ even one gets $a_k^m = -\dfrac{m}{2\pi k} b_k^{m-1}$. Thus, for $m \geq 3$ odd,

$$b_k^m = \frac{m}{2\pi k} a_k^{m-1} = -\frac{m(m-1)}{(2\pi k)^2} b_k^{m-2},$$

so that, recursively, we obtain

$$b_k^m = (-1)^{\frac{m-1}{2}} \frac{m!}{(2\pi k)^{m-1}} b_k^1 = 2(-1)^{\frac{m-1}{2}+1} \frac{m!}{(2\pi k)^m}.$$

On the other side, for $m \geq 2$ even, again recursively, we get

$$a_k^m = -\frac{m}{2\pi k} b_k^{m-1} = -\frac{m}{2\pi k} 2(-1)^{\frac{m-2}{2}+1} \frac{(m-1)!}{(2\pi k)^{m-1}} = (-1)^{\frac{m}{2}-1} 2 \frac{m!}{(2\pi k)^m}. \qquad \square$$

Figures 13.2, 13.3 and 13.4 depict, respectively, $B_1(x)$, $B_2(x)$ and $B_5(x)$ on $[0, 1]$ and some of their Fourier expansions.

Definition 13.8 (*The sup-norm μ_m*) In the sequel we denote by μ_m the sup-norm $\| B_m(x) \|_{L^\infty([0,1])}$ of the polynomial functions $B_m(x)$ on $[0, 1]$. $\qquad \square$

Let us see an important upper estimate μ_m and a relation of the Bernoulli numbers B_m of *even* index m with $\zeta(m)$.

Corollary 13.9 (Fundamental estimates) *The following identities hold:*

$$\mu_m := \| B_m(x) \|_{L^\infty([0,1])} \leq 4 \frac{m!}{(2\pi)^m} \qquad \forall m \in \mathbb{N}_{\geq 1}; \tag{13.9.a}$$

$$\mu_m := \| B_m(x) \|_{L^\infty([0,1])} = | B_m | \qquad \forall m \geq 2 \text{ even}; \tag{13.9.b}$$

$$\zeta(m) = (-1)^{\frac{m}{2}+1} \frac{1}{2} \frac{(2\pi)^m}{m!} B_m \qquad \forall m \geq 2 \text{ even}. \tag{13.9.c}$$

Proof. If $m = 1$ then $\mu_1 = \dfrac{1}{2} \leq \dfrac{4}{\pi} \dfrac{1}{2} = 4 \dfrac{1!}{2\pi}$, due to the fact that $\pi \leq 4$. Assume now that $m \geq 2$. By Lemma 13.7, for each x in $[0, 1]$ we have

Fig. 13.2 $B_1(x)$ and its
Fourier expansions of order
3 and 7

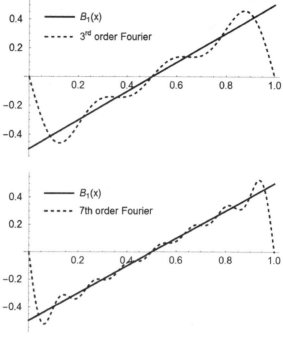

Fig. 13.3 $B_2(x)$ with its 2nd
order Fourier expansion and
its 6th order Fourier
expansion

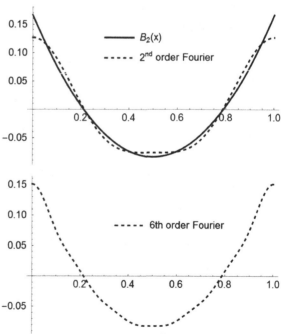

Fig. 13.4 $B_5(x)$ and its
Fourier expansion of order 1:
they almost coincide!

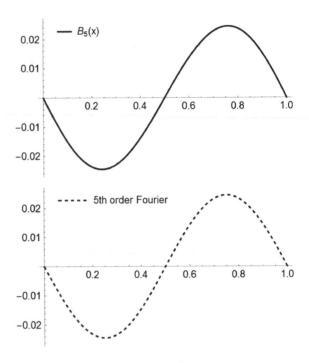

$$|B_m(x)| \leq 2\frac{m!}{(2\pi)^m}\sum_{k=1}^{\infty}\frac{1}{k^m} = 2\frac{m!}{(2\pi)^m}\zeta(m) \leq 4\frac{m!}{(2\pi)^m},$$

since, from (8.55.a), $\zeta(m) \leq 2$. Then we get (13.9.a).

Assume now that $m \geq 2$ is even. Then by (13.7.b)

$$2\frac{m!}{(2\pi)^m}\sum_{k=1}^{\infty}\frac{1}{k^m} = |B_m(0)| = |B_m|;$$

therefore $\mu_m = |B_m(0)| = |B_m|$. Finally, applying (13.7.b) at $x = 0$, for all $m \geq 2$
even we get

$$B_m = B_m(0) = 2(-1)^{\frac{m}{2}+1}\frac{m!}{(2\pi)^m}\sum_{k=1}^{\infty}\frac{1}{k^m} = 2(-1)^{\frac{m}{2}+1}\frac{m!}{(2\pi)^m}\zeta(m). \qquad \square$$

☛ **Corollary 13.10** *The following identity relates* $B_m(1/2)$ *with* B_m:

$$B_m(1/2) = (2^{1-m} - 1)B_m \quad \forall m \in \mathbb{N}_{\geq 1}. \tag{13.10.a}$$

Proof. For $m = 1$ one has $B_1(1/2) = 0 = (2^{1-1} - 1) B_1$. Assume that $m \geq 2$. If m is odd then, since $\sin(k\pi) = 0$ for each $k \geq 1$, it follows from (13.7.c) that $B_m(1/2) = 0$ as well as $B_m = 0$. Assume now that m is even: we deduce from (13.7.b) that

$$B_m(1/2) = 2(-1)^{\frac{m}{2}+1} \frac{m!}{(2\pi)^m} \sum_{k=1}^{\infty} \frac{(-1)^k}{k^m}, \qquad (13.10.b)$$

$$B_m = B_m(0) = 2(-1)^{\frac{m}{2}+1} \frac{m!}{(2\pi)^m} \sum_{k=1}^{\infty} \frac{1}{k^m}. \qquad (13.10.c)$$

We now use a standard way to recover $\sum_{k=1}^{\infty} \frac{(-1)^k}{k^m}$ from $\sum_{k=1}^{\infty} \frac{1}{k^m}$. By regrouping the terms of different parity we get

$$\sum_{k=1}^{\infty} \frac{(-1)^k}{k^m} = -\sum_{k \text{ odd}} \frac{1}{k^m} + \sum_{k \geq 2 \text{ even}} \frac{1}{k^m} = -\sum_{k \text{ odd}} \frac{1}{k^m} + 2^{-m} \sum_{k=1}^{\infty} \frac{1}{k^m};$$

we use here the fact that every even natural is of the form $2k$ for some natural k. Similarly

$$\sum_{k=1}^{\infty} \frac{1}{k^m} = \sum_{k \text{ odd}} \frac{1}{k^m} + \sum_{k \geq 2 \text{ even}} \frac{1}{k^m} = \sum_{k \text{ odd}} \frac{1}{k^m} + 2^{-m} \sum_{k=1}^{\infty} \frac{1}{k^m}.$$

Thus

$$\sum_{k \text{ odd}} \frac{1}{k^m} = (1 - 2^{-m}) \sum_{k=1}^{\infty} \frac{1}{k^m},$$

and therefore

$$\sum_{k=1}^{\infty} \frac{(-1)^k}{k^m} = (-(1 - 2^{-m}) + 2^{-m}) \sum_{k=1}^{\infty} \frac{1}{k^m} = (2^{1-m} - 1) \sum_{k=1}^{\infty} \frac{1}{k^m}. \qquad (13.10.d)$$

Equalities (13.10.b) and (13.10.c) yield the conclusion. $\qquad\square$

Remark 13.11 (The Dirichlet η function at even numbers) In (13.9.c) we found the sum of the harmonic series of even powers in terms of the Bernoulli numbers. We are now able to express the sum of the *alternating* harmonic series of even powers, namely the **eta Dirichlet**[2] function η, defined by

---

[2]Johann Peter Gustav Lejeune Dirichlet (1805–1859).

$$\eta(m) := \sum_{k=1}^{\infty} \frac{(-1)^{k-1}}{k^m}.$$

Indeed, let $m \geq 2$ be *even*. It follows from (13.10.b) that

$$B_m(1/2) = 2(-1)^{\frac{m}{2}+1} \frac{m!}{(2\pi)^m} (-\eta(m)),$$

so that

$$\eta(m) = (-1)^{\frac{m}{2}} \frac{(2\pi)^m}{2\,m!} B_m(1/2) = (-1)^{\frac{m}{2}} \frac{(2\pi)^m}{m!} (2^{-m} - 2^{-1}) B_m$$
$$= (-1)^{\frac{m}{2}} \frac{\pi^m}{m!} (1 - 2^{m-1}) B_m \quad \forall m \geq 2 \text{ even.}$$

Notice that, from (13.10.d), we get the well-known relation between the eta and the zeta function

$$\eta(m) = (1 - 2^{1-m})\zeta(m) \qquad \forall m \geq 2 \text{ even.}$$

We mention that the expression of the eta function for odd values of m is unknown, except for $m = 1$:

$$\eta(1) = 1 - \frac{1}{2} + \frac{1}{3} + \cdots = \log 2.$$

13.2   The Euler–Maclaurin Expansion of Order m

In Sect. 12.2 we have seen the first- and second- order Euler–Maclaurin formulas and some of their notable applications. In this section we write the Euler–Maclaurin expansions of arbitrary order.

Definition 13.12 Let f be of class $\mathscr{C}^{m-1}$ on $[a, b]$, where a, b are **natural numbers**. The **Euler–Maclaurin expansion of order** m of f on $[a, b]$ is

$$\mathrm{EM}_m(f; [a, b]) := \int_a^b f(x)\,dx + \sum_{i=1}^{m} \frac{B_i}{i!} \left[f^{(i-1)} \right]_a^b. \qquad \square$$

In the applications, we will no go further than $m = 8$: the Euler–Maclaurin expansions of order less than 8 can be obtained easily from the one of order 8.

Example 13.13 (The first terms in the Euler–Maclaurin expansion) The Euler–Maclaurin expansion $\mathrm{EM}_8(f; [a, b])$ of order 8 of a function f on an integer interval $[a, b]$ is

$$\int_a^b f(x)\,dx + \left[-\frac{1}{2}f + \frac{1}{12}f' - \frac{1}{720}f^{(3)} + \frac{1}{30\,240}f^{(5)} - \frac{1}{1\,209\,600}f^{(7)} \right]_a^b. \qquad \square$$

13.2.1   The Euler–Maclaurin Formula of Order *m*

We are now ready to generalize the Euler–Maclaurin formula established for functions of class $\mathscr{C}^1$ or $\mathscr{C}^2$ in Theorem 12.27.

☞ *Remark 13.14* Note that in the Euler–Maclaurin expansion of order m of a function there are involved the function's derivatives up to order $m - 1$, and the Bernoulli numbers $B_1, \ldots, B_m$. For each $\ell \geq 1$, the Euler–Maclaurin expansion of order 2ℓ of a function coincides with its Euler–Maclaurin expansion of order $2\ell + 1$: indeed $B_{2\ell+1} = 0$.

Theorem 13.15 (Euler–Maclaurin formula of order m) *Given two integers $a < b$, consider a function $f : [a, b] \to \mathbb{R}$ of class $\mathscr{C}^m$, with $m \geq 1$. Then*

$$\sum_{a \leq k < b} f(k) = \int_a^b f(x)\,dx + \sum_{i=1}^m \frac{B_i}{i!} \left[f^{(i-1)} \right]_a^b + R_m = \mathrm{EM}_m(f; [a, b]) + R_m,$$

$$R_m = \frac{(-1)^{m+1}}{m!} \int_a^b B_m(x - [x])\, f^{(m)}(x)\,dx. \qquad (13.15.a)$$

Moreover, denoted by μ_m the sup-norm $\| B_m(x) \|_{L^\infty([0,1])}$, the following estimates of R_m hold:

$$|R_m| \leq \frac{\mu_m}{m!} \int_a^b |f^{(m)}(x)|\,dx \leq \frac{\mu_m}{m!}(b-a)\| f^{(m)} \|_{L^\infty([a,b])} \leq \frac{4(b-a)}{m!(2\pi)^m}\| f^{(m)} \|_{L^\infty([a,b])}. \qquad (13.15.b)$$

Remark 13.16 Since for each $\ell \geq 1$, the Euler–Maclaurin expansion of order 2ℓ of a function coincides with its Euler–Maclaurin expansion of order $2\ell + 1$, we get

$$\frac{(-1)^{2\ell+1}}{(2\ell)!} \int_a^b B_{2\ell}(x - [x])\, f^{(2\ell)}(x)\,dx = R_{2\ell} =$$

$$= R_{2\ell+1} = \frac{(-1)^{2\ell+2}}{(2\ell+1)!} \int_a^b B_{2\ell+1}(x - [x]) f^{(2\ell+1)}(x)\, dx.$$

When estimating the remainder in the Euler–Maclaurin formula, one of these two descriptions will typically be more useful than the other.

Example 13.17 Example 13.6 and Corollary 13.9 yield the following values of $\mu_m = \| B_m(x) \|_{L^\infty([0,1])}$:

$$\mu_1 = \frac{1}{2} = |B_1|, \quad \mu_2 = \frac{1}{6} = |B_2|, \quad \mu_3 = \frac{1}{12\sqrt{3}} < \frac{1}{20}, \quad \mu_4 = |B_4| = \frac{1}{30},$$

$$\mu_5 < \frac{1}{40}, \quad \mu_6 = |B_6| = \frac{1}{42}, \quad \mu_7 < \frac{1}{38}, \quad \mu_8 = |B_8| = \frac{1}{30}. \qquad \square$$

The following remark will be used whenever the remainder term contains the integral of the modulus of the derivative of a monotonic function.

Remark 13.18 (The remainder if $f^{(m-1)}$ is monotonic) In the relevant applications it is often the case that $f^{(m)}$ has a constant sign, or in other words that $f^{(m-1)}$ is *monotonic*; in this case from Lemma 12.25 we get

$$\int_a^b |f^{(m)}(x)|\, dx = |f^{(m-1)}(b) - f^{(m-1)}(a)|.$$

In particular it follows from (13.15.b) that

$$|R_m| \le \frac{\mu_m}{m!} \left| \left[f^{(m-1)} \right]_a^b \right|. \qquad (13.18.a)$$

If $m = 2\ell$ is even, then $\mu_{2\ell} = |B_{2\ell}|$ (Corollary 13.9) and hence the sum $\sum_{a \le k < b} f(k)$ differs from the Euler–Maclaurin expansion of order 2ℓ on $[a, b]$ for at most the modulus of its last term $\dfrac{B_{2\ell}}{(2\ell)!} \left[f^{(2\ell-1)} \right]_a^b$.

Proof (of Theorem 13.15). We begin by considering the case $a = 0, b = 1$. Let $g : [0, 1] \to \mathbb{R}$ be of class $\mathscr{C}^m$. Given that $B_1'(x) = 1$, integration by parts gives

$$\int_0^1 g(x)\, dx = \int_0^1 g(x)\, B_1'(x)\, dx = [g(x)\, B_1(x)]_0^1 - \int_0^1 g'(x)\, B_1(x)\, dx$$

and since $B_1(0) = -1/2 = B_1$ and $B_1(1) = 1/2 = -B_1$ one obtains

$$\int_0^1 g(x)\,dx = \frac{1}{2}(g(1) + g(0)) - \int_0^1 g'(x)\,B_1(x)\,dx.$$

From $B_2'(x) = 2\,B_1(x)$ one obtains, (again integrating by parts) that

$$\int_0^1 g'(x)\,B_1(x)\,dx = \int_0^1 g'(x)\frac{B_2'(x)}{2}\,dx = \left[g'(x)\frac{B_2(x)}{2}\right]_0^1 - \int_0^1 g''(x)\frac{B_2(x)}{2}\,dx,$$

from which, on recalling that $B_m(0) = B_m(1) = B_m$ for $m \geq 2$, it follows that

$$\int_0^1 g(x)\,dx = \frac{g(0) + g(1)}{2} - \frac{B_2}{2}\left[g'\right]_0^1 + \int_0^1 g''(x)\frac{B_2(x)}{2}\,dx.$$

At this point it is now easy to prove by induction that one has

$$\int_0^1 g(x)\,dx = \frac{g(0) + g(1)}{2} - \sum_{i=2}^{m}(-1)^m\frac{B_i}{i!}\left[g^{(i-1)}\right]_0^1 + (-1)^m\int_0^1 g^{(m)}(x)\frac{B_m(x)}{m!}\,dx$$

or, equivalently,

$$\frac{g(0) + g(1)}{2} = \int_0^1 g(x)\,dx + \sum_{i=2}^{m}(-1)^m\frac{B_i}{i!}\left[g^{(i-1)}\right]_0^1 - (-1)^m\int_0^1 g^{(m)}(x)\frac{B_m(x)}{m!}\,dx.$$

Applying this last relation to $g(x) = f(x + k)$ for $k = a, \ldots, b - 1$, and taking into account that $B_m = 0$ if $m \geq 3$ is odd, one obtains

$$\frac{f(k) + f(k + 1)}{2} = \int_k^{k+1} f(x)\,dx +$$

$$+ \sum_{i=2}^{m}\frac{B_i}{i!}\left[f^{(i-1)}\right]_k^{k+1} - (-1)^m\int_k^{k+1} f^{(m)}(x)\frac{B_m(x - k)}{m!}\,dx.$$

Summing for k that varies from a to $b - 1$ one gets

$$\frac{f(a)}{2} + \sum_{k=a+1}^{b-1} f(k) + \frac{f(b)}{2} =$$

$$= \int_a^b f(x)\,dx + \sum_{i=2}^{m}\frac{B_i}{i!}\left[f^{(i-1)}\right]_a^b - (-1)^m\frac{1}{m!}\int_a^b B_m(x - [x])\,f^{(m)}(x)\,dx,$$

from which it follows that

$$\sum_{a \le k < b} f(k) = \frac{1}{2}f(a) - \frac{1}{2}f(b) + \int_a^b f(x)\,dx +$$

$$+ \sum_{i=2}^{m} \frac{B_i}{i!}\left[f^{(i-1)}\right]_a^b - (-1)^m \frac{1}{m!}\int_a^b B_m(x - [x])\,f^{(m)}(x)\,dx.$$

Recalling that $B_1 = -1/2$, one immediately obtains relation (13.15.a). The estimate of the remainder in (13.15.b) follows directly.   □

Example 13.19 If $f \in \mathscr{C}^5([a, b])$ it follows from (13.15.a), with $m = 5$, that

$$\sum_{a \le k < b} f(k) = \int_a^b f(x)\,dx + \left[-\frac{1}{2}f + \frac{1}{6}\frac{1}{2!}f' - \frac{1}{30}\frac{1}{4!}f^{(3)}\right]_a^b + R_5,$$

and

$$R_5 = \frac{1}{5!}\int_a^b B_5(x - [x])\,f^{(5)}(x)\,dx = R_4 = -\frac{1}{4!}\int_a^b B_4(x - [x])\,f^{(4)}(x)\,dx. \quad □$$

Example 13.20 If $f(x) = x^2$ one has $f'(x) = 2x$, $f'' = 2$, $f^{(3)} = f^{(4)} = 0$, and so, since $R_3 = 0$, one has

$$\sum_{1 \le k < n} k^2 = \int_1^n x^2\,dx - \frac{1}{2}\left[x^2\right]_1^n + \frac{1}{12}\left[2x\right]_1^n = \frac{1}{3}n^3 - \frac{1}{2}n^2 + \frac{1}{6}n. \qquad □$$

More generally, if f is a polynomial function of degree less than or equal to m, then the $(m + 1)$-st derivative of f is zero, and so the remainder R_{m+1} in (13.15.a) is equal to 0 and the m-th derivative of f is a constant.

Corollary 13.21 *Let f be a polynomial function of degree $m \ge 1$. Then*

$$\sum_{a \le k < b} f(k) = \mathrm{EM}_m(f;[a,b]) = \int_a^b f(x)\,dx + \sum_{i=1}^{m} \frac{B_i}{i!}\left[f^{(i-1)}\right]_a^b.$$

In particular, Faulhaber's formula for the sum of the successive m-th powers of the integers (that we saw in Example 8.52) is rather simply expressed via the Bernoulli numbers.

Example 13.22 (Faulhaber's formula) Let $n \in \mathbb{N}$. For each $m \ge 1$ one has

$$\sum_{0\le k<n} k^m = \frac{1}{m+1}\sum_{i=0}^{m}\binom{m+1}{i}\mathrm{B}_i\,n^i = \frac{1}{m+1}\sum_{i=0}^{m}\binom{m+1}{i}\mathrm{B}_i\,n^{m+1-i}.$$

$$(13.22.\mathrm{a})$$

Indeed, if $f(x) = x^m$, for each integer $1 \le i \le m$ one has

$$f^{(i-1)}(x) = m(m-1)\cdots(m-i+2)x^{m-i+1}$$

$$= \frac{1}{m+1}\frac{(m+1)!}{(m+1-i)!}x^{m-i+1}.$$

By Corollary 13.21 one gets

$$\sum_{0\le k<n} k^m = \int_0^n x^m\,dx + \sum_{i=1}^{m}\frac{\mathrm{B}_i}{i!}\big[f^{(i-1)}\big]_0^n$$

$$= \frac{1}{m+1}n^{m+1} + \frac{1}{m+1}\sum_{i=1}^{m}\binom{m+1}{i}\mathrm{B}_i\,n^{m+1-i}$$

$$= \frac{1}{m+1}\sum_{i=0}^{m}\binom{m+1}{i}\mathrm{B}_i n^{m+1-i}. \qquad \square$$

The remainder R_m in the Euler–Maclaurin formula (13.15.a) can be very large, rendering it useless. Considering longer Euler–Maclaurin expansions of a function do not constitute in general a solution. Here is an example illustrating how things can get worse as the order of the expansions increases.

Example 13.23 Consider $f(x) = \dfrac{1}{x^2}$, $x \ge 1$. One can easily compute

$$\sum_{1\le k<10} f(k) = \frac{9778141}{6350400} = 1.53976\ldots.$$

For all $i \ge 1$, the derivative $f^{(i)}$ of f is given by

$$f^{(i)}(x) = (-1)^i\frac{(i+1)!}{x^{i+2}};$$

thus the Euler–Maclaurin expansion of order m of f on $[1, 10]$ is

$$\mathrm{EM}_m\left(\frac{1}{x^2};[1,10]\right) := \int_1^{10} \frac{1}{x^2}\, dx + \sum_{i=1}^{m} \frac{\mathrm{B}_i}{i!}\left[(-1)^{i-1}\frac{i!}{x^{i+1}}\right]_1^{10}$$

$$= \frac{9}{10} + \sum_{i=1}^{m}(-1)^{i-1}\,\mathrm{B}_i\left[\frac{1}{10^{i+1}}-1\right].$$

The values of

$$R_m = \frac{(-1)^{m+1}}{m!}\int_1^{10} \mathrm{B}_m(x-[x])f^{(m)}(x)\, dx = \sum_{1\le k<10} f(k) - \mathrm{EM}_m\left(\frac{1}{x^2};[1,10]\right),$$

for $m = 1,\ldots,12$ seem to oscillate between -0.75 and -0.96; then for $m > 12$ the values become also larger in module. We list here the values of R_m for $m = 1,\ldots,25$:

m	1	2	3	4	5	6	7	8	9	10	11	12	13
R_m	−0.76	−0.92	−0.92	−0.89	−0.89	−0.91	−0.91	−0.88	−0.88	−0.95	−0.95	−0.70	−0.70

m	14	15	16	17	18	19	20	21	22	23	24	25
R_m	−1.87	−1.87	5.22	5.22	−49.76	−49.76	479.54	479.54	−5712.46	−5712.46	80867.54	80867.54

From $m = 26$ the values of $|R_m|$ are greater than 10^6.   □

13.2.2 The Euler–Maclaurin Approximation Formula of Order m

We now extend to arbitrary order the approximation formulas established in Corollary 12.45.

Corollary 13.24 *Let* $f : [a,+\infty[\to \mathbb{R}$ *be of class* $\mathscr{C}^m$, *with* $m \ge 1$ *and* $a \in \mathbb{N}$. *If* n, N *are naturals, with* $a \le n \le N$, *then*

$$\sum_{a\le k<N} f(k) = \sum_{a\le k<n} f(k) + \int_n^N f(x)\, dx + \sum_{i=1}^{m} \frac{\mathrm{B}_i}{i!}\left[f^{(i-1)}\right]_n^N + \varepsilon_m(n,N),$$

$$(13.24.\mathrm{a})$$

where $\varepsilon_m(n,N) = \dfrac{(-1)^{m+1}}{m!}\displaystyle\int_n^N \mathrm{B}_m(x-[x])f^{(m)}(x)\, dx$ *and*

$$|\varepsilon_m(n,N)| \le \frac{\mu_m}{m!}\int_n^N |f^{(m)}(x)|\, dx \le \frac{\mu_m}{m!}\int_n^{+\infty} |f^{(m)}(x)|\, dx, \qquad (13.24.\mathrm{b})$$

where $\mu_m = \|\,\mathrm{B}_m(x)\,\|_{L^\infty([0,1])}$.

Proof. It is enough to apply (13.15.a) to the sum

$$\sum_{n \le k < N} f(k) = \sum_{a \le k < N} f(k) - \sum_{a \le k < n} f(k).$$

$\square$

Example 13.25 In Example 12.53 we found an approximated value of

$$\sum_{1 \le k < 1000^{1000}} \frac{1}{k},$$

with an error within 10^{-3}, by means of the approximation formula of order 2, with $f(x) = \dfrac{1}{x}$, $x \ge 1$. Choosing $n = 10$ in (12.45.b), we got

$$\sum_{1 \le k < 1000^{1000}} \frac{1}{k} \approx \sum_{1 \le k < 10} \frac{1}{k} + \int_{10}^{1000^{1000}} \frac{1}{x}\, dx - \frac{1}{2}\left[\frac{1}{x}\right]_{10}^{1000^{1000}} + \frac{1}{12}\left[-\frac{1}{x^2}\right]_{10}^{1000^{1000}}$$

$$= 6\,908.332495476\ldots.$$

We now wish to apply the approximation formula of order 4, i.e., (13.24.a) with $N = 1000^{1000}$ and $m = 4$. The remainder satisfies

$$|\varepsilon_4(n, 1000^{1000})| \le \frac{1}{720} \int_n^{+\infty} |f^{(4)}(x)|\, dx.$$

Now $f^{(3)}(x) = \dfrac{-6}{x^4}$ is monotonic: it follows from Remark 13.18 that

$$\int_n^{+\infty} |f^{(4)}(x)|\, dx = |f^{(3)}(\infty) - f^{(3)}(n)| = \frac{6}{n^4};$$

thus

$$|\varepsilon_4(n, 1000^{1000})| \le \frac{1}{120 n^4}.$$

Therefore, with $n = 10$, the error is at most equal to $\dfrac{1}{1\,200\,000}$; actually we get

$$\sum_{1 \le k < 1000^{1000}} \frac{1}{k} \approx \sum_{1 \le k < 10} f(k) + \int_{10}^{1000^{1000}} f(x)\, dx + \sum_{i=1}^{4} \frac{B_i}{i!} \left[f^{(i-1)}\right]_{10}^{1000^{1000}}$$

$$= \frac{7129}{2520} + \log\left(1000^{1000}/10\right) + \left[-\frac{1}{2} f + \frac{1}{12} f' - \frac{1}{720} f^{(3)}\right]_{10}^{1000^{1000}}$$

$$= 6908.\mathbf{33249464}31112603,$$

whereas the exact value of the given sum is 6908.**33249464**70385849.... Notice
that, if one just wants an approximation up to 10^{-2}, it is enough that

$$|\varepsilon_4(n, 1000^{1000})| \leq \frac{1}{120\,n^4} \leq 10^{-2},$$

and this occurs for $n \geq (5/6)^{1/4}$, thus for $n \geq 1$. We do not have to make too many
calculations to get the result: for $n = 1$ the 4th order approximation yields

$$\sum_{1 \leq k < 1000^{1000}} \frac{1}{k} \approx \int_1^{1000^{1000}} f(x)\,dx + \sum_{i=1}^4 \frac{B_i}{i!} \left[f^{(i-1)} \right]_1^{1000^{1000}}$$

$$= \frac{7129}{2520} + \log\left(1000^{1000}/10\right) + \left[-\frac{1}{2}f + \frac{1}{12}f' - \frac{1}{720}f^{(3)} \right]_{10}^{1000^{1000}}$$

$$= 6908.330278\ldots.$$

Notice that in this case we did not need any term of the original sum in order to
obtain its approximation!   □

Example 13.26 Let $N \in \mathbb{N}_{\geq 1}$. We wish to compute $\displaystyle\sum_{1 \leq k < N} \sqrt{k}$ with an error up to
10^{-6}. If we set $f(x) = \sqrt{x}$, $x \geq 1$, (13.24.a) with $m = 4$ yields

$$\sum_{1 \leq k < N} f(k) = \sum_{1 \leq k < n} f(k) + \int_n^N f(x)\,dx + \left[-\frac{1}{2}f + \frac{1}{12}f' - \frac{1}{720}f^{(3)} \right]_n^N + \varepsilon_4(n, N),$$

with, due to the fact that $f^{(3)}(x) = \dfrac{3}{8x^{5/2}}$ is decreasing,

$$|\varepsilon_4(n, N)| \leq \frac{1}{720} \left| f^{(3)}(N) - f^{(3)}(n) \right| \leq \frac{1}{720} f^{(3)}(n) = \frac{1}{1\,920 n^{5/2}}.$$

It turns out that $|\varepsilon_4(n, N)| \leq 10^{-6}$ whenever $n \geq \left(\dfrac{10^6}{1\,920} \right)^{2/5} \approx 12.2$. By taking
$n = 13 \leq N$, we thus get

$$\sum_{1 \leq k < N} \sqrt{k} \approx \sum_{1 \leq k < 13} \sqrt{k} + \left[\frac{2}{3}x^{3/2} \right]_n^N + \left[-\frac{1}{2}\sqrt{x} + \frac{1}{12}\frac{x^{-1/2}}{2} - \frac{1}{720}\frac{3x^{-5/2}}{8} \right]_{13}^N$$

$$= -0.20788622\cdots + \frac{2N^{3/2}}{3} - \frac{1}{1920N^{5/2}} - \frac{\sqrt{N}}{2} + \frac{1}{24\sqrt{N}},$$

with an error within 10^{-6}.   □

13.2.3 The Convergence of the Euler–Maclaurin Series ☕

Let $f \in \mathscr{C}^{\infty}([a, b])$. Then we can consider the Euler–Maclaurin expansions

$$\mathrm{EM}_m(f; [a, b]) = \int_a^b f(x)\,dx + \sum_{i=1}^m \frac{\mathrm{B}_i}{i!} \left[f^{(i-1)} \right]_a^b,$$

for each $m \in \mathbb{N}_{\geq 1}$. Thus, it is the natural to question the behavior of the *Euler–Maclaurin series*.

Definition 13.27 (*Euler–Maclaurin series*) Let $f \in \mathscr{C}^{\infty}([a, b])$. The **Euler-Maclaurin series** of f on $[a, b]$ is:

$$\mathrm{EM}_{\infty}(f; [a, b]) := \int_a^b f(x)\,dx + \sum_{i=1}^{\infty} \frac{\mathrm{B}_i}{i!} \left[f^{(i-1)} \right]_a^b. \qquad \square$$

☞   We invite the reader not to confuse the following Euler–Maclaurin series with the Maclaurin formal power series of Definition 7.66.

Here is a formal motivation of the Euler–Maclaurin series.

Example 13.28 (Formal motivation for the Euler–Maclaurin series) The appearance of the Bernoulli numbers in the Euler–Maclaurin formula may seem surprising. However, in reality one can infer such a formula via an argument due to Lagrange. That discussion is once again indicated here for interested readers, although it is offered without a formally rigorous proof. We have seen in Remark 6.29 that the difference operator Δ is the inverse of the sum operator Σ, just as the integral $\int$ is the inverse of the derivation operator D. If f is developable into a power series over $\mathbb{R}$, then for every $n \in \mathbb{Z}$ one has

$$f(n+1) = f(n) + f'(n) + \frac{f''(n)}{2!} + \frac{f'''(n)}{3!} + \cdots,$$

from which it follows that

$$\Delta f(n) = f(n+1) - f(n) = (D/1! + D^2/2! + D^3/3! + \cdots)f(n) = (e^D - 1)f(n),$$

where we use e^D to indicate the operator

$$e^D = 1 + D + D^2/2! + D/3! + \cdots.$$

Thus one has $\Delta = e^D - 1$ and so Σ is the inverse of $e^D - 1$. Now, since by Proposition 8.50

$$\frac{X}{e^X - 1} = \sum_{k=0}^{\infty} \frac{B_k}{k!} X^k,$$

then the following equality holds:

$$D\Sigma = \frac{D}{e^D - 1} = \sum_{k=0}^{\infty} \frac{B_k}{k!} D^k = 1 + \sum_{k=1}^{\infty} \frac{B_k}{k!} D^k.$$

Since the operator D is invertible with inverse $D^{-1} = \int$ we get

$$\Sigma = \frac{1}{D} + \sum_{k=1}^{\infty} \frac{B_k}{k!} D^{k-1} = \int + \sum_{k=1}^{\infty} \frac{B_k}{k!} D^{k-1},$$

from which it follows that

$$\sum_{a \le k < b} f(k) = \mathrm{EM}_{\infty}(f; [a, b]) = \int_a^b f(x)\,dx + \sum_{k=1}^{\infty} \frac{B_k}{k!} \left[f^{(k-1)} \right]_a^b.$$

It seems that Euler and Maclaurin deduced the Euler–Maclaurin formula with a similar piece of reasoning, without, however, determining the most important part constituted by the remainder R_m in (13.15.a): Poisson [28] explicitly calculated the remainder R_m at a much later date.   □

The motivation of the Euler–Maclaurin formula may lead one to believe that the Euler–Maclaurin series of f in $[a, b]$ converges to $\sum_{a \le k < b} f(k)$. In reality things do not quite go as we might wish; the following example based on Example 13.23 shows that such a series may not converge.

Example 13.29 We consider again the function $f(x) = \dfrac{1}{x^2}$, $x \in [1, 10]$, considered in Example 13.23. Since $f^{(i)}(x) = (-1)^i \dfrac{(i+1)!}{x^{i+2}}$ for all $i \ge 1$ it turns out that the Euler–Maclaurin series of f on $[1, 10]$ is

$$\mathrm{EM}_{\infty}\left(\frac{1}{x^2}; [1, 10]\right) = \int_1^{10} \frac{1}{x^2}\,dx + \sum_{i=1}^{\infty} \frac{B_i}{i!} \left[(-1)^{i-1} \frac{i!}{x^{i+1}} \right]_1^{10}.$$

The absolute value of the general term of the series is given by

$$|B_i| \left| \left[\frac{1}{x^{i+1}} \right]_1^{10} \right| = |B_i| \left(1 - \frac{1}{10^{i+1}} \right) \ge \frac{|B_i|}{2};$$

Corollary 8.57 shows that $\lim\limits_{i\to+\infty} |B_{2i}| = +\infty$, proving that the $EM_\infty\left(\dfrac{1}{x^2}; [1, 10]\right)$ does not converge. $\qquad\square$

In Example 13.29 the derivatives of the function blow up too quickly: we see now how some growth conditions on the derivatives of a function ensure the convergence of the Euler–Maclaurin series.

Theorem 13.30 (Convergence of the Euler–Maclaurin series) *Let $f \in \mathscr{C}^\infty([a, b])$, with $a, b \in \mathbb{N}$. Assume that there exist $M \in \mathbb{N}$, $C \geq 0$ and $0 \leq L < 2\pi$ such that*

$$\|f^{(m)}(x)\|_{L^\infty([a,b])} \leq C\,L^m \quad \forall m \geq M. \tag{13.30.a}$$

Then the Euler–Maclaurin series $EM_\infty(f; [a, b])$ converges to $\sum\limits_{a\leq k<b} f(k)$:

$$\sum_{a\leq k<b} f(k) = EM_\infty(f; [a, b]) = \int_a^b f(x)\,dx + \sum_{i=1}^\infty \frac{B_i}{i!}\left[f^{(i-1)}\right]_a^b.$$

Proof. From (13.15.b) for every natural $m \geq M$ we have

$$|R_m| \leq \frac{4}{(2\pi)^m} \int_a^b |f^{(m)}(x)|\,dx$$
$$\leq \frac{4(b-a)}{(2\pi)^m} \|f^{(m)}\|_{L^\infty([a,b])} \leq 4C(b-a)\left(\frac{L}{2\pi}\right)^m.$$

Then since $\lim\limits_{m\to+\infty} \left(\dfrac{L}{2\pi}\right)^m = 0$, we get also $\lim\limits_{m\to+\infty} |R_m| = 0$, proving the claim. $\qquad\square$

Some simple conditions ensure the validity of (13.30.a).

Example 13.31 Let $f \in \mathscr{C}^\infty([a, b])$. Let us illustrate some case in which Condition (13.30.a) is fulfilled.

- The derivatives of f are definitively equibounded, i.e., there exists $M \in \mathbb{N}$ and $C \geq 0$ such that $\|f^{(m)}\|_{L^\infty([a,b])} \leq C$ for $m \geq M$. Indeed in this case (13.30.a) is satisfied with $L = 1$.
- $f(x) = e^{Lx}$ with $0 \leq L < 2\pi$: indeed in this case, for every $m \in \mathbb{N}$, we have

$$\|f^{(m)}\|_{L^\infty([a,b])} = L^m \|e^{Lx}\|_{L^\infty([a,b])}.$$

- f is a polynomial: we saw in Corollary 13.21 that the Euler–Maclaurin series is actually a finite sum.

- $f(x) = e^{L_0 x} P(x)$, where $0 \le L_0 < 2\pi - 1$ and $P(X)$ is a polynomial. Indeed, for any natural m the Leibniz rule for the derivatives of a product (see Theorem 3.26) yields

$$f^{(m)}(x) = \sum_{k=0}^{m} \binom{m}{k} (e^{L_0 x})^{(k)} P(x)^{(m-k)}.$$

Now, for every $k = 0, \ldots, m$ we have

$$\left| (e^{L_0 x})^{(k)} \right| = L_0^k |e^{L_0 x}| \le L_0^k \|e^{L_0 x}\|_{L^\infty([a,b])},$$

whereas, if N is the degree of $P(X)$,

$$\left| P(x)^{(m-k)} \right| \le K := \max \left\{ \|P\|_{L^\infty([a,b])}, \|P'\|_{L^\infty([a,b])}, \ldots, \|P^{(N)}\|_{L^\infty([a,b])} \right\},$$

from which

$$\|f^{(m)}\|_{L^\infty([a,b])} \le K \|e^{L_0 x}\|_{L^\infty([a,b])} \sum_{k=0}^{m} \binom{m}{k} L_0^k = K \|e^{L_0 x}\|_{L^\infty([a,b])} (1 + L_0)^m.$$

$\square$

13.3   Approximation of Integrals

In this section we will present the Euler–Maclaurin approximation formula for integrals and the trapezoidal method for functions of class $\mathscr{C}^m$.

13.3.1   The Euler–Maclaurin Approximation Formula of an Integral

We consider here the Euler–Maclaurin formula from Maclaurin's point of view, namely the approximation of integrals. We saw in Sect. 12.6 how the Euler–Maclaurin formulas of order 1 and 2 allow one to prove the rectangle and the trapezoidal method. We follow here a similar approach based on the general Euler–Maclaurin formula; we will see that the more the function is regular, the more precise is our approximation of its integral.

Given a function $g \in \mathscr{C}([\alpha, \beta])$, with $\alpha < \beta \in \mathbb{R}$ (and not necessarily integers!), set for each $n \in \mathbb{N}_{\ge 1}$

$$x_k = \alpha + h_n k \quad k = 0, \ldots, n-1, \quad h_n := \frac{\beta - \alpha}{n}.$$

In Definition 12.88 we have introduced the sum T_n^g of the areas of the trapezes with vertices

$$(x_k, 0), (x_k, g(x_k)), (x_{k+1}, g(x_{k+1})), (x_{k+1}, 0) \qquad k = 0, \ldots, n-1$$

given by

$$
\begin{aligned}
T_n^g :&= h_n \frac{g(x_0) + g(x_1)}{2} + h_n \frac{g(x_1) + g(x_2)}{2} + \cdots + h_n \frac{g(x_{n-1}) + g(x_n)}{2} \\
&= \frac{h_n}{2} \left(g(x_0) + 2g(x_1) + \cdots + 2g(x_{n-1}) + g(x_n) \right) \\
&= h_n \left(\sum_{0 \le k < n} g(x_k) + \frac{1}{2} [g]_\alpha^\beta \right).
\end{aligned}
$$

The following result follows readily from the order m Euler–Maclaurin formula (13.15.a).

Theorem 13.32 (*m*-th order Euler–Maclaurin approximation of integrals) *Let* $g \in \mathscr{C}^m([\alpha, \beta])$, *with* $\alpha < \beta \in \mathbb{R}$ *and* $m \ge 2$. *For every* $n \ge 1$ *we have*

$$
\int_\alpha^\beta g(x)\,dx = T_n^g - \sum_{i=2}^m \frac{\mathrm{B}_i}{i!} h_n^i \left[g^{(i-1)} \right]_\alpha^\beta + \varepsilon_m(n),
$$

$$
|\varepsilon_m(n)| \le \frac{\mu_m}{m!} \frac{(\beta - \alpha)^m}{n^m} \int_\alpha^\beta |g^{(m)}(x)|\,dx \le \frac{\mu_m}{m!} \frac{(\beta - \alpha)^{m+1}}{n^m} \|g^{(m)}\|_{L^\infty([\alpha, \beta])},
$$

$$(13.32.a)$$

where $\mu_m = \| \mathrm{B}_m(x) \|_{L^\infty([0,1])}$.

Remark 13.33 The remainder in (13.32.a) is $O\left(\dfrac{1}{n^m} \right)$ as $n \to +\infty$: for n sufficiently big it is thus more convenient to raise the values of m in order to get better approximations.

Proof (of Theorem 13.32). Set $f(t) = g(\alpha + th_n)$, $t \in [0, n]$. The change of variables $x = \alpha + th_n$ yields

$$
\int_\alpha^\beta g(x)\,dx = h_n \int_0^n f(t)\,dt.
$$

The Euler–Maclaurin formula (13.15.a) applied to f on $[0, n]$ yields

$$
\int_0^n f(t)\,dt = \sum_{0 \le k < n} f(k) + \frac{1}{2} [f]_0^n - \sum_{i=2}^m \frac{\mathrm{B}_i}{i!} \left[f^{(i-1)} \right]_0^n - R_m(n), \qquad (13.33.a)
$$

with $|R_m(n)| \leq \dfrac{\mu_m}{m!} \displaystyle\int_0^n |f^{(m)}(t)|\, dt$. Now

$$\left[f^{(i-1)} \right]_0^n = h_n^{i-1} \left[g^{(i-1)} \right]_\alpha^\beta \quad i = 2, \ldots, m$$

and

$$\int_0^n |f^{(m)}(t)|\, dt = h_n^{m-1} \int_\alpha^\beta |g^{(m)}(x)|\, dx.$$

By translating (13.33.a) in terms of g we get

$$\frac{1}{h_n} \int_\alpha^\beta g(x)\, dx = \left(\sum_{0 \leq k < n} g(x_k) + \frac{1}{2} [g]_\alpha^\beta \right) - \sum_{i=2}^m \frac{B_i}{i!} h_n^{i-1} \left[g^{(i-1)} \right]_\alpha^\beta - R_m(n),$$

$$\tag{13.33.b}$$

with $|R_m(n)| \leq \dfrac{\mu_m}{m!} h_n^{m-1} \displaystyle\int_\alpha^\beta |g^{(m)}(x)|\, dx$. If we multiply both members of (13.33.b) by h_n, we get (13.32.a) with $\varepsilon_m(n) = -h_n R_m(n)$. Clearly

$$|\varepsilon_m(n)| = h_n |R_m(n)| \leq \frac{\mu_m}{m!} h_n^m \int_\alpha^\beta |g^{(m)}(x)|\, dx = \frac{\mu_m}{m!} \frac{(\beta - \alpha)^m}{n^m} \int_\alpha^\beta |g^{(m)}(x)|\, dx;$$

the last inequality in (13.32.a) follows directly.   □

Remark 13.34 (Integral approximation of order 8) For the convenience of the reader it may be useful to write explicitly the term in (13.32.a) involving Bernoulli polynomials for $m = 8$:

$$\sum_{i=2}^8 \frac{B_i}{i!} h_n^i \left[g^{(i-1)} \right]_\alpha^\beta = \left[-\frac{h_n^2}{12} g' + \frac{h_n^4}{720} g^{(3)} - \frac{h_n^6}{30\,240} g^{(5)} + \frac{h_n^8}{1\,209\,600} g^{(7)} \right]_\alpha^\beta.$$

Example 13.35 In Example 12.93 we dealt with the problem of approximating the integral $\displaystyle\int_0^1 g(x)\, dx$, where $g(x) = e^{x^2}$. By means of (13.32.a) with $m = 2$ and $n = 8$ we got $1.46263\ldots$, whereas the true value of the integral is

$$I := \int_0^1 e^{x^2}\, dx = 1.4626517459071816088\ldots.$$

If we still consider $n = 8$ in formula (13.32.a), with $m = 4$, we get the value

$$I \approx T_8^g + \left[-\frac{1}{12 \times 8^2} g' + \frac{1}{720 \times 8^4} g^{(3)} \right]_0^1 = 1.4626518\ldots,$$

which is thus an improvement of two more exact decimals, with a total of 6 exact decimals. We now wish to obtain the value I, with an error below 10^{-30}, by means of (13.32.a) with $m = 8$. We have

$$g^{(8)}(x) = 13\,440e^{x^2}x^2 + 680e^{x^2} + 256e^{x^2}x^8 + 3\,584e^{x^2}x^6 + 13\,440e^{x^2}x^4,$$

so that $\|g^{(8)}\|_{L^\infty([0,1])} = e(13\,440 + 680 + 256 + 3\,584 + 13\,440) = 32\,400e$. If in (13.32.a) we wish that $|\varepsilon_8(n)| \leq 10^{-30}$, it is sufficient that

$$\frac{\mu_8}{8!}\frac{1}{n^8}32\,400e = \frac{|B_8|}{8!}\frac{1}{n^8}32\,400e = \frac{32\,400e}{1\,209\,600n^8} \leq 10^{-30},$$

or, equivalently,

$$n \geq \left(\frac{32\,400e \times 10^{30}}{1\,209\,600}\right)^{1/8} = 1000\sqrt[4]{2} \times 5^{3/4}\sqrt[8]{\frac{3e}{7}} \approx 4\,052.98.$$

By choosing $n = 4053$ we get the approximated value of I

$$I \approx T_n^8 + \left[-\frac{1}{12n^2}g' + \frac{1}{720n^4}g^{(3)} - \frac{1}{30\,240n^6}g^{(5)} + \frac{1}{1\,209\,600n^8}g^{(7)}\right]_0^1 =$$
$$= 1.46265174590718160880404858685698155128887\ldots,$$

$$(13.35.a)$$

whereas $I = 1.46265174590718160880404858685698155120870\ldots$: 38 decimals are now correct. We actually obtained an approximation up to 10^{-38}. □

13.3.2 The Trapezoidal Method of Order m

Similarly to what we established in Theorem 12.101, we now want to show that it is possible to drop the term $h_n^m\left[g^{(m-1)}\right]_\alpha^\beta$ in (13.32.a), with no essential loss of sup-norm precision in our estimates.

Example 13.36 (A naive estimate of R_{m-1} in terms of $f^{(m)}$) Let $m \geq 2$ be and $f \in \mathscr{C}^m([a,b])$. The Euler–Maclaurin formulas (13.15.a), applied at orders $m-1$ and m, yields

$$\sum_{a \leq k < b} f(k) = \int_a^b f(x)\,dx + \sum_{i=1}^{m-1}\frac{B_i}{i!}\left[f^{(i-1)}\right]_a^b + R_{m-1}$$

$$= \int_a^b f(x)\,dx + \sum_{i=1}^{m}\frac{B_i}{i!}\left[f^{(i-1)}\right]_a^b + R_m,$$

whence the equality

$$R_{m-1} = R_m + \frac{B_m}{m!} \left[f^{(m-1)} \right]_a^b = R_m + \frac{B_m}{m!} \int_a^b f^{(m)}(x) \, dx. \qquad (13.36.a)$$

The estimate of R_m in (13.15.b) yields

$$|R_m| \leq \frac{\mu_m}{m!} (b-a) \| f^{(m)} \|_{L^\infty([a,b])},$$

where $\mu_m = \| B_m(x) \|_{L^\infty([0,1])}$. It follows from (13.36.a) that

$$|R_{m-1}| \leq \frac{\mu_m + |B_m|}{m!} (b-a) \| f^{(m)} \|_{L^\infty([a,b])}. \qquad (13.36.b)$$

Since $|B_m| \leq \mu_m$ we get

$$|R_{m-1}| \leq \frac{2\mu_m}{m!} (b-a) \| f^{(m)} \|_{L^\infty([a,b])}. \qquad (13.36.c)$$

$\square$

We show now that, actually, we can halve the upper estimate in (13.36.c). The following smarter estimate of the remainder in the Euler–Maclaurin formula of order $m-1$ holds if one assumes that f is of class $\mathscr{C}^m$.

Proposition 13.37 (The reduced Euler–Maclaurin formula) *Let $f \in \mathscr{C}^m([a,b])$, with $m \geq 2$, and $a, b \in \mathbb{N}$. Then*

$$\sum_{a \leq k < b} f(k) = EM_{m-1}(f; [a,b]) + R_{m-1}, \quad |R_{m-1}| \leq \frac{\mu_m}{m!} (b-a) \| f^{(m)} \|_{L^\infty([a,b])},$$

$$(13.37.a)$$

where $\mu_m = \| B_m(x) \|_{L^\infty([0,1])}$.

Proof. Let R_m be the remainder in the Euler–Maclaurin formula (13.15.a). As in (13.36.a) we get

$$\begin{aligned}
R_{m-1} &= R_m + \frac{B_m}{m!} \left[f^{(m-1)} \right]_a^b \\
&= \frac{(-1)^{m+1}}{m!} \int_a^b B_m(x - [x]) f^{(m)}(x) \, dx + \frac{B_m}{m!} \int_a^b f^{(m)}(x) \, dx \quad (13.37.b) \\
&= \frac{1}{m!} \int_a^b \left(B_m + (-1)^{m+1} B_m(x - [x]) \right) f^{(m)}(x) \, dx.
\end{aligned}$$

If $m \geq 3$ is odd, then $B_m = 0$ and hence by (13.15.b)

$$|R_{m-1}| = |R_m| \leq \frac{\mu_m}{m!}(b-a)\|f^{(m)}\|_{L^\infty([a,b])}.$$

Assume now that $m = 2\ell$ is even; then from (13.9.b) it follows that the term $B_{2\ell} + (-1)^{2\ell+1} B_{2\ell}(x - [x]) = B_{2\ell} - B_{2\ell}(x - [x])$ has a constant sign. We are thus allowed to use the 2nd Mean Value Theorem 12.97: there exists $\xi \in [a, b]$ satisfying

$$\int_a^b (B_{2\ell} - B_{2\ell}(x - [x]))\, f^{(2\ell)}(x)\, dx = f^{(2\ell)}(\xi) \int_a^b (B_{2\ell} - B_{2\ell}(x - [x]))\, dx.$$

$$(13.37.c)$$

Now, $B'_{2\ell+1}(x) = (2\ell + 1)\, B_{2\ell}(x)$, so that

$$\int_a^b B_{2\ell}(x - [x])\, dx = (b - a) \int_0^1 B_{2\ell}(x)\, dx = \frac{b-a}{2\ell + 1}(B_{2\ell+1}(1) - B_{2\ell+1}(0)) = 0,$$

due to the fact that $B_{2\ell+1}(0) = B_{2\ell+1}(1) = B_{2\ell+1} = 0$. Therefore, thanks to (13.37.c), (13.37.b) yields

$$
\begin{aligned}
|R_{2\ell-1}| &= \frac{|f^{(2\ell)}(\xi)|}{(2\ell)!}\left|\int_a^b B_{2\ell}\, dx\right| = |f^{(2\ell)}(\xi)|(b-a)\frac{|B_{2\ell}|}{(2\ell)!}\\
&\leq \frac{|B_{2\ell}|}{(2\ell)!}(b-a)\|f^{(2\ell)}\|_\infty = \frac{\mu_{2\ell}}{(2\ell)!}(b-a)\|f^{(2\ell)}\|_\infty. \qquad \square
\end{aligned}
$$

$$(13.37.d)$$

Remark 13.38 If $m > 2$ is even then $B_{m-1} = 0$ and $\mu_m = |B_m|$ (see (13.9.b)): therefore $EM_{m-1}(f; [a, b]) = EM_{m-2}(f; [a, b])$ and (13.37.a) becomes:

$$\sum_{a \leq k < b} f(k) = EM_{m-2}(f; [a, b]) + R_{m-1}, \quad |R_{m-1}| \leq \frac{|B_m|}{m!}(b-a)\|f^{(m)}\|_{L^\infty([a,b])}.$$

We can now state, and easily prove, the generalization of the trapezoidal method.

Theorem 13.39 (Trapezoidal method of order m) *Let $g \in \mathscr{C}^m([\alpha, \beta])$, with $\alpha < \beta \in \mathbb{R}$ and $m \geq 2$. For every $n \geq 1$, set $h_n = \dfrac{\beta - \alpha}{n}$, one has:*

$$\int_\alpha^\beta g(x)\, dx = T_n^g - \sum_{i=2}^{m-1} \frac{B_i}{i!} h_n^i \left[g^{(i-1)} \right]_\alpha^\beta + \varepsilon_{m-1}(n),$$

$$|\varepsilon_{m-1}(n)| \le \frac{\mu_m}{m!} \frac{(\beta - \alpha)^{m+1}}{n^m} \| g^{(m)} \|_{L^\infty([\alpha,\beta])}, \qquad (13.39.\mathrm{a})$$

where $\mu_m = \| B_m(x) \|_{L^\infty([0,1])}$.

Remark 13.40 If $m = 2\ell > 4$ is even, then $B_{2\ell-1} = 0$ and (13.39.a) becomes

$$\int_\alpha^\beta g(x)\, dx = T_n^g - \sum_{i=2}^{2\ell-2} \frac{B_i}{i!} h_n^i \left[g^{(i-1)} \right]_\alpha^\beta + \varepsilon_{2\ell-1}(n),$$

$$|\varepsilon_{2\ell-1}(n)| \le \frac{|B_{2\ell}|}{(2\ell)!} \frac{(\beta - \alpha)^{2\ell+1}}{n^{2\ell}} \| g^{(2\ell)} \|_{L^\infty([\alpha,\beta])}.$$

Remark 13.41 If one does not care for the estimates of the remainder terms, the formula (13.39.a) of order m is *exactly* the approximation formula (13.32.a) of order $m - 1$. The difference between the two formulas relies on the estimate of the remainder $\varepsilon_{m-1}(n)$. The advantage of (13.39.a) of order m with respect to (13.32.a) of order $m - 1$ is due to the fact that the further regularity of f ensures in (13.39.a) that $\varepsilon_{m-1}(n) = O\left(\dfrac{1}{n^m} \right)$ as $n \to +\infty$, whereas in (13.32.a) we are just supposed to know that $\varepsilon_{m-1}(n) = O\left(\dfrac{1}{n^{m-1}} \right)$ as $n \to +\infty$. Nevertheless, the formula (13.39.a) of order m is less precise than (13.32.a) of the *same* order m. Indeed the upper bound of the estimate of $\varepsilon_{m-1}(n)$ in (13.39.a) is bigger than that of $\varepsilon_m(n)$ in (13.32.a):

$$\frac{\mu_m}{m!} \frac{(\beta - \alpha)^m}{n^m} \int_\alpha^\beta |g^{(m)}(x)|\, dx \le \frac{\mu_m}{m!} \frac{(\beta - \alpha)^{m+1}}{n^m} \| g^{(m)} \|_{L^\infty([\alpha,\beta])},$$

and the inequality is usually strict, as we noticed in Remark 12.23. Thus, for a given $m \ge 1$, we may approximate the integral $\displaystyle\int_\alpha^\beta g(x)\, dx$ by means of the following formulas, in an *increasing* order of precision:

1. The approximation formula (13.32.a) of order $m - 1$;
2. The trapezoidal method formula (13.39.a) of order m;
3. The approximation formula (13.32.a) of order m.

We prove Theorem 13.39 following the lines of the proof of Theorem 13.32 by replacing the Euler–Maclaurin formula (13.15.a) with the reduced formula established in Proposition 13.37.

Proof (of Theorem 13.39). Let $f(t) = g(\alpha + th_n)$, $t \in [0, n]$. The change of variable $x = \alpha + th_n$ yields

$$\int_\alpha^\beta g(x)\, dx = h_n \int_0^n f(t)\, dt.$$

Now (13.37.a) applied to f on $[0, n]$ gives

$$\int_0^n f(t)\, dt = \sum_{0 \le k < n} f(k) - \sum_{i=1}^{m-1} \frac{B_i}{i!} \big[f^{(i-1)}\big]_0^n - R_{m-1}, \qquad (13.41.a)$$

with

$$|R_{m-1}| \le \frac{\mu_m}{m!} n \|f^{(m)}\|_{L^\infty([a,b])}.$$

We now write (13.41.a) in terms of g. For $i = 1, \ldots, m-1$ we have

$$\big[f^{(i-1)}\big]_0^n = h_n^{i-1} \big[g^{(i-1)}\big]_\alpha^\beta.$$

Moreover

$$\|f^{(m)}\|_{L^\infty([a,b])} = h_n^m \|g^{(m)}\|_{L^\infty([\alpha,\beta])}.$$

Equation (13.41.a) thus becomes

$$\frac{1}{h_n} \int_\alpha^\beta g(x)\, dx = \left(\sum_{0 \le k < n} g(x_k) + \frac{1}{2} [g]_\alpha^\beta \right) - \sum_{i=2}^{m-1} \frac{B_i}{i!} h_n^{i-1} \big[g^{(i-1)}\big]_\alpha^\beta - R_{m-1},$$
$$(13.41.b)$$

with

$$|R_{m-1}| \le \frac{\mu_m}{m!} n \|f^{(m)}\|_{L^\infty([a,b])} = \frac{\mu_m}{m!} n h_n^m \|g^{(m)}\|_{L^\infty([\alpha,\beta])}. \qquad (13.41.c)$$

By multiplying by h_n both members of (13.41.b) we get

$$\int_\alpha^\beta g(x)\, dx = T_n^g - \sum_{i=2}^{m-1} \frac{B_i}{i!} h_n^i \big[g^{(i-1)}\big]_\alpha^\beta - h_n R_{m-1}.$$

Set $\varepsilon_{m-1}(n) = -h_n R_{m-1}$, from (13.41.c) the following estimate holds:

$$|\varepsilon_{m-1}(n)| = h_n |R_{m-1}| \le \frac{\mu_m}{m!} n h_n^{m+1} \|g^{(m)}\|_{L^\infty([\alpha,\beta])} = \frac{\mu_m}{m!} \frac{(\beta - \alpha)^{m+1}}{n^m} \|g^{(m)}\|_{L^\infty([\alpha,\beta])}. \quad \square$$

The special case of Theorem 13.39 with $m = 4$ is known as the **Hermite rule** it follows directly from (13.39.a) by taking into account that $B_2 = \dfrac{1}{6}$, $B_3 = 0$ and

$$\mu_4 = |B_4| = \frac{1}{30}.$$

Corollary 13.42 (Hermite's rule) *Let* $g \in \mathscr{C}^4([\alpha, \beta])$ *with* $\alpha < \beta \in \mathbb{R}$. *For every* $n \geq 1$, *set* $h_n = \dfrac{\beta - \alpha}{n}$, *the following formula holds:*

$$\int_\alpha^\beta g(x)\,dx = T_n^g - \frac{1}{12}h_n^2\left[g'\right]_\alpha^\beta + \varepsilon_3(n), \quad |\varepsilon_3(n)| \leq \frac{(\beta - \alpha)^5}{720\,n^4}\|g^{(4)}\|_{L^\infty([\alpha,\beta])}.$$
$$(13.42.a)$$

Remark 13.43 It is worth noticing that the estimate in Hermite's formula (13.42.a) is one fourth of the upper estimate of the remainder of the most popular Simpson's rule [4].

Example 13.44 We use Hermite's rule (13.42.a) in order to approximate the value of

$$I := \int_1^{10} x \sin\left(\frac{1}{x}\right) dx,$$

with an error at most equal to 10^{-10}. Set $g(x) = x \sin\left(\dfrac{1}{x}\right)$, $1 \leq x \leq 10$. Since

$$g^{(4)}(x) = \frac{1}{x^7}\sin\left(\frac{1}{x}\right) - \frac{8}{x^6}\cos\left(\frac{1}{x}\right) - \frac{12}{x^5}\sin\left(\frac{1}{x}\right),$$

then $\|g^{(4)}\|_{L^\infty([1,10])} \leq 1 + 8 + 12 = 21$. In order to be sure that $|\varepsilon_3(n)| \leq 10^{-10}$ in (13.42.a) it is enough that

$$21 \times \frac{9^5}{720\,n^4} \leq 10^{-10}, \qquad n \geq \sqrt[4]{\frac{21 \times 9^5 \times 10^{10}}{720}} \approx 2037.16.$$

By choosing $n = 2038$, we obtain the following approximate value:

$$I \approx T_{2038}^g - \frac{9^2}{12 \times 2038^2}\left[g'\right]_1^{10} = 8.\mathbf{8527357087656}\ldots,$$

whereas $I = 8.\mathbf{8527357087640}\ldots$ □

13.4 The Asymptotic Formulas of Euler–Maclaurin of Order m

In Sect. 12.5 we studied the first-order and second-order tests for the convergence of the series $\sum_{k=0}^{+\infty} f(k)$. Here we give the corresponding test of order m.

13.4.1 The Euler–Maclaurin Asymptotic Formulas of Order m and the Euler Constant of a Function

Let us recall (see Definition 12.58) that, for a given locally integrable function $f :$ $[a, +\infty[\to \mathbb{R}$ and $n \in \mathbb{N}_{\geq a}$, we set

$$\gamma_n^f := \sum_{a \leq k < n} f(k) - \int_a^n f(x)\,dx, \quad \forall n \in \mathbb{N}.$$

Whenever the sequence $(\gamma_n^f)_n$ converges to a real number, its limit, denoted by γ^f, is called the **Euler constant** of f.

The asymptotic formula of order $m \geq 1$ for a function f holds whenever f is of class at least $\mathscr{C}^m$ and if either its derivative $f^{(m)}$ is summable or its derivative $f^{(m-1)}$ is definitively monotonic. The Euler constant of f exists under the more severe conditions that $f^{(m)}$ is summable and the limits $f^{(i)}(\infty)$, $i = 0, \ldots, m - 2$ are finite.

Theorem 13.45 (Criterion of order m for the existence of the Euler constant and Euler–Maclaurin asymptotic formula) *Let $f : [a, +\infty[\to \mathbb{R}$ be of class $\mathscr{C}^m$ for some $m \geq 1$ and $a \in \mathbb{N}$.*

1. If $f^{(m)} \in L^1([a, +\infty[)$ there exists a constant C_m^f satisfying

$$\sum_{a \leq k < n} f(k) = \int_a^n f(x)\,dx + C_m^f + \sum_{i=1}^{m-1} \frac{B_i}{i!} f^{(i-1)}(n) +$$
$$+ O\left(\int_n^{+\infty} |f^{(m)}(x)|\,dx\right) \quad n \to +\infty. \tag{13.45.a}$$

One has $f^{(m-1)}(\infty) \in \mathbb{R}$; assuming further that, for $i = 1, \ldots, m$, the limit $f^{i-1}(\infty)$ belongs to $\mathbb{R}$, the Euler constant γ^f of f is defined and it is given by

$$\gamma^f = C_m^f + \sum_{i=1}^{m-1} \frac{B_i}{i!} f^{(i-1)}(\infty).$$

*Moreover for $n \in \mathbb{N}_{\geq a}$ we have the following **estimate of order** m of γ^f*

$$\gamma^f = \gamma_n^f + \sum_{i=1}^{m} \frac{B_i}{i!} \left[f^{(i-1)} \right]_n^{\infty} + \varepsilon_m(n), \quad |\varepsilon_m(n)| \leq \frac{\mu_m}{m!} \int_n^{+\infty} |f^{(m)}(x)| \, dx.$$

$$(13.45.b)$$

*1'. If $f^{(m-1)}$ is **bounded** and **definitively monotonic** then $f^{(m)} \in L^1([a, +\infty[)$, and (13.45.a) may be rewritten as*

$$\sum_{a \leq k < n} f(k) = \int_a^n f(x) \, dx + C_m^f + \sum_{i=1}^{m-1} \frac{B_i}{i!} f^{(i-1)}(n) + O\left(f^{(m-1)}(\infty) - f^{(m-1)}(n) \right)$$

$$(13.45.c)$$

as $n \to +\infty$.

Moreover in the estimate (13.45.b) of γ^f we have

$$|\varepsilon_m(n)| \leq \frac{\mu_m}{m!} |f^{(m-1)}(n) - f^{(m-1)}(\infty)|.$$

*2. If $f^{(m-1)}$ is **unbounded** and **definitively monotonic** then*

$$\sum_{a \leq k < n} f(k) = \int_a^n f(x) \, dx + \sum_{i=1}^{m-1} \frac{B_i}{i!} f^{(i-1)}(n) + O\left(f^{(m-1)}(n) \right)$$

$$(13.45.d)$$

as $n \to +\infty$.

Remark 13.46 1. Notice that in (13.45.a) the remainder tends to 0 as $n \to +\infty$:

$$\lim_{n \to +\infty} \int_n^{+\infty} |f^{(m)}(x)| \, dx = \int_a^{+\infty} |f^{(m)}(x)| \, dx - \lim_{n \to +\infty} \int_a^n |f^{(m)}(x)| \, dx = 0.$$

2. At Point 2 of Theorem 13.45 the remainder is a "big O" of the last term $f^{(m-1)}(n)$ of the Euler–Maclaurin expansion.

Proof (of Theorem 13.45). It follows from the Euler–Maclaurin formula (Theorem 13.15) that

$$\sum_{a \leq k < n} f(k) = \int_a^n f(x)\,dx + \sum_{i=1}^m \frac{B_i}{i!} \left[f^{(i-1)} \right]_a^n +$$

$$+ \frac{(-1)^{m+1}}{m!} \int_a^n B_m(x - [x]) f^{(m)}(x)\,dx. \tag{13.46.a}$$

1. Assume that $f^{(m)} \in L^1([a, +\infty[)$: since $B_m(x - [x])$ is bounded on $[a, +\infty[$, then also $B_m(x - [x]) f^{(m)}(x) \in L^1([a, +\infty[)$. From

$$\int_a^n B_m(x - [x]) f^{(m)}(x)\,dx =$$

$$= \int_a^{+\infty} B_m(x - [x]) f^{(m)}(x)\,dx - \int_n^{+\infty} B_m(x - [x]) f^{(m)}(x)\,dx,$$

we get

$$\sum_{a \leq k < n} f(k) = \int_a^n f(x)\,dx + D_m^f + \sum_{i=1}^m \frac{B_i}{i!} f^{(i-1)}(n) + \varepsilon_m(n), \tag{13.46.b}$$

where we set

$$D_m^f = \frac{(-1)^{m+1}}{m!} \int_a^{+\infty} B_m(x - [x]) f^{(m)}(x)\,dx - \sum_{i=1}^m \frac{B_i}{i!} f^{(i-1)}(a),$$

$$\varepsilon_m(n) := \frac{(-1)^m}{m!} \int_n^{+\infty} B_m(x - [x]) f^{(m)}(x)\,dx. \tag{13.46.c}$$

Now

$$|\varepsilon_m(n)| \leq \frac{\mu_m}{m!} \int_n^{+\infty} |f^{(m)}(x)|\,dx = O\left(\int_n^{+\infty} |f^{(m)}(x)|\,dx \right).$$

Moreover, it follows from Lemma 12.57 that $f^{(m-1)}(\infty)$ is finite and that

$$f^{(m-1)}(n) = f^{(m-1)}(\infty) - \int_n^{+\infty} f^{(m)}(x)\,dx = f^{(m-1)}(\infty) + O\left(\int_n^{+\infty} |f^{(m)}(x)|\,dx \right)$$

for $n \to +\infty$. Then, equation (13.46.b) with $C_m^f = D_m^f + \dfrac{B_m}{m!} f^{(m-1)}(\infty)$ yields (13.45.a).

If, moreover, $f^{(i-1)}(\infty) \in \mathbb{R}$ for $i = 1, \ldots, m - 1$ then passing to the limit as $n \to +\infty$ in (13.45.a), we get $\gamma^f = C_m^f + \sum_{i=1}^{m-1} \frac{B_i}{i!} f^{(i-1)}(\infty)$.

From (13.46.a), for every $n \in \mathbb{N}_{\geq a}$ we have

$$\gamma_n^f = \sum_{i=1}^{m} \frac{B_i}{i!} \left[f^{(i-1)} \right]_a^n + \frac{(-1)^{m+1}}{m!} \int_a^n B_m(x - [x]) f^{(m)}(x)\, dx.$$

By passing to the limit as $n \to +\infty$ in the above equality we get

$$\gamma^f := \lim_{n \to +\infty} \gamma_n^f = \sum_{i=1}^{m} \frac{B_i}{i!} \left[f^{(i-1)} \right]_a^\infty + \frac{(-1)^{m+1}}{m!} \int_a^{+\infty} B_m(x - [x]) f^{(m)}(x)\, dx.$$

We thus obtain that

$$\gamma^f - \gamma_n^f = \sum_{i=1}^{m} \frac{B_i}{i!} \left[f^{(i-1)} \right]_n^\infty + \varepsilon_m(n),$$

with

$$\varepsilon_m(n) = \frac{(-1)^{m+1}}{m!} \int_n^\infty B_m(x - [x]) f^{(m)}(x)\, dx.$$

The estimate of $\varepsilon_m(n)$ follows directly from the fact that $\mu_m = \| B_m(x) \|_{L^\infty([0,1])}$.

1'. For sure $f^{(m)} \in L^1([a, +\infty[)$: indeed, if $f^{(m-1)}$ is monotonic on $[b, +\infty[$ for some $b \geq a$, then

$$\lim_{n \to +\infty} \int_a^n \left| f^{(m)}(x) \right| dx = \int_a^b \left| f^{(m)}(x) \right| dx + \left| f^{(m-1)}(\infty) - f^{(m-1)}(b) \right|$$

is finite. The assumptions of Point 1 are thus satisfied; moreover in (13.45.a) we have that for each $n \geq b$

$$\int_n^{+\infty} |f^{(m)}(x)|\, dx = |f^{(m-1)}(\infty) - f^{(m-1)}(n)|,$$

and therefore (13.45.c) follows.

2. Assume that $f^{(m-1)}$ is monotonic on $[b, +\infty[$ for a suitable $b \geq a$. For $n \geq b$ we have

$$\int_b^n \left| f^{(m)}(x) \right| dx = \left| \int_b^n f^{(m)}(x)\, dx \right| = |f^{(m-1)}(n) - f^{(m-1)}(b)|. \qquad (13.46.d)$$

We deduce from (13.46.a) that

$$\sum_{a \le k < n} f(k) = \int_a^n f(x)\,dx + D_m^f + \sum_{i=1}^m \frac{B_i}{i!}\left[f^{(i-1)}\right]_a^n + \frac{(-1)^{m+1}}{m!}\int_b^n B_m(x-[x])f^{(m)}(x)\,dx,$$

where

$$D_m^f = \frac{(-1)^{m+1}}{m!}\int_a^b B_m(x-[x])f^{(m)}(x)\,dx.$$

Now, since $f^{(m-1)}$ diverges at infinity, any constant is $O(f^{(m-1)}(n))$ for $n \to +\infty$; in particular

$$D_m^f - \sum_{i=1}^m \frac{B_i}{i!}f^{(i-1)}(a) = O(f^{(m-1)}(n)).$$

Since $f^{(m)}$ has a constant sign on $[b, +\infty[$ we get

$$\left|\frac{(-1)^{m+1}}{m!}\int_b^n B_m(x-[x])f^{(m)}(x)\,dx\right| \le \frac{\mu_m}{m!}|f^{(m-1)}(n) - f^{(m-1)}(b)| = O(f^{(m-1)}(n))$$

as $n \to +\infty$, whence (13.45.d) follows. □

As applications we compute the Euler–Mascheroni constant and the asymptotic expansion of the harmonic numbers H_n.

Example 13.47 (The Euler–Mascheroni constant γ, part III) By means of the second-order approximation formula (12.65.b) we got, in Example 12.68, with $n = 50$ the first 8 decimals of the Euler–Mascheroni constant γ. By applying, again with $n = 50$, the approximation formula (13.45.b) of order $m = 4$ with $f(x) = \dfrac{1}{x}$, we obtain

$$\gamma \approx \gamma_{50} - \frac{1}{2}[f]_{50}^\infty + \frac{1}{12}[f']_{50}^\infty - \frac{1}{720}[f^{(3)}]_{50}^\infty =$$
$$= \frac{434760950922906578789288 0539}{968451407576873970750000000} - \log 50 = 0.57721566490127\ldots,$$
$$(13.47.a)$$

whereas $\gamma = 0.57721566490153\ldots$. □

Example 13.48 (Asymptotic expansion of H_n) In Example 12.68 we showed that

$$H_n = \sum_{1 \le k < n} \frac{1}{k} + \frac{1}{n} = \log n + \gamma + \frac{1}{2n} + O\left(\frac{1}{n^2}\right) \qquad n \to +\infty,$$

where γ is the Euler–Mascheroni constant. For $m = 9$, (13.45.c) applied to $f(x) = \frac{1}{x}$, $x \geq 1$, and $C_m^f = \gamma$ yield

$$\sum_{1 \leq k < n} f(k) = \int_a^n f(x)\,dx + \gamma + \sum_{i=1}^{8} \frac{B_i}{i!} f^{(i-1)}(n) + O\left(f^{(8)}(n)\right), \quad n \to +\infty.$$

Since $f^{(i)}(x) = (-1)^i \dfrac{i!}{x^{i+1}}$ we get

$$\sum_{1 \leq k < n} \frac{1}{k} = \log n + \gamma + \sum_{i=1}^{8} (-1)^{i-1} \frac{B_i}{i} \frac{1}{n^i} + O\left(\frac{1}{n^9}\right)$$

$$= \log n + \gamma - \frac{1}{2n} - \frac{1}{12n^2} + \frac{1}{120n^4} - \frac{1}{252n^6} + \frac{1}{240n^8} + O\left(\frac{1}{n^9}\right), \quad n \to +\infty.$$

By adding $1/n$ to both members of the above inequality we thus obtain the following asymptotic formula for H_n:

$$H_n = \log n + \gamma + \frac{1}{2n} - \frac{1}{12n^2} + \frac{1}{120n^4} - \frac{1}{252n^6} + \frac{1}{240n^8} + O\left(\frac{1}{n^9}\right), \quad n \to +\infty.$$

The very same reasoning shows that, for every $m \geq 2$ we have

$$H_n = \log n + \gamma + \frac{1}{2n} - \sum_{2 \leq i \leq m} \frac{B_i}{i} \frac{1}{n^i} + O\left(\frac{1}{n^{m+1}}\right), \quad n \to +\infty. \qquad \square$$

13.4.2 Series: The Integral Test and the Approximation Formula of Order m

We present now the following version of order m of the integral test for the convergence of a series stated in Corollaries 12.78 and 12.84.

Corollary 13.49 (The integral test of order m) *Let* $f : [a, +\infty[\to \mathbb{R}$ *be a function of class* $\mathscr{C}^m$ ($m \geq 2$, $a \in \mathbb{N}$) *and such that:*

- *The limits $f^{(i)}(\infty)$, $i = 0, \ldots, m-2$, are finite;*
- *$f^{(m)} \in L^1([a, +\infty[)$.*

The following properties hold:

1. *The series $\sum\limits_{k=a}^{\infty} f(k)$ and the generalized integral $\int_a^{+\infty} f(x)\,dx$ have the same behavior (both convergent, divergent or do not exist);*

2. *If the series $\sum\limits_{k=a}^{\infty} f(k)$ converges, then for every n in $\mathbb{N}_{\geq a}$ we have the following* **approximation of order m**

$$\sum_{k=a}^{\infty} f(k) = \sum_{a \leq k < n} f(k) + \int_n^{+\infty} f(x)\,dx + \sum_{i=1}^{m} \frac{B_i}{i!} \left[f^{(i-1)} \right]_n^{\infty} + \varepsilon_m(n),$$

$$|\varepsilon_m(n)| \leq \frac{\mu_m}{m!} \int_n^{+\infty} |f^{(m)}(x)|\,dx, \quad (13.49.\mathrm{a})$$

where $\mu_m = \| B_m(x) \|_{L^{\infty}([0,1])}$.

Remark 13.50 Notice that, if $f^{(m-1)}$ is *monotonic* on $[b, +\infty[$ for a suitable $b \geq a$, then the fact that $f^{(m-1)}(\infty) \in \mathbb{R}$ is by Lemma 12.57 equivalent to the property that $f^{(m)} \in L^1([a, +\infty[)$; moreover in (13.49.a), thanks to Lemma 12.25, we have

$$|\varepsilon_m(n)| \leq \frac{\mu_m}{m!} \left| f^{(m-1)}(n) - f^{(m-1)}(\infty) \right|, \quad n \geq b. \quad (13.50.\mathrm{a})$$

Proof. 1. From Theorem 13.45 the sequence

$$(\gamma_n^f)_n = \left(\sum_{a \leq k < n} f(k) - \int_a^n f(x)\,dx \right)_n$$

converges to $\gamma^f \in \mathbb{R}$. Therefore the sequences $\left(\sum\limits_{a \leq k < n} f(k) \right)_n$ and $\left(\int_a^n f(x)\,dx \right)_n$ are both convergent, divergent or do not admit limit. Now, if $\left(\int_a^n f(x)\,dx \right)_n$ does not admit limit or diverges, then also $\lim\limits_{x \to +\infty} \int_a^x f(t)\,dt$ does not exist or diverges.

If $\left(\int_a^n f(x)\,dx\right)_n$ converges, then $f(\infty) = 0$ and hence also $\lim\limits_{x \to +\infty} \int_a^x f(t)\,dt$ converges (see Problem 12.15).

2. It follows from (13.24.a) that, for all $a \le n \le N$, we have

$$\sum_{a \le k < N} f(k) = \sum_{a \le k < n} f(k) + \int_n^N f(x)\,dx + \sum_{i=1}^m \frac{B_i}{i!} \left[f^{(i-1)} \right]_n^N + \varepsilon_m(n, N),$$

with $\varepsilon_m(n, N) = \dfrac{(-1)^{m+1}}{m!} \displaystyle\int_n^N B_m(x - [x]) f^{(m)}(x)\,dx$. We know from Point 1 that the generalised integral of f exists on $[n, +\infty[$: by passing to the limit as $N \to +\infty$ we get

$$\sum_{k=a}^{+\infty} f(k) = \sum_{a \le k < n} f(k) + \int_n^{+\infty} f(x)\,dx + \sum_{i=1}^m \frac{B_i}{i!} \left[f^{(i-1)} \right]_n^{+\infty} + \varepsilon_m(n),$$

with $\varepsilon_m(n) = \dfrac{(-1)^{m+1}}{m!} \displaystyle\int_n^{+\infty} B_m(x - [x]) f^{(m)}(x)\,dx$. The conclusion follows. □

Example 13.51 (The series $\zeta(3)$ and Apéry's constant) In Example 12.87 we computed the approximate value of Apery's constant $\zeta(3)$ up to an error of 10^{-6}; in that case we used the approximation formula of order 2, namely (12.84.a). Now, imagine you are in the middle of the sea with this book, a pocket calculator and no access to software or internet. For some reason (the Riemann ζ function almost certainly has some useful applications in navigation) you need to calculate Apery's constant up to an error of 10^{-15}. In order to minimize the number of summands you are led to solve the problem by means of some high order Euler–Maclaurin approximation formula: observe that the derivatives of $f(x) = 1/x^3, x \ge 1$ become smaller and smaller. Let us try the formula of order 8: it follows from Remark 13.50 that, for $n \ge 1$, the remainder $\varepsilon_8(n)$ in (13.49.a) satisfies the inequality

$$|\varepsilon_8(n)| \le \frac{|B_8|}{8!} \left| f^{(7)}(n) - f^{(7)}(\infty) \right| = \frac{1}{30 \times 8!} \left| f^{(7)}(n) \right|.$$

Since $f^{(i)}(x) = \dfrac{(-1)^i \, (i+2)!}{2} \dfrac{1}{x^{i+3}}$ then $|\varepsilon_8(n)| \le \dfrac{1}{60 \times 8!} \dfrac{9!}{n^{10}} = \dfrac{3}{20n^{10}}$; therefore $|\varepsilon_8(n)|$ is less than 10^{-15} whenever $n \ge \left(\dfrac{3}{20}\right)^{1/10} 10^{3/2} \approx 26.16$. By taking $n = 27$ we get the following approximation:

$$\zeta(3) \approx \sum_{1 \le k < 27} f(k) + \int_{27}^{+\infty} f(x)\,dx +$$

$$+ \left[-\frac{1}{2}f + \frac{1}{12}f' - \frac{1}{720}f^{(3)} + \frac{1}{30\,240}f^{(5)} - \frac{1}{1\,209\,600}f^{(7)} \right]_{27}^{\infty}$$

$$= \sum_{1 \le k < 27} \frac{1}{k^3} + \frac{1}{2 \times 27^2} +$$

$$+ \left[-\frac{1}{2x^3} - \frac{1}{4x^4} + \frac{1}{12x^6} - \frac{1}{12x^8} + \frac{567}{655x^{10}} \right]_{27}^{\infty}$$

$$= \frac{2561097446355634460847287110189}{2131858131361319942957376000000} + \frac{1}{2 \times 27^2} + \frac{1}{2 \times 27^3} +$$

$$+ \frac{1}{4 \times 27^4} - \frac{1}{12 \times 27^6} + \frac{1}{12 \times 27^8} - \frac{567}{655 \times 27^{10}}$$

$$= 1.\mathbf{20205690315959428}263\ldots.$$

Notice that $\zeta(3) = 1.\mathbf{2020569031595942}854\ldots.$ □

The following example shows how to deal with (13.49.a), if one cannot precisely compute the integral $\int_{n}^{+\infty} f(x)\,dx$.

Example 13.52 We want to compute the sum of the convergent series $\sum_{k=1}^{\infty} \frac{\sin \sqrt{k}}{k^2}$, whose partial sums are depicted in Fig. 13.5, with an error below 10^{-3}.

Set $f(x) = \frac{\sin \sqrt{x}}{x^2}$, $x \ge 1$. From (13.49.a), with $m = 4$, for every $n \ge 1$ we get

Fig. 13.5 The values of the sum $\sum_{k=1}^{n} \frac{\sin \sqrt{k}}{k^2}$ as n varies from 1 to 300

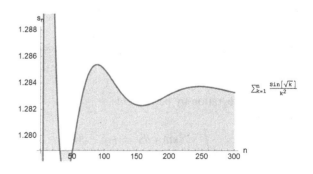

$$\sum_{k=1}^{\infty} f(k) = \sum_{1 \le k < n} f(k) + \int_{n}^{+\infty} f(x)\,dx + \left[-\frac{1}{2}f + \frac{1}{12}f' - \frac{1}{720}f^{(3)} \right]_{n}^{\infty} + \varepsilon_4(n),$$

$$|\varepsilon_4(n)| \le \frac{1}{720} \int_{n}^{+\infty} |f^{(4)}(x)|\,dx.$$

(13.52.a)

Now (here a CAS helped us quite a lot)

$$f^{(4)}(x) = \frac{((x-207)x + 1920)\sin\left(\sqrt{x}\right) + \sqrt{x}(22x - 975)\cos\left(\sqrt{x}\right)}{16x^6}.$$

By means of some very rough estimates it follows that for $x \ge 1$

$$|f^{(4)}(x)| \le \frac{((x+207)x + 1920) + \sqrt{x}(22x + 975)}{16x^6}$$

$$= \frac{22x^{3/2} + x^2 + 207x + 975\sqrt{x} + 1920}{16x^6}$$

$$\le \frac{(22 + 1 + 207 + 975 + 1\,920)x^2}{16x^6} = \frac{3\,125}{16x^4}.$$

We thus get

$$|\varepsilon_4(n)| \le \frac{1}{720} \int_{n}^{+\infty} \frac{3\,125}{16x^4}\,dx = \frac{625}{6912n^3} \le \frac{1}{n^3}.$$

(13.52.b)

We are not able to compute explicitly the integral $\int_{n}^{+\infty} \dfrac{\sin\sqrt{x}}{x^2}\,dx$ that appears in (13.52.a); in view of (13.52.b) we wish to estimate its value with an error that is at most $O\left(\dfrac{1}{n^3}\right)$ for $n \to +\infty$. The change of variables $y = \sqrt{x}$ yields

$$\int_{n}^{+\infty} \frac{\sin\sqrt{x}}{x^2}\,dx = 2\int_{\sqrt{n}}^{+\infty} y^{-3}\sin y\,dy.$$

(13.52.c)

Repeated integration by parts yields

$$\int y^{-3}\sin y\,dy = -y^{-3}\cos y - 3\int y^{-4}\cos y\,dy,$$

$$\int y^{-4}\cos y\,dy = y^{-4}\sin y + 4\int y^{-5}\sin y\,dy,$$

$$\int y^{-5}\sin y\,dy = -y^{-5}\cos y - 5\int y^{-6}\cos y\,dy,$$

$$\int y^{-6}\cos y\,dy = -y^{-6}\sin y + 6\int y^{-7}\sin y\,dy.$$

It follows that

$$\int \frac{\sin y}{y^3}\,dy = -y^{-3}\cos y - 3y^{-4}\sin y + 12y^{-5}\cos y + 60y^{-6}\sin y + 360\int y^{-7}\sin y\,dy.$$

We deduce from (13.52.c) that

$$\int_n^{+\infty}\frac{\sin\sqrt{x}}{x^2}\,dx = 2\frac{\cos\sqrt{n}}{n^{3/2}} + 6\frac{\sin\sqrt{n}}{n^2} - 24\frac{\cos\sqrt{n}}{n^{5/2}} - 120\frac{\sin\sqrt{n}}{n^3} + \sigma(n),$$

with

$$|\sigma(n)| \le 720\int_{\sqrt{n}}^{+\infty} y^{-7}\,dy = \frac{120}{n^3}.$$

Thus (13.52.a) becomes

$$\sum_{k=1}^{\infty} f(k) = \sum_{1\le k<n} f(k) + \left[-\frac{1}{2}f + \frac{1}{12}f' - \frac{1}{720}f^{(3)}\right]_n^{\infty} +$$

$$+ 2\frac{\cos\sqrt{n}}{n^{3/2}} + 6\frac{\sin\sqrt{n}}{n^2} - 24\frac{\cos\sqrt{n}}{n^{5/2}} - 120\frac{\sin\sqrt{n}}{n^3} + \sigma(n) + \varepsilon_4(n),$$

with

$$|\sigma(n) + \varepsilon_4(n)| \le \frac{120}{n^3} + \frac{1}{n^3} = \frac{121}{n^3}.$$

Now $\dfrac{121}{n^3}$ is less or equal than 10^{-3} whenever $n \ge 121^{1/3}\times 10 \approx 49.5$. For $n = 50$ we thus get, with an error certainly within 10^{-3},

$$\sum_{k=1}^{\infty} f(k) \approx \sum_{1\le k<50} f(k) + \left[-\frac{1}{2}f + \frac{1}{12}f' - \frac{1}{720}f^{(3)}\right]_{50}^{\infty} +$$

$$+ 2\frac{\cos\sqrt{50}}{50^{3/2}} + 6\frac{\sin\sqrt{50}}{50^2} - 24\frac{\cos\sqrt{50}}{50^{5/2}} - 120\frac{\sin\sqrt{50}}{50^3}$$

$$= 1.28266\ldots. \qquad\qquad \square$$

13.5   Problems

Problem 13.1 Prove that the Bernoulli polynomial functions $B_m(x)$ with $m \geq 2$ have the following properties:

1. If m is divisible by 4 then there exists a real number a such that $a < 1/2$ and

$$\begin{cases} B_m(x) < 0 \text{ on } [0, a[; \\ B_m(x) > 0 \text{ on }]a, 1 - a[; \\ B_m(x) < 0 \text{ on }]1 - a, 1]. \end{cases}$$

2. If $m - 1$ is divisible by 4 then

$$\begin{cases} B_m(x) < 0 \text{ on }]0, 1/2[; \\ B_m(x) > 0 \text{ on }]1/2, 1[. \end{cases}$$

3. If $m - 2$ is divisible by 4 then there exists $a < 1/2$ such that

$$\begin{cases} B_m(x) > 0 \text{ on } [0, a[; \\ B_m(x) < 0 \text{ on }]a, 1 - a[; \\ B_m(x) > 0 \text{ on }]1 - a, 1]. \end{cases}$$

4. If $m - 3$ is divisible by 4 then

$$\begin{cases} B_m(x) > 0 \text{ on }]0, 1/2[; \\ B_m(x) < 0 \text{ on }]1/2, 1[. \end{cases}$$

Problem 13.2 Find the Euler–Maclaurin expansions of order 2 and 4 of $f(x) = 1/x^3$ on $[1, 100]$, and give in both cases an estimate of their difference with $S :=$
$$\sum_{1 \leq k < 100} \frac{1}{k^3}.$$

Problem 13.3 Let $f : [a, +\infty[\rightarrow \mathbb{R}$ a function of class $\mathscr{C}^m$ for some $m \geq 1$, and $a \in \mathbb{N}$. Suppose moreover that the integral $\displaystyle\int_a^{+\infty} |f^{(m)}(x)| \, dx$ exists and is finite and that the series $\displaystyle\sum_{k=a}^{\infty} f(k)$ converges. Prove that there exists a constant c_m^f, depending only on m and on f, such that

$$\int_a^n f(x) \, dx - \left(c_m^f - \sum_{i=1}^m \frac{B_i}{i!} f^{(i-1)}(n) \right) \rightarrow 0 \qquad \text{for } n \rightarrow +\infty.$$

Chapter 14
Cauchy and Riemann Sums, Factorials, Ramanujan Numbers and Their Approximations

Abstract After a short recall of the basic asymptotic relations "big O" and "small o", we consider the Cauchy sums of the form $\displaystyle\sum_{a \leq k < n} g\left(\frac{k}{n^\alpha}\right)$, where $\alpha > 0$; in the case where $\alpha = 1$ these are strictly related to the celebrated Riemann sums. After having learned how to approximate such sums, we apply the results to the approximation of sums of the form $\displaystyle\sum_{0 \leq k < n} e^{-k^2/n}$ this involves of course the Gauss integral. The second part is dedicated to the Ramanujan Q-distribution $Q(n, k) = \dfrac{n!}{(n-k)!n^k}$. Dealing with uniform estimates on families or sequences, this analysis is by far the most difficult in the book. The Ramanujan Q-distribution appears in many applications: for instance if $n = 365$ then $Q(365, k)$ is the probability that at least two people among k are born the same day. One of the goals of the chapter is to approximate $Q(n, k)$ as n goes to infinity. In addition, we will meet sums of the Ramanujan distribution $Q(n, k)$ as k varies. For the reader who is mainly interested in the applications we recommend looking at the main claims, without delving into the proofs.

14.1 A Revision of the "Big O" and the "Small o" Asymptotic Relations

We recall here the basic properties of the asymptotic relations among functions and sequences: the readers, depending on their background, can either skip it or come back to it later, if necessary.

Definition 14.1 Let $a \in \mathbb{R} \cup \{\pm\infty\}$ and let f, g be two real valued functions defined in a **deleted neighborhood** of a, i.e., a neighborhood of a minus $\{a\}$. (The neighborhoods of $+\infty$ are the subsets of the real line that contain open intervals of the type $]c, +\infty[$, while those of $-\infty$ are the subsets of the real line that contain open intervals of the type $] - \infty, c[$).

C. Mariconda and A. Tonolo, *Discrete Calculus*,
UNITEXT - La Matematica per il 3+2 103, DOI 10.1007/978-3-319-03038-8_14

1. f is "**small** o" of g for x that tends to a if there exists a function σ defined in a neighborhood W of a, with $\lim_{x \to a} \sigma(x) = 0$, such that $f = \sigma g$; in this case one writes $f(x) \in o(g(x))$ or $f(x) = o(g(x))$ for $x \to a$. If $g \neq 0$ in a suitable deleted neighborhood of a, then $f(x) = o(g(x))$ for $x \to a$ if and only if

$$\lim_{x \to a} \frac{f(x)}{g(x)} = 0.$$

2. f and g are **asymptotic** for x that tends to a if $f(x) = g(x) + o(g(x))$ for $x \to a$; in this case one writes $f(x) \sim g(x)$ for $x \to a$. If $g \neq 0$ in a suitable deleted neighborhood of a, one has that $f(x) \sim g(x)$ for $x \to a$ if and only if

$$\lim_{x \to a} \frac{f(x)}{g(x)} = 1.$$

3. f is said to be "**big** O" of g for x that tends to a if there exists a deleted neighborhood U of a and a constant $C > 0$ such that

$$|f(x)| \leq C|g(x)| \quad \forall x \in U;$$

in that case one writes $f(x) \in O(g(x))$ or $f(x) = O(g(x))$ for $x \to a$. $\square$

Remark 14.2 Traditionally it is common to write $f(x) = o(g(x))$ or $f(x) = O(g(x))$. It would, however, be preferable to use the notation with the symbol "$\in$"[1] in place of the equal sign "$=$" given that use of $o(f(x))$ or $O(f(x))$ does not indicate a single function but rather a set of functions, and the notation using the equal sign may well result in an error. Note that if $g(x) = o(f(x))$ and $h(x) = o(f(x))$ for $x \to a$ one has $g(x) - h(x) = o(f(x))$ for $x \to a$ and not, as the equal sign seems to suggest, $g(x) - h(x) = 0$: for example, $x^2 = o(x)$, $x^3 = o(x)$ and $x^2 - x^3 = o(x)$ for $x \to 0$, but $x^2 - x^3 \neq 0$ in every deleted neighborhood of zero. We generally do not adopt this "more reasonable" notation out of respect for a well-consolidated traditional terminology. Nevertheless, following the "$\in$" notation, we prefer to write $g_1(x) + o(f_1(x)) \subseteq g_2(x) + o(f_2(x))$ rather than $g_1(x) + o(f_1(x)) = g_2(x) + o(f_2(x))$ (and analogously for the big O) when the relation is not symmetric, as in

$$x + o(x^3) \subseteq \sin x + o(x^2) \quad x \to 0.$$

Example 14.3 If $\lim_{x \to a} \dfrac{f(x)}{g(x)}$ exists and is finite, then one has

$$\lim_{x \to a} \frac{|f(x)|}{|g(x)|} = c \in \mathbb{R}_{\geq 0};$$

---

[1] This is the choice made in the book [1].

therefore there exists a deleted neighborhood V of a such that

$$\frac{|f(x)|}{|g(x)|} \in]c - 1, c + 1[\text{ for every } x \in V.$$

Thus, $|f(x)| \le (c + 1)|g(x)|$ for each $x \in V$ and so

$$f(x) = O(g(x)) \text{ for } x \to a.$$

☞ In particular, if $f(x) = o(g(x))$ or $f(x) \sim g(x)$ then $f(x) = O(g(x))$ for $x \to a$.

□

Example 14.4 A function is $O(1)$ for $x \to a$ if and only if it is bounded in a deleted neighborhood of a. □

In what follows we will make use of these simple properties, whose proofs follow immediately from the definitions.

Proposition 14.5 *Let f, g, h be real valued functions of a real variable defined on a deleted neighborhood of $a \in \mathbb{R} \cup \{\pm\infty\}$.*

1. *If $f(x) = O(g(x))$ for $x \to a$ and $\lim\limits_{x \to a} g(x) = 0$ then*

$$\lim_{x \to a} f(x) = 0.$$

2. *If $f(x) = o(g(x))$ for $x \to a$ then*

$$f(x)h(x) = o(g(x))h(x) \subseteq o(g(x)h(x)) \qquad x \to a.$$

3. *If $f(x) \sim g(x)$ for $x \to a$ then*

$$f(x)h(x) \sim g(x)h(x) \qquad x \to a.$$

4. *If $f(x) = O(g(x))$ for $x \to a$ then*

$$f(x)h(x) = O(g(x))h(x) \subseteq O(g(x)h(x)) \qquad x \to a.$$

In the asymptotic estimates that we will discuss we will make use of the following simple expansions, which we recall here for completeness.

Proposition 14.6 *Let $n \in \mathbb{N}$.*

- *Expansion of e^x with remainder in the form "small o":*

$$e^x = 1 + x + \frac{x^2}{2!} + \cdots + \frac{x^n}{n!} + o(x^n) \quad x \to 0.$$

- *Expansion of $\log(1 + x)$ with remainder in the form "small o":*

$$\log(1 + x) = x - \frac{x^2}{2} + \cdots + (-1)^{n-1}\frac{x^n}{n} + o(x^n) \quad x \to 0.$$

- *Expansion of e^x with remainder in the form "big O":*

$$e^x = 1 + x + \frac{x^2}{2!} + \cdots + \frac{x^{n-1}}{(n-1)!} + O(x^n) \quad x \to 0.$$

- *Expansion of $\log(1 + x)$ with remainder in the form "big O":*

$$\log(1 + x) = x - \frac{x^2}{2} + \cdots + (-1)^{n-1}\frac{x^{n-1}}{n-1} + O(x^n) \quad x \to 0.$$

We limit ourselves to the observation that the expansions in the big O version are a consequence of the more refined small o versions; indeed, one obviously has $x^n + o(x^n) \subseteq O(x^n)$. The advantage of using the big O version consists in not having to pay attention to the coefficient of degree n which is neglected; such a series expansion is often sufficient for the applications we make in our discussions.

One also has quite analogous notions that apply to sequences, where one makes obvious appropriate minor adaptations of the above properties.

Definition 14.7 Let $(a_n)_n$, $(b_n)_n$ be two sequences.

1. One says that a_n is "small o" of b_n for $n \to +\infty$ if there exists a sequence $(c_n)_n$ converging to 0 such that

$$a_n = c_n b_n \quad \forall n \in \mathbb{N};$$

in that case one writes $a_n = o(b_n)$. If $(b_n)_n$ is definitively non-zero this is equivalent to

$$\lim_{n \to +\infty} \frac{a_n}{b_n} = 0.$$

2. One says that a_n and b_n are asymptotic for $n \to +\infty$ if

$$a_n = b_n + o(b_n) \quad n \to +\infty;$$

in this case one writes $a_n \sim b_n$. If $(b_n)_n$ is definitively non-zero this is equivalent to

$$\lim_{n \to +\infty} \frac{a_n}{b_n} = 1.$$

3. One says that a_n is "big O" of b_n for $n \to +\infty$ if there exist a constant $C > 0$ and $n_0 \in \mathbb{N}$ such that

$$|a_n| \le C|b_n| \quad \forall n \ge n_0;$$

in this case one writes $a_n = O(b_n)$. $\qquad \square$

As we have already seen for functions, the asymptotic relations behave well with regard to products of sequences:

Proposition 14.8 *Let a_n, b_n, c_n be three sequences.*

1. *If $a_n = o(b_n)$ for $n \to +\infty$, then $a_n c_n = o(b_n)c_n = o(b_n c_n)$ for $n \to +\infty$.*
2. *If $a_n \sim b_n$ for $n \to +\infty$, then $a_n c_n \sim b_n c_n$ for $n \to +\infty$.*
3. *If $a_n = O(b_n)$ for $n \to +\infty$, then $a_n c_n = O(b_n)c_n \subseteq O(b_n c_n)$ for $n \to +\infty$.*

Example 14.9 Let a_n, b_n be two sequences and suppose that $b_n \to 0$ for $n \to +\infty$. By Proposition 14.6 one then has

$$e^{a_n+b_n} = e^{a_n}e^{b_n} = e^{a_n}(1 + b_n + O(b_n^2)) = e^{a_n} + b_n e^{a_n} + O(b_n^2 e^{a_n})$$

for $n \to +\infty$. $\qquad \square$

Proposition 14.10 (Substitution of the independent variable) *Let f be a function defined in a deleted neighborhood of 0 such that for some $\alpha > 0$*

$$f(t) = O(t^\alpha) \qquad t \to 0.$$

If $a_n \to 0$ for $n \to +\infty$, then

$$f(a_n) = O(|a_n|^\alpha) \qquad n \to +\infty.$$

Proof. Let $\delta > 0$ be such that $|f(t)| \le M|t^\alpha|$ for $|t| \le \delta$. If $\bar{n}$ is such that $|a_n| \le \delta$ for $n \ge \bar{n}$, for such n one has $f(a_n) \le M|a_n|^\alpha$. $\qquad \square$

Example 14.11 Let us verify that for every $m \geq 1$

$$e^{1/n} = 1 + 1/n + \cdots + \frac{(1/n)^{m-1}}{(m-1)!} + O\left(\frac{1}{n^m}\right) \qquad n \to +\infty.$$

In view of Proposition 14.6 one has that

$$e^x - 1 - x - \cdots - \frac{x^{m-1}}{(m-1)!} = O(x^m) \qquad x \to 0.$$

By Proposition 14.10 one then has

$$e^{1/n} - 1 - 1/n - \cdots - \frac{(1/n)^{m-1}}{(m-1)!} = O\left(\frac{1}{n^m}\right) \qquad n \to +\infty. \qquad \square$$

The following example is elementary, but we report it here since we will meet it in the proof of some estimates involving binomials.

Example 14.12 Let a_n, b_n be two sequences such that, for some $\alpha > 0$, $a_n = b_n(1 + O(1/n^\alpha))$ for $n \to +\infty$. Then $a_n \sim b_n$ for $n \to +\infty$; indeed, for $n \to +\infty$ one has

$$a_n = b_n + O(b_n/n^\alpha) \subseteq b_n + o(b_n). \qquad \square$$

14.2 Approximation of Cauchy and Riemann Sums

Let $a \in \mathbb{N}$, $\alpha > 0$ a real number and $g : [0, +\infty[\to \mathbb{R}$ a function of class $\mathscr{C}^1$ or $\mathscr{C}^2$. In this section we deal with the asymptotic estimates for $n \to +\infty$ of

- *Cauchy sums*: namely sums of the type

$$\sum_{a \leq k < n} g\left(\frac{k}{n^\alpha}\right);$$

- *Riemann sums*: namely sums of the type

$$\sum_{a \leq k < n} \frac{1}{n} g\left(\frac{k}{n}\right).$$

The problem is complicated both because sums do not preserve the asymptotic relation and because the number of summands increases with n.

14.2.1 The Case $0 < \alpha < 1$

This case $0 < \alpha < 1$ will be the only one used in our future applications. Therefore, omitting the reading of the other two cases will not undermine the understanding of the remainder of the chapter.

Proposition 14.13 Let $\alpha \in]0, 1[$ and $a \in \mathbb{N}$. Let $g : [0, +\infty[\to \mathbb{R}$ be a function of class $\mathscr{C}^1$, with both g and $|g'|$ integrable in the generalized sense on $[0, +\infty[$. Then

$$\sum_{a \leq k < n} g\left(\frac{k}{n^\alpha}\right) = n^\alpha \int_0^{+\infty} g(t)\, dt + o(n^\alpha) \qquad n \to +\infty. \tag{14.13.a}$$

In particular, if $\int_0^{+\infty} g(t)\, dt \neq 0$, one has

$$\sum_{a \leq k < n} g\left(\frac{k}{n^\alpha}\right) \sim n^\alpha \int_0^{+\infty} g(t)\, dt \qquad n \to +\infty.$$

Proof. Given n, we set $f(x) = g(x/n^\alpha)$ for each $x \in [a, n]$. The Euler–Maclaurin formula (12.27.a) applied to f yields

$$\sum_{a \leq k < n} f(k) = \int_a^n f(x)\, dx - \frac{f(n) - f(a)}{2} + R_1$$
$$= \int_a^n g\left(\frac{x}{n^\alpha}\right) dx + \frac{1}{2}\left(g\left(\frac{a}{n^\alpha}\right) - g\left(\frac{n}{n^\alpha}\right)\right) + R_1, \tag{14.13.b}$$

where

$$|R_1| \leq \frac{1}{2}\int_a^n |f'(x)|\, dx = \frac{1}{2}\int_a^n \left|g'\left(\frac{x}{n^\alpha}\right)\right| \frac{1}{n^\alpha}\, dx$$
$$= \frac{1}{2}\int_{a/n^\alpha}^{n^{1-\alpha}} |g'(t)|\, dt \leq \frac{1}{2}\int_0^{+\infty} |g'(t)|\, dt,$$

which is finite since g' is summable. One notes that g is bounded on its domain. Indeed for each x one has $g(x) = g(0) + \int_0^x g'(t)\, dt$ from which one sees that

$$|g(x)| \leq |g(0)| + \int_0^{+\infty} |g'(t)| \, dt =: M$$

and therefore the term of (14.13.b) that follows the integral is bounded by a constant:

$$\frac{1}{2} \left| g\left(\frac{n}{n^\alpha}\right) - g\left(\frac{a}{n^\alpha}\right) \right| + |R_1| \leq M + \frac{1}{2} \int_0^{+\infty} |g'(t)| \, dt = O(1) \subseteq o(n^\alpha) \quad n \to +\infty.$$
$$(14.13.c)$$

The change of variable $t = x/n^\alpha$ yields

$$\int_a^n g\left(\frac{x}{n^\alpha}\right) dx = n^\alpha \int_{a/n^\alpha}^{n^{1-\alpha}} g(t) \, dt. \qquad (14.13.d)$$

Now

$$\int_{a/n^\alpha}^{n^{1-\alpha}} g(t) \, dt = \int_0^{+\infty} g(t) \, dt - \int_0^{a/n^\alpha} g(t) \, dt - \int_{n^{1-\alpha}}^{+\infty} g(t) \, dt.$$

Since $0 < \alpha < 1$ one has

$$\int_0^{a/n^\alpha} g(t) \, dt = o(1) = \int_{n^{1-\alpha}}^{+\infty} g(t) \, dt \quad n \to +\infty,$$

from which it follows that

$$n^\alpha \int_{a/n^\alpha}^{n^{1-\alpha}} g(t) \, dt = n^\alpha \int_0^{+\infty} g(t) \, dt + o(n^\alpha) \qquad n \to +\infty;$$

the desired conclusion is now immediate. □

Remark 14.14 The hypotheses of Proposition 14.13 are satisfied if, for example, the function $g : [0, +\infty[\to]0, +\infty[$ is of class $\mathscr{C}^1$, definitively decreasing, and integrable in the generalized sense on $[0, +\infty[$.

Indeed, the generalized integrability implies $\lim_{x \to +\infty} g(x) = 0$; moreover if g is decreasing on $[b, +\infty[$, then

$$\int_a^{+\infty} |g'(x)| \, dx = \int_a^b |g'(x)| \, dx + \int_b^{+\infty} |g'(x)| \, dx$$

$$= \int_a^b |g'(x)| \, dx - \int_b^{+\infty} g'(x) \, dx = \int_a^b |g'(x)| \, dx + g(b) < +\infty.$$

Moreover, since $g(x) > 0$ for every $x \geq 0$, one also has $\int_0^{+\infty} g(t) \, dt \neq 0$.

14.2.2 The Case α = 1: Riemann Sums ☕

Let us now consider the case $\alpha = 1$. The following result is a well-known consequence of the fact that the integral of a continuous function g defined on $[0, 1]$ is for any $a \in \mathbb{N}$ approximated by the sequence of Riemann sums

$$\left(\frac{1}{n} \sum_{a \leq k < n} g\left(\frac{k}{n}\right) \right)_n .$$

The proof of next proposition is based on the trapezoidal method; in Problem 14.4 we ask the reader to look for a proof based on the expansions of Euler–Maclaurin.

Proposition 14.15 (Riemann Sums) *Let $a \in \mathbb{N}$.*

1. If $g : [0, 1] \to \mathbb{R}$ is a function of class $\mathscr{C}^1$, then

$$\sum_{a \leq k < n} g\left(\frac{k}{n}\right) = n \int_0^1 g(x)\, dx + O(1) \qquad n \to +\infty. \tag{14.15.a}$$

2. If $g : [0, 1] \to \mathbb{R}$ is a function of class $\mathscr{C}^2$, then

$$\sum_{a \leq k < n} g\left(\frac{k}{n}\right) = n \int_0^1 g(x)\, dx - \frac{1}{2} [g]_0^1 - a g(0) + O\left(\frac{1}{n}\right) \qquad n \to +\infty. \tag{14.15.b}$$

Proof. Fix $n > a$ and set $f(x) = g\left(\dfrac{x}{n}\right)$, one has

$$\sum_{a \leq k < n} g\left(\frac{k}{n}\right) = \sum_{a \leq k < n} f(k) \quad \forall n > a.$$

1. The first-order Euler–Maclaurin reduced formula (12.98.a) gives

$$\sum_{a \leq k < n} f(k) = \int_a^n f(x)\, dx + R_0(n), \quad |R_0(n)| \leq \frac{n-a}{2} \|f'\|_\infty.$$

In terms of g we get

$$\sum_{a \le k < n} g\left(\frac{k}{n}\right) = \int_a^n g\left(\frac{x}{n}\right) dx + R_0(n),$$

with $|R_0(n)| \le \dfrac{n-a}{2n}\|g'\|_\infty = O(1)$ as $n \to +\infty$. The change of variables $t = \dfrac{x}{n}$ yields

$$\sum_{a \le k < n} g\left(\frac{k}{n}\right) = n\int_{a/n}^1 g(t)\,dt + O(1) = n\int_0^1 g(t)\,dt + O(1) \quad n \to +\infty,$$

due to the fact that $\displaystyle\lim_{n \to +\infty} n\int_0^{a/n} g(t)\,dt = ag(0).$

2. The second-order Euler–Maclaurin reduced formula (12.98.b) gives

$$\sum_{0 \le k < n} f(k) = \int_0^n f(x)\,dx - \frac{1}{2}(f(n) - f(0)) + R_1(n), \quad |R_1(n)| \le \frac{n}{12}\|f''\|_\infty.$$

In terms of g we get

$$\sum_{0 \le k < n} g\left(\frac{k}{n}\right) = \int_0^n g\left(\frac{x}{n}\right) dx - \frac{1}{2}(g(1) - g(0)) + R_1(n),$$

with $|R_1(n)| \le \dfrac{n}{12n^2}\|g''\|_\infty = O\left(\dfrac{1}{n}\right)$ as $n \to +\infty$: indeed $f''(x) = \dfrac{1}{n^2}g''\left(\dfrac{x}{n}\right)$.
The change of variables $t = \dfrac{x}{n}$ yields

$$\sum_{0 \le k < n} g\left(\frac{k}{n}\right) = n\int_0^1 g(t)\,dt - \frac{1}{2}(g(1) - g(0)) + O\left(\frac{1}{n}\right) \qquad (14.15.c)$$

as $n \to +\infty$. Now we take into account the fact that the sums start from a. Notice that

$$\sum_{a \le k < n} g\left(\frac{k}{n}\right) = \sum_{0 \le k < n} g\left(\frac{k}{n}\right) - \sum_{0 \le k < a} g\left(\frac{k}{n}\right). \qquad (14.15.d)$$

Since, by the Mean Value Theorem,

$$|g(t) - g(0)| \le t\|g'\|_\infty \quad \forall t \in [0, 1],$$

we have

$$\forall 0 \le k < a \quad g\left(\frac{k}{n}\right) = g(0) + O\left(\frac{1}{n}\right) \quad n \to +\infty.$$

By adding a finite number a of terms as above, we get

$$\sum_{0 \le k < a} g\left(\frac{k}{n}\right) = ag(0) + O\left(\frac{1}{n}\right) \qquad n \to +\infty.$$

The conclusion now follows immediately from (14.15.c) and (14.15.d). □

Example 14.16 Applying (14.15.a) one obtains

$$\sum_{0 \le k < n} \frac{1}{1 + k/n} \sim n \int_0^1 \frac{1}{1 + x} dx = n \log 2 \qquad n \to +\infty.$$

For example, when $n = 100$ one has $100 \log 2 \approx 69.31$. Relation (14.15.b) yields

$$\sum_{0 \le k < n} \frac{1}{1 + k/n} = n \log 2 + \frac{1}{4} + O\left(\frac{1}{n}\right) \qquad n \to +\infty.$$

For example, with $n = 100$ we get

$$\sum_{0 \le k < 100} \frac{1}{1 + k/100} \approx 100 \log 2 + \frac{1}{4} \approx 69.56.$$

The real sum has value close to 69.57: using relation (14.15.b) in place of (14.15.a) we have thus succeeded in gaining in precision. □

Example 14.17 Let us consider the sum

$$\sum_{5 \le k < 100} e^{k/100}.$$

The approximation furnished by relation (14.15.a) yields

$$\sum_{5 \le k < 100} e^{k/100} \approx 100 \int_0^1 e^t \, dt = 100(e^1 - e^0) \approx 171.83,$$

while that furnished by (14.15.b) gives

$$\sum_{5 \le k < 100} e^{k/100} \approx 100 \int_0^1 e^t \, dt - \frac{1}{2}(e^1 - e^0) - 5e^0 \approx 165.97.$$

This latter approximation is closer to the true value of the sum which is approximately 165.87. □

14.2.3   The Case $\alpha > 1$ ☕

We conclude our study of Cauchy sums with the case $\alpha > 1$.

Proposition 14.18 *Let* $\alpha > 1$, $a \in \mathbb{N}$ *and* $g : [0, 1] \to \mathbb{R}$ *be a function.*

1. If g is of class $\mathscr{C}^1$ *then*

$$\sum_{a \leq k < n} g\left(\frac{k}{n^\alpha}\right) = g(0)(n - a) + O\left(\frac{1}{n^{\alpha-2}}\right) \qquad n \to +\infty. \qquad (14.18.a)$$

2. If g is of class $\mathscr{C}^2$ *then*

$$\sum_{a \leq k < n} g\left(\frac{k}{n^\alpha}\right) = g(0)(n - a) + \frac{g'(0)}{2n^{\alpha-2}} - \frac{g'(0)}{2n^{\alpha-1}} + \frac{g'(0)(a - a^2)}{2n^\alpha} +$$
$$+ O\left(\frac{1}{n^{2\alpha-3}}\right) \qquad n \to +\infty. \qquad (14.18.b)$$

Remark 14.19 Notice that, if g is of class $\mathscr{C}^2$, then (14.18.b) implies (14.18.a). Let us also point out that, since for all $\alpha > 1$, $\alpha - 2 < 2\alpha - 3$, then

$$\frac{1}{n^{\alpha-2}} \neq O\left(\frac{1}{n^{2\alpha-3}}\right) \qquad n \to +\infty,$$

thus the term $\dfrac{g'(0)}{2n^{\alpha-2}}$ does really count in (14.18.b). Instead, the relevance of the terms $-\dfrac{g'(0)}{2n^{\alpha-1}}, \dfrac{g'(0)(a - a^2)}{2n^\alpha}$ in (14.18.b) depends, respectively, upon the fact that $\alpha > 2$ or $\alpha > 3$.

Proof (of Proposition 14.18). Given the integer n, we put $f(x) = g(x/n^\alpha)$.
1. The Euler–Maclaurin reduced formula (12.98.a) applied to f on $[a, n]$ yields

$$\sum_{a \leq k < n} f(k) = \int_a^n f(x)\,dx + R_0(n), \qquad (14.19.a)$$

where
$$|R_0(n)| \leq \frac{1}{2}(n-a)\|f'\|_\infty.$$

The change of variable $t = \dfrac{x}{n^\alpha}$ gives

$$\int_a^n f(x)\,dx = \int_a^n g(x/n^\alpha)\,dx = n^\alpha \int_{a/n^\alpha}^{n^{1-\alpha}} g(t)\,dt. \qquad (14.19.\text{b})$$

Now, using the fact that $|g(t) - g(0)| \leq t\|g'\|_\infty$ (Mean Value Theorem), we get

$$\int_{a/n^\alpha}^{n^{1-\alpha}} (g(t) - g(0))\,dt \leq \|g'\|_\infty \int_{a/n^\alpha}^{n^{1-\alpha}} t\,dt = \|g'\|_\infty \frac{n^2 - a^2}{2n^{2\alpha}},$$

we have

$$\int_{a/n^\alpha}^{n^{1-\alpha}} g(t)\,dt = \int_{a/n^\alpha}^{n^{1-\alpha}} g(0)\,dt + \int_{a/n^\alpha}^{n^{1-\alpha}} (g(t) - g(0))\,dt$$

$$= g(0)\frac{n-a}{n^\alpha} + O\left(\frac{1}{n^{2\alpha-2}}\right), \quad n \to +\infty.$$

Multiplying both terms of the equality by n^α we obtain

$$\int_a^n f(x)\,dx = \int_a^n g(x/n^\alpha)\,dx = g(0)(n-a) + O\left(\frac{1}{n^{\alpha-2}}\right) \quad n \to +\infty.$$

$$(14.19.\text{c})$$

Moreover, since $f'(x) = \dfrac{1}{n^\alpha} g'\left(\dfrac{x}{n^\alpha}\right)$, we get

$$|R_0(n)| \leq \frac{n-a}{2n^\alpha}\|g'\|_\infty = O\left(\frac{1}{n^{\alpha-1}}\right) \subseteq O\left(\frac{1}{n^{\alpha-2}}\right),$$

and (14.18.a) follows.

2. Assume now that g is of class $\mathscr{C}^2$. Taking into account (14.19.b), the reduced formula (12.98.b) applied to f on the interval $[a, n]$ yields

$$\sum_{a \leq k < n} g\left(\frac{k}{n^\alpha}\right) = \int_a^n f(x)\,dx - \frac{f(n) - f(a)}{2} + R_1(n)$$

$$(14.19.\text{d})$$

$$= n^\alpha \int_{a/n^\alpha}^{n^{1-\alpha}} g(t)\,dt + \frac{1}{2}\left(g\left(\frac{a}{n^\alpha}\right) - g\left(\frac{n}{n^\alpha}\right)\right) + R_1(n),$$

with
$$|R_1(n)| \leq \frac{n-a}{12n^{2\alpha}}\|g''\|_\infty = O\left(\frac{1}{n^{2\alpha-1}}\right) \quad n \to +\infty. \qquad (14.19.\text{e})$$

Since both a/n^α and $n^{1-\alpha} = n/n^\alpha$ tend to zero at infinity, by mean of the second-order Maclaurin formula

$$g(t) = g(0) + g'(0)t + O(t^2) \quad t \to 0,$$

and by Proposition 14.10, we get

$$g\left(\frac{a}{n^\alpha}\right) = g(0) + g'(0)\frac{a}{n^\alpha} + O\left(\frac{1}{n^{2\alpha}}\right) \quad n \to +\infty,$$

$$g\left(\frac{n}{n^\alpha}\right) = g(0) + g'(0)\frac{n}{n^\alpha} + O\left(\frac{1}{n^{2\alpha-2}}\right) \quad n \to +\infty.$$

Thus

$$g\left(\frac{a}{n^\alpha}\right) - g\left(\frac{n}{n^\alpha}\right) = g'(0)\frac{a-n}{n^\alpha} + O\left(\frac{1}{n^{2\alpha-2}}\right) \quad n \to +\infty. \qquad (14.19.\text{f})$$

Again by mean of the second-order Maclaurin formula, we obtain

$$n^\alpha \int_{a/n^\alpha}^{n^{1-\alpha}} g(t)\,dt = n^\alpha \int_{a/n^\alpha}^{n^{1-\alpha}} \left(g(0) + g'(0)t + O(t^2)\right)\,dt$$

$$= g(0)(n-a) + g'(0)\frac{n^2 - a^2}{2n^\alpha} + O\left(\frac{1}{n^{2\alpha-3}}\right)$$

$$= g(0)(n-a) + \frac{g'(0)}{2n^{\alpha-2}} - \frac{a^2 g'(0)}{2n^\alpha} + O\left(\frac{1}{n^{2\alpha-3}}\right) \quad n \to +\infty.$$

$$(14.19.\text{g})$$

Rewriting (14.19.d) by taking (14.19.e)–(14.19.g) into account, we deduce

$$\sum_{a\le k<n} g\left(\frac{k}{n^\alpha}\right) = g(0)(n-a) + \frac{g'(0)}{2n^{\alpha-2}} - \frac{a^2 g'(0)}{2n^\alpha} + g'(0)\frac{a-n}{2n^\alpha} + O\left(\frac{1}{n^{2\alpha-3}}\right)$$

$$= g(0)(n-a) + \frac{g'(0)}{2n^{\alpha-2}} - \frac{g'(0)}{2n^{\alpha-1}} + \frac{g'(0)(a-a^2)}{2n^\alpha} + O\left(\frac{1}{n^{2\alpha-3}}\right) \quad n \to +\infty.$$

$$\square$$

Example 14.20 Let $g(x) = 3x + 5$. With the choice of $\alpha = 50$, a CAS easily yields

$$\sum_{7\le k<n} g\left(\frac{k}{n^{50}}\right) = \frac{10n^{51} - 70n^{50} + 3n^2 - 3n - 126}{2n^{50}}$$

$$= 5n - 35 + \frac{3}{2n^{48}} - \frac{3}{2n^{49}} - \frac{63}{n^{50}},$$

which of course implies (14.18.b). In particular we have

$$\sum_{7 \le k < n} g\left(\frac{k}{n^{50}}\right) = 5n - 35 + O\left(\frac{1}{n^{48}}\right) \quad n \to +\infty,$$

which is exactly (14.18.a). □

Example 14.21 Let $g(x) = 3x \log(x^2 + 1) - 67 \cos(x) + 2x$. Then $g(0) = -67$ and $g'(0) = 2$ so that by (14.18.b) we have

$$\sum_{0 \le k < n} g\left(\frac{k}{n^{10}}\right) = -67n + \frac{1}{n^8} - \frac{1}{n^9} + O\left(\frac{1}{n^{17}}\right) \quad n \to +\infty.$$

The approximation turns out to be here excellent even for small values of n. For instance, for $n = 5$ we get

$$\sum_{0 \le k < 5} g\left(\frac{k}{5^{10}}\right) = -334.\mathbf{99999795}199\ldots,$$

whereas

$$-67 \times 5 + \frac{1}{5^8} - \frac{1}{5^9} = -334.\mathbf{99999795}200\ldots. \qquad \square$$

Since, for $\alpha > 1$, $\dfrac{1}{n^{\alpha-2}} \in o(n)$ as $n \to +\infty$, as an immediate consequence of (14.18.a) we obtain the following asymptotic result.

Corollary 14.22 *Let $\alpha > 1$, $a \in \mathbb{N}$ and $g : [0, 1] \to \mathbb{R}$ be a function of class $\mathscr{C}^1$. If $g(0) \neq 0$ then*

$$\sum_{a \le k < n} g\left(\frac{k}{n^{\alpha}}\right) \sim g(0)(n - a) \qquad n \to +\infty. \tag{14.22.a}$$

Moreover, for $\alpha \ge 2$ we get

$$\forall \alpha \ge 2 \quad \sum_{a \le k < n} g\left(\frac{k}{n^{\alpha}}\right) = g(0)(n - a) + O(1) \quad n \to +\infty. \tag{14.22.b}$$

Example 14.23 We provide an asymptotic estimate for $\displaystyle\sum_{0\le k<n} e^{-k/n^2}$. The sum under

consideration is equal to $\displaystyle\sum_{0\le k<n} g\left(\frac{k}{n^2}\right)$ where one has set $g(x) = e^{-x}$. By (14.22.a)

such a sum is asymptotic to $ng(0) = n$ for $n \to +\infty$. For example, when $n = 10$, carrying out the calculations, one finds that

$$\sum_{0\le k<10} e^{-k/n^2} \approx 9.56,$$

while 10 is the value of the approximation.   □

Remark 14.24 The result furnished by Proposition 14.18 justifies the intuitive idea

that, given $\alpha > 1$, summing the terms of the type $g\left(\dfrac{k}{n^\alpha}\right)$ for $k = 0, \ldots, n-1$, all

of which are "close to" $g(0)$, one obtains something close to $ng(0)$. However, it may escape one's intuition that the validity of this result depends on the assumption that g belongs to $\mathscr{C}^1([0, 1])$. Let us consider, for example, $g(x) = \sqrt{x}$ for $x \ge 0$, and $\alpha = 2$. Observe that this function belongs to $\mathscr{C}^1(]0, 1])$, is defined in 0, but

$g'(x) = \dfrac{1}{2\sqrt{x}}$ is not defined in 0. One then has

$$\sum_{0\le k<n} g\left(\frac{k}{n^\alpha}\right) = \sum_{0\le k<n} \frac{\sqrt{k}}{n}.$$

Now, by (12.103.a), from the continuity of g it follows that

$$\sum_{0\le k<n} \sqrt{\frac{k}{n}} \sim n \int_0^1 \sqrt{x}\,dx \sim \frac{2n}{3} \quad n \to +\infty$$

and therefore

$$\sum_{0\le k<n} \frac{\sqrt{k}}{n} = \frac{1}{\sqrt{n}} \sum_{0\le k<n} \sqrt{\frac{k}{n}} \sim \frac{2\sqrt{n}}{3}. \tag{14.24.a}$$

If relation (14.18.a) holds, then from (14.22.b)

$$\sum_{0\le k<n} \frac{\sqrt{k}}{n} = 0 \times n + O(1) = O(1) \quad n \to +\infty$$

is bounded, thus contradicting the relation (14.24.a): therefore, in this case, relation (14.18.a) does not hold.

14.3   The Gauss Integral

It will be useful to recall the value of the integral of Gauss, which gives rise to the famous *gaussian*, the density of the normal variable in probabilistic calculus.

Lemma 14.25 (The Gaussian integral) *The function e^{-x^2} is integrable in the generalized sense on the interval $[0, +\infty[$ and is*

$$\int_0^{+\infty} e^{-x^2}\, dx = \frac{\sqrt{\pi}}{2}.$$

Moreover, for every sequence (a_n) of positive terms one has

$$\int_0^{a_n} e^{-x^2}\, dx = \frac{\sqrt{\pi}}{2} + O\left(e^{-a_n^2}\right) \qquad n \to +\infty.$$

Proof. On setting $I = \displaystyle\int_0^{+\infty} e^{-x^2}\, dx$, the reduction formulas for multiple integrals yield

$$\int_{[0,+\infty[\times[0,+\infty[} e^{-x^2-y^2}\, dxdy = \int_0^{+\infty} e^{-x^2}\, dx \int_0^{+\infty} e^{-y^2}\, dy = I^2.$$

On the other hand, the change to polar coordinates also yields

$$I^2 = \int_{[0,+\infty[\times[0,+\infty[} e^{-x^2-y^2}\, dxdy = \int_0^{+\infty}\int_0^{\pi/2} \rho e^{-\rho^2}\, d\theta d\rho =$$

$$= \frac{\pi}{2} \int_0^{+\infty} \rho e^{-\rho^2}\, d\rho = \frac{\pi}{2}\left[-\frac{1}{2}e^{-\rho^2}\right]_0^{+\infty} = \frac{\pi}{4},$$

from which, since $I > 0$, $I = \dfrac{\sqrt{\pi}}{2}$.

Furthermore one notes that if $(a_n)_n$ is a sequence of positive terms one then has

$$\left(\int_{a_n}^{+\infty} e^{-x^2}\, dx\right)^2 = \int_{[a_n,+\infty[\times[a_n,+\infty[} e^{-x^2-y^2}\, dxdy \le \int_{a_n}^{\infty}\int_0^{\pi/2} \rho e^{-\rho^2}\, d\theta d\rho =$$

$$= \frac{\pi}{2}\left[-\frac{1}{2}e^{-\rho^2}\right]_{a_n}^{+\infty} = \frac{\pi}{4}e^{-a_n^2}.$$

Hence

$$\int_0^{a_n} e^{-x^2}\, dx = \int_0^{+\infty} e^{-x^2}\, dx - \int_{a_n}^{+\infty} e^{-x^2}\, dx = \frac{\sqrt{\pi}}{2} + O(e^{-a_n^2/2}) \quad n \to +\infty.$$

□

The following big O comparison test furnishes a criterion which will be useful to us in what follows.

Example 14.26 Let g be continuous on $[0, +\infty[$. If $g(x) = O(1/x^2)$ for $x \to +\infty$ then $g \in L^1([0, +\infty[)$. Indeed, one has $|g(x)| \le M/x^2$ for every x greater than a suitable $a \in \mathbb{R}$; thus, for every $b > a$ one has

$$\int_0^b |g(x)|\, dx = \int_0^a |g(x)|\, dx + \int_a^b |g(x)|\, dx \le \int_0^a |g(x)|\, dx + M \int_a^b \frac{1}{x^2}\, dx.$$

Now

$$\int_a^{+\infty} \frac{1}{x^2}\, dx := \lim_{b \to +\infty} \int_a^b \frac{1}{x^2}\, dx = \lim_{b \to +\infty} \left(-\frac{1}{b} + \frac{1}{a} \right) = \frac{1}{a}$$

exists and is finite; therefore the increasing function $b \mapsto \int_0^b |g(x)|\, dx$ is bounded and so its limit for $b \to +\infty$ is finite.   □

The following asymptotic expansions will be used in the sequel; they are based on Proposition 14.13.

Corollary 14.27 *Let $c > 0$. One then has:*

$$\sum_{0 \le k < n} e^{-k^2/cn} \sim \frac{\sqrt{c\pi}}{2} n^{1/2} \qquad n \to +\infty; \qquad (14.27.a)$$

$$\sum_{0 \le k < n} k e^{-k^2/cn} \sim \frac{c}{2} n \qquad n \to +\infty; \qquad (14.27.b)$$

$$\sum_{0 \le k < n} k^2 e^{-k^2/cn} \sim \frac{c^{3/2}\sqrt{\pi}}{4} n^{3/2} \qquad n \to +\infty; \qquad (14.27.c)$$

$$\sum_{0 \le k < n} k^3 e^{-k^2/cn} \sim \frac{c^2}{2} n^2 \qquad n \to +\infty. \qquad (14.27.d)$$

Proof. 1. Having set $g(x) = e^{-x^2/c}$, the sum under consideration coincides with

$$\sum_{0 \le k < n} g\left(\frac{k}{n^{1/2}}\right).$$

Now, g is decreasing and of class $\mathscr{C}^1$; then since for any $k \in \mathbb{N}$

$$g(x) = e^{-x^2/c} = o(1/x^k) \quad x \to +\infty,$$

by Example 14.26 it is integrable in a generalized sense on $[0, +\infty[$. In view of Remark 14.14 and Proposition 14.13 one then has

$$\sum_{0 \le k < n} e^{-k^2/cn} = \sum_{0 \le k < n} g\left(\frac{k}{n^{1/2}}\right) = n^{1/2} \int_0^{+\infty} g(x)\,dx + o(\sqrt{n}) \qquad n \to +\infty.$$

By Lemma 14.25 one immediately obtains

$$\int_0^{+\infty} g(x)\,dx = \int_0^{+\infty} e^{-x^2/c}\,dx = \sqrt{c} \int_0^{+\infty} e^{-t^2}\,dt = \frac{\sqrt{c\pi}}{2},$$

whence (14.27.a).

2. Set $g(x) = xe^{-x^2/c}$, the sum under consideration coincides with

$$\sqrt{n} \sum_{0 \le k < n} g\left(\frac{k}{n^{1/2}}\right).$$

Now $g'(x) = \left(1 - 2\frac{x^2}{c}\right)e^{-x^2/c} < 0$ for $x > \sqrt{c/2}$; since $g(x) = xe^{-x^2/c} = o(1/x^k)$ for every $k \in \mathbb{N}$, by Example 14.26 it is integrable in the generalized sense on $[0, +\infty[$.

By Remark 14.14 and Proposition 14.13 one then has

$$\sum_{0 \le k < n} g\left(\frac{k}{n^{1/2}}\right) = n^{1/2} \int_0^{+\infty} g(x)\,dx + o(n^{1/2}) \qquad n \to +\infty$$

$$= n^{1/2} \left[-\frac{c}{2}e^{-x^2/c}\right]_0^{+\infty} + o(n^{1/2}) \qquad n \to +\infty$$

$$= \frac{c\sqrt{n}}{2} + o(n^{1/2}) \qquad n \to +\infty,$$

from which (14.27.b) follows.

3. The sum under consideration coincides with

$$n \sum_{0 \le k < n} g\left(\frac{k}{n^{1/2}}\right),$$

where $g(x) = x^2 e^{-x^2/c}$. One verifies that the function g is definitively decreasing; integrating by parts and using Lemma 14.25 one obtains

$$\int_0^{+\infty} g(x)\,dx = \frac{1}{4} c^{3/2} \sqrt{\pi}.$$

Thus (14.27.c) is then an easy consequence of Remark 14.14 and Proposition 14.13.

4. Having set $g(x) = x^3 e^{-x^2/c}$ the sum under consideration is

$$n^{3/2} \sum_{0 \le k < n} g\left(\frac{k}{n^{1/2}}\right).$$

Since g is definitively decreasing and integrable in a generalized sense on $[0, +\infty[$, by Remark 14.14 and Proposition 14.13, one has

$$\sum_{0 \le k < n} k^3 e^{-k^2/cn} \sim Cn^2 \qquad n \to +\infty,$$

where

$$C = \int_0^{+\infty} x^3 e^{-x^2/c}\,dx.$$

Setting $t = x^2/c$ in the integral one obtains

$$C = \int_0^{+\infty} (ct)^{3/2} e^{-t} t^{-1/2} \frac{c^{1/2}}{2}\,dt = \frac{c^2}{2} \int_0^{+\infty} te^{-t}\,dt$$
$$= \frac{c^2}{2} \left([-te^{-t}]_0^{+\infty} + \int_0^{+\infty} e^{-t}\,dt \right) = \frac{c^2}{2},$$

from which (14.27.d) follows. □

In (14.27.a) we have seen that, if $c > 0$, the sum $\sum_{0 \le k < n} e^{-k^2/cn}$ is asymptotic to $\sqrt{c\pi n}/2$ for $n \to +\infty$. In reality we obtain the same conclusion if we limit ourselves to summing the integers k up to $[n^r]$, for a given r satisfying $1/2 < r \le 1$, in the place of "up to n".[2]

---

[2]For $r = 1/2$ the formula continues to hold, but we do not give its proof which makes use of a different path of reasoning.

Proposition 14.28 *Let $c > 0$ and $r \in]1/2, 1]$. Then*

$$\sum_{0 \le k < [n^r]} e^{-k^2/cn} \sim \frac{\sqrt{c\pi}}{2} n^{1/2} \qquad n \to +\infty. \tag{14.28.a}$$

Proof. We fix $n \in \mathbb{N}$ and apply the Euler–Maclaurin formula (12.27.a) to the monotonic function $f(x) = e^{-x^2/cn}$ on the interval $[0, [n^r]]$: one obtains

$$\sum_{0 \le k < [n^r]} e^{-k^2/cn} = \int_0^{[n^r]} f(x)\,dx - \frac{1}{2}\left[e^{-x^2/cn}\right]_0^{[n^r]} + R_1(n),$$

where

$$|R_1(n)| \le \frac{1}{2}\left|\left[e^{-x^2/cn}\right]_0^{[n^r]}\right|.$$

Analyzing the terms of the expansion, one has, after setting $t = x/\sqrt{cn}$,

$$\int_0^{[n^r]} f(x)\,dx = \int_0^{[n^r]} e^{-x^2/cn}\,dx = \sqrt{cn}\int_0^{[n^r]/\sqrt{cn}} e^{-t^2}\,dt;$$

$$\left|\left[e^{-x^2/cn}\right]_0^{[n^r]}\right| = 1 - e^{-[n^r]^2/cn} \le 1,$$

from which it follows that

$$\sum_{0 \le k < [n^r]} e^{-k^2/cn} = \sqrt{cn}\int_0^{[n^r]/\sqrt{cn}} e^{-t^2}\,dt + O(1) \qquad n \to +\infty.$$

As a consequence of Lemma 14.25 one has

$$\int_0^{[n^r]/\sqrt{cn}} e^{-t^2}\,dt = \frac{\sqrt{\pi}}{2} + O(e^{-[n^r]^2/cn}) = \frac{\sqrt{\pi}}{2} + O(e^{-n^{2r}/cn}) \qquad n \to +\infty$$

and so

$$\sum_{0 \le k < [n^r]} e^{-k^2/cn} = \sqrt{cn}\left(\frac{\sqrt{\pi}}{2} + O(e^{-n^{2r}/cn})\right) + O(1) \qquad n \to +\infty.$$

Now $\sqrt{cn}\,O(e^{-n^{2r}/cn}) = O(\sqrt{cn}\,e^{-n^{2r}/cn}) \subseteq O(1)$ for $n \to +\infty$, given that $r > \frac{1}{2}$, and hence

$$\sum_{0 \le k < [n^r]} e^{-k^2/cn} = \frac{\sqrt{cn\pi}}{2} + O(1) \qquad n \to +\infty. \qquad \square$$

Remark 14.29 One notes that the estimate (14.27.a) is no longer valid if $r < 1/2$: indeed since $e^{-k^2/cn} < 1$ one has

$$\sum_{0 \le k < n^r} e^{-k^2/cn} \le \sum_{0 \le k < n^r} 1 \le n^r.$$

It follows that

$$\frac{\displaystyle\sum_{0 \le k < n^r} e^{-k^2/cn}}{n^r} \le 1;$$

if the relation (14.27.a) held, passing to the limit for $n \to +\infty$ one would obtain

$$+\infty = \lim_{n \to +\infty} \frac{\sqrt{c\pi n}}{2n^r} = \lim_{n \to +\infty} \frac{\displaystyle\sum_{0 \le k < n^r} e^{-k^2/cn}}{n^r} \le 1.$$

14.4   Families of Sequences and Their Estimates

☞   In this section we furnish the tools needed to understand what follows. We suggest that the reader attempts to understand the definitions and essential examples, disregarding, on first reading, the individual technical propositions, whilst reserving, of course, the right to return to them when they are invoked in later sections.

14.4.1   Families of Sequences

Let us introduce some supplementary notions that will allow us to handle sequences depending on two natural parameters.

Definition 14.30 Let $\alpha, \beta \in \mathbb{N}$. A **family of sequences**

$$(a_{k,n})_{n \ge \alpha}, \quad k \ge \beta,$$

is the sequence of sequences

$$((a_{\beta,n})_{n \ge \alpha}, (a_{\beta+1,n})_{n \ge \alpha}, (a_{\beta+2,n})_{n \ge \alpha}, \ldots). \qquad (14.30.a)$$

This means that to each $k \geq \beta$ corresponds a sequence $(a_{k,n})_{n \geq \alpha}$. We will occasionally omit the α, and sometimes the β, if they are obvious (most of the time α equals 0 or 1) or not essential for our purposes. □

Remark 14.31 Do not be fooled by the names of the indices. For instance $\left(\dfrac{k}{n}\right)_{n \geq 1}$, $k \geq 1$, is the sequence (of sequences)

$$\left(\left(\frac{1}{n}\right)_{n \geq 1}, \left(\frac{2}{n}\right)_{n \geq 1}, \left(\frac{3}{n}\right)_{n \geq 1}, \cdots\right),$$

whereas the family $\left(\dfrac{k}{n}\right)_{k \geq 1}$, $n \geq 1$, is the sequence (of sequences)

$$\left((k)_{k \geq 1}, \left(\frac{k}{2}\right)_{k \geq 1}, \left(\frac{k}{3}\right)_{k \geq 1}, \cdots\right).$$

14.4.2 Uniform Estimates of Families of Sequences

The following examples motivate the definition of a *uniform* estimate of a family of sequences.

Example 14.32 Given $k \in \mathbb{N}$, one has

$$\frac{k}{n^2} = o\left(\frac{1}{n^{3/2}}\right) \qquad n \to +\infty.$$

Indeed, $\displaystyle\lim_{n \to +\infty} \frac{k/n^2}{1/n^{3/2}} = 0$. Moreover, for each given $\overline{k} \in \mathbb{N}$, one has

$$\lim_{n \to +\infty} \frac{\displaystyle\sum_{k \leq \overline{k}} k/n^2}{1/n^{3/2}} = 0$$

and hence

$$\sum_{k \leq \overline{k}} \frac{k}{n^2} = o\left(\frac{1}{n^{3/2}}\right) \qquad n \to +\infty.$$

If instead we allow k to vary with n (that is, $k = k(n)$ is a function of n), then the preceding assertion may well be false. For example

$$\sum_{0 \le k < n} \frac{k}{n^2} = \frac{(n-1)n}{2n^2} \ne o\left(\frac{1}{n^{3/2}}\right) \qquad n \to +\infty. \qquad \square$$

Example 14.33 Suppose that we wish to give an estimate of $\displaystyle\sum_{0 \le k < n} e^{k/n^2}$. The problem is easily resolved by (14.18.b): the considered sum is equal to $n + \dfrac{1}{2} + O\left(\dfrac{1}{n}\right)$ for $n \to +\infty$. Here we wish to obtain the same result with another approach, based on the asymptotic estimate of the exponential function around the origin $e^t = 1 + t + O(t^2)$ for $t \to 0$. From this estimate, since $(k/n^2)_n$ is a sequence converging to 0 for each given k, by Proposition 14.10 one obtains that $e^{k/n^2} = 1 + (k/n^2) + O(k^2/n^4)$ for $n \to +\infty$. Therefore

$$\sum_{0 \le k < n} e^{k/n^2} = \sum_{0 \le k < n} \left(1 + (k/n^2) + O(k^2/n^4)\right) \quad n \to +\infty;$$

now, how behaves the sum of the $O(k^2/n^4)$ as k varies between 1 e n? We will give an answer to this problem in Example 14.44. $\square$

Definition 14.34 (*Uniform estimate*) Let $(a_{k,n})_n$ and $(b_{k,n})_n$ be two families of sequences and let $(k_n)_n$ be a sequence of positive terms. We say that $a_{k,n} = O(b_{k,n})$ for $n \to +\infty$, **uniformly** for $k \le k_n$, and write

$$a_{k,n} = O(b_{k,n}) \quad k \overset{\text{unif.}}{\le} k_n, \; n \to +\infty,$$

if there exist M and $\bar{n}$ such that

$$\forall n \ge \bar{n} \qquad |a_{k,n}| \le M|b_{k,n}| \; \forall k \le k_n. \qquad (14.34.\text{a})$$
$$\square$$

☞ *Remark 14.35* What matters here in (14.34.a) is the fact that the constant M not only does not depend on n once the inequality $n \ge \bar{n}$ holds, but indeed it neither depends on k, provided that $k \le k_n$.

Example 14.36 • $\dfrac{k}{k+n} = O\left(\dfrac{1}{\sqrt{n}}\right) \quad k \overset{\text{unif.}}{\le} \sqrt{n}, \; n \to +\infty.$
Indeed if $k \le \sqrt{n}$ then

$$\left| \frac{k}{k+n} \right| = \frac{k}{k+n} \le \frac{\sqrt{n}}{k+n} \le \frac{\sqrt{n}}{n} = \frac{1}{\sqrt{n}} \quad \forall n.$$

- $\dfrac{k}{k+n} \ne O\left(\dfrac{1}{\sqrt{n}}\right) \quad k \overset{\text{unif.}}{\le} \dfrac{n}{2}, \ n \to +\infty.$

Indeed assume, by contradiction, that there are $M > 0$ and $\bar{n} \in \mathbb{N}$ such that

$$\forall n \ge \bar{n} \quad \frac{k}{k+n} \le M\frac{1}{\sqrt{n}} \quad \forall k \le n/2.$$

Then, for $k = \left[\dfrac{n}{2}\right]$, we get $\dfrac{\sqrt{n}\left[\dfrac{n}{2}\right]}{\left[\dfrac{n}{2}\right]+n} \le M$ for all $n \ge \bar{n}$, a contradiction since

$$\lim_{n \to +\infty} \frac{\sqrt{n}\left[\dfrac{n}{2}\right]}{\left[\dfrac{n}{2}\right]+n} = +\infty. \qquad \square$$

Example 14.37 One has $\dfrac{k}{n^2} = O\left(\dfrac{1}{n^{3/2}}\right)$ for $n \to +\infty$, uniformly for $k \le \sqrt{n}$:
indeed, if $k \le \sqrt{n}$, then $\left|\dfrac{k}{n^2}\right| = \dfrac{k}{n^2} \le \dfrac{1}{n^{3/2}}$ for every $n \ge 1$. $\qquad \square$

Example 14.38 The estimate $\dfrac{k}{n} = O\left(\dfrac{\sqrt{k}}{\sqrt{n}}\right)$ for $n \to +\infty$, holds uniformly for $k \le n$: indeed, in such a case

$$\left|\frac{k}{n}\right| = \frac{k}{n} = \frac{\sqrt{k}\,\sqrt{k}}{\sqrt{n}\,\sqrt{n}} \le \frac{\sqrt{k}}{\sqrt{n}}. \qquad \square$$

We now state some simple properties of the uniform estimates that will be used later on. Among other things we see that if the estimates $a_{k,n} = O(b_{k,n})$ for $n \to +\infty$ are uniform for $k \le k_n$, then the comparison remains valid for sums of at most k_n terms.

Proposition 14.39 Let $(a_{k,n})_n$, $(b_{k,n})_n$, $(c_{k,n})_n$ be three families of sequences, $(m_n)_n$, $(k_n)_n$ two sequences of positive terms, and $a_{k,n} = O(b_{k,n})$ for $n \to +\infty$, uniformly for $k \le k_n$. The following properties hold:

1. Product: $a_{k,n}\,c_{k,n} = O(b_{k,n}\,c_{k,n})$ for $n \to +\infty$, uniformly for $k \le k_n$;
2. Transitivity: if $b_{k,n} = O(c_{k,n})$ for $n \to +\infty$, uniformly for $k \le k_n$ then $a_{k,n} = O(c_{k,n})$ for $n \to +\infty$, uniformly for $k \le k_n$;

3. Composition of the index k: *if $m_n \leq k_n$, then one has the following relation between sequences*

$$a_{m_n,n} = O(b_{m_n,n}) \qquad n \to +\infty;$$

☞ 4. Sum of a variable number of terms: *if $m_n \leq k_n$, then one has*

$$\sum_{k \leq m_n} a_{k,n} = O\left(\sum_{k \leq m_n} |b_{k,n}|\right) \qquad n \to +\infty.$$

Proof. Let M_1 and $\bar{n}_1$ be such that $|a_{k,n}| \leq M_1 |b_{k,n}|$ for $k \leq k_n$ and $n \geq \bar{n}_1$; for such k and n one obviously has $|a_{k,n} c_{k,n}| \leq M_1 |b_{k,n} c_{k,n}|$, from which follows Point 1. Next, if one has $|b_{k,n}| \leq M_2 |c_{k,n}|$ for $n \geq \bar{n}_2$ and $k \leq k_n$, it follows immediately, for $n \geq \max\{\bar{n}_1, \bar{n}_2\}$ and $k \leq k_n$, that

$$|a_{k,n}| \leq M_1 M_2 |c_{k,n}|,$$

which establishes Point 2. For $m_n \leq k_n$ and $n \geq \bar{n}_1$ one then has $|a_{m_n,n}| \leq M_1 |b_{m_n,n}|$, from which Point 3 follows; finally $\sum_{k \leq m_n} |a_{k,n}| \leq M_1 \sum_{k \leq m_n} |b_{k,n}|$ gives Point 4. □

14.4.3   Uniform Limits of Families of Sequences

The concept of *uniform convergence* of a family of sequences is closely related to the notion of a uniform estimate.

Definition 14.40 (*Uniform convergence*) Let $(a_{k,n})_n$ be a family of sequences and let $(k_n)_n$ be a sequence of positive terms. We say that $(a_{k,n})_n \to \ell \in \mathbb{R}$ for $n \to +\infty$, uniformly for $k \leq k_n$ if for every $\varepsilon > 0$ there exists a natural number $\bar{n}$ such that

$$\forall n \geq \bar{n}, \quad \forall k \leq k_n \quad |a_{k,n} - \ell| \leq \varepsilon.$$

Analogously we say that $(a_{k,n})_n$ diverges to $+\infty$ (resp. $-\infty$), uniformly for $k \leq k_n$, if for every $M \in \mathbb{R}$ there exists $\bar{n}$ such that $a_{k,n} \geq M$ (resp. $a_{k,n} \leq M$) for $n \geq \bar{n}$ e $k \leq k_n$. We will write

$$\lim_{\substack{\text{unif. } k \leq k_n \\ n \to +\infty}} a_{k,n} = \ell$$

whenever $(a_{k,n})_n$ tends to $\ell \in \mathbb{R} \cup \{\pm\infty\}$, uniformly for $k \leq k_n$.   □

Example 14.41 It follows immediately from the definition of uniform convergence that if $(a_{k,n})_n$ converges to ℓ for $n \to +\infty$, uniformly for $k \leq k_n$ then, for every sequence of integers m_n with $m_n \leq k_n$, one has $\lim_{n \to +\infty} a_{n,m_n} = \ell$.   □

Example 14.42 The family of sequences $(k/n)_n$ converges to 0 for $n \to +\infty$ for every given value of k; it converges uniformly to 0 for $k \leq n^r$ if and only if $r < 1$: indeed, if $k \leq n^r$ one has $0 \leq k/n \leq 1/n^{1-r}$, and the latter converges to 0 for $n \to +\infty$ if and only if $r < 1$.   □

Proposition 14.43 (Substitution of the independent variable) *Let f be a function defined on a neighborhood of 0 such that for some $\alpha > 0$*

$$f(t) = O(t^{\alpha}) \qquad t \to 0.$$

Suppose that $\lim_{\substack{\text{unif. } k \leq k_n \\ n \to +\infty}} a_{k,n} = 0$ *(k and k_n); then*

$$f(a_{k,n}) = O(|a_{k,n}|^{\alpha}) \qquad k \overset{\text{unif.}}{\leq} k_n, \ n \to +\infty.$$

Proof. Let $\delta > 0$ be such that $|f(t)| \leq M|t^{\alpha}|$ for $|t| \leq \delta$. If $\bar{n}$ is such that $|a_{k,n}| \leq \delta$ for $n \geq \bar{n}$ and $k \leq k_n$, then for such k, n one has $f(a_{k,n}) \leq M|a_{k,n}|^{\alpha}$.   □

We now investigate how the notions just introduced allow one to obtain the same estimate as that of Example 14.33 without getting one's hands dirty.

Example 14.44 Let us give an estimate of $\displaystyle\sum_{0 \leq k < n} e^{k/n^2}$ for $n \to +\infty$, using the asymptotic estimate $e^t = 1 + t + O(t^2)$, for $t \to 0$, of the exponential function. Clearly $\lim_{\substack{\text{unif. } k \leq n \\ n \to +\infty}} (k/n^2)_n = 0$; by Proposition 14.43, one has

$$e^{k/n^2} = 1 + k/n^2 + O(k^2/n^4) \qquad k \overset{\text{unif.}}{\leq} n, \ n \to +\infty.$$

Since $k^2/n^4 = O(1/n^2)$ for $n \to +\infty$, uniformly for $k \leq n$, one deduces from Point 2 of Proposition 14.39 that

$$e^{k/n^2} = 1 + k/n^2 + O(1/n^2) \qquad k \overset{\text{unif.}}{\leq} n, \ n \to +\infty.$$

From Point 4 of Proposition 14.39 one then obtains

$$\sum_{0 \le k < n} e^{k/n^2} = n + \frac{n(n-1)}{2n^2} + O\left(\frac{n}{n^2}\right)$$

$$= n + \frac{1}{2} + O\left(\frac{1}{n}\right) \qquad n \to +\infty. \qquad \square$$

The following example will be used in our subsequent discussion.

Example 14.45 Let $0 < r < 1$. By Proposition 14.6 one has $\log(1+t) = t - t^2/2 + O(t^3)$ for $t \to 0$; by Proposition 14.43 and Point 2 of Proposition 14.39 one obtains

$$\log\left(1 + \frac{k}{n}\right) = \frac{k}{n} - \frac{k^2}{2n^2} + O\left(\frac{k^3}{n^3}\right) \qquad k \overset{\text{unif.}}{\le} n^r,\ n \to +\infty$$

$$= \frac{k}{n} - \frac{k^2}{2n^2} + O\left(\frac{1}{n^{3-3r}}\right) \qquad k \overset{\text{unif.}}{\le} n^r,\ n \to +\infty.$$

The reader can also verify in analogous fashion that

$$\log\left(1 - \frac{k}{n}\right) = -\frac{k}{n} - \frac{k^2}{2n^2} + O\left(\frac{k^3}{n^3}\right) \qquad k \overset{\text{unif.}}{\le} n^r,\ n \to +\infty$$

$$= -\frac{k}{n} - \frac{k^2}{2n^2} + O\left(\frac{1}{n^{3-3r}}\right) \qquad k \overset{\text{unif.}}{\le} n^r,\ n \to +\infty. \qquad \square$$

Proposition 14.46 (Composition of sequences with family of sequences) *Let* $(a_n)_n$ *and* $(b_n)_n$ *be two sequences such that* $a_n = O(b_n)$ *for* $n \to +\infty$, *and* $(x_{k,n})_n$ *be a family of sequences with values in* $\mathbb{N}$ *such that*

$$\lim_{\substack{\text{unif. } k \le k_n \\ n \to +\infty}} x_{k,n}(k \text{ and } k_n) = +\infty.$$

Then $a_{x_{k,n}} = O(b_{x_{k,n}})$ *for* $n \to +\infty$, *uniformly for* $k \le k_n$.

Proof. Let M and n_1 be such that $|a_n| \le M|b_n|$ for $n \ge n_1$. Then if n_2 is such that $x_{k,n} \ge n_1$ for $n \ge n_2$ and $k \le k_n$, for such n and k one obviously has $|a_{x_{k,n}}| \le M|b_{x_{k,n}}|$. $\qquad \square$

We will need the following technical result.

Lemma 14.47 *Let* $a_n = O(1/n)$ *for* $n \to +\infty$ *and let* $0 < r < 1$*. Then*

$$a_{n\pm k} = O(1/n) \qquad k \overset{\text{unif.}}{\le} n^r, \ n \to +\infty.$$

Proof. If $k \le n^r$ one has

$$n \pm k \ge n - n^r \to +\infty \qquad n \to +\infty$$

and so $n \pm k \to +\infty$ for $n \to +\infty$, uniformly for $k \le n^r$. Applying Proposition 14.46 with $x_{k,n} = n \pm k$ and $b_n = 1/n$ one deduces that

$$a_{n\pm k} = O\left(\frac{1}{n \pm k}\right) \qquad n \to +\infty,$$

uniformly for $k \le n^r$. Since $\dfrac{1}{n \pm k} = O\left(\dfrac{1}{n - n^r}\right)$ uniformly for $k \le n^r$, by Point 2 of Proposition 14.39 one obtains that

$$a_{n\pm k} = O\left(\frac{1}{n - n^r}\right) \qquad k \overset{\text{unif.}}{\le} n^r, \ n \to +\infty.$$

Now, $1/(n - n^r) \sim 1/n$ for $n \to +\infty$; hence $1/(n - n^r) = O(1/n)$ for $n \to +\infty$ and therefore the desired conclusion follows by Proposition 14.39. $\qquad\square$

14.5 Uniform Asymptotic Estimate of Binomials

In Proposition 12.71 we have determined a version of Stirling formula up to a multiplicative constant. Precisely in (12.71.b) we have obtained

$$n! \sim e^C n^n \sqrt{n}\, e^{-n} \qquad n \to +\infty. \tag{14.47.a}$$

In Corollary 12.73 we established that $e^C = \sqrt{2\pi}$ by means of Wallis formula. This give us the possibility to prove the following estimate:

Theorem 14.48 (Estimate of the binomials) *Let* $0 < r < 2/3$*. Having set*

$$\alpha = \min\{1, 2 - 3r\} = \begin{cases} 1 & \text{if } r \le 1/3, \\ 2 - 3r & \text{if } r \ge 1/3, \end{cases}$$

one has

$$\binom{2n}{n \pm k} = \frac{2^{2n}}{\sqrt{\pi n}} e^{-k^2/n} \left(1 + O\left(\frac{1}{n^\alpha}\right)\right) \qquad k \overset{\text{unif.}}{\leq} n^r, \ n \to +\infty.$$
$$(14.48.\text{a})$$

In particular for every sequence $k_n \leq n^r$ one has

$$\binom{2n}{n \pm k_n} \sim \frac{2^{2n}}{\sqrt{\pi n}} e^{-k_n^2/n} \qquad n \to +\infty. \qquad (14.48.\text{b})$$

Proof. In order to reduce the problem to approximation of sums it is convenient to consider the logarithm

$$\log \binom{2n}{n \pm k} = \log(2n)! - \left(\log(n-k)! + \log(n+k)!\right). \qquad (14.48.\text{c})$$

From (14.47.a) one obtains

$$\log n! = \sum_{k=1}^{n} \log k = n \log n - n + \frac{\log n}{2} + C + O\left(\frac{1}{n}\right) \qquad n \to +\infty.$$
$$(14.48.\text{d})$$

Consequently, one has

$$\log(2n)! = \sum_{k=1}^{2n} \log k = 2n \log(2n) - 2n + \frac{\log(2n)}{2} + C + O\left(\frac{1}{n}\right)$$
$$(14.48.\text{e})$$
$$= 2n \log n + 2n \log 2 - 2n + \frac{\log n}{2} + \frac{\log 2}{2} + C + O\left(\frac{1}{n}\right)$$

for $n \to +\infty$. Since $r < 2/3 < 1$, from Lemma 14.47 and (14.48.d) one has that for $n \to +\infty$, uniformly for $k \leq n^r$,

$$\log(n-k)! = (n-k)\log(n-k) - (n-k) + \frac{\log(n-k)}{2} + C + O\left(\frac{1}{n}\right)$$
$$= (n-k)\log\left(n\left(1 - \frac{k}{n}\right)\right) - (n-k) + \frac{\log\left(n\left(1 - \frac{k}{n}\right)\right)}{2} + C + O\left(\frac{1}{n}\right)$$
$$= (n-k)\log n + (n-k)\log\left(1 - \frac{k}{n}\right) - n + k + \frac{\log n}{2} +$$
$$+ \frac{1}{2}\log\left(1 - \frac{k}{n}\right) + C + O\left(\frac{1}{n}\right). \qquad (14.48.\text{f})$$

In Example 14.45 we obtained that

$$\log\left(1 \pm \frac{k}{n}\right) = \pm\frac{k}{n} - \frac{k^2}{2n^2} + O\left(\frac{k^3}{n^3}\right) \qquad k \overset{\text{unif.}}{\le} n^r, \ n \to +\infty. \quad (14.48.\text{g})$$

Thus, one has, for $n \to +\infty$, uniformly for $k \le n^r$,

$$(n-k)\log\left(1 - \frac{k}{n}\right) = (n-k)\left(-\frac{k}{n} - \frac{k^2}{2n^2} + O\left(\frac{k^3}{n^3}\right)\right)$$

$$= -k - \frac{k^2}{2n} + O\left(\frac{k^3}{n^2}\right) + \frac{k^2}{n} + \frac{k^3}{2n^2} + O\left(\frac{k^4}{n^3}\right)$$

$$= -k + \frac{k^2}{2n} + O\left(\frac{k^3}{n^2}\right) + O\left(\frac{k^4}{n^3}\right).$$

Therefore, by (14.48.g) we get

$$(n-k)\log\left(1 - \frac{k}{n}\right) + \frac{1}{2}\log\left(1 - \frac{k}{n}\right) = -k + \frac{k^2}{2n} + O\left(\frac{k^3}{n^2}\right) + O\left(\frac{k^4}{n^3}\right) +$$

$$- \frac{k}{2n} - \frac{k^2}{4n^2} + O\left(\frac{k^3}{n^3}\right) \qquad k \overset{\text{unif.}}{\le} n^r, \ n \to +\infty.$$

$$(14.48.\text{h})$$

Substituting in (14.48.f), one obtains

$$\log(n-k)! = (n-k)\log n - n + k + \frac{\log n}{2} + C - k + \frac{k^2}{2n} - \frac{k}{2n} - \frac{k^2}{4n^2} +$$

$$+ O\left(\frac{1}{n}\right) + O\left(\frac{k^3}{n^2}\right) + O\left(\frac{k^3}{n^3}\right) + O\left(\frac{k^4}{n^3}\right)$$

$$= n\log n - k\log n - n + \frac{\log n}{2} + C + \frac{k^2}{2n} - \frac{k}{2n} + O\left(\frac{1}{n}\right) + O\left(\frac{k^2}{n^2}\right) +$$

$$+ O\left(\frac{k^3}{n^2}\right) + O\left(\frac{k^4}{n^3}\right) + O\left(\frac{k^3}{n^3}\right) \qquad k \overset{\text{unif.}}{\le} n^r, \ n \to +\infty.$$

In analogous fashion one finds that

$$\log(n+k)! = n\log n + k\log n - n + \frac{\log n}{2} + C + \frac{k^2}{2n} + \frac{k}{2n} + O\left(\frac{1}{n}\right) + O\left(\frac{k^2}{n^2}\right) +$$

$$+ O\left(\frac{k^3}{n^2}\right) + O\left(\frac{k^4}{n^3}\right) + O\left(\frac{k^3}{n^3}\right) \qquad k \overset{\text{unif.}}{\le} n^r, \ n \to +\infty.$$

Thus one gets

$$\log(n-k)! + \log(n+k)! = 2n\log n - 2n + \log n + 2C + \frac{k^2}{n} + O\left(\frac{1}{n}\right) + O\left(\frac{k^2}{n^2}\right) +$$

$$+ O\left(\frac{k^3}{n^2}\right) + O\left(\frac{k^4}{n^3}\right) + O\left(\frac{k^3}{n^3}\right) \qquad k \overset{\text{unif.}}{\leq} n^r, \ n \to +\infty.$$

Now, if $k \leq n^r$ one obtains

$$\frac{k^2}{n^2} \leq n^{2r-2}, \quad \frac{k^3}{n^2} \leq n^{3r-2}, \quad \frac{k^4}{n^3} \leq n^{4r-3}, \quad \frac{k^3}{n^3} \leq n^{3r-3}.$$

Since $0 < r < 2/3 < 1$, one has $3r - 2 \geq 4r - 3$, $3r - 2 \geq 3r - 3$, and $3r - 2 \geq 2r - 2$. Consequently we get

$$\log(n-k)! + \log(n+k)! = 2n\log n - 2n + \log n + \frac{k^2}{n} + 2C +$$

$$+ O\left(\frac{1}{n}\right) + O\left(\frac{1}{n^{2-3r}}\right) \qquad k \overset{\text{unif.}}{\leq} n^r, \ n \to +\infty.$$

Having set $\alpha = \min\{1, 2 - 3r\}$, one finds that

$$O\left(\frac{1}{n}\right) + O\left(\frac{1}{n^{2-3r}}\right) = O\left(\frac{1}{n^{\alpha}}\right).$$

Therefore, for $n \to +\infty$, uniformly for $k \leq n^r$,

$$\log(n-k)! + \log(n+k)! = 2n\log n - 2n + \log n + \frac{k^2}{n} + 2C + O\left(\frac{1}{n^{\alpha}}\right).$$

$$(14.48.\text{i})$$

Bearing in mind (14.48.e) and (14.48.i), the relation (14.48.c) then yields

$$\log\binom{2n}{n \pm k} = 2n\log n + 2n\log 2 - 2n + \frac{\log n}{2} + \frac{\log 2}{2} + C + O\left(\frac{1}{n}\right) +$$

$$- 2n\log n + 2n - \log n - \frac{k^2}{n} - 2C + O\left(\frac{1}{n^{\alpha}}\right)$$

$$= 2n\log 2 - \frac{\log n}{2} + \frac{\log 2}{2} - C - \frac{k^2}{n} + O\left(\frac{1}{n^{\alpha}}\right),$$

uniformly for $k \leq n^r$. Applying the exponential function to both terms of the preceding equation one obtains

$$\binom{2n}{n \pm k} = \frac{2^{2n}\sqrt{2}}{e^C \sqrt{n}} e^{-k^2/n} e^{O(1/n^\alpha)} \qquad k \overset{\text{unif.}}{\leq} n^r, \ n \to +\infty.$$

Since $r < 2/3$, one has $\alpha = \min\{1, 2 - 3r\} > 0$; then $1/n^\alpha \to 0$ for $n \to +\infty$. Thanks to Proposition 14.43, from $e' = 1 + O(t)$ for $t \to 0$ one has

$$e^{O(1/n^\alpha)} = 1 + O\left(\frac{1}{n^\alpha}\right) \qquad n \to +\infty.$$

Therefore one gets that

$$\binom{2n}{n \pm k} = \frac{2^{2n}\sqrt{2}}{e^C \sqrt{n}} e^{-k^2/n} \left(1 + O\left(\frac{1}{n^\alpha}\right)\right) \qquad k \overset{\text{unif.}}{\leq} n^r, \ n \to +\infty.$$

The conclusion follows since, from Corollary 12.73, $e^C = \sqrt{2\pi}$. □

🔖 *Remark 14.49* (*A local version of the Central Limit Theorem*) The estimate (14.48.a) has a strong connection with the Central Limit Theorem. Indeed, consider the binomial random variable X_n with parameters $\left(2n, \frac{1}{2}\right)$. The Central Limit Theorem asserts that, in distribution, the variable tends to the normal random variable N_n of same mean n and variance $\frac{n}{2}$, whose density at $x \in \mathbb{R}$ is given by

$$f_{N_n}(x) := \frac{1}{\sqrt{\pi n}} e^{-(x-n)^2/n}.$$

If $k \leq n$ the probability that X_n equals $n + k$ is given by

$$P(X_n = n + k) = \binom{2n}{n + k} \frac{1}{2^{2n}},$$

so that, by (14.48.a) we get

$$P(X_n = n \pm k) \sim \frac{1}{\sqrt{\pi n}} e^{-k^2/n} = f_{N_n}(n \pm k) \quad n \to +\infty.$$

Thus it turns out that the density of the normal variable N_n approximates the discrete density of the binomial variable X_n. This fact is known as a local version of the Central Limit Theorem.

14.6 Q-Ramanujan Distribution and Function, and Their Uniform Estimate

The Ramanujan[3] Q-distribution and Q-function are very much used in analysing algorithms.

Definition 14.50 The **Ramanujan Q-distribution** is defined by

$$Q(n, k) = \frac{n!}{(n-k)!n^k} \quad (0 \le k \le n);$$

the **Ramanujan Q-function** is defined by

$$Q(n) = \sum_{k=0}^{n} Q(n, k) = \sum_{k=0}^{n} \frac{n!}{(n-k)!n^k}. \qquad \square$$

The quantities defined above arise naturally in combinatorial calculus and in probability theory.

Example 14.51 (*The birthdays problem*) In a group of $k \le 365$ people the probability that no two of them are born on the same day (of a year having 365 days) is

$$Q(365, k) = \frac{365(365-1)\cdots(365-(k-1))}{365^k}.$$

Indeed (see also Example 2.30), labelled with I_k the k people and with I_{365} the days of the year, we can describe the dates of birth by a k-sequence of I_{365}, indicating in the i-th place the date of birth of the person i: there are 365^k such sequences. Among them, the k-sequences that correspond to two by two distinct dates are

$$365 \times (365-1) \times \cdots \times (365-(k-1)).$$

Let us now pretend to admit people to a birthday party. The "average" number of people to be admitted in order that at least two of them have the same birthday is given by the number $Q(365)$.

The proof of this fact is reserved to those having some familiarity with the basic notions of random variables. Let X be the random variable that counts the number of people to be admitted to the party in order for two of them to have the same birthday (not necessarily in the same year, however). Since X assumes integer values, its expected value is given by the sum of the series

---

[3] Srinivasa Ramanujan (1887–1920).

$$E[X] = 0 \times P(X = 0) + 1 \times P(X = 1) + 2 \times P(X = 2) + \cdots$$
$$= P(X = 1) + P(X = 2) + P(X = 3) + \cdots$$
$$+ P(X = 2) + P(X = 3) + \cdots$$
$$+ P(X = 3) + \cdots$$
$$\cdots\cdots\cdots$$
$$= P(X > 0) + P(X > 1) + P(X > 2) + \cdots ;$$

the latter series actually stops at $P(X > 365)$ in view of the fact that $P(X > k) = 0$ if $k \geq 366$. Now, for every $k \leq 365$, $P(X > k)$ is the probability that in a group of k people no two of them are born on the same day, and therefore $P(X > k) = Q(365, k)$; it follows that

$$E[X] = Q(365, 0) + \cdots + Q(365, 365) = Q(365). \qquad \square$$

Example 14.52 The probability that a word of k letters formed by chance from an alphabet of $n \geq k$ letters does not contain some letter twice is given by

$$Q(n, k) = \frac{n!}{(n - k)! n^k}.$$

In the same manner of Example 14.51, the average number of letters to list from an alphabet of n letters in order that a repetition should appear is given by

$$Q(n) = \sum_{k=0}^{n} Q(n, k) = \sum_{k=0}^{n} \frac{n!}{(n - k)! n^k}. \qquad \square$$

We end the section, and the book, proving an uniform asymptotic estimate of the Ramanujan distribution and the Ramanujan function.

Theorem 14.53 (Uniform asymptotic estimates of Ramanujan Q's) *Let* $0 < r < 2/3$. *Then*

$$Q(n, k) = e^{-k^2/2n} \left(1 + O(k/n) + O(k^3/n^2)\right) \qquad k \overset{\text{unif.}}{\leq} n^r, \ n \to +\infty.$$
$$(14.53.a)$$

In particular if k_n is a sequence such that $k_n \leq n^r$ one has

$$Q(n, k_n) \sim e^{-k_n^2/2n} \qquad n \to +\infty. \qquad (14.53.b)$$

Furthermore,

$$Q(n) \sim \sqrt{\pi n/2} \qquad n \to +\infty. \qquad (14.53.c)$$

Remark 14.54 We make a comment on why we need both O's in formula (14.53.a). If k is given, then one certainly has that $O(k^3/n^2) \subseteq O(k/n)$ for $n \to +\infty$; however, this fact no longer holds if one makes k vary with n: thus, for example, if $k_n = n^{2/3}$ one has $k_n^3/n^2 = 1$ while $k_n/n = 1/n^{1/3} \to 0$ for $n \to +\infty$ so that in this case $O(k_n^3/n^2) = O(1)$ but $O(k_n^3/n^2) \not\subseteq O(k_n/n)$ for $n \to +\infty$. Note that $(k^3/n^2)/(k/n) = k^2/n \to 0$ for $n \to +\infty$, uniformly for $k \leq n^r$ provided that $r < 1/2$; in this case by Proposition 14.39 one has that $O(k^3/n^2) \subseteq O(k/n)$ for $n \to +\infty$ uniformly for $k \leq n^r$.

Proof (of Theorem 14.53). For $k = 0$ one has $Q(n, 0) = 1 = e^{-0/n}$; thus, in what follows we may suppose $k \geq 1$. Given that

$$Q(n, k) = \frac{n!}{(n-k)! n^k} = \frac{(n-1)(n-2)\cdots(n-(k-1))}{n^{k-1}}$$

$$= \left(1 - \frac{1}{n}\right)\left(1 - \frac{2}{n}\right)\cdots\left(1 - \frac{k-1}{n}\right), \qquad (14.54.a)$$

one obtains

$$\log Q(n, k) = \log\left(1 - \frac{1}{n}\right) + \log\left(1 - \frac{2}{n}\right) + \cdots + \log\left(1 - \frac{k-1}{n}\right).$$

Recalling that $\log(1 - t) = -t + O(t^2)$ for $t \to 0$ (see Proposition 14.6), by Example 14.45 it follows

$$\log Q(n, k) = -\left(\frac{1}{n} + \cdots + \frac{k-1}{n}\right) + O\left(\frac{1}{n^2} + \cdots + \frac{(k-1)^2}{n^2}\right)$$

for $n \to +\infty$, uniformly for $k \leq n^r$. Since $1 + \cdots + (k-1) = k(k-1)/2$ and

$$1 + 2^2 + \cdots + (k-1)^2 \leq (k-1)(k-1)^2 = (k-1)^3,$$

one finds that

$$\log Q(n, k) = -\frac{k(k-1)}{2n} + O\left(\frac{k^3}{n^2}\right) \qquad k \overset{\text{unif.}}{\leq} n^r, \; n \to +\infty.$$

Passing to the exponential function one obtains

$$Q(n,k) = e^{-k(k-1)/2n} e^{O(k^3/n^2)} = e^{-k^2/2n} e^{k/2n} e^{O(k^3/n^2)} \qquad k \overset{\text{unif.}}{\leq} n^r, \ n \to +\infty,$$

where we use $e^{O(k^3/n^2)}$ to indicate a family of sequences of the type $(e^{a_{k,n}})_n$, parameterised by k, with $a_{k,n} = O(k^3/n^2)$ for $n \to +\infty$, uniformly for $k \leq n^r$. Forasmuch as $0 < r < 2/3$, we have

$$\frac{k}{n} \leq \frac{n^r}{n} = \frac{1}{n^{1-r}} \to 0, \qquad \frac{k^3}{n^2} \leq \frac{1}{n^{2-3r}} \to 0 \qquad k \overset{\text{unif.}}{\leq} n^r, \ n \to +\infty.$$

Since $e^t = 1 + O(t)$ for $t \to 0$, by Proposition 14.43 one has $e^{k/n} = 1 + O(k/n)$ and $e^{O(k^3/n^2)} = 1 + O(k^3/n^2)$ for $n \to +\infty$, uniformly for $k \leq n^r$. Therefore

$$Q(n,k) = e^{-k^2/2n} \left(1 + O\left(\frac{k}{n}\right)\right)\left(1 + O\left(\frac{k^3}{n^2}\right)\right) \qquad k \overset{\text{unif.}}{\leq} n^r, \ n \to +\infty.$$

Since $k^4/n^3 \leq k^3/n^2$, one then has

$$\left(1 + O\left(\frac{k}{n}\right)\right)\left(1 + O\left(\frac{k^3}{n^2}\right)\right) = 1 + O\left(\frac{k}{n}\right) + O\left(\frac{k^3}{n^2}\right) + O\left(\frac{k^4}{n^3}\right)$$

$$= 1 + O\left(\frac{k}{n}\right) + O\left(\frac{k^3}{n^2}\right) \qquad k \overset{\text{unif.}}{\leq} n^r, \ n \to +\infty,$$

from which (14.53.a) follows.

By Point 3 of Proposition 14.39 and from the fact that if $k_n \leq n^r$, then k_n/n and k_n^3/n^2 both tend to zero for $n \to +\infty$ one obtains easily (14.53.b) from (14.53.a).

To prove relation (14.53.c), we fix $r \in]1/2, 2/3[$ in such a way as to make valid both (14.53.a) and (14.28.a).

(A) *Estimates of* $\displaystyle\sum_{[n^r] \leq k \leq n} Q(n,k)$. In view of (14.50) it follows immediately that the function $k \mapsto Q(n,k)$ is decreasing and, in particular, that $Q(n,k) \leq Q(n,[n^r])$ for every $[n^r] \leq k \leq n$. By relation (14.53.b) with $k_n = [n^r]$, one has

$$Q(n,[n^r]) \sim e^{-[n^r]^2/2n} \qquad n \to +\infty$$

and consequently there exists $\bar{n}$ such that $Q(n,[n^r]) \leq 2e^{-[n^r]^2/2n}$ for $n \geq \bar{n}$. Thus, for $n \geq \bar{n}$ one has

$$\sum_{[n^r] \leq k \leq n} Q(n,k) \leq \sum_{[n^r] \leq k \leq n} Q(n,[n^r]) \leq nQ(n,[n^r]) \leq 2ne^{-[n^r]^2/2n}.$$

Since $1/2 < r < 2/3 < 1$, one has $2r - 1 > 0 < 1 - r$ and hence

$$ne^{-[n^r]^2/2n} \le ne^{-(n^r-1)^2/2n} = \frac{n}{e^{n^{2r-1}/2}e^{1/2n}e^{n^{1-r}}} \to 0 \qquad \text{for } n \to +\infty;$$

therefore

$$\lim_{n \to +\infty} \sum_{[n^r] \le k \le n} Q(n, k) = 0. \tag{14.54.b}$$

(B) Estimate of $\displaystyle\sum_{0 \le k < [n^r]} Q(n, k)$. From (14.53.a) and Point 4 of Proposition 14.39, one gets

$$\sum_{0 \le k < [n^r]} Q(n, k) = \sum_{0 \le k < [n^r]} e^{-k^2/2n} + C_n, \tag{14.54.c}$$

with

$$|C_n| \le M \left(\frac{1}{n} \sum_{0 \le k < [n^r]} ke^{-k^2/2n} + \frac{1}{n^2} \sum_{0 \le k < [n^r]} k^3 e^{-k^2/2n} \right),$$

for sufficiently large n and a suitable $M > 0$. From relations (14.27.b)–(14.27.d) one has

$$\sum_{0 \le k < [n^r]} ke^{-k^2/2n} \sim n \quad \text{and} \quad \sum_{0 \le k < [n^r]} k^3 e^{-k^2/2n} \sim 2n^2 \qquad n \to +\infty.$$

Then, for suitable constants c_1 and c_2, it is

$$\sum_{0 \le k < [n^r]} ke^{-k^2/2n} \le c_1 n \quad \text{and} \quad \sum_{0 \le k < [n^r]} k^3 e^{-k^2/2n} \le c_2 n^2.$$

Therefore, $|C_n| \le M(c_1 + c_2)$ is bounded. By Proposition 14.28

$$\sum_{0 \le k < [n^r]} e^{-k^2/2n} \sim \sqrt{\pi n/2} \qquad n \to +\infty;$$

consequently by (14.54.c) one deduces that

$$\sum_{0 \le k < [n^r]} Q(n, k) \sim \sqrt{\pi n/2} \qquad n \to +\infty. \tag{14.54.d}$$

Finally, if $1/2 < r < 2/3$, one has

$$Q(n) = \sum_{k=0}^{n} Q(n, k) = \sum_{0 \le k < [n^r]} Q(n, k) + \sum_{[n^r] \le k \le n} Q(n, k),$$

and so by (14.54.b)–(14.54.d) it follows

$$Q(n) \sim \sqrt{\pi n/2} \qquad n \to +\infty.$$ □

Example 14.55 The number k of people necessary to have a probability greater than 50% that there are at least two people who have the same birthday is 23. Indeed, by Example 14.51 the probability is equal to $1 - Q(365, k)$ and, carrying out the calculations, one finds that $Q(365, 22) \approx 0.52$ while $Q(365, 23) \approx 0.49$. Approximating $Q(365, k)$ as in (14.53.a), one finds that

$$Q(365, k) \approx e^{-k^2/730}$$

which is less than $1/2$ for $k \geq \sqrt{730 \log 2} \approx 23$. Again by Example 14.51, the average number of people to allow admission to the party in order to have at least two people born on the same day is equal to $Q(365)$. One has $Q(365) \approx 24.62$; using relation (14.53.c) one gets

$$Q(365) \approx \sqrt{\frac{365\pi}{2}} \approx 23.94.$$ □

Example 14.56 We pick a card from a deck of 52 playing cards and we reinsert it in the deck. After shuffling the deck, we pick another card and we reinsert it again in the deck. How many times, on average, must we repeat this procedure in order for a repetition to appear? The answer is

$$Q(52) = \sum_{k=0}^{52} \frac{52!}{(52 - k)! 52^k} \approx 9.72.$$

Using relation (14.53.c) one gets

$$Q(52) \approx \sqrt{\pi \times \frac{52}{2}} \approx 9.04.$$ □

14.7   Problems

Problem 14.1 Provide asymptotic estimates for the following Cauchy sums:

1. $\sum_{1 \leq k < n} e^{-k/n^2}$;

2. $\sum_{1 \leq k < n} e^{-k/\sqrt{n}}$;

3. $\displaystyle\sum_{1\le k<n} \frac{n}{(k+n^{1/2})(k+2n^{1/2})}$;

4. $\displaystyle\sum_{1\le k<n} \frac{n^4}{(k+n^2)(k+2n^2)}$.

Problem 14.2 Let $(a_{k,n})_n$ and $(b_{k,n})_n$ be two families of sequences, with $b_{k,n} \ne 0$ for all k, n, and such that $\displaystyle\lim_{n\to+\infty} \frac{a_{k,n}}{b_{k,n}} = 0$ uniformly for $k \le k_n$, where k_n is a sequence of positive terms. Prove that $O(a_{k,n}) = O(b_{k,n})$ for $n \to +\infty$, uniformly for $k \le k_n$, or, in other words, that if $c_{k,n} = O(a_{k,n})$ for $n \to +\infty$ uniformly for $k \le k_n$ then $c_{k,n} = O(b_{k,n})$ for $n \to +\infty$ uniformly for $k \le k_n$.

Problem 14.3 (*Ramanujan R-distribution*) Ramanujan R-distribution is

$$\forall n, k \in \mathbb{N}, \quad k \le n, \quad R(n,k) = \frac{n!n^k}{(n+k)!}.$$

Show that

$$\frac{n!n^k}{(n+k)!} = e^{-k^2/(2n)} \left(1 + O\left(\frac{1}{n}\right)\right) \quad n \to +\infty.$$

🖐 **Problem 14.4** Prove Proposition 14.15 using the Euler–Maclaurin expansion of first rank (that is, of order 1) to replace the role played by the trapezoidal method.

🖐 **Problem 14.5** Give asymptotic estimates for both the first- and second- order for the following Riemann sums:

1. $\displaystyle\sum_{1\le k<n} e^{-k/n}$;

2. $\displaystyle\sum_{1\le k<n} \frac{n^2}{(k+n)(k+2n)}$.

🖐 **Problem 14.6** Furnish an estimate of $\displaystyle\sum_{0\le k<n} \frac{1}{n^2+k^2}$ by means of (14.15.a) and (14.15.b). Compare, for $n = 10$, the approximations with the actual value of the sum.

🖐 **Problem 14.7** Estimate the following sums by means of (14.15.b):

1. $\displaystyle\sum_{0\le k<n} \frac{1}{1+k^2/n^2}$;

2. $\displaystyle\sum_{0\le k<n} \frac{1}{1+k^3/n^3}$.

Appendix

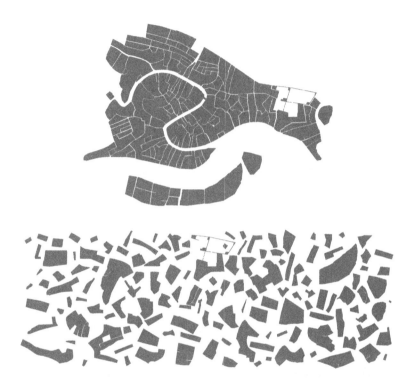

© Springer International Publishing Switzerland 2016
C. Mariconda and A. Tonolo, *Discrete Calculus*,
UNITEXT - La Matematica per il 3+2 103, DOI 10.1007/978-3-319-03038-8

Formulas and Tables

Basic Combinatorial Definitions ($k, n \in \mathbb{N}_{\geq 1}$)

Symbol	Number of	Value
$S(n, k)$	k-sequences without repetition of I_n	$\dfrac{n!}{(n-k)!} = n^{\underline{k}}$
$C(n, k) = \dbinom{n}{k}$	k-collections without repetition of I_n	$\dfrac{n!}{k!(n-k)!}$
$S((n, k))$	k-sequences of I_n or n-sharings of I_k	n^k
$C((n, k))$	k-collections of I_n or n-compositions of k	$\dbinom{(n-1)+k}{k} = \dbinom{(n-1)+k}{(n-1)}$
$S(n, k; (k_1, \ldots, k_n))$	k-sequences or n-sharings of I_n with occupancy $(k_1, \ldots, k_n)$	$\dfrac{k!}{k_1! \cdots k_n!}$
$P(a_1, \ldots, a_k)$	permutations of the k-sequence $(a_1, \ldots, a_k)$ with occupancy $(k_1, \ldots, k_n)$	$\dfrac{k!}{k_1! \cdots k_n!}$
$S(n, k; [k_1, \ldots, k_n])$	k-sequences or n-sharings of I_n with occupancy $[k_1, \ldots, k_n]$	$S(n, k; (k_1, \ldots, k_n)) \times P(k_1, \ldots, k_n)$
$C(n, k; [k_1, \ldots, k_n])$	k-collections of I_n or n-compositions di k with occupancy $[k_1, \ldots, k_n]$	$P(k_1, \ldots, k_n)$

© Springer International Publishing Switzerland 2016
C. Mariconda and A. Tonolo, *Discrete Calculus*,
UNITEXT - La Matematica per il 3+2 103, DOI 10.1007/978-3-319-03038-8

Symbol	Number of	Value
D_n	derangements of $(1,\dots,n)$	$n!\left(1-\dfrac{1}{1!}+\dfrac{1}{2!}-\dfrac{1}{3!}+\cdots+\dfrac{(-1)^n}{n!}\right)$
$D_n(n_1,\dots,n_k)$	derangements of I_k with occupancy $(n_1,\dots,n_k)$	$\displaystyle\sum_{m=0}^{n}(-1)^m\sum_{\substack{j_1+\cdots+j_k=m\\0\le j_1\le n_1\\ \vdots \\ 0\le j_k\le n_k}}\binom{n_1}{j_1}\cdots\binom{n_k}{j_k}\dfrac{(n-m)!}{(n_1-j_1)!\cdots(n_k-j_k)!}$
$\begin{Bmatrix}n\\k\end{Bmatrix}$ $n\ge k\ge 1$	k-partitions of I_n Stirling number of the II type	$\dfrac{1}{k!}\displaystyle\sum_{i=0}^{k}(-1)^i\binom{k}{i}(k-i)^n$ $=\displaystyle\sum_{n_1+\cdots+n_k=n-k}1^{n_1}2^{n_2}\cdots k^{n_k}$
$\Pi(k,n;[n_1,\dots,n_k])$	k-partitions of I_n with occupancy $[n_1,\dots,n_k]$	$\dfrac{1}{k!}S(k,n;[n_1,\dots,n_k])=$ $=\dfrac{1}{k!}\times\dfrac{n!}{n_1!n_2!\cdots n_k!}\times P(n_1,\dots,n_k)$
	k-cycles of I_n	$\dfrac{1}{k}S(n,k)=\dfrac{n!}{k\times(n-k)!}$
$\begin{bmatrix}n\\k\end{bmatrix}$ $n\ge k\ge 1$	k-partitions into cycles I_n Stirling number of the II type	$\displaystyle\sum_{\substack{I\subseteq I_{n-1}\\|I|=n-k}}\prod_{i\in I}i$ $=\dfrac{n!}{k!}\displaystyle\sum_{\substack{n_1+\cdots+n_k=n\\n_i\ge1}}\dfrac{1}{n_1\cdots n_k}$
$\Pi_{cyc}(k,n;[n_1,\dots,n_k])$	k-partitions into cycles of I_n with occupancy $[n_1,\dots,n_k]$	$\dfrac{1}{k!}\times\dfrac{n!}{n_1!\cdots n_k!}\times P(n_1,\dots,n_k)$

Combinatorial Identities ($k, m, n \in \mathbb{N}$)

$n \geq 0$	$$\binom{n}{0} + \cdots + \binom{n}{n} = 2^n$$
$n_1, \ldots, n_k \in \mathbb{N}$	$$\sum_{\substack{0 \leq \ell_1 \leq n_1 \\ 0 \leq \ell_k \leq n_k}} \binom{n_1}{\ell_1} \cdots \binom{n_k}{\ell_k} = 2^{n_1 + \cdots + n_k}$$
$n \geq 0$	$$\binom{n}{0}^2 + \binom{n}{1}^2 + \binom{n}{2}^2 + \cdots + \binom{n}{n-1}^2 + \binom{n}{n}^2 = \binom{2n}{n}$$
$n, k \geq 0$	$$\binom{n}{0} + \binom{n+1}{1} + \binom{n+2}{2} + \cdots + \binom{n+k}{k} = \binom{n+k+1}{k}$$
$n \geq 0, 0 \leq k \leq n$	$$\binom{0}{k} + \binom{1}{k} + \cdots + \binom{n}{k} = \binom{n+1}{k+1}$$
$n \geq m \geq 1$	$$m! \left\{ {n \atop m} \right\} = \sum_{k \in \mathbb{N}} \left\langle {n \atop k} \right\rangle \binom{k}{n-m} = \sum_{k=n-m}^{n-1} \left\langle {n \atop k} \right\rangle \binom{k}{n-m}$$
$n \geq 1, m \geq 0$	$$m^n = \sum_{k=0}^{n-1} \left\langle {n \atop k} \right\rangle \binom{m+k}{n}$$
$k, m, n \geq 0$	$$\sum_{j=0}^{k} \binom{m}{j} \binom{n}{k-j} = \binom{m+n}{k}$$

Alternating Sign Binomial Identities ($i, k, j, m, n, q \in \mathbb{N}$)

$0 \leq i \leq n \in \mathbb{N}$	$\displaystyle\sum_{k=i}^{n}(-1)^k \binom{n}{k}\binom{k}{i} = (-1)^n \delta_{i,n}, \quad \delta_{i,n} := \left\{ \begin{array}{l} 1,\ i = n, \\ 0,\ i \neq n. \end{array} \right.$
$a, b, c \in \mathbb{N}$	$\displaystyle\sum_{c=0}^{b}(-1)^c \binom{a+b}{b-c} C((a,c)) = 1$
$0 \leq j \leq n$	$\displaystyle\sum_{k=j}^{n}(-1)^k \binom{n+1}{k+1}\binom{k}{j} = (-1)^j$
$q \geq n \geq 1$ $0 \leq k < n$	$\displaystyle\sum_{i=0}^{n}(-1)^i \binom{n}{i}(q-i)^k = 0$
$Q(X)$ polynomial $0 \leq \deg Q(X) < m$	$\displaystyle\sum_{i=0}^{m}(-1)^i \binom{m}{i} Q(i) = 0$
$n \geq 1$	$\displaystyle \mathrm{H}_n = \sum_{k=1}^{n}(-1)^{k-1}\frac{1}{k}\binom{n}{k}$

Triangles of Classes of Numbers

☞ The empty boxes correspond to the value 0.

Binomial Triangle

n	$\binom{n}{0}$	$\binom{n}{1}$	$\binom{n}{2}$	$\binom{n}{3}$	$\binom{n}{4}$	$\binom{n}{5}$	$\binom{n}{6}$	$\binom{n}{7}$	$\binom{n}{8}$	$\binom{n}{9}$	$\binom{n}{10}$
0	1										
1	1	1									
2	1	2	1								
3	1	3	3	1							
4	1	4	6	4	1						
5	1	5	10	10	5	1					
6	1	6	15	20	15	6	1				
7	1	7	21	35	35	21	7	1			
8	1	8	28	56	70	56	28	8	1		
9	1	9	36	84	126	126	84	36	9	1	
10	1	10	45	120	210	252	210	120	45	10	1

Triangle of the Stirling Numbers of the 1st Type

n	$\left[{n\atop 0}\right]$	$\left[{n\atop 1}\right]$	$\left[{n\atop 2}\right]$	$\left[{n\atop 3}\right]$	$\left[{n\atop 4}\right]$	$\left[{n\atop 5}\right]$	$\left[{n\atop 6}\right]$	$\left[{n\atop 7}\right]$	$\left[{n\atop 8}\right]$	$\left[{n\atop 9}\right]$	$\left[{n\atop 10}\right]$
0	1										
1	0	1									
2	0	1	1								
3	0	2	3	1							
4	0	6	11	6	1						
5	0	24	50	35	10	1					
6	0	120	274	225	85	15	1				
7	0	720	1764	1624	735	175	21	1			
8	0	5040	13068	13132	6769	1960	322	28	1		
9	0	40320	109584	118124	67284	22449	4536	546	36	1	
10	0	362880	1026576	1172700	723680	269325	63273	9450	870	45	1

Triangle of the Stirling Numbers of the 2nd Type

n	$\left\{n \atop 0\right\}$	$\left\{n \atop 1\right\}$	$\left\{n \atop 2\right\}$	$\left\{n \atop 3\right\}$	$\left\{n \atop 4\right\}$	$\left\{n \atop 5\right\}$	$\left\{n \atop 6\right\}$	$\left\{n \atop 7\right\}$	$\left\{n \atop 8\right\}$	$\left\{n \atop 9\right\}$	$\left\{n \atop 10\right\}$
0	1										
1	0	1									
2	0	1	1								
3	0	1	3	1							
4	0	1	7	6	1						
5	0	1	15	25	10	1					
6	0	1	31	90	65	15	1				
7	0	1	63	301	350	140	21	1			
8	0	1	127	966	1701	1050	266	28	1		
9	0	1	255	3025	7770	6951	2646	462	36	1	
10	0	1	511	9330	34105	42525	22827	5880	750	45	1

Triangle of the Eulerian Numbers

n	$\left\langle n \atop 0\right\rangle$	$\left\langle n \atop 1\right\rangle$	$\left\langle n \atop 2\right\rangle$	$\left\langle n \atop 3\right\rangle$	$\left\langle n \atop 4\right\rangle$	$\left\langle n \atop 5\right\rangle$	$\left\langle n \atop 6\right\rangle$	$\left\langle n \atop 7\right\rangle$	$\left\langle n \atop 8\right\rangle$	$\left\langle n \atop 9\right\rangle$
1	1									
2	1	1								
3	1	4	1							
4	1	11	11	1						
5	1	26	66	26	1					
6	1	57	302	302	57	1				
7	1	120	1191	2416	1191	120	1			
8	1	247	4293	15619	15619	4293	247	1		
9	1	502	14608	88234	156190	88234	14608	502	1	
10	1	1013	47840	455192	1310354	1310354	455192	47840	1013	1

Discrete Primitives

$m \leq n$	$\displaystyle\sum_{m \leq k < n} g(k) = [G(k)]_{k=m}^{n} \iff \Delta G = g$
$m \neq -1, k \in \mathbb{N}$	$\displaystyle\sum k^{\underline{m}} = \frac{1}{m+1} k^{\underline{m+1}} + \mathbb{R}$
$k \in \mathbb{N}$	$\displaystyle\sum k^{\underline{-1}} = \mathrm{H}_k + \mathbb{R}$
$\ell, k \in \mathbb{N}$	$\displaystyle k^{\ell} = \sum_{i=0}^{\ell} \left\{ {\ell \atop i} \right\} k^{\underline{i}}$
$\ell, k \in \mathbb{N}$	$\displaystyle k^{\underline{\ell}} = \sum_{i=0}^{\ell} \left[{\ell \atop i} \right] (-1)^{l+i} k^{i}$
$\ell, k \in \mathbb{N}$	$\displaystyle\sum k^{\ell} = \sum_{i=0}^{\ell} \left\{ {\ell \atop i} \right\} \frac{1}{i+1} k^{\underline{i+1}} + \mathbb{R}$
$a \neq 1$	$\displaystyle\sum a^{k} = \frac{1}{a-1} a^{k} + \mathbb{R}$
$m < n$	$\displaystyle\sum_{m \leq k < n} u(k)\Delta v(k) = [u(k)v(k)]_{k=m}^{n} - \sum_{m \leq k < n} \Delta u(k)\theta\, v(k)$

OGF's Rules and Properties

Sequence	OGF	Operation
$(a_n + b_n)_n$	$\mathrm{OGF}\,(a_n)_n + \mathrm{OGF}\,(b_n)_n$	OGF of a sum/sum of OGF
$(a_n)_n \star (b_n)_n$	$\mathrm{OGF}\,(a_n)_n \cdot \mathrm{OGF}\,(b_n)_n$	OGF of a convolution/product of OGF
$((n+1)a_{n+1})_n$	$\mathrm{OGF}\,(a_n)'_n$	derivative of a OGF $(a_n)_n$
$(a_{n+1})_n$	$\dfrac{1}{X}\,(\mathrm{OGF}(a_n)_n - a_0)$	OGF of the shift

EGF's Rules and Properties

Sequence	EGF	Operation
$(a_n + b_n)_n$	$\mathrm{EGF}\,(a_n)_n + \mathrm{OGF}\,(b_n)_n$	EGF of a sum/sum of EGF
$(a_n)_n \diamond (b_n)_n$	$\mathrm{EGF}\,(a_n)_n \cdot \mathrm{EGF}\,(b_n)_n$	EGF of a binom. convolution/product of OGF
$(a_{n+1})_n$	$\mathrm{EGF}\,(a_n)'_n$	EGF of a shift/derivative of a EGF
$(a_{n+1}/(n+1))_n$	$\dfrac{1}{X}\,(\mathrm{EGF}(a_n)_n - a_0)$	EGF of $(a_{n+1}/(n+1))_n$

Frequent Use OGFs

$$\sum_{n=0}^{\infty} X^n = \frac{1}{1 - X}$$	$\mathrm{OGF}(1)_n$
$$\sum_{n=0}^{\infty} \binom{a}{k} X^k = (1 + X)^a \quad (a \in \mathbb{R})$$	$\mathrm{OGF}\binom{a}{k}_k$
$$(1 + X)^{1/2} = \sum_{n=0}^{\infty} \binom{1/2}{k} X^k$$ $$(1 + X)^{1/2} = 1 + \sum_{k=1}^{\infty} (-1)^{k-1} \frac{2}{4^k} \operatorname{Cat}_{k-1} X^k$$	$\mathrm{OGF}\binom{1/2}{k}_k$
$$\sum_{n=0}^{\infty} \binom{k + n}{k} X^n = \frac{1}{(1 - X)^{k+1}} \quad (m \in \mathbb{N})$$	$\mathrm{OGF}\binom{k + n}{n}_n$
$$\sum_{n=k}^{\infty} \binom{n}{k} X^n = \frac{X^k}{(1 - X)^{k+1}} \quad (m \in \mathbb{N})$$	$\mathrm{OGF}\binom{n}{k}_n$
$$\sum_{k=0}^{\infty} \binom{1/m}{k} X^n = (1 + X)^{1/k} \quad (m \in \mathbb{N})$$	$\mathrm{OGF}\binom{1/m}{k}_k$
$$\sum_{n=1}^{\infty} (-1)^{n-1} \frac{X^n}{n} = \log(1 + X)$$	$\mathrm{OGF}\left(\frac{(-1)^{n-1}}{n}\right)_n$
$$\sum_{n=1}^{\infty} \frac{X^n}{n} = -\log(1 - X)$$	$\mathrm{OGF}\left(\frac{1}{n}\right)_n$

$$\sum_{n=1}^{\infty} H_n X^n = \frac{\log(1-X)}{X-1}$$	OGF harmonic numbers H_n
$$\sum_{n=1}^{\infty} n(H_n - 1)X^n = -X\frac{\log(1-X)}{(X-1)^2}$$	
$$\sum_{k=0}^{\infty} (H_{n+k} - H_n)\binom{n+k}{k}X^k = -\frac{\log(1-X)}{(1-X)^{n+1}} \quad (n \in \mathbb{N})$$	
$$\sum_{n=0}^{\infty}\left(\sum_{k=0}^{n} a_k\right)X^n = \frac{1}{1-X}\sum_{n=0}^{\infty} a_n X^n$$	OGF partial sums
$$\sum_{n=0}^{\infty} \text{Cat}_n X^n = \frac{1-(1-4X)^{1/2}}{2X}$$	OGF Catalan numbers Cat_n
$$\sum_{k=1}^{\infty} k^n X^k = \frac{X}{(1-X)^{n+1}}\sum_{k=0}^{n-1}\left\langle{n\atop k}\right\rangle X^k$$	OGF powers $(k^n)_k$
$$\sum_{k=0}^{n}\left[{n\atop k}\right]X^k = X(X+1)(X+2)\cdots(X+(n-1)) \quad (n \geq 1)$$	OGF Stirling 1st $\left[{n\atop k}\right]_k$
$$\sum_{n=0}^{\infty}\left\{{n\atop k}\right\}X^n = \frac{X^k}{(1-X)(1-2X)\cdots(1-kX)} \quad (k \geq 1)$$	OGF Stirling 2nd $\left\{{n\atop k}\right\}_n$

Frequent Use EGFs

$\sum_{n=0}^{\infty} \dfrac{X^n}{n!} = e^X$	$\mathrm{EGF}(1)_n$
$\sum_{n=0}^{\infty}(-1)^n \dfrac{X^{2n}}{(2n)!} = \cos X$	$\cos X$
$\sum_{n=0}^{\infty}(-1)^n \dfrac{X^{2n+1}}{(2n+1)!} = \sin X$	$\sin X$
$\sum_{n=0}^{\infty} \dfrac{X^{2n}}{(2n)!} = \cosh X$	$\cosh X$
$\sum_{n=0}^{\infty} \dfrac{X^{2n+1}}{(2n+1)!} = \sinh X$	$\sinh X$
$\sum_{n=0}^{\infty}\left[\begin{matrix}n\\k\end{matrix}\right]\dfrac{X^n}{n!} = \dfrac{1}{k!}\left[\log\left(\dfrac{1}{1-X}\right)\right]^k$ $(k \geq 1)$	EGF Stirling 1st $\left[\begin{matrix}n\\k\end{matrix}\right]_n$
$\sum_{n=0}^{\infty}\left\{\begin{matrix}n\\k\end{matrix}\right\}\dfrac{X^n}{n!} = \dfrac{(e^X-1)^k}{k!}$ $(k \geq 1)$	EGF Stirling 2nd $\left\{\begin{matrix}n\\k\end{matrix}\right\}_n$
$\sum_{n=0}^{\infty}\mathfrak{B}_n \dfrac{X^n}{n!} = e^{(e^X-1)}$	EGF Bell $\mathfrak{B}_n$
$\sum_{n=0}^{\infty} B_n \dfrac{X^n}{n!} = \dfrac{X}{e^X-1}$	EGF Bernoulli B_n
$\sum_{n=1}^{\infty}(-1)^{n-1}\dfrac{1}{n}\dfrac{X^n}{n!} = \mathrm{Ein}(X)$	$\mathrm{EGF}\left(\dfrac{(-1)^{n-1}}{n}\right)_n$
$\sum_{n=1}^{\infty}\dfrac{1}{n}\dfrac{X^n}{n!} = -\mathrm{Ein}(-X)$	$\mathrm{EGF}\left(\dfrac{1}{n}\right)_n$
$\sum_{n=1}^{\infty} H_n \dfrac{X^n}{n!} = e^X \mathrm{Ein}(X)$	EGF harmonic numbers H_n

Recurrence Relations

Basis-Solutions to Homogeneous Relations

Consider the homogeneous linear recurrence relation of order r

$$c_0 x_n + c_1 x_{n-1} + \cdots + c_r x_{n-r} = 0, \quad n \geq r. \quad (c_0 c_r \neq 0) \qquad (R_o)$$

$$P_{\text{car}}(X) := c_0 X^r + c_1 X^{r-1} + \cdots + c_r$$

1. Let $\lambda_1, \ldots, \lambda_m$ be the distinct complex roots of the characteristic polynomial and $\mu_1, \ldots, \mu_m$ their multiplicity. The general complex solution of the recurrence (R_o) is given by the linear combinations with complex coefficients of the $r = \mu_1 + \cdots + \mu_m$ sequences

$$(\lambda_j^n)_n, \ldots, (n^{\mu_j - 1} \lambda_j^n)_n \quad j = 1, \ldots, m.$$

2. Suppose that the coefficients $c_0, \ldots, c_r$ are real numbers. Let

   - $\rho_1(\cos \alpha_1 \pm i \sin \alpha_1), \ldots, \rho_h(\cos \alpha_h \pm i \sin \alpha_h)$ the pairs of the non real complex conjugate roots of the characteristic polynomial and with $\mu_1, \ldots, \mu_h$ their multiplicity;
   - $\lambda_1, \ldots, \lambda_\ell$ be the real roots of the characteristic polynomial and $\mu_1', \ldots, \mu_\ell'$ their multiplicity.

   Then, the general real solution of the recurrence (R_o) is the set of all linear combinations with real coefficients of the $r = 2\mu_1 + \cdots + 2\mu_h + \mu_1' + \cdots + \mu_\ell'$ sequences

$$(\rho_j^n \cos(n\alpha_j))_n, \ldots, (n^{\mu_j - 1} \rho_j^n \cos(n\alpha_j))_n \quad j = 1, \ldots, h,$$

$$(\rho_j^n \sin(n\alpha_j))_n, \ldots, (n^{\mu_j - 1} \rho_j^n \sin(n\alpha_j))_n \quad j = 1, \ldots, h,$$

$$(\lambda_j^n)_n, \ldots, (n^{\mu_j' - 1} \lambda_j^n)_n \quad j = 1, \ldots, \ell.$$

Particular Solutions to a Non Homogeneous Linear Equation

Complex Solutions

Let $Q(X)$ be a polynomial, $q \in \mathbb{C}$, $c_0, \ldots, c_r \in \mathbb{C}$. Consider the linear recurrence relation of order r

$$c_0 x_n + c_1 x_{n-1} + \cdots + c_r x_{n-r} = Q(n) q^n, \quad n \geq r \quad (c_0 c_r \neq 0). \quad (R)$$

The above recurrence relation have a particular solution of the type

$$\left(n^{\mu} \widetilde{Q}(n) q^n \right)_n$$

where

- $\widetilde{Q}(X)$ is a polynomial of degree less or equal than the degree of $Q(X)$;
- μ is the multiplicity of q as root of the characteristic polynomial ($\mu = 0$ if q is not a root).

Real Solutions

Suppose we have a linear recurrence relation with *real coefficients* of one of the following types

$$c_0 x_n + c_1 x_{n-1} + \cdots + c_r x_{n-r} = Q(n) \rho^n \cos(n\gamma), \quad n \geq r, \quad (R_1)$$

$$c_0 y_n + c_1 y_{n-1} + \cdots + c_r y_{n-r} = Q(n) \rho^n \sin(n\gamma), \quad n \geq r, \quad (R_2)$$

with $c_0, \ldots, c_r \in \mathbb{R}$ ($c_0 c_r \neq 0$), $\gamma, \rho \in \mathbb{R}$ and $Q(X)$ is a polynomial with real coefficients. Then one obtains a particular real solution of (R_1) (resp. (R_2)) considering the real (resp. imaginary) part of a particular solution of the recurrence

$$c_0 z_n + c_1 z_{n-1} + \cdots + c_r z_{n-r} = Q(n) q^n, \quad n \geq r, \ q := \rho(\cos \gamma + i \sin \gamma).$$

Divide and Conquer Relations

1. Let $(a_n)_n$, $n = b^k$, be a solution to the divide and conquer recurrence

$$y_n = \lambda y_{n/b} + cn^\rho, \quad n = b^k, k \geq 1.$$

Then

$$a_n = \begin{cases} O(n^\rho) & \text{if } \lambda < b^\rho, \\ O(n^\rho \log n) & \text{if } \lambda = b^\rho, \\ O(n^{\log_b \lambda}) & \text{if } \lambda > b^\rho, \end{cases} \quad \text{for } n = b^k, k \to +\infty.$$

2. Let $(a_n)_n$, $n = b^k$, be a solution to the divide and conquer recurrence

$$y_n = \lambda y_{n/b} + c \log_b n \quad n = b^k, k \geq 1.$$

Then

$$a_n = \begin{cases} O(\log n) & \text{if } \lambda < 1, \\ O(\log^2 n) & \text{if } \lambda = 1, \\ O(n^{\log_b \lambda}) & \text{if } \lambda > 1, \end{cases} \quad \text{for } n = b^k, k \to +\infty.$$

OGF of a Linear Recurrence

Consider the recurrence relation

$$c_0 x_n + c_1 x_{n-1} + \cdots + c_r x_{n-r} = h_n, \quad n \geq r \quad (c_0 c_r \neq 0). \qquad (R)$$

The formal series $A(X) = \sum_{n=0}^{\infty} a_n X^n$ is the OGF of a solution of the linear recurrence (R) if and only if there exists a polynomial $S(X)$ satisfying

$$A(X) = \frac{S(X) + \sum_{n=r}^{\infty} h_n X^n}{c_0 + c_1 X + \cdots + c_r X^r}, \quad \deg S(X) < r.$$

In that case, necessarily, $S(X)$ is given by

$$S(X) = [X^{\leq r-1}] \left(c_0 + c_1 X + \cdots + c_{r-1} X^{r-1} \right) A(X)$$

$$= [X^{\leq r-1}] \left(\sum_{n=0}^{r-1} c_n X^n \right) \left(\sum_{n=0}^{r-1} a_n X^n \right).$$

Symbolic Calculus

The OGF of the combinatorial class $\mathrm{SEQ}^{\neg p}\,\Gamma$ of sequences in $\mathrm{SEQ}\,\Gamma$ that *do not contain p as a subsequence* is

$$\mathrm{SEQ}^{\neg p}\,\Gamma(X) = \frac{C_p(X)}{X^{\ell(p)} + (1 - |\Gamma|X)C_p(X)}.$$

Set for convenience $c_0 = c_{\ell(p)} := 1$, the sequence $(a_n)_n$ of the fibres of $\mathrm{SEQ}^{\neg p}\,\Gamma$ satisfies the recurrence

$$a_n = \sum_{i=1}^{\ell(p)} \left(|\Gamma|c_{i-1} - c_i\right)a_{n-i} \quad \forall n \geq \ell(p),$$

with initial data $a_i = |\Gamma|^i, i = 0, \ldots, \ell(p) - 1$.

The expected length of a randomly selected sequence of Γ in which appears the pattern p is

$$|\Gamma|^{\ell(p)}C_p(1/|\Gamma|).$$

Let $p \neq q$ be two patterns of the same length. The odds in favor of the appearing of q before p are equal to

$$\frac{C_p(1/|\Gamma|) - C_{(p,q)}(1/|\Gamma|)}{C_q(1/|\Gamma|) - C_{(q,p)}(1/|\Gamma|)}. \qquad \text{(Conway equation)}$$

Bernoulli Polynomials

$B_0(X) = 1$, $B_1(X) = -\dfrac{1}{2} + X$	$B_5(X) = -\dfrac{X}{6} + \dfrac{5X^3}{3} - \dfrac{5X^4}{2} + X^5$
$B_2(X) = \dfrac{1}{6} - X + X^2$	$B_6(X) = \dfrac{1}{42} - \dfrac{X^2}{2} + \dfrac{5X^4}{2} - 3X^5 + X^6$
$B_3(X) = \dfrac{X}{2} - \dfrac{3X^2}{2} + X^3$	$B_7(X) = \dfrac{X}{6} - \dfrac{7X^3}{6} + \dfrac{7X^5}{2} - \dfrac{7X^6}{2} + X^7$
$B_4(X) = -\dfrac{1}{30} + X^2 - 2X^3 + X^4$	$B_8(X) = -\dfrac{1}{30} + \dfrac{2X^2}{3} - \dfrac{7X^4}{3} + \dfrac{14X^6}{3} - 4X^7 + X^8$

Euler–Maclaurin Formulas

The Basic Euler–Maclaurin Formulas

First-order formulas f monotonic	$$\sum_{a \leq k < b} f(k) = \int_a^b f(t)\, dt - \frac{1}{2}\, [f]_a^b + R_1, \quad |R_1| \leq \frac{1}{2}|f(a) - f(b)|$$
Bernoulli numbers	$$B_0 = 1,\ B_1 = -\frac{1}{2},\ B_2 = \frac{1}{6},\ B_3 = 0,\ B_4 = -\frac{1}{30},\ B_5 = 0,$$ $$B_6 = \frac{1}{42},\ B_7 = 0,\ B_8 = -\frac{1}{30},\ B_9 = 0,\ B_{10} = \frac{5}{66}$$
Sup-norm of **the Bernoulli polynomials**	$$\mu_m := \| B_m(x) \|_{L^\infty([0,1])} \leq 4\frac{m!}{(2\pi)^m}$$ $$\mu_1 = \frac{1}{2},\ \mu_m = |B_m|\ \ \forall\, m \text{ even}$$
Formula of order m $f \in \mathscr{C}^m([a, b])$ Remainder R_m $f^{(m-1)}$ monotonic	$$\sum_{a \leq k < n} f(k) = \int_a^b f(x)\, dx + \sum_{i=1}^m \frac{B_i}{i!}\left[f^{(i-1)} \right]_a^b + R_m$$ $$|R_m| \leq \frac{\mu_m}{m!} \int_a^b |f^{(m)}(x)|\, dx$$ $$|R_m| \leq \frac{\mu_m}{m!} \left| [f^{(m-1)}]_a^b \right|$$
Estimate of partial sums $f \in \mathscr{C}^m([a, b])$	$$\sum_{a \leq k < N} f(k) = \sum_{a \leq k < n} f(k) + \int_n^N f(x)\, dx + \sum_{i=1}^m \frac{B_i}{i!}\left[f^{(i-1)} \right]_n^N + \varepsilon_m(n, N)$$ $$|\varepsilon_m(n, N)| \leq \frac{\mu_m}{m!} \int_n^N |f^{(m)}(x)|\, dx$$

Asymptotic Versions of the Euler–Maclaurin Formula

First-order asymptotic formula f monotonic and bounded	$$\sum_{a \le k < n} f(k) = \int_a^n f(t)\,dt + O(f(\infty) - f(n))$$
First-order asymptotic formula f monotonic and unbounded	$$\sum_{a \le k < n} f(k) = \int_a^n f(t)\,dt + O(f(n))$$
Asymptotic formula of order $m \ge 2$ $f \in \mathscr{C}^m$ and $f^{(m)} \in L^1$	$$\sum_{a \le k < n} f(k) = \int_a^n f(x)\,dx + C_m^f + \sum_{i=1}^m \frac{\mathrm{B}_i}{i!} f^{(i-1)}(n) +$$ $$+\, O\left(\int_n^{+\infty} |f^{(m)}(x)|\,dx \right)$$
Asymptotic formula of order $m \ge 2$ $f \in \mathscr{C}^m$, $f^{(m-1)}$ bounded and definitively monotonic	$$\sum_{a \le k < n} f(k) = \int_a^n f(x)\,dx + C_m^f + \sum_{i=1}^{m-1} \frac{\mathrm{B}_i}{i!} f^{(i-1)}(n) +$$ $$+\, O\left(f^{(m-1)}(n) - f^{(m-1)}(\infty) \right)$$
Asymptotic formula of order $m \ge 2$ $f \in \mathscr{C}^m$, $f^{(m-1)}$ unbounded and definitively monotonic	$$\sum_{a \le k < n} f(k) = \int_a^n f(x)\,dx \sum_{i=1}^{m-1} \frac{\mathrm{B}_i}{i!} f^{(i-1)}(n) + O\left(f^{(m-1)}(n) \right)$$

Euler Constant of a Function

Euler number of f	$$\gamma_n^f = \sum_{a \leq k < n} f(k) - \int_a^n f(x)\,dx$$ $$\gamma^f = \lim_{n \to +\infty} \gamma_n^f$$
f monotonic and bounded	$$\gamma^f = \gamma_n^f - \frac{1}{2}\,[f\,]_n^\infty + \varepsilon_1(n)$$ $$|\varepsilon_1(n)| \leq \frac{1}{2}\,|f(\infty) - f(n)|$$
$f \in \mathscr{C}^m,\ f^{(m)} \in L^1,$ $f(\infty), \ldots, f^{(m-2)}(\infty) \in \mathbb{R}$ convergent series	$$\gamma^f = \gamma_n^f + \sum_{i=1}^{m} \frac{\mathrm{B}_i}{i!}\left[f^{(i-1)}\right]_n^\infty + \varepsilon_m(n)$$ $$\sum_{k=a}^{\infty} f(k) = \sum_{a \leq k < n} f(k) + \int_n^{+\infty} f(x)\,dx + \sum_{i=1}^{m} \frac{\mathrm{B}_i}{i!}\left[f^{(i-1)}\right]_n^\infty + \varepsilon_m(n)$$ $$|\varepsilon_m(n)| \leq \frac{\mu_m}{m!}\int_n^{+\infty} |f^{(m)}(x)|\,dx$$

Approximation of Integrals

$h_n := \dfrac{\beta - \alpha}{n},\ x_k := \alpha + h_n k$	$T_n^g := h_n\left(\dfrac{g(x_0) + g(x_1)}{2} + \dfrac{g(x_1) + g(x_2)}{2} + \cdots + \dfrac{g(x_{n-1}) + g(x_n)}{2}\right)$ $D_n^g := h_n \sum_{0 \leq k < n} g(x_k)$
First-order approximation $g \in \mathscr{C}^1([\alpha, \beta])$	$$\int_\alpha^\beta g(x)\,dx = T_n^g + \varepsilon_1(n)$$ $$|\varepsilon_1(n)| \leq \frac{(\beta - \alpha)^2}{2n}\,\|g'\|_\infty$$
Rectangle method $g \in \mathscr{C}^1([\alpha, \beta])$	$$\int_\alpha^\beta g(x)\,dx = D_n^g + \varepsilon_1'(n)$$ $$|\varepsilon_1'(n)| \leq \frac{(\beta - \alpha)^2}{2n}\,\|g'\|_\infty$$

Approximation of order $m \geq 2$ $g \in \mathscr{C}^m([\alpha, \beta])$	$\int_\alpha^\beta g(x)\,dx = T_n^g - \sum_{i=2}^m \dfrac{B_i}{i!} h_n^i \left[g^{(i-1)} \right]_\alpha^\beta + \varepsilon_m(n)$ $\|\varepsilon_m(n)\| \leq \dfrac{\mu_m}{m!} \dfrac{(\beta - \alpha)^{m+1}}{n^m} \|g^{(m)}\|_\infty$
Trapezoidal method of order $m \geq 2$ $g \in \mathscr{C}^m([\alpha, \beta])$	$\int_\alpha^\beta g(x)\,dx = T_n^g - \sum_{i=2}^{m-2} \dfrac{B_i}{i!} h_n^i \left[g^{(i-1)} \right]_\alpha^\beta + \varepsilon_{m-1}(n)$ $\|\varepsilon_{m-1}(n)\| \leq \dfrac{\mu_m}{m!} \dfrac{(\beta - \alpha)^{m+1}}{n^m} \|g^{(m)}\|_\infty$

Riemann and Cauchy Sums

Riemann sums	
$g \in \mathscr{C}^1$	$\displaystyle\sum_{a \leq k < n} g\left(\frac{k}{n}\right) \sim n \int_0^1 g(x)\,dx \quad n \to +\infty$
$g \in \mathscr{C}^2$	$\displaystyle\sum_{a \leq k < n} g\left(\frac{k}{n}\right) \sim n \int_0^1 g(x)\,dx - \frac{1}{2}\, [g]_0^1 - ag(0) + O\left(\frac{1}{n}\right) \quad n \to +\infty$
Cauchy sums	
$0 < \alpha < 1$ $g \in \mathscr{C}^1,\ \int_0^\infty |g| + |g'| < +\infty$	$\displaystyle\sum_{a \leq k < n} g\left(\frac{k}{n^\alpha}\right) = n^\alpha \int_0^{+\infty} g(t)\,dt + o(n^\alpha) \quad n \to +\infty$
$\alpha > 1$ $g \in \mathscr{C}^1$	$\displaystyle\sum_{a \leq k < n} g\left(\frac{k}{n^\alpha}\right) = ng(0) + O(1) \quad n \to +\infty$
$\alpha > 1$ $g \in \mathscr{C}^2$	$\displaystyle\sum_{a \leq k < n} g\left(\frac{k}{n^\alpha}\right) = g(0)(n - a) + \frac{g'(0)}{2n^{\alpha-2}} - \frac{g'(0)}{2n^{\alpha-1}} + \frac{g'(0)(a - a^2)}{2n^\alpha} + O\left(\frac{1}{n^{2\alpha-3}}\right) \quad n \to +\infty$

Approximation of Combinatorial Numbers

Factorial and Binomial

$$n! \sim \sqrt{2\pi n}\left(\frac{n}{e}\right)^n \qquad n \to +\infty \quad \textbf{(Stirling's Formula)}$$

$$\log n! = n \log n - n + \frac{\log n}{2} + \log\left(\sqrt{2\pi}\right) + O\left(\frac{1}{n}\right) \qquad n \to +\infty$$

$$\binom{2n}{n \pm k} = \frac{2^{2n}}{\sqrt{\pi n}} e^{-k^2/n}\left(1 + O\left(\frac{1}{n}\right)\right) \qquad n \to +\infty$$

Ramanujan Q's

$$Q(n,k) := \frac{n!}{(n-k)! n^k} = e^{-k^2/2n}\left(1 + O\left(\frac{1}{n}\right)\right) \qquad n \to +\infty$$

$$Q(n) := \sum_{k=0}^{n} Q(n,k) \sim \sqrt{\pi n/2} \qquad n \to +\infty$$

References

1. Apéry, R.: Interpolation de fractions continues et irrationalitè de certaines constantes. Mathematics, CTHS: Bull. Sec. Sci. III. Bib. Nat. Paris, pp. 37–53 (1981)
2. Apostol, T.M.: Calculus. Vol. I: One-variable Calculus, with an Introduction to Linear Algebra. 2nd edn. Blaisdell Publishing Co. Ginn and Co., Waltham, Mass.-Toronto, Ont.-London (1967)
3. Apostol, T.M.: An elementary view of Euler's summation formula. Am. Math. Monthly **106**, 409–418 (1999)
4. Atkinson, K.E.: An Introduction to Numerical Analysis, 2nd edn. Wiley, New York (1989)
5. Benjamin, A.T., Quinn, J.J.: Proofs that really count. The Dolciani Mathematical Expositions, vol. 27. Mathematical Association of America, Washington, DC (2003)
6. Berndt, B.C.: Ramanujan's notebooks. Part I. Springer, New York (1985)
7. Bernoulli, J.: Ars Conjectandi. Thurneysen Brothers, Basel (1713)
8. De Marco, G.: Analisi 1. Primo corso di analisi matematica. Teoria ed esercizi. Collana di matematica. Testi e manuali. Zanichelli, Bologna (1996)
9. De Marco, G.: Analisi due. Teoria ed esercizi. Collana di matematica. Testi e manuali. Zanichelli, Bologna (1999)
10. De Marco, G., De Zotti, M., Mariconda, C.: Euler-Maclaurin formulas for BV functions. submitted (2016)
11. Devaney, R.L.: An Introduction to Chaotic Dynamical Systems. Addison-Wesley studies in nonlinearity, 2nd edn. Addison-Wesley Publishing Company, Advanced Book Program, Redwood City (1989)
12. Devaney, R.L.: A First Course in Chaotic Dynamical Systems. Addison-Wesley studies in non linearity. Addison-Wesley Publishing Company, Advanced Book Program, Reading (1992)
13. Diaconis, P., Graham, R.: Magical Mathematics: The Mathematical Ideas that Animate Great Magic Tricks. Princeton University Press, Princeton (2012)
14. Even, S., Gillis, J.: Derangements and Laguerre polynomials. Math. Proc. Camb. Philos. Soc. **79**, 135–143 (1976)
15. Faulhaber, J.: Academia Algebrae, Darinnien die miraculosische Inventiones zu den höchsten Cossen weiters continuirt und profitiert werden. Johann Ulrich Schönigs, Augsburg (1631)
16. Felix, D.: Optimal PenneyAnte strategy via correlation polynomial identities. Electron. J. Comb., 13: Research Paper 35, 15 pp. (electronic) (2006)
17. Gardner, M.: On the paradoxical situations that arise from nontransitive relations. Sci. Am. **231**, 120–125 (1974)
18. Gilbreath, N.: Magnetic collectors. Link. Ring **38**, 60 (1958)
19. Goldstein, P.T., Doth. Phil Goldstein, 1st edn. (1987)

© Springer International Publishing Switzerland 2016
C. Mariconda and A. Tonolo, *Discrete Calculus*,
UNITEXT - La Matematica per il 3+2 103, DOI 10.1007/978-3-319-03038-8

20. Graham, R.L., Knuth, D.E., Patashnik, O.: Concrete Mathematics, 2nd edn. Addison-Wesley Publishing Company, Reading, MA, USA (1994)
21. Guibas, L.J., Odlyzko, A.M.: String overlaps, pattern matching, and nontransitive games. J. Comb. Theory Ser. A **30**, 183–208 (1981)
22. Hardy, G.H., Ramanujan, S.: Asymptotic formulæ for the distribution of integers of various types. Proc. London Math. Soc **XVI**, 112–132 (1917)
23. Karamata, J.: Théorèmes sur la sommabilité exponentielle et d'autres sommabilitś rattachant. Mathematica (Cluj) **9**, 164–178 (1935)
24. Knuth, D.E.: The Art of Computer Programming: Vol. 3: Sorting and Searching, 2nd edn. Addison-Wesley, Reading (1998)
25. Maurer, R.: Problem E 2404. Am. Math. Mon. **80**, 316 (1973)
26. Moscovich, I., Stewart, I.: The Big Book of Brain Games: 1000 Playthinks: Puzzles, Paradoxes. Illusions and Games. Workman Publishing Company, New York (2001)
27. Pevzner, P.A.: Computational Molecular Biology: An Algorithmic Approach. MIT Press, Cambridge (2000)
28. Poisson, S.D.: Mémoire sur le calcul numérique des intégrales définies. Mémoires de lAcadémie Royale des Sciences de lInstitut de France **6**, 571–602 (1823)
29. Ramanujan, S.: On the sum of the square roots of the first n natural numbers. J. Indian Math. Soc. **7**, 173–175 (1915)
30. Reinhold, R.: Theory of Complex Functions. Graduate Texts in Mathematics, vol. 122. Springer, New York (1991)
31. Ross, S.: A First Course in Probability, 2nd edn. Macmillan Co., New York; Collier Macmillan Ltd., London (1984)
32. Rota, G.-C.: Combinatorial snapshots. Math. Intell. **21**, 8–14 (1999)
33. Walter, R.: Principles of Mathematical Analysis. International Series in Pure and Applied Mathematics, 3rd edn. McGraw-Hill Book Co., New York-Auckland-Düsseldorf (1976)
34. Rudin, W.: Real and Complex Analysis, 3rd edn. McGraw-Hill Book Co., New York (1987)
35. Sedgewick, R., Flajolet, P.: An Introduction to the Analysis of Algorithms. Addison-Wesley Longman Publishing Co., Inc, Boston, MA, USA (1996)
36. Stanley, R.P.: Enumerative Combinatorics. Volume 62 of Cambridge Studies in Advanced Mathematics, vol. 2. Cambridge University Press, Cambridge (1999)
37. Stirling, J.: Methodus differentialis: sive tractatus de summatione et interpolatione serierum infinitarum. Typis Gul. Bowyer, London (1730)
38. Whittaker, E.T., Watson, G.N.: A course of modern analysis. Cambridge Mathematical Library. Cambridge University Press, Cambridge (1996)

Index

Symbols

$()$, 421

$\mathfrak{S}_k(A_1, \ldots, A_n)$, 84

Θ, 139

$\approx$, 470

$\binom{a}{k}$, $a \in \mathbb{R}$, 239

$\binom{n}{k}$, $n \in \mathbb{N}$, 31

χ_A, 3

$\langle a_1, a_2, \ldots, a_k \rangle$, 123

$\cong$, 423

$\diamond$, 200

Δ, 158

$\sum_k f(k)$, $\sum f$, 165

$\lim\limits_{\substack{\text{unif. } k \leq k_n \\ n \to +\infty}}$, 604

$\sum\limits_{i=0}^{\infty} A_i(X)$, 210

η, 545

$\left\langle \begin{matrix} n \\ k \end{matrix} \right\rangle$, 138

γ, 493, 495, 571

γ^f, 489

γ_n^f, 488, 527

$\left[X^n \right] A(X)$, 194

$\left[X^{>n} \right] A(X)$, 194

$\left[X^{\leq n} \right] A(X)$, 194

$\left[\begin{matrix} n \\ k \end{matrix} : [n_1, \ldots, n_k] \right]$, 136

$[f]_a^b$, 472

$\left\{ \begin{matrix} n \\ k \end{matrix} : [n_1, \ldots, n_k] \right\}$, 115

$\overset{\text{unif.}}{\leq}$, 602

$\mathbb{C}[[X]]$, 194

$\mathbb{N}_{\geq 1}$, 2

$\mathbb{N}_{\geq k}$, 2

$\mathbb{R}[[X]]$, 194

$\mathbb{R}((X))$, 222

$\mathscr{R}_k$, 266

$\mathbb{1}$, 158, 420

$\mathfrak{B}_n$, 112

μ_m, 542

ν, 419

$\oplus$, 421

θ, 158

$\sim$, 583

$\|f\|_{L^\infty(I)}$, $\|f\|_\infty$, 470

$\star$, 200

$\left\{ \begin{matrix} n \\ k \end{matrix} \right\}$, 105

$\mathscr{U}$, 425

$\mathscr{R}$, 454

$\mathscr{T}$, 450

$\left[\begin{matrix} n \\ k \end{matrix} \right]$, 128

ζ, 304

$\zeta(2)$, 509

$\zeta(3)$, 507, 510, 574

$a_n = O(b_n)$, 583

$a_n = o(b_n)$, 582

$a_n \sim b_n$, 583

$f(X)$, 226

$f(\infty)$, 483

© Springer International Publishing Switzerland 2016

C. Mariconda and A. Tonolo, *Discrete Calculus*,

UNITEXT - La Matematica per il 3+2 103, DOI 10.1007/978-3-319-03038-8

Printed in the United States
By Bookmasters